Human Physiology

Second Edition

Human Physiology

Stuart Ira Fox, Ph.D.
Pierce College

Wm. C. Brown Publishers
Dubuque, Iowa

Cover Illustration

© William B. Westwood
The cover shows an artistic representation of the theme of neural and endocrine regulation. The large cell in the center is a heart cell (note the myofibrils) from the SA node that is receiving neurotransmitter molecules—norepinephrine and acetylcholine—from autonomic nerve endings and hormones (epinephrine and norepinephrine) from blood capillaries. The neurotransmitters are represented by yellow spheres and the hormone molecules are represented by scarlet spheres. Although the molecules of neurotransmitters and hormones are about the same size, the hormone molecules are represented by larger spheres in order to depict their interaction with receptor proteins in the membrane of the heart cell. The sizes of both neurotransmitters and hormones in relation to the size of the cell have been exaggerated for artistic purposes.

Book Team

Edward G. Jaffe
Executive Editor

Lynne M. Meyers
Developmental Editor

Kevin Campbell
Production Editor

Julie E. Anderson
Designer

Carla D. Arnold
Permissions Editor

Shirley Charley
Photo Research Editor

Matt Shaughnessy
Product Manager

wcb group

Wm. C. Brown
Chairman of the Board

Mark C. Falb
President and Chief Executive Officer

wcb

Wm. C. Brown Publishers, College Division

G. Franklin Lewis
Executive Vice-President, General Manager

E. F. Jogerst
Vice-President, Cost Analyst

George Wm. Bergquist
Editor in Chief

Beverly Kolz
Director of Production

Chris C. Guzzardo
Vice-President, Director of Marketing

Bob McLaughlin
National Sales Manager

Craig S. Marty
Manager, Marketing Research

Julie A. Kennedy
Production Editorial Manager

Marilyn A. Phelps
Manager of Design

Faye M. Schilling
Photo Research Manager

The credits section for this book begins on page 688 and is considered an extension of the copyright page.

Library of Congress Catalog Card Number: 86-71661

ISBN 0-697-08338-1

Printed in the United States of America.
10 9 8 7 6 5 4 3 2 1

To Professor Robert J. Lyon, mentor and friend

Brief Contents

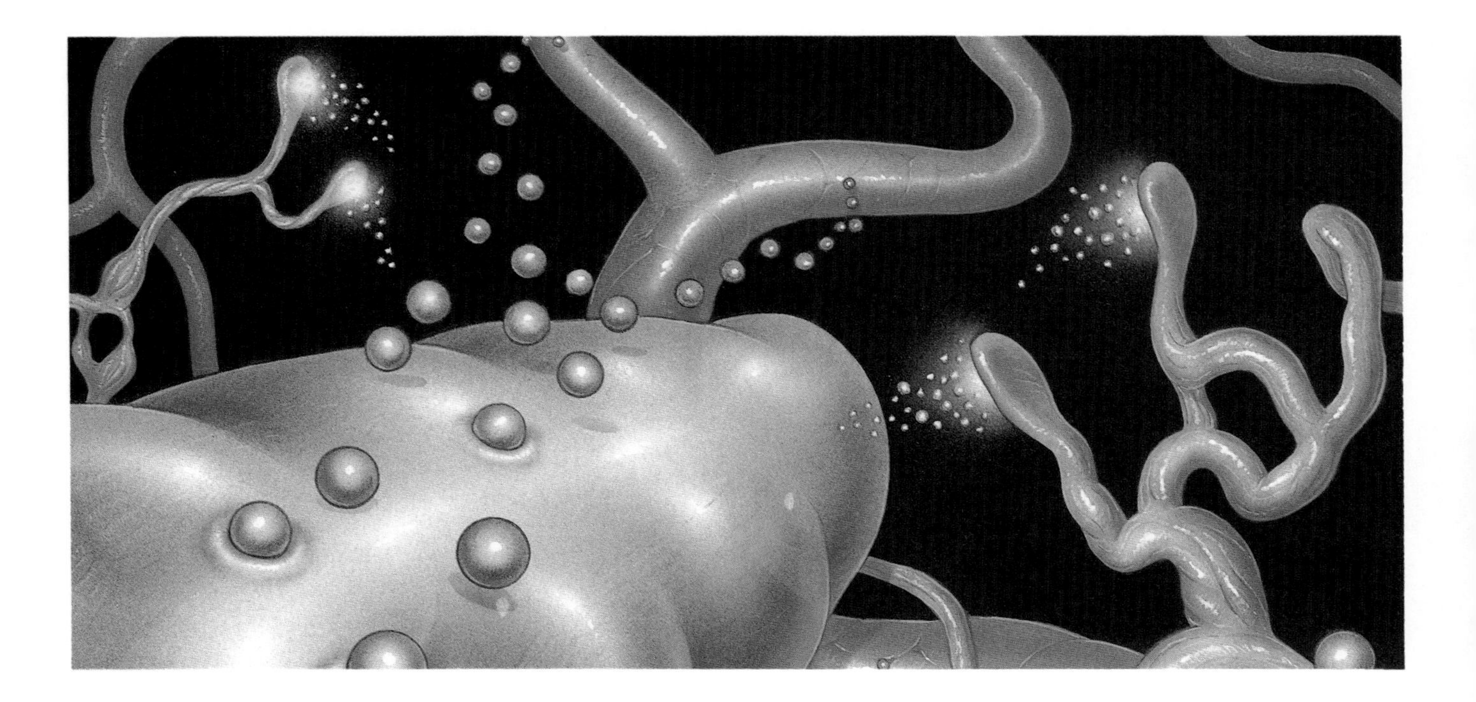

Expanded Contents

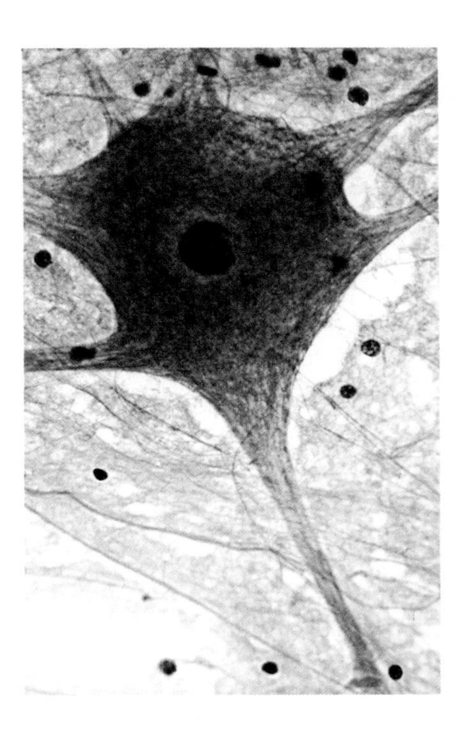

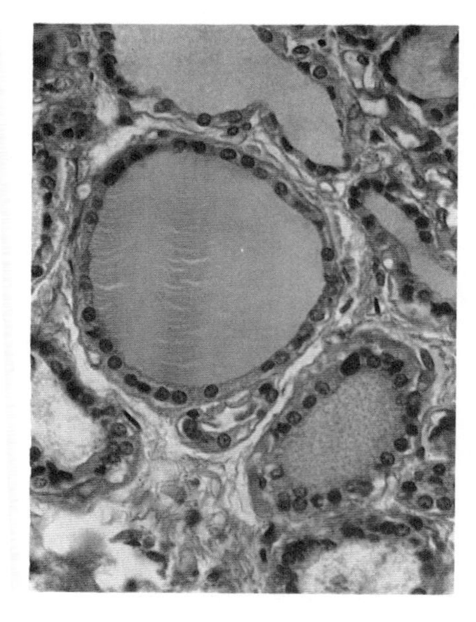

12 The Autonomic Nervous System and Its Control of Visceral Function 322

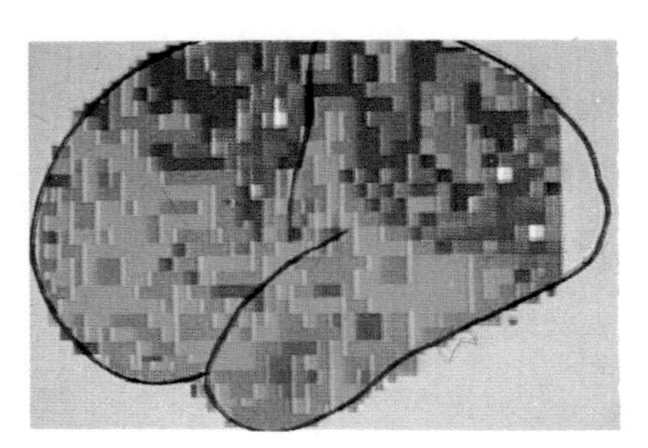

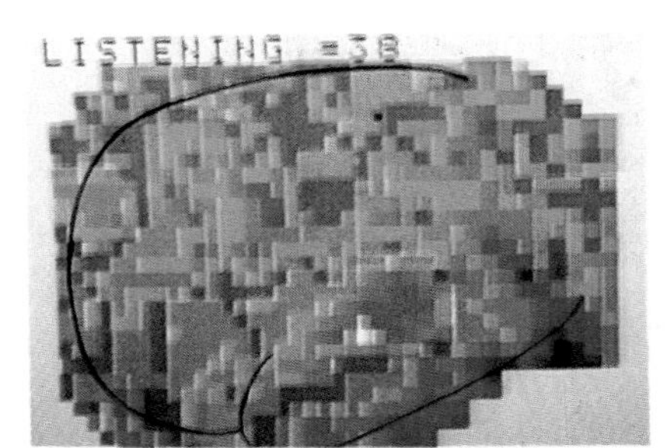

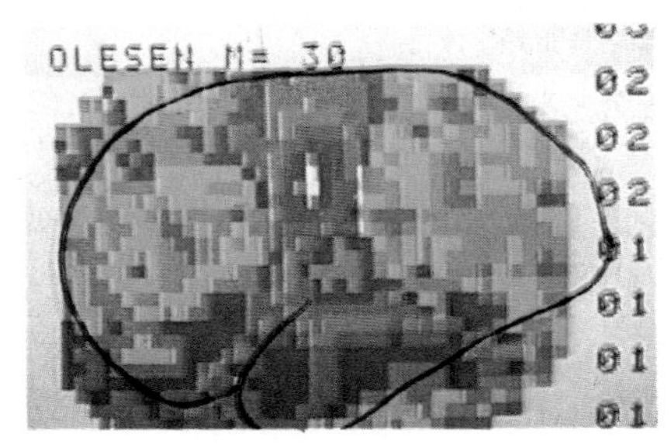

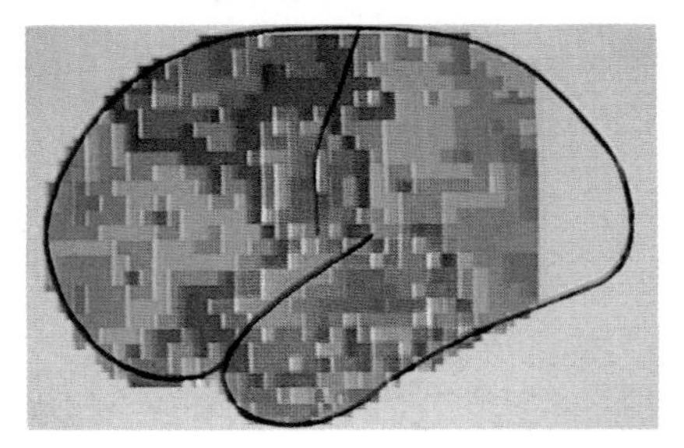

13 Heart and Circulation 346

14 Cardiac Output, Blood Flow, and Blood Pressure 384

15 Respiratory Physiology 426

16 Physiology of the Kidneys 476

17 The Immune System 512

18 The Digestive System 550

19 Regulation of Metabolism 588

20 Reproduction 618

Preface

Human physiology is a subject that is fascinating and useful to almost everyone. In addition to the inherent attraction of the subject, it is also a recommended or required course for students majoring in many and diverse areas. This is so for a very good reason: physiology is the major scientific foundation of medicine and all other technologies related to human health and physical performance. The scope of topics that can be covered in a human physiology course is thus large, yet each topic must be covered in sufficient detail to provide a firm basis for future expansion and application. The scientific study of body function need not diminish the student's initial fascination with "how the body works." On the contrary, the understanding of physiological mechanisms can instill a deeper appreciation for the awesome complexity and beauty of the body.

This text is designed to serve the needs of students taking an undergraduate human physiology course. Since most of these students do not have extensive science backgrounds, the beginning chapters introduce and review basic chemical and biological concepts that are needed to understand physiology. Physiological mechanisms are explained fully and in a straightforward style, and they are illustrated with flowcharts and reviewed in outline summaries whenever possible. Every effort has been made to help students integrate related concepts and mechanisms, and to see the relationships between anatomical structures and their functions. Health applications are discussed often to heighten interest, to improve understanding of physiological concepts, and to aid the transfer of knowledge to the areas of the student's career interests. The ability of the text to serve all of these functions is aided by extensive, but not obtrusive, use of a variety of pedagogical devices.

Pedagogical Aids

The straightforward writing style, clear explanations, and logical organization of the text help students learn the material presented. The ability of students to learn is further aided by a variety of pedagogical tools, including:

1. A list of behavioral objectives at the beginning of the chapter.
2. An outline, presented at the beginning of the chapter, of the topics covered.
3. Extensive use of tables, line drawings, flowcharts, photomicrographs, and beautifully rendered illustrations, many in full color.
4. Boldface and italics to highlight important concepts and scientific terms.
5. Study activities following every major heading within the chapter, so that students have a framework to organize the information presented as they progress through the chapter.
6. Boxed essays containing clinical applications within the text of the chapter.
7. Review activities at the end of each chapter, consisting of sample essay questions and objective questions.

8. A complete summary at the end of each chapter.
9. A list of recommended readings at the end of each chapter.
10. A glossary at the end of the book.
11. Answers to the objective questions in the chapter review activities are presented in the appendix.

Second Edition

All chapters in this text were rewritten and revised to include the latest scientific advances in knowledge about physiology and to present this knowledge in as clear a form as possible. Examples of such information that is new or updated in the second edition include, but are by no means limited to: hormone actions and interactions, synaptic plasticity, atrial natriuretic hormone, functions and uses of interleukin-2, macrophage-lymphocyte interactions, chorionic villus biopsy, renal transport mechanisms, transport of cholesterol and atherosclerosis, AIDS, causes of muscle fatigue, and brain mechanisms involved in memory and language ability.

In addition to the rewriting and updating of the text, the second edition has a number of new organizational features:

1. A new chapter on the CNS and brain function is presented (chapter 9).
2. Respiration is covered in one chapter (chapter 15) rather than in two, so that the functions of the respiratory system can be presented in a more integrated fashion.
3. Cardiovascular physiology is covered in two chapters (chapters 13 and 14) rather than in three, in order to achieve a more integrated coverage of this subject.
4. An introduction to the endocrine system is presented earlier in the text (chapter 8), immediately following the introduction to the nervous system, so that the major regulatory systems can be treated together. Endocrine control of metabolism and reproduction is covered at the end of the text, as in the first edition.
5. Boxed essays of clinical applications are presented within each chapter. In this way the flow of information about normal physiology is not interrupted, and the applications of physiological principles are highlighted.
6. Essay questions have been added to the objective questions for self-study at the end of each chapter.
7. Selected readings are presented at the end of each chapter rather than in an appendix.
8. Many new and redrawn figures are included in the second edition, and the use of full-color figures has been greatly expanded.

Supplementary Materials

Supplementary materials accompany this text and are designed to help instructors to organize and plan their course presentation and to more successfully integrate laboratory and lecture material.

A Laboratory Guide to Human Physiology: Concepts and Clinical Applications, Fourth Edition, by Stuart I. Fox. The laboratory guide has been updated and revised to include a variety of new pedagogical devices, such as a statement of the main idea, or concept, of each exercise and the learning objectives of each exercise. Boxed essays describing some aspects of the clinical significance of the procedures and concepts are now included. Laboratory reports at the end of each exercise contain spaces for data and an organized framework of self-quiz objective and essay questions. As in the previous editions of the lab guide, a wide variety of exercises, with concise explanations and clearly written procedures, is available to provide the maximum possible instructional support to the teaching and learning of physiology.

An *Instructor's Manual*, by Stuart I. Fox, is available to help instructors adapt the text to their own courses. This manual contains chapter objectives, outlines and summaries, study activities, answers to essay and objective questions in the chapter review activities, and guides to use of supplementary audiovisual materials. The Instructor's Manual also contains a test bank of additional objective questions that can be used by instructors to help construct examinations. An Instructor's Manual for the laboratory guide is also available.

A set of *acetate transparencies* is available to instructors who adopt this text. These transparencies are taken from selected illustrations in the text and are ideal supplements to classroom lecture.

wcb *QuizPak*, a student self-testing program that operates on an Apple® II*e* or IIc or IBM® PC microcomputer, is available to instructors who adopt this text.

wcb *TestPak*, a free, computerized testing service for generating examinations, is available to instructors who adopt this text.

Acknowledgements

Human Physiology, Second Edition, could not have been written without the support and encouragement of my colleagues at Los Angeles City College and Pierce College. I would like to thank Professors Robert J. Lyon, Lester S. Schneider, Larry Thouin, and James Rikel for their professional advice and example.

Many of the new figures and organizational changes in this second edition are the result of interaction with my friend, Dr. Kent M. Van De Graaff of Brigham Young University, who is coauthor of our combined text *Concepts of Human Anatomy and Physiology*, Wm. C. Brown Publishers, 1986.

I am deeply grateful to the editorial staff and production people at Wm. C. Brown, who have consistently labored to help produce a text that adheres to the highest values of college textbook publishing. It is a personal and professional joy to work with these people. I am also indebted to Elizabeth Blake for her dedicated and conscientious efforts in copyediting the manuscript.

The editors at Wm. C. Brown assembled a panel of accomplished physiology professors to review the manuscript. I am very grateful for their suggestions, criticisms, and encouragements. The reviewers were: Larry O. Miller, Moorpark College; F. W. Munz, University of Oregon; Ernest Schwab, Loma Linda University; Richard F. Weick, University of Western Ontario; and Gary L. Whitson, University of Tennessee at Knoxville.

Finally, but most importantly, I wish to extend my deepest appreciation to my wife Ellen and daughter Laura for their encouragement, patience, and support.

Human Physiology

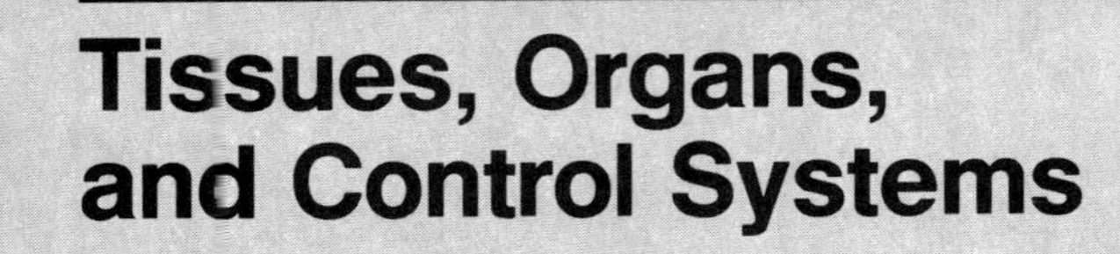

1 Tissues, Organs, and Control Systems

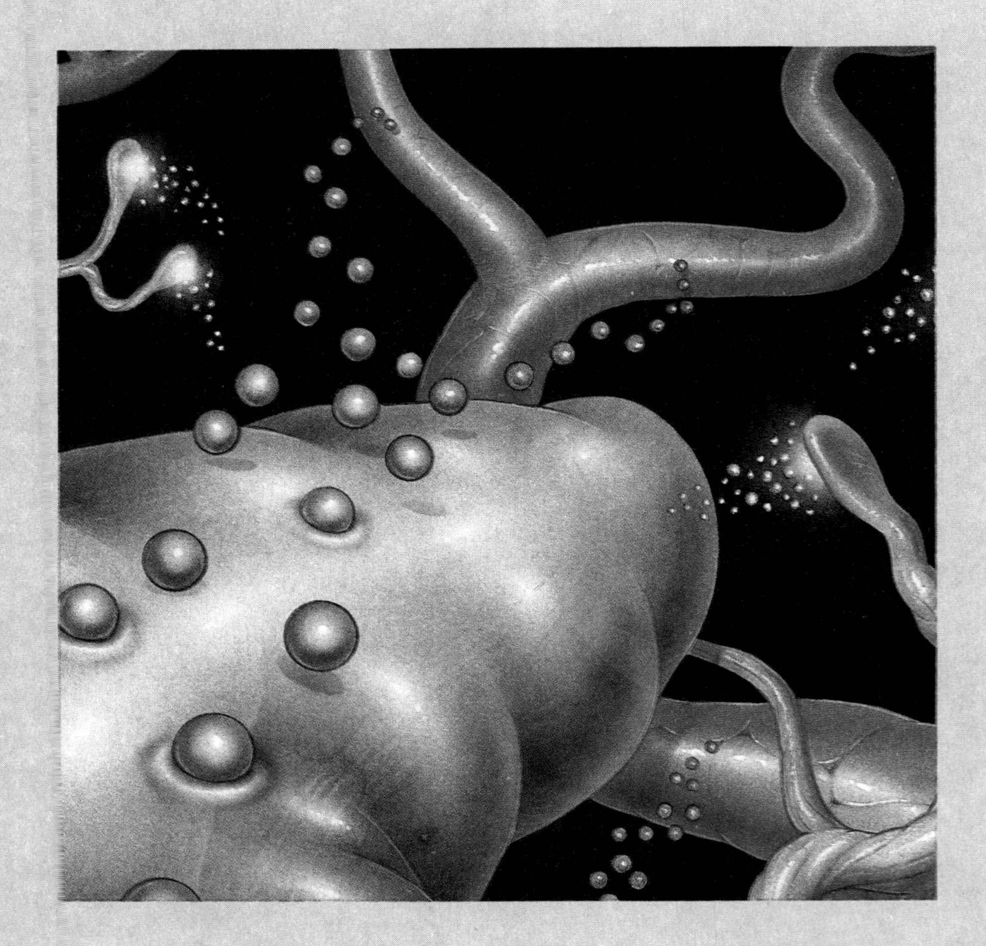

Objectives

By studying this chapter, you should be able to

1. describe, in a general way, the topics studied in physiology and the importance of physiology in modern medicine
2. describe the characteristics and explain the importance of the scientific method
3. list the four primary tissues and their subtypes and describe the distinguishing features of each primary tissue
4. relate the structure of each primary tissue to its functions
5. describe how the primary tissues are organized into organs, using the skin as an example
6. define homeostasis and describe how this concept is used in physiology and medicine
7. explain the nature of negative feedback loops and how these mechanisms act to maintain homeostasis
8. explain how antagonistic effectors help to maintain homeostasis
9. explain the nature of positive feedback loops and how these function in the body
10. explain the function of receptor proteins in the regulation of the nervous and endocrine systems
11. explain how negative feedback inhibition helps to regulate the secretion of hormones

Outline

Introduction

Physiology is the study of biological function—of how the body works, from cell to tissue, tissue to organ, organ to system, and of how the organism as a whole accomplishes particular tasks essential for life. In the study of physiology, the emphasis is on *mechanisms*—with questions that begin with the word *how* and answers that involve cause-and-effect sequences. These sequences can be woven into larger and larger stories that include descriptions of the structures involved (anatomy) and that overlap with the sciences of chemistry and physics.

The separate facts and relationships of these cause-and-effect sequences are derived empirically from experimental evidence. Explanations that seem logical are not necessarily true; they are only as valid as the data on which they are based, and they can change as new techniques are developed and further experiments are performed.

One standard technique for investigating the function of an organ is to observe what happens when it is surgically removed from an experimental animal or when its function is altered in a specific way. This study is often aided by "experiments of nature"—diseases—that involve specific damage to the function of an organ. The study of disease processes has thus aided our understanding of normal function, and the study of normal physiology has provided much of the scientific basis of modern medicine. This relationship is recognized by the Nobel Prize committee who award prizes in the category "Physiology or Medicine."

Scientific Method

All of the information in this text has been gained by the application of the scientific method. Although many different techniques are involved in the scientific method, all share three attributes: (1) confidence that the natural world, including ourselves, is ultimately explainable in terms we can understand; (2) descriptions and explanations of the natural world that are honestly based on observations and that could be modified or refuted by other observations; and (3) humility, that is, the willingness to accept the fact that we could be wrong. If further study should yield conclusions that refute all or part of an idea, the idea must be accordingly modified. In short, the scientific method is based on a confidence in our rational ability, honesty, and humility. Practicing scientists may not always display these attributes, but the validity of the large body of scientific knowledge that has been accumulated—as shown by the technological applications and the predictive value of scientific hypotheses—are ample testimony to the fact that the scientific method works.

Suppose you measured the resting pulse rate of one athlete and one sedentary person and found that the athlete had a lower resting pulse rate. If you were following the scientific method, you could state that the first person had a lower resting pulse than the second at the time you made the measurements, but no generalization would be scientifically justified by this data. If you repeated these measurements at different times and found that the differences remained, you could construct a scientific hypothesis that athletes have lower resting pulse rates than sedentary people. This hypothesis is scientific because it is *testable;* you could measure the pulse rates of 100 athletes and 100 sedentary people and see if statistically significant differences were obtained. If they were, you would be justified in saying that athletes, on the average, have lower resting pulse rates than sedentary people *based on your data.* You must still be open to the fact that you could be wrong (your measurement techniques may have been biased, or your sample of people may not have been representative of the general population) or that there could be alternative explanations for your results. Before your discovery would become generally accepted and be included in textbooks, other scientists would consistently have to replicate your results—for scientific theories are based on *reproducible* data.

It is quite possible that when others attempt to replicate your experiment their results will be slightly different from your own. They may then construct scientific hypotheses that the differences in resting pulse rate also depend on factors such as the nature of the exercise performed by the athlete or on nutrition or on genetic influences. When other scientists attempt to test these hypotheses, they will likely encounter new problems, requiring new explanatory hypotheses, which must be tested by additional experiments.

In this way, a large body of highly specialized information is gradually accumulated and a more generalized explanation can be formulated. This explanation will almost always be different from preconceived notions. People who follow the scientific method will then appropriately modify their concepts, realizing that their new ideas will probably have to be changed again in the future as additional experiments are performed.

New concepts that emerge from the scientific method are not always welcome; people often prefer comfortable, older ideas, particularly when these are part of a large, cherished belief structure. In this situation, the knowledge gained by the scientific method may be rejected, and the scientists involved villified, persecuted, or even murdered. Because the open communication of knowledge is a scientific tradition, however, those ideas that are closest to the truth have survived and eventually supplanted their competitors in the minds of most educated people.

Figure 1.1. Three skeletal muscle fibers showing the characteristic cross-striations.

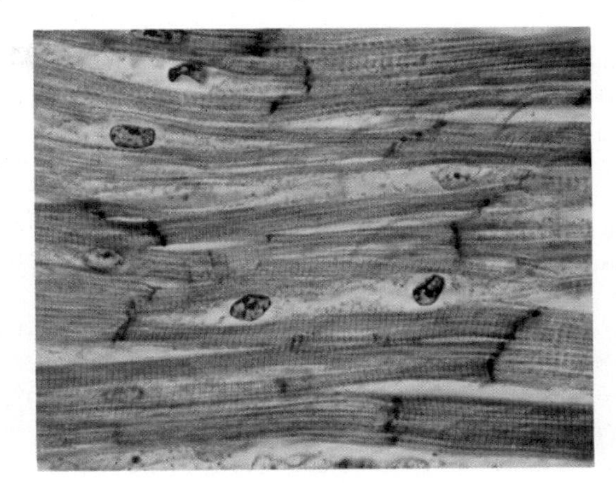

Figure 1.2. Human cardiac muscle. Notice the striated appearance and dark-staining intercalated discs.

1. *Define the science of physiology, and describe the relationship of this science to medicine.*
2. *Describe the characteristics of the scientific method.*

The Primary Tissues

Although physiology is the study of function, it is difficult to properly understand the function of the body without some knowledge of its anatomy, particularly at a microscopic level. The anatomy of specific organs will be discussed together with their functions in later chapters. In this section, the common "fabric" of all organs is described.

Cells that have similar functions are grouped into categories called *tissues.* The entire body is composed of only four types of tissues. These **primary tissues** include (1) muscle, (2) nervous, (3) epithelial, and (4) connective tissues. Groupings of these four primary tissues into anatomical and functional units are called *organs.* Organs, in turn, may be grouped together by common functions into *systems.*

Muscle

Muscle tissue is specialized for contraction. There are three types of muscles: (1) **skeletal muscle,** (2) **cardiac muscle,** and (3) **smooth muscle.** Skeletal muscle is often called *voluntary muscle,* because we have conscious control of its contraction without special training. Both skeletal and cardiac muscles are **striated;** they have striations, or stripes, that extend across the width of the muscle cell (figs. 1.1 and 1.2), and for this reason they have similar

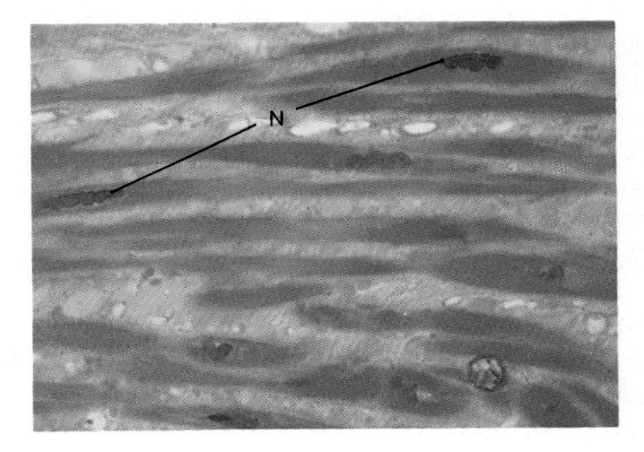

Figure 1.3. Photomicrograph of smooth muscle cells. Notice that these cells lack striations and contain single, centrally located nuclei.

mechanisms of contraction. Smooth muscle (fig. 1.3) lacks these cross-striations and has a different mechanism of contraction.

Skeletal Muscle. Skeletal muscles are generally attached by means of tendons to bones at both ends, so that contraction produces movements of the skeleton. There are, however, exceptions to this pattern; the tongue, superior portion of the esophagus, anal sphincter, and diaphragm are also composed of skeletal muscle.

Figure 1.4. The neuron. (*a*) A photomicrograph of nerve tissue. (*b*) A simplified diagram of a neuron and its principal parts. The arrows indicate the direction of the nerve impulse.

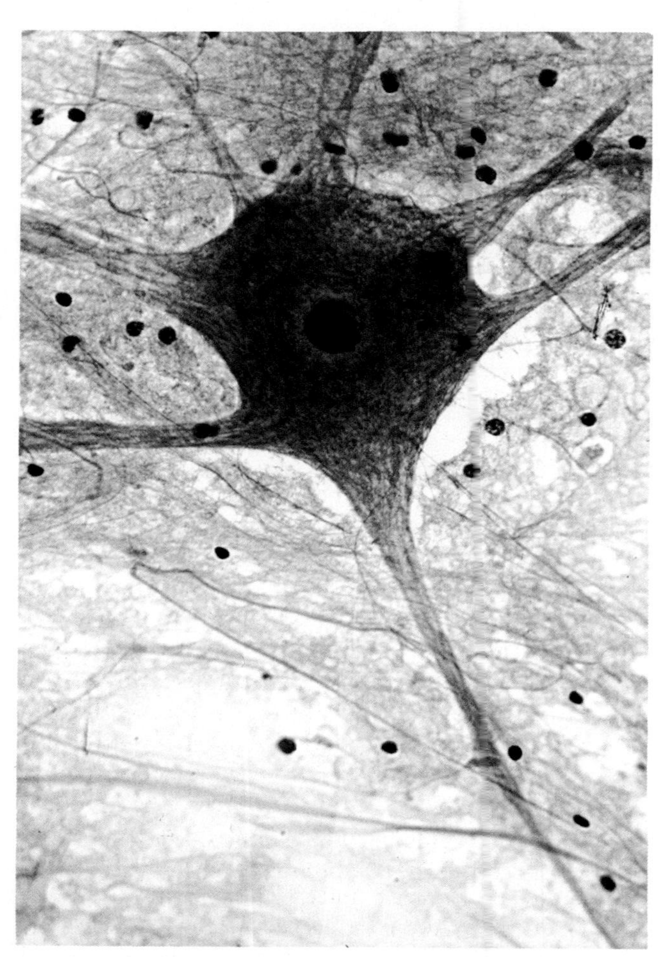

(a)

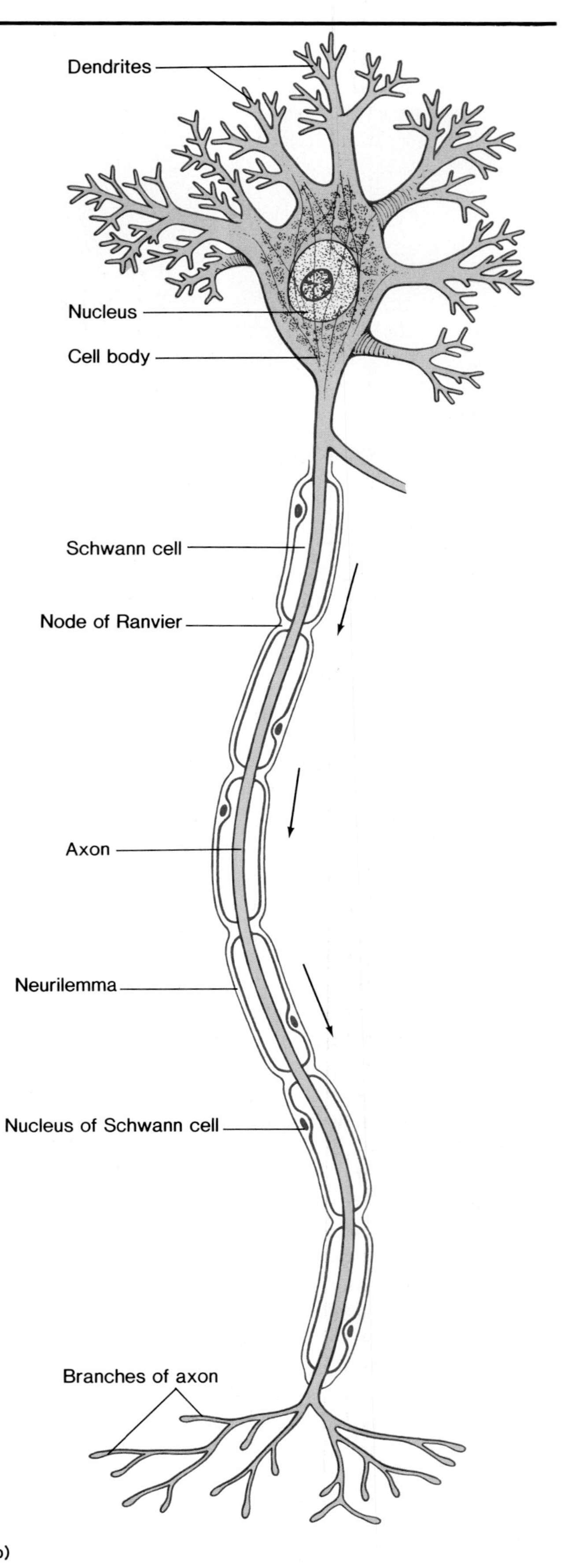

(b)

Since skeletal muscle cells are long and thin they are called **fibers,** or **myofibers** (*myo* = muscle). Despite their specialized structure and function, each myofiber contains structures common to all cells (nuclei, mitochondria, and other organelles, described in chapter 3).

Within a skeletal muscle the muscle fibers are arranged in bundles, and within these bundles the fibers extend in parallel from one end to the other of the bundle. The parallel arrangement of muscle fibers (seen in fig. 1.1) allows each fiber to be controlled individually: one can thus contract fewer or more muscle fibers and, in this way, vary the strength of the whole muscle's contraction. The ability to vary, or "grade," the strength of skeletal muscle contraction is obviously needed for proper control of skeletal movements.

Cardiac Muscle. Although cardiac muscle is striated, it has a very different appearance from skeletal muscle. Cardiac muscle is found only in the heart, where the **myocardial cells** are short, branched, and intimately interconnected to form a continuous fabric. Special areas of contact between adjacent cells stain darkly to show *intercalated discs* (fig. 1.2), which are characteristic of heart muscle.

The intercalated discs couple myocardial cells together mechanically and electrically. Unlike skeletal muscles, therefore, the heart cannot produce a graded contraction by varying the number of cells stimulated to contract. Because of the way it is constructed, the stimulation of one myocardial cell results in the stimulation of all other cells in the mass and a whole-hearted contraction.

Smooth Muscle. As implied by the name, smooth muscle cells (fig. 1.3) do not have the cross-striations characteristic of skeletal and cardiac muscle. Smooth muscle is found in the digestive tract, blood vessels, bronchioles (small air passages in the lungs), and in the urinary and reproductive systems. Circular arrangements of smooth muscle in these organs produce constriction of the *lumen* (cavity) when the muscle cells contract. The digestive tract also contains longitudinally arranged layers of smooth muscle. Rhythmic contractions of circular and longitudinal layers of muscle produce *peristalsis,* a process that pushes food from one end of the digestive tract to the other.

Nervous Tissue

Nervous tissue consists of nerve cells, or **neurons,** which are specialized for the generation and conduction of electrical events, and of **neuroglia,** which provide anatomical and functional support to the neurons.

Each neuron consists of three parts (fig. 1.4): (1) a *cell body,* which contains the nucleus and serves as the metabolic center of the cell; (2) *dendrites* (literally, "branches"), which are highly branched cytoplasmic extensions of the cell body that receive input from other neurons or from receptor cells; and (3) an *axon,* which is a single cytoplasmic extension of the cell body, which can be quite long (up to a few feet in length) and is specialized for conducting nerve impulses from the cell body to another neuron or an effector (muscle or gland) cell.

The neuroglia, composed of *neuroglial cells,* do not conduct impulses but instead serve to bind neurons together, modify the extracellular environment of the nervous system, and influence the nourishment and electrical activity of neurons. Neuroglial cells are about five times more abundant than neurons in the nervous system and, unlike neurons, maintain a limited ability to divide by mitosis throughout life.

Epithelial Tissue

Epithelial tissue consists of cells that form **membranes,** which cover and line the body surfaces, and of **glands** that are derived from these membranes (fig. 1.5). There are two categories of glands. *Exocrine glands* (*exo* = outside) secrete chemicals through a duct that leads to the outside of the membrane and thus to the outside of the body. *Endocrine glands* (*endo* = within) secrete chemicals called *hormones* into the blood.

Epithelial membranes are classified according to the number of their layers and the shape of the cells in the upper layer (table 1.1). Epithelial cells that are flattened in shape are *squamous,* those that are taller than they are wide are *columnar,* and those that are as wide as they are tall are *cuboidal* (fig. 1.6). Those epithelial membranes that are only one cell layer in thickness are known as *simple* membranes; those that are composed of a number of layers are *stratified* membranes.

A simple squamous membrane is adapted for diffusion and filtration; such a membrane lines all blood vessels, where it is known as an *endothelium.* A simple cuboidal epithelium lines the ducts of exocrine glands and part of the tubules of the kidney. A simple columnar epithelium lines the lumen of the stomach and intestine; this epithelium contains specialized unicellular glands, called *goblet cells,* which secrete mucus and are dispersed among the columnar epithelial cells. The columnar epithelial cells in the uterine (fallopian) tubes of females and in the respiratory passages contain numerous *cilia* (oarlike structures described in chapter 3), which can move in a coordinated fashion and aid the functions of these organs.

Figure 1.5. The formation of exocrine and endocrine glands from epithelial membranes.

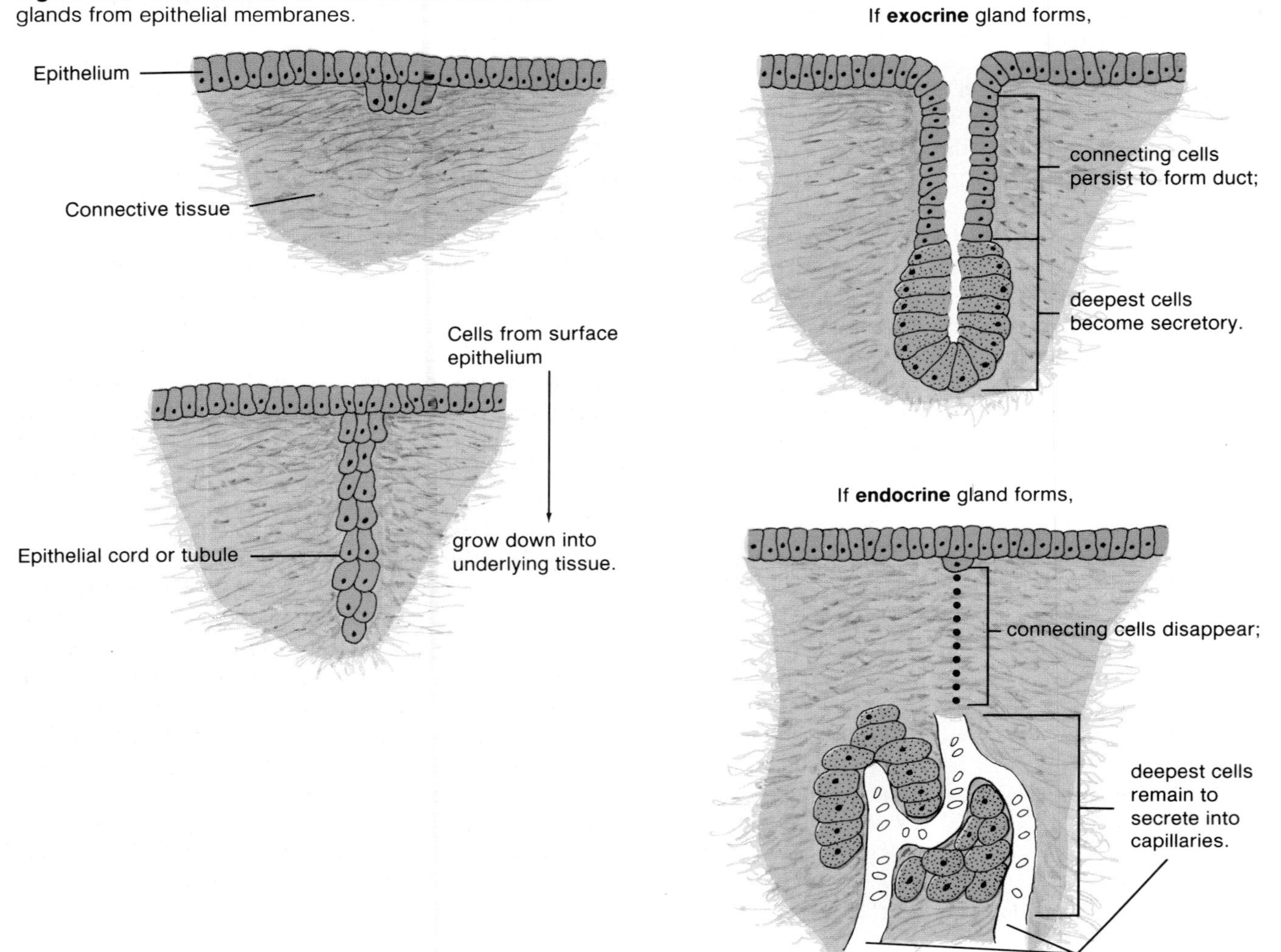

Table 1.1 Summary of epithelial tissues.

Type	Structure and Function	Location
Simple Epithelia	Single layer of cells; diffusion and filtration	Covering visceral organs, linings of lumina and body cavities
Simple squamous epithelium	Single layer of flattened, tightly bound cells; diffusion and filtration	Capillary walls, air sacs of lungs, covering visceral organs, linings of body cavities
Simple cuboidal epithelium	Single layer of cube-shaped cells; excretion, secretion, or absorption	Surface of ovaries; linings of kidney tubules, salivary ducts, and pancreatic ducts
Simple columnar epithelium	Single, nonciliated layer of tall, columnar-shaped cells; protection, secretion, and absorption	Lining of digestive tract
Simple ciliated columnar epithelium	Single, ciliated layer of columnar-shaped cells; transportive role through ciliary motion	Lining the lumen of the uterine tubes

Source: Van De Graaff, Kent M., *Human Anatomy.* © 1984 Wm. C. Brown Publishers, Dubuque, Iowa. All Rights Reserved. Reprinted by permission.

Figure 1.6. (*a*) Simple squamous, (*b*) simple cuboidal, and (*c*) simple columnar epithelial membranes. The tissue beneath each membrane is connective tissue.

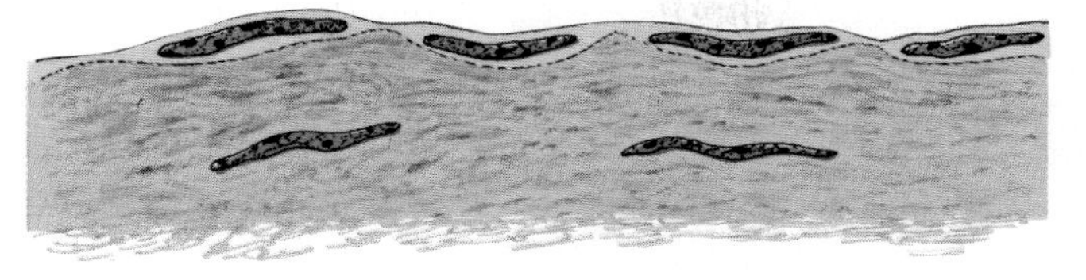

(a)

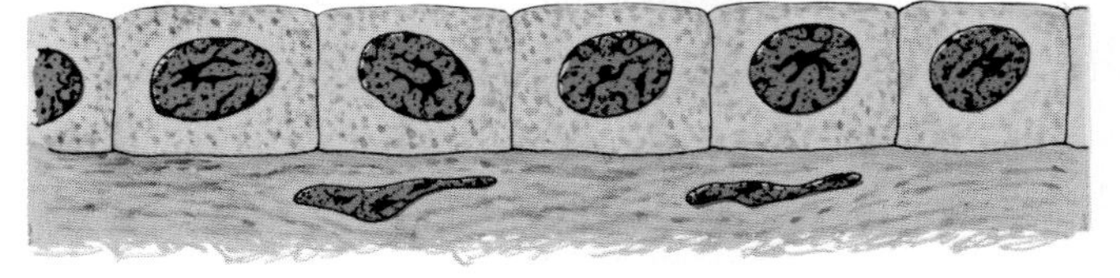

(b)

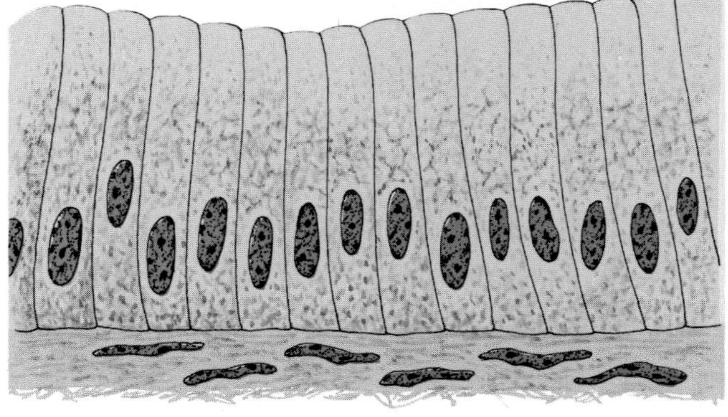

(c)

The epithelial covering of the esophagus that provides protection for this organ is a stratified squamous epithelium (fig. 1.7). This is a *nonkeratinized* membrane, and all layers of this epithelium consist of living cells. The *epidermis* of the skin, in contrast, is *keratinized,* or *cornified* (fig. 1.8). Since the epidermis is dry and exposed to the potentially desiccating effects of the air, the surface is covered with dead cells that are filled with a water-resistant protein known as *keratin.* This protective layer is constantly flaked off from the surface of the skin and therefore must be constantly replaced by the division of cells in the deeper layers of the epidermis.

The constant loss and renewal of cells is characteristic of epithelial membranes. The entire epidermis is completely replaced every two weeks; the stomach lining is renewed every two to three days. Examination of the cells that are lost, or "exfoliated," from the surface of the female genital tract is a common procedure in gynecology (as in the Pap smear).

In order to form a strong membrane that is effective as a barrier at the body surfaces, epithelial cells are very closely packed and are joined together by structures collectively called **junctional complexes.** There is no room for blood vessels between adjacent epithelial cells. The epithelium must therefore receive nourishment from the tissue beneath, which has large intercellular spaces that can accommodate blood vessels and nerves. This underlying tissue is called *connective tissue.* Epithelial membranes are attached to the underlying connective tissue by a layer of proteins and polysaccharides known as the **basement membrane,** which can only be observed under the microscope using specialized staining techniques.

Table 1.1 (Cont.)

Type	Structure and Function	Location
Pseudostratified ciliated columnar epithelium	Single layer of ciliated, irregularly shaped cells, many goblet cells; protection, secretion, ciliary movement	Lining of respiratory passageways
Stratified Epithelia	Two or more layers of cells; protection, strengthening, or distension	Epidermal layer of skin; linings of body openings, ducts, urinary bladder
Stratified squamous epithelium (keratinized)	Numerous layers, contains keratin, outer layers flattened and dead; protection	Epidermis of skin
Stratified squamous epithelium (nonkeratinized)	Numerous layers, lacks keratin, outer layers moistened and alive; protection and pliability	Linings of oral and nasal cavities, vagina, and anal canal
Stratified cuboidal epithelium	Usually two layers of cube-shaped cells; strengthen luminal walls	Larger ducts of sweat glands, salivary glands, and pancreas
Transitional epithelium	Numerous layers of rounded, nonkeratinized cells; distension	Luminal walls of ureters and urinary bladder

Figure 1.7. The stratified squamous nonkeratinized epithelial membrane of the vagina.

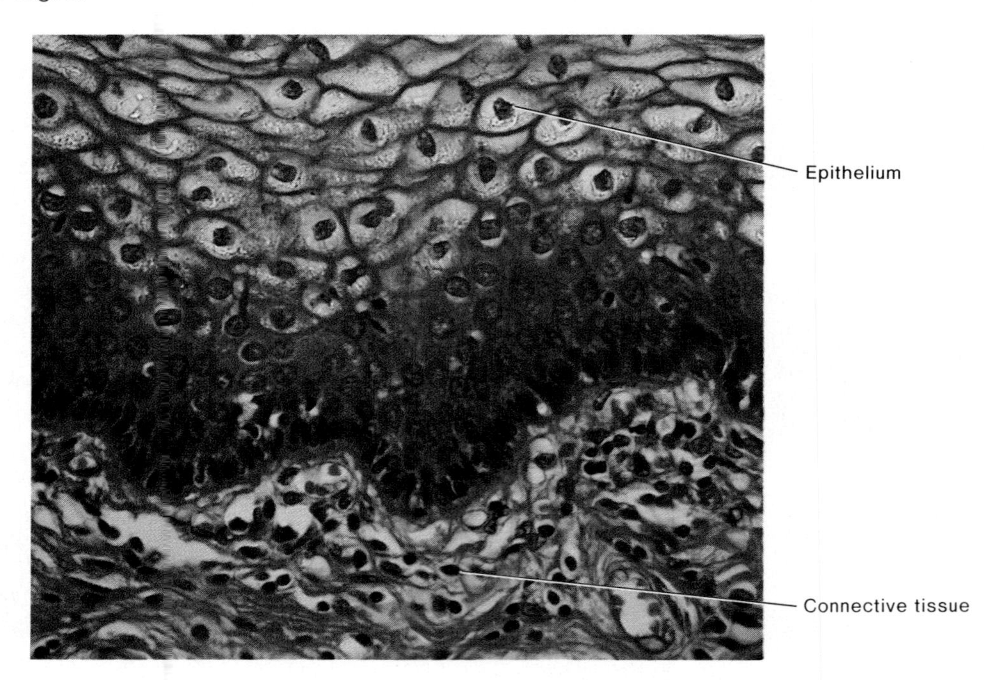

Figure 1.8. A section of skin showing the loose connective tissue dermis beneath the cornified epidermis. Loose connective tissue contains scattered collagen fibers in a matrix of protein-rich fluid. The intercellular spaces also contain cells and blood vessels.

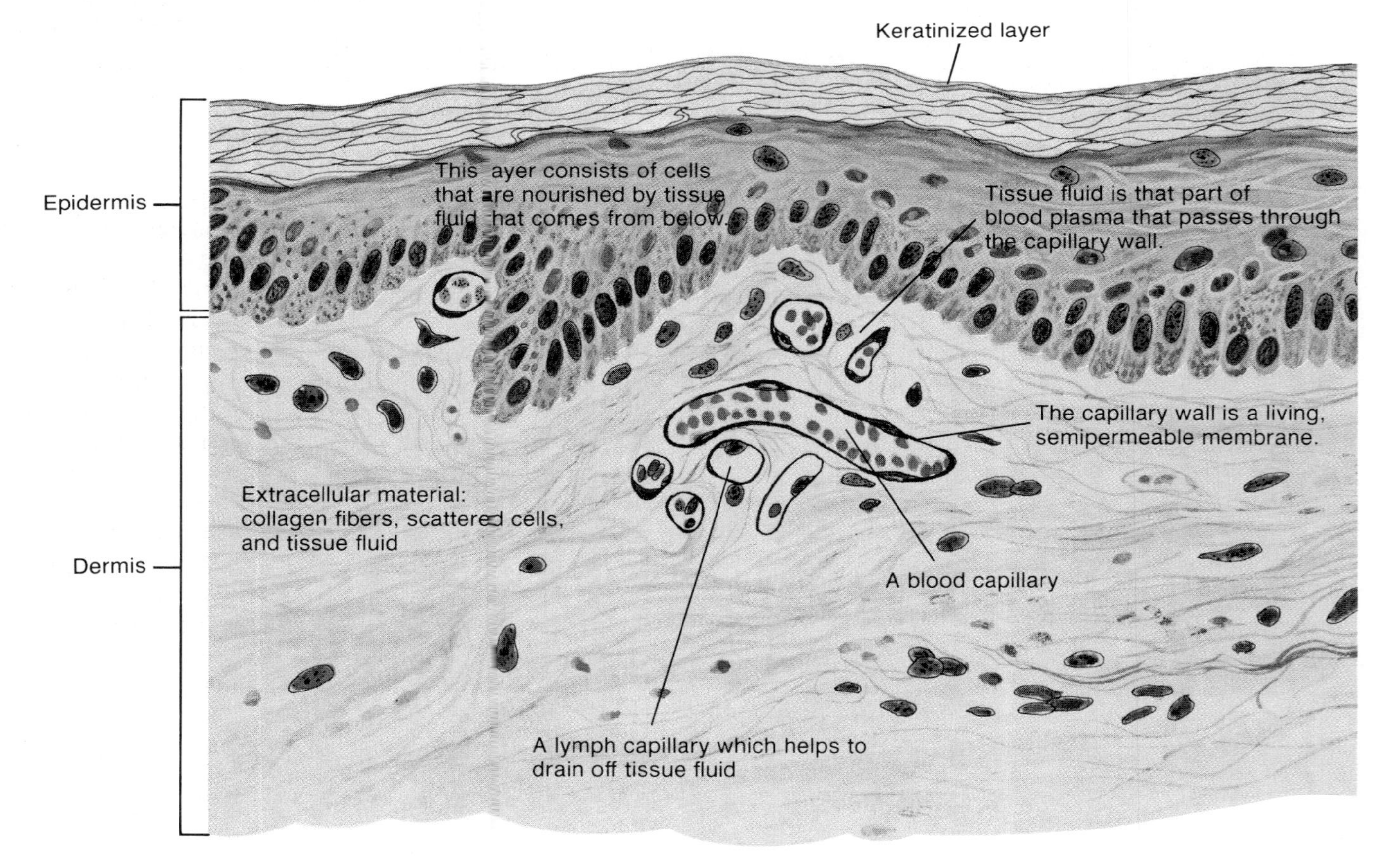

Figure 1.9. A photomicrograph of dense irregular connective tissue. Note the tightly packed, irregularly arranged collagen proteins.

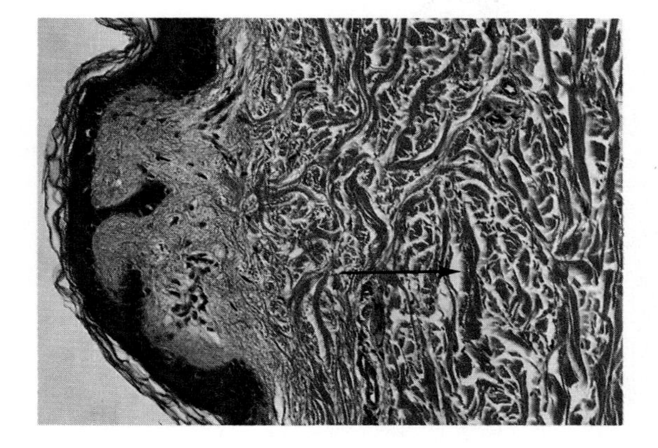

Figure 1.10. A photomicrograph of a tendon showing a dense, regular arrangement of collagen fibers.

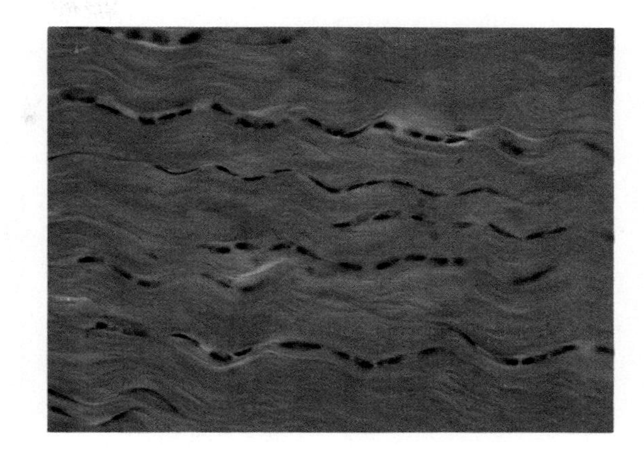

Table 1.2 Summary of connective tissue proper.

Type	Structure and Function	Location
Loose connective (areolar) tissue	Predominantly fibroblast cells with lesser amounts of collagen and elastin cells; binds organs, holds tissue fluids, diffusion	Surrounding nerves and vessels, between muscles, beneath the skin
Dense fibrous connective tissue	Densely packed collagen fibers; provides strong, flexible support	Tendons, ligaments, sclera of eye, deep skin layers
Elastic connective tissue	Predominantly irregularly arranged elastic fibers; supports, provides framework	Large arteries, lower respiratory tract, between vertebrae
Reticular connective tissue	Reticular fibers forming supportive network; stores, phagocytic	Lymph nodes, liver, spleen, thymus, bone marrow
Adipose connective tissue	Adipose cells; protects, stores fat, insulates	Hypodermis of skin, surface of heart, omentum, around kidneys, back of eyeball, surrounding joints

Source: Van De Graaff, Kent M., *Human Anatomy.*

Connective Tissue

Connective tissue is characterized by large amounts of extracellular material in the spaces between the connective tissue cells. This extracellular material may be of various types and arrangements and, on this basis, several types of connective tissues are recognized: (1) connective tissue proper; (2) cartilage; (3) bone; and (4) blood. Blood is usually classified as connective tissue because about half its volume is composed of an extracellular fluid known as *plasma.*

Connective tissue proper includes a variety of subtypes. An example of *loose connective tissue* (or *areolar tissue*) is the *dermis* of the skin (fig. 1.8). This connective tissue consists of scattered fibrous proteins called *collagen* and tissue fluid, which provides abundant space for the entry of blood and lymphatic vessels and nerve fibers. Another type of connective tissue proper is *dense fibrous connective tissue,* which contains densely packed fibers of collagen that may be in an irregular or a regular arrangement. Dense irregular connective tissue contains a meshwork of collagen fibers and forms the tough capsules and sheaths around organs (fig. 1.9). Tendons, which connect muscle to bone, and ligaments, which connect bones together at joints, are examples of dense regular connective tissue. This tissue contains a dense arrangement of collagen fibers that are parallel to each other (fig. 1.10). The characteristics of these and other types of connective tissue proper are summarized in table 1.2.

Cartilage consists of cells, called *chondrocytes,* surrounded by a semisolid ground substance that imparts elastic properties to the tissue. Cartilage is a type of supportive and protective tissue commonly called "gristle." It forms the precursor to many bones in the body and persists at the articular surfaces on the bones at all movable joints.

Figure 1.11. A diagram illustrating how a bone grows in width. Cells within the outer connective tissue covering of the bone (the periosteum) add new bone lamellae around blood vessels within the periosteum. This produces new haversian systems with a central canal containing the blood vessel and lined by the same connective tissue (which is now an endosteum).

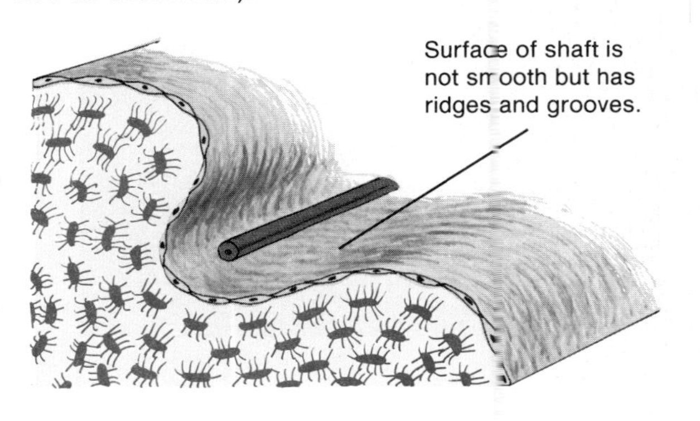

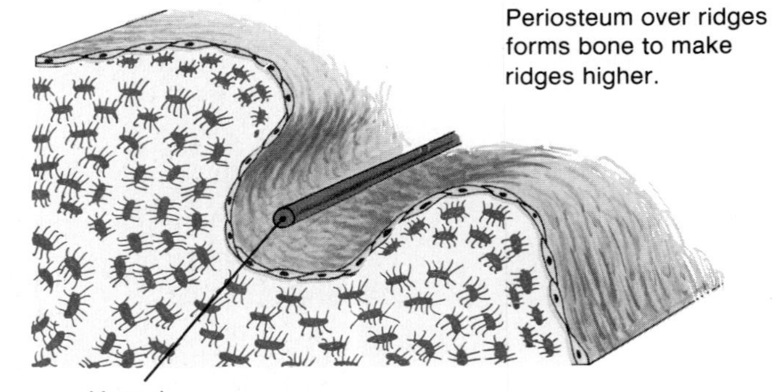

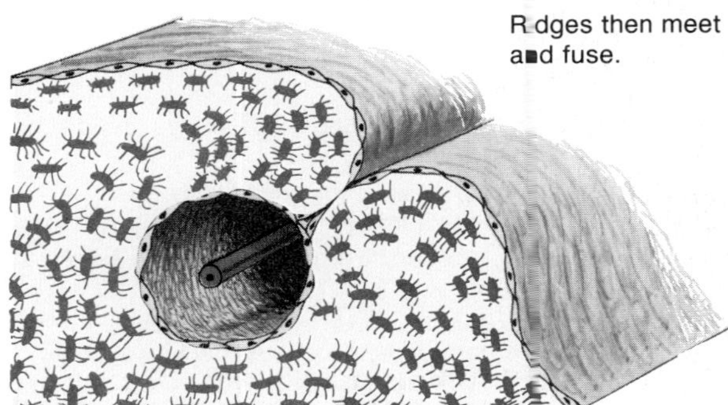

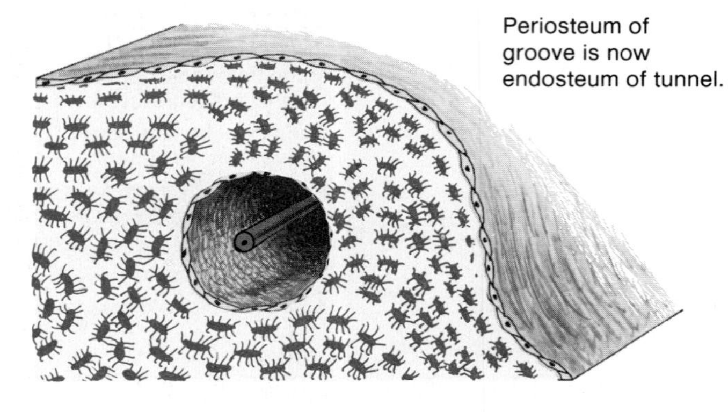

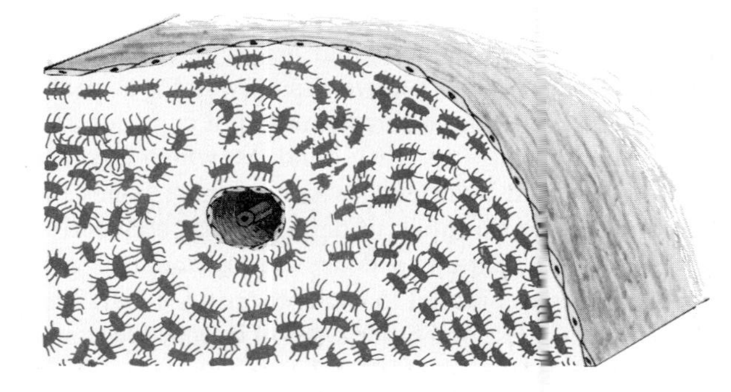

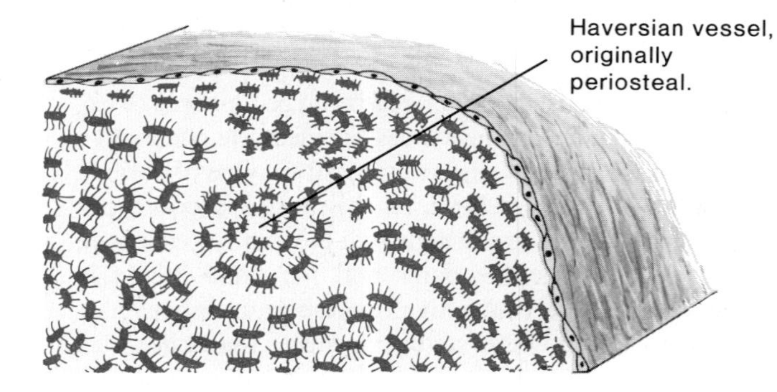

Figure 1.12. A cross section of a tooth showing pulp, dentin, and enamel. The cells that form dentin (odontoblasts) are located in the pulp, and their processes extend into the dentin-forming tubules. When these processes die, the tubules fill with air and appear dark. The odontoblasts in the pulp can form new reparative dentin to seal off the pulp from the dead tracts.

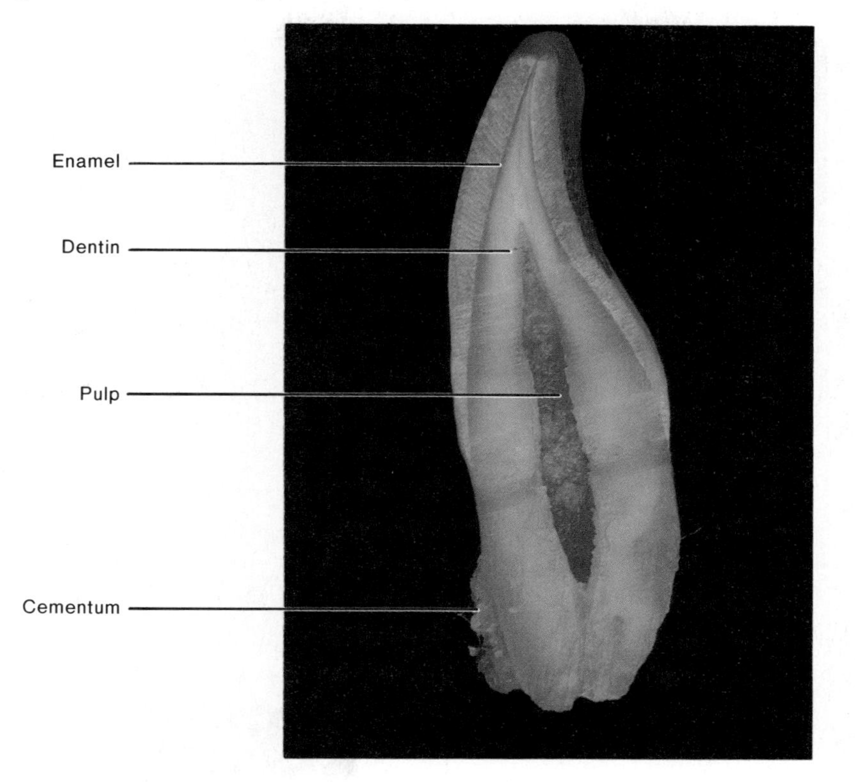

Bone is produced as concentric layers, or *lamellae,* of calcified material laid around blood vessels. The bone-forming cells, or *osteoblasts,* surrounded by their calcified products, become trapped within cavities (called *lacunae*). The trapped cells, which are now called *osteocytes,* remain alive because they are nourished by "lifelines" of cytoplasm that extend from the cells to the blood vessels in *canaliculi* (little canals). The blood vessels lie within central canals, surrounded by concentric rings of bone lamellae with their trapped osteocytes; these units are called *haversian systems* (fig. 1.11).

The *dentin* of a tooth (fig. 1.12) is similar in composition to bone, but the cells that form this calcified tissue are located in the pulp (composed of loose connective tissue). These cells send cytoplasmic extensions, called *dentinal tubules,* into the dentin. Dentin, like bone, is thus a living tissue which can be remodeled in response to stresses. The cells that form the outer *enamel* of a tooth, in contrast, are lost as the tooth erupts. Enamel is a highly calcified material, harder than bone or dentin, that cannot be regenerated; artificial "fillings" are therefore required to patch holes in the enamel.

1. *List the four primary tissues, and describe the distinguishing features of each type.*
2. *Name the three types of muscle tissue, and distinguish between them.*
3. *Describe the different types of epithelial tissues, and indicate their locations in the body.*
4. *Describe the different types of connective tissues, and explain how they differ from each other in their content of extracellular material.*

Figure 1.13. A diagram showing the structure of the skin. Notice that all four types of primary tissues are present.

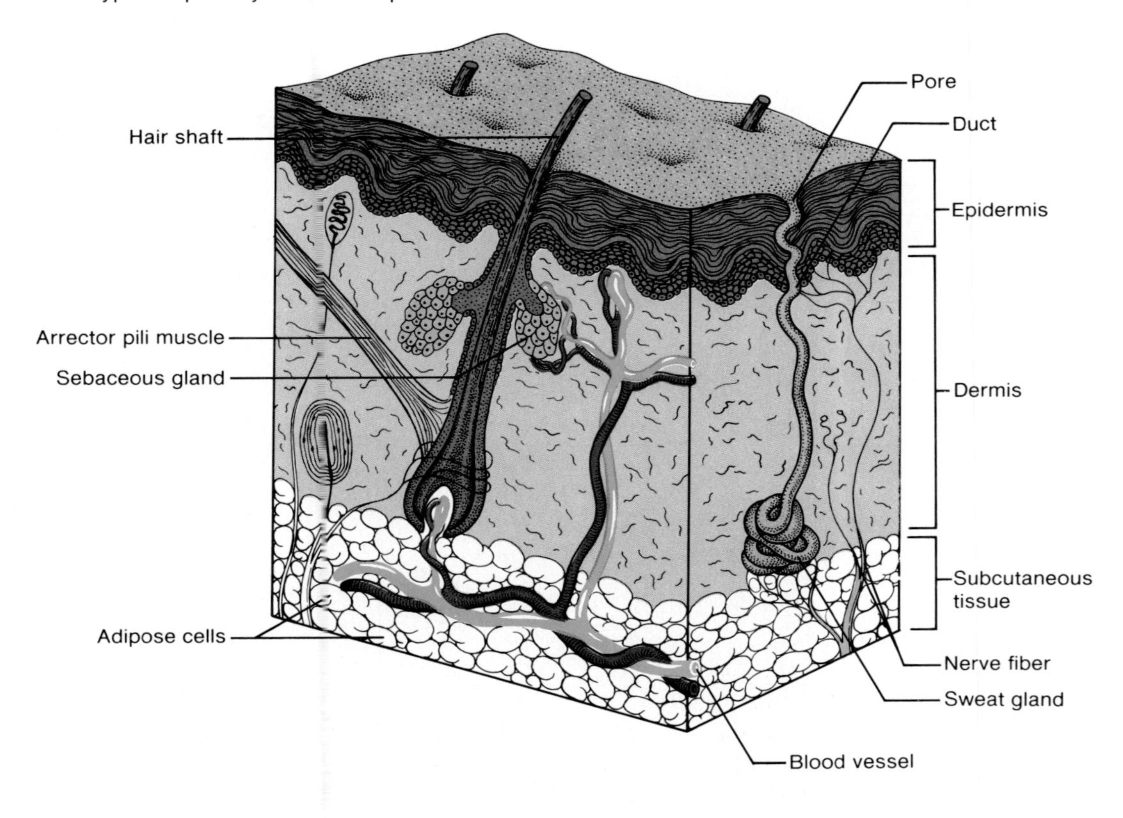

Organs and Systems

An organ is a structure composed of at least two, and usually all four, types of primary tissues. Each primary tissue contributes to the structure and function of each organ. The largest organ in the body, in terms of its surface area, is the skin. The numerous functions of the skin serve to illustrate how primary tissues cooperate in the service of organ physiology.

An Example of an Organ: The Skin

The cornified epidermis protects the skin (fig. 1.13) against water loss and against invasion by disease-causing organisms. Invaginations of the epithelium into the underlying connective tissue dermis creates the exocrine glands of the skin. These include hair follicles (which secrete the proteins that compose hair), sweat glands, and sebaceous glands. The secretion of sweat glands cools the body by evaporation and produces odors that, at least in lower animals, serve as sexual attractants. Sebaceous glands secrete oily sebum into hair follicles, where it is transported to the surface of the skin (unless these ducts are blocked, in which case blackheads are produced). Sebum lubricates the cornified surface of the skin, helping to prevent it from drying and cracking (as in chapped lips—there are no sebaceous glands in the lips; one must, therefore, moisten the lips periodically with the tongue).

The skin is nourished by blood vessels within the dermis. In addition to blood vessels, the dermis contains wandering white blood cells and other types of cells that protect against invading disease-causing organisms, as well as nerve fibers and fat cells. Most of the fat cells, however, are grouped together to form the *hypodermis* (a layer beneath the dermis). Although fat cells are a type of connective tissue, masses of fat deposits throughout the body—such as subcutaneous fat—are referred to as **adipose tissue.**

Figure 1.14. The pacinian corpuscle is a receptor for deep pressure. It consists of epithelial cells and connective tissue proteins that form concentric layers around the ending of a sensory nerve.

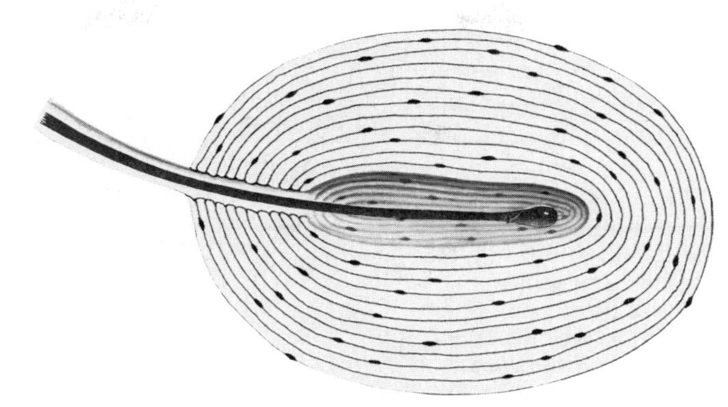

Sensory nerve endings within the dermis mediate the cutaneous sensations of touch, pressure, heat, cold, and pain. Some of these sensory stimuli directly affect the sensory nerve endings. Others act via sensory structures derived from nonneural primary tissues. The pacinian corpuscles in the dermis of the skin (fig. 1.14), for example, monitor sensations of pressure. Motor nerve fibers in the skin stimulate effector organs, resulting in, for example, the secretions of exocrine glands and contractions of the arrector pili muscles, which attach to hair follicles and surrounding connective tissue (producing goose bumps). The degree of constriction or dilation of cutaneous blood vessels—and therefore the rate of blood flow—is also regulated by motor nerve fibers.

The epidermis itself is a dynamic structure that can respond to environmental stimuli. The rate of its cell division—and consequently the thickness of the cornified layer—increases under the stimulus of constant abrasion. This produces calluses. The skin also protects itself against the dangers of ultraviolet light by increasing its production of *melanin* pigment, which absorbs ultraviolet light while producing a tan. In addition, the skin is an endocrine gland that produces and secretes vitamin D (derived from cholesterol under the influence of ultraviolet light), which functions as a hormone.

The architecture of most organs is similar to that of the skin. Most are covered by an epithelium immediately over a connective tissue layer. The connective tissue contains blood vessels, nerve endings, scattered cells for fighting infection, and possibly glandular tissue as well. If the organ is hollow—as in the digestive tract or in blood vessels—the lumen is also lined with an epithelium immediately over a connective tissue layer. The presence, type, and distribution of muscular and nervous tissue varies in different organs.

Systems

Organs that are located in different regions of the body and that perform related functions are grouped into **systems.** These include the nervous system, endocrine system, cardiovascular system, respiratory system, excretory system, reproductive system, digestive system, and immune system. By means of numerous regulatory mechanisms, these systems work together to maintain the life and health of the entire organism.

1. *Describe the location of each type of primary tissue in the skin.*
2. *Describe the functions of nerve, muscle, and connective tissue in the skin.*
3. *Describe the functions of the epidermis, and explain why this tissue is called "dynamic."*

Homeostasis and Feedback Control

Over a century ago the French physiologist Claude Bernard observed that the *milieu intérieur* ("internal environment") remains remarkably constant despite changing conditions in the external environment. In a book entitled *The Wisdom of the Body* (published in 1932), Walter

Table 1.3 Approximate normal ranges for measurements of some blood values.

Measurement	Normal Range	Measurement	Normal Range
Arterial pH	7.35–7.43	Urea	12–35 mg/100 ml
Bicarbonate	21.3–28.5 mEq/L	Amino acids	3.3–5.1 mg/100 ml
Sodium	136–151 mEq/L	Protein	6.5–8.0 g/100 ml
Calcium	4.6–5.2 mEq/L	Total lipids	350–850 mg/100 ml
Oxygen content	17.2–22.0 ml/100 ml	Glucose	75–110 mg/100 ml

Figure 1.15. A rise in some factor of the internal environment (↑X) is detected by a sensor. The sensor activates an effector (generally, nerves or hormones), which causes a decrease in this factor (↓X). In this way the factor returns to its initial level and homeostasis is maintained. Completion of the negative feedback loop is shown by a dotted arrow and a negative sign. The circled numbers indicate the sequence of events.

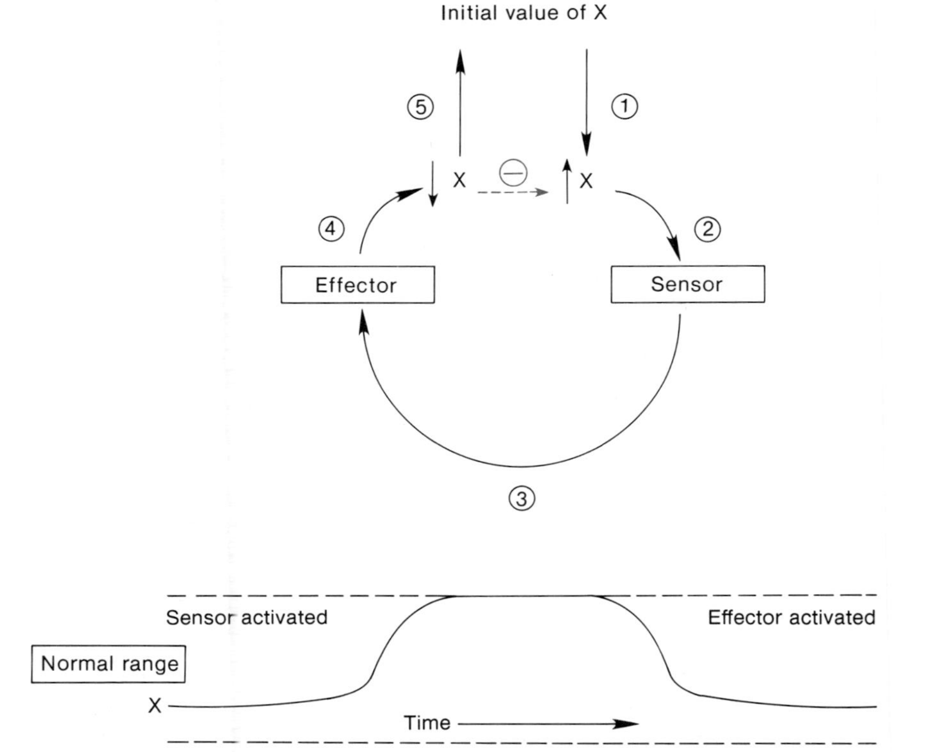

Cannon coined the term **homeostasis** to describe this internal constancy. Cannon further suggested that mechanisms of physiological regulation exist for one purpose—the maintenance of internal constancy.

The concept of homeostasis has been of inestimable value in the study of anatomy and physiology, because it allows diverse regulatory mechanisms to be understood in terms of their "why" as well as their "how." The concept of homeostasis also provides a major foundation for medical diagnostic procedures. When a particular measurement of the internal environment, such as blood measurements (table 1.3), deviates significantly from the normal range of values, it can be concluded that homeostasis is not maintained and the person is sick. A number of such measurements, combined with clinical observations, may allow the particular defective mechanism to be identified.

Negative Feedback Loops

In order for internal constancy to be maintained, the body must have *sensors* that are able to detect deviations from the normal range of particular internal conditions and *effectors* that are able to reduce and ultimately to reverse

Figure 1.16. A negative feedback loop in which a decrease in some factor of the internal environment is compensated by the actions of an effector. Compare this diagram with that shown in figure 1.15. The circled numbers indicate the sequence of events.

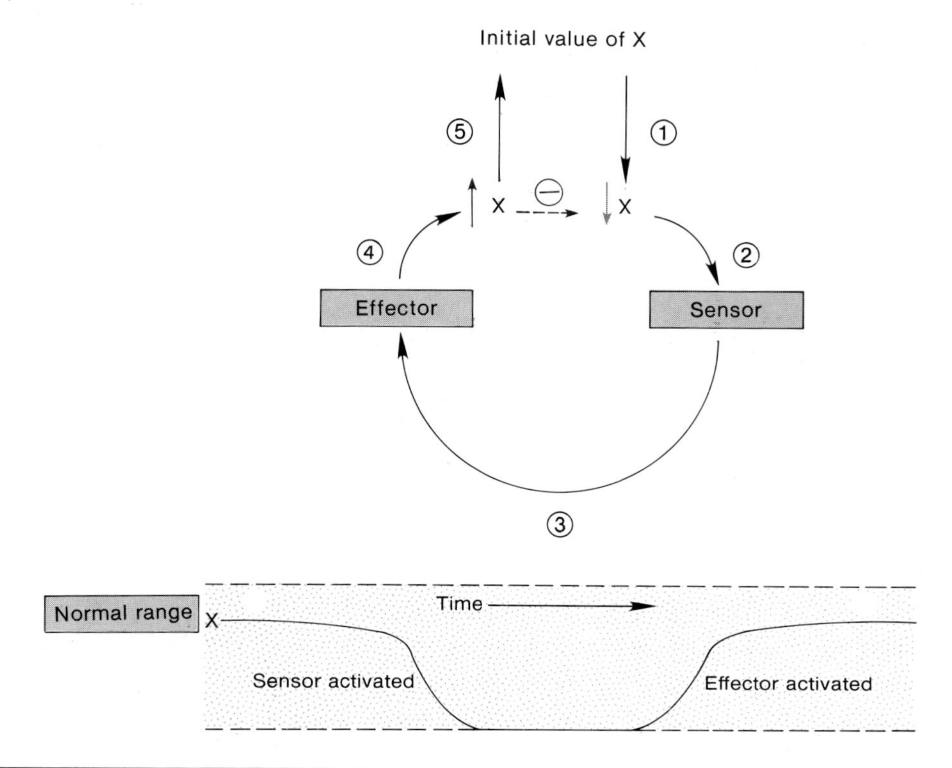

Figure 1.17. Negative feedback loops (indicated by negative signs) maintain a state of dynamic constancy within the internal environment.

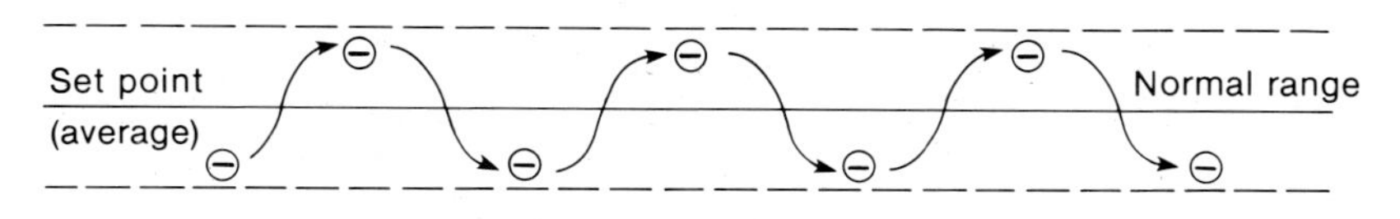

these deviations. A detectable change in the internal environment, thus, results in changes in the reverse direction. Internal constancy is maintained by such **negative feedback loops** (fig. 1.15).

An increase in a value (indicated as X in figure 1.15) above its normal range is detected by a sensor; the sensor in turn activates an effector that causes X to decrease. This negative feedback loop prevents X from rising too far above its normal range. It is important to realize, however, that even when X is in its normal range there is some activity of the effector. Even under normal, resting conditions, to be more specific, there is always some spontaneous nerve activity and some hormone secretion. If an increase in the activity of an effector causes a decrease in X (as shown in figure 1.15), a decrease in the activity of the same effector will cause an increase in X. Changes from the normal range in either direction can be compensated by reverse changes in effector activity (fig. 1.16).

Homeostasis is best conceived as a state of **dynamic constancy,** rather than simply a state of absolute constancy. The values of particular measurements of the internal environment fluctuate above and below the average, or *set point,* of these measurements. Values are, therefore, at their set point only in passing (fig. 1.17), and regulatory mechanisms are constantly active to greater or lesser degrees in ongoing negative feedback loops.

Antagonistic Effectors. Most factors in the internal environment are controlled by several effectors, which often have antagonistic actions. Control by antagonistic effectors is sometimes described as "push-pull." This affords a finer degree of control than could be achieved by simply switching one effector on and off. Normal body temperature, for example, is maintained about a set point of 37° C by the antagonistic effects of sweating, shivering, and other mechanisms (fig. 1.18).

Figure 1.18. A simplified scheme by which body temperature is maintained within the normal range (with a set point of 37° C) by two antagonistic mechanisms—shivering and sweating. Shivering is induced when the body temperature falls too low and gradually subsides as the temperature rises. Sweating occurs when the body temperature is too high and diminishes as the temperature falls. Most aspects of the internal environment are regulated by the antagonistic actions of different effector mechanisms.

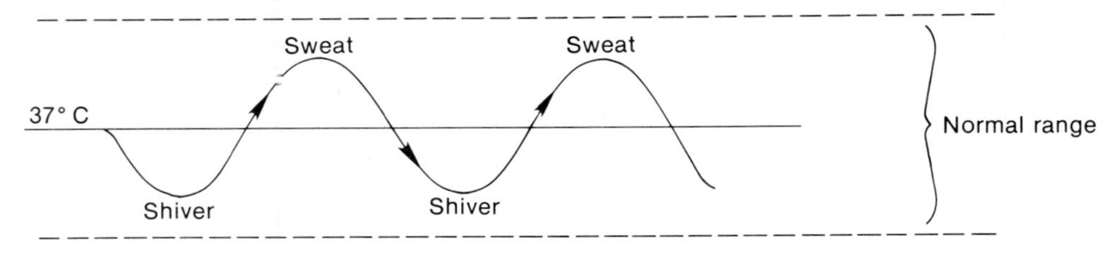

The blood concentrations of glucose, calcium, and other substances are regulated by negative feedback loops that involve hormones which promote opposite effects. While insulin, for example, lowers blood glucose, other hormones raise the blood glucose concentration. The heart rate, similarly, is controlled by nerve fibers that produce opposite effects: stimulation of one group of nerve fibers increases heart rate, and stimulation of another group slows the heart rate.

Positive Feedback

Constancy of the internal environment is maintained by effectors, which act to compensate the change that served as the stimulus for their activation; in short, by negative feedback loops. A thermostat, for example, maintains a constant temperature by increasing heat production when it is cold and decreasing heat production when it is warm. The opposite occurs during **positive feedback**—in this case, the action of effectors *amplifies* those changes that stimulated the effectors. A thermostat that works by positive feedback, for example, would increase heat production in response to a rise in temperature.

It is clear that homeostasis must ultimately be maintained by negative rather than by positive feedback mechanisms. The effectiveness of some negative feedback loops, however, is increased by positive feedback mechanisms that amplify the actions of a negative feedback response. Blood clotting, for example, occurs as a result of a sequential activation of clotting factors; the activation of one clotting factor results in activation of many in a positive feedback, avalanchelike, manner. In this way, a single change is amplified to produce a blood clot. Formation of the clot, however, can prevent further loss of blood and thus represents the completion of a negative feedback loop.

Neural and Endocrine Regulation

Homeostasis is maintained by two general categories of regulatory mechanisms: (1) those that are **intrinsic,** or "built-in," to the organ; these are usually due to the effects of chemicals that act within the organs that produce them; and (2), those that are **extrinsic;** that is, regulation of an organ by the nervous and endocrine systems.

The endocrine system functions closely with the nervous system in regulating and integrating body processes and in maintaining homeostasis. The nervous system regulates body activities through the action of electrochemical nerve impulses, which are transmitted by the axons of nerve cells (neurons). Nerve impulses generally produce rapid but brief responses. In contrast, the glands of the endocrine system secrete chemical regulators (hormones) into the blood, which produce slower but more long lasting responses in their target organs.

Common Aspects of Neural and Endocrine Regulation. It may be thought that neural regulation is electrical and therefore distinct from the chemical nature of endocrine control systems. This idea is incorrect. Electrical nerve impulses are in fact chemical events produced by the diffusion of ions through the neuron cell membrane. Interestingly, the action of some hormones (such as insulin) is accompanied by ion diffusion and electrical changes in the target cells. Also, most nerve fibers stimulate the cells they innervate through the release of a chemical **neurotransmitter.** Neurotransmitters differ from hormones in that they do not travel in the blood but instead diffuse only a very short distance from the nerve ending to the target cell. In other respects, however, the actions of neurotransmitters and hormones are very similar.

Indeed, many polypeptide hormones, including those secreted by the pituitary gland and by the digestive tract, have been discovered in the brain. In certain locations in

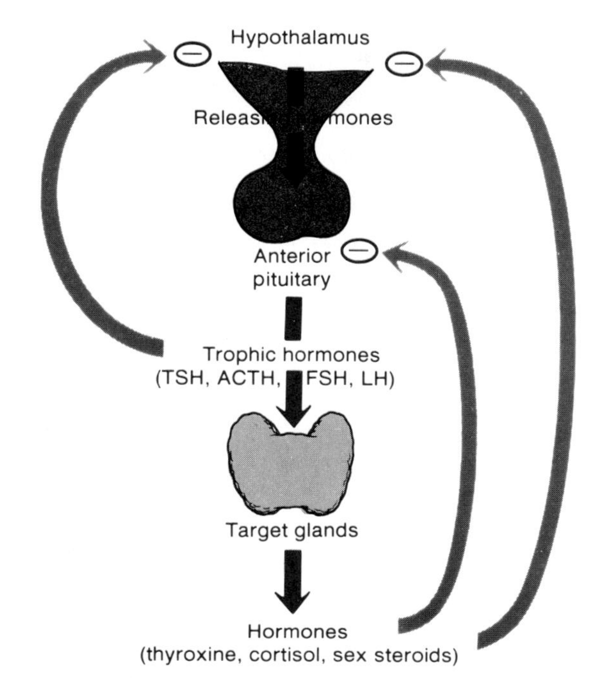

Figure 1.19. The regulation of hormone secretion by the hypothalamus, anterior pituitary, and target glands (thyroid, adrenal cortex, and gonads). Hormones secreted by the target glands exert negative feedback inhibition on the release of hormones from the hypothalamus and anterior pituitary.

the brain, some of these compounds are produced and secreted as hormones (the brain is an endocrine gland). In other brain locations, some of these compounds apparently serve as neurotransmitters. The discovery of some of these polypeptides in unicellular organisms, which of course lack a nervous and endocrine system, suggests that these regulatory molecules appeared early in evolution and became incorporated into the function of nerve and endocrine tissue as these systems evolved. This fascinating theory would help to explain, for example, why insulin, a polypeptide hormone produced in the pancreas of vertebrates, is found in neurons of invertebrates (which lack a separate endocrine system).

Regardless of whether a particular chemical is acting as a neurotransmitter or as a hormone, in order for it to function in physiological regulation: (1) target cells must have specific **receptor proteins** that combine with the chemical; (2) the combination of the regulator molecule with its receptor proteins must cause a specific sequence of changes in the target cells; and (3) there must be a mechanism to rapidly turn off the action of the regulator; without an "off switch," physiological control is impossible. This last process involves rapid removal and/or chemical inactivation of the regulator molecules.

Feedback Control of Hormone Secretion

Hormones are secreted in response to specific chemical stimuli. The presence of proteins in the stomach, for example, stimulates the secretion of gastrin, and a rise in plasma glucose concentration stimulates insulin secretion from the islets of Langerhans in the pancreas. Hormones are also secreted in response to neurotransmitters released from nerve endings and to stimulation by other hormones.

The nervous system helps to regulate hormone secretion in a variety of ways. Autonomic nerves, for example, stimulate the secretion of epinephrine from the adrenal medulla and also participate in the regulation of insulin and glucagon secretion from the islets of Langerhans in the pancreas. Hormonal secretion by the adrenal cortex, thyroid, and gonads is stimulated by hormones secreted by the anterior pituitary gland. Secretions of the anterior pituitary, in turn, are regulated by hormones released by neurons in the hypothalamus of the brain. This control is not one-way; many neural centers in the brain are affected by the actions of hormones (fig. 1.19).

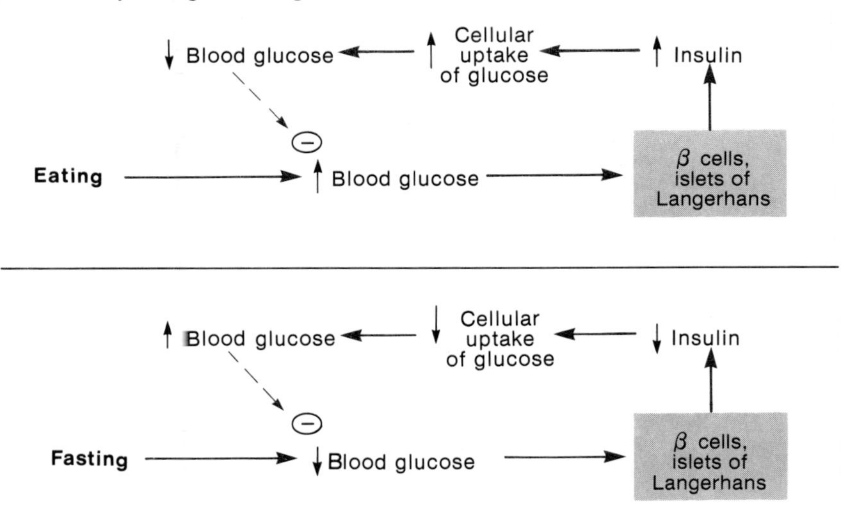

Figure 1.20. The negative feedback control of insulin secretion by changes in the blood glucose concentration. Dashed arrows and negative signs indicate that negative feedback loops compensate the initial changes in blood glucose concentrations produced by eating or fasting.

The secretion of hormones is controlled by inhibitory as well as by stimulatory influences. The effect of a given hormone's action can inhibit its own secretion. The secretion of insulin—which acts to lower the plasma glucose concentration—is stimulated by a rise in glucose concentration, for example, and is inhibited by a fall in blood glucose. The lowering of blood glucose levels by insulin thus has an inhibitory feedback effect on further insulin secretion. This closed loop control system is called **negative feedback inhibition** (see fig. 1.20).

1. *Define homeostasis, and describe how this concept can be used to understand physiological control mechanisms.*
2. *Describe the meaning of the term negative feedback, and explain how it contributes to homeostasis. Illustrate this concept by drawing a negative feedback loop.*
3. *Describe positive feedback, and explain how this process functions in the body.*
4. *Describe the functions of receptor proteins, and explain how they contribute to the function of the nervous and endocrine systems.*
5. *Explain how the secretion of hormones is controlled by the nervous system.*
6. *Explain how the secretion of a hormone is controlled by negative feedback inhibition. Illustrate this process with an example.*

Summary

Introduction p. 4

I. Physiology is the study of how cells, tissues, and organs function.
  A. In the study of physiology, cause-and-effect sequences are emphasized.
  B. Knowledge of physiological mechanisms is deduced from data obtained experimentally.

II. All of the information in this book has been gained by applications of the scientific method. This method has three essential characteristics.
  A. It is assumed that the subject under study can ultimately be explained in terms we can understand.
  B. Descriptions and explanations are honestly based on observations of the natural world and can be changed when warranted by new observations.
  C. Humility is an important characteristic of the scientific method; the scientist must be willing to change his or her theories when warranted by the weight of the evidence.

The Primary Tissues, Organs, and Systems p. 5

I. The body is composed of four primary tissues: muscular, nervous, epithelial, and connective tissues.
  A. There are three types of muscle tissue: skeletal, cardiac, and smooth muscle.
    1. Skeletal and cardiac muscle are striated.
    2. Smooth muscle is found in the walls of the internal organs.
  B. Nervous tissue is composed of neurons and neuroglial cells.
    1. Neurons are specialized for the generation and conduction of electrical impulses.
    2. Neuroglial cells provide anatomical and functional support to the neurons.
  C. Epithelial tissue includes membranes and glands.
    1. Epithelial membranes cover and line the body surfaces, and their cells are tightly joined by junctional complexes.
    2. Epithelial membranes may be simple or stratified, and their cells may be squamous, cuboidal, or columnar.
    3. Exocrine glands, which secrete into ducts, and endocrine glands, which lack ducts and secrete hormones into the blood, are derived from epithelial membranes.
  D. Connective tissue is characterized by large intercellular spaces that contain extracellular material.
    1. Connective tissue proper includes subtypes such as loose, dense fibrous, adipose, and others.
    2. Cartilage, bone, and blood are classified as connective tissues because their cells are widely spaced with abundant extracellular material between them.

II. Organs are units of structure and function that are composed of at least two, and usually all four, primary tissues.
  A. The skin is a good example of an organ.
    1. The epidermis is a stratified squamous keratinized epithelium, which serves to protect underlying structures and also produces vitamin D.
    2. The dermis is an example of loose connective tissue.
    3. Hair follicles, sweat glands, and sebaceous glands are exocrine glands found in the dermis.
    4. Sensory and motor nerve fibers enter the spaces within the dermis to innervate sensory organs and smooth muscles.
    5. The arrector pili muscles that attach to the hair follicles are composed of smooth muscle.
  B. Organs that are located in different regions of the body and perform related functions are grouped into systems; these include the cardiovascular system, digestive system, endocrine system, and others.

Homeostasis and Feedback Control p. 15

I. Homeostasis refers to the dynamic constancy of the internal environment.
  A. Homeostasis is maintained by mechanisms that act through negative feedback loops.
    1. A negative feedback loop requires a sensor that can detect a change in the internal environment, and an effector that can be activated by the sensor.
    2. In a negative feedback loop the effector acts to cause changes in the internal environment that compensate the initial deviations that activated the sensor.
  B. Positive feedback loops serve to amplify changes and may be part of the action of an overall negative feedback mechanism.
  C. The nervous and endocrine systems provide extrinsic regulation of other body systems and act to maintain homeostasis.
    1. Many nerve fibers release chemicals called neurotransmitters; endocrine glands secrete chemicals called hormones into the blood.
    2. Both neurotransmitters and hormones combine with specific receptor proteins and regulate specific target cells.
  D. The secretion of hormones is stimulated by specific chemicals, including neurotransmitters from nerve fibers, and is inhibited by negative feedback mechanisms.

Review Activities

Objective Questions

Match the following:

1. Glands are derived from
2. Cells are joined closely together in
3. Cells are separated by large extracellular spaces in
4. Blood vessels and nerves are usually located within

(a) nervous tissue
(b) connective tissue
(c) muscular tissue
(d) epithelial tissue

Multiple Choice

5. The wall of a blood vessel is composed of
   (a) epithelial tissue
   (b) muscle tissue
   (c) connective tissue
   (d) all of these
6. Sweat is secreted by exocrine glands. This means that
   (a) it is produced by epithelial cells
   (b) it is a hormone
   (c) it is secreted into a duct
   (d) it is produced outside the body
7. Which of the following statements about homeostasis is *true?*
   (a) The internal environment is maintained absolutely constant.
   (b) Negative feedback mechanisms act to correct deviations from a normal range within the internal environment.
   (c) Homeostasis is maintained by switching effector actions on and off.
   (d) All of these are true.
8. In a negative feedback loop the effector organ produces changes that are
   (a) similar in direction to that of the initial stimulus
   (b) opposite in direction to that of the initial stimulus
   (c) unrelated to the initial stimulus
9. A hormone called *parathyroid hormone* acts to help raise the blood-calcium concentration. According to the principles of negative feedback, an effective stimulus for parathyroid hormone secretion would be
   (a) a fall in blood calcium
   (b) a rise in blood calcium
10. Which of the following consists of dense, parallel arrangements of collagen fibers?
    (a) skeletal muscle tissue
    (b) nervous tissue
    (c) tendons
    (d) dermis
11. The act of breathing raises the blood oxygen level, lowers the blood carbon dioxide concentration, and raises the blood pH. According to the principles of negative feedback, sensors that regulate breathing should respond to
    (a) a rise in blood oxygen
    (b) a rise in blood pH
    (c) a rise in blood carbon dioxide concentration
    (d) all of the above

Essay Questions

1. Describe the structure of different epithelial membranes, and explain how their structures relate to their functions.
2. What are the similarities between the dermis of the skin, bone, and blood? What are the major structural differences between these tissues?
3. Explain the role of antagonistic negative feedback processes in the maintenance of homeostasis.
4. Explain, using examples, how the secretion of a hormone is controlled by the effects of that hormone's actions.
5. Describe the marriage of neural and endocrine regulation in the control of the anterior pituitary gland.

Selected Readings

Adolph, E. F. 1968. *Origins of physiological regulations.* New York: Academic Press.

Bloom, W. B., and D. W. Fawcett. 1975. *A textbook of histology.* 10th ed. Philadelphia: W. B. Saunders Co.

Cannon, W. B. 1932. *The wisdom of the body.* New York: Norton and Company.

Di Fiore, M. S. H. 1974. *An atlas of histology.* 4th ed. Philadelphia: Lea and Febiger.

Jacobs, S., and P. Cuatrecasas. 1977. Cell receptors in disease. *New England Journal of Medicine* 297:1383.

Jones, R. W. 1973. *Principles of biological regulation: an introduction to feedback systems.* New York: Academic Press.

Kessel, R. G., and R. H. Kardon. 1979. *Tissues and organs: a text-atlas of scanning electron microscopy.* San Francisco: W. H. Freeman and Co.

Snyder, S. H. October 1985. The molecular basis for communication between cells. *Scientific American.*

Soderberg, U. 1964. Neurophysiological aspects of homeostasis. *Annual Review of Physiology* 26:271.

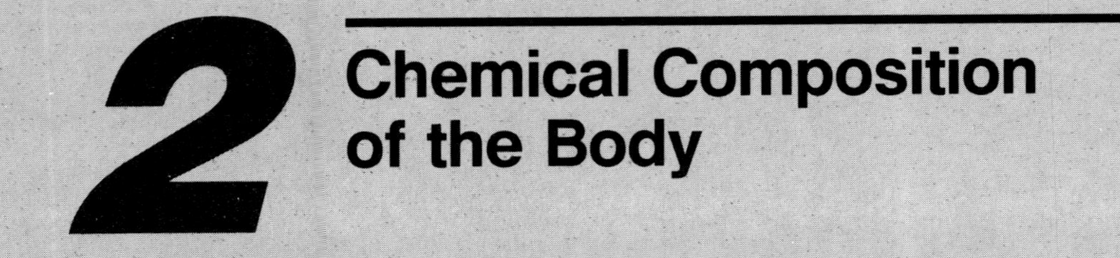

2 Chemical Composition of the Body

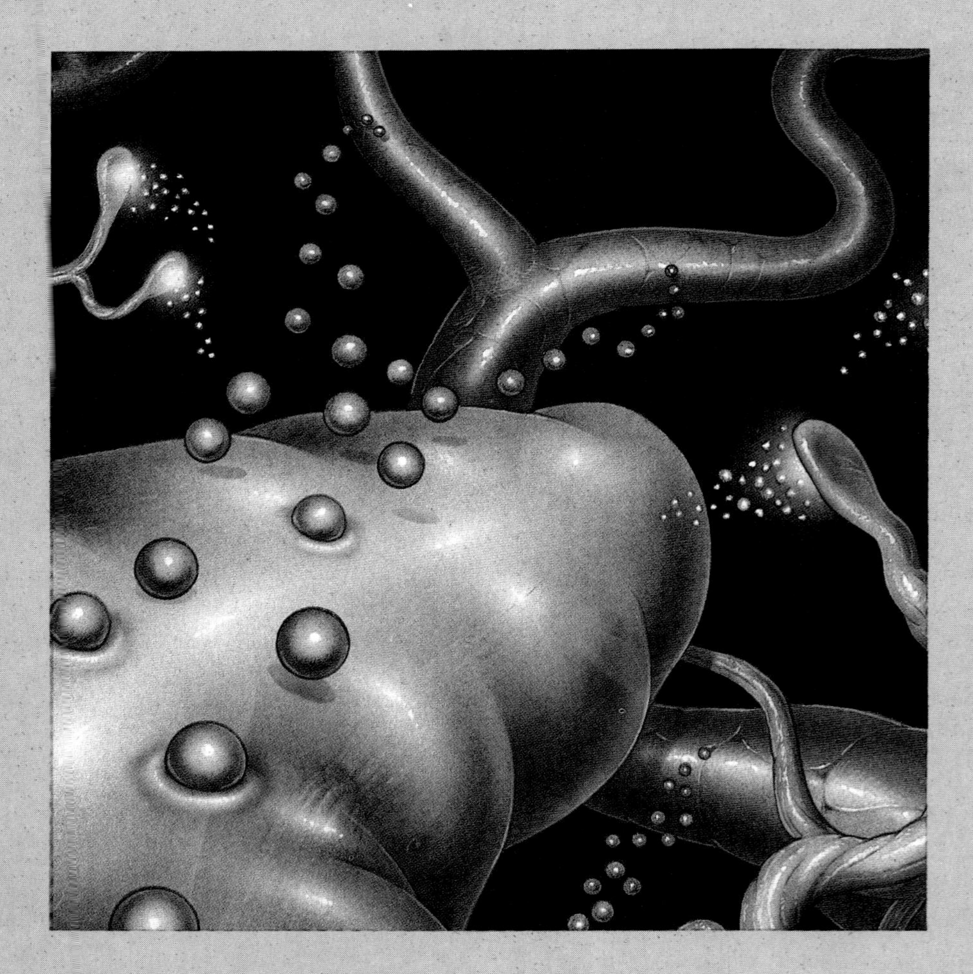

Objectives

By studying this chapter, you should be able to

1. describe the structure of an atom and explain the meaning of the terms *atomic mass unit* and *atomic number*
2. explain how covalent bonds are formed and distinguish between nonpolar and polar covalent bonds
3. explain how ions and ionic bonds are formed and describe the nature of hydrogen bonds
4. describe the structure of a water molecule and explain why some compounds are hydrophilic and others are hydrophobic
5. define the terms *acid* and *base* and explain the meaning of the pH scale
6. describe the different types of carbohydrates and give examples of each type
7. explain the mechanisms and significance of dehydration synthesis and hydrolysis reactions
8. explain the common characteristic of lipids and describe the different categories of lipids
9. describe how peptide bonds are formed and the different orders of protein structure
10. list some of the functions of proteins and explain why proteins provide specificity in their functions
11. describe the structure of DNA and explain the law of complementary base pairing
12. describe the structure of RNA and explain how its structure differs from that of DNA
13. describe, in a general way, the functions of DNA, RNA, and protein in the cell

Outline

Table 2.1 Atoms commonly present in organic molecules.

Atom	Symbol	Atomic Number	Atomic Weight	Orbital 1	Orbital 2	Orbital 3	Number of Chemical Bonds
Hydrogen	H	1	1.01	1	0	0	1
Carbon	C	6	12.01	2	4	0	4
Nitrogen	N	7	14.01	2	5	0	3
Oxygen	O	8	16.00	2	6	0	2
Phosphorus	P	15	30.97	2	8	5	5
Sulfur	S	16	32.06	2	8	6	2

Atoms, Ions, and Chemical Bonds

The anatomical structures and physiological processes of the body are based, to a large degree, on the properties and interactions of atoms, ions, and molecules. Water is the major solvent in the body and contributes 65% to 75% of the total weight of an average adult. Of this amount 30% to 40% is within the body cells (in the *intracellular compartment*); the remainder is in the *extracellular compartment,* including the blood and tissue fluids. Dissolved in this water are many organic molecules (carbon-containing molecules such as carbohydrates, lipids, proteins, and nucleic acids) and inorganic molecules and ions (atoms with a net charge). Before describing the structure and function of organic molecules within the body, some basic chemical concepts, terminology, and symbols will be introduced.

Atoms

Atoms are much too small to be seen individually, even with the most powerful electron microscope. Through the efforts of many generations of scientists, however, the structure of an atom is now well understood. At the center of an atom is its *nucleus.* The nucleus contains two types of particles—*protons,* which have a positive charge, and *neutrons,* which are noncharged. The sum of the protons and neutrons in a nucleus is the **atomic mass unit,** which is approximately equal to the **atomic weight** (table 2.1).

Orbiting the positively charged nucleus are negatively charged particles called *electrons.* The number of electrons in an atom is equal to the number of its protons; each atom, therefore, has a net charge of zero. The number of protons (or electrons) in an atom is called its **atomic number.**

The exact location of an electron at a given time cannot be predicted. One can only predict that, with a given probability, the electron will be within a given volume of space around the nucleus. The *orbital* of an electron describes the outer boundary of this volume of space. This orbital can be drawn as a circle around the nucleus, much like the orbit of a satellite (like the moon) around a planet.

The first orbital can only contain two electrons. If an atom has more than two electrons (as do all atoms except hydrogen and helium), the additional electrons must occupy orbitals that are farther removed from the nucleus than the first. The next outer orbital can contain a maximum of eight electrons. The orbitals are filled from the innermost to the outermost. Carbon—with six electrons—thus has two electrons in its first orbital and four electrons in the second orbital (fig. 2.1). It is always the electrons in the outermost orbital that participate in chemical reactions.

Chemical Bonds

Molecules are formed when two or more atoms are joined together by sharing or some other interaction of the electrons in their outer orbitals. Such sharing of electrons or other attractive forces produces *chemical bonds* (fig. 2.2). The number of bonds that each atom can have is determined by the number of electrons in its outer orbital. Hydrogen, for example, must obtain only one more electron—and can thus form one chemical bond—to complete the first orbital of two electrons. Carbon, in contrast, must obtain four more electrons—and can thus form four chemical bonds—to complete the second orbital of eight electrons (fig. 2.3).

Covalent Bonds. In **covalent bonds,** atoms are joined together by the sharing of electrons—equally in nonpolar covalent bonds, unequally in polar covalent bonds. Covalent bonds that are formed between identical atoms—as in oxygen gas (O_2) and hydrogen gas (H_2)—are the strongest because their electrons are equally shared. Since the electrons are equally distributed between the two atoms, these molecules are said to be *nonpolar.* When covalent bonds are formed between two different atoms, however, the electrons may be pulled more toward one atom than the other. The side of the molecule toward which the electrons are pulled is electrically negative in comparison to the other end. Such a molecule is said to be *polar* (has a positive and negative "pole"). Atoms of oxygen, nitrogen, and phosphorus strongly attract electrons—are *electronegative*—when they bond with other atoms.

Figure 2.1. Diagrams of the hydrogen and carbon atoms. The electron orbitals on the left are represented by dots indicating probable positions of the electrons. The orbitals on the right are represented by concentric circles.

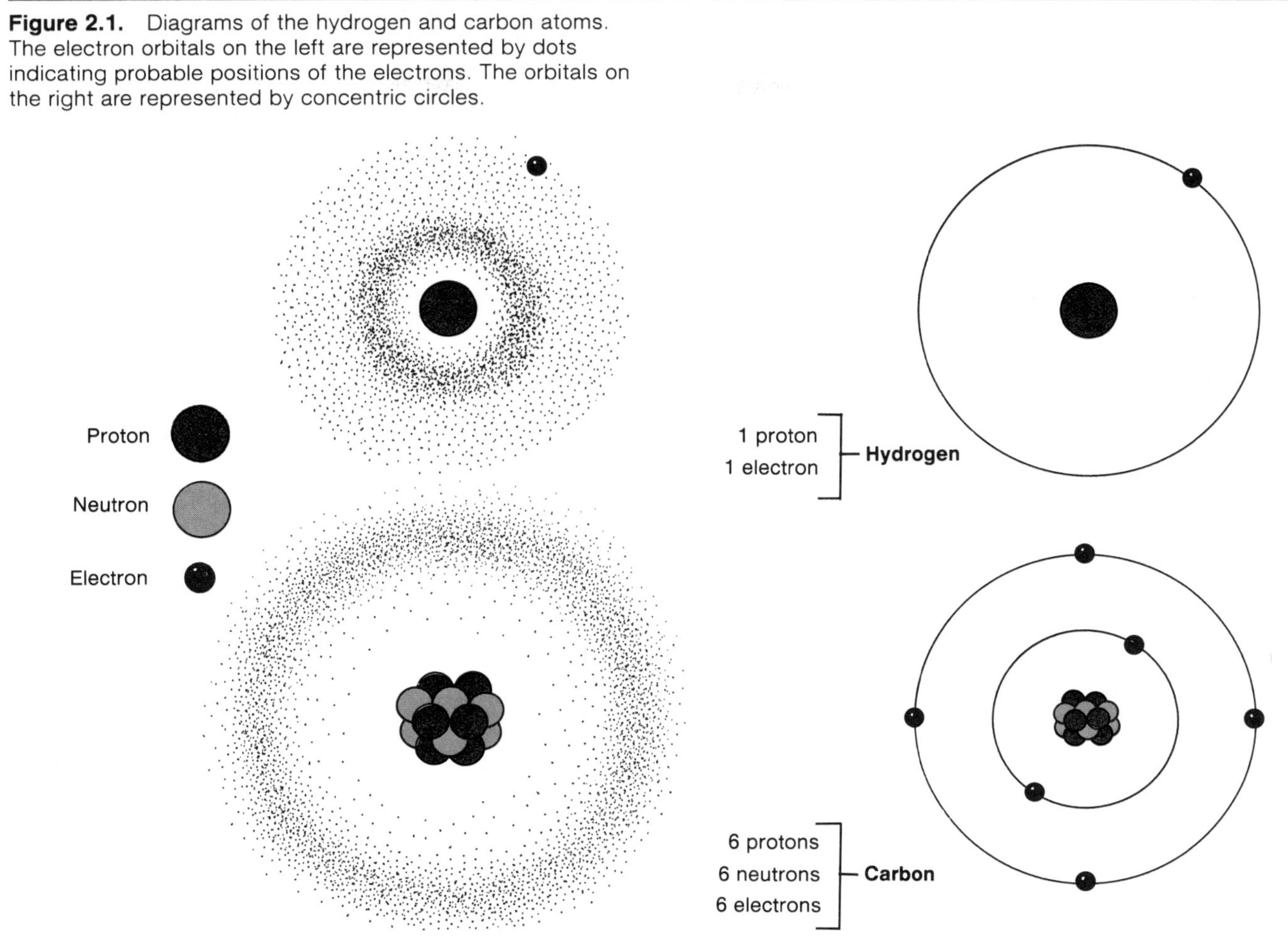

Figure 2.2. The hydrogen molecule, showing the covalent bonds between hydrogen atoms formed by the equal sharing of electrons.

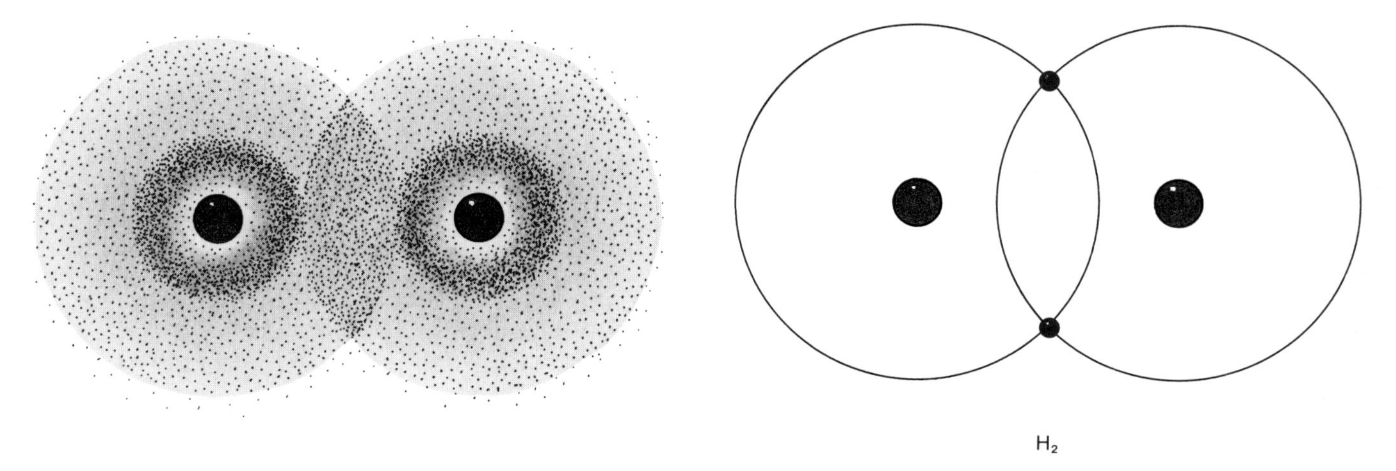

Figure 2.3. The molecules methane and ammonia represented in three different ways. Notice that a bond between two atoms consists of a pair of shared electrons (the electrons from the outer orbital of each atom).

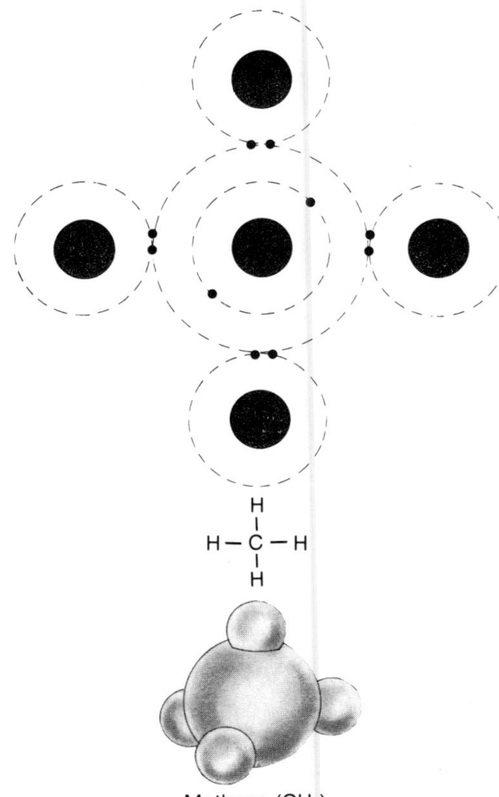

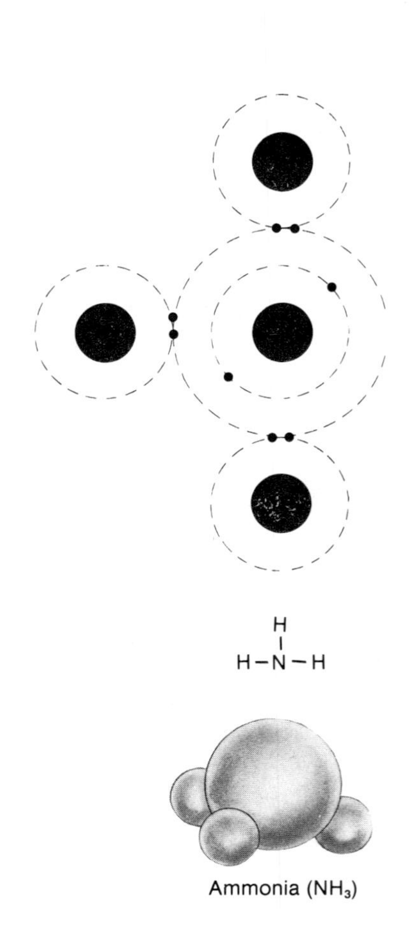

Water is the most abundant molecule in the body and serves as the solvent of body fluids. Water is a good solvent because it is polar; the oxygen atom pulls electrons from the two hydrogens toward its side of the water molecule, so that the oxygen side is negatively charged compared to the hydrogen side of the molecule (fig. 2.4). The significance of the polar nature of water in its function as a solvent is discussed in the next section.

Ionic Bonds. In **ionic bonds** the electrons are not shared at all. Instead, they are transferred from one atom to another. Common table salt, sodium chloride (NaCl), serves as an example. Sodium, with a total of eleven electrons, has two in its first orbital, eight in its second orbital, and only one in its third orbital.

Chloride, conversely, is only one electron short of completing its outer orbital of eight electrons. The lone electron in sodium's outer orbital is attracted to chloride's outer orbital. This creates a *chloride ion*—represented by the symbol Cl^-—with one extra electron from sodium's

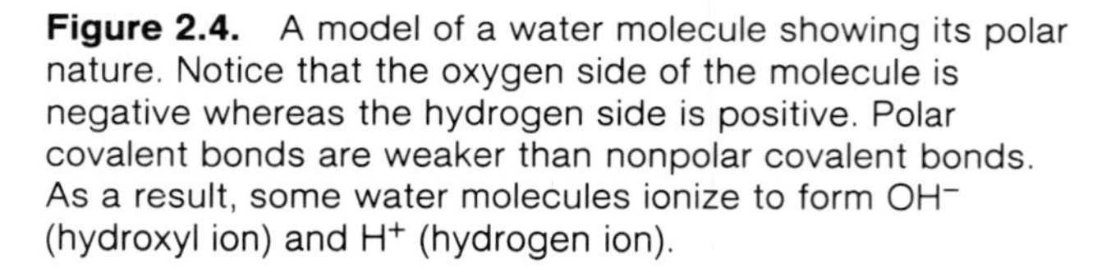

Figure 2.4. A model of a water molecule showing its polar nature. Notice that the oxygen side of the molecule is negative whereas the hydrogen side is positive. Polar covalent bonds are weaker than nonpolar covalent bonds. As a result, some water molecules ionize to form OH^- (hydroxyl ion) and H^+ (hydrogen ion).

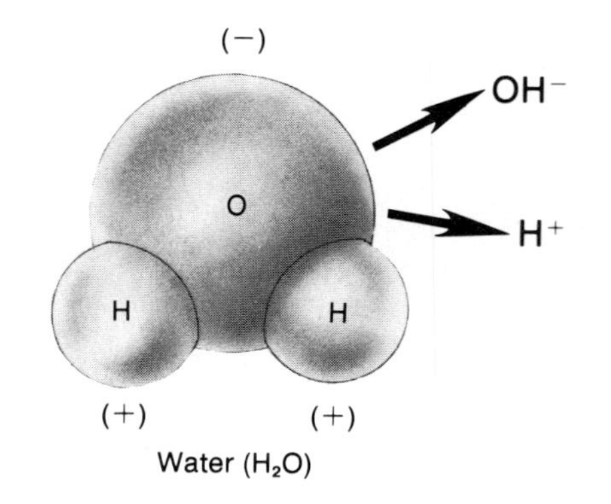

Figure 2.5. The ionization of sodium chloride to produce sodium and chloride ions. The positive sodium and the negative chloride ions attract each other to produce the ionic compound sodium chloride (NaCl).

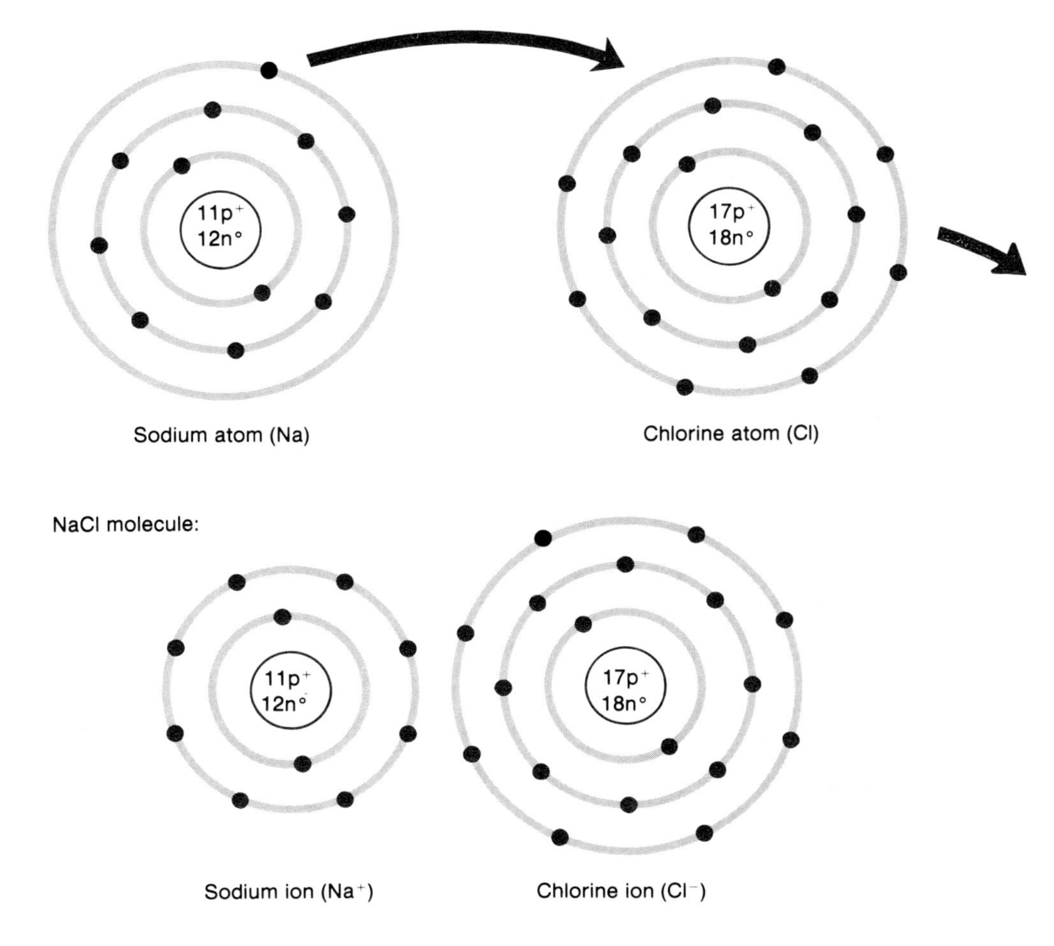

outer orbital, and a *sodium ion* (Na^+), which lost an electron to chloride's outer orbital. The attraction between these two ions produces the *ionic compound* NaCl (fig. 2.5).

Positively charged ions are called *cations,* and negatively charged ions are called *anions.* These terms are derived from the way ions move in an electric field, in which there is a positive pole (the anode) and a negative pole (the cathode). Since positively charged ions move toward the cathode, they are called cations. Negatively charged ions move toward the anode and are thus known as anions.

Ionic bonds are weaker than polar covalent bonds and, therefore, break more easily when dissolved in water. Sodium chloride almost completely ionizes (separates into ions) when dissolved. Each of the ions released (Na^+ and Cl^-) attracts polar water molecules; the negative ends of water molecules are attracted to the Na^+, and the positive ends are attracted to the Cl^- (fig. 2.6). The water molecules that surround these ions in turn attract other molecules of water to form *hydration spheres* around each ion.

The formation of hydration spheres makes an ion or a molecule soluble in water. Glucose, amino acids, and many other organic molecules are water-soluble because hydration spheres can form around atoms of oxygen, nitrogen, and phosphorus, which are joined by polar covalent bonds to other atoms. Such molecules are said to be **hydrophilic.** In contrast, molecules composed primarily of nonpolar covalent bonds, such as the hydrocarbon chains of fat molecules, have few charges and, thus, cannot form hydration spheres. They are insoluble in water and, in fact, actually avoid water—they are **hydrophobic.**

Figure 2.6. The negatively charged oxygen-ends of water molecules are attracted to the positively charged Na^+, whereas the positively charged hydrogen-ends of water molecules are attracted to the negatively charged Cl^-. Other water molecules are attracted to this first concentric layer of water, forming hydration spheres around the sodium and chloride ions.

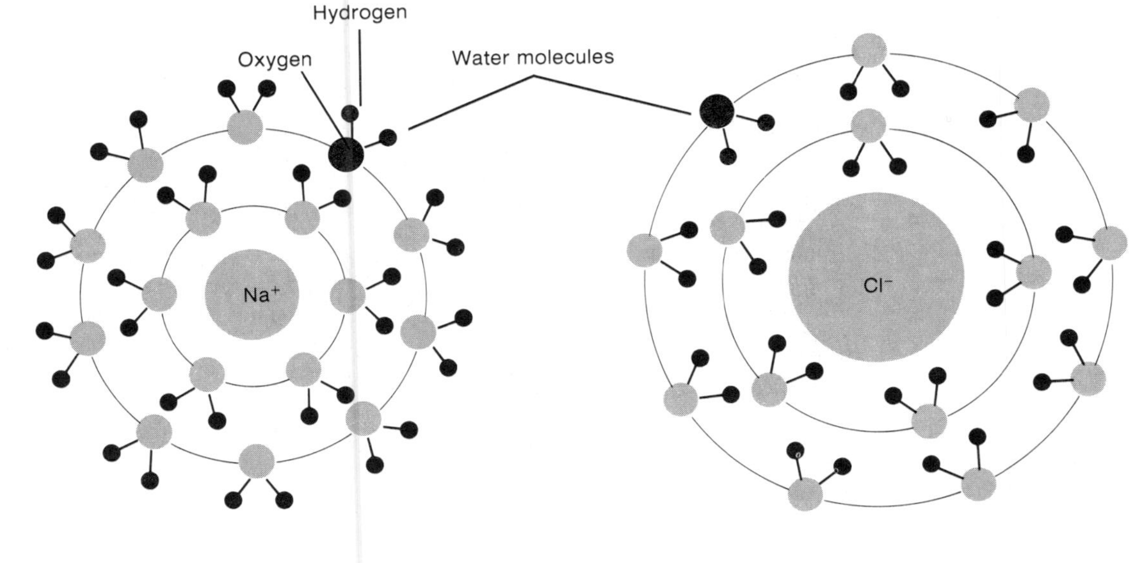

Figure 2.7. The oxygen atoms of water molecules are weakly joined together by the attraction of the electronegative oxygen for the positively charged hydrogen. These weak bonds are called *hydrogen bonds*.

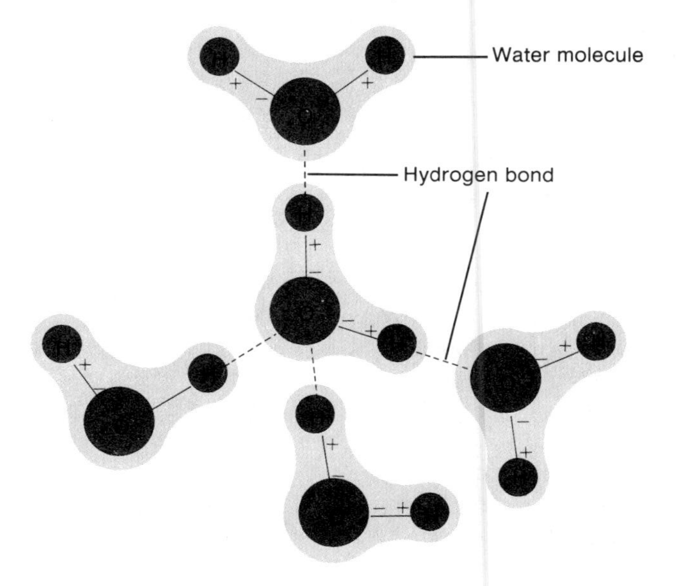

Hydrogen Bonds. Hydrogen bonds are weaker than ionic bonds. When hydrogen forms a polar covalent bond with an atom of oxygen or nitrogen, the hydrogen gains a slight positive charge as its electron is pulled toward the electronegative atom. Since the hydrogen has a slight positive charge, it will have a weak attraction for a second electronegative atom. This weak attraction is called a **hydrogen bond.** Hydrogen bonds are often shown with dotted lines (fig. 2.7) to distinguish them from stronger covalent bonds, which are shown with solid lines.

Water molecules are bonded to each other by weak hydrogen bonds (fig. 2.7). These bonds are responsible for many of the physical properties of water, including its *surface tension* and its ability to be pulled as a column through narrow channels in a process called *capillary action.*

Acids, Bases, and the pH Scale

The bonds joining hydrogen atoms to oxygen atoms in water molecules are, as previously discussed, polar covalent bonds. Although these bonds are stronger than ionic bonds, they are sufficiently weak to allow some water molecules to break apart, or dissociate, as a hydrogen atom completely loses its electron to oxygen. When this occurs, the hydrogen nucleus—consisting of a single proton—is released as a hydrogen ion (H^+).

The dissociation of hydrogen from water results in the production of two ions—H^+ (a hydrogen ion) and OH^- (a hydroxyl ion). Very few water molecules dissociate in

Table 2.2 Common acids and bases.

Acid	Symbol	Base	Symbol
Hydrochloric acid	HCl	Sodium hydroxide	$NaOH$
Phosphoric acid	H_3PO_4	Potassium hydroxide	KOH
Nitric acid	HNO_3	Calcium hydroxide	$Ca(OH)_2$
Sulfuric acid	H_2SO_4	Ammonium hydroxide	NH_4OH
Carbonic acid	H_2CO_3		

Source: Fox, Stuart Ira, *A Laboratory Guide to Human Physiology, Concepts and Clinical Applications,* 3d ed. © 1976, 1980, 1984 Wm. C. Brown Publishers, Dubuque, Iowa. All Rights Reserved. Reprinted by permission.

this way, but those that do produce equal amounts of H^+ and OH^-. The concentration of each of these ions in pure water is equal to only 10^{-7} molar. (The term *molar* is a unit of concentration, described in chapter 6; for hydrogen, one molar equals one gram per liter.) A solution with this H^+ concentration—that is, a solution produced by the dissociation of water molecules, where the H^+ concentration is equal to the OH^- concentration—is said to be **neutral.**

A solution that contains a higher H^+ concentration than that of water is called *acidic,* and a solution with a lower H^+ concentration than water is called *basic.* An **acid** is defined as a molecule that can release H^+ (protons) into a solution. A **base** is defined as a molecule that lowers the H^+ concentration of a solution. Most bases release OH^- into a solution, which combines with free H^+ to form water and thus lowers the H^+ concentration. Examples of common acids and bases are shown in table 2.2.

pH. The H^+ concentration of a solution is usually indicated in pH units on a pH scale that runs from 0 to 14. The pH number is equal to the logarithm of one over the H^+ concentration:

$$pH = \log \frac{1}{[H^+]}$$

where $[H^+]$ = molar H^+ concentration.

Pure water has a H^+ concentration of 10^{-7} molar and, thus, has a pH of 7 (neutral). Because of the logarithmic relationship, a solution with ten times the hydrogen ion concentration (10^{-6} M) has a pH of 6, whereas a solution with one-tenth the H^+ concentration (10^{-8} M) has a pH of 8. The pH number is easier to write than the molar H^+ concentration, but it is admittedly confusing because it is *inversely related* to the H^+ concentration: a solution with a higher H^+ concentration has a lower pH number; one with a lower H^+ concentration has a higher pH number. A strong acid with a high H^+ concentration of 10^{-2} molar, for example, has a pH of 2, whereas a solution with only 10^{-10} molar H^+ has a pH of 10. **Acidic solutions,** therefore, have a pH of less than 7 (that of pure water), whereas **basic solutions** have a pH of between 7 and 14 (table 2.3).

Table 2.3 The pH scale.

	H^+ Concentration (molar)	pH	OH^- Concentration (molar)
	1.0	0	10^{-14}
	0.1	1	10^{-13}
	0.01	2	10^{-12}
Acids	0.001	3	10^{-11}
	0.0001	4	10^{-10}
	10^{-5}	5	10^{-9}
	10^{-6}	6	10^{-8}
Neutral	10^{-7}	7	10^{-7}
	10^{-8}	8	10^{-6}
	10^{-9}	9	10^{-5}
	10^{-10}	10	0.0001
Bases	10^{-11}	11	0.001
	10^{-12}	12	0.01
	10^{-13}	13	0.1
	10^{-14}	14	1.0

Buffers. A *buffer* is a system of molecules and ions that acts to prevent changes in H^+ concentration and thus that serves to stabilize the pH of a solution. In blood plasma, for example, the pH is stabilized by the following reversible reaction involving the bicarbonate ion (HCO_3^-) and carbonic acid (H_2CO_3):

$$HCO_3^- + H^+ \rightleftharpoons H_2CO_3$$

The double arrows indicate that the reaction could go either to the right or to the left; the net direction depends on the concentration of molecules and ions on each side. If an acid (such as lactic acid) should release H^+ into the solution, for example, the following reaction would be promoted:

$$HCO_3^- + H^+ \longrightarrow H_2CO_3$$

The above reaction serves to decrease the effect of added H^+ on the pH of the blood. Bicarbonate, in fact, is the major buffer of the blood. Under the opposite condition, when the concentration of free H^+ in the blood is falling, the reaction previously described can be reversed:

$$H_2CO_3 \longrightarrow H^+ + HCO_3^-$$

The dissociation of carbonic acid yields free H^+, which helps to prevent an increase in pH. Bicarbonate ions and carbonic acid thus act as a *buffer pair* to prevent either decreases or increases in pH, respectively. This buffering action normally maintains the blood pH at a very stable 7.40 ± 0.05.

Figure 2.8. Two carbon atoms joined by a single covalent bond (*above*) or by a double covalent bond (*below*). In both cases each carbon atom shares four pairs of electrons (has four bonds) to complete the eight electrons required to fill its outer orbital.

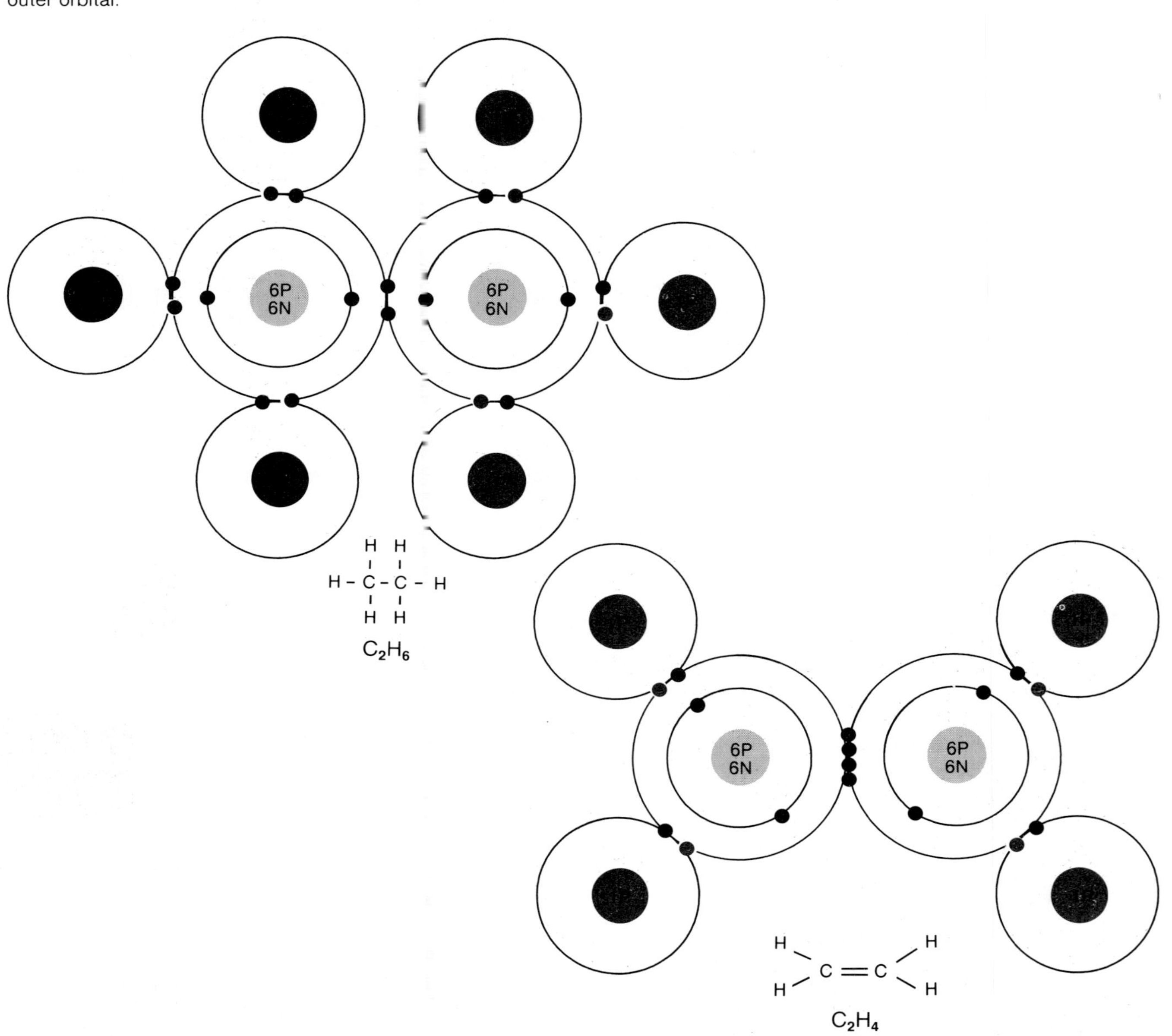

Organic Molecules

Organic molecules are those that contain the atom *carbon.* Since the carbon atom has four electrons in its outer orbital, it must share four additional electrons by covalent bonding with other atoms to fill its outer orbital with eight electrons. The unique bonding requirements of carbon enable it to join with other carbon atoms to form chains and rings, while still allowing the carbon atoms to bond with hydrogen and other atoms.

Most organic molecules in the body contain hydrocarbon chains and rings as well as other atoms bonded to carbon. Two adjacent carbon atoms in a chain or ring may share one or two pairs of electrons. If the two carbon atoms share one pair of electrons, they are said to have a *single covalent bond;* this leaves each carbon atom free to bond to as many as three other atoms. If the two carbon atoms share two pairs of electrons, they have a *double covalent bond,* and each carbon atom can only bond to a maximum of two additional atoms (fig. 2.8).

Figure 2.9. Hydrocarbons that are (*a*) linear, (*b*) cyclic, and (*c*) aromatic rings.

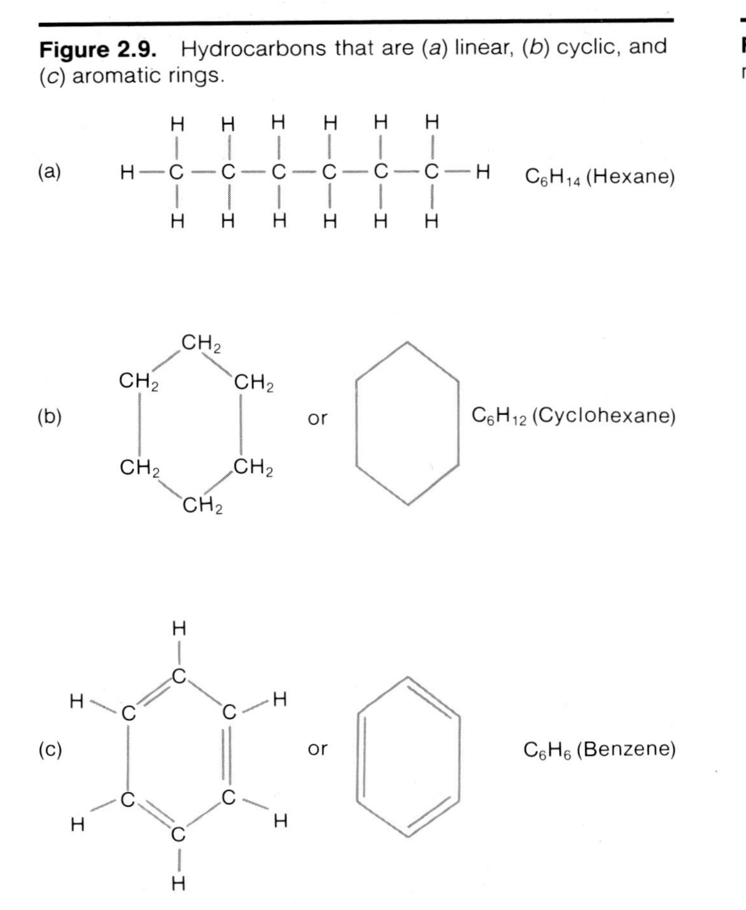

The ends of some hydrocarbons are joined together to form rings. In the shorthand structural formulas of these molecules, the carbon atoms are not shown but are understood to be located at the corners of the ring. Some of these cyclic molecules have a double bond between two adjacent carbon atoms. Benzene and related molecules are shown as a six-sided ring with alternating double bonds. Such compounds are called *aromatic*. Since all of the carbons in an aromatic ring are equivalent, double bonds can be shown between any two adjacent carbons in the ring (fig. 2.9).

The hydrocarbon chain or ring of many organic molecules provides a relatively inactive molecular "backbone," to which more reactive groups of atoms are attached. Known as *functional groups* of the molecule, these reactive groups usually contain atoms of oxygen, nitrogen, phosphorus, or sulfur and are, in large part, responsible for the unique chemical properties of the molecule (fig. 2.10).

Classes of organic molecules can be named according to their functional groups. *Ketones*, for example, have a carbonyl group within the carbon chain. An organic molecule is an *alcohol* if it has a hydroxyl group at one end of the chain. All *organic acids* (such as acetic acid, citric acids, and others) have a *carboxyl* group (fig. 2.11).

Figure 2.10. Various functional groups of organic molecules.

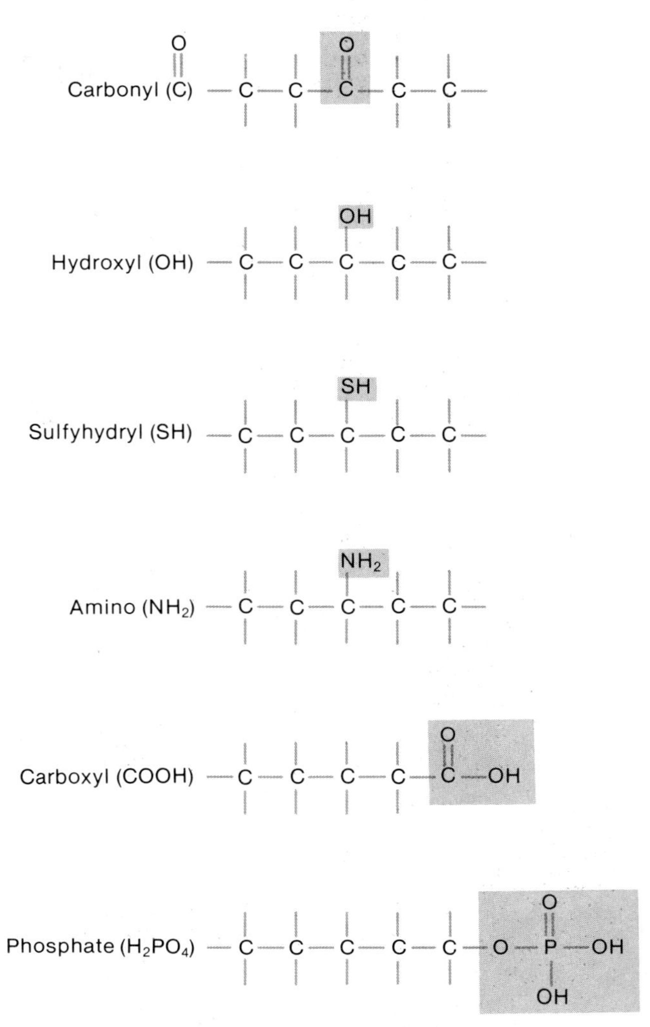

Figure 2.11. Categories of organic molecules based on functional groups.

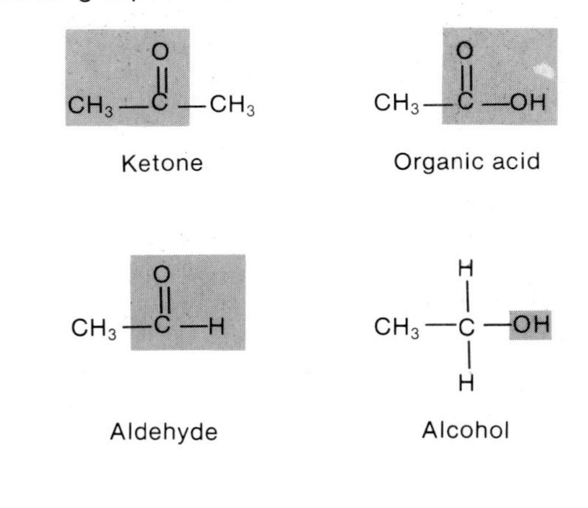

Stereoisomers. Two molecules may have exactly the same atoms arranged in exactly the same sequence, yet may differ with respect to the spatial orientation of key functional groups. Such molecules are called *stereoisomers* of each other. The isomers that have a key functional group represented on the right side of the molecule are called **D-isomers** (for *dextro,* or right-handed). Molecules that are represented by structures showing functional groups on the left side are called **L-isomers** (for *levo,* or left-handed).

The two stereoisomers are mirror images of each other—they cannot be superimposed. These subtle differences in structure are extremely important biologically, because enzymes—which interact with such molecules in a stereo-specific way in chemical reactions—cannot combine with the "wrong" stereoisomer. The enzymes of all cells (human and others) can only combine with L-amino acids and D-sugars, for example. The opposite stereoisomers (D-amino acids and L-sugars) cannot be used by the body.

1. *Describe the structure of an atom, and define the terms* atomic weight *and* atomic number.
2. *List the types of chemical bonds in order of decreasing strength, and describe how these bonds are formed.*
3. *Write the definition of an acid and the equation for the definition of pH. Describe the relationship between pH and the* H^+ *concentration.*
4. *Describe the structure of water, and explain why ions and molecules with many polar covalent bonds are water-soluble.*
5. *Describe the structure of a carbon atom, and explain how this structure allows carbon atoms to bond together to form chains and rings.*
6. *Define the term* stereoisomer, *and explain the physiological significance of this concept.*

Carbohydrates and Lipids

Carbohydrates and lipids are similar in many ways. Both groups of molecules consist primarily of the atoms carbon, hydrogen, and oxygen, and both serve as major sources of energy in the body (comprising most of the calories consumed in food). Carbohydrates and lipids differ, however, in some important aspects of their chemical structures and physical properties. Such differences significantly affect the functions of these molecules in the body.

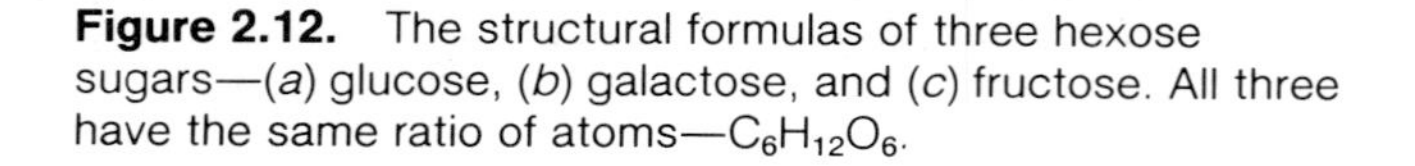

Figure 2.12. The structural formulas of three hexose sugars—(*a*) glucose, (*b*) galactose, and (*c*) fructose. All three have the same ratio of atoms—$C_6H_{12}O_6$.

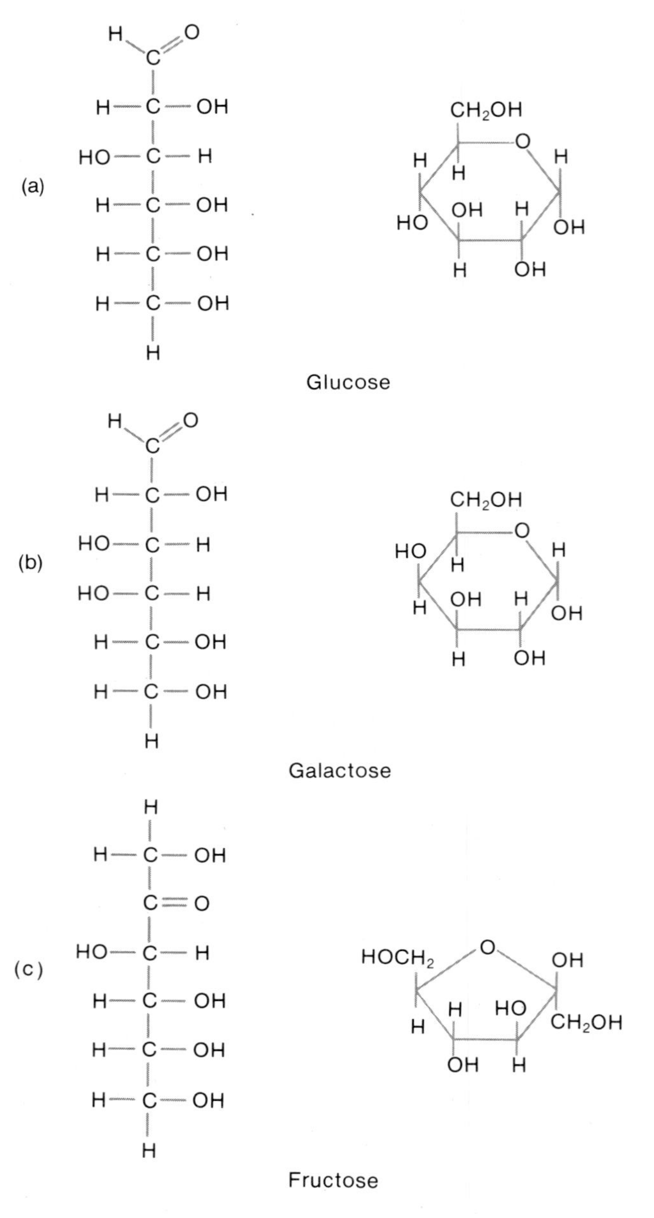

Carbohydrates

Carbohydrates are organic molecules that contain carbon, hydrogen, and oxygen in the ratio described by their name—*carbo* (carbon) and *hydrate* (water, H_2O). The general formula of a carbohydrate molecule is thus CH_2O; the molecule contains twice the number of hydrogen atoms as it contains carbon or oxygen atoms.

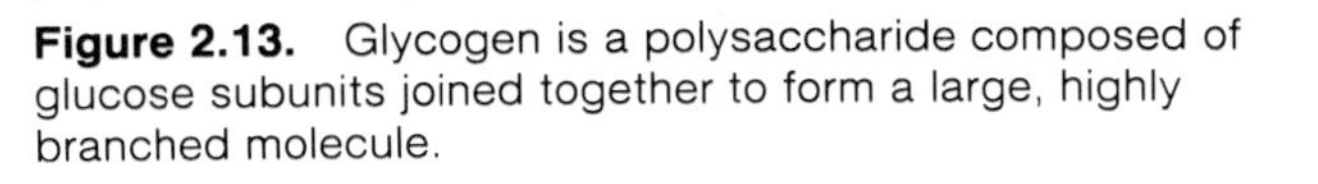

Figure 2.13. Glycogen is a polysaccharide composed of glucose subunits joined together to form a large, highly branched molecule.

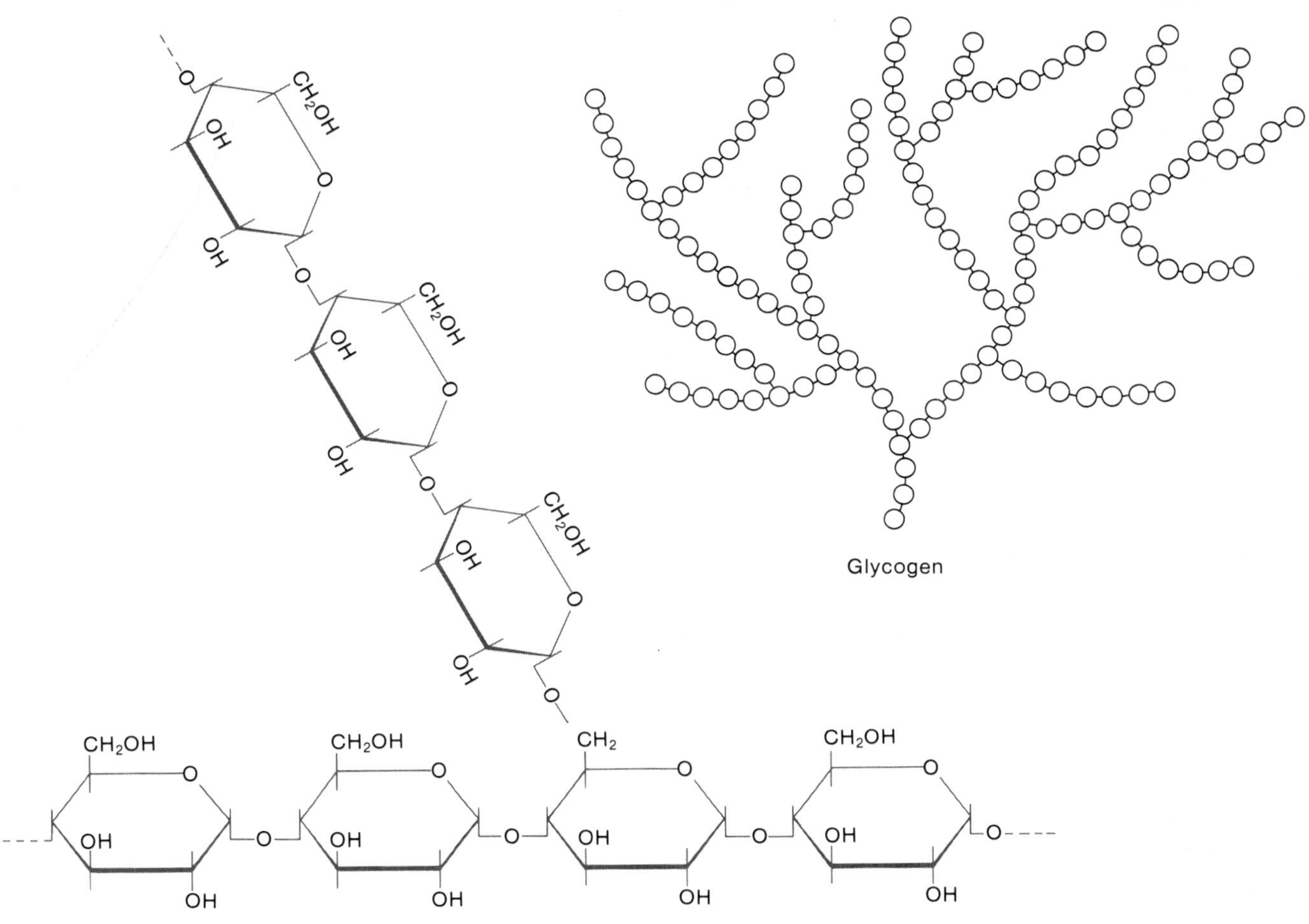

Monosaccharides, Disaccharides, and Polysaccharides. Carbohydrates include simple sugars, or **monosaccharides,** and longer molecules that contain a number of monosaccharides joined together. The suffix *-ose* denotes a sugar molecule; the term *hexose,* for example, refers to a six-carbon monosaccharide with the formula $C_6H_{12}O_6$. This formula is adequate for some purposes, but it does not distinguish between related hexose sugars, which are *structural isomers* of each other. The structural isomers glucose, fructose, and galactose, for example, are monosaccharides that have the same ratio of atoms arranged in slightly different ways (fig. 2.12).

Two monosaccharides can be joined covalently to form a **disaccharide,** or double sugar. Common disaccharides include table sugar, or *sucrose* (composed of glucose and fructose), milk sugar, or *lactose* (composed of glucose and galactose), and malt sugar, or *maltose* (composed of two glucose molecules). When many monosaccharides are joined together, the resulting molecule is called a **polysaccharide.** *Starch,* for example, is a polysaccharide found in many plants, which is formed by the bonding together of thousands of glucose subunits. Animal starch **(glycogen),** found in the liver and muscles, likewise consists of repeating glucose molecules but differs from plant starch in that it is more highly branched (fig. 2.13).

Dehydration Synthesis and Hydrolysis. In the formation of disaccharides and polysaccharides, the separate subunits (monosaccharides) are bonded together covalently by a type of reaction called **dehydration synthesis,** or **condensation.** In this reaction, which requires the participation of specific enzymes (chapter 4), a hydrogen atom

Figure 2.14. Dehydration synthesis of two disaccharides, (*a*) maltose and (*b*) sucrose. Notice that as the disaccharides are formed a molecule of water is produced.

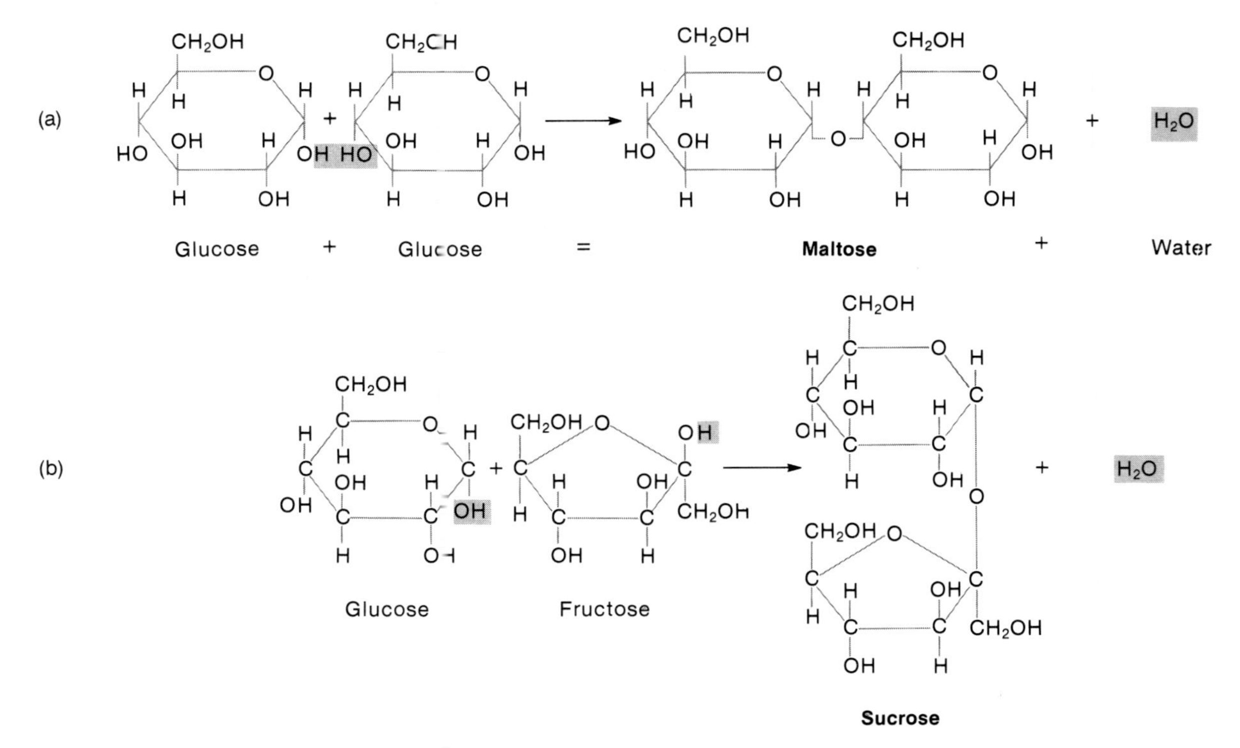

Figure 2.15. The hydrolysis of starch (*a*) into disaccharides (maltose) and (*b*) into monosaccharides (glucose). Notice that as the covalent bond between the subunits breaks, a molecule of water is split. In this way the hydrogen and hydroxyl from water is added to the ends of the released subunits.

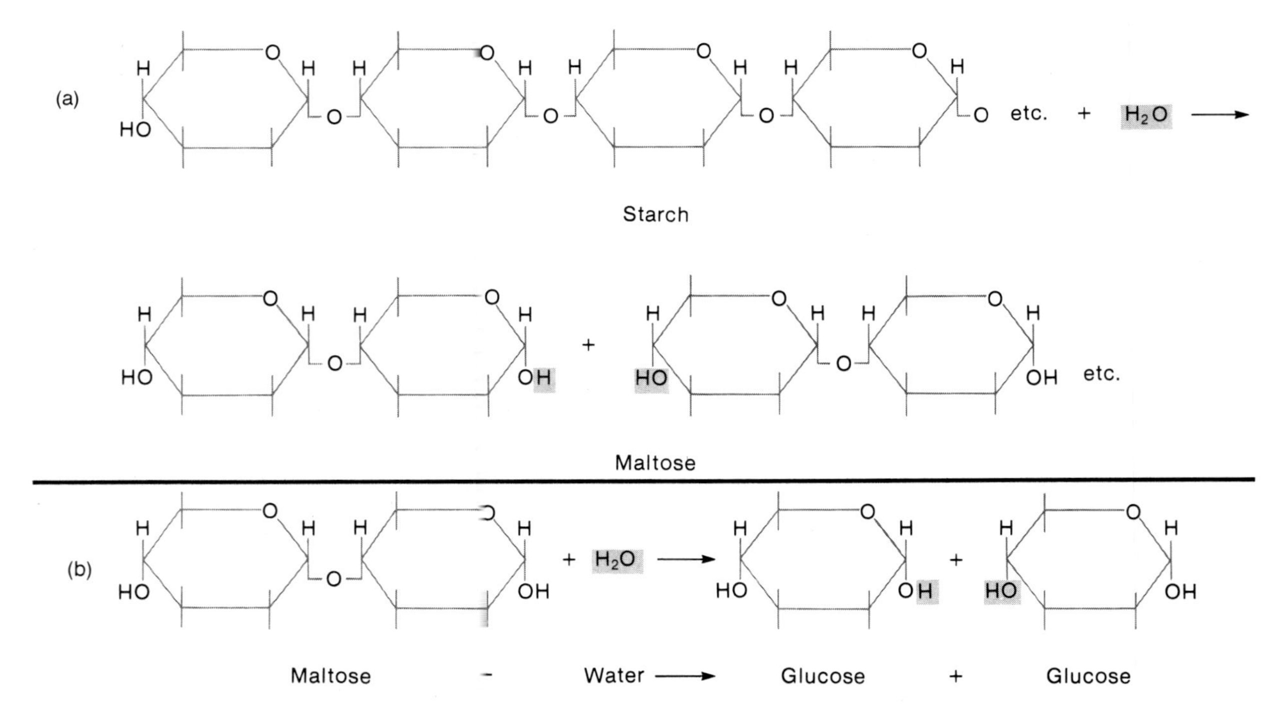

Figure 2.16. Structural formulas for (*a*) saturated and (*b*) unsaturated fatty acids.

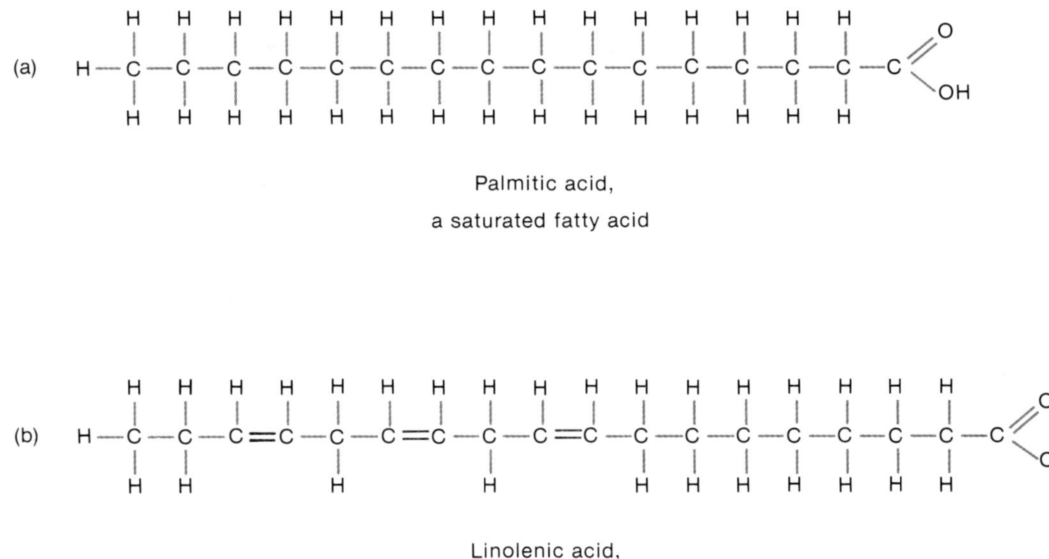

is removed from one monosaccharide and a hydroxyl group (OH) is removed from another. As a covalent bond is formed between the two monosaccharides, water (H_2O) is produced. Dehydration synthesis reactions are illustrated in figure 2.14.

When a person eats disaccharides and polysaccharides or when the stored glycogen in the liver and muscles is to be used by tissue cells, the covalent bonds that join monosaccharides into disaccharides and polysaccharides must be broken. These *digestion* reactions occur by means of **hydrolysis.** Hydrolysis is the reverse of dehydration synthesis. A water molecule is split, as implied by the word *hydrolysis,* and the resulting hydrogen atom is added to one of the free glucose molecules as the hydroxyl group is added to the other (fig. 2.15).

When a potato is eaten, the starch within it is hydrolyzed into separate glucose molecules within the intestine. This glucose is absorbed into the blood and carried to the tissues. Some tissue cells may use this glucose for energy. Liver and muscles, however, can store excess glucose in the form of glycogen by dehydration synthesis reactions in these cells. During fasting or prolonged exercise, the liver can add glucose to the blood through hydrolysis of its stored glycogen.

Dehydration synthesis, or condensation, and hydrolysis reactions do not occur spontaneously; they require the action of specific enzymes. Similar reactions, in the presence of other enzymes, build and break down lipids, proteins, and nucleic acids. In general, therefore, hydrolysis reactions digest molecules into their subunits, and dehydration synthesis reactions build larger molecules by the bonding together of their subunits.

Lipids

The category of molecules known as lipids includes several types of molecules that differ greatly in chemical structure. These diverse molecules are all in the lipid category by virtue of a common physical property—they are all *insoluble in polar solvents* such as water. This is because lipids consist primarily of hydrocarbon chains and rings, which are nonpolar and, thus, hydrophobic. Although lipids are insoluble in water, they can be dissolved in nonpolar solvents such as ether, benzene, and related compounds.

Triglycerides. Triglycerides are a subcategory of lipids that includes fat and oil. These molecules are formed by the condensation of one molecule of *glycerol* (a three-carbon alcohol) with three molecules of *fatty acids.* Each fatty acid molecule consists of a nonpolar hydrocarbon chain with a carboxylic acid group (abbreviated COOH) on one end. If the carbon atoms within the hydrocarbon chain are joined by single covalent bonds, so that each carbon atom can also bond to two hydrogen atoms, the fatty acid is said to be *saturated.* If there are a number of double covalent bonds within the hydrocarbon chain, so that each carbon atom can only bond to one hydrogen atom, the fatty acid is said to be *unsaturated.* Triglycerides that contain saturated fatty acids are called **saturated fats;** those that contain unsaturated fatty acids are **unsaturated fats** (fig. 2.16).

Figure 2.17. Dehydration synthesis of a triglyceride molecule from a glycerol and three fatty acids. A molecule of water is produced as an ester bond forms between each fatty acid and the glycerol. Sawtooth lines represent carbon chains, which are symbolized by an *R*.

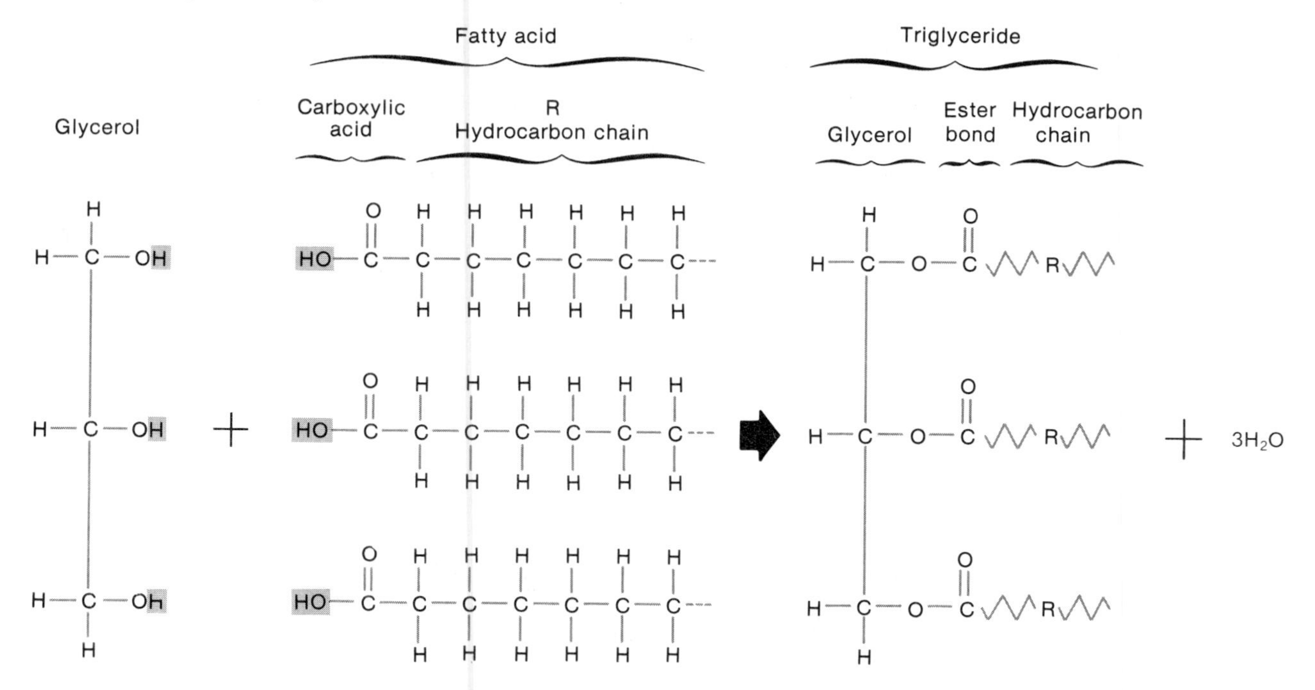

Within the adipose cells of the body, triglycerides are formed as the carboxylic acid ends of fatty acid molecules condense with the hydroxyl groups of a glycerol molecule (fig. 2.17). Since the hydrogen atoms from the carboxyl ends of fatty acid molecules form water molecules during dehydration synthesis, fatty acids that are combined with glycerol can no longer release H^+ and function as acids. For this reason, triglycerides are described as *neutral fats*.

Ketone Bodies. Hydrolysis of triglycerides within adipose tissue releases *free fatty acids* into the blood. Free fatty acids can be used as an immediate source of energy by many organs; they can also be converted by the liver into derivatives called *ketone bodies.* These include four-carbon-long acidic molecules (acetoacetic acid and β-hydroxybutyric acid) and acetone (the active ingredient in nail polish remover). A rapid breakdown of fat, such as occurs during dieting and in uncontrolled diabetes mellitus, results in elevated levels of ketone bodies in the blood, a condition called **ketosis.** If there are sufficient amounts of ketone bodies in the blood to lower the blood pH, the condition is called **ketoacidosis.**

Figure 2.18. The structure of lecithin, a typical phospholipid (*above*), and its more simplified representation (*below*).

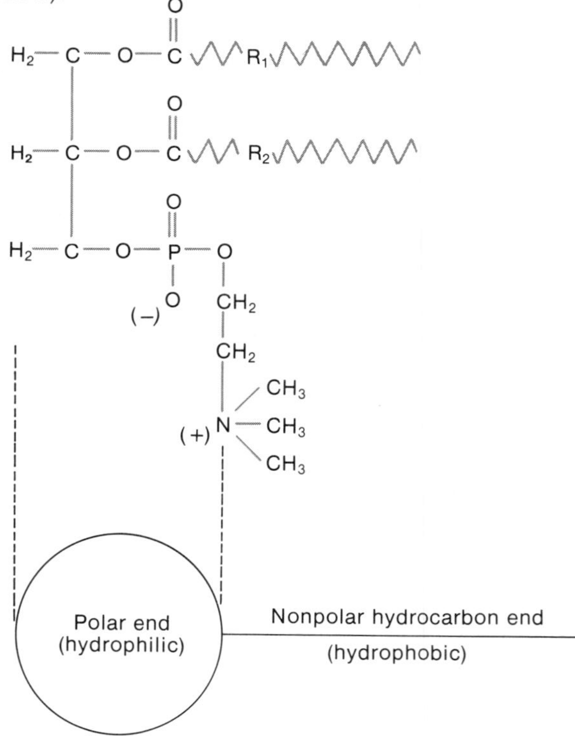

Figure 2.19. The formation of a micelle structure by phospholipids such as lecithin. The straight lines represent the hydrophobic fatty acid parts of the molecule, and the circles represent the polar phosphate part of the molecule. The detailed structure of lecithin is shown in one part of the micelle.

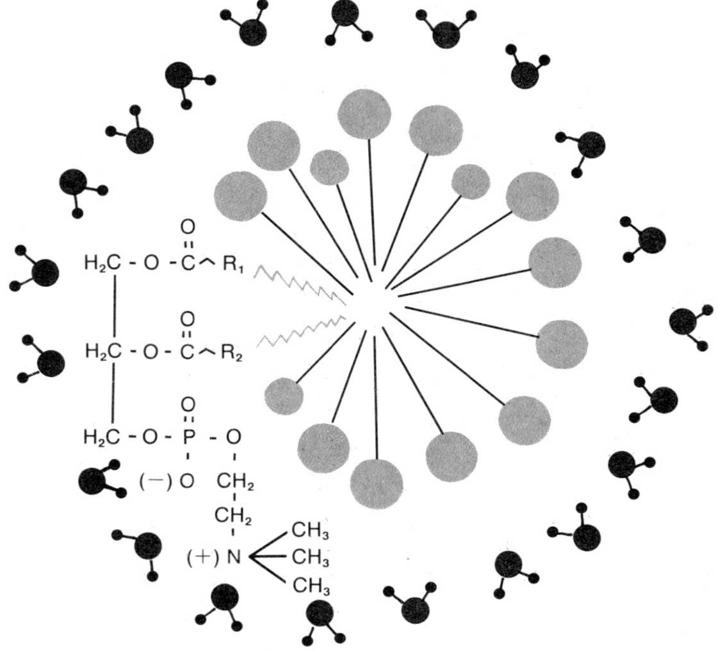

Phospholipids. The class of lipids known as *phospholipids* contains a number of different categories, which have in common the fact that they are lipids that contain a phosphate group. The most common type of phospholipid molecule has this structure: the three-carbon alcohol molecule glycerol is attached to two fatty acid molecules; the third carbon atom of the glycerol molecule is attached to a phosphate group, and the phosphate group in turn is bonded to other molecules. If the phosphate group is attached to a nitrogen-containing choline molecule, the phospholipid molecule thus formed is known as *lecithin* (fig. 2.18). Figure 2.18 shows a simple way of illustrating the structure of a phospholipid—the parts of the molecule capable of ionizing and thus becoming charged are shown as a circle, whereas the nonpolar parts of the molecule are represented by lines.

Since the nonpolar ends of phospholipids are hydrophobic, they tend to group together when mixed in water; this allows the hydrophilic parts (which are polar) to face the surrounding water molecules (fig. 2.19). Such aggregates of molecules are called **micelles.** The dual nature of phospholipid molecules (part polar, part nonpolar) allows them to form the major component of the cell membrane as well as to alter the interaction of water molecules and thus decrease the surface tension of water. This latter function of phospholipids, which makes them *surfactants* (surface-active-agents), prevents collapse of the lungs.

Steroids. The structure of steroid molecules is quite different from that of triglycerides or phospholipids, and yet steroids are still included in the lipid category of molecules because they are nonpolar and insoluble in water. All steroid molecules have the same basic structure; three six-carbon rings are joined to one five-carbon ring (fig. 2.20). However, different kinds of steroids have different functional groups attached to this basic structure, and they vary in the number and position of the double covalent bonds between the carbon atoms in the rings.

Cholesterol is an important molecule in the body because it serves as the precursor (parent molecule) for the steroid hormones produced by the gonads and adrenal cortex. The testes and ovaries (collectively called the *gonads*) secrete **sex steroids,** which include estradiol and progesterone from the ovaries and testosterone from the testes. The adrenal cortex secretes the **corticosteroids,** including hydrocortisone and aldosterone, among others.

Figure 2.20. Cholesterol and some steroid hormones derived from cholesterol.

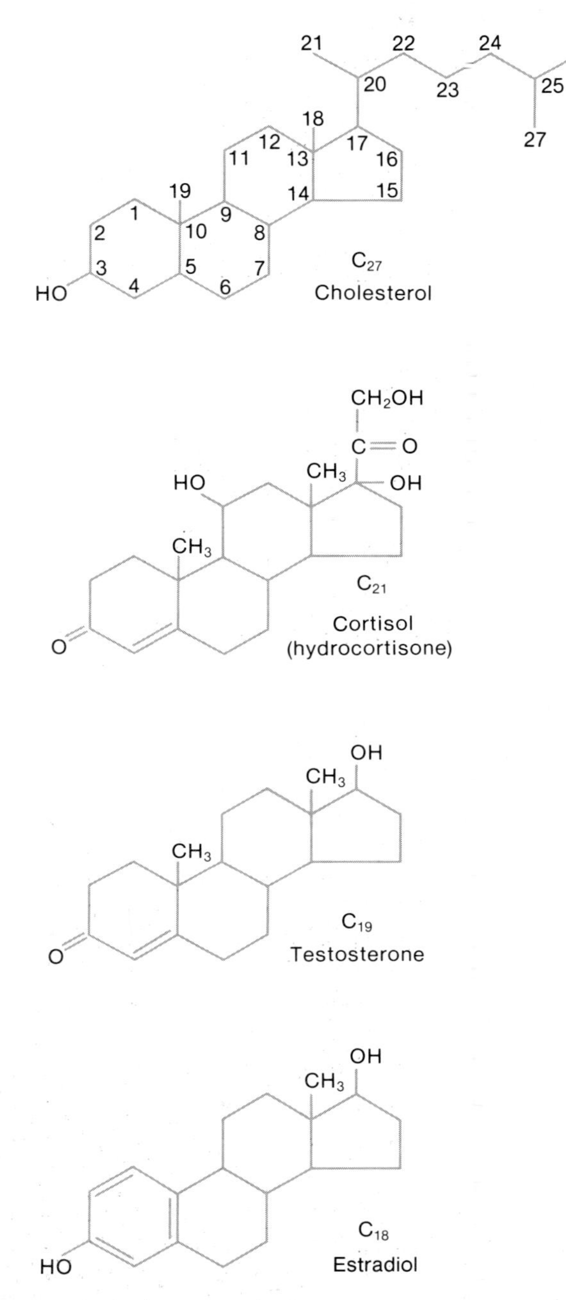

Figure 2.21. Structural formulas of some prostaglandins.

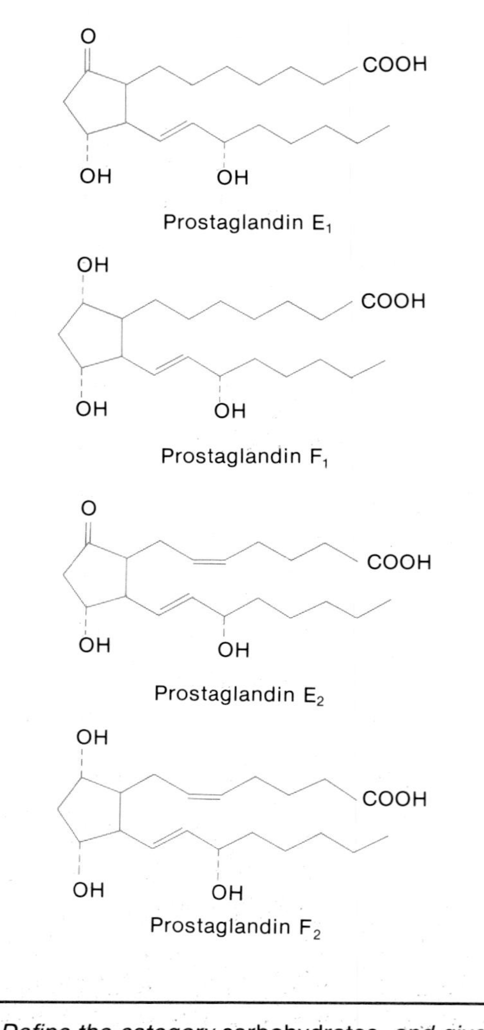

Prostaglandins. Prostaglandins are a type of fatty acid (with a cyclic hydrocarbon group), which have a variety of regulatory functions. Although their name is derived from the fact that they were originally found in the semen as a secretion of the prostate gland, it has since been shown that they are produced by and active in almost all tissues. Prostaglandins are implicated in the regulation of blood vessel diameter, ovulation, uterine contraction during labor, inflammation reactions, blood clotting, and many other functions. Some of the different types of prostaglandins are shown in figure 2.21.

1. *Define the category* carbohydrates, *and give examples of monosaccharides, disaccharides, and polysaccharides.*
2. *Describe dehydration synthesis and hydrolysis and where these reactions occur in the body, and give examples of these reactions in the metabolism of carbohydrates.*
3. *Define the different categories of lipids, and explain the structure and biological functions of these molecules.*

Proteins and Nucleic Acids

Proteins and nucleic acids are large molecules formed of amino acids and nucleotides, respectively. Proteins provide much of the structure of body cells and are important for many physiological activities, and nucleic acids provide the basis for the genetic control of cellular activity. Proteins and nucleic acids are giant molecules, or *macromolecules.* The macromolecule **DNA (deoxyribonucleic acid)** contains the genetic code; the macromolecules **RNA**

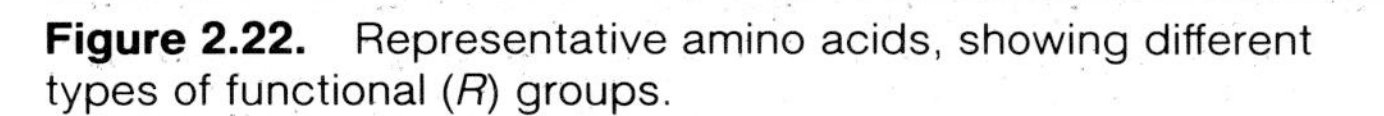

Figure 2.22. Representative amino acids, showing different types of functional (*R*) groups.

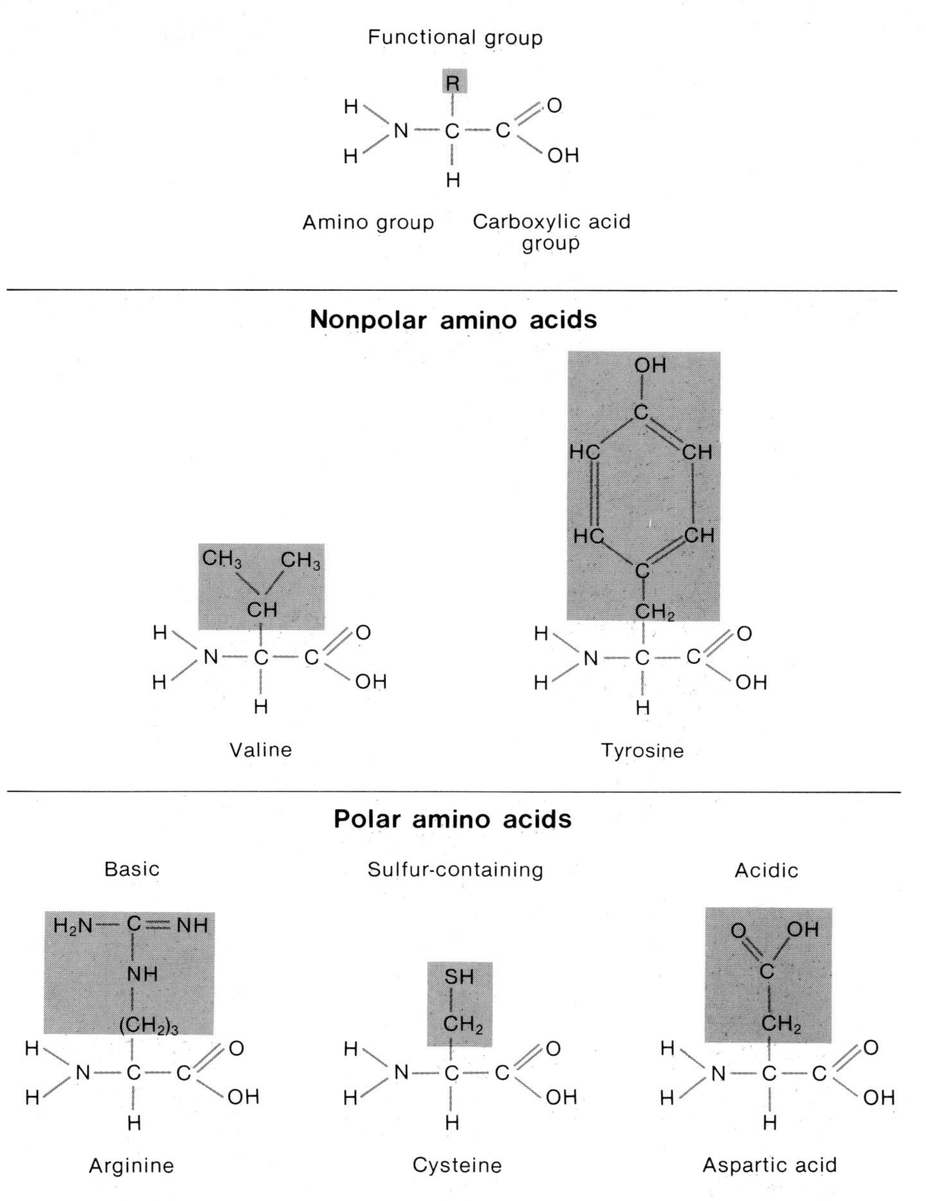

(**ribonucleic acid**) and **protein** provide the means by which this code can be expressed within the cells of the body. In order to understand these genetic control mechanisms (chapter 3), a knowledge of the structure of proteins and nucleic acids is required.

Proteins

Proteins consist of long chains of subunits called **amino acids.** As the name implies, each amino acid contains an *amino group* (NH_2) on one end of the molecule and a *carboxylic acid group* (COOH) on another end. There are approximately twenty different amino acids, with different structures and chemical properties that are used to build proteins. These differences are due to differences in the *functional groups* of these amino acids. *R* is the abbreviation for *functional group* in the general formula for an amino acid (fig. 2.22). The *R* symbol actually stands for the word *residue,* but it can be thought of as indicating the "rest of the molecule."

When amino acids are joined together by dehydration synthesis, the hydrogen from the amino end of one amino acid combines with the hydroxyl group of the carboxylic acid end of another amino acid. As a covalent bond is formed between the two amino acids, water is produced (fig. 2.23). The bond between adjacent amino acids is called a **peptide bond,** and the compound formed is called a *peptide.* When many amino acids are joined in this way, a chain of amino acids, or **polypeptide,** is produced.

Figure 2.23. The formation of peptide bonds by a dehydration synthesis reaction between amino acids.

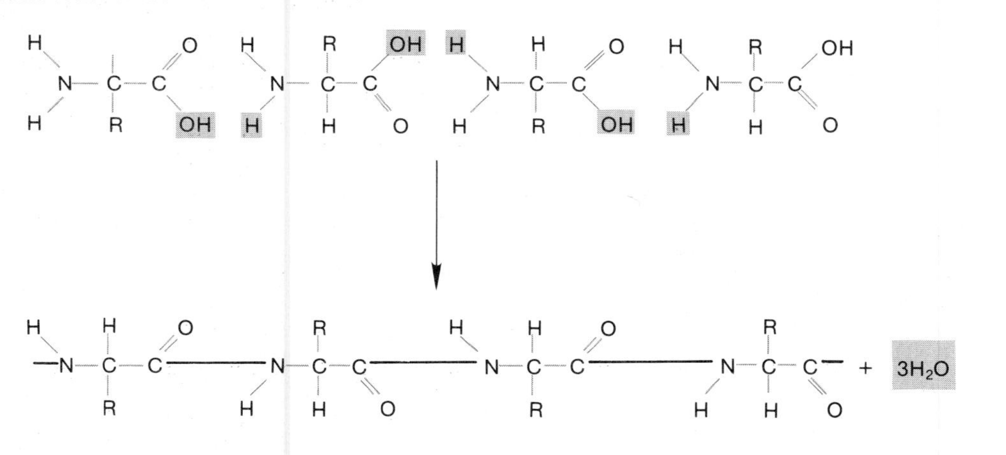

Figure 2.24. A polypeptide chain, showing (*a*) its primary structure and (*b*) secondary structure.

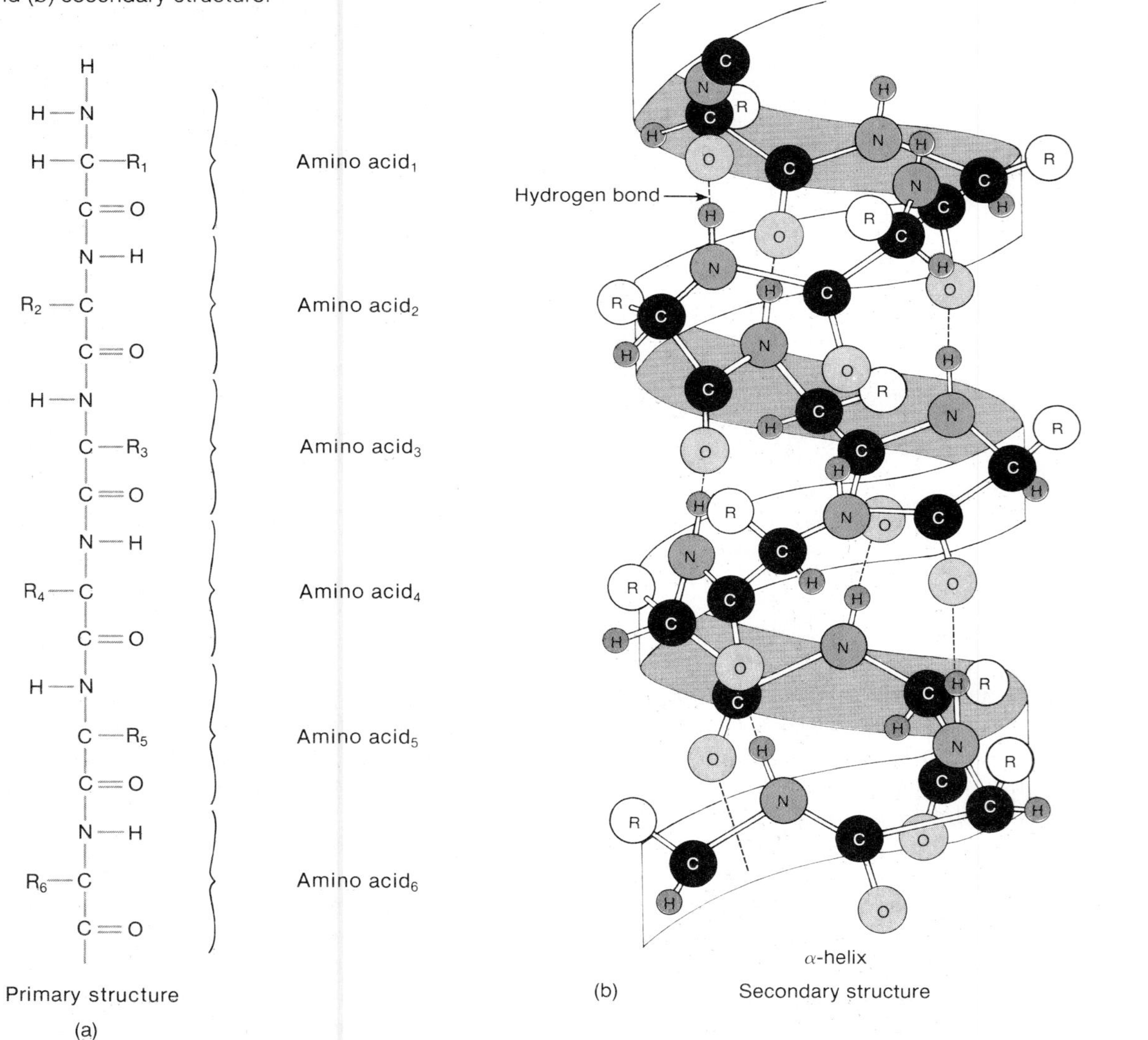

Figure 2.25. The tertiary structure of a protein. (*a*) interactions between functional (*R*) groups of amino acids result in (*b*) the formation of complex three-dimensional shapes of proteins.

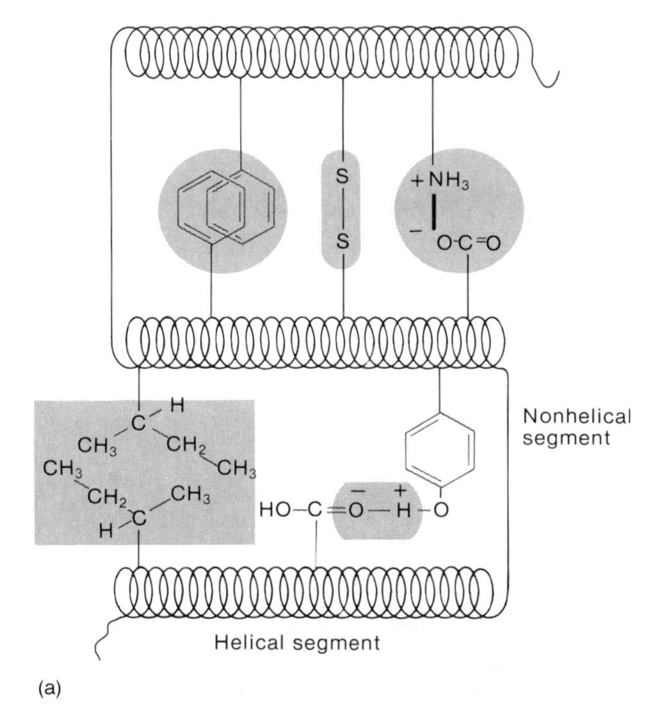

Carboxyl end

Amino end

(b)

The lengths of polypeptide chains vary greatly. A hormone called *thyrotrophin-releasing hormone,* for example, is only three amino acids long, whereas *myosin,* a muscle protein, contains about forty-five hundred amino acids. When the length of a polypeptide chain becomes very long (greater than about a hundred amino acids), the molecule is called a **protein.**

Protein Structure. The structure of a protein can be described at four different levels. At the first level, the sequence of amino acids in the protein, called the **primary structure** of the protein, is described. Each type of protein has a different primary structure. All of the billions of *copies* of a given type of protein in the body, however, have the same structure, because the structure of a given protein is coded by the genes. The primary structure of a protein is illustrated in figure 2.24.

Weak interactions (such as hydrogen bonds) between functional (*R*) groups of amino acids in nearby positions in the polypeptide chain cause this chain to twist into a *helix.* The extent and location of the helical structure is different for each protein because of differences in amino acid composition. A description of the helical structure of a protein is termed its **secondary structure** (fig. 2.24).

Most polypeptide chains bend and fold on themselves to produce complex three-dimensional shapes, called the **tertiary structure** of the proteins. Each type of protein has its own characteristic tertiary structure. This is because the folding and bending of the polypeptide chain is produced by chemical interactions between particular amino acids that are located in different regions of the chain.

Most of the tertiary structure of proteins is formed and stabilized by weak chemical interactions (such as hydrogen bonds) between widely spaced amino acids. The tertiary structure of some proteins, however, is made more stable by covalent bonds between sulfur atoms (called *disulfide bonds* and abbreviated S-S) in the functional (*R*) groups of amino acids known as cysteines (fig. 2.25). These strong covalent bonds are the exception. Since most of the tertiary structure is stabilized by weak bonds, this structure can easily be disrupted by high temperature or by changes in pH. Irreversible changes in the tertiary structure of proteins produced by this means are referred to as *denaturation* of the proteins.

Table 2.4 Composition of selected proteins in the body.

Protein	Number of Polypeptide Chains	Nonprotein Component	Function
Hemoglobin	4	Heme pigment	Carries oxygen in the blood
Myoglobin	2	Heme pigment	Stores oxygen in muscle
Insulin	2	None	Hormone-regulating metabolism
Luteinizing hormone	1	Carbohydrate	Hormone that stimulates gonads
Fibrinogen	1	Carbohydrate	Involved in blood clotting
Mucin	1	Carbohydrate	Forms mucus
Blood group proteins	1	Carbohydrate	Produces blood types
Lipoproteins	1	Lipids	Transports lipids in blood

Denatured proteins retain their primary structure (the peptide bonds are not broken) but have altered chemical properties. Cooking a pot roast, for example, alters the texture of the meat proteins—it doesn't result in an amino acid soup. Denaturation is most dramatically demonstrated by frying an egg. Egg-albumin proteins are soluble in their native state, in which they form the clear, viscous fluid of a raw egg. When denatured by cooking, these proteins change shape, cross-bond with each other, and by this means form an insoluble white precipitate—the egg white.

Some proteins (such as hemoglobin and insulin) are composed of a number of polypeptide chains covalently bonded together. This is the **quaternary structure** of these proteins. Insulin, for example, is composed of two polypeptide chains, one that is twenty-one amino acids long, the other that is thirty amino acids long. Hemoglobin (the protein in red blood cells that carries oxygen) is composed of four separate polypeptide chains. The composition of various body proteins is shown in table 2.4.

Many proteins in the body are normally found combined, or *conjugated,* with other types of molecules. *Glycoproteins* are proteins conjugated with carbohydrates. Examples of such molecules include certain hormones and some proteins found in the cell membrane. *Lipoproteins* are proteins conjugated with lipids. These are found in cell membranes and in the plasma (the fluid portion of the blood). Proteins conjugated with pigment molecules are *chromoproteins.* These include hemoglobin, which transports oxygen in red blood cells, and the cytochromes, which are needed for oxygen utilization and energy production within cells.

Figure 2.26. A photomicrograph of collagen fibers.

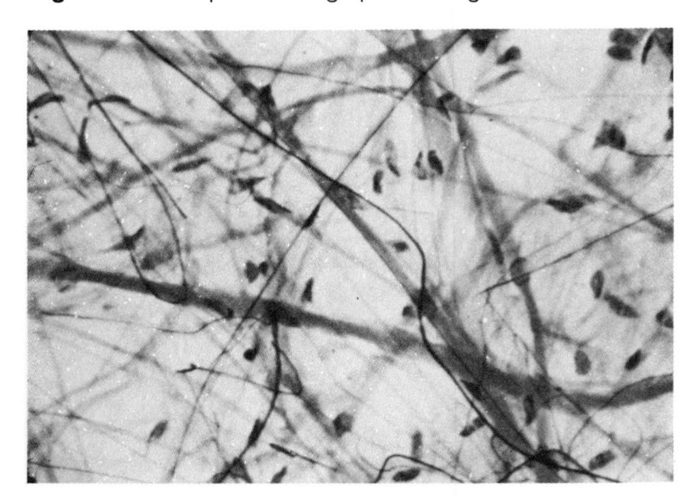

Functions of Proteins. Because of their tremendous structural diversity, proteins can serve a wider variety of functions than any other type of molecule in the body. Many proteins, for example, contribute significantly to the structure of different tissues and in this way play a passive role in the functions of these tissues. Examples of such *structural proteins* include collagen (fig. 2.26) and keratin. Collagen is a fibrous protein that provides tensile strength to connective tissues, such as tendons and ligaments. Keratin is found in the outer layer of dead cells in the epidermis, where it serves to prevent water loss through the skin.

Many proteins serve a more active role in the body where specialized structure and function are required. *Enzymes* and *antibodies,* for example, are proteins—no

Figure 2.27. The general structure of a nucleotide and the formation of sugar-phosphate bonds between nucleotides to form a polymer.

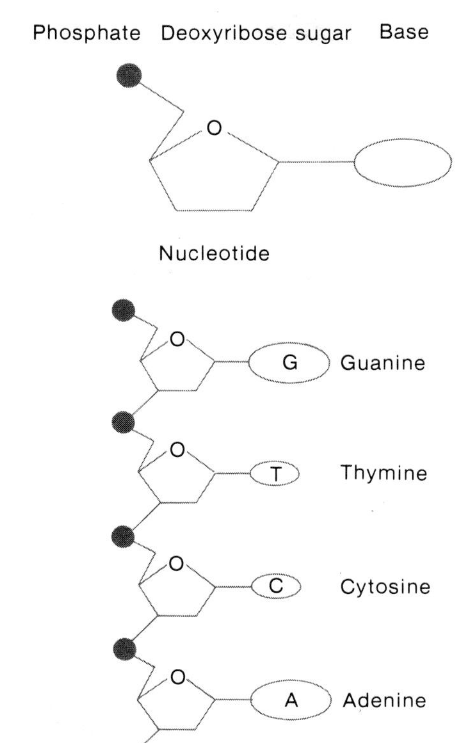

other type of molecule could provide the vast array of different structures needed for these functions. Proteins in cell membranes serve as *receptors* for specific regulator molecules (such as hormones) and as *carriers* that transport specific molecules across the membrane. Proteins provide the diversity of shape and chemical properties for the specificity required by these functions.

Nucleic Acids

Nucleic acids include the macromolecules of **DNA** and **RNA,** which are critically important in genetic regulation, and the subunits from which these molecules are formed. These subunits are known as *nucleotides.*

Nucleotides are used as subunits in the formation of long polynucleotide chains. The nucleotides themselves, however, are composed of subunits. Each nucleotide is composed of three parts—a five-carbon sugar, a phosphate group bonded to one end of the sugar, and a *nitrogenous base* bonded to the other end of the sugar (fig. 2.27). The nucleotide bases are cyclic nitrogen-containing molecules with either one ring of carbons (the *pyrimidines*) or two rings (the *purines*).

Deoxyribonucleic Acid. The structure of DNA serves as the basis for the genetic code. One might, therefore, expect DNA to have an extremely complex structure. Actually, although DNA is the largest molecule in the cell, it has a simpler structure than that of most proteins. This simplicity of structure deceived some of the early scientists into believing that the protein content of chromosomes, rather than their DNA content, provided the basis for the genetic code.

Sugar molecules in the nucleotides of DNA are a type of pentose (five-carbon) sugar called **deoxyribose** (hence the name for this nucleic acid). Each deoxyribose sugar can be covalently bonded to one of four possible bases. These bases include the two purines (adenine and guanine) and the two pyrimidines (cytosine and thymine). There are thus four different types of nucleotides that can be used to produce the long DNA chains.

Figure 2.28. The four nitrogenous bases in deoxyribonucleic acid (DNA). Notice that hydrogen bonds can form between guanine and cytosine and between thymine and adenine.

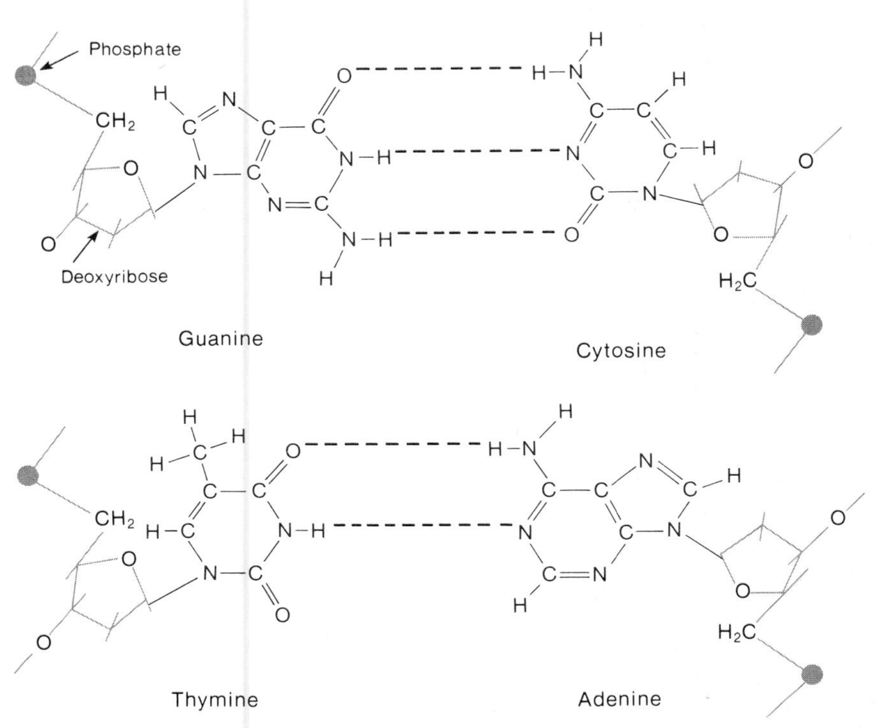

When nucleotides combine to form a chain, the phosphate group of one condenses with the deoxyribose sugar of another nucleotide. This forms a sugar-phosphate chain as water is removed in dehydration synthesis. Since the nitrogenous bases are attached to the sugar molecules, the sugar-phosphate chain looks like a "backbone" from which the bases project. Each of these bases can form hydrogen bonds with other bases, which are in turn joined to a different chain of nucleotides. Such hydrogen bonding between bases thus produces a *double-stranded* DNA molecule; the two strands are like a staircase, with the paired bases as steps (fig. 2.28).

Actually, the two chains of DNA twist about each other to form a **double helix**—the molecule is like a spiral staircase (fig. 2.29). It has been shown that the number of purine bases in DNA is equal to the number of pyrimidine bases. The reason for this is explained by the **law of complementary base pairing;** adenine can only pair with thymine (through two hydrogen bonds), whereas guanine can only pair with cytosine (through three hydrogen bonds). Knowing this rule, we could predict the base sequence of one DNA strand if we knew the sequence of bases in the complementary strand.

Although we can predict which base is opposite a given base in DNA, we cannot predict which bases are above or below that position within a single polynucleotide chain. Although there are only four bases, the number of possible base sequences along a stretch of several thousand nucleotides (the length of a gene) is almost infinite. Despite the almost infinite possible variety of sequences, almost all of the billions of copies of a particular gene in a person are identical.

Ribonucleic Acid. The genetic information contained in DNA functions to direct the activities of the cell through its production of another type of nucleic acid—*RNA (ribonucleic acid).* Like DNA, RNA consists of long chains of nucleotides joined together by sugar-phosphate bonds.

Figure 2.29. The double helix structure of DNA.

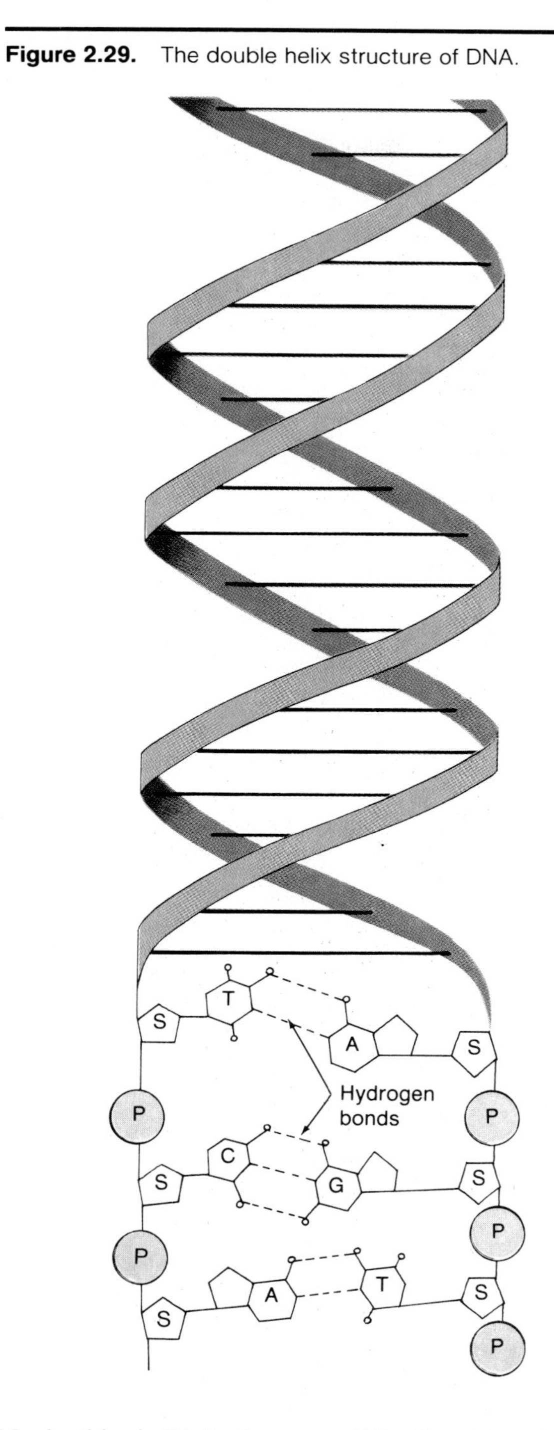

Nucleotides in RNA, however, differ from those in DNA (fig. 2.30) in three ways: (1) a **ribonucleotide** contains the sugar *ribose* (instead of deoxyribose); (2) the base *uracil* is found in place of thymine; and (3) RNA is composed of a single polynucleotide strand (it is not double-stranded like DNA).

Figure 2.30. Differences between the nucleotides and sugars in DNA and RNA.

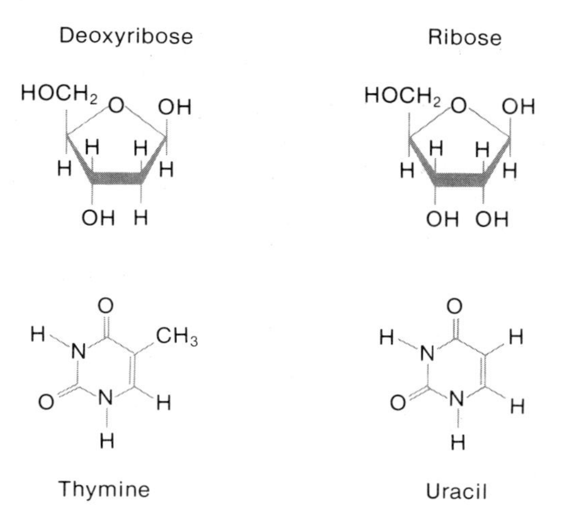

There are three types of RNA molecules that function in the cytoplasm of cells: *messenger RNA (mRNA), transfer RNA (tRNA),* and *ribosomal RNA (rRNA).* All three types are made within the cell nucleus by using information contained in DNA as a guide. The synthesis of RNA and its role in gene expression are discussed in chapter 3.

1. *Write the general formula for an amino acid, and describe how amino acids differ from each other.*
2. *Describe the structure of proteins, list some of the functions of different proteins, and explain how this structure grants specificity to the actions of proteins.*
3. *Define the law of complementary base pairing, and explain how this law allows us to predict the sequence of bases on one DNA strand if we know the sequence of bases on the other strand.*
4. *Compare the structure of DNA with that of RNA.*

Summary

Atoms, Ions, and Chemical Bonds p. 26

I. Covalent bonds are formed by atoms that share electrons; these are the strongest type of chemical bonds.
   A. Electrons are equally shared in nonpolar covalent bonds and unequally shared in polar covalent bonds.
   B. Atoms of oxygen, nitrogen, and phosphorus strongly attract electrons and become electrically negative compared to the other atoms sharing electrons with them.

II. Ionic bonds are formed by atoms that transfer electrons; these weak bonds join atoms together in an ionic compound.
   A. If one atom in this compound takes the electron from another atom, it gains a net negative charge and the other becomes positively charged.
   B. Ionic bonds easily break when the ionic compound is dissolved in water; dissociation of the ionic compound yields charged atoms called ions.

III. When hydrogen is bonded to an electronegative atom, it gains a slight positive charge and is weakly attracted to another electronegative atom; this weak attraction is a hydrogen bond.

IV. Acids donate hydrogen ions to solution, whereas bases lower the hydrogen ion concentration of a solution.
   A. The pH scale is a negative function of the logarithm of the hydrogen ion concentration.
   B. In a neutral solution the concentration of H^+ is equal to the concentration of OH^-, and the pH is 7.
   C. Acids raise the H^+ concentration and thus lower the pH below 7; bases lower the H^+ concentration and thus raise the pH above 7.

V. Organic molecules contain atoms of carbon joined together by covalent bonds; atoms of nitrogen, oxygen, phosphorus, or sulfur may be present as specific functional groups in the organic molecule.

Carbohydrates and Lipids p. 34

I. Carbohydrates contain carbon, hydrogen, and oxygen, usually in a ratio of 1:2:1.
   A. Carbohydrates consist of simple sugars (monosaccharides), disaccharides, and polysaccharides (such as glycogen).
   B. Covalent bonds between monosaccharides are formed by dehydration synthesis, or condensation; bonds are broken by hydrolysis reactions.

II. Lipids are organic molecules that are insoluble in polar solvents such as water.
   A. Triglycerides (fat and oil) consist of three fatty acid molecules joined to a molecule of glycerol.
   B. Ketone bodies are smaller derivatives of fatty acids.
   C. Phospholipids (such as lecithin) are phosphate-containing lipids that have a polar group, which is hydrophilic; the rest of the molecule is hydrophobic.
   D. Steroids (including the hormones of the adrenal cortex and gonads) are lipids with a characteristic five-ring structure.
   E. Prostaglandins are a family of cyclic fatty acids, which serve a variety of regulatory functions.

Proteins and Nucleic Acids p. 40

I. Proteins are composed of long chains of amino acids bonded together by covalent peptide bonds.
   A. Each amino acid contains an amino group, a carboxyl group, and a functional group that is different for each of the more than twenty different amino acids.
   B. The polypeptide chain may be twisted into a helix (secondary structure) and bent and folded to form the tertiary structure of the protein.
   C. Proteins that are composed of two or more polypeptide chains are said to have a quaternary structure.
   D. Proteins may be combined with carbohydrates, lipids, or other molecules.
   E. Because of their great variety of possible structures, proteins serve a wider variety of specific functions than any other type of molecule.

II. Nucleic acids include the macromolecules DNA and RNA, and their nucleotide subunits.
   A. Each nucleotide consists of a pentose sugar (ribose or deoxyribose), a phosphate, and a nitrogenous base.
   B. There are four nitrogenous bases, and thus four possible nucleotides used to make either DNA or RNA.
   C. DNA bases include adenine, guanine, cytosine, and thymine; in RNA, uracil is substituted for thymine.
   D. DNA consists of two polynucleotide strands, in the form of a double helix, with each strand joined to the other by hydrogen bonding between complementary bases.
   E. Complementary base pairing in DNA occurs because adenine can only form hydrogen bonds to thymine (and vice versa), and guanine can only pair with cytosine (and vice versa).
   F. RNA consists of only one polynucleotide strand.
   G. Messenger RNA (mRNA), transfer RNA (tRNA), and ribosomal RNA (rRNA) are made in the nucleus according to the instructions given in the base sequence of DNA.

Review Activities

Objective Questions

1. Which of the following statements about atoms is *true?*
   (a) They have more protons than electrons.
   (b) They have more electrons than protons.
   (c) They are electrically neutral.
   (d) They have as many neutrons as they have electrons.
2. The bond between oxygen and hydrogen in a water molecule is a(n)
   (a) hydrogen bond
   (b) polar covalent bond
   (c) nonpolar covalent bond
   (d) ionic bond
3. Which of the following is a nonpolar covalent bond?
   (a) The bond between two carbons.
   (b) The bond between sodium and chloride.
   (c) The bond between two water molecules.
   (d) The bond between nitrogen and hydrogen.
4. Solution A has a pH of 2, and solution B has a pH of 10. Which of the following statements about these solutions is *true?*
   (a) Solution A has a higher H^+ concentration than solution B.
   (b) Solution B is basic.
   (c) Solution A is acidic.
   (d) All of these are true.
5. Glucose is a
   (a) disaccharide
   (b) polysaccharide
   (c) monosaccharide
   (d) phospholipid
6. Digestion reactions occur by means of
   (a) dehydration synthesis
   (b) hydrolysis
7. Carbohydrates are stored in the liver and muscles in the form of
   (a) glucose
   (b) triglycerides
   (c) glycogen
   (d) cholesterol
8. Lecithin is a
   (a) carbohydrate
   (b) protein
   (c) steroid
   (d) phospholipid
9. Which of the following lipids have regulatory roles in the body?
   (a) steroids
   (b) prostaglandins
   (c) triglycerides
   (d) both *a* and *b*
   (e) both *b* and *c*
10. The tertiary structure of a protein is *directly* determined by
    (a) the genes
    (b) the primary structure of the protein
    (c) enzymes that "mold" the shape of the protein
    (d) the position of peptide bonds
11. If four bases in one DNA strand are A (adenine), G (guanine), C (cytosine), and T (thymine), the complementary bases in the opposite strand will be
    (a) T,C,G,A
    (b) C,G,A,T
    (c) A,G,C,T
    (d) U,C,G,A
12. Which of the following statements about RNA is *true?*
    (a) It is made in the nucleus.
    (b) It contains the base uracil.
    (c) It is double stranded.
    (d) Both *a* and *b* are true.
    (e) Both *b* and *c* are true.

Essay Questions

1. Compare and contrast nonpolar covalent bonds, polar covalent bonds, and ionic bonds.
2. Give the definition of an acid and base, and explain how these influence the pH of a solution.
3. Using dehydration synthesis and hydrolysis reactions, explain the relationships between starch in an ingested potato, liver glycogen, and blood glucose.
4. "All fats are lipids, but not all lipids are fats." Explain why this statement is true.
5. Explain the relationship between the primary structure of a protein and its secondary and tertiary structures.

Selected Readings

Darnell, J. E., Jr. October 1985. RNA. *Scientific American.*

Demers, L. M. September 1984. The effects of prostaglandins. *Diagnostic Medicine,* p. 37.

Doolittle, R. F. October 1985. Proteins. *Scientific American.*

Felsenfeld, G. October 1985. DNA. *Scientific American.*

Hakomori, Sen-itiroh. May 1986. Glycosphingolipids. *Scientific American.*

Jackson, R. W., and A. M. Gotto. 1974. Phospholipids in biology and medicine. *New England Journal of Medicine* 290:24.

Kuehl, F. A., Jr., and R. W. Egan. 1980. Prostaglandins, arachidonic acid, and inflammation. *Science* 210:978.

Prockop, D. J., and N. A. Guzman. 1977. Collagen disease and the biosynthesis of collagen. *Hospital Practice* 12:61.

Sharon, N. May 1974. Glycoproteins. *Scientific American.*

Sharon, N. November 1980. Carbohydrates. *Scientific American.*

Weinberg, R. A. October 1985. The molecules of life. *Scientific American.*

3 Cell Structure and Genetic Control

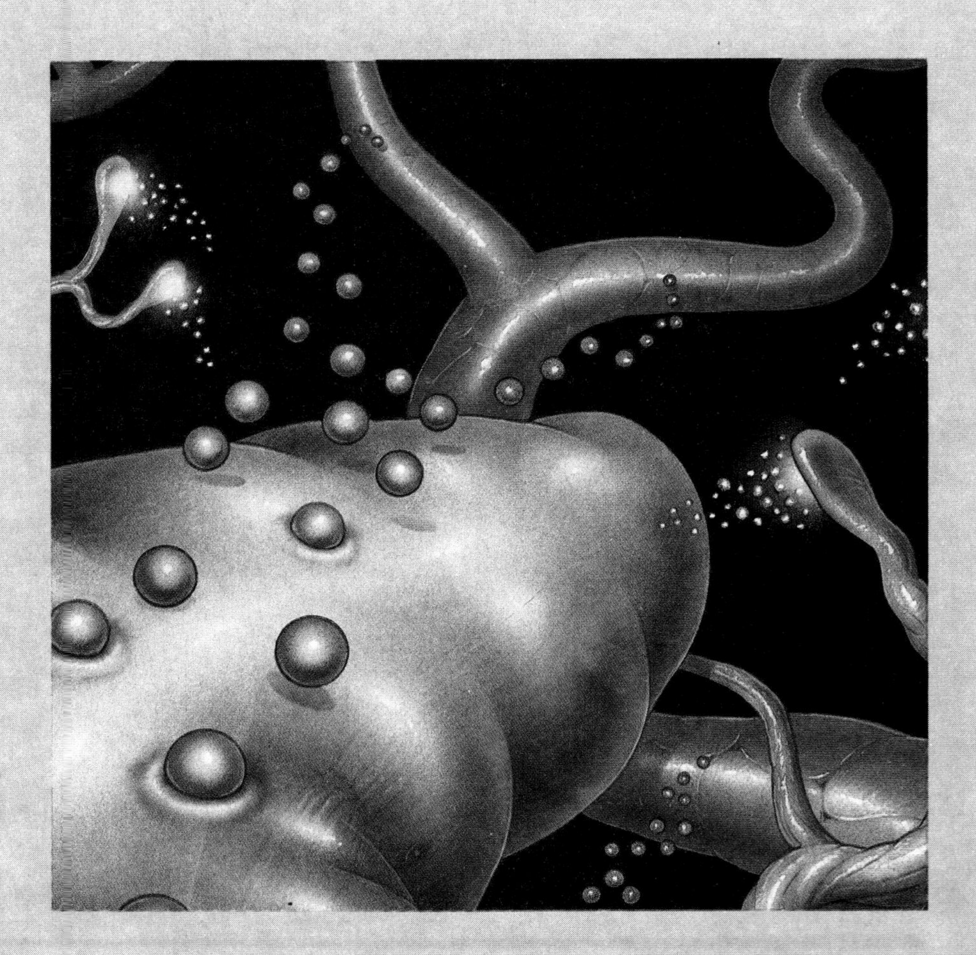

Objectives

By studying this chapter, you should be able to

1. describe the structure of the cell membrane and the cytoskeleton
2. explain the processes of endocytosis and exocytosis and describe the mechanisms of cellular movements
3. describe the structure and functions of lysosomes and mitochondria
4. describe the structure of the cell nucleus and the appearance and composition of chromatin
5. explain the semiconservative replication of DNA
6. describe the phases of the cell cycle
7. list the phases of mitosis and explain the events that occur in each phase
8. define the terms *hypertrophy, hyperplasia,* and *atrophy* and explain their significance
9. compare meiosis with mitosis and explain the separate functions of each process
10. define the terms *genetic transcription* and *genetic translation* and explain the significance of these processes
11. explain how the different types of RNA cooperate in the synthesis of proteins
12. explain the roles of the rough endoplasmic reticulum and Golgi apparatus in the synthesis and secretion of proteins

Outline

Table 3.1 Structure and function of cellular components.

Component	Structure	Function
Cell (plasma) membrane	Membrane composed of phospholipid and protein molecules	Gives form to cell and controls passage of materials in and out of cell
Cytoplasm	Fluid, jellylike substance in which organelles are suspended	Serves as matrix substance in which chemical reactions occur
Endoplasmic reticulum	System of interconnected membrane-forming canals and tubules	Supporting framework within cytoplasm; transports materials and provides attachment for ribosomes
Ribosomes	Granular particles composed of protein and RNA	Synthesize proteins
Golgi apparatus	Cluster of flattened, membranous sacs	Synthesizes carbohydrates and packages molecules for secretion; secretes lipids and glycoproteins
Mitochondria	Membranous sacs with folded inner partitions	Release energy from food molecules and transform energy into usable ATP
Lysosomes	Membranous sacs	Digest foreign molecules and worn and damaged cells
Peroxisomes	Spherical membranous vesicles	Contain certain enzymes; form hydrogen peroxide
Centrosome	Nonmembranous mass of two rodlike centrioles	Helps organize spindle fibers and distribute chromosomes during mitosis
Vacuoles	Membranous sacs	Store and excrete various substances within the cytoplasm
Fibrils and microtubules	Thin, hollow tubes	Support cytoplasm and transport materials within the cytoplasm
Cilia and flagella	Minute cytoplasmic extensions from cell	Move particles along surface of cell or move cell
Nuclear membrane	Membrane surrounding nucleus, composed of protein and lipid molecules	Supports nucleus and controls passage of materials between nucleus and cytoplasm
Nucleolus	Dense, nonmembranous mass composed of protein and RNA molecules	Forms ribosomes
Chromatin	Fibrous strands composed of protein and DNA molecules	Controls cellular activity for carrying on life processes

Source: Van De Graaff, Kent M., *Human Anatomy*. © 1984 Wm. C. Brown Publishers, Dubuque, Iowa. All Rights Reserved. Reprinted by permission.

Cell Structure and Movements

The cell is so small and so simple in appearance when viewed with the ordinary (light) microscope that it is difficult to conceive that each cell is a living entity unto itself. Equally amazing is the fact that the physiology of our organs and systems derive from the complex functions of the cells of which they are composed. Complexity of function demands complexity of structure even at the subcellular level.

As the basic functional unit of the body, each cell is a highly organized molecular factory. Cells come in a great variety of shapes and sizes. This great variation, which is also apparent in the subcellular structures within different cells, reflects the variation of functions of different cells in the body. All cells, however, share certain characteristics—such as the fact that they are surrounded by a cell membrane—and most cells possess the structures listed in table 3.1. Thus, although no single cell can be considered "typical," the general structure of cells can be shown with a single illustration (fig. 3.1).

Cell Membrane

Because both the intracellular and extracellular environments (or "compartments") are aqueous, a barrier must be present to prevent the loss of cellular molecules such as enzymes, nucleotides, and others that are water soluble. Since this barrier cannot itself be composed of water-soluble molecules, it makes sense that the cell membrane is composed of lipids.

The cell membrane and indeed all of the membranes surrounding organelles within the cell are composed primarily of phospholipids. Phospholipids, as described in

Figure 3.1. A generalized cell and the principal organelles.

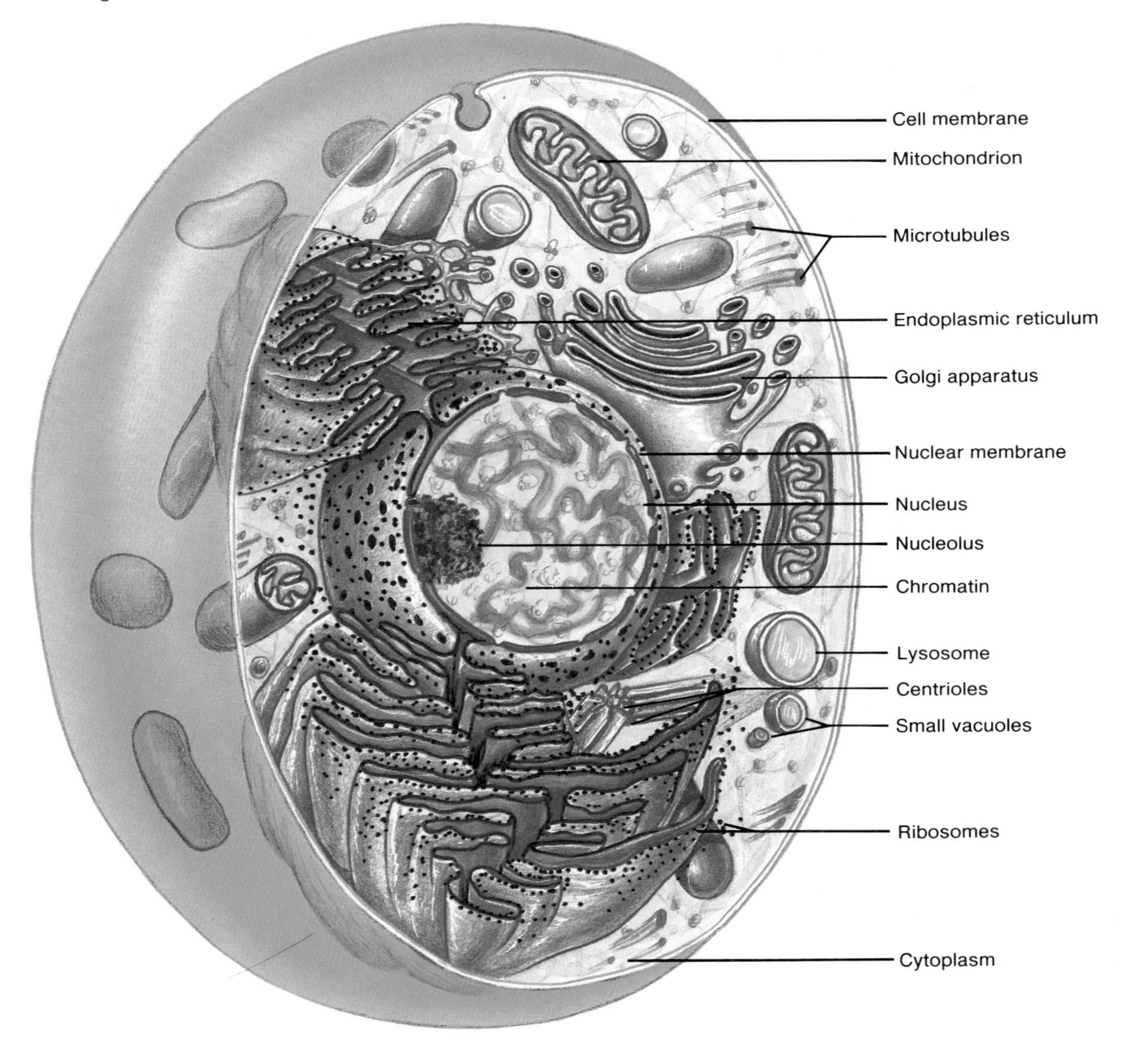

chapter 2, are polar on the end that contains the phosphate group and nonpolar (and hydrophobic) throughout the rest of the molecule. Since there is an aqueous environment on each side of the membrane, the hydrophobic parts of the molecules "huddle together" in the center of the membrane, leaving the polar ends exposed to water on both surfaces. This results in the formation of a double layer of phospholipids in the cell membrane.

The hydrophobic core of the membrane restricts the passage of water and water-soluble molecules and ions. Certain of these polar compounds, however, do pass through the membrane. The specialized functions and selective transport properties of the membrane are believed to be due to its protein content. Proteins are found partially submerged on each side of the membrane; some proteins span the membrane completely from one side to the other. Since the membrane is not solid—phospholipids and proteins are free to move laterally—the proteins within

Figure 3.2. The fluid-mosaic model of the cell membrane. The membrane consists of a double layer of phospholipids, with the phosphates (*open circles*) oriented outward and the hydrophobic hydrocarbons (*wavy lines*) oriented toward the center. Proteins may completely or partially span the membrane. Carbohydrates are attached to the outer surface.

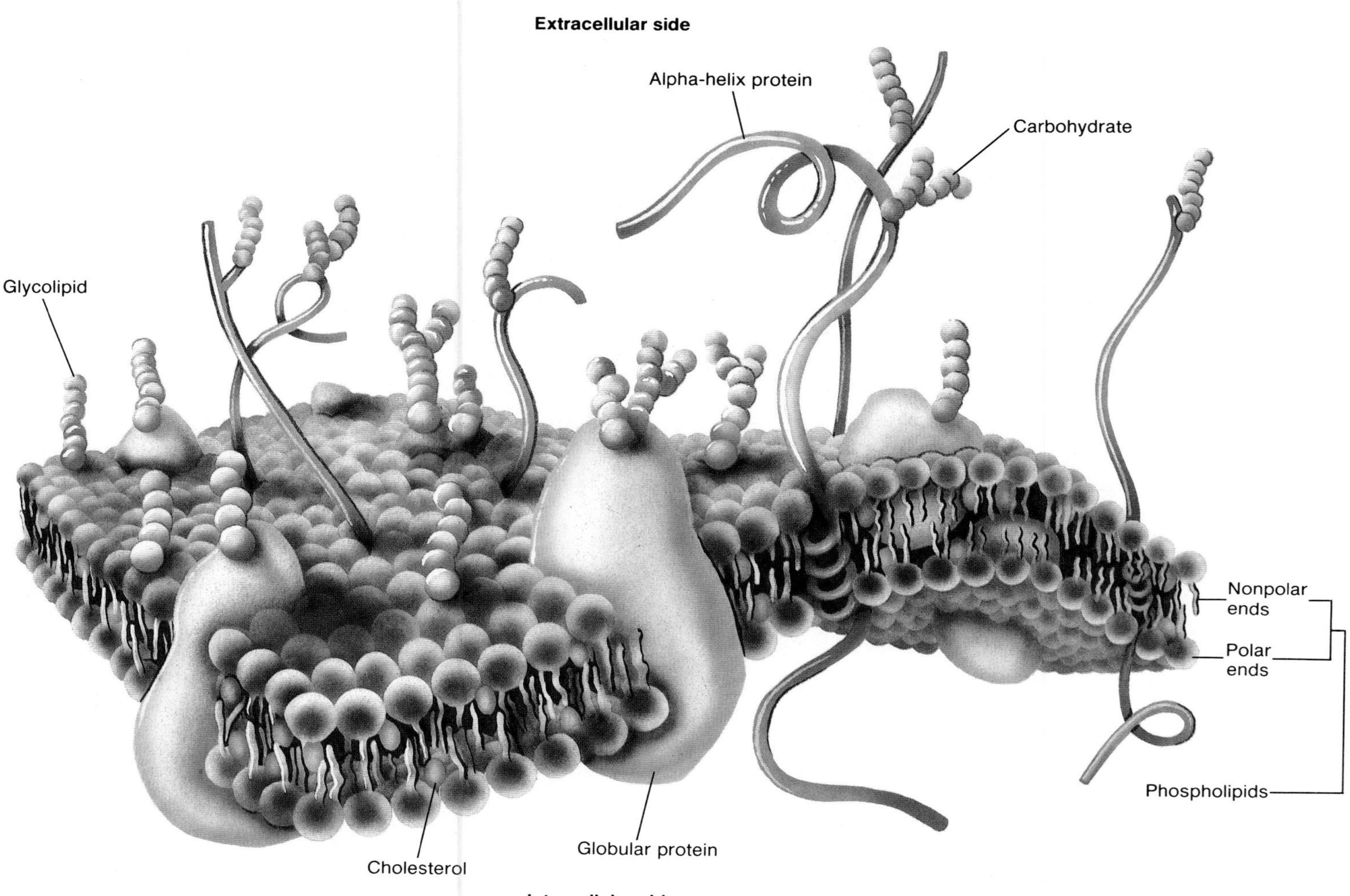

the phospholipid "sea" are not uniformly distributed, but rather present a mosaic pattern. This structure is known as the **fluid-mosaic model** of membrane structure (fig. 3.2).

The proteins found in the cell membrane serve a variety of functions, including (1) structural support; (2) transport of molecules across the membrane; (3) enzymatic control of chemical reactions at the cell surface; (4) receptors for hormones and other regulatory molecules that arrive at the outer surface of the membrane; and (5) cellular "markers" (antigens), which identify the blood and tissue type.

The cell membranes of all higher organisms contain cholesterol. The ratio of cholesterol to phospholipids, as well as the ratio of different types of phospholipids, determines the flexibility of the cell membrane. When there is an inherited defect in these ratios, the flexibility of the cell may be reduced. This could result, for example, in the inability of red blood cells to flex at the middle when passing through narrow blood channels (thereby causing the occlusion of these small vessels).

Figure 3.3. An immunofluorescence photograph of microtubules forming the cytoskeleton of a cell. Microtubules are visualized with the aid of antibodies against tubulin, the major protein component of the microtubules.

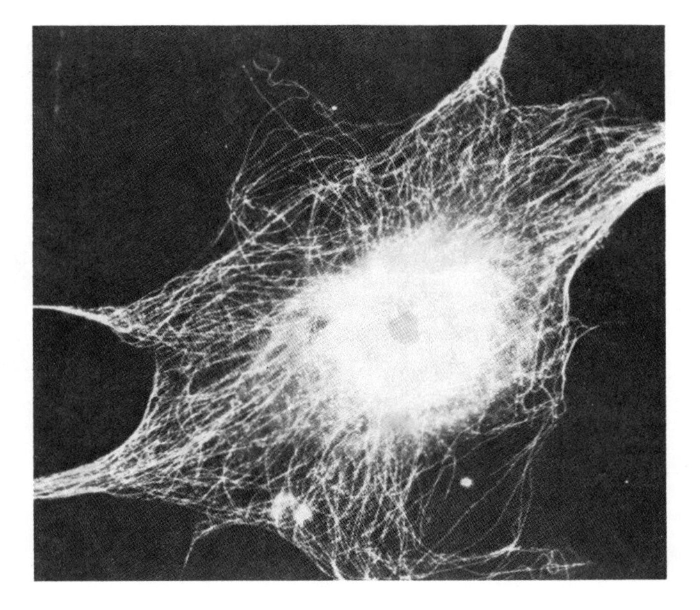

In addition to lipids and proteins, the cell membrane also contains carbohydrates, which are primarily attached to the outer surface of the membrane as glycoproteins and glycolipids. These surface carbohydrates have many negative charges and, as a result, affect the interaction of regulatory molecules with the membrane. The negative charges at the surface also affect interactions between cells—they help keep red blood cells apart, for example. Stripping the carbohydrates from the outer red blood cell surface results in their more rapid destruction by the liver, spleen, and bone marrow.

Cytoskeleton

The jellylike matrix within a cell (exclusive of that within the nucleus) is known as **cytoplasm.** When viewed in a microscope without special techniques, the cytoplasm appears to be uniform and unstructured. According to recent evidence, however, the cytoplasm is not a homogenous solution; it is, rather, a highly organized structure in which protein fibers—in the form of *microtubules* and *microfilaments*—are arranged in a complex latticework. These can be seen by fluorescence microscopy with the aid of antibodies against the proteins that compose these structures (fig. 3.3). The interconnected microfilaments and microtubules are believed to provide structural organization for cytoplasmic enzymes and support for various organelles.

The latticework of microfilaments and microtubules is thus said to function as a **cytoskeleton** (fig. 3.4). The structure of this "skeleton" is not rigid; it has been shown to be capable of quite rapid reorganization. Contractile proteins—including actin and myosin, which are responsible for muscle contraction—may be able to shorten the length of some microfilaments. The cytoskeleton may thus represent the cellular "musculature." Microtubules, for example, form the *spindle apparatus* that pulls chromosomes away from each other in cell division; they also form the central parts of cilia and flagella.

Cellular Movements

Some body cells—including certain white blood cells and macrophages in connective tissues and microglial cells in the brain—are able to move like an amoeba (a single-celled animal). This "amoeboid" movement is performed by the extension of parts of the cytoplasm to form *pseudopods,* which attach to a substrate and pull the cell along. These cells—as well as liver cells, which are not capable of amoeboid movement—use pseudopods to surround and engulf particles of organic matter (such as bacteria). This process is a type of cellular "eating" called **phagocytosis.**

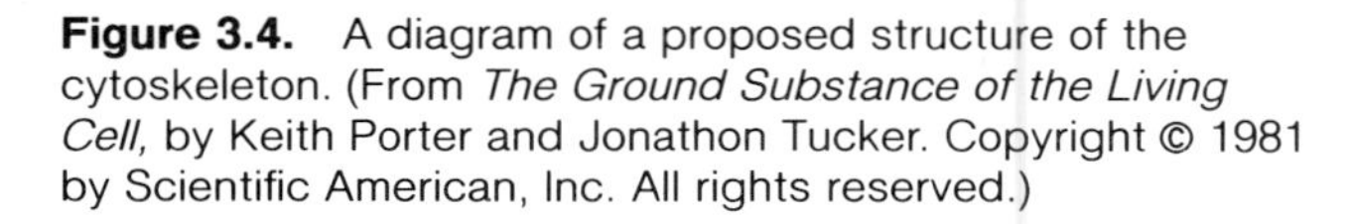

Figure 3.4. A diagram of a proposed structure of the cytoskeleton. (From *The Ground Substance of the Living Cell,* by Keith Porter and Jonathon Tucker. Copyright © 1981 by Scientific American, Inc. All rights reserved.)

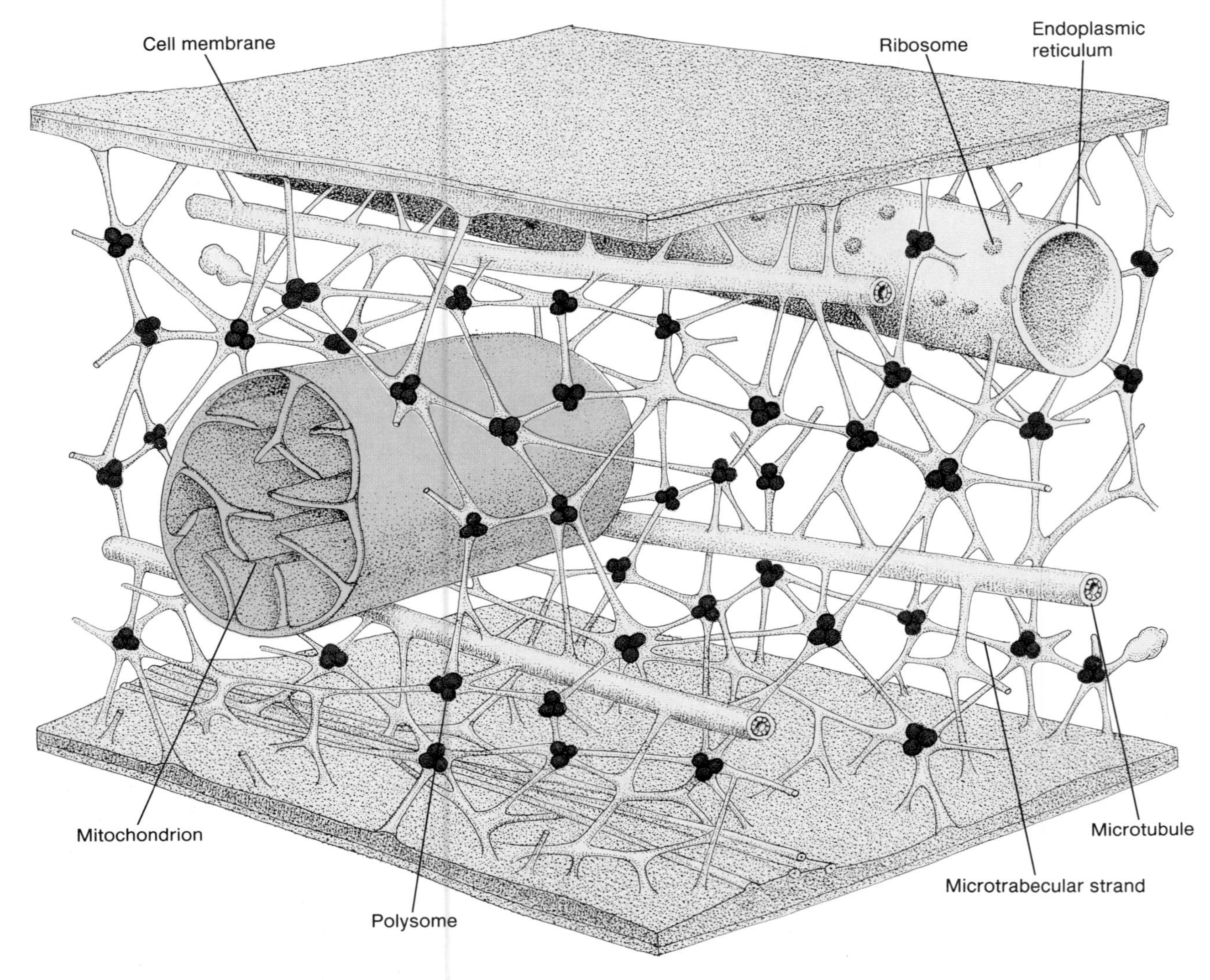

Phagocytic cells surround their victim with pseudopods, which join together and fuse (fig. 3.5). After the inner membrane of the pseudopods forms a continuous membranous barrier around the ingested particle it pinches off from the cell membrane, so that the ingested particle is contained in a *food vacuole* within the cell. The ingested particle will subsequently be digested by enzymes contained in a different organelle (the lysosome).

Pinocytosis is a related process performed by many cells. Instead of forming pseudopods, the cell membrane invaginates to produce a deep, narrow furrow. The membrane near the surface of this furrow then fuses, and a small vacuole, containing extracellular fluid, is pinched off and enters the cell.

Endocytosis and Exocytosis. The processes by which part of the extracellular environment are brought into a cell by invagination of the cell membrane are called **endocytosis.** There are three types of endocytosis: phagocytosis, pinocytosis (as previously described), and *receptor-mediated endocytosis.* In the latter case, the interaction of very specific molecules in the extracellular fluid with specific membrane receptor proteins causes the membrane to invaginate, fuse, and pinch off to form a *vesicle*—a small vacuole (fig. 3.6). Vesicles formed in this way contain extracellular fluid and molecules that could not have passed by other means into the cell.

Proteins and other molecules produced within the cell that are destined for export (secretion) are packaged inside of the cell within vesicles by an organelle known as the Golgi apparatus. In the process of **exocytosis,** these

Figure 3.5. Scanning electron micrographs of phagocytosis, showing the formation of pseudopods and the entrapment of the prey within a food vacuole.

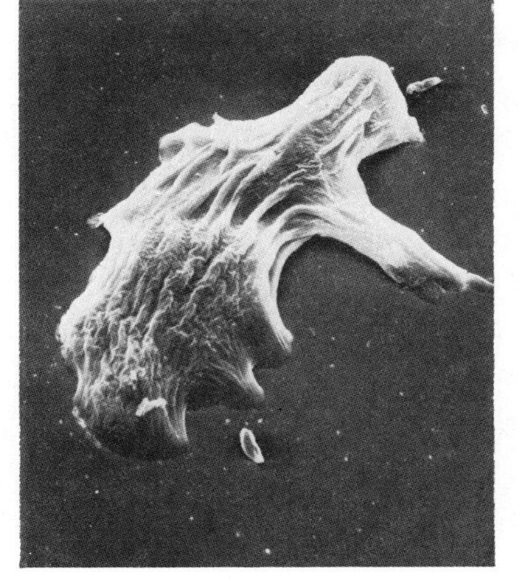

(a)

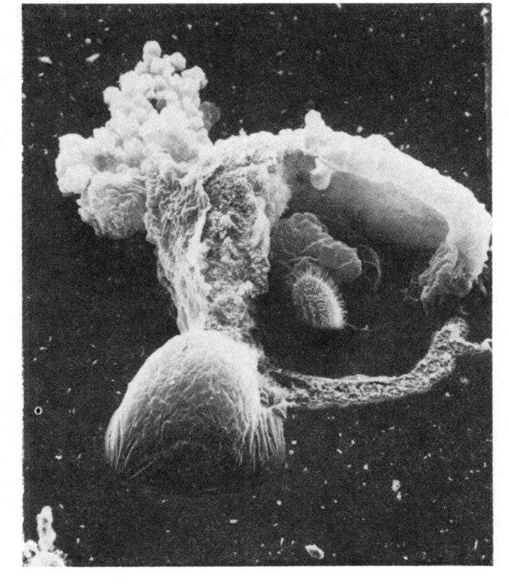

(b)

Figure 3.6. Stages (1–4) of endocytosis, where specific bonding of extracellular particles to membrane receptor proteins is believed to occur.

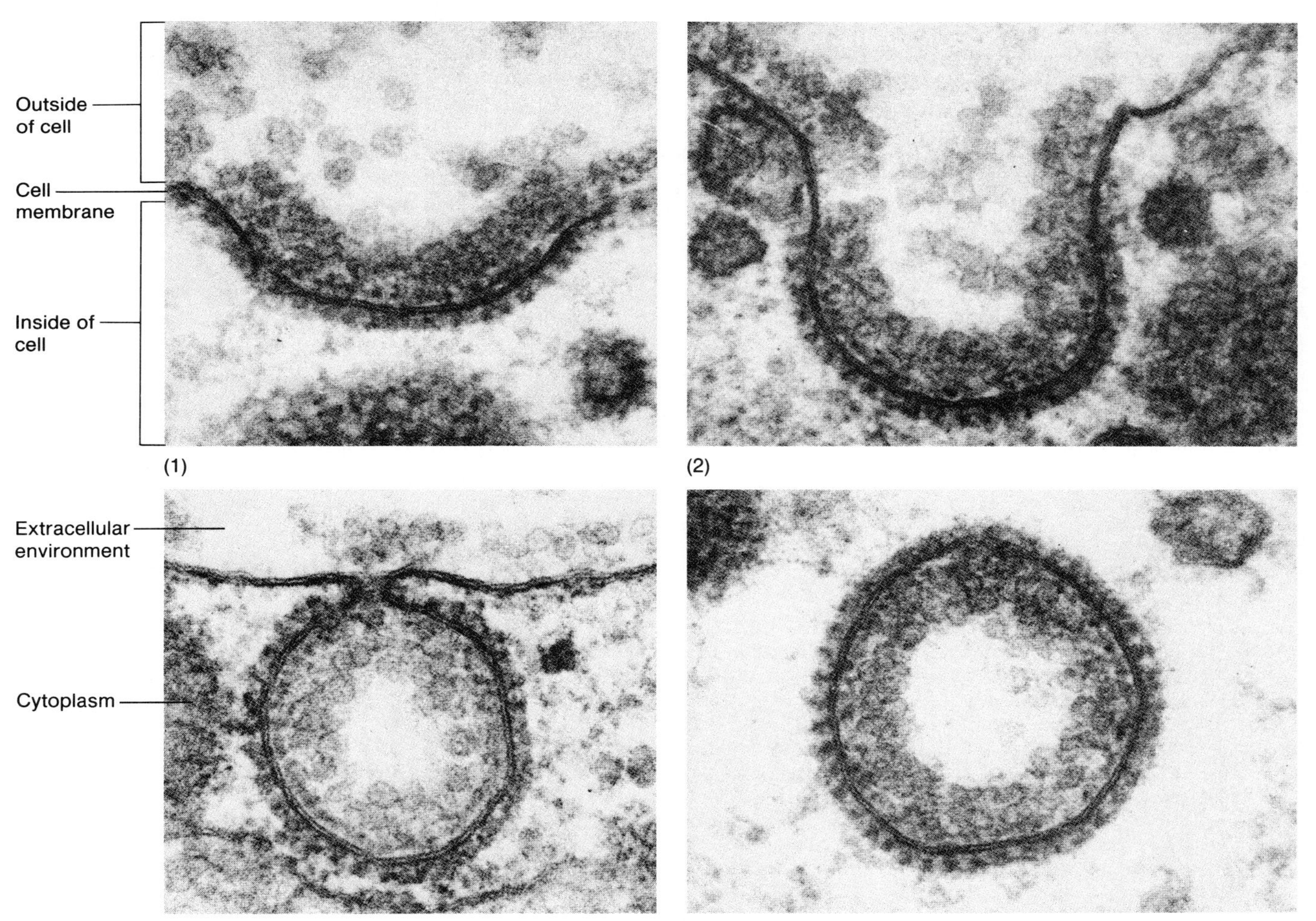

Figure 3.7. Electron micrographs of cilia, showing (*a*) longitudinal and (*b*) cross sections. Notice the characteristic "9 + 2" arrangement of microtubules in the cross sections.

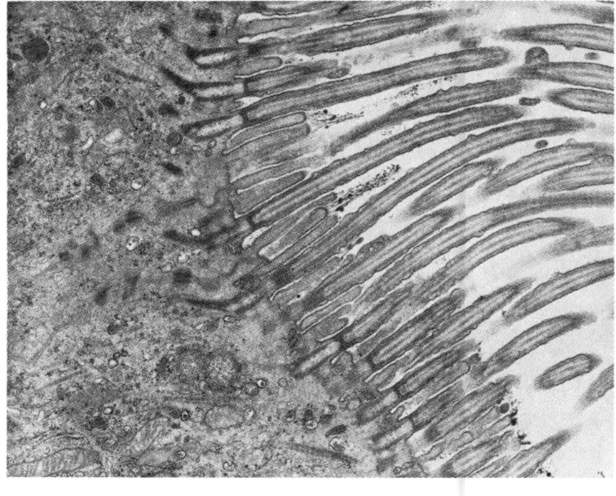

(a)

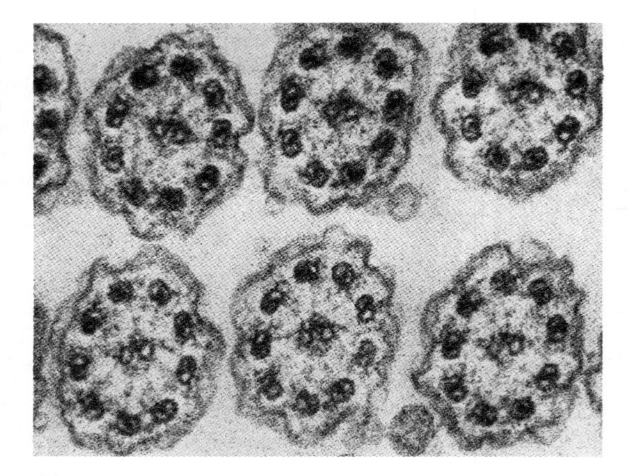

(b)

secretory vesicles fuse with the cell membrane and release their contents into the extracellular environment. This process adds new membrane material which replaces that which was lost from the cell membrane during endocytosis.

Endocytosis and exocytosis account for only part of the two-way traffic between the intracellular and extracellular compartments. Most of this traffic is due to membrane transport processes, the movement of molecules and ions through the cell membrane (chapter 6).

Cilia and Flagella. **Cilia** are tiny hairlike structures that protrude from the cell and, like the coordinated action of oarsmen in a boat, stroke in unison. Cilia in the human body are found on the apical surface (the surface facing the lumen, or cavity) of epithelial cells in the respiratory and female genital tracts. In the respiratory system, the cilia transport strands of mucus, which are then conveyed by ciliary action to a region (the pharynx) where the mucus can either be swallowed or expectorated. In the female genital tract, ciliary movements in the epithelial lining are believed to draw the egg (ovum) into the uterine tube and move it toward the uterus.

Sperm are the only cells in the human body that have **flagella.** The flagellum is a single, whiplike structure that propels the sperm through its environment. Both cilia and flagella are composed of microtubules arranged in a characteristic way. Nine pairs of microtubules in the periphery of a cilium or a flagellum surround a single pair of microtubules in the center (fig. 3.7).

1. *Draw the fluid-mosaic model of the cell membrane, and describe the structure of the membrane.*
2. *Describe the structure and possible functions of the cytoskeleton.*
3. *List the parts of a generalized cell, and distinguish between cilia and flagella.*
4. *Draw a figure showing phagocytosis; explain the events that occur during exocytosis.*

Lysosomes and Mitochondria

Lysosomes and mitochondria are cellular organelles involved in enzymatic activities. Lysosomes contain digestive enzymes capable of digesting organic material brought into them. Mitochondria contain enzymes and other molecules needed for obtaining energy from the combustion of food molecules.

Lysosomes

After a phagocytic cell has engulfed the proteins, polysaccharides, and lipids present in a particle of "food" (such as a bacterium), these molecules are still kept isolated from the cytoplasm by the membranes surrounding the food vacuole. The large molecules of proteins, polysaccharides, and lipids must first be digested into their smaller subunits (amino acids, monosaccharides, and so on) before they can cross the vacuole membrane and enter the cytoplasm.

The digestive enzymes of a cell are isolated from the cytoplasm and concentrated within membrane-bound organelles called **lysosomes** (fig. 3.8). A *primary lysosome*

Figure 3.8. Electron micrograph showing primary lysosomes (Lys_1) and secondary lysosomes (Lys_2). Mitochondria (Mi), Golgi apparatus (GA), and the nuclear envelope (NE) are also seen.

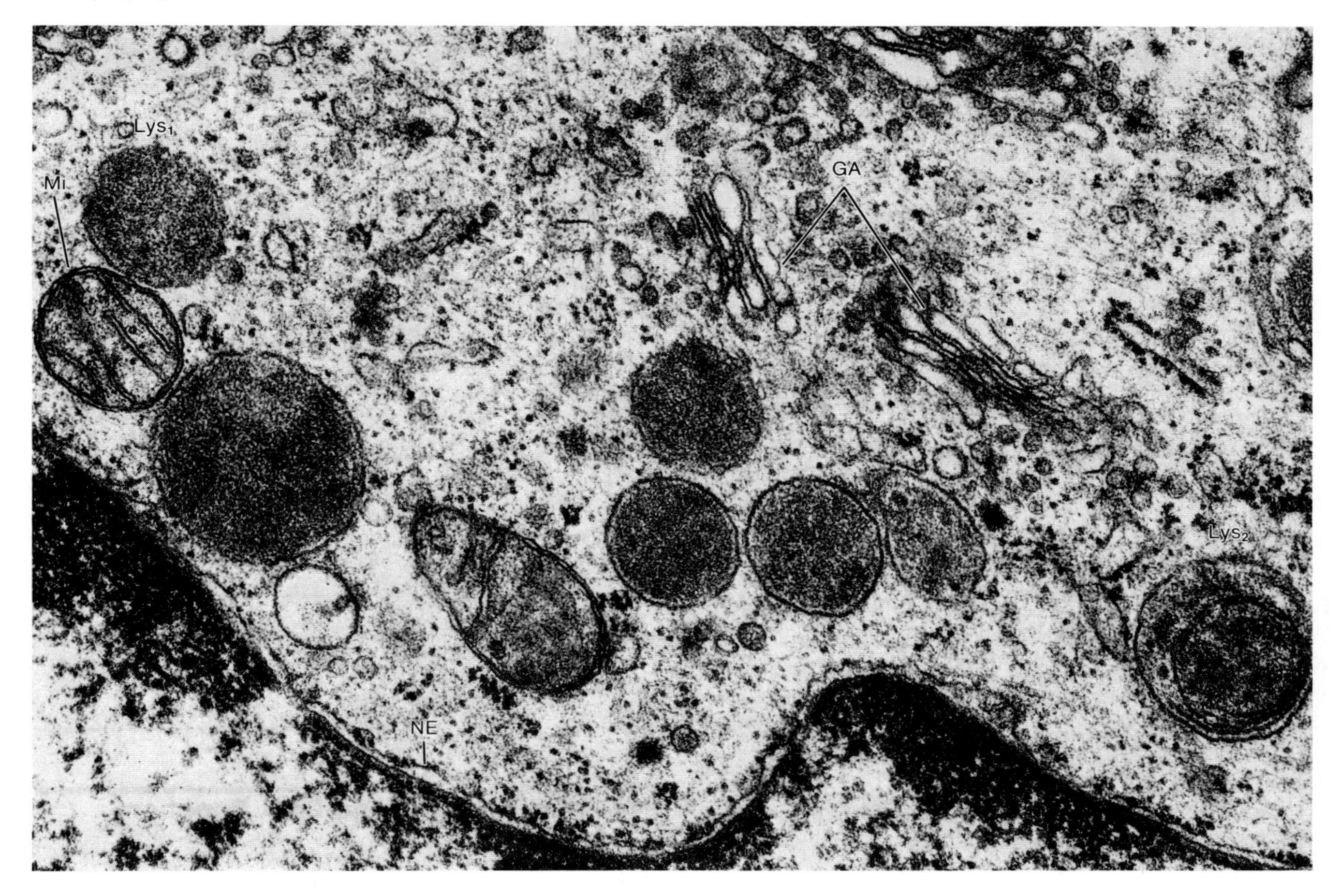

is one which contains only digestive enzymes (about forty different species) within an environment that is considerably more acidic than the surrounding cytoplasm. A primary lysosome may fuse with a food vacuole (or with another cellular organelle) to form a *secondary lysosome* in which worn-out organelles and the products of phagocytosis can be digested. Thus, a secondary lysosome contains partially digested remnants of other organelles and ingested organic material. A lysosome that contains undigested wastes is called a *residual body.* Residual bodies may eliminate their wastes by exocytosis, or the wastes may accumulate within the cell as the cell ages.

Partly digested membranes of various organelles and other cellular debris are often observed within secondary lysosomes. This is a result of **autophagy,** a process that destroys worn-out organelles so that they can be continuously replaced. Lysosomes are thus aptly referred to as the "digestive system" of the cell.

Lysosomes have also been called "suicide bags," because a break in their membranes would release their digestive enzymes and thus destroy the cell. This happens normally as part of *programmed cell death,* in which the destruction of tissues is part of embryological development. It also occurs in white blood cells during an inflammation reaction.

Most, if not all, molecules in the cell have a limited life-span. They are continuously destroyed and must be continuously replaced. Glycogen and some complex lipids in the brain, for example, are digested normally at a particular rate by lysosomes. If a person, because of some genetic defect, does not have the proper amount of these lysosomal enzymes, the resulting abnormal accumulation of glycogen and lipids could destroy the tissues.

Figure 3.9. A mitochondrion (*a*). The outer membrane (OM) and the infoldings of the inner membrane—the cristae (Cr)—are clearly seen. The fluid in the center is the matrix (Ma). The structure of a mitochondrion is illustrated in (*b*).

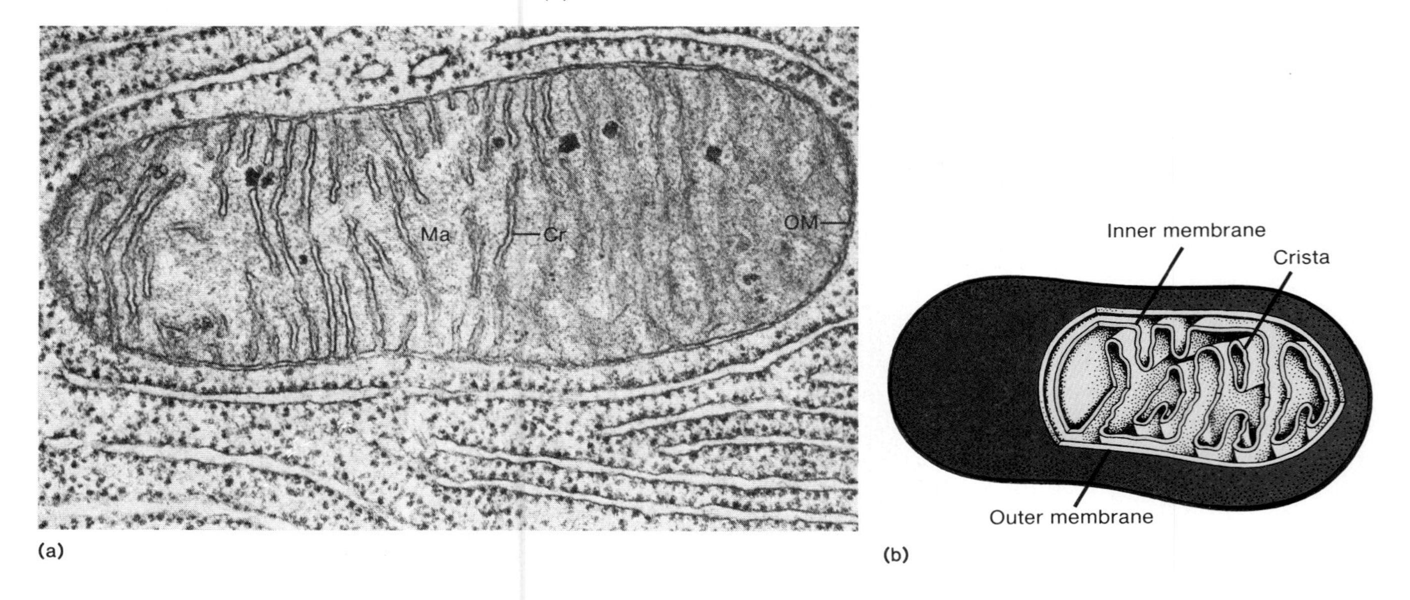

Mitochondria

All cells in the body, with the exception of mature red blood cells, have a hundred to a few thousand organelles called **mitochondria.** Mitochondria serve as sites for the production of most of the cellular energy (chapter 5). For this reason, mitochondria are sometimes called the "powerhouses" of the cell.

Mitochondria vary in size and shape, but all have the same basic structure (fig. 3.9). Each is surrounded by an *outer membrane* that is separated by a narrow space from an *inner membrane*. The inner membrane has many folds, called *cristae,* which extend into the central area (or *matrix*) of the mitochondrion. The cristae and the matrix provide different compartments in the mitochondrion and have different roles in the generation of cellular energy.

Mitochondria are able to migrate through the cytoplasm of a cell, and it is believed that they are able to reproduce themselves. Indeed, mitochondria contain their own DNA! This is a more primitive form of DNA than that found within the cell nucleus. For these and other reasons, many scientists believe that mitochondria evolved from separate organisms, related to bacteria, which entered animal cells and remained in a state of symbiosis.

1. *Describe the roles of lysosomes, and explain the importance of autophagy.*
2. *Describe the structure of a mitochondrion and the function it performs.*

Nuclear Structure and DNA Function

The nucleus of a cell contains DNA in its chromosomes and serves as the cell's control center. During cell division DNA replicates itself, and duplicated copies are distributed to the daughter cells. When a cell is not dividing, DNA serves as the information source for the formation of RNA and hence of proteins.

Most cells in the body have a single nucleus, although some—such as skeletal muscle cells—are multinucleate. The nucleus is surrounded by a *nuclear envelope* composed of an inner and an outer membrane; these two membranes fuse together to form thin sacs with openings called *nuclear pores* (figs. 3.10 and 3.11). These pores allow RNA to exit the nucleus (where it is formed) and enter the cytoplasm but prevent DNA from leaving the nucleus.

Many granulated threads in the nuclear fluid can be seen with an electron microscope. These threads are called *chromatin* and consist of a combination of DNA and protein. There are two forms of chromatin. Thin, extended chromatin—or *euchromatin*—appears to be the active form of DNA in a nondividing cell. Regions of condensed, "blotchy"-appearing chromatin, known as *heterochromatin,* are believed to contain inactive DNA.

Figure 3.10. An electron micrograph of a freeze-fractured nuclear envelope, showing the nuclear pores.

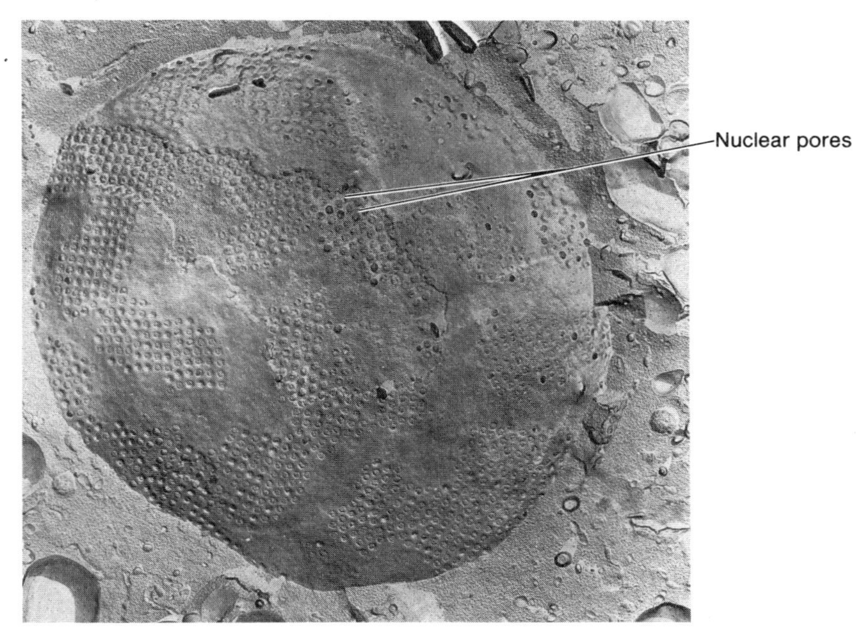

Figure 3.11. The nucleus of a liver cell showing the nuclear envelope, heterochromatin, and nucleolus.

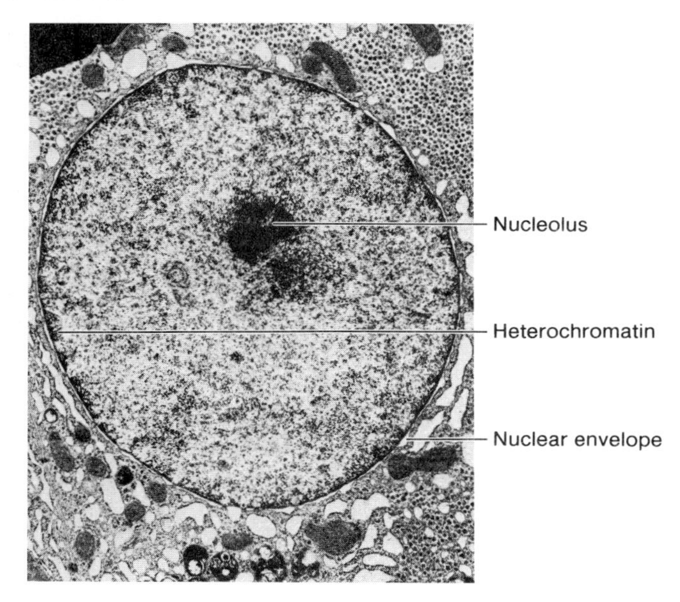

Figure 3.12. The replication of DNA. Each new double helix is composed of one old and one new strand. The base sequences of each of the new molecules is identical to that of the parent DNA because of complementary base pairing.

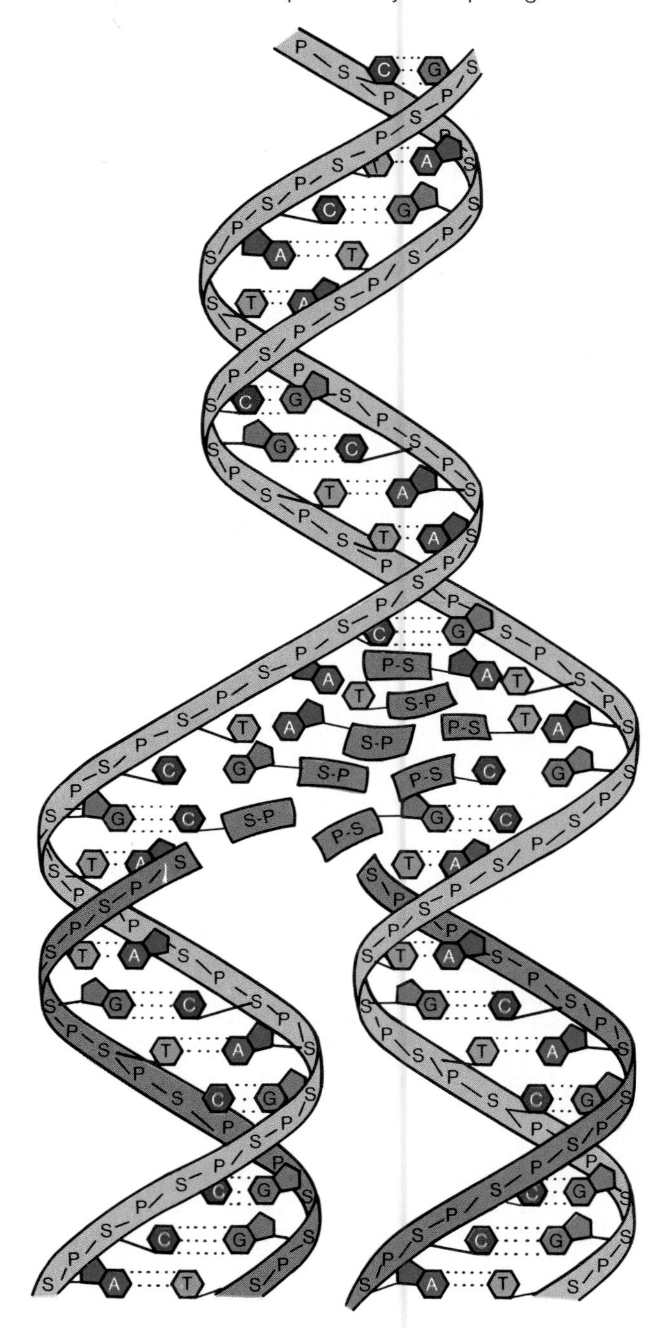

Figure 3.13. The life cycle of a cell.

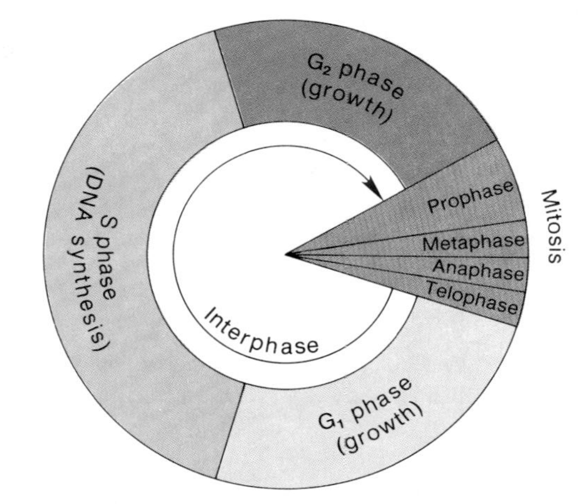

One or more dark areas within each nucleus can be seen. These regions, which are not surrounded by membranes, are called *nucleoli*. The DNA within nucleoli contains genes that code for the production of ribosomal RNA (rRNA), an essential component of ribosomes.

DNA Replication

When a cell is going to divide, each DNA molecule replicates itself, and each of the identical DNA copies thus produced is distributed to the two daughter cells. Replication of DNA requires the action of a specific enzyme known as *DNA polymerase*. This enzyme moves along the DNA molecule, breaking the weak hydrogen bonds between complementary bases as it travels. As a result, the bases of each of the two DNA strands become free to bond to new complementary bases (which are part of nucleotides) that are available within the surrounding environment.

Because of the rules of complementary base pairing, the bases of each original strand will bond to the appropriate free nucleotides: adenine bases pair with thymine-containing nucleotides; guanine bases pair with cytosine-containing nucleotides, and so on. In this way, two new molecules of DNA, each containing two complementary strands, are formed. The DNA polymerase enzyme links the phosphate groups and deoxyribose sugar groups together to form a second polynucleotide chain in each DNA that is complementary to the first DNA strands. Thus two new double-helix DNA molecules are produced that contain the same base sequence as the parent molecule (fig. 3.12).

When DNA replicates, therefore, each copy is composed of one new strand and one strand from the original DNA molecule. Replication is said to be **semiconservative** (half of the original DNA is "conserved" in each of the new DNA molecules). Through this mechanism, the sequence of bases in DNA—which is the basis of the genetic code—is preserved from one cell generation to the next.

Cell Growth and Division

Unlike the life of an organism, which can be pictured as a linear progression from birth to death, the life of a cell follows a cyclical pattern. Each cell is produced as a part of its "parent" cell; when the daughter cell divides, it in turn becomes two new cells. In a sense, then, each cell is potentially immortal as long as its progeny can continue to divide. Some cells in the body divide frequently; the epidermis of the skin, for example, is renewed approximately every two weeks, and the stomach lining is renewed about every two or three days. Other cells, however, such as nerve and striated muscle cells in the adult, do not divide at all. All cells in the body, of course, live only as long as the person lives (some cells live longer than others, but eventually all cells die when vital functions cease).

The Cell Cycle. The nondividing cell is in a part of its life cycle known as **interphase** (fig. 3.13). The chromosomes are in their extended form (as euchromatin), and their genes actively direct the synthesis of RNA. Through their direction of RNA synthesis, genes control the metabolism of the cell. During this time the cell may be

Figure 3.14. A photograph of homologous pairs of chromosomes from a human male cell at metaphase of mitosis (homologous chromosomes have been paired and numbered according to convention).

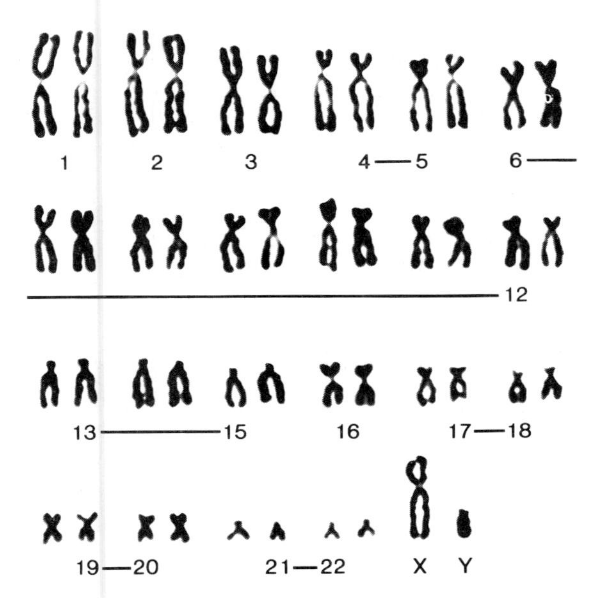

growing, so this part of interphase is known as the *G_1 phase* (*G* stands for *growth*). Although sometimes described as "resting," cells in the G_1 phase perform the physiological functions characteristic of the tissue in which they are found.

If a cell is going to divide, it replicates its DNA in a part of interphase known as the *S phase* (*S* stands for *synthesis*). The mechanisms that cause transformation of a cell from the G_1 to the S phase are not known. Some experiments suggest that two nucleotides—cyclic adenosine monophosphate (cAMP) and cyclic guanosine monophosphate (cGMP)—may regulate this transformation. An increase in the intracellular concentration of cAMP and a decrease in cGMP can stimulate cell division. Reverse changes in the concentration of the nucleotides inhibit cell division. Besides helping to regulate cell division, these cyclic nucleotides have a wide variety of other regulatory functions in the cell (as described in later chapters).

Once DNA has replicated, the chromatin condenses to form short, thick, rodlike structures. This is the more familiar form of chromosomes because they are easily seen in the ordinary (light) microscope. It should be remembered that this form of the chromatin represents a "packaged" state of DNA—not the extended, threadlike form that is active in directing the metabolism of the cell during the G_1 phase.

The matched pairs of chromosomes are called **homologous chromosomes.** One member of each homologous pair is derived from a chromosome inherited from the father, and the other member is a copy of one of the chromosomes inherited from the mother. Homologous chromosomes do not have identical DNA base sequences; one member of the pair may code for blue eyes, for example, and the other for brown eyes. There are twenty-two homologous pairs of *autosomal chromosomes* and one pair of *sex chromosomes,* described as X and Y. Females have two X chromosomes, whereas males have one X and one Y chromosome (fig. 3.14).

Mature nerve and muscle cells do not replicate at all; neurons are thus particularly susceptible to damage by alcohol and other drugs. Epithelial cells, in contrast, have very rapid cell cycles that help to replace the continuous loss of cells. *Cancers* have rapid rates of cell division but not necessarily more rapid than normal tissue. The fast growth of some cancers is thus not due simply to a rapid rate of cell division but is rather due to the fact that the rate of cell division is much greater than the rate of cell death. The observation that many normal tissues are at least as rapid in their rates of cell divisions as cancer makes the chemotherapy of cancer much more difficult than it would be otherwise.

Figure 3.15. (*a*) An electron micrograph of a centriole magnified about 120,000 times. (*b*) The diagram shows the arrangement of the bundles of three microtubules in a centriole.

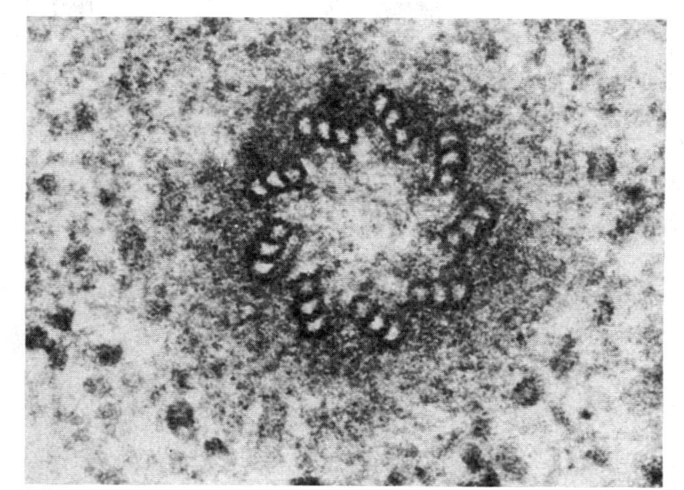

(a)

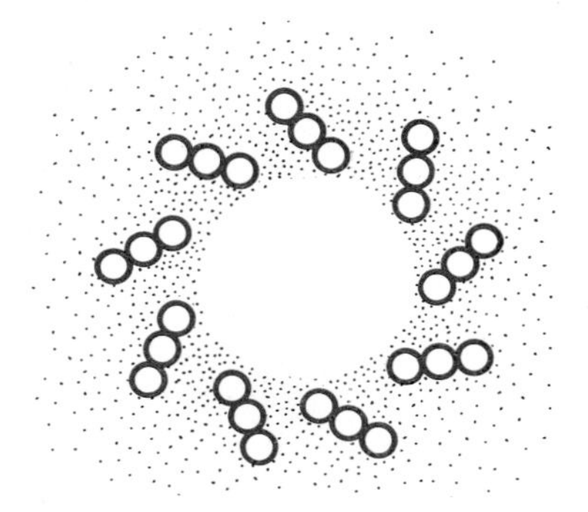

(b)

Centrioles. **Centrioles** are small, rodlike structures located near the nucleus. There are two centrioles, positioned at right angles to each other, within a spherical nonmembranous mass called a **centrosome.** Each centriole is composed of nine evenly spaced bundles of microtubules, with three microtubules per bundle (fig. 3.15).

Centrosomes are found only in those cells that are capable of division. During the process of cell division, the centrioles take up positions on opposite sides of the nucleus. The centrioles are then involved in the production of, and are attached to, **spindle fibers,** which are also composed of microtubules. The spindle fibers and centrioles help to pull the duplicated chromosomes to opposite poles of the cell during cell division. Cells that lack centrioles, such as mature muscle and nerve cells, cannot divide.

Mitosis. Following the S phase of the cell cycle, each chromosome consists of two strands called **chromatids,** which are joined together by a *centromere*. The two chromatids within a chromosome contain identical DNA base sequences because each is produced by the semiconservative replication of DNA. Each chromatid, therefore, contains a complete double helix DNA molecule that is a copy of the single DNA molecule existing prior to replication. Each chromatid will become a separate chromosome once cell division has been completed.

Following a second resting phase (G_2), which is usually shorter than G_1, the cell proceeds through the various stages of cell division, or **mitosis** (the *M phase* of the cell cycle). In mitosis, the chromosomes line up single file along the equator of the cell. Spindle fibers from the centrioles form and attach to the centromere of each chromosome (fig. 3.16).

Figure 3.16. The stages of mitosis.

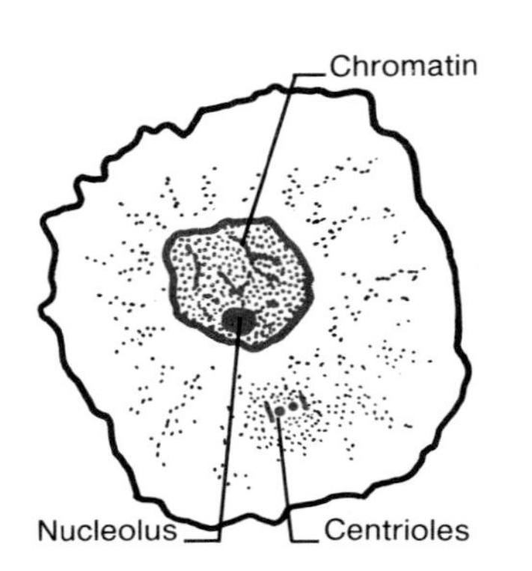

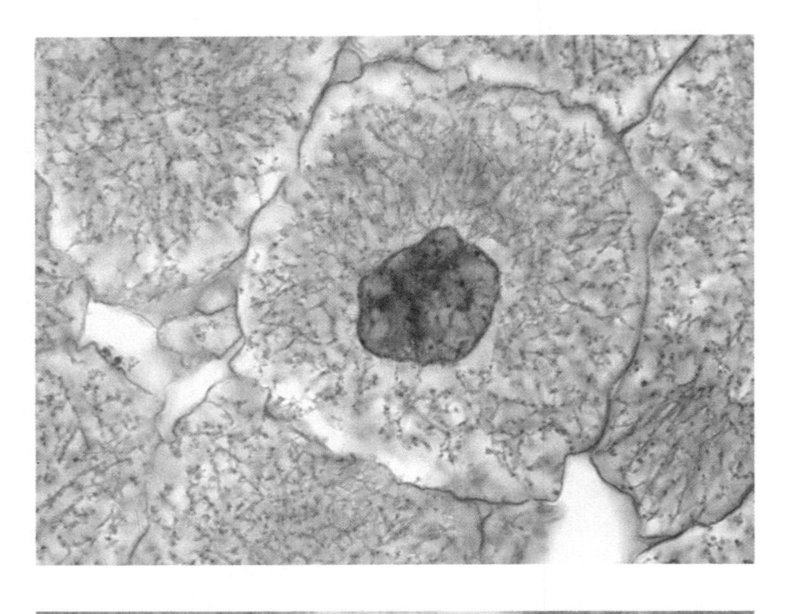

(a) Interphase

The chromosomes are in an extended form and seen as chromatin in the electron microscope.
The nucleus is visible.

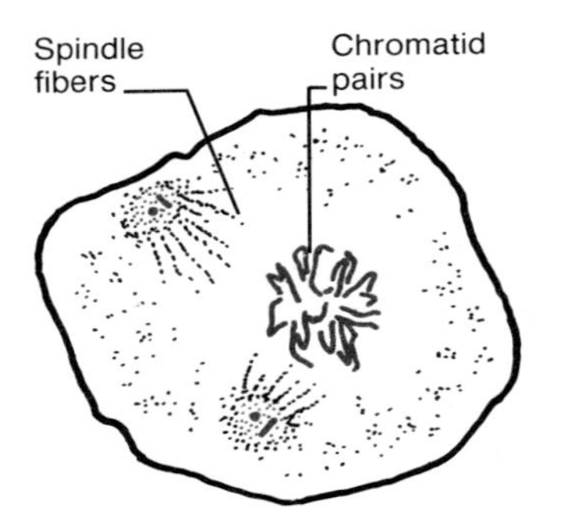

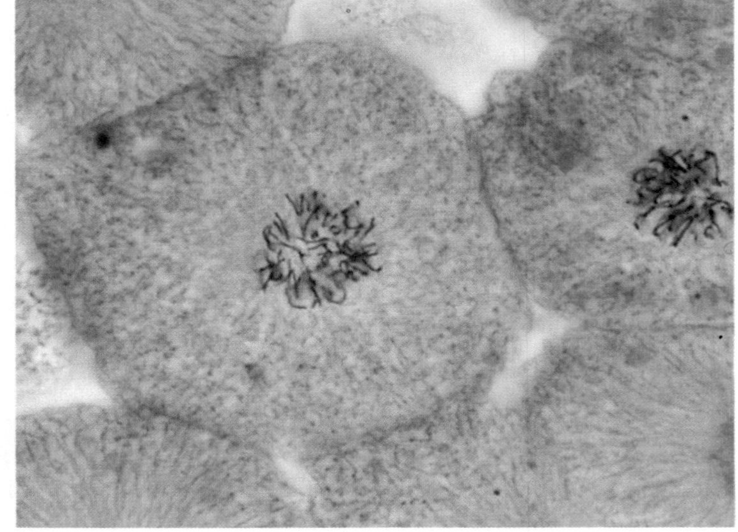

(b) Prophase

The chromosomes are seen and observed to consist of two chromatids joined together by a centromere.
The centrioles move apart towards opposite poles of the cell.
Spindle fibers are produced and extend from each centriole.
The nuclear membrane starts to disappear.
The nucleolus is no longer visible.

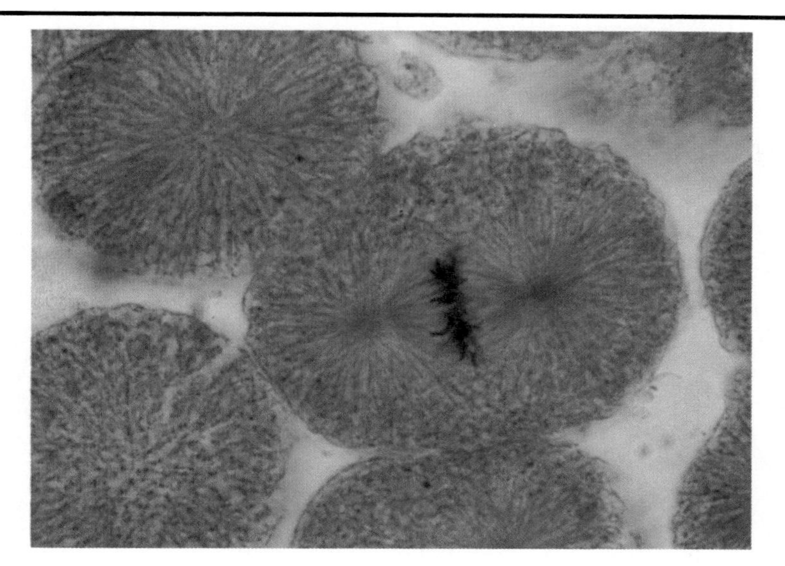

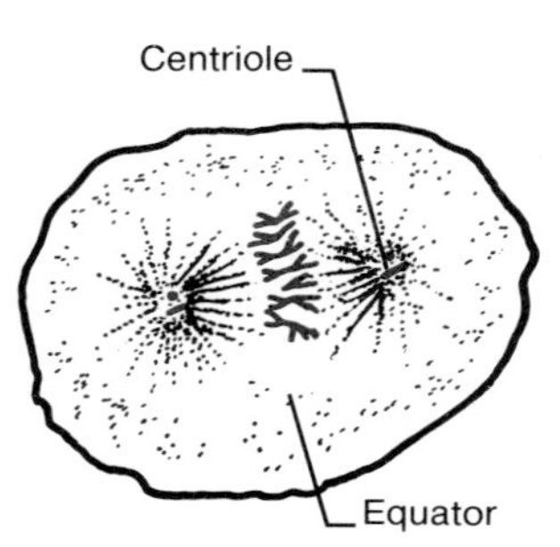

(c) Metaphase

The chromosomes are lined up at the equator of the cell.
The spindle fibers from each centriole are attached to the centromeres of the chromosomes.
The nuclear membrane has disappeared.

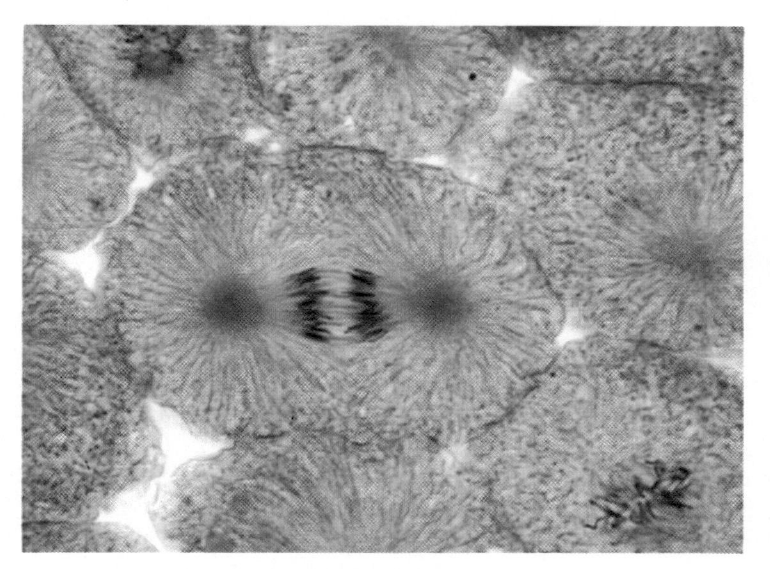

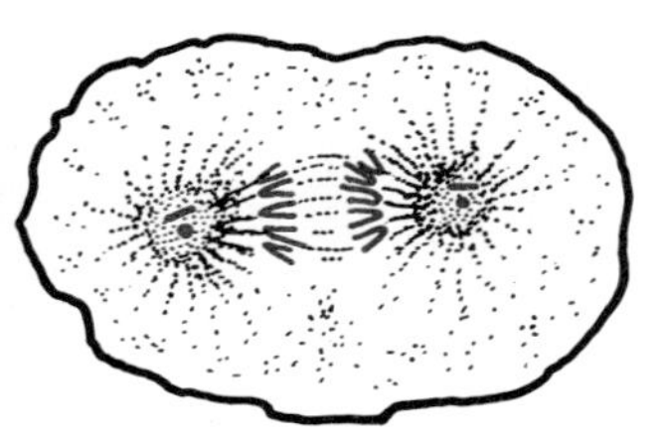

(d) Anaphase

The centromeres split, and the sister chromatids separate as each is pulled to an opposite pole.

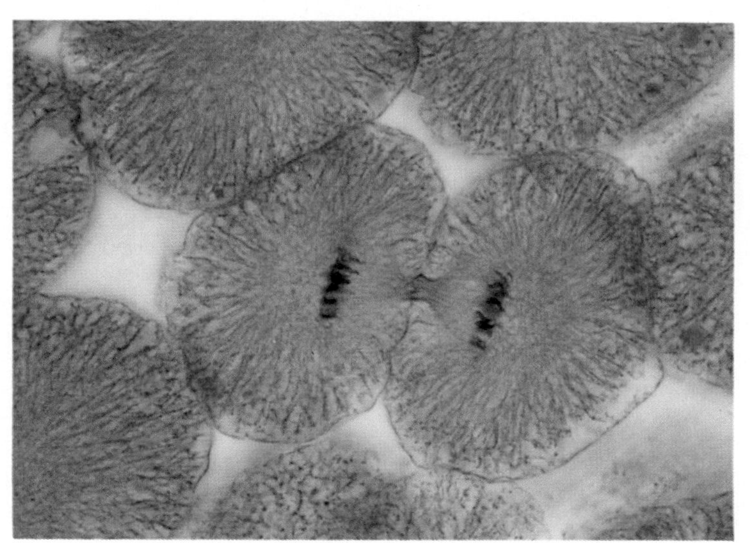

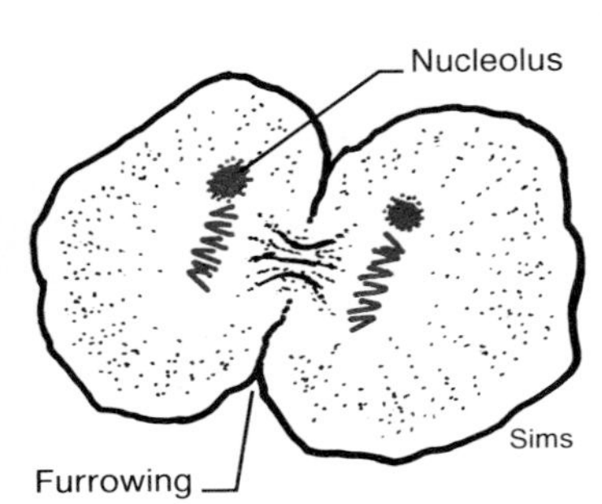

(e) Telophase

The chromosomes become longer, thinner, and less distinct.
New nuclear membranes form.
The nucleolus reappears.
Cell division (cytokinesis) is nearly complete.

Table 4.2 Examples of the diagnostic value of some enzymes found in plasma.

Enzyme	Diseases Associated with Abnormal Plasma Enzyme Concentrations
Alkaline phosphatase	Obstructive jaundice, Paget's disease (osteitis deformans), carcinoma of bone
Acid phosphatase	Benign hypertrophy of prostate, cancer of prostate
Amylase	Pancreatitus, perforated peptic ulcer
Aldolase	Muscular dystrophy
Creatine kinase (or creatine phosphokinase-CPK)	Muscular dystrophy, myocardial infarction
Lactate dehydrogenase (LDH)	Myocardial infarction, liver disease, renal disease, pernicious anemia
Transaminases (GOT and GPT)	Myocardial infarction, hepatitis, muscular dystrophy

Table 4.3 Some clinical uses of isoenzyme measurements.

Enzyme	Isoenzyme Form Number	Disease Associated with Abnormal Elevation of Isoenzyme
Creatine phosphokinase (CPK)	1 2	Muscular dystrophy Myocardial infarction
Lactic acid dehydrogenase (LDH)	1 5	Myocardial infarction Liver disease (such as hepatitis)

Control of Enzyme Activity

The activity of an enzyme, as measured by the rate at which its substrates are converted to products, is influenced by a variety of factors, including (1) the temperature and pH of the solution; (2) the concentration of cofactors and coenzymes, which are needed by many enzymes as "helpers" for their catalytic activity; (3) the concentration of enzyme and substrate molecules in the solution; and (4) the stimulatory and inhibitory effects of some products of enzyme action on the activity of the enzymes that helped form these molecules.

Figure 4.3. The effect of temperature on enzyme activity, as measured by the rate of the enzyme-catalyzed reaction under standardized conditions.

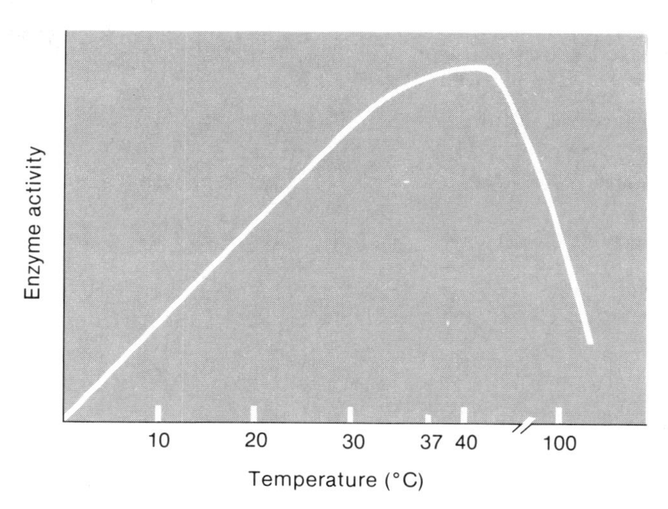

Effects of Temperature and pH

An increase in temperature, as previously described, will increase the rate of non-enzyme-catalyzed reactions because a higher proportion of reactant molecules will have the activation energy required. A similar relationship between temperature and reaction rate occurs in enzyme-catalyzed reactions. At a temperature of 0° C the reaction rate is unmeasurably slow. As the temperature is raised above 0° C the reaction rate increases but only up to a point. At a few degrees above body temperature (which is 37° C) the reaction rate reaches a plateau; further increases in temperature actually *decrease* the rate of the reaction (fig. 4.3). This decrease is due to the fact that the tertiary structure of enzymes becomes altered at higher temperatures.

A similar relationship is observed when the rate of an enzymatic reaction is measured at different pH values. Each enzyme characteristically has its peak activity in a very narrow pH range, which is the **pH optimum** for the

Figure 4.4. The effect of pH on the activity of three digestive enzymes.

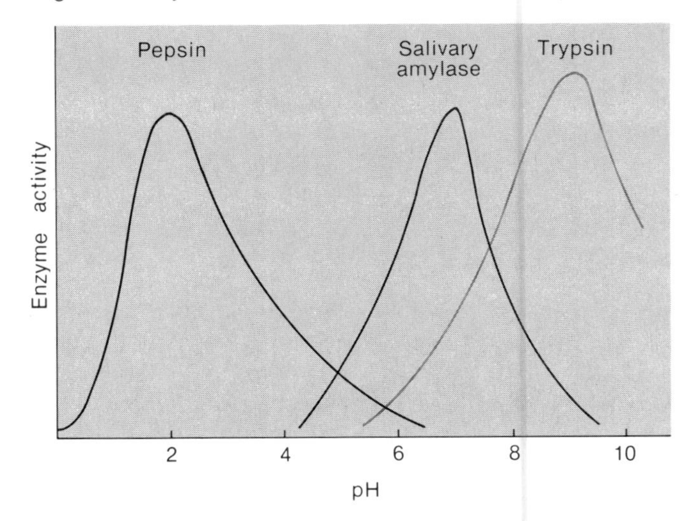

Table 4.4 The pH optima of selected enzymes.		
Enzyme	**Reaction Catalyzed**	**pH Optimum**
Pepsin (stomach)	Digestion of protein	2.0
Acid phosphatase (prostate)	Removal of phosphate group	5.5
Salivary amylase (saliva)	Digestion of starch	6.8
Lipase (pancreatic juice)	Digestion of fat	7.0
Alkaline phosphatase (bone)	Removal of phosphate group	9.0
Trypsin (pancreatic juice)	Digestion of protein	9.5
Monoamine oxidase (nerve endings)	Removal of amine group from norepinephrine	9.8

enzyme. If the pH is changed from this optimum, the reaction rate decreases (fig. 4.4). This decreased enzyme activity is due to changes in the conformation of the enzyme and in the charges of the R groups of the amino acids lining the active sites.

The pH optimum of an enzyme usually reflects the pH of the body fluid in which the enzyme is found. The acidic pH optimum of the protein-digesting enzyme *pepsin,* for example, allows it to be active in the strong hydrochloric acid of gastric juice. Similarly, the neutral pH optimum of *salivary amylase* and the alkaline pH optimum of *trypsin* in pancreatic juice allow these enzymes to digest starch and protein, respectively, in other parts of the digestive tract.

Although the pH of other body fluids shows less variation than the fluids of the digestive tract, significant differences exist between the pH optima of different enzymes found throughout the body (see table 4.4). Some of these differences can be exploited for diagnostic purposes. Disease of the prostate, for example, may be associated with elevated blood levels of a prostatic phosphatase with an acidic pH optimum (descriptively called acid phosphatase). Bone disease, on the other hand, may be associated with elevated blood levels of alkaline phosphatase, which has a higher pH optimum than the similar enzyme released from the diseased prostate.

Cofactors and Coenzymes

Many enzymes are completely inactive when they are isolated in a pure state. Evidently some of the ions and smaller organic molecules that were removed in the purification procedure are needed for enzyme activity. These ions and smaller organic molecules are called *cofactors* and *coenzymes.*

Cofactors are metal ions such as Ca^{++}, Mg^{++}, Mn^{++}, Cu^{++}, and Zn^{++}. Some enzymes with a cofactor requirement do not have a properly shaped active site in the absence of the cofactor. In these enzymes, the attachment of cofactors causes a conformational change in the protein that allows it to combine with its substrate. The cofactors of other enzymes participate in the temporary bonds between the enzyme and its substrate when the enzyme-substrate complex is formed (fig. 4.5).

Coenzymes are organic molecules that are derived from water-soluble vitamins, such as niacin and riboflavin. Coenzymes participate in enzyme-catalyzed reactions by transporting hydrogen atoms and small molecules from one enzyme to another. Examples of the actions of cofactors and coenzymes in specific reactions will be given in the context of their roles in cellular metabolism later in this chapter.

Figure 4.5. The roles of cofactors in enzyme function. In (*a*) the cofactor changes the conformation of the active site, allowing a better fit between the enzyme and its substrates. In (*b*) the cofactor participates in the temporary bonding between the active site and the substrates.

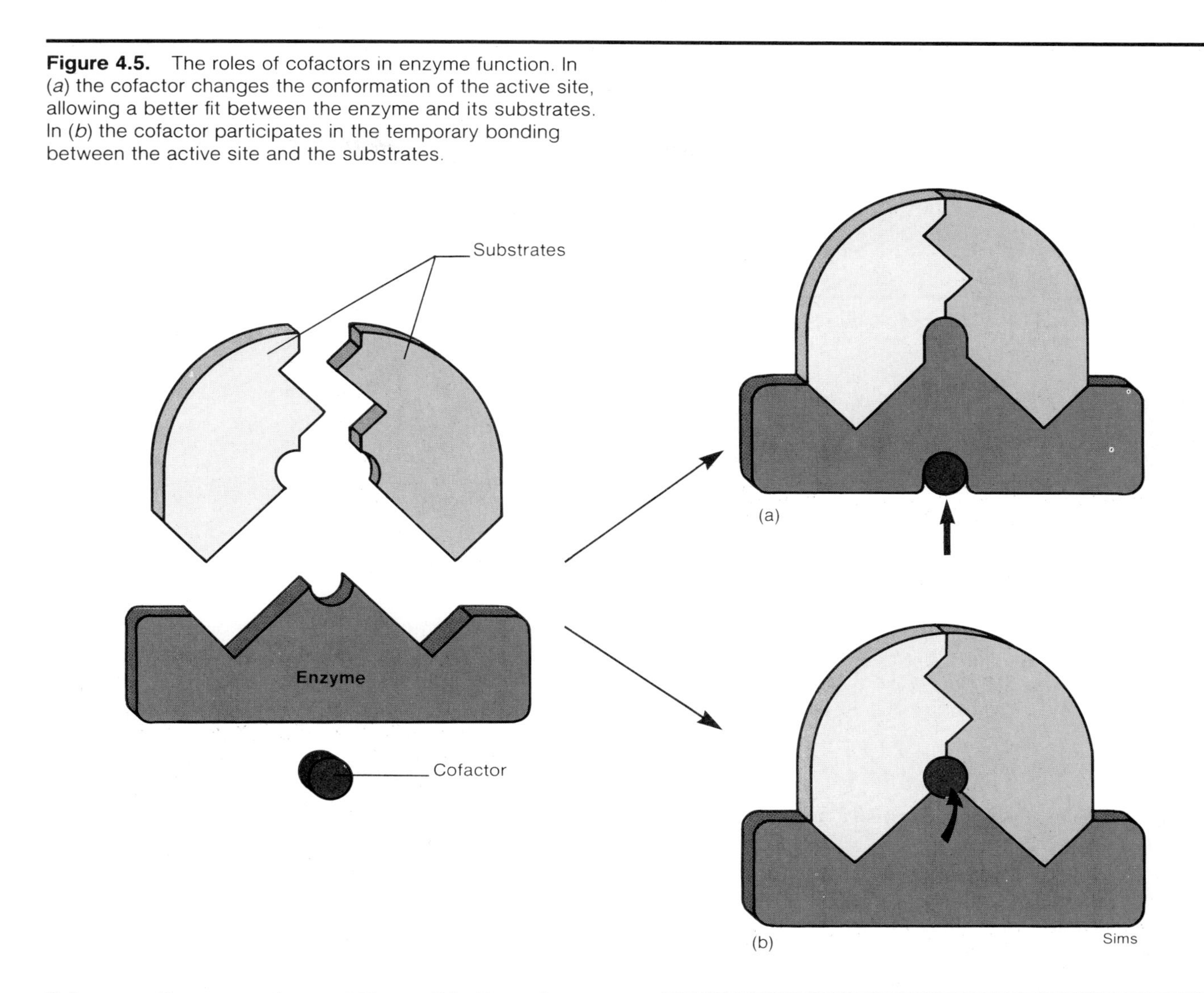

Substrate Concentration and Reversible Reactions

The rate at which an enzymatic reaction converts substrates into products depends on the enzyme concentration and on the concentration of substrates. When the enzyme concentration is at a given level, the rate of product formation will increase as the substrate concentration increases. Eventually, however, a point will be reached where additional increases in substrate concentration do not result in comparable increases in reaction rate. When the relationship between substrate concentration and reaction rate reaches a plateau, the enzyme is said to be *saturated.* If one thinks of enzymes as workers and substrates as jobs, there is 100% employment when the enzyme is saturated; further availability of jobs (substrate) cannot further increase employment (conversion of substrate to product). This is illustrated in figure 4.6.

Figure 4.6. The effect of substrate concentration on the reaction rate of an enzyme-catalyzed reaction. When the reaction rate is maximum, the enzyme is said to be *saturated.*

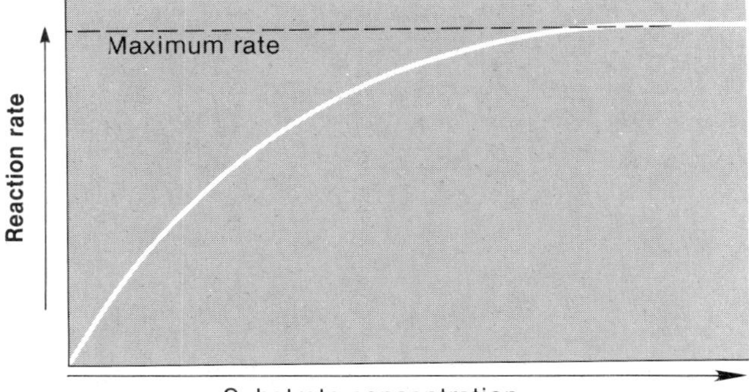

Figure 4.7. A metabolic pathway, where the product of one enzyme becomes the substrate of the next in a multi-enzyme system.

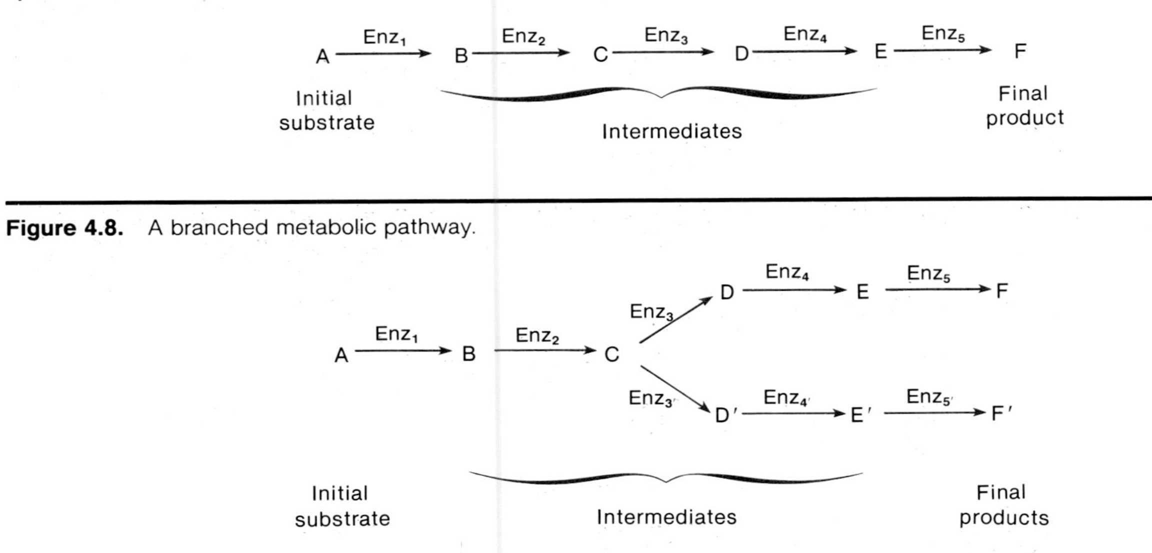

Figure 4.8. A branched metabolic pathway.

Some enzymatic reactions within a cell are reversible, with both the forward and backward reactions catalyzed by the same enzyme. The enzyme *carbonic anhydrase,* for example, is named because it can catalyze the following reaction:

$$H_2CO_3 \rightarrow H_2O + CO_2$$

The same enzyme, however, can also catalyze the reverse reaction:

$$H_2O + CO_2 \rightarrow H_2CO_3$$

The two reactions can be more conveniently illustrated by a single equation:

$$H_2O + CO_2 \rightleftharpoons H_2CO_3$$

The direction of the reversible reaction depends, in part, on the relative concentrations of the molecules to the left and right of the arrows. If the concentration of CO_2 is very high (as it is in the tissues), the reaction will be driven to the right. If the concentration of CO_2 is low and that of H_2CO_3 is high (as it is in the lungs), the reaction will be driven to the left. The ability of reversible reactions to be driven from the side of the equation where the concentration is higher to the side where the concentration is lower is known as the **law of mass action.**

Although some enzymatic reactions are not directly reversible, the net effects of the reactions can be reversed by the action of different enzymes. The enzymes that convert glucose to pyruvic acid, for example, are different from those that reverse the pathway and produce pyruvic acid from glucose. Likewise, the formation and breakdown of glycogen (a polymer of glucose) are catalyzed by different enzymes.

Metabolic Pathways

The many thousands of different types of enzymatic reactions within a cell do not occur independently of each other. They are, rather, all linked together by intricate webs of interrelationships, the total pattern of which constitutes cellular metabolism. A part of this web that begins with an *initial substrate,* progresses through a number of *intermediates,* and ends with a *final product* is known as a **metabolic pathway.**

The enzymes in a metabolic pathway cooperate in a manner analogous to workers on an assembly line where each contributes a small part to the final product. In this process, the product of one enzyme in the line becomes the substrate of the next enzyme, and so on (fig. 4.7).

Few metabolic pathways are completely linear. Most are branched so that one intermediate at the branch point can serve as a substrate for two different enzymes. Two different products can thus be formed that serve as intermediates of two divergent pathways (fig. 4.8).

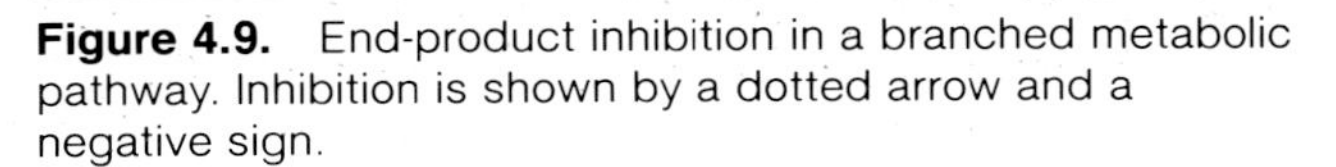

Figure 4.9. End-product inhibition in a branched metabolic pathway. Inhibition is shown by a dotted arrow and a negative sign.

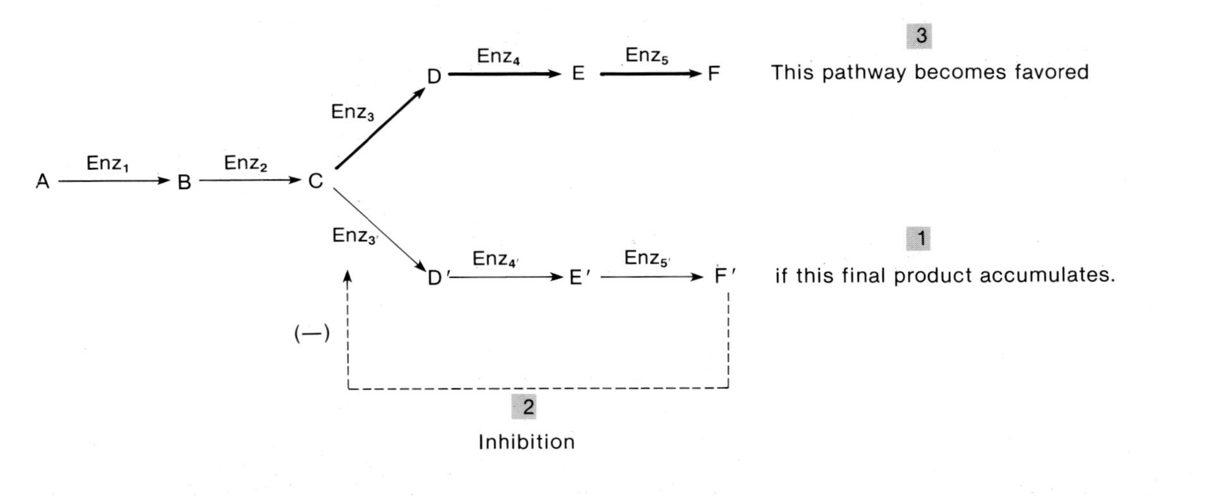

Figure 4.10. The effects of an inborn error of metabolism on a branched metabolic pathway.

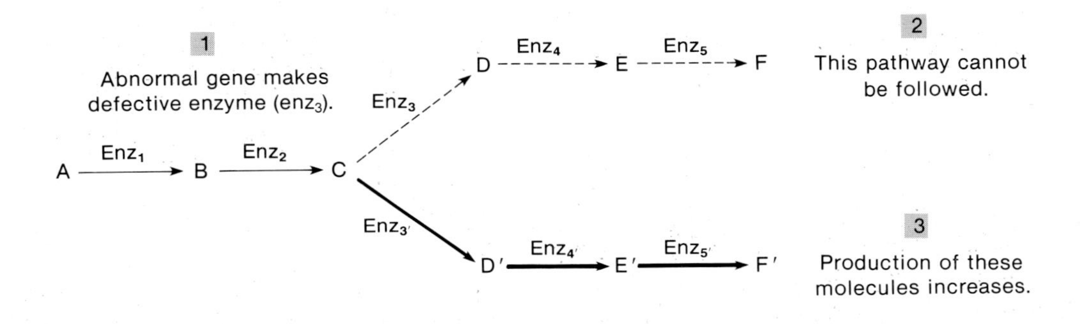

End-Product Inhibition. The activities of enzymes at the branch points of metabolic pathways are often regulated by a process called **end-product inhibition.** In this process, one of the final products of a divergent pathway inhibits the branch point enzyme that began the path toward the production of this inhibitor. This inhibition prevents that final product from accumulating excessively and results in a shift toward the final product of the alternate divergent pathway (fig. 4.9).

The mechanism by which a final product inhibits an earlier enzymatic step in its pathway is known as **allosteric inhibition.** The allosteric inhibitor combines with a part of the enzyme located away from the active site, causing the active site to change shape so that it can no longer combine properly with its substrate.

Inborn Errors of Metabolism. Each enzyme in a metabolic pathway is coded by a different gene. An inherited defect in one of these genes may result in a disease known as an "inborn error of metabolism." In these diseases there is an *increased* amount of intermediates formed *prior* to the defective step and a *decrease* in the intermediates and final products formed *after* the defective enzymatic step. If the defective enzyme is active at a step after a branch point in a pathway, the intermediates and final products of the divergent pathway will increase as a result of the block in the alternate pathway (fig. 4.10).

Figure 4.11. Metabolic pathways for the degradation of the amino acid phenylalanine. Defective enzyme$_1$ produces phenylketonuria (PKU), defective enzyme$_2$ produces alcaptonuria (not a clinically significant condition), and defective enzyme$_3$ produces albinism.

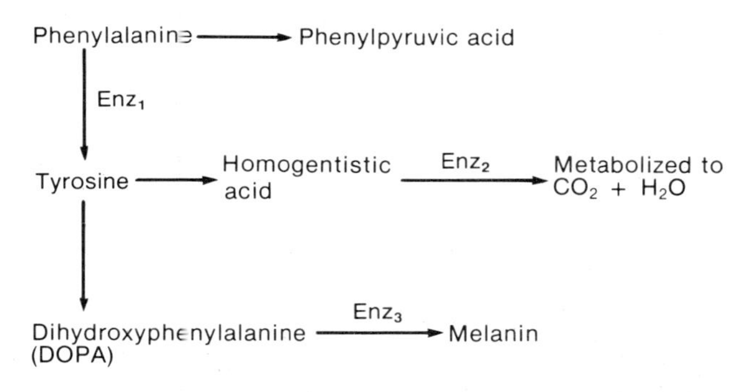

Table 4.5 Selected examples of inborn errors in the metabolism of amino acids, carbohydrates, and lipids.

Metabolic Defect	Disease	Abnormality	Clinical Result
Amino acid metabolism	Phenylketonuria (PKU)	Increase in phenylalanine	Mental retardation, epilepsy
	Albinism	Lack of melanin	Susceptibility to skin cancer
	Maple-syrup disease	Increase in leucine, isoleucine, and valine	Degeneration of brain, early death
	Homocystinuria	Accumulation of homocystine	Mental retardation, eye problems
Carbohydrate metabolism	Lactose intolerance	Lactose not utilized	Diarrhea
	Glucose-6-phosphatase deficiency (Gierke's disease)	Accumulation of glycogen in liver	Liver enlargement, hypoglycemia
	Glycogen phosphorylase deficiency (McArdle syndrome)	Accumulation of glycogen in muscle	Muscle fatigue and pain
Lipid metabolism	Gaucher's disease	Lipid accumulation (glucocerebroside)	Liver and spleen enlargement, brain degeneration
	Tay-Sachs disease	Lipid accumulation (ganglioside G_{M_2})	Brain degeneration, death by age 5
	Hypercholestremia	High blood cholesterol	Atherosclerosis of coronary and large arteries

The branched metabolic pathway that begins with phenylalanine as the initial substrate is subject to a number of inborn errors of metabolism (fig. 4.11). When the enzyme that converts this amino acid to the amino acid *tyrosine* is defective, the final products of a divergent pathway accumulate and can be detected in the blood and urine. This disease—*phenylketonuria (PKU)*—can result in severe mental retardation and a shortened life-span. Although no inborn error of metabolism is common, PKU occurs so frequently and is so easy to detect that all newborn babies are tested for this defect. If this disease is detected early, brain damage can be prevented by placing the child on an artificial diet low in the amino acid phenylalanine.

One of the conversion products of phenylalanine is a molecule known as *DOPA,* which is an acronym for dihydroxyphenylalanine. DOPA is a precursor of the pigment molecule *melanin.* An inherited defect in the enzyme that catalyzes the formation of melanin from DOPA results in an albino. Besides PKU and albinism, there are a large number of other inborn errors of amino acid metabolism, as well as errors in carbohydrate and lipid metabolism (table 4.5).

1. *Draw graphs to represent the effects of changes in temperature, pH, enzyme and substrate concentration, cofactors, and coenzymes on the rate of enzymatic reactions. Explain the mechanisms responsible for the appearance of these graphs.*
2. *Draw a flow chart of a metabolic pathway (using arrows and letters such as* A, B, C, *etc.) with one branch point.*
3. *Describe a reversible reaction, and explain how the law of mass action affects this reaction.*
4. *Define end-product inhibition, and use your diagram of a branched metabolic pathway to explain how this process will affect the concentration of different intermediates.*
5. *Suppose, due to an inborn error of metabolism, that the enzyme that catalyzes the third reaction in your pathway (question no. 2) is defective. Describe the effects this would have on the concentrations of the intermediates in your pathway. Give two "real-life" examples of inborn errors of metabolism.*

Bioenergetics

Bioenergetics refers to the flow of energy in living systems. Organisms maintain their highly ordered structure and life-sustaining activities through the constant expenditure of energy obtained ultimately from the environment. The energy flow in living systems obeys the first and second laws of a branch of physics known as *thermodynamics*.

According to the **first law of thermodynamics,** energy can be transformed, but it can neither be created nor destroyed. This is sometimes called the law of *conservation of energy*. As a result of energy transformations, according to the **second law of thermodynamics,** the universe and its parts (including living systems) have increased amounts of *entropy*. Entropy is related to the degree of disorganization of a system. Only energy that is in an organized state—called *free energy*—can be used to do work. Thus, since entropy increases in every energy transformation, the amount of free energy available to do work decreases. As a result of the increased entropy described by the second law, systems tend to go from states of higher to states of lower free energy.

Matter is a form of energy (from Einstein's $E = mc^2$). Atoms that are organized into complex organic molecules, such as glucose, have more free energy (less entropy) than six separate molecules each of carbon dioxide and water.

Figure 4.12. A simplified diagram of photosynthesis. Some of the sun's radiant energy is captured by plants and used to produce glucose from carbon dioxide and water. As the product of this endergonic reaction, glucose has a higher free energy content than the initial reactants.

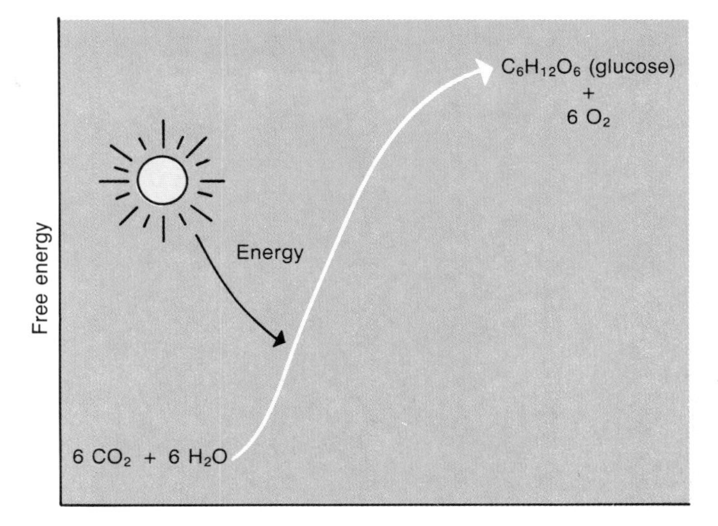

Therefore, in order to convert carbon dioxide and water to glucose, energy must be added. Plants perform this feat using energy from the sun in the process of *photosynthesis* (fig. 4.12).

Endergonic and Exergonic Reactions

Chemical reactions that require an input of energy are known as **endergonic reactions.** Since energy is added to make these reactions "go," the products of endergonic reactions must contain more free energy than the reactants. A portion of the energy added, in other words, is contained within the product molecules. This follows from the fact that energy cannot be created or destroyed (first law of thermodynamics) and from the fact that a more organized state of matter contains more free energy (less entropy) than a less organized state (as described by the second law).

The fact that glucose contains more free energy than carbon dioxide and water can be easily proven by the combustion of glucose to CO_2 and H_2O. This reaction releases energy in the form of heat. Reactions that convert molecules with more free energy to molecules with less—and, therefore, release energy as they proceed—are called **exergonic reactions.**

Figure 4.13. Since glucose contains more energy than carbon dioxide and water, the combustion of glucose is an exergonic reaction. The same amount of energy is released if glucose is broken down stepwise within the cell.

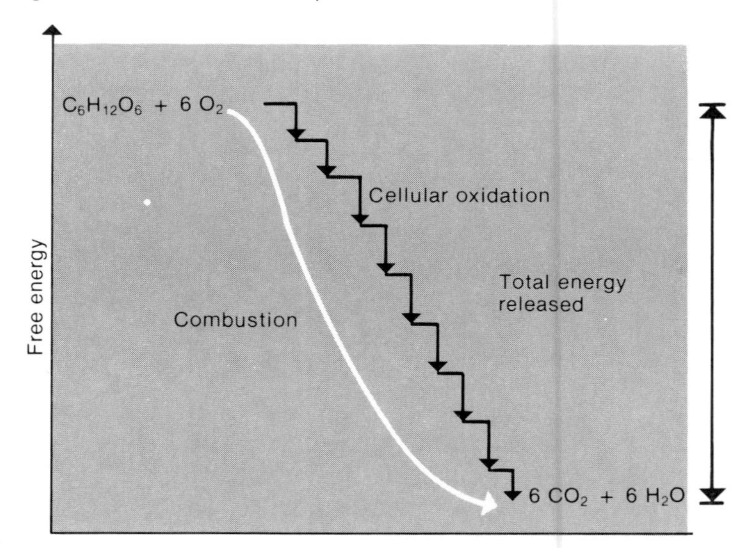

As illustrated in figure 4.13, the amount of energy released by an exergonic reaction is the same whether the energy is released in a single combustion reaction or in the many small, enzymatically controlled steps that occur in tissue cells. The energy that the body obtains from the consumption of particular foods can, therefore, be measured as the amount of heat energy released when these foods are combusted.

Heat is measured in units called *calories*. One calorie is defined as the amount of heat required to raise the temperature of one cubic centimeter of water one degree Celsius. The caloric value of food is usually indicated in kilocalories (one kilocalorie equals a thousand calories), which is commonly expressed with a capital letter—*C*alories.

Coupled Reactions: ATP

In order to remain alive a cell must maintain its highly organized, low entropy state at the expense of free energy in its environment. Accordingly, the cell contains many enzymes that catalyze exergonic reactions, using substrates that come ultimately from the environment. The energy released by these exergonic reactions is used to drive the energy-requiring processes (endergonic reactions) in the cell. Since the cell cannot use heat energy to drive energy-requiring processes, chemical bond energy that is released in the exergonic reactions must be directly transferred to chemical bond energy in the products of endergonic reactions. Energy-liberating reactions are thus *coupled* to energy-requiring reactions. This relationship is like two meshed gears; the turning of one (the energy-releasing, exergonic gear) causes turning of the other (the energy-requiring, endergonic gear—fig. 4.14).

The energy released by most exergonic reactions in the cell is used, either directly or indirectly, to drive *one* endergonic reaction (fig. 4.15): the formation of **adenosine triphosphate** (**ATP**) from adenosine diphosphate (ADP) and inorganic phosphate (abbreviated P_i).

The formation of ATP requires the input of a fairly large amount of energy. Since this energy must be conserved (first law of thermodynamics), the bond that is produced by joining P_i to ADP must contain a part of this energy. Thus, when enzymes reverse this reaction and convert ATP to ADP and P_i, a large amount of energy is released. Energy released from the breakdown of ATP is used to power the energy-requiring processes in all cells. As the **universal energy carrier,** ATP serves to couple more efficiently the energy released by the breakdown of food molecules to the energy required by the diverse endergonic processes in the cell (fig. 4.16).

Coupled Reactions: Oxidation-Reduction

When an atom or a molecule gains electrons, it is said to become **reduced;** when it loses electrons, it is said to become **oxidized.** Reduction and oxidation are always coupled reactions: an atom or a molecule cannot become oxidized unless it donates electrons to another, which therefore becomes reduced. The atom or molecule that donates electrons to another is a *reducing agent,* and the one that accepts electrons from another is an *oxidizing agent*. It should be noted that an atom or a molecule may function as an oxidizing agent in one reaction and as a reducing agent in another reaction; it may gain electrons from one atom or molecule and pass them on to another in a series of coupled oxidation-reduction reactions—like a bucket brigade.

Notice that the term *oxidation* does not imply that oxygen participates in the reaction. This term is derived from the fact that oxygen has a great tendency to accept electrons, that is, to act as a strong oxidizing agent. This property of oxygen is exploited by cells; oxygen acts as the final electron acceptor in a chain of oxidation-reduction reactions that provides energy for ATP production.

Figure 4.14. A model of the coupling of exergonic and endergonic reactions. The drive shaft (representing the energy of activation) turns the exergonic gear, which turns the endergonic gear. The reactants of the exergonic reaction (represented by the larger gear) have more free energy than the products of the endergonic reaction because the coupling is not 100% efficient—some energy is lost as heat.

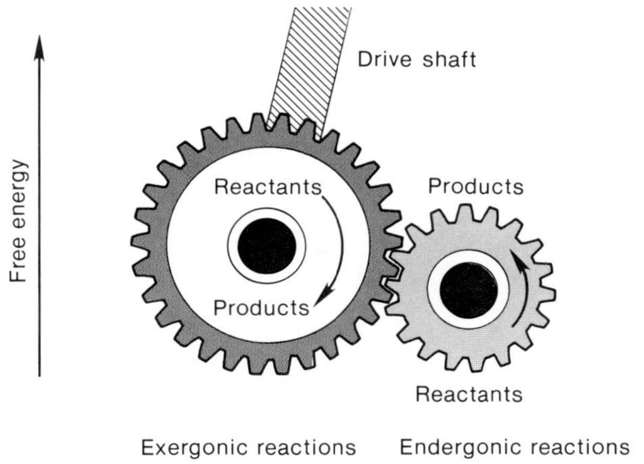

Drive shaft
Free energy
Reactants
Products
Products
Reactants
Exergonic reactions
Endergonic reactions

Figure 4.15. The formation and structure of adenosine triphosphate (ATP).

Adenosine diphosphate (ADP)

+

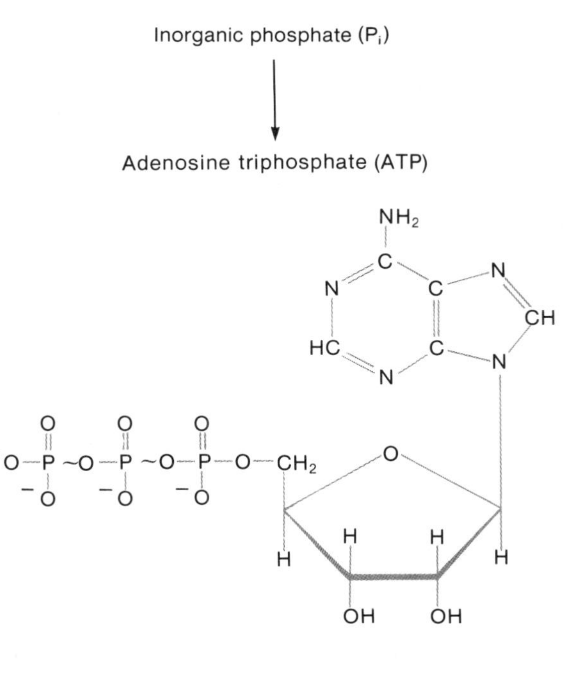

Figure 4.16. A model of ATP as the universal energy carrier of the cell. Exergonic reactions are shown as gears with arrows going down (reactions produce decrease in free energy); endergonic reactions are shown as gears with arrows going up (reactions produce increase in free energy).

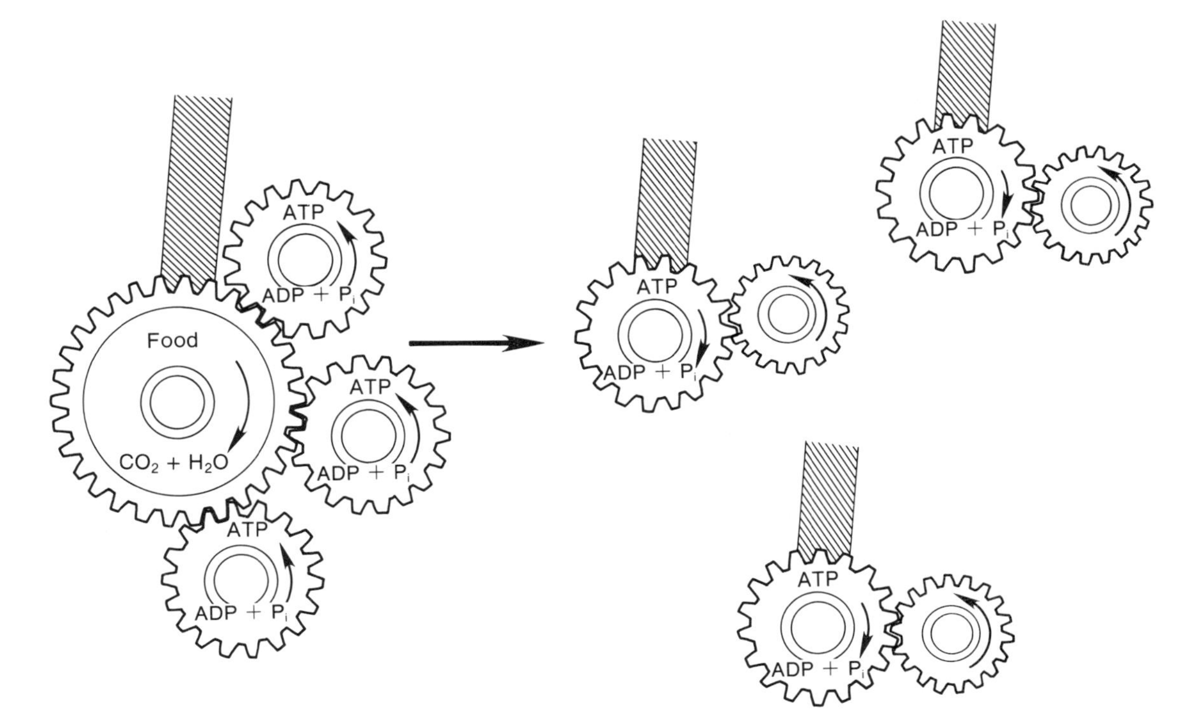

Figure 4.17. Structures of (*a*) the oxidized form of NAD (nicotinamide adenine dinucleotide) and (*b*) the reduced form of FAD (flavin adenine dinucleotide). Note the two additional hydrogen atoms (shown in color) that reduce FAD.

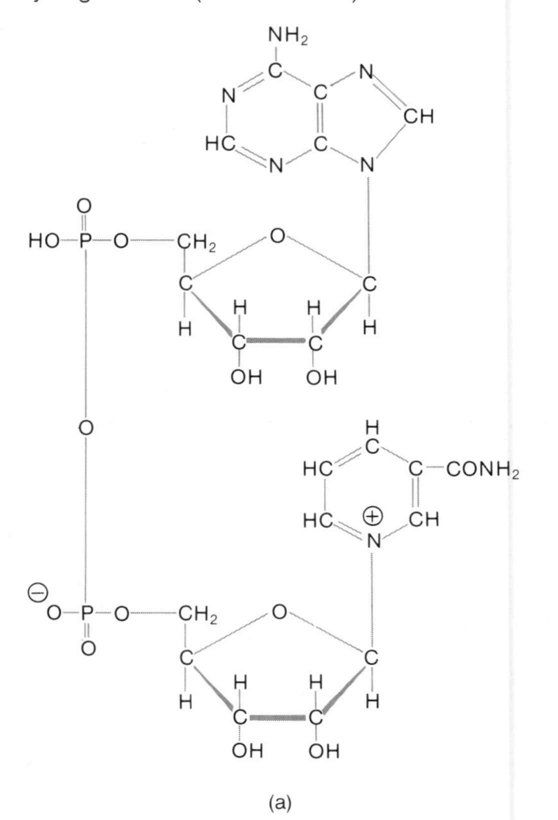

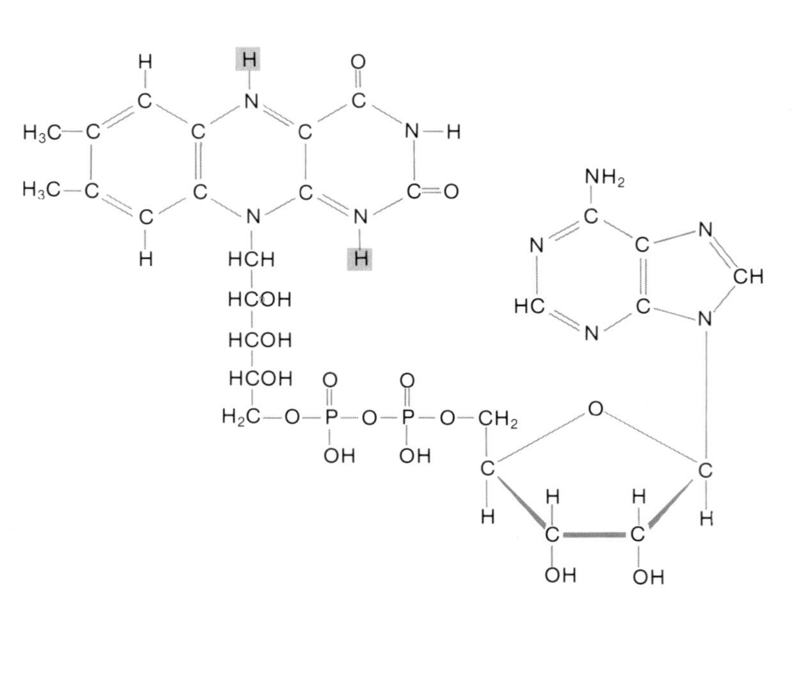

Oxidation-reduction reactions in cells often involve the transfer of hydrogen atoms rather than of free electrons. Since a hydrogen atom contains one electron (and one proton in the nucleus) a molecule that loses hydrogen becomes oxidized, and one that gains hydrogen becomes reduced. In many oxidation-reduction reactions, pairs of electrons—either as free electrons or as a pair of hydrogen atoms—are transferred from the reducing agent to the oxidizing agent.

Two molecules that serve important roles in the transfer of hydrogens are **nicotinamide adenine dinucleotide (NAD),** which is derived from the vitamin niacin (vitamin B_3), and **flavin adenine dinucleotide (FAD),** which is derived from the vitamin riboflavin (vitamin B_2). These molecules are *hydrogen carriers* because they accept hydrogens (becoming reduced) in one cellular location and donate hydrogens (becoming oxidized in the process) at a different cellular location (fig. 4.17).

Each FAD can accept two electrons and can bind two protons. The reduced form of FAD can, therefore, combine with the equivalent of two hydrogen atoms and can be written as $FADH_2$. Each NAD can also accept two electrons but can only bind one proton. The reduced form of NAD may, therefore, be shown as $NADH + H^+$ (the H^+ represents a free proton). In order to represent these two hydrogen carriers in a way that is both consistent and correct, the oxidized forms will subsequently be shown as NAD_{ox} and FAD_{ox}, and the reduced forms will be shown as NAD_{red} and FAD_{red} (fig. 4.18).

1. *Describe the first and second laws of thermodynamics, and use these laws to explain why the chemical bonds in glucose represent a source of potential energy and how cells can obtain this energy.*
2. *Define the terms* exergonic reaction *and* endergonic reaction. *Use these terms to explain the function of ATP in cells.*
3. *Using the symbols X-H_2 and Y, draw a coupled oxidation-reduction reaction. Identify the molecule that is reduced and the one that is oxidized, and tell which one is the reducing agent and which is the oxidizing agent.*
4. *Describe the functions of NAD, FAD, and oxygen (in terms of oxidation-reduction reactions), and explain the meaning of the symbols NAD_{red} and FAD_{red}.*

Figure 4.18. NAD_{ox} becomes reduced by the addition of two electrons from hydrogen atoms removed from an organic molecule (*X*). In this reaction, NAD acts as an oxidizing agent. In another cellular location, NAD_{red} can donate these two electrons to a different organic molecule (*Y*). (If *Y* could also bind to protons [H^+], it would receive two complete hydrogen atoms [H_2].) The molecule *Y* is thus reduced by this reaction in which NAD serves as a reducing agent.

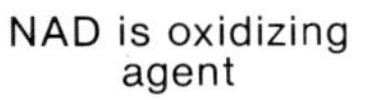

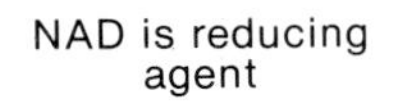

Summary

Enzymes as Catalysts p. 82

I. Enzymes are biological catalysts.
   A. Catalysts increase the rate of chemical reactions.
      1. A catalyst is not altered by the reaction.
      2. Catalysts do not change the final result of a reaction.
   B. Catalysts lower the activation energy of chemical reactions.
      1. The activation energy is the amount of energy needed by the reactant molecules to participate in a reaction.
      2. In the absence of a catalyst, only a small proportion of the reactants have the activation energy.
      3. By lowering the activation energy, enzymes allow a larger proportion of the reactants to participate in the reaction, thus increasing the reaction rate.

II. All enzymes are proteins.
   A. Protein enzymes have specific three-dimensional shapes, which are determined by the amino acid sequence and, ultimately, by the genes.
   B. The reactants in an enzyme-catalyzed reaction—called the substrates of the enzyme—fit into a specific pocket in the enzyme called the active site.
   C. By forming an enzyme-substrate complex, substrate molecules are brought into proper orientation and existing bonds are weakened; this allows new bonds to be more easily formed.

Control of Enzyme Activity p. 85

I. The activity of an enzyme is affected by a variety of factors.
   A. The rate of enzyme-catalyzed reactions increases with increasing temperature, up to a maximum.
      1. This is because increasing the temperature increases the energy in the total population of reactant molecules, thus increasing the proportion of reactants that have the activation energy.
      2. At a few centigrade degrees above body temperature, however, most enzymes start to denature, and the rate of the reactions at which they catalyze therefore decreases.
   B. Each enzyme has optimal activity at a characteristic pH—called the pH optimum for that enzyme.
      1. This is because the pH affects the shape and charges within the active site.
      2. The pH optima of different enzymes can be quite different—pepsin has a pH optimum of 2, for example, while trypsin is most active at a pH of 9.
   C. Many enzymes require metal ions in order to be active—these ions are therefore said to be cofactors for the enzymes.
   D. Many enzymes require smaller organic molecules for activity. These smaller organic molecules are called coenzymes.
      1. Many coenzymes are derived from water-soluble vitamins.
      2. Coenzymes transport hydrogen atoms and small substrate molecules from one enzyme to another.
   E. The rate of enzymatic reactions increases when either the substrate concentration or the enzyme concentration is increased.
      1. If the enzyme concentration is constant, the rate of the reaction increases as the substrate concentration is raised, up to a maximum rate.
      2. When the rate of the reaction does not increase upon further addition of substrate, the enzyme is said to be saturated.

II. Metabolic pathways involve a number of enzyme-catalyzed reactions.
   A. A number of enzymes usually cooperate to convert an initial substrate to a final product by way of several intermediates.
   B. Metabolic pathways are produced by multi-enzyme systems in which the product of one enzyme becomes the substrate of the next.

C. If an enzyme is defective due to an abnormal gene, the intermediates formed after the step catalyzed by the defective enzymes decrease, and the intermediates formed prior to the defective step accumulate.
   1. Diseases that result from defective enzymes are called inborn errors of metabolism.
   2. Accumulation of intermediates often results in damage to the organ that contains the defective enzyme.

D. Many metabolic pathways are branched so that one intermediate can serve as the substrate for two different enzymes.

E. The activity of a particular pathway can be regulated by end-product inhibition.
   1. In end-product inhibition, one of the products of the pathway inhibits the activity of a key enzyme.
   2. This is an example of allosteric inhibition, in which the product combines with its specific site on the enzyme, changing the conformation of the active site.

Bioenergetics p. 91

I. The flow of energy in the cell is called bioenergetics.
   A. According to the first law of thermodynamics, energy can neither be created nor destroyed but only transformed from one form to another.
   B. According to the second law of thermodynamics, all energy transformation reactions result in an increase in entropy (disorder).
      1. As a result of the increase in entropy, there is a decrease in free (usable) energy.
      2. Atoms that are organized into large organic molecules thus contain more free energy than more disorganized, smaller molecules.
   C. In order to produce glucose from carbon dioxide and water, energy must be added as sunlight.
      1. This process is called photosynthesis.
      2. Reactions that require the input of energy to produce molecules with higher free energy than the reactants are called endergonic reactions.
   D. The combustion of glucose to carbon dioxide and water releases energy in the form of heat.
      1. A reaction that releases energy and thus forms products that contain less free energy than the reactants is called an exergonic reaction.
      2. The same total amount of energy is released when glucose is converted into carbon dioxide and water within cells, even though this process occurs in many small steps.
   E. The exergonic reactions that convert food molecules into carbon dioxide and water in cells are coupled to endergonic reactions that form ATP.
      1. Some of the chemical-bond energy in glucose is, therefore, transferred to the "high-energy" bonds of adenosine triphosphate (ATP).
      2. The breakdown of ATP into adenosine diphosphate (ADP) and inorganic phosphate results in the liberation of energy.
      3. The energy liberated by the breakdown of ATP is used to power all of the energy-requiring processes of the cell—ATP is thus the "universal energy carrier" of the cell.

II. Oxidation-reduction reactions have several characteristics.
   A. A molecule is said to be oxidized when it loses electrons and to be reduced when it gains electrons.
   B. A reducing agent is thus an electron donor, and an oxidizing agent is an electron acceptor.
   C. Although oxygen is the final electron acceptor in the cell, other molecules can act as oxidizing agents.
   D. A single molecule can be an electron acceptor in one reaction and an electron donor in another.
      1. NAD and FAD can become reduced by accepting electrons from hydrogen atoms removed from other molecules.
      2. NAD_{red} and FAD_{red}, in turn, donate these electrons to other molecules in other locations within the cells.
      3. Oxygen is the final electron acceptor (oxidizing agent); since protons (H^+) follow the electrons, oxygen becomes reduced to H_2O.

Review Activities

Objective Questions

1. Which of the following statements about enzymes is *true?*
   (a) All proteins are enzymes.
   (b) All enzymes are proteins.
   (c) Enzymes are changed by the reactions they catalyze.
   (d) The active sites of enzymes have little specificity for substrates.
2. Which of the following statements about enzyme-catalyzed reactions is *true?*
   (a) The rate of reaction is independent of temperature.
   (b) The rate of all enzyme-catalyzed reactions is decreased when the pH is lowered from 7 to 2.
   (c) The rate of reaction is independent of substrate concentration.
   (d) Under given conditions of substrate concentration, pH, and temperature, the rate of product formation varies directly with enzyme concentration, up to a maximum, at which point the rate cannot be further increased.
3. Which of the following statements about lactate dehydrogenase is *true?*
   (a) It is a protein.
   (b) It oxidizes lactic acid.
   (c) It reduces another molecule (pyruvic acid).
   (d) All of these are true.
4. In a metabolic pathway
   (a) the product of one enzyme becomes the substrate of the next
   (b) the substrate of one enzyme becomes the product of the next
5. In an inborn error of metabolism
   (a) a genetic change results in the production of a defective enzyme
   (b) intermediates produced before the defective step accumulate
   (c) alternate pathways are taken by intermediates at branch points located before the defective step
   (d) All of these are true.
6. Which of the following represents an *endergonic* reaction?
   (a) $ADP + P_i \rightarrow ATP$
   (b) $ATP \rightarrow ADP + P_i$
   (c) glucose $+ O_2 \rightarrow CO_2 + H_2O$
   (d) $CO_2 + H_2O \rightarrow$ glucose
   (e) both *a* and *d*
   (f) both *b* and *c*
7. Which of the following statements about ATP is *true?*
   (a) The bond joining ADP and the third phosphate is a high-energy bond.
   (b) The formation of ATP is coupled to energy-liberating reactions.
   (c) The conversion of ATP to ADP and P_i provides energy for biosynthesis, cell movement, and other cellular processes that require energy.
   (d) ATP is the "universal energy carrier" of cells.
   (e) All of these are true.
8. When oxygen is combined with two hydrogens to make water,
   (a) oxygen is reduced
   (b) the molecule that donated the hydrogens becomes oxidized
   (c) oxygen acts as a reducing agent
   (d) both *a* and *b* apply
   (e) both *a* and *c* apply

Essay Questions

1. Explain the relationship between the chemical structure and the function of an enzyme, and describe how various conditions may alter both the structure and the function of an enzyme.
2. Explain how the rate of enzymatic reactions may be regulated by the relative concentrations of substrates and products.
3. Explain how end-product inhibition represents a form of negative feedback regulation.
4. Using the first and second laws of thermodynamics, explain how ATP is formed and how it serves as the universal energy carrier.
5. The coenzymes NAD and FAD can "shuttle" hydrogens from one reaction to another. Explain how this process serves to couple oxidation and reduction reactions.

Selected Readings

Baker, J. J. W., and G. E. Allen. 1982. *Matter, energy, and life.* 4th ed. Mass.: Addison-Wesley.

Berry, H. K. March 1984. The spectrum of metabolic disorders. *Diagnostic Medicine,* p. 39.

Dyson, R. D. 1978. *Essentials of cell biology.* 2d ed. Boston: Allyn & Bacon.

Hinkle, P., and R. E. McCarty. March 1979. How cells make ATP. *Scientific American.*

Lehninger, A. L. September 1961. How cells transform energy. *Scientific American.*

Lehninger, A. L. 1982. *Principles of biochemistry.* New York: Worth.

Sheeler, P., and D. E. Bianchi. 1980. *Cell biology: structure, biochemistry, and function.* New York: John Wiley & Sons.

Stryker, L. 1981. *Biochemistry,* 2d ed. New York: Freeman.

5 Cell Respiration and Metabolism

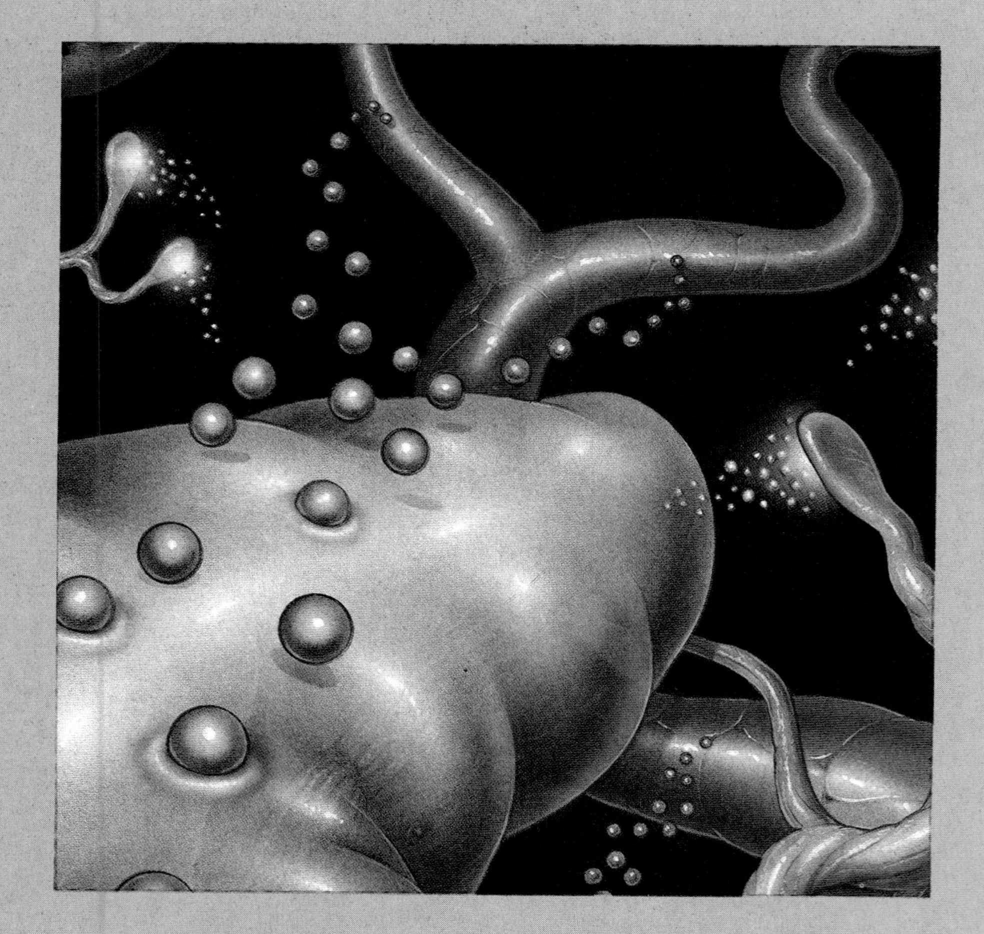

Objectives

By studying this chapter, you should be able to

1. describe the steps of glycolysis and explain its significance
2. describe how lactic acid is formed in anaerobic respiration and explain the physiological significance of the anaerobic pathway
3. define the term *oxygen debt* and explain the contribution of lactic acid to the oxygen debt
4. define the term *gluconeogenesis* and describe the Cori cycle
5. describe the pathway for the aerobic respiration of glucose through the steps of the Krebs cycle
6. explain the functional significance of the Krebs cycle in relation to the electron transport system
7. describe the electron transport system and oxidative phosphorylation
8. describe the role of oxygen in aerobic respiration
9. compare anaerobic and aerobic respiration in terms of initial substrates, final products, cellular locations, and the total number of ATP molecules produced per glucose respired
10. explain how the relative proportions of anaerobic and aerobic respiration in skeletal muscles are influenced by the maximal oxygen uptake
11. define the terms *lipolysis* and *β-oxidation* and describe how these processes function in cellular energy production
12. explain how ketone bodies are formed
13. describe the processes of oxidative deamination and transamination of amino acids and explain how these processes can contribute to energy production
14. explain, in terms of the metabolic pathways involved, how excess calories in the form of carbohydrates or protein can be converted to fat
15. describe the changes in skeletal muscle metabolism during exercise

Outline

Glycolysis and Anaerobic Respiration

In cellular respiration, chemical reactions that transform food molecules into simpler molecules liberate energy, some of which is used to produce ATP. The complete combustion of a molecule requires the presence of oxygen; some energy is liberated, however, during incomplete combustion in the absence of oxygen. This anaerobic respiration can serve as an important source of energy in some organs under special circumstances.

All of the reactions in the body that involve energy transformation are collectively termed **metabolism.** Metabolism may be divided into two categories: *anabolism* and *catabolism*. Catabolic reactions release energy, usually by the breakdown of larger organic molecules into smaller molecules. Anabolic reactions require the input of energy and include the synthesis of large, energy-storage molecules such as glycogen, fat, and protein.

The catabolic reactions that break down glucose, fatty acids, and amino acids serve as the primary sources of energy for the cellular synthesis of ATP. These metabolic pathways are known collectively as *cellular respiration.* When oxygen serves as the final electron acceptor, these processes are called **aerobic cell respiration.** The final products of aerobic respiration are carbon dioxide, water, and energy (a part of which is trapped in the chemical bonds of ATP). The overall equation for aerobic respiration, therefore, is identical to the equation that describes combustion (fuel + $O_2 \rightarrow CO_2 + H_2O$ + energy).

Notice that the term *respiration* refers to chemical reactions that liberate energy for the production of ATP. The oxygen used in aerobic respiration by tissue cells is obtained from the blood; the blood, in turn, becomes oxygenated in the lungs by the process of breathing. Breathing (also called *ventilation* or *external respiration*) is thus needed for, but is different from, aerobic respiration.

Unlike combustion, the conversion of glucose to carbon dioxide and water within the cells occurs in small, enzymatically catalyzed steps. Oxygen is only used at the last step (this will be described in a later section). Since a small amount of the chemical bond energy of glucose is released at early steps in the metabolic pathway, some cells in the body can obtain energy for ATP production in the temporary absence of oxygen. This process is called **anaerobic respiration.**

Glycolysis

Both the anaerobic and the aerobic respiration of glucose begin with a metabolic pathway known as **glycolysis.** Glycolysis (*glyco* = sugar; *lysis* = break) is the metabolic pathway by which glucose—a six-carbon (hexose) sugar—is converted into two molecules of *pyruvic acid.* Each pyruvic acid molecule contains three carbons, three oxygens, and four hydrogens. The number of carbon and oxygen atoms in one molecule of glucose—$C_6H_{12}O_6$—can thus be accounted for in the two pyruvic acid molecules. Since the two pyruvic acids together account for only eight hydrogens, however, it is clear that four hydrogen atoms are removed from the intermediates in glycolysis. These hydrogen atoms are used to reduce two molecules of NAD_{ox} to two molecules of NAD_{red}.

Glycolysis is exergonic, and a portion of the energy that is released is used to drive the endergonic reaction $ADP + P_i \rightarrow ATP$. At the end of the glycolytic pathway there is a net gain of two ATP per glucose molecule, as indicated in the overall equation for glycolysis:

$$\text{Glucose} + 2NAD_{ox} + 2ADP + 2P_i \rightarrow$$
$$2 \text{ pyruvic acid} + 2NAD_{red} + 2ATP$$

Although the overall equation for glycolysis is exergonic, glucose must be "activated" at the beginning of the pathway before energy can be obtained. This activation requires the addition of two phosphate groups derived from two molecules of ATP. Energy from the reaction $ATP \rightarrow ADP + P_i$ is therefore consumed at the beginning of glycolysis. This is shown as an "up-staircase" in figure 5.1. At later steps in glycolysis, however, four molecules of ATP are produced (and two molecules of NAD are reduced) as energy is liberated (the "down-staircase" in figure 5.1). The two molecules of ATP used in the beginning, therefore, represent an energy investment; the net gain of two ATP and two NAD_{red} by the end of the pathway represent an energy "profit."

The overall equation for glycolysis obscures the fact that this is a metabolic pathway consisting of nine separate steps. The individual steps in this pathway are shown in figure 5.2, and the enzymes that catalyze these steps are listed in table 5.1.

Objectives

By studying this chapter, you should be able to

1. describe diffusion and explain its physical basis
2. explain how nonpolar molecules, inorganic ions, and water can diffuse through a cell membrane
3. describe the factors that influence the rate of diffusion through cell membranes
4. define osmosis and describe the conditions required for osmosis to occur
5. describe the meanings of the terms *osmolality* and *osmotic pressure* and explain how these factors relate to osmosis
6. define the terms *isotonic, hypertonic,* and *hypotonic* and explain their physiological significance
7. describe the characteristics of carrier-mediated transport
8. describe the facilitated diffusion of glucose through cell membranes, and give examples of where this occurs in the body
9. define active transport and explain how active transport may be accomplished
10. explain the active transport of Na^+ and K^+ by the Na^+/K^+ pump and the physiological significance of this activity
11. explain how an equilibrium potential is produced when only one ion is able to diffuse through a cell membrane
12. explain why nerve and muscle cells have a resting membrane potential that is slightly different than the potassium equilibrium potential
13. explain the role of the Na^+/K^+ pump in the maintenance of the resting membrane potential

Outline

Diffusion and Osmosis

The cell (plasma) membrane separates the intracellular environment from the extracellular environment. Proteins, nucleotides, and other molecules needed for the structure and function of the cell cannot penetrate, or "permeate," the membrane. The cell membrane is, however, **selectively permeable** to certain molecules and many ions; this allows two-way traffic in nutrients and wastes needed to sustain metabolism and provides electrical currents created by the movements of ions through the membrane.

The mechanisms involved in the transport of molecules and ions through the cell membrane may be divided into two categories: (1) transport that requires the action of specific *carrier proteins* in the membrane (*carrier-mediated transport*); and (2) transport through the membrane that is not carrier-mediated. Carrier-mediated transport includes *facilitated diffusion* and *active transport;* non-carrier-mediated transport consists of the *simple diffusion* of ions, lipid-soluble molecules, and water through the membrane. The diffusion of water (solvent) through a membrane is called *osmosis.*

Diffusion—whether it requires the presence of carriers or not and whether it involves the movement of solute or solvent through a membrane—is driven by the thermal energy present in the transported substances. Metabolic energy is not required for these transport processes, which, indeed, can occur across artificial membranes or across the membranes of dead cells. Active transport, in contrast, requires an active cellular metabolism, because this form of membrane transport does not occur without the energy supplied by the conversion of ATP to ADP and P_i. As will be described in a later section, metabolic energy is needed in active transport because this process moves molecules and ions "uphill," from regions of lower to regions of higher concentrations.

Diffusion

Molecules in a gas and molecules and ions dissolved in a solution are in a constant state of random motion as a result of their thermal (heat) energy. This random motion, called **diffusion,** tends to make the gas or solution evenly mixed, or diffusely spread out, within a given volume. Whenever a *concentration difference,* or *concentration gradient,* exists between two parts of a solution, therefore, random molecular motion tends to abolish the gradient and to make the molecules uniformly distributed (fig. 6.1).

Figure 6.1. Net diffusion occurs when there is a concentration difference (or concentration gradient) between two regions of a solution (*a*) provided that the membrane separating these regions is permeable to the diffusing substance. Diffusion tends to equalize the concentration of these solutions (*b*) and thus to abolish the concentration differences.

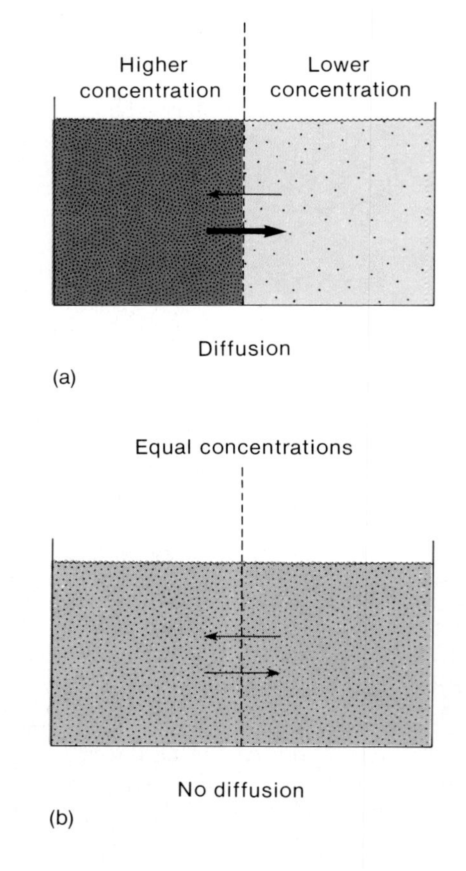

In terms of the second law of thermodynamics, the concentration difference represents an unstable state of high organization (low entropy), which changes to produce a uniformly distributed solution with maximum disorganization (high entropy).

As a result of random molecular motion, molecules in the part of the solution with a higher concentration will enter the area of lower concentration. Molecules will also move in the opposite direction, but not as frequently. As a result, there will be a *net movement* from the region of higher to the region of lower concentration until the concentration difference is abolished. This net movement is called **net diffusion.** Net diffusion is a physical process that occurs whenever there is a concentration difference; when the concentration difference exists across a membrane, diffusion becomes a type of membrane transport.

Actually, the Na^+/K^+ pump does more than simply work against the ion leaks, since it transports *three* Na^+ ions out of the cell for every *two* K^+ ions that it moves in, its action helps generate a potential difference across the membrane (fig. 6.21). As a result of all of these activities, a real cell has (1) a relatively constant intracellular concentration of Na^+ and K^+; and (2) a constant membrane potential (in the absence of stimulation) in nerves and muscles of −65 mV to −85 mV.

1. *Define the term* membrane potential, *and describe how it is measured.*
2. *Describe how an equilibrium potential is produced when potassium is the only diffusible cation.*
3. *Explain why the resting membrane potential is different from the potassium equilibrium potential.*
4. *Explain the role of the Na^+/K^+ pump in the generation and maintenance of the resting membrane potential.*

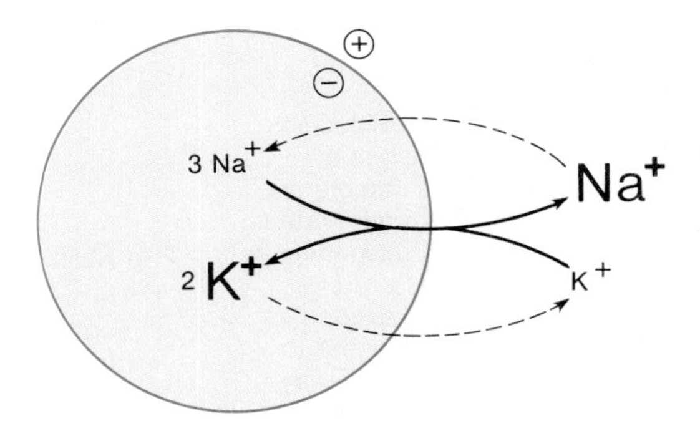

Figure 6.21. The concentrations of Na^+ and K^+ both inside and outside the cell do not change as a result of diffusion (*dotted arrows*) because of active transport (*solid arrows*) by the Na^+/K^+ pump. Since the pump transports three Na^+ for every two K^+, the pump itself helps to create a charge separation (a potential difference, or voltage) across the membrane.

Summary

Diffusion and Osmosis p. 124

I. Diffusion is the net movement of molecules or ions from regions of high to regions of low concentration.
   A. This is a type of passive transport—energy is provided by the thermal energy of the molecules, not by cellular metabolism.
   B. Net diffusion stops when the concentration is equal on both sides of the membrane.

II. The rate of diffusion is dependent on a variety of factors.
   A. The rate of diffusion depends on the concentration difference on the two sides of the membrane.
   B. The rate depends on the permeability of the cell membrane to the diffusing substance.
   C. The rate of diffusion through a membrane is also directly proportional to the surface area of the membrane, which can be increased by such adaptations as microvilli.

III. Simple diffusion is the type of passive transport in which small molecules and inorganic ions, such as Na^+ and K^+, move through membrane.
   A. Inorganic ions pass through specific channels in the membrane.
   B. Lipids such as steroid hormones, and dipolar water molecules, can pass by simple diffusion directly through the phospholipid layers of the membrane.

IV. Osmosis is the simple diffusion of solvent (water) through a membrane that is more permeable to the solvent than it is to the solute.
   A. Water moves from the solution that is more dilute to the solution that has a higher solute concentration.
   B. Osmosis depends on a difference in total solute concentration, not on the chemical nature of the solute.
      1. The concentration of total solute (in moles) per kilogram (liter) of water is measured in osmolality units.
      2. The solution with the higher osmolality has the higher osmotic pressure.
      3. Water moves by osmosis from the solution of lower osmolality and osmotic pressure to the solution of higher osmolality and osmotic pressure.
   C. Solutions that have the same osmotic pressure as plasma (such as 0.9% NaCl and 5% glucose) are said to be isotonic.
      1. Solutions with a lower osmotic pressure are hypotonic; those with a higher osmotic pressure are hypertonic.
      2. Cells in a hypotonic solution gain water and swell; those in a hypertonic solution lose water and shrink (crenate).
   D. The osmolality and osmotic pressure of the plasma is maintained within a normal range by the action of osmoreceptors in the hypothalamus.
      1. Increased osmolality stimulates the osmoreceptors; decreased osmolality inhibits the electrical activity of the osmoreceptors.
      2. Stimulation of the osmoreceptors causes thirst and the secretion of antidiuretic hormone (ADH) from the pituitary.
      3. ADH stimulates water retention by the kidneys, which serves to maintain a normal blood volume and osmolality.

Carrier-Mediated Transport p. 132

I. The passage of glucose, amino acids, and other substances through the cell membrane is mediated by carrier proteins in the cell membrane.
   A. Carrier-mediated transport exhibits the properties of specificity, competition, and saturation.
   B. The transport rate of molecules such as glucose reaches a maximum when the carriers are saturated—this maximum rate is called the transport maximum, or T_m.
II. The transport of molecules such as glucose from the side of higher to the side of lower concentration by means of membrane carriers is called facilitated diffusion.
   A. Like simple diffusion, this is passive transport—cellular energy is not required.
   B. Unlike simple diffusion, facilitated diffusion is specific, exhibits competition, and can be saturated.
III. The active transport of molecules and ions across a membrane requires the expenditure of cellular energy (ATP).
   A. In active transport, carriers move molecules or ions from the side of lower to the side of higher concentration.
   B. One example of active transport is the action of the Na^+/K^+ pump.
      1. Sodium is more concentrated on the outside of the cell, whereas potassium is more concentrated on the inside of the cell.
      2. The Na^+/K^+ pump helps to maintain these concentration differences by transporting Na^+ out of the cell and K^+ into the cell.
      3. Three Na^+ are transported out for every two K^+ ions that are transported into the cell.

The Membrane Potential p. 136

I. The cytoplasm of the cell contains negatively charged organic ions (anions) that cannot leave the cell—they are "fixed anions."
   A. These fixed anions attract K^+, which is the inorganic cation that can most easily pass through the cell membrane.
   B. As a result of this electrical attraction, the concentration of K^+ within the cell is greater than the concentration of K^+ in the extracellular fluid.
   C. If K^+ were the only diffusible ion, the concentrations of K^+ on the inside and outside of the cell would reach an equilibrium.
      1. At this point, the rate of K^+ entry (due to electrical attraction) would equal the rate of K^+ exit (due to diffusion).
      2. At this equilibrium, there would still be a higher concentration of negative charges within the cell (due to the fixed anions) than outside the cell.
      3. At this equilibrium, the inside of the cell would be ninety millivolts (90 mV) negative compared to the outside of the cell—this is called the K^+ equilibrium potential (E_K).
   D. The true membrane potential is less than E_K—it is usually -65 mV to -85 mV.
      1. This is because some Na^+ can also enter the cell.
      2. Na^+ is more highly concentrated outside than inside the cell, and the inside of the cell is negative—these forces attract Na^+ into the cell.
      3. The rate of Na^+ entry is generally slow because the membrane is usually not very permeable to Na^+.
II. The slow rate of Na^+ entry is accompanied by a slow rate of K^+ exit from the cell.
   A. These changes are corrected by the active transport Na^+/K^+ pump, which maintains constant concentrations and a constant resting membrane potential.
   B. There are numerous Na^+/K^+ pumps in all cells of the body that require a constant expenditure of energy.
   C. The Na^+/K^+ pump itself contributes to the membrane potential because it pumps more Na^+ out than it pumps K^+ in (by a ratio of three to two).

Review Activities

Objective Questions

1. The movement of water across a cell membrane occurs by
   (a) active transport
   (b) facilitated diffusion
   (c) simple diffusion (osmosis)
   (d) all of the above
2. Which of the following statements about the facilitated diffusion of glucose is *true*?
   (a) There is a net movement from the region of low to the region of high concentration.
   (b) Carrier proteins in the cell membrane are required for this transport.
   (c) This transport requires energy obtained from ATP.
   (d) This is an example of co-transport.
3. If a poison such as cyanide stops the production of ATP, which of the following transport processes would cease?
   (a) the movement of Na^+ out of a cell
   (b) osmosis
   (c) the movement of K^+ out of a cell
   (d) all of the above

4. Red blood cells crenate in
   (a) a hypotonic solution
   (b) an isotonic solution
   (c) a hypertonic solution
5. Plasma has an osmolality of about 300 mOsm. Isotonic saline has an osmolality of
   (a) 150 mOsm
   (b) 300 mOsm
   (c) 600 mOsm
   (d) none of the above
6. A 0.5 m NaCl solution and a 1.0 m glucose solution
   (a) have the same osmolality
   (b) have the same osmotic pressure
   (c) are isotonic to each other
   (d) are all of the above
7. The diffusible ion that is most important in the establishment of the membrane potential is
   (a) K^+
   (b) Na^+
   (c) Ca^{++}
   (d) Cl^-
8. An increase in blood osmolality
   (a) can occur as a result of dehydration
   (b) causes a decrease in blood osmotic pressure
   (c) is accompanied by a decrease in ADH secretion
   (d) all of the above
9. In hyperkalemia, the membrane potential
   (a) increases
   (b) decreases
   (c) is not changed
10. Which of the following statements about the Na^+/K^+ pump is *true*?
   (a) Na^+ is actively transported into the cell.
   (b) K^+ is actively transported out of the cell.
   (c) An equal number of Na^+ and K^+ ions are transported with each cycle of the pump.
   (d) The pumps are constantly active in all cells.

Essay Questions

1. Describe the conditions required to produce osmosis, and explain why osmosis occurs under these conditions.
2. Explain how simple diffusion can be distinguished from facilitated diffusion and how active transport can be distinguished from passive transport.
3. Compare the theoretical membrane potential that occurs at K^+ equilibrium with the true resting membrane potential. Explain the reasons for the differences between these.
4. Explain how the Na^+/K^+ pump contributes to the resting membrane potential.

Selected Readings

Christensen, H. N. 1975. *Biological transport,* 2nd ed. Menlo Park, CA: W. A. Benjamen, Inc.

Dyson, R. D. 1978. *Essentials of cell biology,* 2d ed. Boston: Allyn & Bacon.

Finean, J. B., Coleman, R., and R. H. Mitchell. 1978. *Membranes and their cellular functions,* 2d ed. New York: John Wiley & Sons, Inc.

Johnson, L. G. 1983. *Biology.* Dubuque: Wm. C. Brown Company Publishers.

Kaplan, J. H. 1985. Ion movements through the sodium pump. *Annual Review of Physiology* 47:535.

Keeton, W. T. 1980. *Biological science,* 3rd ed. New York: W. W. Norton & Co.

Kotyk, A., and K. Janacek. 1975. *Cell membrane transport: principles and techniques,* 2nd. ed. New York: Plenum Publishing Corp.

Mader, S. S. 1982. *Inquiry into life,* 3rd. ed. Dubuque: Wm. C. Brown Company Publishers.

Sheeler, P., and D. E. Bianchi. 1980. *Cell biology: structure, biochemistry, and function.* New York: John Wiley & Sons.

Wheeler, T. J., and P. C. Hinkle. 1985. The glucose transporter of mammalian cells. *Annual Review of Physiology* 47:503.

7 The Nervous System: *Organization, Electrical Activity, and Synaptic Transmission*

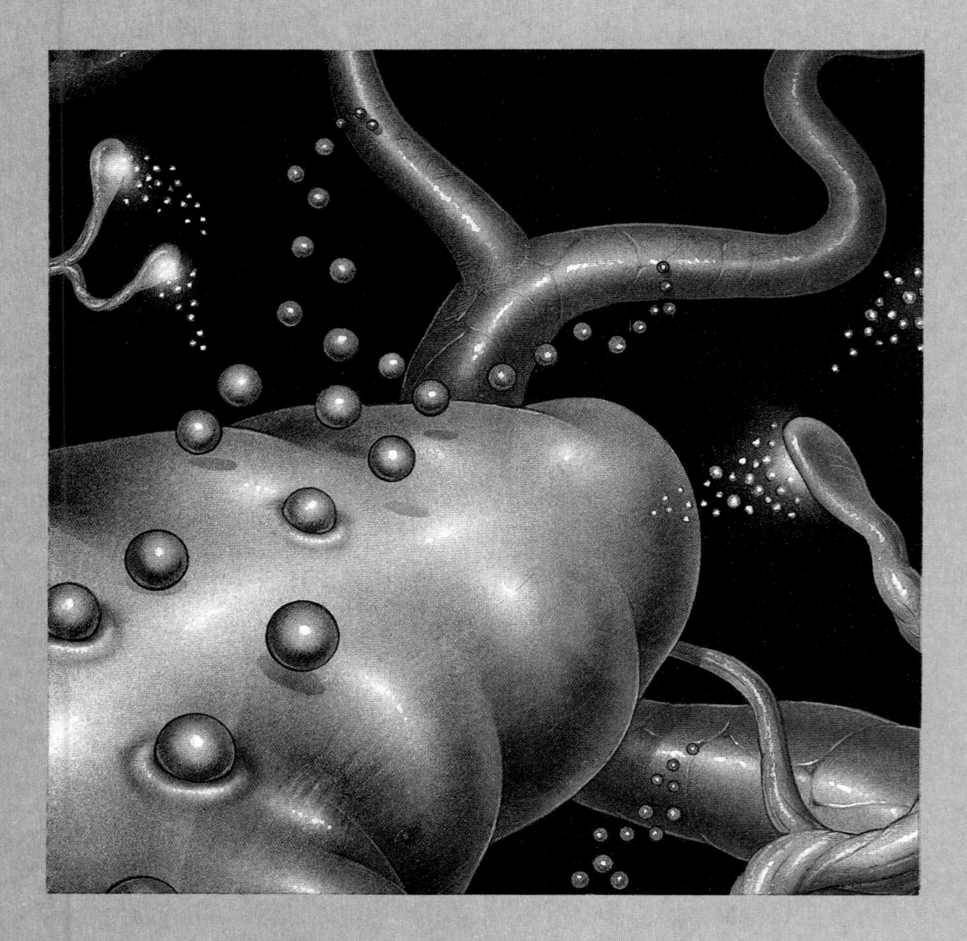

Objectives

By studying this chapter, you should be able to

1. describe the parts of a neuron and their functional significance
2. describe the anatomical and functional divisions of the nervous system
3. describe the locations and functions of the different types of neuroglial cells
4. describe the blood-brain barrier and explain its significance
5. describe the sheath of Schwann and explain how this sheath helps the regeneration of cut peripheral nerve fibers
6. explain how a myelin sheath is formed
7. define depolarization, repolarization, and hyperpolarization
8. explain the actions of voltage-regulated Na^+ and K^+ gates and the events that occur during the production of an action potential
9. describe the properties of action potentials and explain the significance of the all-or-none law and the refractory periods
10. explain how action potentials are regenerated along a myelinated and a nonmyelinated axon and explain why saltatory conduction improves the conduction speed of nerve impulses
11. describe the events that occur between the electrical excitation of an axon and the release of neurotransmitter
12. describe how ACh stimulates a postsynaptic cell
13. describe the characteristics of an EPSP and compare these characteristics to those of an action potential
14. compare the mechanisms that inactivate ACh and catecholamine neurotransmitters
15. explain the role of cyclic AMP in the action of catecholamine neurotransmitters
16. explain the nature and significance of the inhibitory effects of glycine and GABA in the central nervous system
17. explain how EPSPs and IPSPs can interact and explain the significance of spatial and temporal summation
18. describe the processes of presynaptic inhibition and post-tetanic potentiation

Outline

Figure 7.1. The structure of two kinds of neurons. (*a*) A motor neuron; (*b*) a sensory neuron.

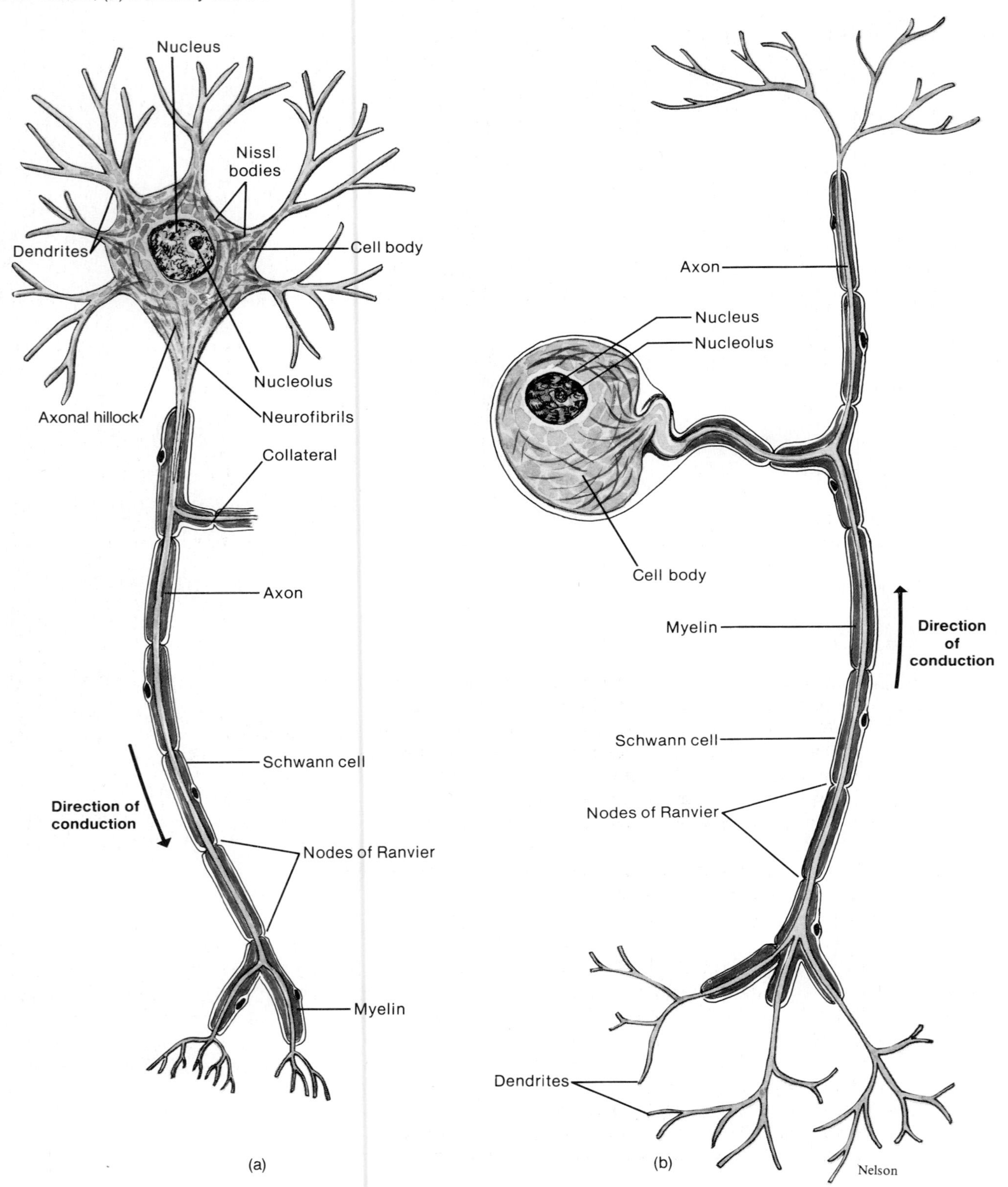

Table 7.1 Anatomical terms used in describing the nervous system.

Term	Definition
Central nervous system (CNS)	Brain and spinal cord
Peripheral nervous system (PNS)	Nerves and ganglia
Sensory nerve fiber	Nerve that transmits impulses from sensory receptor into CNS (afferent fiber)
Motor nerve fiber	Nerve that transmits impulses from CNS to an effector organ (e.g., muscle efferent fiber)
Nerve	Cablelike collection of nerve fibers; may be "mixed" (contain both sensory and motor fibers)
Somatic motor nerve	Nerve that stimulates contraction of skeletal muscles
Autonomic motor nerve	Nerve that stimulates contraction (or inhibits contraction) of smooth muscle and cardiac muscle and secretion of glands
Ganglion	Collection of neuron cell bodies located outside CNS
Nucleus	Groupings of neuron cell bodies within CNS
Tract	Collections of nerve fibers that interconnect regions of CNS

Organization and Histology of the Nervous System

The nervous system is divided into the **central nervous system (CNS),** which includes the brain and spinal cord, and the **peripheral nervous system (PNS),** which includes the *cranial nerves* arising from the brain and the *spinal nerves* arising from the spinal cord.

The nervous system is composed of only two principal types of cells, neurons and neuroglia. **Neurons** are the basic structural and functional units of the nervous system. They are specialized to respond to physical and chemical stimuli, conduct electrochemical impulses, and release specific chemical regulators. Through these activities, neurons perform such functions as the perception of sensory stimuli, learning, memory, and the control of muscles and glands. Neurons cannot divide by mitosis, although some neurons can regenerate a severed portion or sprout small new branches under some conditions.

Neuroglia, or **glial** (*glia* = glue) **cells,** are supportive cells in the nervous system that aid the function of neurons. Glial cells are about five times more abundant than neurons and have limited mitotic abilities (brain tumors that occur in adults are usually composed of glial cells rather than neurons).

Neurons

Although neurons vary considerably in size and shape, they generally have three principal regions: (1) a cell body; (2) dendrites; and (3) an axon (figs. 7.1 and 7.2).

The **cell body,** or **perikaryon** (*peri* = around; *karyon* = nucleus), is the enlarged portion of the neuron, which contains the nucleus and serves as the "nutritional center" of the neuron where macromolecules are produced. The perikaryon also contains granular, densely

Figure 7.2 A photomicrograph of neurons from the anterior column of gray matter of the spinal cord (120X).

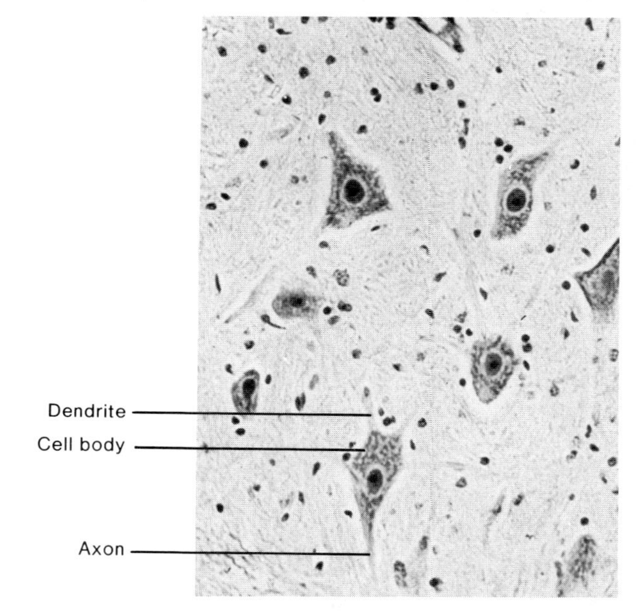

staining material known as *Nissl bodies,* which are not found in the dendrites or axon. The Nissl bodies are composed of granular (rough) endoplasmic reticulum, an organelle involved in protein synthesis. The cell bodies within the CNS are frequently clustered into groups called *nuclei* (not to be confused with the nucleus of a cell). Cell bodies in the PNS usually occur in clusters called *ganglia* (table 7.1).

Table 7.2 Comparison of axoplasmic flow with axonal transport.

Axoplasmic flow	Axonal transport
Transport rate comparatively slow (1–2mm/day)	Transport rate comparatively fast (200–400mm/day)
Molecules transported only from cell body	Molecules transported from cell body to axon endings and in reverse direction
Bulk movement of proteins in axoplasm, including microfilaments and tubules	Transport of specific proteins, mainly of membrane proteins and acetylcholinesterase
Transport accompanied by peristaltic waves of axon membrane	Transport dependent on cagelike microtubule structure within axon and on actin and Ca^{++}

Table 7.3 Some types of neuroglial cells and their functions.

Neuroglia	Functions
Schwann cells	Surround axons of all peripheral nerve fibers, forming neurilemmal sheath, or sheath of Schwann; wrap around many peripheral fibers to form myelin sheaths
Oligodendrocytes	Form myelin sheaths around central axons, producing "white matter" of CNS
Astrocytes	Perivascular foot processes, covering capillaries within brain and contributing to the blood-brain barrier
Microglia	Amoeboid cells within CNS that are phagocytic
Ependyma	Form epithelial lining of brain cavities (ventricles) and central canal of spinal cord; cover tufts of capillaries to form choroid plexus—structures that produce cerebrospinal fluid

Dendrites (*dendron* = tree branch) are branched processes that extend from the cytoplasm of the cell body. Dendrites serve as a receptive area that transmits electrical impulses to the cell body. The **axon,** or **nerve fiber,** is a longer process that conducts impulses away from the cell body. Axons vary in length from only a millimeter to as long as a meter or more (for axons that extend from the CNS to the foot) in length. The origin of the axon near the cell body is called the *axon hillock,* and side branches that may extend from the axon are called *axon collaterals.*

Proteins and other molecules are transported through the axon at faster rates than could be achieved by simple diffusion. This rapid movement is produced by two different mechanisms: axoplasmic flow and axonal transport (table 7.2). **Axoplasmic flow,** the slower of the two, results from rhythmic waves of contraction that push the cytoplasm from the axon hillock to the nerve endings. **Axonal transport,** which is more rapid and more selective, may occur in a reverse (retrograde) as well as a forward (orthograde) direction. Indeed, retrograde transport may be responsible for the movement of herpes virus, rabies virus, and tetanus toxin from the nerve terminals into cell bodies.

Neuroglia

Unlike other organs that are "packaged" in connective tissue derived from mesoderm (the middle layer of embryonic tissue), the supporting neuroglial cells of the nervous system are derived from the same embryonic tissue layer (ectoderm) that produces neurons. There are six categories of neuroglial cells: (1) **Schwann cells,** which form myelin sheaths around peripheral axons; (2) **oligodendrocytes,** which form myelin sheaths around axons of the CNS; (3) **microglia,** which are phagocytic cells that migrate through the CNS and remove foreign and degenerated material; (4) **astrocytes,** which help regulate the passage of molecules from the blood to the brain; (5) **ependyma,** which line the ventricles of the brain and the central canal of the spinal cord; and (6) **satellite cells,** which support neuron cell bodies within the ganglia of the PNS (table 7.3).

Sheath of Schwann and Myelin Sheath. Some axons in the CNS and PNS are surrounded by a myelin sheath and are known as *myelinated* axons. Other axons do not have a myelin sheath and are *unmyelinated.*

Unmyelinated axons in the PNS are surrounded by a living sheath of Schwann cells, known as the **sheath of Schwann.** The outer surface of this layer of Schwann cells is encased in a glycoprotein *basement membrane,* often called the *neurilemma,* which is analogous to the basement membrane that underlies epithelial membranes. The

Figure 7.3. The formation of a myelin sheath in a peripheral axon. The myelin sheath is formed by successive wrappings of the Schwann cell membranes, leaving most of the Schwann cell cytoplasm outside the myelin. The sheath of Schwann cells is thus located outside the myelin sheath.

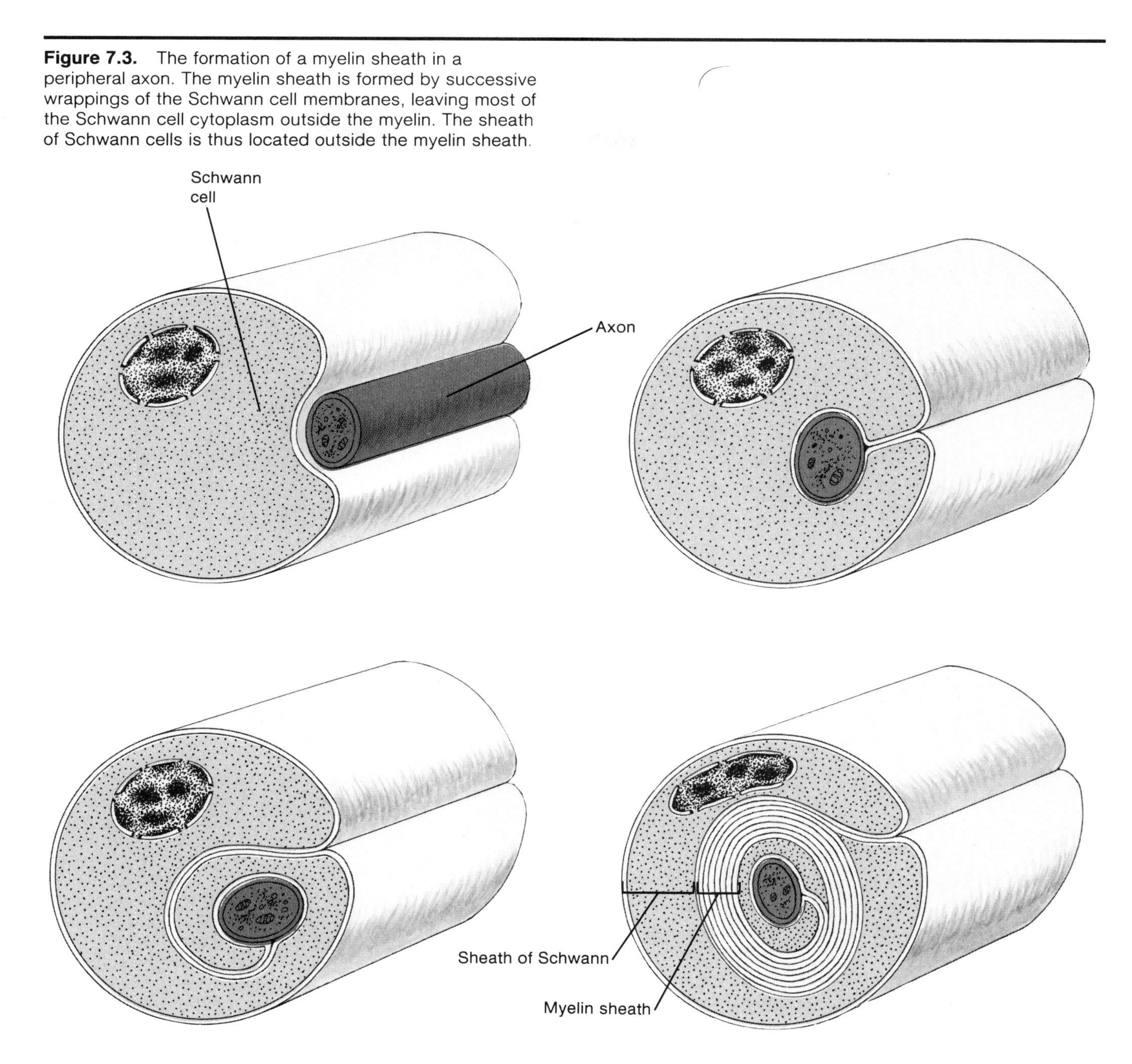

unmyelinated axons of the CNS, in contrast, lack a sheath of Schwann (because Schwann cells are only found in the PNS) and also lack a continuous basement membrane.

Axons that are less than two micrometers in diameter are usually unmyelinated. Larger axons are generally surrounded by a **myelin sheath,** which is composed of successive wrappings of the cell membrane of Schwann cells or oligodendrocytes. In the process of myelin formation, Schwann cells (in the PNS) wrap around the axon so that their cytoplasm becomes squeezed towards the outermost region (fig. 7.3), like toothpaste rolled from the bottom of a tube. Each Schwann cell only wraps about 1 mm of axon, leaving gaps of exposed axon between the adjacent Schwann cells. These gaps in the myelin sheath are known as the **nodes of Ranvier.** The successive wrappings of Schwann cell membrane provide insulation around the axon, leaving only the nodes of Ranvier exposed to produce nerve impulses.

Figure 7.4. An electron micrograph of unmyelinated and myelinated axons

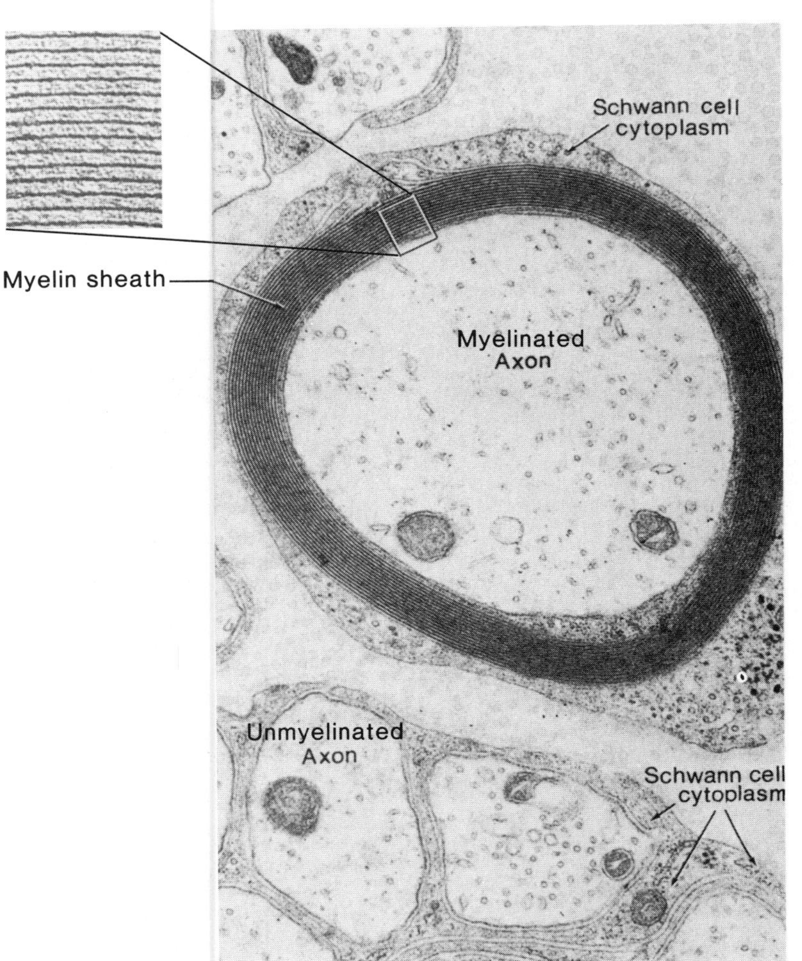

The Schwann cells remain alive as their cytoplasm is squeezed to the outside of the myelin sheath. As a result, myelinated axons of the PNS, like their unmyelinated counterparts, are surrounded by a living sheath of Schwann and a continuous basement membrane (fig. 7.4).

The myelin sheaths of the CNS are formed by oligodendrocytes. Unlike a Schwann cell, which forms a myelin sheath around only one axon, each oligodendrocyte has extensions that form myelin sheaths around several axons. Myelinated axons of the CNS, as a result, are not surrounded by a continuous basement membrane. The myelin sheaths around axons of the CNS give this tissue a white color; areas of the CNS that contain a high concentration of axons thus form the *white matter.* The *grey matter* of the CNS is composed of high concentrations of cell bodies and dendrites, which lack myelin sheaths.

Multiple sclerosis (MS) is a relatively common neurological disease in persons between the ages of twenty and forty. MS is a chronic, degenerating, remitting, and relapsing disease that progressively destroys the myelin sheaths of neurons in multiple areas of the CNS. Initially, lesions form on the myelin sheaths and soon develop into hardened *scleroses* (*skleros* = hardened). Destruction of the myelin sheaths prohibits the normal conduction of impulses, resulting in a progressive loss of functions. Because myelin degeneration is widely spread, MS has a greater variety of symptoms than any other neurological disease. This characteristic, coupled with remissions, frequently causes misdiagnosis of this disease.

Figure 7.5. The process of neuron regeneration. (*a*) If a neuron is severed through a myelinated axon, the proximal portion may survive, but (*b*) the distal portion degenerates. The myelin sheath provides a pathway for (*c*) the regeneration of an axon and (*d*) innervation is restored.

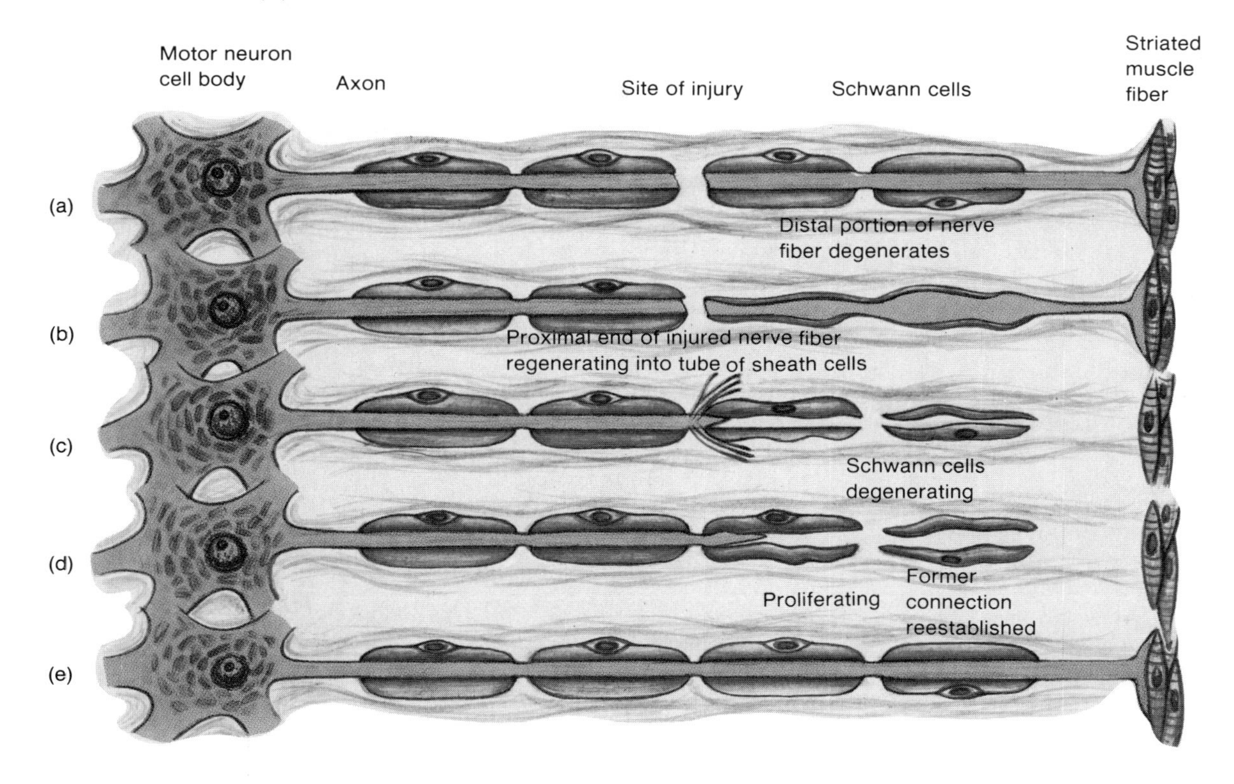

Regeneration of a Cut Axon. When an axon in a peripheral nerve is cut, the distal portion of the axon that was severed from the cell body degenerates and is phagocytosed by Schwann cells. The Schwann cells, surrounded by the basement membrane, then form a *regeneration tube* (fig. 7.5), as the part of the axon that is connected to the cell body begins to grow and exhibit amoeboid movement. The Schwann cells of the regeneration tube are believed to secrete chemicals that attract the growing axon tip, and the regeneration tube helps to guide the regenerating axon to its proper destination. Even a severed major nerve may be surgically reconnected and the function of the nerve largely reestablished if the surgery is performed before tissue death.

Injury in the CNS stimulates growth of axon collaterals, but central axons have a much more limited ability to regenerate than peripheral axons. This is believed to be primarily due to the absence of a continuous basement membrane, so that a regeneration tube cannot be formed. There is currently much research devoted to the study of nerve growth factors and to surgical methods of compensating for the lack of a regeneration tube by guiding the growth of axons in the CNS by artificial techniques.

Astrocytes and the Blood-Brain Barrier. Astrocytes (*aster* = star) are large, stellate cells with numerous cytoplasmic processes that radiate outwards. They are the most abundant of the neuroglial cells in the CNS, constituting up to 90% of the nervous tissue in some areas of the brain.

Capillaries in the brain, unlike those of most other organs, do not have pores between adjacent endothelial cells (the cells that compose the walls of capillaries). Unlike other organs, therefore, the brain cannot obtain molecules from the blood plasma by a nonspecific filtering process. Instead, molecules within brain capillaries must

Figure 7.6. A photomicrograph showing the perivascular feet of astrocytes (a type of neuroglial cell), which cover most of the surface area of brain capillaries.

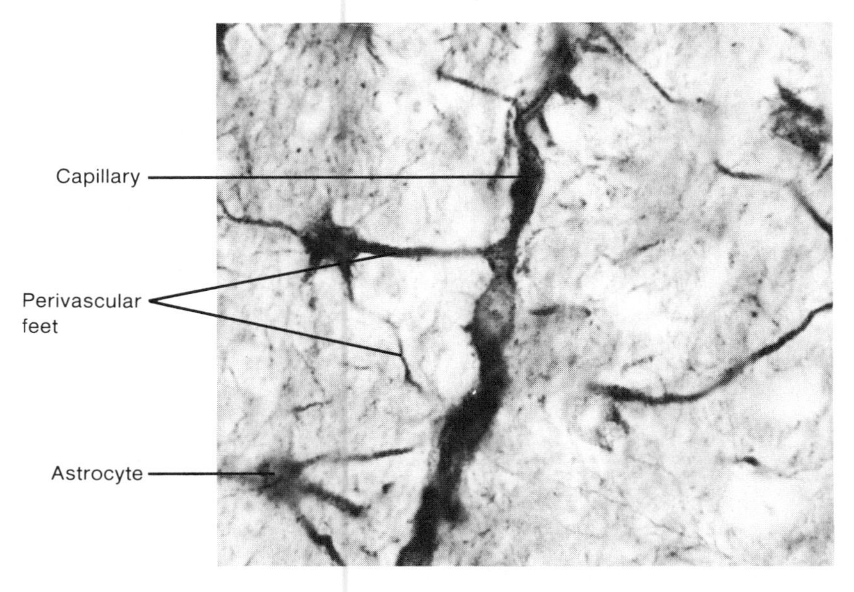

Figure 7.7. The relationship between sensory and motor fibers of the peripheral nervous system (*PNS*) and the central nervous system (*CNS*).

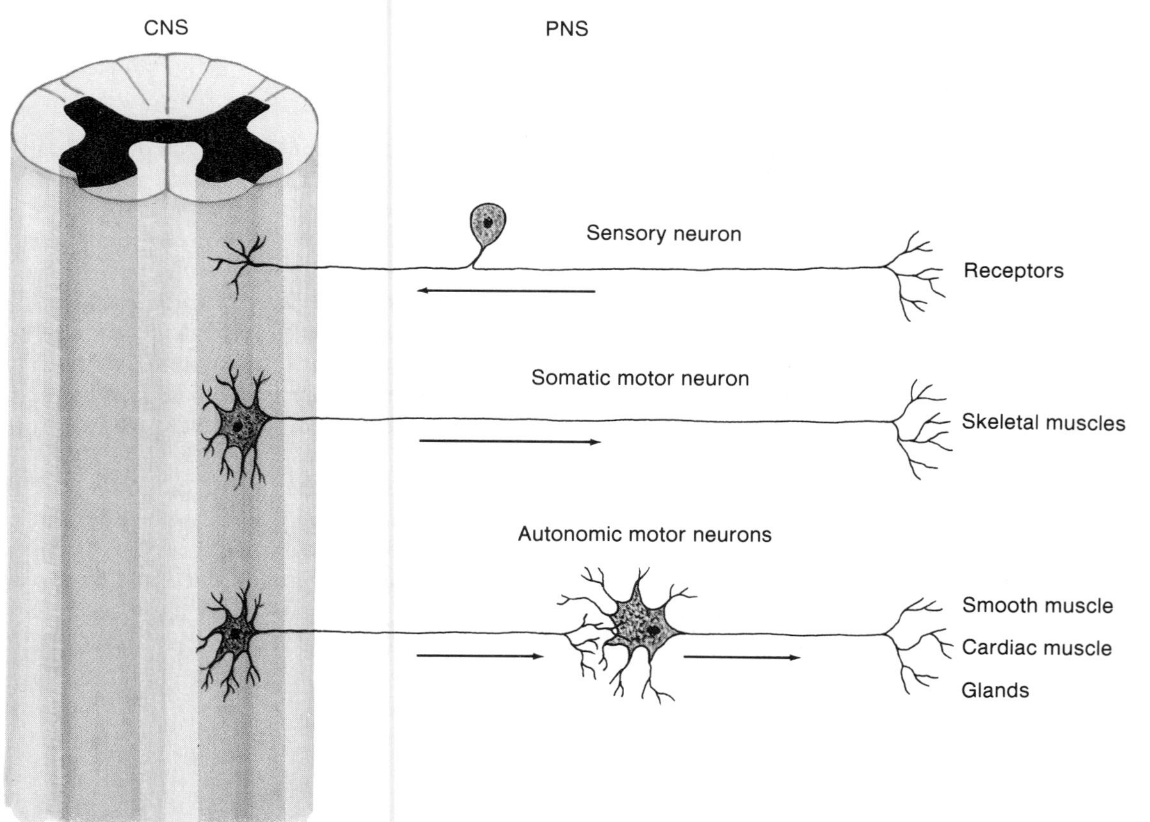

be moved through the endothelial cells by active transport, endocytosis, and exocytosis. This imposes a very selective **blood-brain barrier.** Astrocytes contribute to this blood-brain barrier because the brain capillaries are surrounded by extensions of astrocytes known as *perivascular feet* (fig. 7.6). Before molecules in the blood can enter neurons in the CNS they may have to pass through both the endothelial cells of the capillaries and the astrocytes.

The blood-brain barrier presents difficulties in the chemotherapy of brain diseases because drugs that could enter other organs may not be able to enter the brain. In the treatment of Parkinson's disease, for example, patients who need a chemical called dopamine in the brain must be given a precursor molecule called levodopa (L-dopa). This is because dopamine cannot cross the blood-brain barrier, whereas L-dopa can enter the neurons and be changed to dopamine in the brain.

Classification of Neurons and Nerves

Neurons may be classified according to their structure or function. The functional classification is based on the direction that they conduct impulses. **Sensory,** or **afferent, neurons** conduct impulses from sensory receptors into the CNS. **Motor,** or **efferent, neurons** (fig. 7.7) conduct impulses out of the CNS to effector organs (muscles and glands). **Association neurons,** or **interneurons,** are located entirely within the CNS and serve the associative, or integrative, functions of the nervous system.

There are two types of motor neurons: somatic and autonomic. **Somatic motor neurons** provide both reflex and voluntary control of skeletal muscles. **Autonomic motor neurons** innervate the involuntary effectors—smooth muscle, cardiac muscle, and glands. The cell bodies of the autonomic neurons that innervate these organs are located outside the CNS in autonomic ganglia (fig. 7.7). There are two subdivisions of autonomic neurons: *sympathetic* and *parasympathetic*. Autonomic motor neurons, together with their central control centers, comprise the *autonomic nervous system,* which will be discussed in chapter 12.

The structural classification of neurons is based on the number of processes that extend from the cell body of the neuron (fig. 7.8). **Bipolar neurons** have two processes, one at either end; this type is found in the retina of the eye. **Multipolar neurons** have several dendrites and one axon extending from the cell body; this is the most common type of neuron (motor neurons are good examples of this type). A **pseudounipolar neuron** has a single short process that divides like a T to form a longer process. Sensory neurons are pseudounipolar—one end of the process formed by the T receives sensory stimuli and produces nerve impulses; the other end of the T delivers these impulses to synapses within the brain or spinal cord. The cell bodies of these sensory neurons are located outside the CNS in the dorsal root ganglia of spinal and cranial nerves.

Figure 7.8. Three different types of neurons.

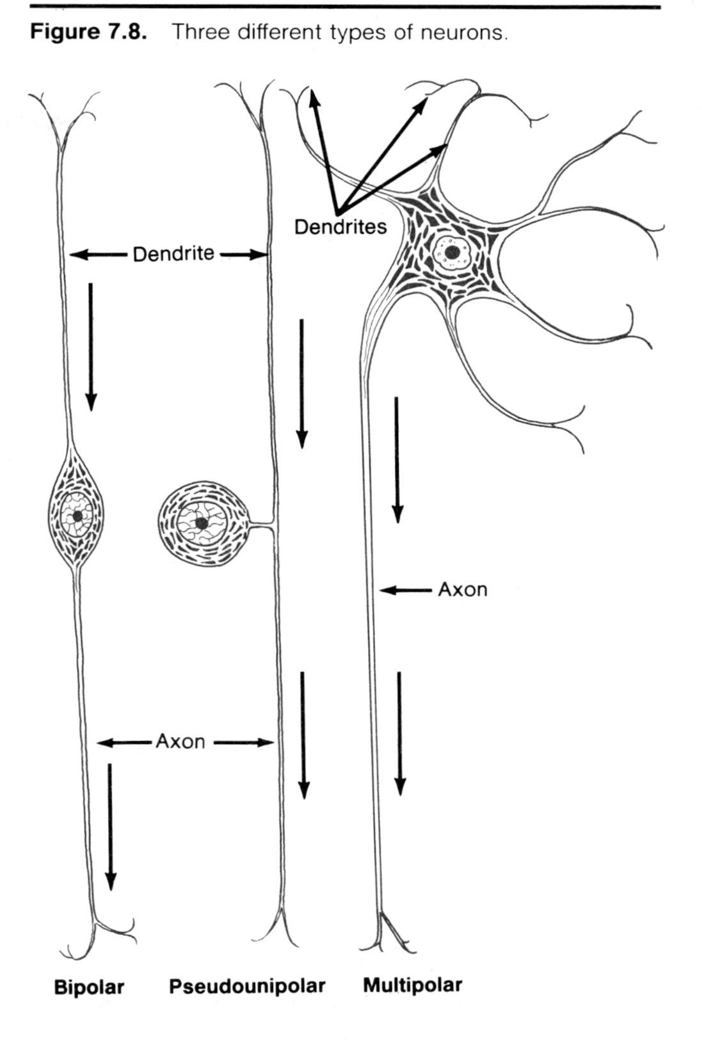

A **nerve** is a collection of axons outside the CNS. Most nerves are composed of both motor and sensory fibers and are thus called *mixed nerves.* Some of the cranial nerves, however, serve only sensory neurons. These are the nerves that serve the special senses of sight, hearing, taste, and smell.

1. *Describe the major parts of a neuron and their functions.*
2. *Distinguish between the structure of a myelinated and an unmyelinated axon, and describe how a myelin sheath is formed.*
3. *Describe how a peripheral nerve fiber is regenerated if cut, and explain why an axon of the CNS cannot regenerate as well.*
4. *Explain the nature of the blood-brain barrier, and describe its structure.*
5. *Explain the functional and structural classifications of neurons.*

Figure 7.9. The difference in potential (in millivolts) between an intracellular and extracellular recording electrode is displayed on an oscilloscope screen. The resting membrane potential (*rmp*) of the axon may be reduced (depolarization) or increased (hyperpolarization).

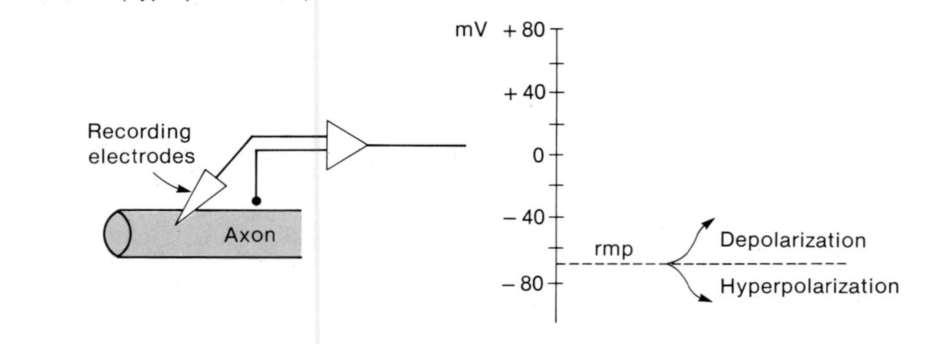

Electrical Activity in Nerve Fibers

Although all cells in the body maintain a potential difference across their cell membranes, only nerve and muscle cells can alter their membrane potential in response to appropriate stimulation. These alterations are achieved by varying the membrane permeability to different ions. Tissues that can respond to stimuli in this manner are said to be *irritable.*

An increase in membrane permeability to a specific ion results in the diffusion of that ion down its concentration gradient, either into or out of the cell. These *ion currents* occur only across limited patches of membrane (located fractions of a millimeter apart), where specific ion channels are located. Changes in the potential difference across the membrane at these points can be measured by the voltage developed between two electrodes—one placed inside the cell, the other placed outside the cell membrane at the region being recorded. The voltage between these two *recording electrodes* can be visualized by connecting these electrodes to an oscilloscope (fig. 7.9).

In an oscilloscope, electrons from a cathode ray "gun" are sprayed across a fluorescent screen, producing a line of light. Changes in the potential difference between the two recording electrodes produce a deflection of this line. The oscilloscope can be calibrated in such a way that an upward deflection of this line indicates that the inside of the membrane has become less negative (or more positive) compared to the outside of the membrane. A downwards deflection of the line, conversely, indicates that the inside of the cell has become more negative.

If both recording electrodes were placed outside of the cell, the potential difference between the two would be zero (there would be no charge separation). When one of the two electrodes penetrates the cell membrane, the oscilloscope shows that the intracellular electrode is electrically negative with respect to the extracellular electrode. If appropriate stimulation causes positive charges to flow into the cell, the line would deflect upward; this change is called **depolarization,** or hypopolarization, because the potential difference between the two recording electrodes is reduced. If the inside of the membrane becomes more negative as a result of stimulation, the line on the oscilloscope would deflect downwards; this is called **hyperpolarization.**

Ion Gating in Axons

As discussed in chapter 6, the intracellular potassium concentration is much higher than the concentration of potassium in the extracellular fluid. The sodium concentration, conversely, is higher in extracellular fluid than it is within the cell. These concentration differences are a result of (1) electrical attraction by negatively charged proteins and organic phosphates that are fixed within the cell; (2) the greater permeability of the resting cell membrane to K^+ than to Na^+; and (3) active transport by the Na^+/K^+ pump. Because of the uneven distribution of ions, the inside of a neuron is about 65 millivolts negative (−65 mV) compared to the extracellular fluid (the exact value of this membrane potential is different in different cells).

The permeability of the membrane to Na^+, K^+, and other ions can be regulated by parts of the ion channels through the membrane called **gates.** Gates are believed to be composed of polypeptide chains that can open or close a membrane channel according to specific conditions.

Figure 7.10. Depolarization of an axon has two effects: (*1*) Na^+ gates open and Na^+ diffuses into the cell and (*2*) after a brief period, K^+ gates open and K^+ diffuses out of the cell. An inward diffusion of Na^+ causes further depolarization—this causes further opening of Na^+ gates in a positive feedback (+) fashion. The opening of K^+ gates and outward diffusion of K^+ make the inside of the cell more negative and thus have a negative feedback effect (−) on the initial depolarization.

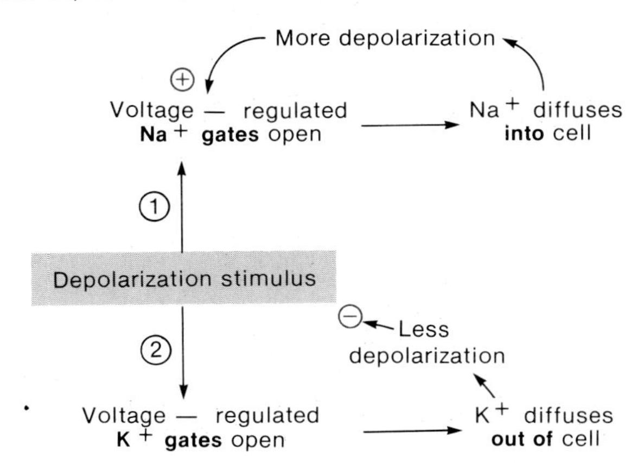

When the gates of specific ion channels are closed, the membrane is not very permeable to that ion, and when the gates are opened, the permeability to that ion can be greatly increased.

It is believed that there are two types of channels for K^+; one type lacks gates and is always open, whereas the other type has gates that are closed in the resting cell. Channels for Na^+, in contrast, always have gates, and these gates are closed in the resting cell. The resting cell is thus more permeable to K^+ than to Na^+ (some Na^+ does leak into the cell, as described in chapter 6; this leakage may occur in a nonspecific manner through open K^+ channels).

Depolarization of a small region of an axon can be experimentally induced by a pair of stimulating electrodes. If a pair of recording electrodes are placed in the same region (one electrode within the axon and one outside), an upward deflection of the oscilloscope line will be observed as a result of this depolarization. If a certain level of depolarization is achieved (from −65 mV to −55 mV for example,) by this artificial stimulation, a sudden and very rapid change in the membrane potential will be observed. This is due to the fact that *depolarization to a threshold level causes the Na^+gates to open.* Now the permeability properties of the membrane are changed, and Na^+ diffuses down its concentration gradient into the cell.

A fraction of a second after the Na^+ gates open, they close again. At this time, the depolarization stimulus causes the K^+ gates to open. This makes the membrane more permeable to K^+ than it is at rest, and K^+ diffuses down its concentration gradient out of the cell. The K^+ gates will then close and the permeability properties of the membrane will return to what they were at rest.

Notice that whether the gates for the Na^+ and K^+ channels are open or closed depends on the membrane potential. The gated channels are closed at the resting membrane potential of −65 mV, but they open when the membrane is depolarized to a certain threshold level. Since the opening and closing of these gates is regulated by the membrane voltage, the gates are said to be **voltage regulated.**

Action Potentials

When the axon membrane has been depolarized to a threshold level—by stimulating electrodes, in the previous example—the Na^+ gates open and the membrane becomes permeable to Na^+. This permits Na^+ to enter the axon by diffusion, which further depolarizes the membrane (makes the inside less negative, or more positive). Since the Na^+ gates of the axon are voltage regulated, this further depolarization makes the membrane even more permeable to Na^+, so that even more Na^+ can enter the cell and open even more voltage-regulated Na^+ gates. A *positive feedback loop* (fig. 7.10) is thus created, which causes the rate of Na^+ entry and depolarization to accelerate in an explosive fashion.

Figure 8.5. The mechanism of the action of T_3 (triiodothyronine) on the target cells.

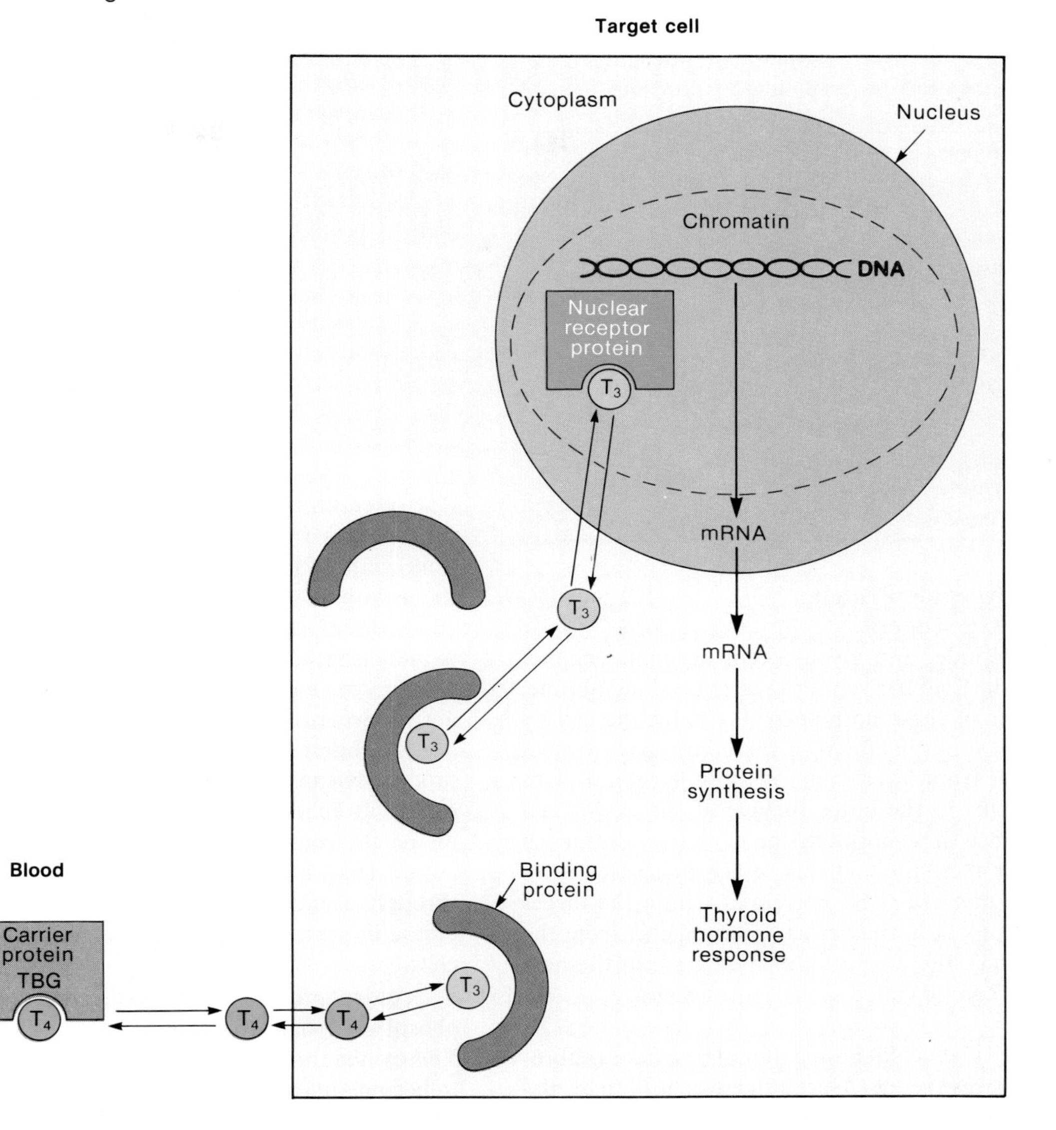

Thyroxine. The major hormone secreted by the thyroid gland is thyroxine, or **tetraiodothyronine (T_4).** Like steroid hormones, thyroxine travels in the blood attached to carrier proteins (primarily attached to *thyroxine-binding globulin,* or *TBG*). The thyroid also secretes a small amount of **triiodothyronine,** or **T_3.** The carrier proteins have a higher affinity for T_4 than for T_3, however, and as a result, the amount of unbound (or "free") T_3 is about ten times greater than the amount of free T_4 in the plasma.

Approximately 99.96% of the thyroxine in the blood is attached to carrier proteins in the plasma; the rest is free. Only the free thyroxine and T_3 can enter target cells; the protein-bound thyroxine serves as a "reservoir" of this hormone in the blood (this is why it takes a couple of weeks after surgical removal of the thyroid for the symptoms of hypothyroidism to develop). Once the free thyroxine passes into the target cell cytoplasm, it is enzymatically converted into T_3. Thyroxine, therefore, is not the chemical form of the hormone that is active in the target cells.

Inactive T_3 receptor proteins are already in the nucleus attached to chromatin. These receptors are inactive until T_3 enters the nucleus from the cytoplasm. The attachment of T_3 to the chromatin-bound receptor proteins activates genes and results in the production of new mRNA and new proteins. This sequence of events is summarized in figure 8.5.

Table 8.5 Hormones that activate adenylate cyclase and use cAMP as a second messenger and hormones that use other second messengers.

Hormones that Stimulate Adenylate Cyclase	Hormones that Do Not Stimulate Adenylate Cyclase
Adrenocorticotrophic hormone (ACTH)	Catecholamines (α-adrenergic)
Calcitonin	Growth hormone (GH)
Epinephrine (β-adrenergic)	Insulin
Follicle-stimulating hormone (FSH)	Oxytocin
Glucagon	Prolactin
Luteinizing hormone (LH)	Somatomedin
Parathyroid hormone	Somatostatin
Thyrotrophin-releasing hormone (TRH)	
Thyroid-stimulating hormone (TSH)	
Antidiuretic hormone	

Membrane Receptor Proteins and Second Messengers

Amine, polypeptide, and glycoprotein hormones cannot pass through the lipid barrier of the target cell membrane. Although some of these hormones may enter the cell by pinocytosis, most of the effects of these hormones are believed to result from interaction of these hormones with receptor proteins in the outer surface of the target cell membrane. Since these hormones do not have to enter the target cells to exert their effects, other molecules must mediate the actions of these hormones within the target cells. If you think of hormones as "messengers" from the endocrine glands, the intracellular mediators of the hormone's action can be called **second messengers.**

Cyclic AMP. Cyclic adenosine monophosphate (abbreviated cAMP) was the first "second messenger" to be discovered and is the best understood. The hormonal effects of epinephrine (adrenalin) and norepinephrine are due to the actions of cAMP, and the effects of norepinephrine released as a neurotransmitter by sympathetic nerve endings are also due to cAMP production. It was later discovered that the effects of many (but not all—see table 8.5) polypeptide and glycoprotein hormones are also mediated by cAMP.

The bonding of these hormones to their membrane receptor proteins activates an enzyme called **adenylate cyclase.** This enzyme is built into the cell membrane and, when activated, it catalyzes the following reaction:

$$ATP \rightarrow cAMP + PP_i$$

Table 8.6 Sequence of events that occurs with cyclic AMP as a second messenger.

1. The hormones combine with their receptors on the outer surface of target cell membranes.
2. Hormone-receptor interaction stimulates activation of adenylate cyclase on the cytoplasmic side of the membranes.
3. Activated adenylate cyclase catalyzes the conversion of ATP to cyclic AMP (cAMP) within the cytoplasm.
4. Cyclic AMP activates protein kinase enzymes that were already present in the cytoplasm in an inactive state.
5. Activated cAMP-dependent protein kinase transfers phosphate groups to (phosphorylates) other enzymes in the cytoplasm.
6. The activity of specific enzymes is either increased or inhibited by phosphorylation.
7. Altered enzyme activity mediates the target cell's response to the hormone.

Adenosine triphosphate (ATP) is thus converted into cAMP plus two inorganic phosphates (*pyrophosphate,* abbreviated PP_i). As a result of the interaction of the hormone with its receptor and the activation of adenylate cyclase, therefore, the intracellular concentration of cAMP is increased. Cyclic AMP activates a previously inactive enzyme in the cytoplasm called **protein kinase.** The inactive form of this enzyme consists of two subunits: a catalytic subunit and an inhibitory subunit. The enzyme is produced in an inactive form and becomes active only when cAMP attaches to the inhibitory subunit and causes it to dissociate from the catalytic subunit, which then becomes active (fig. 8.6). The hormone, in summary—acting through an increase in cAMP production—causes an increase in protein kinase enzyme activity within its target cells.

Active protein kinase catalyzes the attachment of phosphate groups to different proteins in the target cells. This causes some enzymes to become activated, and others to become inactivated. Cyclic AMP, acting through protein kinase, thus modulates the activity of enzymes that are already present in the target cell. This alters the metabolism of the target tissue in a manner characteristic of the actions of that specific hormone (table 8.6).

Like all biologically active molecules, cAMP must be rapidly inactivated for it to function effectively as a second messenger in hormone action. This function is served by an enzyme called **phosphodiesterase** within the target cells, which hydrolyzes cAMP into inactive fragments. Through the action of phosphodiesterase, the stimulatory effect of a hormone that uses cAMP as a second messenger depends upon the continuous generation of new cAMP molecules, and thus depends upon the level of secretion of the hormone.

Figure 8.6. Cyclic AMP (cAMP) as a second messenger in the action of catecholamine, polypeptide, and glycoprotein hormones.

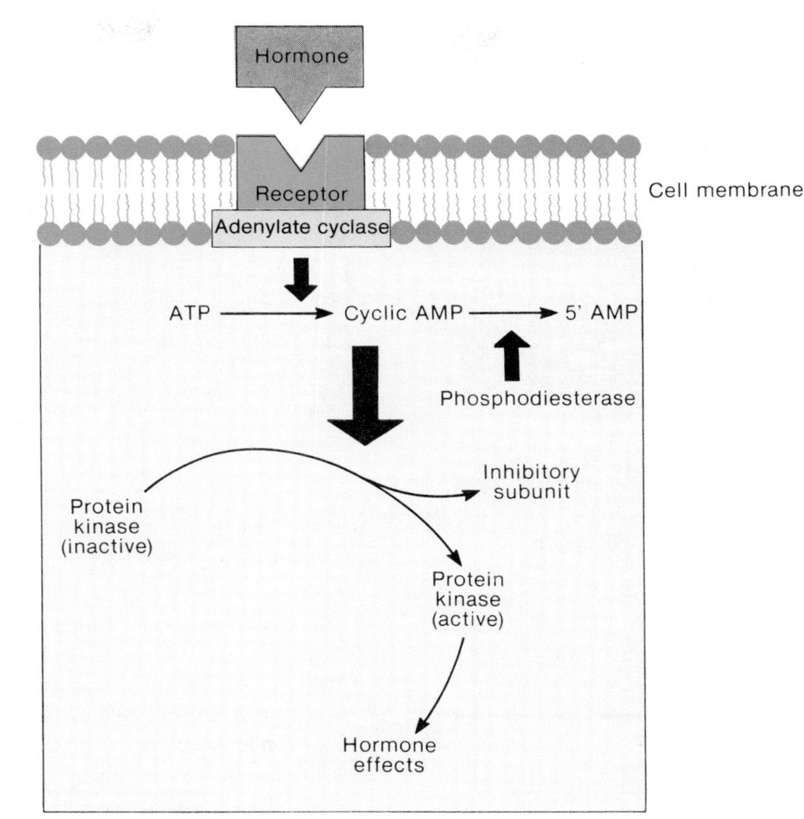

Drugs that inhibit the activity of phosphodiesterase thus prevent the breakdown of cAMP and result in increased concentrations of cAMP within the target cells. The drug **theophylline,** for example, is used clinically to raise cAMP levels within bronchiolar smooth muscle. This duplicates and enhances the effect of epinephrine on the bronchioles (producing dilation) in people who suffer from asthma.

Cyclic GMP and Ca^{++} as Second Messengers. Liver cells respond to epinephrine by an increase in cAMP production and a breakdown of stored glycogen; the action of insulin, in contrast, produces an increase in the production of glycogen. Clearly, since the action of epinephrine is mediated by cAMP, the action of insulin in liver cells must be mediated by a different second messenger. The identity of the second messenger(s) for insulin action is not yet established.

Some of the hormones that do not use cAMP as a second messenger may act via production of **cyclic guanosine monophosphate (cGMP).** Indeed, cAMP and cGMP produced in response to the action of different hormones may act antagonistically within the target cells. The control of mitotic cell division and the cell cycle, for example, appears to be related to the ratio of cAMP to cGMP. In addition to these two "cyclic nucleotides," **calcium ion (Ca^{++})** may also function as a second messenger in the action of certain hormones.

Active transport carriers help to maintain a very steep concentration gradient for Ca^{++}, which is at a much higher concentration in the extracellular fluid than it is in the cytoplasm of cells. Transient inhibition of this Ca^{++} pump produces a sudden inflow of Ca^{++} that can serve as a second messenger. Analogous events occur in muscles and nerves, where calcium stimulates contraction and the release of neurotransmitters, respectively, in response to electrical stimulation.

In some tissue cells, Ca^{++} binds to and activates a cellular protein called **calmodulin.** The activation of calmodulin by a sudden influx of Ca^{++} is analogous to the activation of protein kinase by cAMP. Recent experiments, for example, suggest that insulin may exert at least some of its effects via a transient influx of Ca^{++} that activates calmodulin.

Figure 8.7. The hydrolysis of glycogen (glycogenolysis) in the liver and the skeletal muscles in response to cyclic AMP (cAMP) and Ca^{++} as second messengers. Both protein kinase and calmodulin are proteins activated by the second messengers.

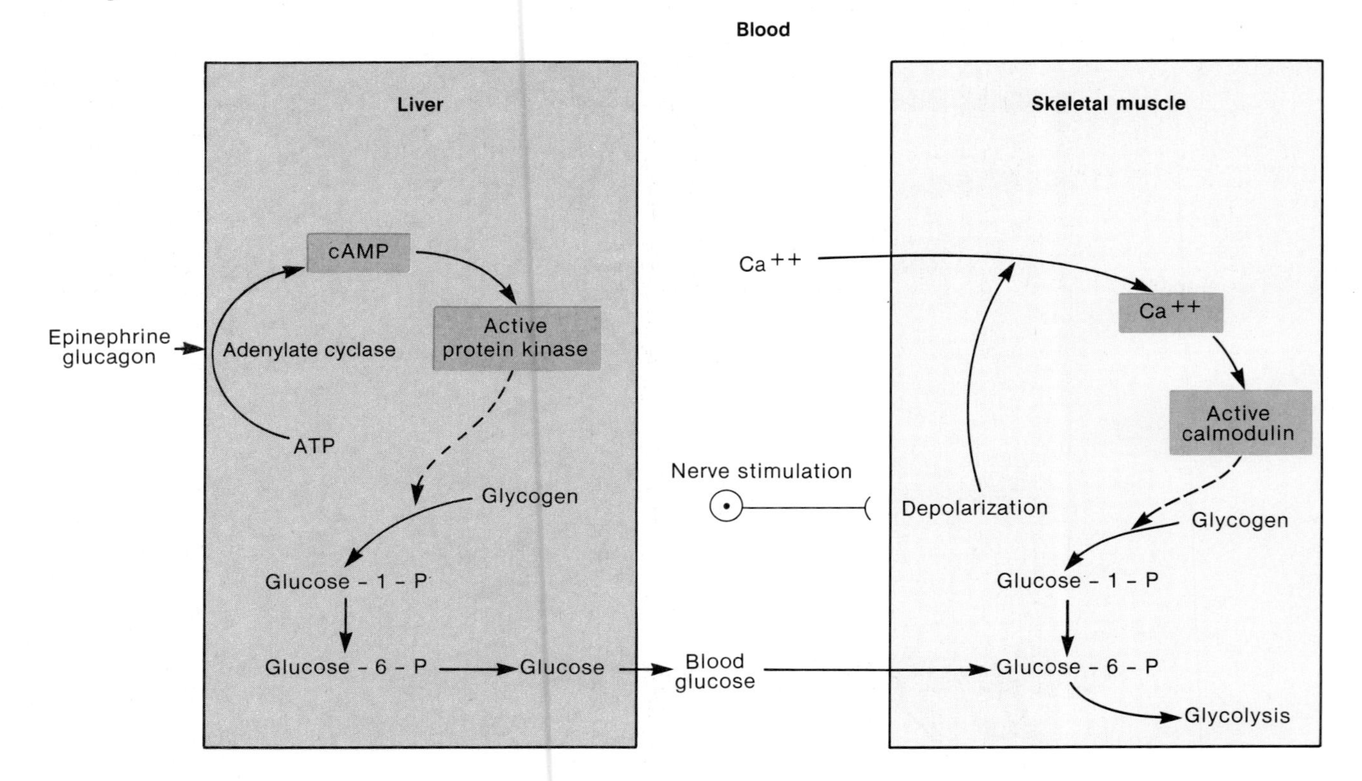

Activated calmodulin, like activated protein kinase, can affect various target cell enzymes. Figure 8.7 illustrates how a single process—the hydrolysis of glycogen—can be stimulated in one case by the hormonal activation of the cAMP/protein kinase regulatory system and in another case by the neural stimulation that activates the Ca^{++}/calmodulin system. Glycogen breakdown in the liver in response to epinephrine is mediated by cAMP and protein kinase; glycogen breakdown in the skeletal muscles during nerve stimulation appears to be mediated by Ca^{++} inflow and the activation of calmodulin. The effects of cAMP and of Ca^{++} in different tissues are summarized in table 8.7.

A given cell may respond to a variety of different hormones that produce similar as well as antagonistic effects. The response of a target cell to these different hormones may depend upon interactions between different second messengers. Calmodulin, for example, has been found to stimulate adenylate cyclase (and thus to raise cAMP levels in target cells) under some circumstances and to stimulate phosphodiesterase activity (and thus to lower cAMP) in other instances. These effects, together with the antagonism between cAMP and cGMP, suggest that the effects of various hormones in a target cell may depend on complex and poorly understood interactions between different second messengers.

1. *Using diagrams, describe how steroid hormones and thyroxine exert their effects on their target cells.*
2. *Using a diagram, describe how cyclic AMP is produced within a target cell in response to hormone stimulation. List three hormones that use cAMP as a second messenger.*
3. *Explain how cAMP functions as a second messenger in the action of some hormones.*
4. *Explain how Ca^{++} can have regulatory functions in different tissues and why it may also be considered to be a second messenger in the action of some hormones.*

Table 8.7 Comparison of the effects of cyclic AMP (cAMP) and Ca^{++} as second messengers in different tissues.

Second Messenger	Mechanism	Effects: Nerves	Effects: Muscles	Effects: Hormones
cAMP	Activation of cAMP-dependent protein kinase	Opens ion channels in postsynaptic membrane	Phosphorylation of myosin in skeletal muscle in response to depolarization (significance controversial)	Activation or inactivation of target cell enzymes in response to many polypeptide and glycoprotein hormones
Ca^{++}	Activation of Ca^{++}-dependent regulatory protein (calmodulin)	Mediates release of neurotransmitter in response to depolarization	Phosphorylation of myosin activates actomyosin ATPase in smooth muscles and stimulates contraction	May mediate the effects of insulin on target cells; may increase cAMP by stimulation of adenylate cyclase or decrease cAMP by stimulation of phosphodiesterase

Figure 8.8. The structure of the pituitary gland as seen in sagittal view.

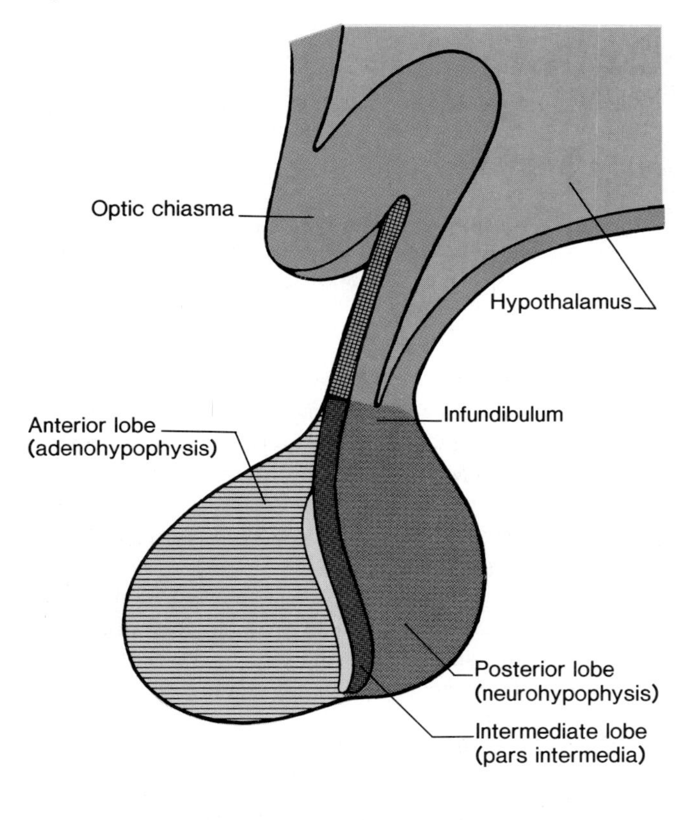

Pituitary Gland

The pituitary gland, or hypophysis, is located on the inferior aspect of the brain in the region of the diencephalon, and is attached to the hypothalamus by a stalklike structure called the infundibulum (fig. 8.8). The pituitary is a rounded, pea-shaped gland measuring about 1.3 cm (0.5 in.) in diameter. It is covered by the dura mater and is supported by the sella turcica of the sphenoid bone. The circle of Willis surrounds the highly vascular pituitary gland, providing it with a rich blood supply.

The pituitary gland is structurally and functionally divided into an anterior lobe, or **adenohypophysis,** and a posterior lobe called the **neurohypophysis.** These two parts have different embryonic origins. The adenohypophysis is derived from epithelial tissue, whereas the neurohypophysis is formed as a downgrowth of the brain. The adenohypophysis consists of three parts: (1) the *pars distalis* is the bulbar portion; (2) the *pars tuberalis* is the thin extension in contact with the infundibulum; and (3) the *pars intermedia* is between the anterior and posterior parts of the pituitary. These parts are illustrated in figure 8.8.

The neurohypophysis is the neural part of the pituitary gland. It consists of the bulbar *pars nervosa,* which is in contact with the pars intermedia and the pars distalis of the adenohypophysis, and the *infundibulum,* which is the connecting stalk to the hypothalamus. Nerve fibers extend through the infundibulum along with minute neuroglia-like cells, called *pituicites.*

Pituitary Hormones

There are eight important hormones secreted by the pituitary gland. The first six in the following list are secreted by the **anterior pituitary** (the pars distalis of the adenohypophysis). The last two hormones in the list are produced in the hypothalamus, transported through axons in the infundibulum to the posterior pituitary, and secreted by the **posterior pituitary** (the pars nervosa of the neurohypophysis) into the blood.

The hormones secreted by the pars distalis are called **trophic hormones.** The term *trophic* means "food." This term is used because high amounts of the anterior pituitary hormones make their target glands hypertrophy. These hormones are secreted by the pars distalis:

1. **Growth hormone (GH, or somatotrophin).** This hormone stimulates growth in all body organs. It also promotes the movement of amino acids into tissue cells and the incorporation of these amino acids into tissue proteins.
2. **Thyroid-stimulating hormone (TSH, or thyrotrophin).** This hormone stimulates the thyroid gland to produce and secrete thyroxine (tetraiodothyronine, or T_4).
3. **Adrenocorticotrophic hormone (ACTH, or corticotrophin).** This hormone stimulates the adrenal cortex to secrete the glucocorticoids, such as hydrocortisone (cortisol).
4. **Follicle-stimulating hormone (FSH, or folliculotrophin).** This hormone stimulates the growth and secretion of ovarian follicles in the ovaries of females and the production of sperm in the testes of males.
5. **Luteinizing hormone (LH, or luteotrophin).** This hormone and FSH are collectively called **gonadotrophic hormones.** In females, LH stimulates ovulation and the conversion of the ovulated ovarian follicle into an endocrine structure called a corpus luteum. In males, LH (which is sometimes also called interstitial cell-stimulating hormone, or ICSH) stimulates the secretion of male sex hormones (mainly testosterone) from the interstitial cells of Leydig in the testes.
6. **Prolactin.** This hormone is secreted by both males and females, but its function is only well understood in females where it stimulates the production of milk in the mammary glands of women after the birth of their babies.

Inadequate growth hormone secretion during childhood causes *pituitary dwarfism.* Hyposecretion of growth hormone in an adult produces a rare condition called *pituitary cachexia (Simmonds' disease).* One of the symptoms of this disease is premature aging caused by tissue atrophy. Oversecretion of growth hormone during childhood, in contrast, causes *gigantism.* Excessive growth hormone secretion in an adult does not cause further growth in length because the epiphyseal discs have ossified. Hypersecretion of growth hormone in an adult causes *acromegaly,* in which the person's appearance gradually changes as a result of thickening of bones and the growth of soft tissues, particularly in the face, hands, and feet.

The posterior pituitary, or pars nervosa, secretes only two hormones, both of which are produced in the hypothalamus and merely stored in the posterior lobe of the pituitary:

1. **Antidiuretic hormone (ADH, or vasopressin).** Antidiuretic hormone stimulates the kidneys to retain water so that less water is excreted in the urine and more water is retained in the blood. This hormone also causes vasoconstriction in experimental animals, although the significance of this effect in humans is controversial.
2. **Oxytocin.** This hormone has no known function in males, but in females it is known to stimulate contractions of the uterus during labor and contractions of the mammary gland alveoli and ducts, which result in the milk-ejection reflex during lactation.

Oxytocin, like ADH, is produced by neurons in the hypothalamus and travels down axons in the infundibulum to the pars nervosa, where it is stored and secreted in response to nerve impulses from the hypothalamus.

Injections of oxytocin may be given to a woman during labor if she is having difficulties in parturition. Increased amounts of oxytocin assist uterine contractions and generally speed up delivery. Oxytocin administration after parturition causes the uterus to regress in size and squeezes the blood vessels, thus minimizing the danger of hemorrhage.

Hypothalamic Control of the Neurohypophysis

The posterior pituitary (pars nervosa of the neurohypophysis) secretes two hormones: antidiuretic hormone (ADH) and oxytocin. These two hormones, however, are actually produced in neuron cell bodies of the *supraoptic nuclei* and *paraventricular nuclei* of the hypothalamus. These nuclei within the hypothalamus are thus endocrine glands; the hormones they produce are transported along axons of the **hypothalamo-hypophyseal tract** (fig. 8.9) to the posterior pituitary, which stores and later secretes these hormones. The posterior pituitary is thus more a storage organ than a true gland.

The secretion of ADH and oxytocin from the posterior pituitary is controlled by **neuroendocrine reflexes.** In nursing mothers, for example, the stimulus of sucking acts via sensory nerve impulses to the hypothalamus to stimulate the reflex secretion of oxytocin. The secretion of ADH is stimulated by osmoreceptor neurons in the hypothalamus in response to a rise in blood osmotic pressure; its secretion is inhibited by sensory impulses from stretch receptors in the left atrium of the heart in response to a rise in blood volume.

Figure 8.9. The posterior pituitary, or neurohypophysis, stores and secretes hormones (vasopressin and oxytocin) produced in neuron cell bodies within the supraoptic and paraventricular nuclei of the hypothalamus. These hormones are transported to the posterior pituitary by nerve fibers of the hypothalamo-hypophyseal tract.

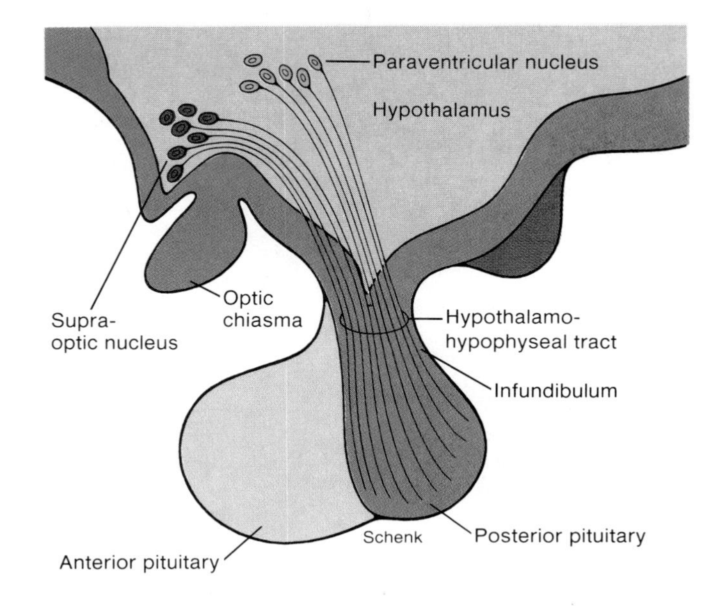

Table 8.8 Hormones secreted by the anterior pituitary.

Hormone	Target Tissue	Stimulated by Hormone	Regulation of Secretion
ACTH (adrenocorticotrophic hormone)	Adrenal cortex	Secretion of glucocorticoids	Stimulated by CRH (corticotrophin-releasing hormone); inhibited by glucocorticoids
TSH (thyroid-stimulating hormone)	Thyroid gland	Secretion of thyroid hormones	Stimulated by TRH (thyrotrophin-releasing hormone); inhibited by thyroid hormones
GH (growth hormone)	Most tissue	Protein synthesis and growth; lipolysis and increased blood glucose	Inhibited by somatostatin; stimulated by growth hormone-releasing hormone
FSH (follicle-stimulating hormone) and LH (luteinizing hormone)	Gonads	Gamete production and sex steroid hormone secretion	Stimulated by GnRH (gonadotrophin-releasing hormone); inhibited by sex steroids
Prolactin	Mammary glands and other sex accessory organs	Milk production Controversial actions in other organs	Inhibited by PIH (prolactin-inhibiting hormone)
LH (luteinizing hormone)	Gonads	Sex hormone secretion: ovulation and corpus luteum formation	Stimulated by GnRH

Hypothalamic Control of the Adenohypophysis

At one time the anterior pituitary was called the "master gland" because it secretes hormones that regulate some other endocrine glands (see fig. 8.10 and table 8.8). Adrenocorticotrophic hormone (ACTH), thyroid-stimulating hormone (TSH), and the gonadotrophic hormones (FSH and LH) stimulate the adrenal cortex, thyroid, and gonads, respectively, to secrete their hormones. The anterior pituitary hormones also have a "trophic" effect on their target glands, in that the structure and health of the target glands depend on adequate stimulation by anterior pituitary hormones. The anterior pituitary, however, is not really the master gland, because secretion of its hormones is in turn controlled by hormones secreted by the hypothalamus.

Figure 8.10. The hormones secreted by the anterior pituitary and the target organs for those hormones.

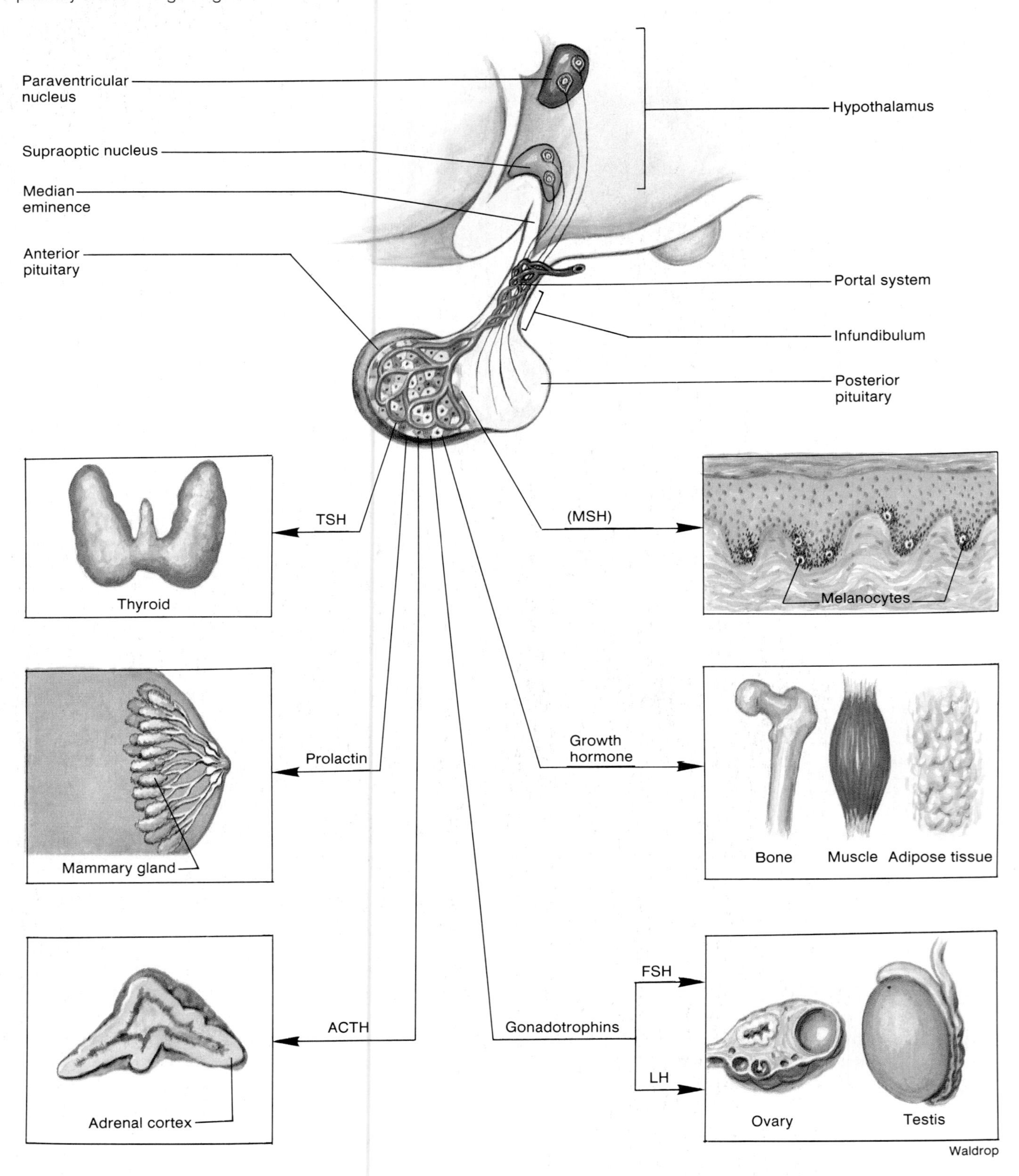

Table 8.9 Some hypothalamic hormones involved in the control of the anterior pituitary.

Hypothalamic Hormone	Structure	Effect on Anterior Pituitary	Action of Anterior Pituitary Hormone
Corticotrophin-releasing hormone (CRH)	41 amino acids	Stimulates secretion of adrenocorticotrophic hormone (ACTH)	Stimulates secretions of adrenal cortex
Gonadotrophin-releasing hormone (GnRH)	10 amino acids	Stimulates secretion of follicle-stimulating hormone (FSH) and luteinizing hormone (LH)	Stimulates gonads to produce gametes (sperm and ova) and secrete sex steroids
Prolactin-inhibiting hormone (PIH)	Controversial (may be dopamine)	Inhibits prolactin secretion	Stimulates production of milk in mammary glands
Somatostatin	14 amino acids	Inhibits secretion of growth hormone	Stimulates anabolism and growth in many organs
Thyrotrophin-releasing hormone (TRH)	3 amino acids	Stimulates secretion of thyroid-stimulating hormone (TSH)	Stimulates secretion of thyroid gland
Growth hormone–releasing hormone (GRH)	44 amino acids	Stimulates growth hormone secretion	Stimulates anabolism and growth in many organs

Releasing and Inhibiting Hormones. Since axons do not enter the anterior pituitary, hypothalamic control of the anterior pituitary is achieved through hormonal rather than neural regulation. Neurons in the hypothalamus produce releasing and inhibiting hormones, which are transported to axon endings in the basal portion of the hypothalamus. This region, known as the *median eminence,* contains blood capillaries that are drained by venules in the stalk of the pituitary.

The venules that drain the median eminence deliver blood to a second capillary bed in the anterior pituitary. Since this second capillary bed receives venous rather than arterial blood, the vascular link between the median eminence and the anterior pituitary comprises a portal system. (This is analogous to the hepatic portal system that delivers venous blood from the intestine to the liver.) The vascular link between the hypothalamus and the anterior pituitary is thus called the **hypothalamo-hypophyseal portal system.**

Neurons of the hypothalamus secrete polypeptide hormones into this portal system that regulate the secretions of the anterior pituitary (table 8.9). Thyrotrophin-releasing hormone (**TRH**) stimulates the secretion of TSH, and corticotrophin-releasing hormone (**CRH**) stimulates the secretion of ACTH from the anterior pituitary. A single releasing hormone, gonadotrophin-releasing hormone, or **GnRH,** appears to stimulate the secretion of both gonadotrophic hormones (FSH and LH) from the anterior pituitary. The secretion of prolactin and of growth hormone from the anterior pituitary is regulated by hypothalamic inhibitory hormones, known as **PIH** (prolactin-inhibiting hormone) and **somatostatin,** respectively. Recently, a specific hypothalamic releasing hormone that stimulates growth hormone secretion has also been identified as a polypeptide consisting of forty-four amino acids. Experiments suggest that a releasing hormone for prolactin may also exist, but no such specific releasing hormone has yet been discovered (although there is evidence that TRH may stimulate the secretion of prolactin as well as TSH).

Feedback Control of the Adenohypophysis

In view of its secretion of releasing and inhibiting hormones, the hypothalamus might be considered the "master gland." The chain of command, however, is not linear; the hypothalamus and anterior pituitary are controlled by the effects of their own actions. In the endocrine system, to use an analogy, the general takes orders from the private. The hypothalamus and anterior pituitary are not master glands because their secretions are controlled by the target glands they regulate.

Anterior pituitary secretion of ACTH, TSH, and the gonadotrophins (FSH and LH) is controlled by **negative feedback inhibition** from the target gland hormones. Secretion of ACTH is inhibited by a rise in corticosteroid secretion, for example, and TSH is inhibited by a rise in the secretion of thyroxine from the thyroid. These negative feedback relationships are easily demonstrated by removal of the target glands. Castration (surgical removal of the gonads), for example, produces a rise in the secretion of FSH and LH. In a similar manner, removal of the adrenals or the thyroid would result in an abnormal increase in ACTH or TSH secretion from the anterior pituitary.

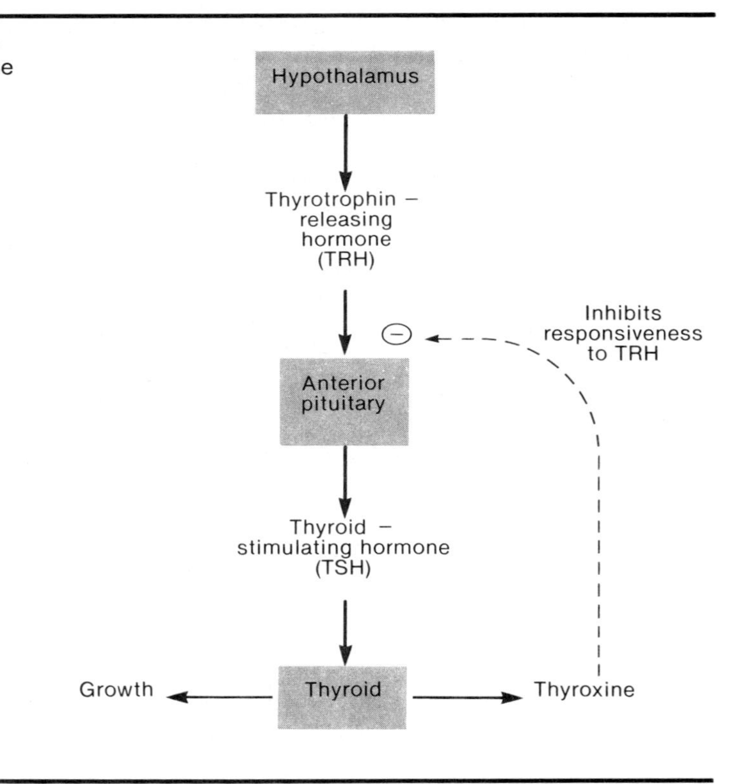

Figure 8.11. The secretion of thyroxine from the thyroid is stimulated by the thyroid-stimulating hormone (TSH) from the anterior pituitary. The secretion of TSH is stimulated by the thyrotrophin-releasing hormone (TRH) secreted from the hypothalamus into the hypothalamo-hypophyseal portal system. This stimulation is balanced by the negative feedback inhibition of thyroxine, which decreases the responsiveness of the anterior pituitary to stimulation by TRH.

Figure 8.12. Negative feedback control of gonadotrophin secretion. The possible short negative feedback loop involving the inhibition of GnRH secretion by the retrograde transport of gonadotrophins is shown by a dotted arrow.

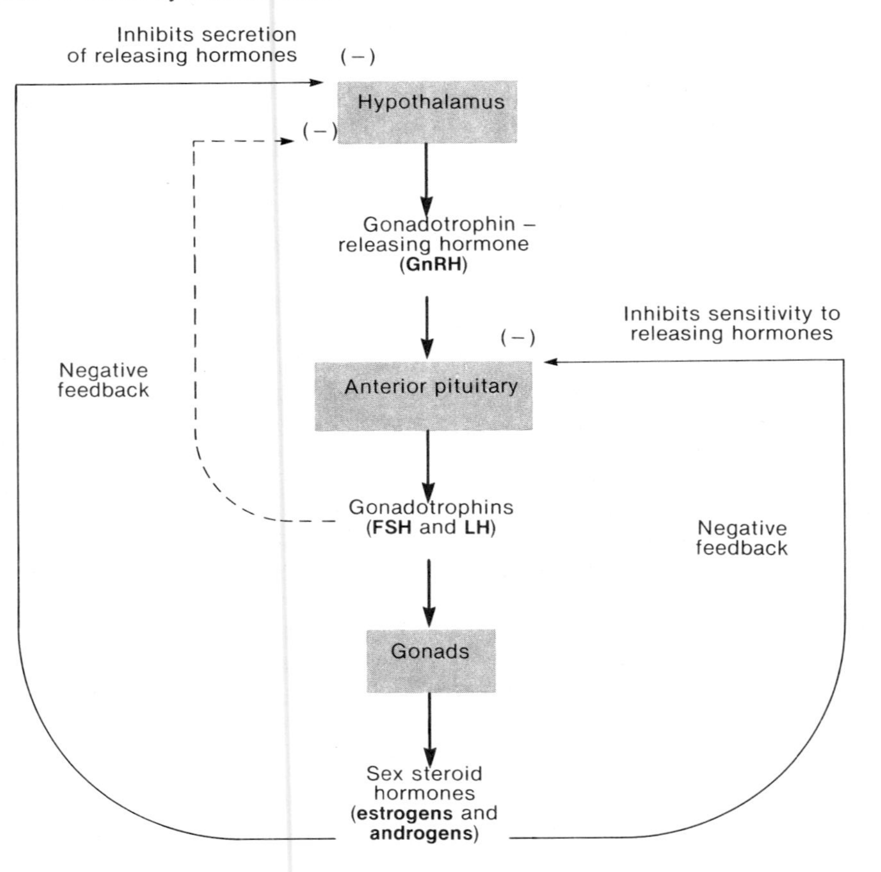

These effects demonstrate that under normal conditions the target glands exert an inhibitory effect on the anterior pituitary. This inhibitory effect can occur at two levels: (1) the target gland hormones could act on the hypothalamus and inhibit the secretion of releasing hormones, and (2) the target gland hormones could act on the anterior pituitary and inhibit its response to the releasing hormones. Thyroxine, for example, appears to inhibit the response of the anterior pituitary to TRH and thus acts to reduce TSH secretion (fig. 8.11). Sex steroids, in contrast, reduce the secretion of gonadotrophins by inhibiting both GnRH secretion and the ability of the anterior pituitary to respond to stimulation by GnRH (see fig. 8.12).

Recent evidence suggests that there may be retrograde transport of blood from the anterior pituitary to the hypothalamus. This may permit a *short feedback loop* where a particular trophic hormone inhibits the secretion of its releasing hormone from the hypothalamus. A high secretion of TSH, for example, may inhibit further secretion of TRH by this means.

In addition to negative feedback control of the anterior pituitary, there is one example where a hormone from a target organ actually stimulates the secretion of an anterior pituitary hormone. This occurs towards the middle of the menstrual cycle when the rising secretion of estradiol from the ovaries stimulates the anterior pituitary to secrete a "surge" of LH, which results in ovulation. This case is commonly referred to as a *positive feedback* effect to distinguish it from the more usual negative feedback inhibition of target gland hormones on anterior pituitary hormone secretion. Interestingly, higher levels of estradiol at a later stage of the menstrual cycle exert the opposite effect—negative feedback inhibition—on LH secretion.

Higher Brain Function and Pituitary Secretion

The feedback effect of estradiol on the secretion of gonadotrophic hormones is believed to be exerted at the level of the pituitary gland and hypothalamus. Since the hypothalamus receives neural input from "higher brain centers," however, it is not surprising that the pituitary-gonad axis can be affected by emotions, so that intense emotions may alter the timing of ovulation or menstruation. The influences of higher brain centers on the pituitary-gonad axis also helps to explain the "dormitory effect," in which researchers have noted a tendency for the menstrual cycles to synchronize in girls who room together.

The effect of stress on the pituitary-adrenal axis is another good example of the influence of higher brain centers on pituitary function. Stressors, as described later in this chapter, produce an increase in CRH secretion from the hypothalamus, which in turn results in elevated ACTH and corticosteroid secretion. In addition, the influence of higher brain centers produces *circadian* ("about a day") *rhythms* in the secretion of many anterior pituitary hormones. The secretion of growth hormone, for example, is highest during sleep and decreases during wakefulness, although its secretion is also stimulated following a meal by the absorption of particular amino acids.

---

1. *Describe the embryonic origins of the adenohypophysis and neurohypophysis, and list the parts of each. Indicate which of these parts are also called the "anterior pituitary" and "posterior pituitary."*
2. *List the hormones secreted by the posterior pituitary. Describe the site of origin of these hormones and the mechanisms by which their secretions are regulated.*
3. *List the hormones secreted by the anterior pituitary, and describe how the hypothalamus controls the secretion of each hormone.*
4. *Draw a negative feedback loop showing the control of ACTH secretion. Explain how this system would be affected by (a) an injection of ACTH; (b) surgical removal of the pituitary; (c) an injection of corticosteroids; and (d) surgical removal of the adrenal glands.*

Adrenal Glands

Each adrenal gland, like the pituitary, is actually composed of two different glands that are located together in the same organ (fig. 8.13). The inner part, or **adrenal medulla,** is derived from modified postganglionic sympathetic neurons and secretes catecholamine hormones (epinephrine and norepinephrine) into the blood upon stimulation by the sympathetic nervous system. The **adrenal cortex,** or outer part of the adrenals, is derived from a different embryonic tissue, secretes different hormones, and is under a different control system than the adrenal medulla.

Functions of the Adrenal Cortex

The adrenal cortex secretes steroid hormones called **corticosteroids,** or **corticoids,** for short. There are three functional categories of corticosteroids: (1) **mineralocorticoids,** which regulate Na^+ and K^+ balance; (2) **glucocorticoids,** which regulate the metabolism of glucose and other organic molecules; and (3) **sex steroids,** which are weak androgens (and lesser amounts of estrogens) that supplement the sex steroids secreted by the gonads. These hormones are secreted by the different zones of the adrenal cortex (fig. 8.13).

Figure 8.13. The structure of the adrenal gland showing the three zones of the adrenal cortex.

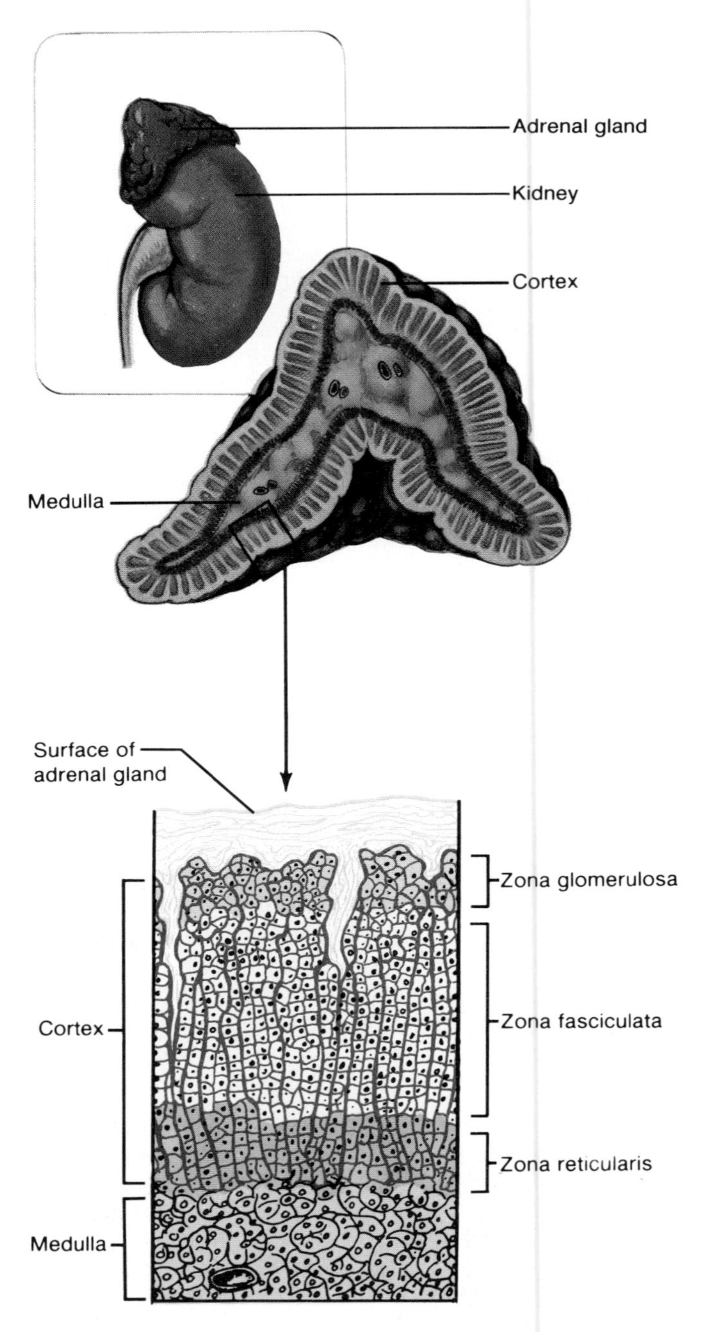

Adrenogenital syndrome is caused by the hypersecretion of adrenal sex hormones, particularly the androgens. Adrenogenital syndrome in young children causes premature puberty and enlarged genitals, especially the penis in a male and the clitoris in a female. An increase in body hair and a deeper voice are other characteristics. This condition in a mature woman can cause the growth of a beard.

Aldosterone is the most potent mineralocorticoid. The mineralocorticoids are produced in the zona glomerulosa (fig. 8.14). The predominant glucocorticoid in humans is cortisol (hydrocortisone), which is secreted by the zona fasciculata and perhaps also by the zona reticularis. The secretion of aldosterone by the zona glomerulosa is stimulated by angiotensin II and by blood K^+. The secretion of cortisol by the zona fasciculata is stimulated by ACTH (fig. 8.15).

Hypersecretion of corticosteroids results in **Cushing's syndrome.** This is generally caused by a tumor of the adrenal cortex or by oversecretion of ACTH from the adenohypophysis. Cushing's syndrome is characterized by changes in carbohydrate and protein metabolism, hyperglycemia, hypertension, and muscular weakness. Metabolic problems give the body a puffy appearance and can cause structural changes characterized as "buffalo hump" and "moon face." Similar effects are also seen when people with chronic inflammatory diseases receive prolonged treatment with corticosteroids, which are given to reduce inflammation and inhibit the immune response.

Addison's disease is caused by inadequate secretion of both glucocorticoids and mineralocorticoids, which results in hypoglycemia, sodium and potassium imbalance, dehydration, hypotension, rapid weight loss, and generalized weakness. A person with this condition who is not treated with corticosteroids will die within a few days because of the severe electrolyte imbalance and dehydration.

Stress and the Adrenal Gland. In 1936, Hans Selye discovered that injections of cattle ovaries into rats (1) stimulated growth of the adrenal cortex; (2) caused atrophy of the lymphoid tissue of the spleen, lymph nodes, and thymus; and (3) produced bleeding peptic ulcers. At first he thought that these ovarian extracts contained a specific hormone that caused these effects. He later discovered that injections of a variety of substances, including foreign chemicals such as formaldehyde, could produce the same effects. Indeed, the same pattern of effects occurred when he placed rats in cold environments or when he dropped them into water and made them swim until they were exhausted.

Figure 8.14. Simplified pathways for the synthesis of steroid hormones in the adrenal cortex. The adrenal cortex produces steroids that regulate Na^+ and K^+ balance (mineralocorticoids), steroids that regulate glucose balance (glucocorticoids), and small amounts of sex steroid hormones.

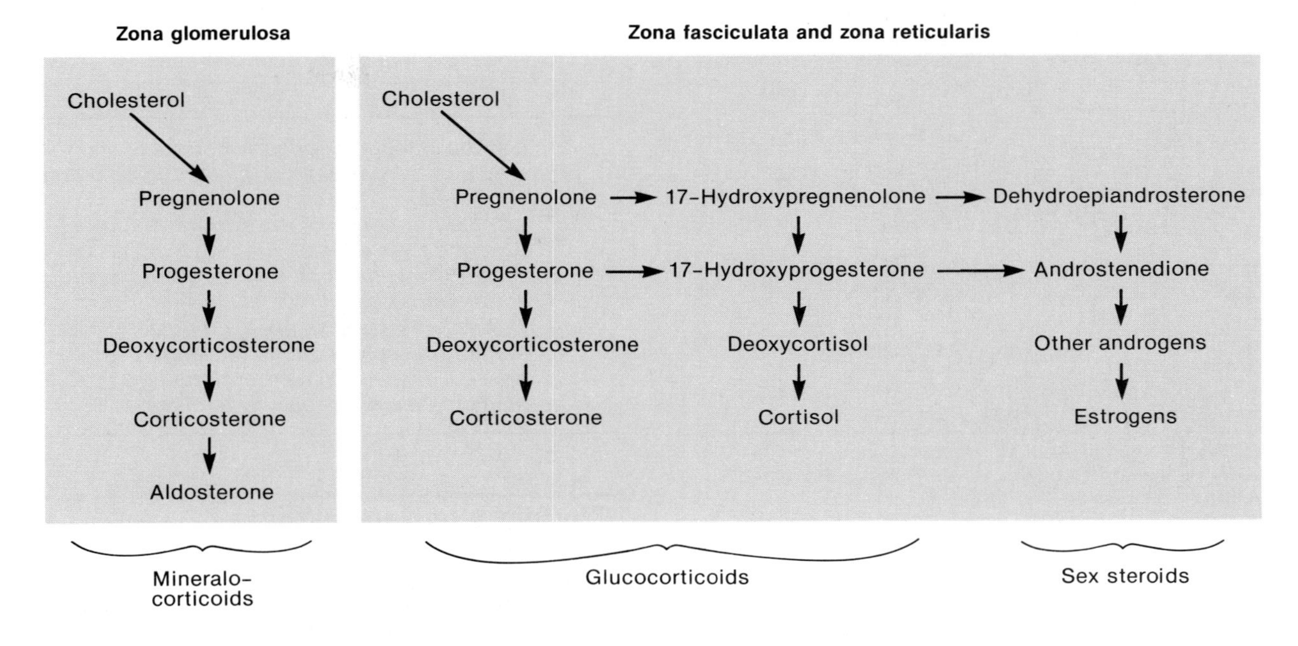

Figure 8.15. The activation of the pituitary-adrenal axis by nonspecific stress.

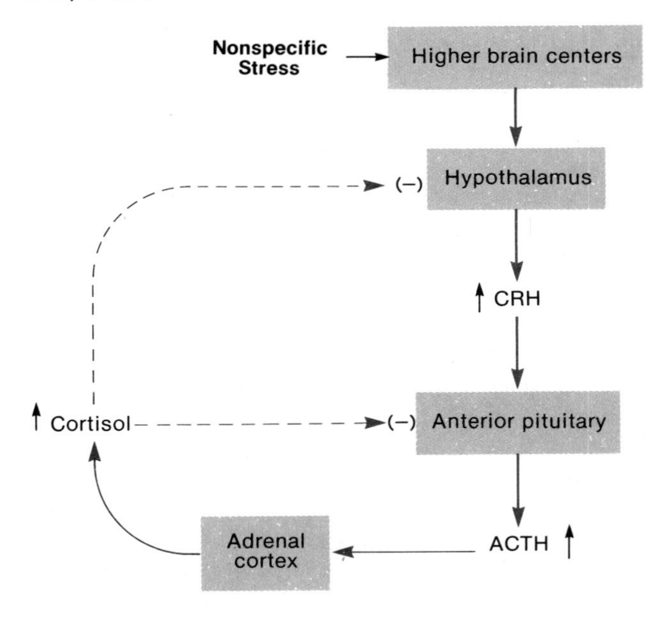

The specific pattern of effects produced by these procedures suggested that these effects were the result of something that the procedures shared in common. Selye reasoned that all of the procedures were *stressful* and that the pattern of changes he observed represented a specific response to any stressful agent. He later discovered that all forms of stress produce these effects because all stressors stimulate the pituitary-adrenal axis. Under stressful conditions, there is increased secretion of ACTH and thus increased secretion of corticosteroids from the adrenal cortex.

Since stress is so very difficult to define, many people prefer to define stress operationally as any stimulus that activates the pituitary-adrenal axis. Using this criterion, it has been found that pleasant changes in one's life—such as marriage, a recent promotion, and so on—can be as stressful as unpleasant changes. On this basis, Selye has stated that there is "a nonspecific response of the body to readjust itself following any demand made upon it."

Selye termed this nonspecific response the **general adaptation syndrome (GAS).** Stress, in other words, produces GAS. There are three stages in the response to stress: (1) the *alarm reaction,* when the adrenal glands are activated; (2) the *stage of resistance,* in which readjustment occurs; and (3) if the readjustment is not complete, the *stage of exhaustion* may follow, leading to sickness and possibly death.

Table 8.10 Comparison of the hormones from the adrenal medulla.

Epinephrine	Norepinephrine
Elevates blood pressure because of increased cardiac output and peripheral vasoconstriction	Elevates blood pressure because of generalized vasoconstriction
Accelerates respiratory rate and dilates respiratory passageways	Similar effect but to a lesser degree
Increases efficiency of muscular contraction	Similar effect but to a lesser degree
Increases rate of glycogen breakdown into glucose, so level of blood glucose rises	Similar effect but to a lesser degree
Increases rate of fatty acid released from fat, so level of blood fatty acids rises	Similar effect but to a lesser degree
Increases release of ACTH and TSH from the adenohypophysis of the pituitary gland	No effect

Source: Van De Graaff, Kent M., *Human Anatomy.*

Functions of the Adrenal Medulla

The chromaffin cells of the adrenal medulla secrete **epinephrine** and **norepinephrine** in an approximate ratio of 4:1, respectively. These hormones are classified as amines (more specifically, as catecholamines) and are derived from the amino acid tyrosine.

The effects of these hormones are similar to those caused by stimulation of the sympathetic nervous system, except that the hormonal effect lasts about ten times longer. The hormones from the adrenal medulla increase cardiac output and heart rate, dilate coronary blood vessels, increase mental alertness, increase the respiratory rate, and elevate metabolic rate. A comparison of the effects of epinephrine and norepinephrine are presented in table 8.10.

The adrenal medulla is innervated by sympathetic nerve fibers. The impulses are initiated from the hypothalamus via the spinal cord when the sympathetic nervous system is stimulated. Many stressors, therefore, activate the adrenal medulla as well as the adrenal cortex. Activation of the adrenal medulla together with the sympathetic nervous system prepares the body for greater physical performance—the *fight-or-flight* response.

Excessive stimulation of the adrenal medulla can result in depletion of the body's energy reserves, and high levels of corticosteroid secretion from the adrenal cortex can significantly impair the immune system. It is reasonable to expect, therefore, that prolonged stress can result in increased susceptibility to disease. Indeed, many studies show that prolonged stress results in an increased incidence of cancer and other diseases.

1. *List the categories of corticosteroids and the zones of the adrenal cortex that secrete these hormones.*
2. *List the hormones of the adrenal medulla, and describe their effects.*
3. *Explain how the secretions of the adrenal cortex and adrenal medulla are regulated.*
4. *Describe how stress affects the secretions of the adrenal cortex and medulla, and explain how excessive adrenal hormones may produce an increased susceptibility to disease.*

Thyroid and Parathyroids

The thyroid gland secretes thyroxine and triiodothyronine, which participate in the regulation of energy metabolism and are critically important for proper growth and development. The thyroid also secretes calcitonin, which may antagonize the action of parathyroid hormone in the regulation of calcium and phosphate balance.

The thyroid gland is positioned just below the larynx (fig. 8.16). This gland consists of two lobes that lie on either side of the trachea and are connected anteriorly by a broad *isthmus.* The thyroid is the largest of the endocrine glands, weighing between 20 and 25 g.

On a microscopic level, the thyroid gland consists of many spherical hollow sacs called **thyroid follicles** (fig. 8.17). These follicles are lined with a simple cuboidal epithelium composed of *principal cells,* which synthesize the principal thyroid hormones. The interior of the follicles contains *colloid,* which is a protein-rich fluid. Between the follicles are epithelial cells called *parafollicular cells,* which produce a hormone called calcitonin (or thyrocalcitonin).

Production and Action of Thyroid Hormones

The thyroid follicles actively accumulate iodide (I^-) from the blood and secrete it into the colloid. Once the iodide is in the colloid, it is oxidized to iodine and attached to specific amino acids (tyrosines) within the polypeptide chain of a protein called **thyroglobulin.** The attachment of one iodine to tyrosine produces *monoiodotyrosine (MIT);* the attachment of two iodines produces *diiodotyrosine (DIT).*

Figure 8.16. The thyroid gland. (*a*) Its relationship to the larynx and trachea. (*b*) A scan of the thyroid gland twenty-four hours after the intake of radioactive iodine.

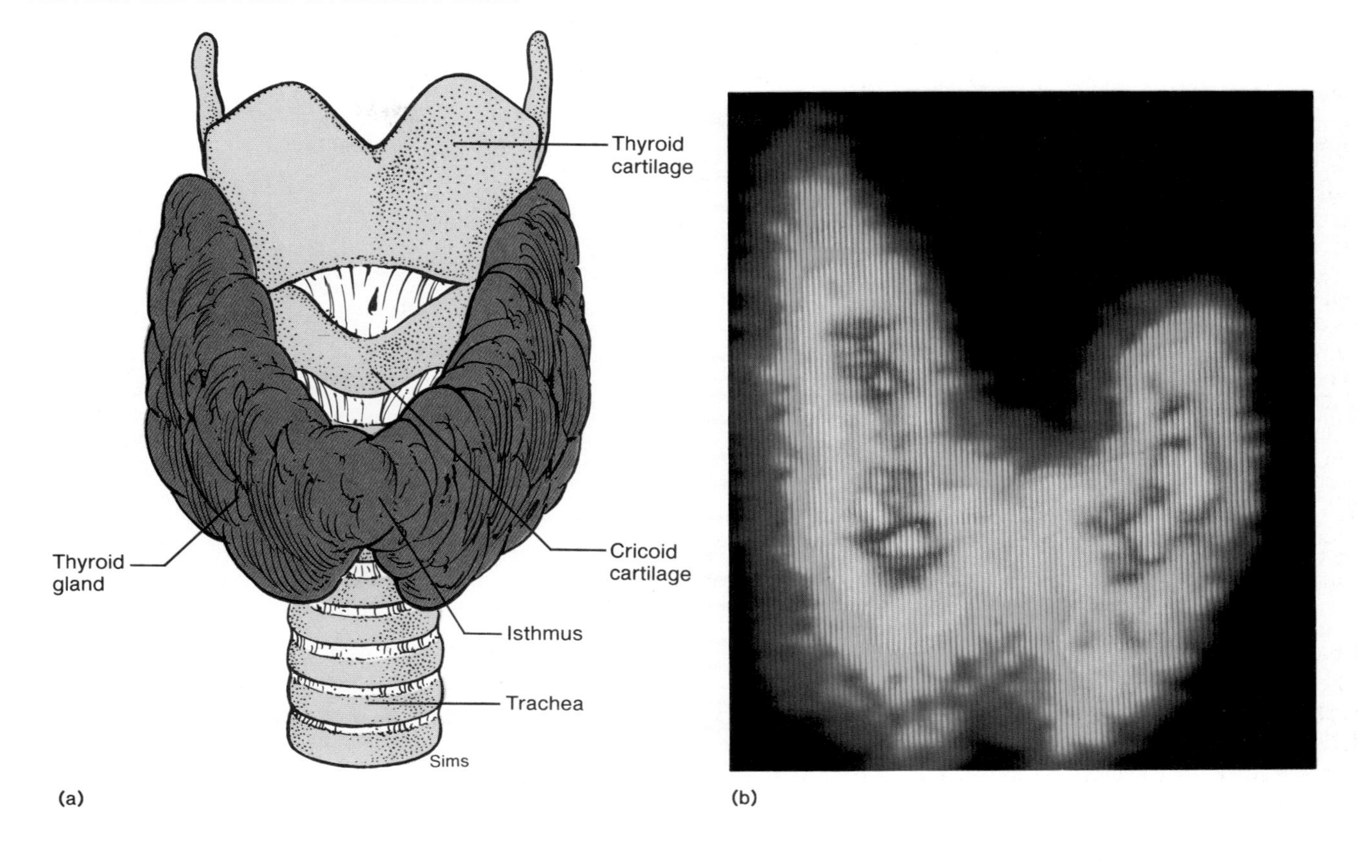

Figure 8.17. A photomicrograph (magnification ×250) of a thyroid gland, showing numerous thyroid follicles. Each follicle consists of follicular cells surrounding the fluid known as colloid, which contains thyroglobulin.

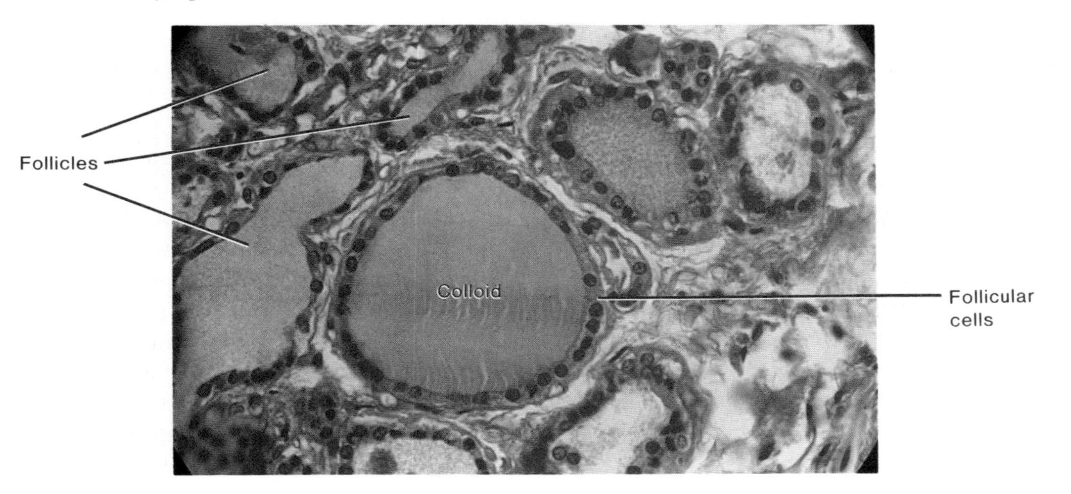

Figure 8.18. Stages in the formation and secretion of thyroid hormones. Iodide is actively accumulated by the follicular cells. In the colloid it is converted into iodine and attached to tyrosine amino acids within the thyroglobulin protein. Pinocytosis of iodinated thyroglobulin, coupling of MIT and DIT, and the release of thyroid hormones are stimulated by TSH from the anterior pituitary.

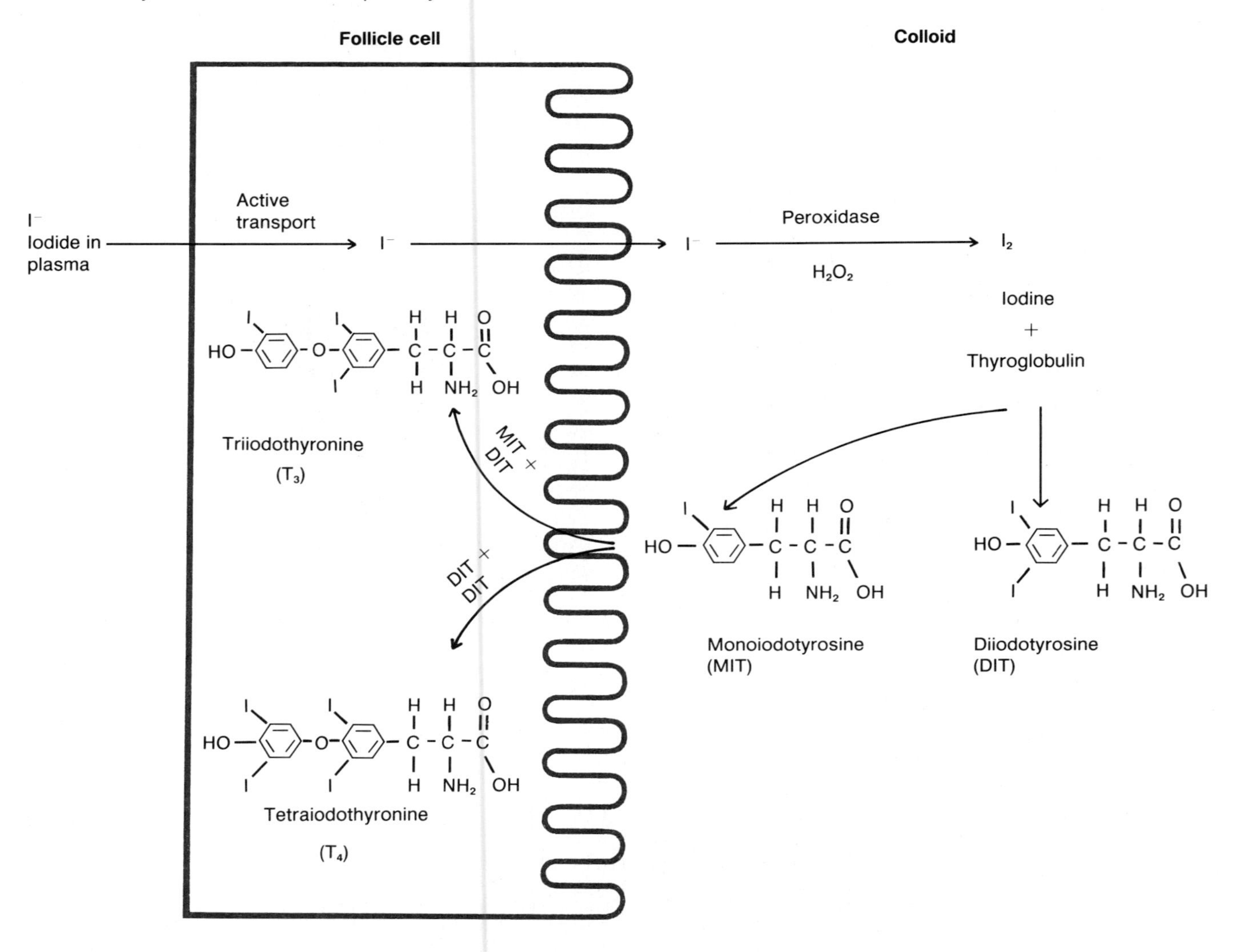

Within the colloid, enzymes modify the structure of MIT and DIT and couple them together (fig. 8.18). When two DIT molecules that are appropriately modified are coupled together, a molecule of **tetraiodothyronine, T_4,** or **thyroxine,** is produced. The combination of one MIT with one DIT forms **triiodothyronine, T_3.** Note that within the colloid T_4 and T_3 are still attached to thyroglobulin.

As previously described, most of the thyroxine in blood is attached to carrier proteins. Only the very small percentage of thyroxine that is free in the plasma can enter the target cells, where it is converted to triiodothyronine and attached to nuclear receptor proteins. Through the activation of genes, thyroid hormones stimulate protein synthesis, promote maturation of the nervous system, and increase the rate of energy utilization by the body.

The parafollicular cells of the thyroid secrete a hormone called **calcitonin** (also called **thyrocalcitonin**). Under certain conditions, calcitonin promotes a decrease in blood calcium levels and thus antagonizes the effects of parathyroid hormone and vitamin D_3.

Diseases of the Thyroid. Thyroid-stimulating hormone (TSH) from the anterior pituitary stimulates the thyroid to secrete thyroxine and exerts a trophic effect on the thyroid gland. This trophic effect is dramatically revealed in people who develop an **iodine-deficiency (endemic) goiter** (fig. 8.19). In the absence of sufficient dietary iodine the thyroid cannot produce adequate amounts of T_4 and T_3. The resulting lack of negative feedback inhibition causes

Figure 8.19. A simple or endemic goiter is caused by insufficient iodine in the diet.

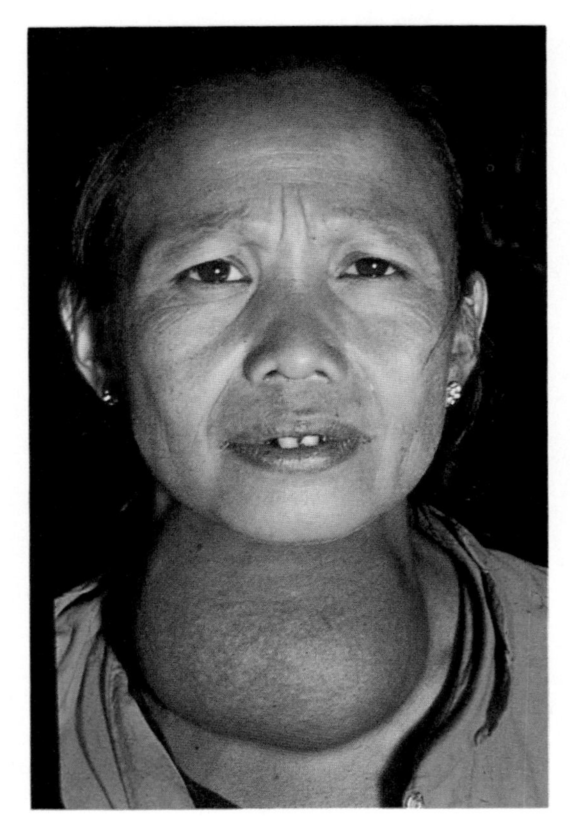

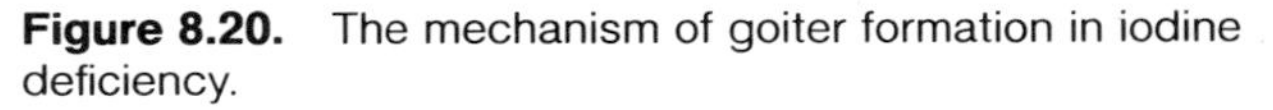

Figure 8.20. The mechanism of goiter formation in iodine deficiency.

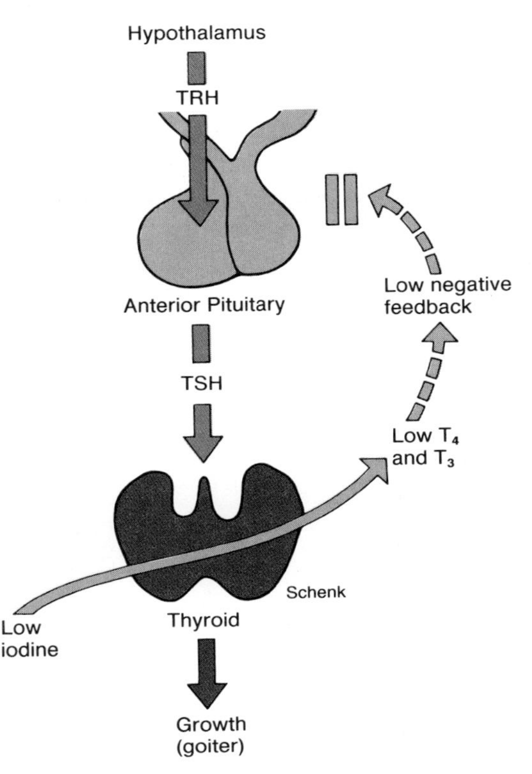

Figure 8.21. Hyperthyroidism is characterized by an increased metabolic rate, weight loss, muscular weakness, and nervousness. The eyes may also protrude.

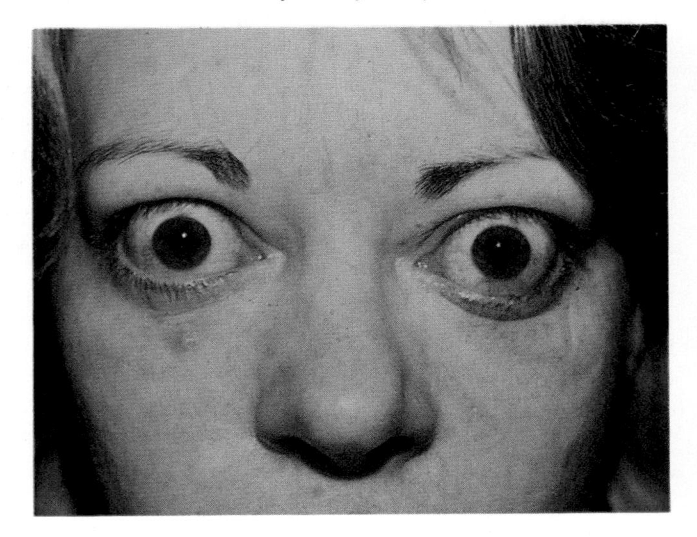

abnormally high levels of TSH secretion, which in turn stimulate the abnormal growth of the thyroid (a goiter). These events are summarized in figure 8.20.

The infantile form of **hypothyroidism** is known as *cretinism.* An affected child usually appears normal at birth because thyroxine is received from the mother through the placenta. The clinical symptoms of cretinism are stunted growth, thickened facial features, abnormal bone development, mental retardation, low body temperature, and general lethargy. If cretinism is diagnosed early, it can be successfully treated by administering thyroxine.

Hypothyroidism in an adult causes *myxedema.* This disorder affects body fluids, causing edema and increasing blood volume, hence increasing blood pressure. A person with myxedema has a low metabolic rate, lethargy, and a tendency to gain weight. This condition is treated with thyroxine or with triiodothyronine, which are taken orally (as pills).

Graves' disease, also called **toxic goiter,** involves growth of the thyroid associated with hypersecretion of thyroxine. This hyperthyroidism is produced by antibodies that act like TSH and stimulate the thyroid; it is an autoimmune disease. As a consequence of high levels of thyroxine secretion, the metabolic rate and heart rate increase, there is loss of weight, and the autonomic nervous system induces excessive sweating. In about half of the cases, *exophthalmos* (bulging of the eyes) also develops (fig. 8.21) because of edema in the tissues of the eye sockets and swelling of the extrinsic eye muscles.

Figure 8.22. A posterior view of the parathyroid glands.

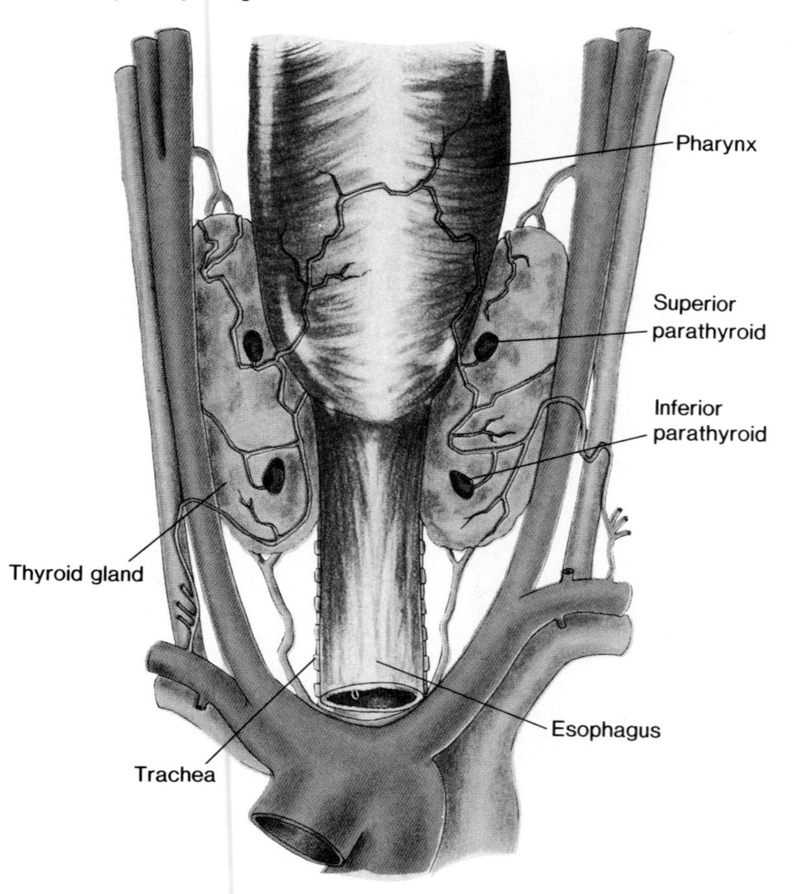

Radiotherapy of Thyroid Cancer. Thyroid cancer, particularly when it occurs in younger adults, has a more optimistic prognosis than most other forms of cancer. It is extremely slow growing and usually spreads only into the lymph nodes of the neck (although other sites can be invaded). Surgical treatment involves removal of the thyroid and of the affected cervical lymph nodes. This surgical treatment is followed by swallowing solutions containing radioactive iodine (^{131}I), which is selectively transported into cells of the thyroid gland and into cancerous thyroid cells that have traveled ("metastasized") to other regions of the body. These cells are then killed by the radioactively labeled iodine.

Thyroid-stimulating hormone (TSH), secreted from the anterior pituitary, stimulates the active accumulation of iodine into thyroid cells. Such active transport of iodine is obviously required for the thyroid cells to produce and secrete thyroid hormone (thyroxine). Therefore, in order for the ingested radioactive iodine to be effective in the treatment of thyroid cancer, the blood levels of TSH must be raised to high levels. This is accomplished by removing most of the patient's thyroid gland.

When the patient's thyroid gland is surgically removed the blood levels of thyroxine gradually decline, and—through the decrease in negative feedback inhibition that results—the secretion of TSH increases. A period of about four to five weeks is required after removal of the thyroid for the TSH levels to rise sufficiently. At this point the high secretion of TSH that results from low thyroxine levels can stimulate accumulation of radioactive iodine into thyroid cells and promote the destruction of both normal and metastasized cells.

Figure 8.23. Actions of parathyroid hormone. An increased level of parathyroid hormone causes the bones to release calcium, the kidneys to conserve calcium loss through the urine, and the absorption of calcium through the intestinal wall. Negative feedback of increased calcium levels in the blood inhibits the secretion of this hormone.

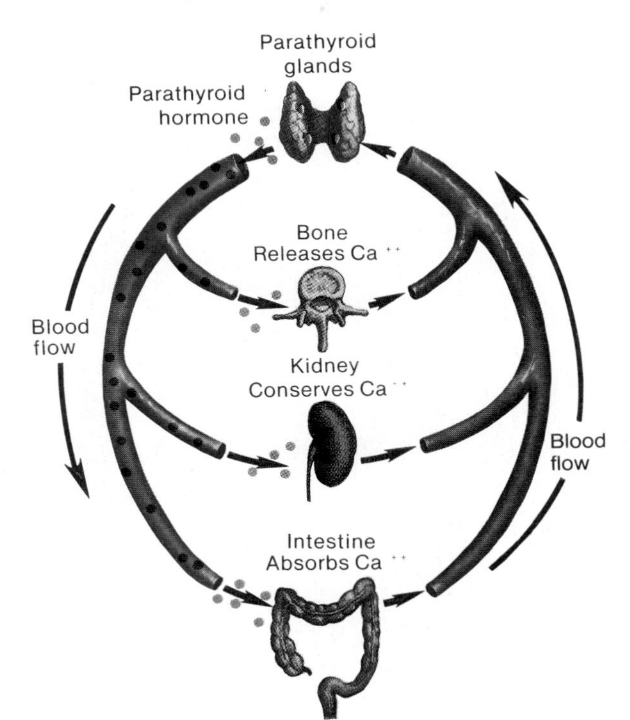

Parathyroid Glands

The small, flattened parathyroid glands are embedded in the posterior surfaces of the lateral lobes of the thyroid gland (see fig. 8.22). There are usually four parathyroid glands: a *superior* and an *inferior* pair. Each parathyroid gland is a small yellowish brown body measuring 3–8 mm (0.1–0.3 in.) in length, 2–5 mm (.07–0.2 in.) in width, and about 1.5 mm (0.05 in.) in depth.

On a microscopic level, the parathyroids are composed of two types of epithelial cells. The cells that synthesize parathyroid hormone are called *chief cells* and are scattered among *oxyphil cells*. Oxyphil cells support the chief cells and are believed to produce reserve quantities of parathyroid hormone.

The parathyroid glands secrete one hormone called **parathyroid hormone (PTH).** This hormone promotes a rise in blood calcium levels by acting on the bones, kidneys, and intestine (fig. 8.23). Regulation of calcium balance is described in more detail in chapter 19.

1. *Describe the structure of the thyroid gland, and list the effect of thyroid hormones.*
2. *Describe how thyroid hormones are produced and how their secretion is regulated, and explain the consequences of an inadequate dietary intake of iodine.*
3. *Describe how the secretion of parathyroid hormone is regulated, and explain the actions of parathyroid hormone.*

Figure 8.24. The pancreas and the associated islets of Langerhans.

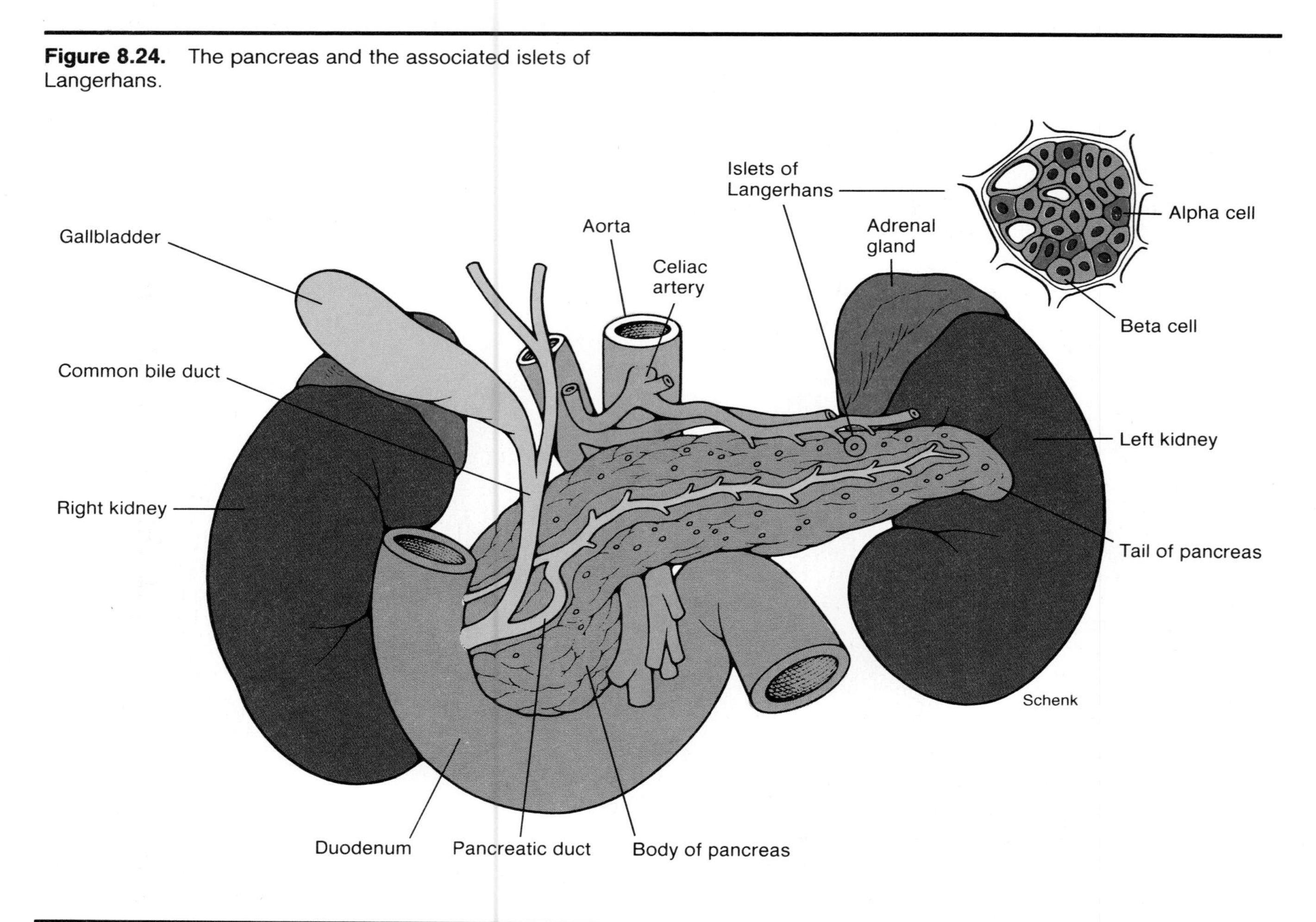

Pancreas and Other Endocrine Glands

The islets of Langerhans in the pancreas secrete two hormones, insulin and glucagon, which are critically involved in the regulation of metabolic balance in the body. Additionally, many other organs secrete hormones that help regulate digestion, metabolism, growth, immune function, and reproduction.

Islets of Langerhans

The pancreas is both an endocrine and an exocrine gland. The gross structure of this gland and its exocrine functions in digestion are described in chapter 18. The endocrine portion of the pancreas consists of scattered clusters of cells called the **islets of Langerhans.** These endocrine structures are most common in the body and tail of the pancreas (fig. 8.24).

Histologically, the most conspicuous cells in the islets are the *alpha* and *beta* cells (fig. 8.25). The alpha cells secrete the hormone **glucagon,** and the beta cells secrete **insulin.**

Figure 8.25. The histology of the islets of Langerhans.

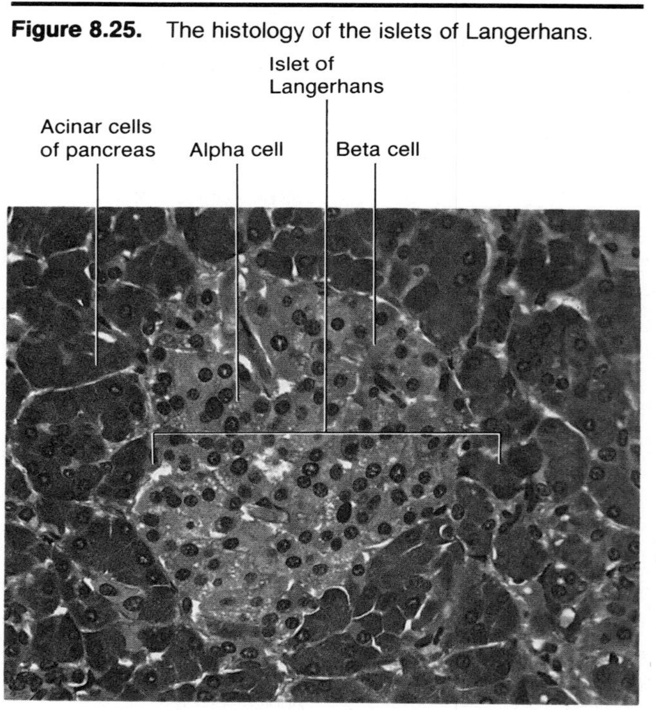

Alpha cells secrete glucagon in response to a fall in the blood glucose concentrations. Glucagon stimulates the liver to convert glycogen to glucose (**glycogenolysis**), which causes the blood glucose level to rise. This effect represents the completion of a negative feedback loop. Glucagon also stimulates the hydrolysis of stored fat (**lipolysis**) and the consequent release of free fatty acids into the blood. This effect helps to provide energy substrates for the body during fasting, when blood glucose levels decrease. Glucagon, together with other hormones, also stimulates the conversion of fatty acids to ketone bodies, which can be secreted by the liver into the blood and used by many organs as an energy source.

Diabetes mellitus is characterized by fasting hyperglycemia and the presence of glucose in the urine. There are two forms of this disease. *Type I,* or *juvenile-onset* diabetes mellitus, is caused by destruction of the beta cells and the resulting lack of insulin secretion. *Type II,* or *maturity-onset* diabetes mellitus (which is the more common form), is caused by decreased tissue sensitivity to the effects of insulin, so that larger amounts of insulin are required to produce a normal effect. Both types of diabetes mellitus are also associated with abnormally high levels of glucagon secretion.

Beta cells secrete insulin in response to a rise in the blood glucose concentrations. Insulin promotes the entry of glucose into tissue cells, and the conversion of this glucose into energy storage molecules of glycogen and fat. Insulin also aids the entry of amino acids into cells and the production of cellular protein. The actions of insulin and glucagon are thus antagonistic. After a meal, insulin secretion is increased and glucagon secretion is decreased; fasting, in contrast, causes a rise in glucagon and a fall in insulin secretion.

People who have a genetic predisposition for type II diabetes mellitus often first develop **reactive hypoglycemia.** In this condition, the rise in blood glucose that follows the ingestion of carbohydrates stimulates excessive secretion of insulin, which in turn causes the blood glucose levels to fall below the normal range. This can result in weakness, changes in personality, and mental disorientation.

Pineal Gland

The small, cone-shaped pineal gland is located in the roof of the third ventricle near the corpora quadrigemina, where it is encapsulated by the meninges covering the brain. The pineal gland of a child weighs about 0.2 g and is 5–8 mm (0.2–0.3 in.) long and 9 mm wide. The gland begins to regress in size at about age seven and in the adult appears as a thickened strand of fibrous tissue. Histologically, the pineal gland consists of specialized parenchymal and neuroglial cells. Although the pineal gland lacks direct nervous connections to the rest of the brain, it is highly innervated by the sympathetic nervous system from the superior cervical ganglion.

The principal hormone of the pineal is **melatonin.** The secretion of melatonin is inhibited by light and is therefore maximal at night. It has long been suspected that this hormone inhibits the pituitary-gonad axis. Melatonin secretion is highest in children of ages one to five and decreases thereafter, reaching its lowest levels at the end of puberty, where concentrations are 75% lower than during early childhood. The secretion of melatonin is, therefore, believed by some researchers to play an important role in the onset of puberty.

Figure 8.26. The thymus is a bilobed organ within the mediastinum of the thorax.

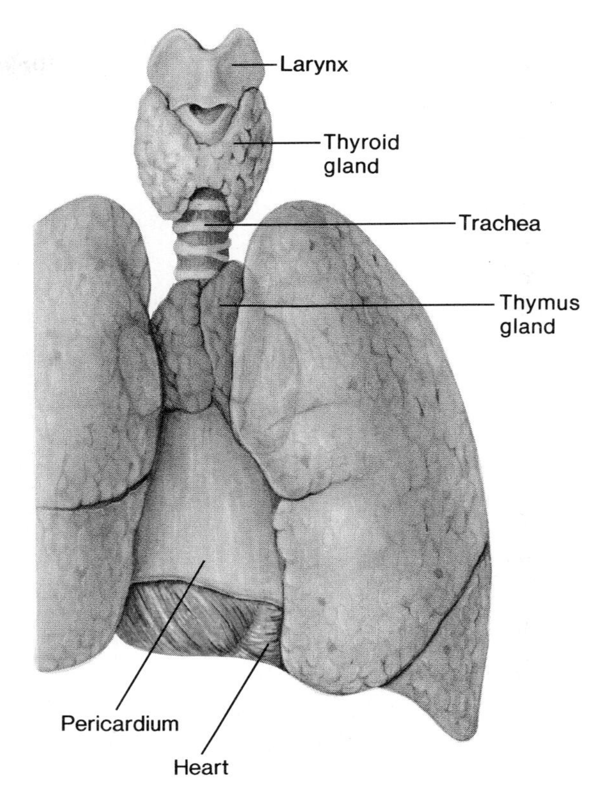

Thymus

The thymus is a bilobed organ positioned in the upper mediastinum, in front of the aorta and behind the manubrium of the sternum (fig. 8.26). Although the size of the thymus varies considerably from person to person, it is relatively large in newborns and children and sharply regresses in size after puberty. Besides decreasing in size, the thymus of adults becomes infiltrated with strands of fibrous and fatty connective tissue.

Table 8.11 Summary of the physiological effects of gastrointestinal hormones.

Secreted by	Hormone	Effects
Stomach	Gastrin	Stimulates parietal cells to secrete HCl Stimulates chief cells to secrete pepsinogen Maintains structure of gastric mucosa
Small intestine	Secretin	Stimulates water and bicarbonate secretion in pancreatic juice Potentiates actions of cholecystokinin on pancreas
Small intestine	Cholecystokinin (CCK)	Stimulates contraction of the gallbladder Stimulates secretion of pancreatic juice enzymes Potentiates action of secretin on pancreas Maintains structure of exocrine pancreas (acini)
Small intestine	Gastric inhibitory peptide (GIP)	Inhibits gastric emptying Inhibits gastric acid secretion Stimulates secretion of insulin from endocrine pancreas (islets of Langerhans)

Source: Fox, Stuart Ira, *A Laboratory Guide to Human Physiology, Concepts and Clinical Applications,* 3d ed.

The thymus serves as the site of production of **T cells** *(thymus-dependent cells),* which are the lymphocytes involved in cell-mediated immunity. In addition to providing T cells, the thymus secretes a number of hormones that are believed to stimulate T cells after they leave the thymus.

Gastrointestinal Tract

The stomach and small intestine secrete a number of hormones that act on the gastrointestinal tract itself and on the pancreas and gallbladder (table 8.11). The effects of these hormones act together with regulation by the autonomic nervous system to coordinate the activities of different regions of the digestive tract and the secretions of pancreatic juice and bile.

Gonads and Placenta

The gonads (**testes** and **ovaries**) secrete **sex steroids.** These include male sex hormones, or **androgens,** and female sex hormones—**estrogens** and **progestogens.** The principal hormones in each of these categories are *testosterone, estradiol-17β,* and *progesterone,* respectively.

The testes consist of two compartments: *seminiferous tubules,* which produce sperm, and *interstitial tissue* between the convolutions of the tubules. Within the interstitial tissue are *Leydig cells,* which secrete testosterone and lesser amounts of weaker androgens (as well as small amounts of estradiol-17β). Testosterone is needed for the development and maintenance of the male genitalia (penis and scrotum) and male sex accessory organs (prostate, seminal vesicles, epididymides, and ductus deferens), as well as for the development of male secondary sexual characteristics.

During the first half of the menstrual cycle, estrogen is secreted by many small structures within the ovary called *ovarian follicles.* These follicles contain the egg cell, or *ovum,* and *granulosa cells* that secrete estrogen. By about mid-cycle, one of these follicles grows very large and, in the process of ovulation, extrudes its ovum from the ovary. The empty follicle, under the influence of luteinizing hormone (LH) from the anterior pituitary, then becomes a new endocrine structure called a *corpus luteum.* The corpus luteum secretes progesterone as well as estradiol-17β.

The **placenta** is the organ responsible for nutrient and waste exchange between the fetus and mother. The placenta is also an endocrine gland; it secretes large amounts of estrogens and progesterone, as well as a number of polypeptide and protein hormones that are similar to some hormones secreted by the anterior pituitary gland. These latter hormones include *human chorionic gonadotrophin (hCG),* which is similar to LH, and *somatomammotrophin,* which is similar in action to growth hormone and prolactin.

---

1. *Describe the structure of the endocrine pancreas, and indicate the sites of origin of insulin and glucagon.*
2. *Describe how insulin and glucagon secretion is affected by eating and by fasting, and explain the actions of these two hormones.*
3. *Describe the location of the pineal gland and the possible function of melatonin.*
4. *Describe the location and function of the thymus gland.*
5. *Explain how the hormones secreted by the gonads and placenta are categorized, and describe which hormones of these categories are secreted by each gland.*

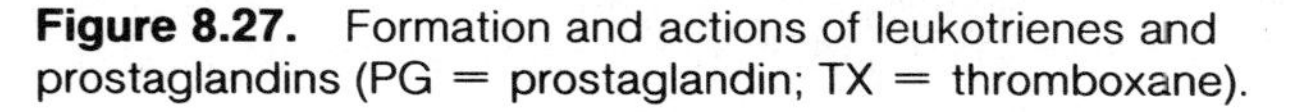

Figure 8.27. Formation and actions of leukotrienes and prostaglandins (PG = prostaglandin; TX = thromboxane).

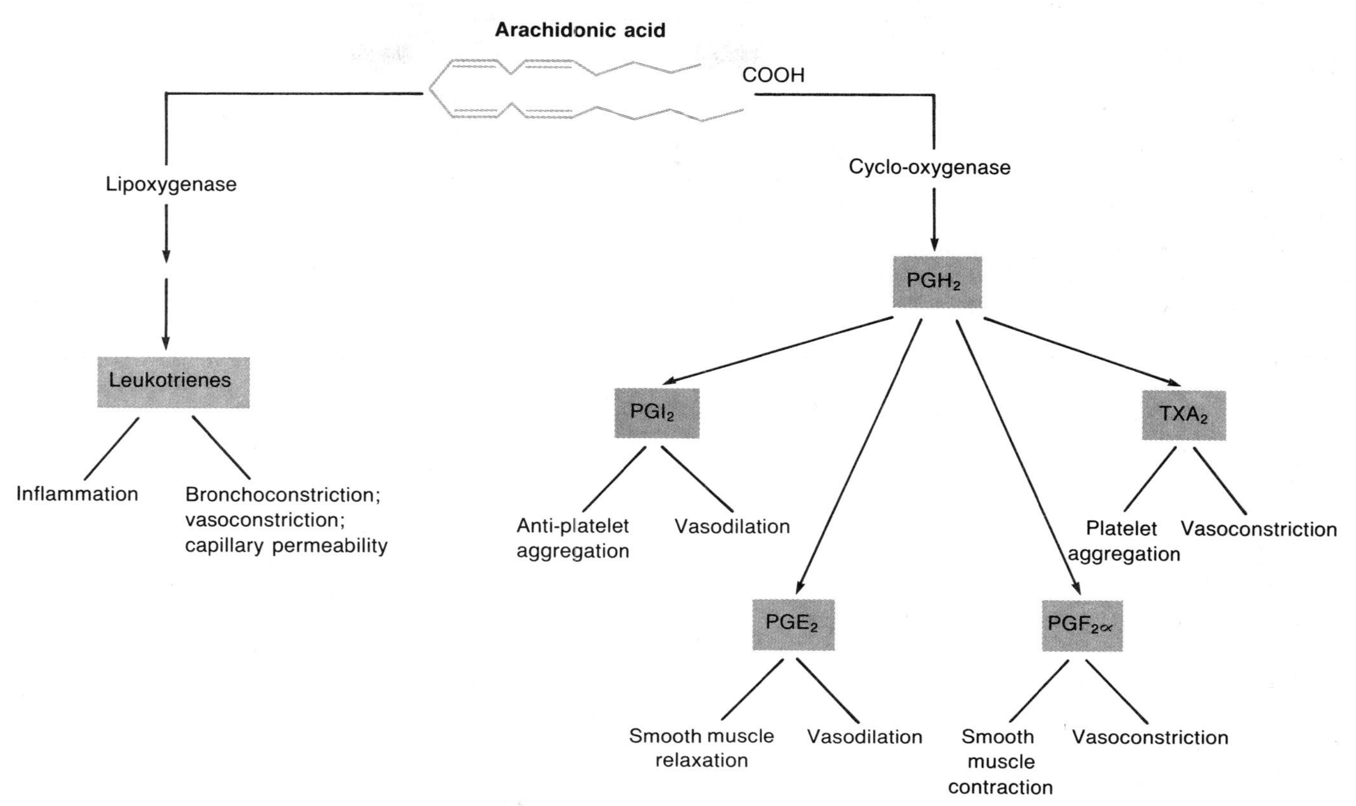

Prostaglandins

In chapters 7 and 8, two types of regulatory molecules have been considered—neurotransmitters and hormones. These two classes of regulatory molecules cannot simply be defined by differences in chemical structure, since the same molecule (such as norepinephrine) may be in both categories, but they must rather be defined by function. Neurotransmitters are released by axons, travel across a narrow synaptic cleft, and affect a postsynaptic cell. Hormones are secreted into the blood by an endocrine gland and, through transport in the blood, come to influence the activities of one or more target organs.

There is yet another class of regulatory molecules, which are distinguished by the fact that they are produced in many different organs and are generally *active within the organ in which they are produced.* Molecules of this type are called *autocrine,* or *paracrine,* regulators. The major group of such molecules are the **prostaglandins.** Since prostaglandins do not generally function as hormones, they are not properly considered a part of the endocrine system. Nevertheless, as a result of their regulatory function and their close association with hormonal control mechanisms, prostaglandins will be considered in this chapter as a corollary of endocrine regulation.

A prostaglandin is a twenty-carbon-long fatty acid that contains a five-membered carbon ring. This molecule is derived from the precursor molecule *arachidonic acid,* which can be released from phospholipids in the cell membrane under hormonal or other stimulation. Arachidonic acid can then enter one of two possible metabolic pathways. In one case, the arachidonic acid may be converted by the enzyme *cyclo-oxygenase* into a prostaglandin, which can then be changed by other enzymes into other prostaglandins. In the other case, arachidonic acid may be converted by the enzyme *lipoxygenase* into **leukotrienes,** which are compounds that are closely related to the prostaglandins (fig. 8.27).

Prostaglandins are produced in almost every organ and have been implicated in a wide variety of regulatory functions. The study of prostaglandin function can be confusing, in part because a given prostaglandin may have opposite effects in different organs. Prostaglandins of the E series (PGE), for example, cause relaxation of smooth muscle in the bladder, bronchioles, intestine, and uterus, but the same molecules cause contraction of vascular smooth muscle. A different prostaglandin, designated $PGF_{2\alpha}$, has exactly the opposite effects.

The antagonistic effects of prostaglandins on blood clotting makes good physiological sense. Blood platelets, which are required for blood clotting, produce *thromboxane A_2*. This prostaglandin promotes clotting by stimulating platelet aggregation and vasoconstriction. The endothelial cells of blood vessels, in contrast, produce a different prostaglandin, known as PGI_2, or *prostacyclin,* which has the opposite effects—it inhibits platelet aggregation and causes vasodilation. These antagonistic effects help to promote clotting but to insure that clots do not normally form on the walls of intact blood vessels.

The following are some of the regulatory functions proposed for prostaglandins in different organs and systems:

1. **Inflammation.** Prostaglandins promote many aspects of the inflammatory process, including the development of pain and fever. Drugs that inhibit prostaglandin synthesis help to alleviate these symptoms.
2. **Reproductive system.** Prostaglandins may play a role in ovulation and corpus luteum function in the ovaries and in contraction of the uterus. Excessive prostaglandin production may be involved in premature labor, endometriosis, premenstrual tension, and other gynecological disorders.
3. **Gastrointestinal tract.** The stomach and intestines produce prostaglandins, which are believed to inhibit gastric secretions and influence intestinal motility and fluid absorption. Since prostaglandins inhibit gastric secretion, drugs that suppress prostaglandin production may make a patient more susceptible to peptic ulcers.
4. **Respiratory system.** Some prostaglandins cause constriction whereas others cause dilation of blood vessels in the lungs and of bronchiolar smooth muscle. The leukotrienes are potent bronchoconstrictors, and these compounds, together with some prostaglandins, may cause respiratory distress and contribute to bronchoconstriction in asthma.
5. **Blood vessels.** Some prostaglandins are vasoconstrictors and others are vasodilators. In a fetus, PGE_2 is believed to promote dilation of the ductus arteriosus, a short vessel that connects the pulmonary artery with the aorta. After birth the ductus arteriosus normally closes as a result of a rise in blood oxygen when the baby breathes. If the ductus remains patent (open), however, it can be closed by the administration of drugs that inhibit prostaglandin synthesis.
6. **Blood clotting.** Thromboxane A_2, produced by blood platelets, promotes platelet aggregation and vasoconstriction. Prostacyclin, produced by vascular endothelial cells, inhibits platelet aggregation and promotes vasodilation.
7. **Kidneys.** Prostaglandins are produced in the medulla of the kidneys and cause vasodilation, resulting in increased renal blood flow and increased excretion of water and electrolytes in the urine.

Aspirin is the most widely used member of a class of drugs known as *nonsteroidal anti-inflammatory drugs.* Aspirin and other drugs of this class produce their effects because they specifically inhibit the cyclo-oxygenase enzyme that is needed for prostaglandin synthesis. Since prostaglandins promote inflammation, therefore, aspirin helps to reduce inflammation, pain, and fever. Acting through the inhibition of prostaglandin synthesis, which is needed for proper blood clotting, aspirin significantly increases the clotting time for eight days after administration (platelets are replaced every eight days). For this reason aspirin may be used to treat venous embolisms and arterial thromboses and has been used in patients with myocardial ischemia in the hope that it would prevent the formation of blood clots in the coronary vessels.

---

1. *Describe the chemical nature of prostaglandins, and explain why these molecules are not considered to be hormones.*
2. *Describe the metabolic pathway by which prostaglandins are produced, and describe the different types of functions performed by different prostaglandins.*
3. *Explain how aspirin works, and describe the effects of aspirin on different body processes.*

Summary

Hormones: Actions and Interactions p. 178

I. Precursors of active hormones may be called either prohormones or prehormones.
   A. Prohormones are relatively inactive precursor molecules made in the endocrine cells.
   B. Prehormones are the normal secretions of an endocrine gland that must be converted to other derivatives by target cells to be active.

II. The effects of a hormone in the body depend on its concentration.
   A. Abnormally high amounts of a hormone can result in effects that are not normally found.
   B. Target tissues can become desensitized by high hormone concentrations.

III. Hormones can interact in permissive, synergistic, or antagonistic ways.

Mechanisms of Hormone Action p. 183

I. Steroid and thyroid hormones enter their target cells.
   A. Thyroid hormones attach to chromatin-bound receptors located in the nucleus.
   B. Steroid hormones bond to cytoplasmic receptor proteins and translocate to the nucleus.
   C. Attachment of the hormone-receptor protein complex to the chromatin activates genes and thereby stimulates RNA and protein synthesis.

II. Amine, polypeptide, and glycoprotein hormones bond to receptor proteins on the outer surface of the target cell membrane.
   A. In many cases, this leads to the intracellular production of cyclic AMP, which serves as a second messenger in the action of these hormones.
   B. In other cases, cyclic GMP, or Ca^{++} may serve as a second messenger in the action of the hormone.

Pituitary Gland p. 189

I. The pituitary secretes eight hormones.
   A. The anterior pituitary secretes growth hormone, thyroid stimulating hormone, adrenocorticotrophic hormone, follicle-stimulating hormone, luteinizing hormone, and prolactin.
   B. The posterior pituitary secretes antidiuretic hormone (also called vasopressin) and oxytocin.

II. Secretions of the posterior pituitary are controlled by the hypothalamo-hypophyseal nerve tract.

III. Secretions of the anterior pituitary are controlled by hypothalamic hormones that stimulate or inhibit secretions of the anterior pituitary.
   A. Hypothalamic hormones include TRH, CRH, GnRH, PIH, somatostatin, and a growth hormone–releasing hormone.
   B. These hormones are carried to the anterior pituitary by the hypothalamo-hypophyseal portal system.

IV. Secretions of the anterior pituitary are also regulated by the feedback (usually negative feedback) of hormones from the target glands.

V. Higher brain centers, acting through the hypothalamus, can influence pituitary secretion.

Adrenal Glands p. 195

I. The adrenal cortex secretes mineralocorticoids (mainly aldosterone), glucocorticoids (mainly hydrocortisone), and sex steroids (primarily weak androgens).
   A. The glucocorticoids help regulate energy balance; they also can inhibit inflammation and suppress immune function.
   B. The pituitary-adrenal axis is stimulated by stress as part of the general adaptation syndrome.

II. The adrenal medulla secretes epinephrine and lesser amounts of norepinephrine, which complement the action of the sympathetic nervous system.

Thyroid and Parathyroids p. 198

I. The thyroid follicles secrete tetraiodothyronine (T_4, or thyroxine) and lesser amounts of triiodothyronine (T_3).
   A. These hormones are formed within the colloid of the thyroid follicles.
   B. The parafollicular cells of the thyroid secrete the hormone calcitonin, which may act to lower blood calcium levels.

II. The parathyroids are small structures embedded within the thyroid gland; the parathyroids secrete a hormone that promotes a rise in blood calcium levels.

Pancreas and Other Endocrine Glands p. 204

I. Beta cells in the islets secrete insulin; alpha cells secrete glucagon.
   A. Insulin lowers blood glucose and stimulates the production of glycogen, fat, and protein.
   B. Glucagon raises blood glucose by stimulating the breakdown of liver glycogen; glucagon also promotes lipolysis and the formation of ketone bodies.
   C. The secretion of insulin is stimulated by a rise in blood glucose following meals; the secretion of glucagon is stimulated by a fall in blood glucose during periods of fasting.

II. The pineal gland, located on the roof of the third ventricle of the brain, secretes melatonin; this hormone may play a role in regulating reproductive function.

III. The thymus is located in front of the aorta in the mediastinum; it is the site of the production of T cell lymphocytes and secretes a number of hormones that may help regulate the immune system.

IV. The gastrointestinal tract secretes a number of hormones that help regulate functions of the digestive system.

V. The gonads secrete sex steroid hormones.
   A. Leydig cells in the interstitial tissue of the testes secrete testosterone and other androgens.
   B. Granulosa cells of the ovarian follicles secrete estrogen.
   C. The corpus luteum of the ovaries secretes progesterone as well as estrogen.

VI. The placenta secretes estrogen, progesterone, and a variety of polypeptide hormones that have actions similar to some anterior pituitary hormones.

Prostaglandins p. 207

I. Prostaglandins are special, twenty-carbon-long fatty acids produced by many different organs that usually have regulatory functions within the organ in which they are produced.

Review Activities

Objective Questions

1. Hypothalamic releasing hormones
   (a) are secreted into capillaries in the median eminence
   (b) are transported by portal veins to the anterior pituitary
   (c) stimulate the secretion of specific hormones from the anterior pituitary
   (d) all of the above
2. The hormone primarily responsible for setting the basal metabolic rate and for promoting the maturation of the brain is
   (a) cortisol
   (b) ACTH
   (c) TSH
   (d) thyroxine
3. Which of the following statements about the adrenal cortex is *true?*
   (a) It is not innervated by nerve fibers.
   (b) It secretes some androgens.
   (c) The zona granulosa secretes aldosterone.
   (d) The zona fasciculata is stimulated by ACTH.
   (e) All of the above are true.
4. The hormone insulin
   (a) is secreted by alpha cells in the islets of Langerhans
   (b) is secreted in response to a rise in blood glucose
   (c) stimulates the production of glycogen and fat
   (d) both *a* and *b*
   (e) both *b* and c

Match the hormone with the primary agent that stimulates its secretion.

5. Epinephrine	(a) TSH
6. Thyroxine	(b) ACTH
7. Corticosteroids	(c) Growth hormone
8. ACTH	(d) Sympathetic nerves
	(e) CRH

9. Steroid hormones are secreted by
   (a) the adrenal cortex
   (b) the gonads
   (c) the thyroid
   (d) both *a* and *b*
   (e) both *b* and *c*
10. The secretion of which of the following hormones would be *increased* in a person with endemic goiter?
    (a) TSH
    (b) thyroxine
    (c) triiodothyronine
    (d) all of the above
11. Which of the following hormones use cAMP as a second messenger?
    (a) testosterone
    (b) corticol
    (c) insulin
    (d) epinephrine
12. Which of the following terms best describes the type of interaction between the effects of insulin and glucagon?
    (a) synergistic
    (b) permissive
    (c) antagonistic
    (d) cooperative

Essay Questions

1. Explain how the regulation of the neurohypophysis and adrenal medulla are related to their embryonic origins.
2. Compare steroid and polypeptide hormones in terms of their mechanism of action in target organs.
3. Explain the significance of the term *trophic* in regard to the actions of anterior pituitary hormones.
4. Suppose a drug blocks the conversion of T_4 to T_3. Explain what the effects of this drug would be on (*a*) TSH secretion, (*b*) thyroxine secretion, and (*c*) the size of the thyroid gland.
5. Explain why the phrase "master gland" is sometimes used to describe the anterior pituitary, and why this term is misleading.
6. Suppose a person's immune system made antibodies against insulin receptor proteins. Describe the possible effect of this condition on carbohydrate and fat metabolism.

Selected Readings

Austin, L. A., and H. Heath, III. 1981. Calcitonin: Physiology and pathophysiology. *New England Journal of Medicine.* 304:269.

Axelrod, J., and T. D. Reisine. 1984. Stress hormones: their interaction and regulation. *Science* 224:452.

Baxter, J. D., and W. J. Funder. 1979. Hormone receptors. *New England Journal of Medicine.* 300:117.

Brownstein, M. J., et al. 1980. Synthesis, transport and release of posterior pituitary hormones. *Science* 207:373.

Carmichael, S. W., and H. Winkler. August 1985. The adrenal chromaffin cell. *Scientific American.*

Demers, L. M. September 1984. The effects of prostaglandins. *Diagnostic Medicine,* p. 37.

Frohman, L. A. 1975. Neurotransmitters as regulators of endocrine functions. *Hospital Practice* 10:54.

Ganong, W. F., L. C. Alpert, and T. C. Lee. 1974. ACTH and the regulation of adrenocortical secretion. *New England Journal of Medicine* 290:1006.

Gershengorn, M. C. 1986. Mechanism of thyrotrophin releasing hormone stimulation of pituitary hormone secretion. *Annual Review of Physiology* 48:515.

Gillie, R. B. June 1971. Endemic goiter. *Scientific American.*

Goldsmith, R. S. 1969. Hyperparathyroidism. *New England Journal of Medicine* 281:367.

Guillemin, R., and R. Burgus. November 1972. The hormones of the hypothalamus. *Scientific American.*

Katzenellenbogen, B. S. 1980. Dynamics of steroid hormone receptor action. *Annual Review of Physiology* 42:17.

Krieger, D. T. 1984. Brain peptides: what, where, and why? *Science* 222:975.

McEwen, B. S. July 1976. Interactions between hormones and nerve tissue. *Scientific American.*

Martin, C. R. *Endocrine physiology.* 1985. New York: Oxford University Press.

O'Malley, B., and W. T. Shrader. February 1976. The receptors of steroid hormones. *Scientific American.*

Rasmussen, H. 1986. The calcium messenger system. *New England Journal of Medicine* 314: First part, p. 1094; Second part, p. 1164.

Reichlin, S., et. al. 1976. Hypothalamic hormones. *Annual Review of Physiology* 39:389.

Roth, J. 1980. Insulin receptors in diabetes. *Hospital Practice* 15:98.

Roth, J., and S. I. Taylor. 1982. Receptors for peptide hormones: alterations in diseases of humans. *Annual Review of Physiology* 44:639.

Schally, A. V. 1978. Aspects of the hypothalamic control of the pituitary gland. *Science* 202:18.

Selye, H. 1973. The evolution of the stress concept. *American Scientist* 61:692.

Turner, C. D. 1976. *General endocrinology.* 6th ed. Philadelphia: W. B. Saunders Co.

Ville, D. B. 1975. *Human endocrinology: A developmental approach.* Philadelphia: W. B. Saunders Co.

Williams, R. 1974. *Textbook of endocrinology.* 5th ed. Philadelphia: W. B. Saunders Co.

9 Central Nervous System

Objectives

By studying this chapter you should be able to

1. describe the organization of the spinal cord and how sensory and motor pathways are conveyed in the spinal cord
2. describe the organization of the cerebrum and the primary roles of its lobes
3. explain the structure and function of the sensory and motor cortex
4. describe different electroencephalogram patterns
5. describe the location and principal functions of the thalamus and hypothalamus and the principal structures of the midbrain and hindbrain and their functions
6. explain how skeletal movements are controlled by the CNS and distinguish between the structures and pathways of the pyramidal and extrapyramidal systems
7. describe the structure and location of the limbic system and explain the role of the limbic system and hypothalamus in emotion and motivation
8. explain the lateralization of functions in the right and left cerebral hemispheres
9. identify the structures involved in the control of speech and explain their interrelationships
10. describe the different types of aphasias that result from damage to specific regions of the brain
11. distinguish between short-term and long-term memory and identify the roles of the hippocampus and cerebral cortex in memory
12. explain how memory may be encoded in the brain and how damage to specific brain regions affects memory

Outline

Figure 9.1. The CNS consists of the brain and the spinal cord, both of which are covered with meninges and bathed in cerebrospinal fluid.

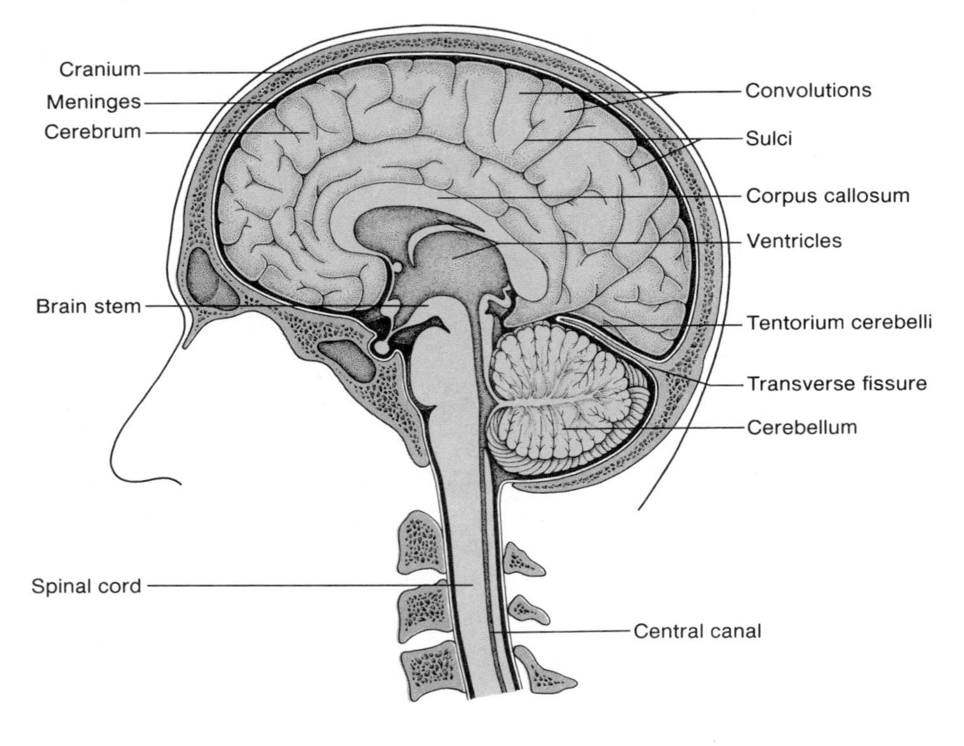

Organization of the Central Nervous System

The **central nervous system** (**CNS**), consisting of the brain and spinal cord (fig. 9.1), receives input from *sensory neurons* and directs the activity of *motor neurons,* which innervate muscles and glands. The *association neurons* within the brain and spinal cord are in a position, as their name implies, to associate appropriate motor responses with sensory stimuli and, thus, to maintain homeostasis in the internal environment and the continued existence of the organism in a changing external environment.

The interposition of a trillion neurons between stimulus input and motor output allows the CNS to create subjective perceptions and to coordinate extremely complex motor behaviors. These behaviors may be "hard-wired," as instinct, in the CNS, or they may be composed of responses that have been learned from past experience and can be changed as a consequence of future experiences. Perceptions, learning, memory, emotions, and the "self-awareness" that forms the basis of consciousness are creations of the brain.

Spinal Cord and Peripheral Nerves

Sensory and motor neurons comprise the **peripheral nervous system** (**PNS**), which provides the avenues of communication between the CNS and the internal and external environments of the body. The PNS consists of *nerves* (collections of axons) and *ganglia* (collections of neuron cell bodies). There are twelve pairs of cranial nerves and thirty-one pairs of spinal nerves. The spinal nerves are grouped into eight cervical, twelve thoracic, five lumbar, five sacral, and one coccygeal (fig. 9.2). These names are derived from the region of the vertebral column from which they arise.

Figure 9.2. The distribution of the spinal nerves.

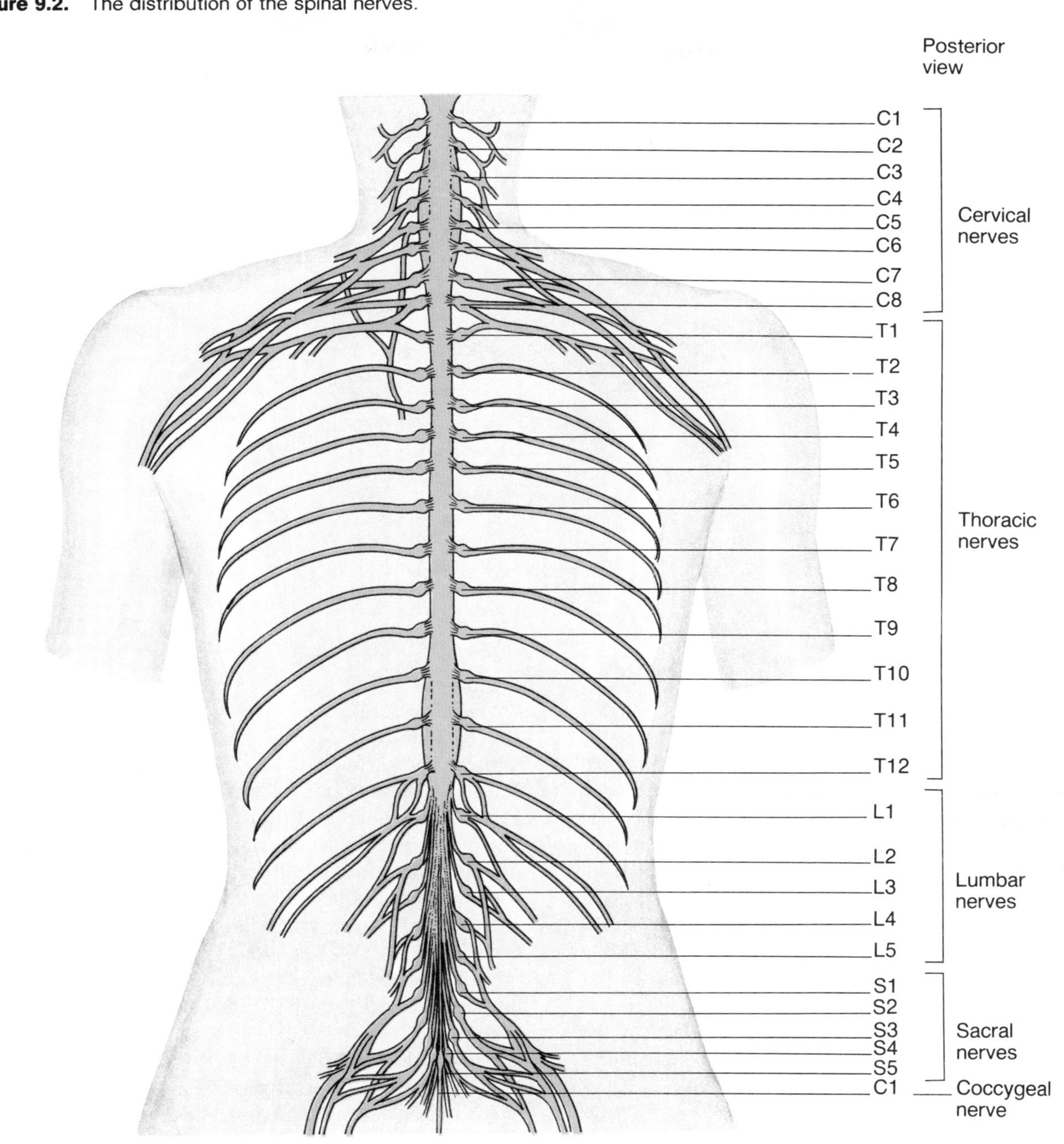

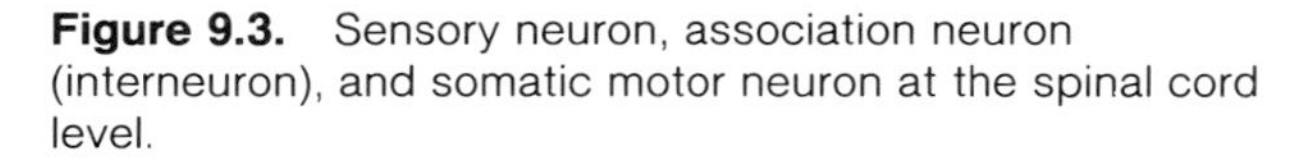

Figure 9.3. Sensory neuron, association neuron (interneuron), and somatic motor neuron at the spinal cord level.

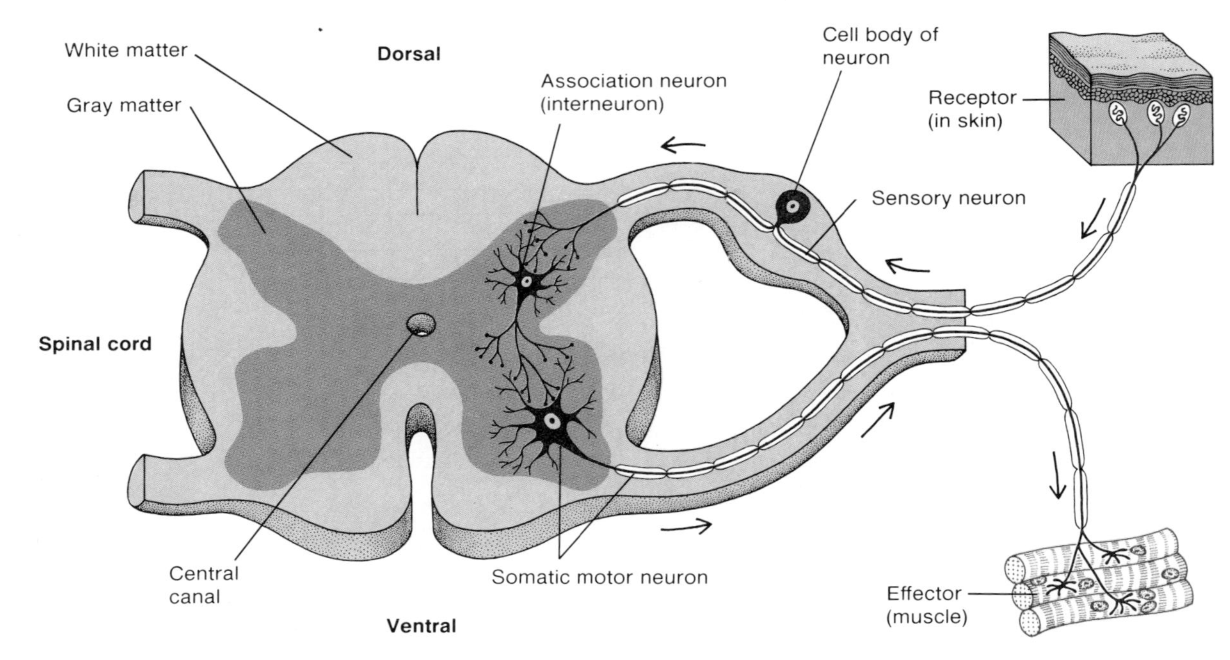

A spinal nerve is a mixed nerve composed of sensory and motor fibers. These fibers are grouped together in the nerve, but split apart near the attachment to the spinal cord, forming a *dorsal root* composed of sensory fibers and a *ventral root* composed of motor fibers (fig. 9.3). The dorsal root contains an enlargement called the *dorsal root ganglion,* where the cell bodies of sensory neurons are located. The cell bodies of motor neurons and of association neurons are located within the **grey matter** of the spinal cord. The grey matter of the spinal cord is surrounded by **white matter,** which is composed of myelinated axons that form *tracts* (collections of axons in the CNS). The arrangement of grey and white matter is reversed in the brain, where the white matter is beneath the outer layers of grey matter.

Cerebrum

The **cerebrum** (fig. 9.4) is the largest portion of the brain, accounting for about 80% of its mass, and is believed to be the brain region primarily responsible for higher mental functions. The cerebrum consists of *right* and *left hemispheres,* which are connected internally by a large fiber tract called the *corpus callosum* (see fig. 9.1).

The cerebrum consists of an outer **cerebral cortex,** which is 2–4 mm of grey matter, and underlying white matter. The surface of the cerebrum is folded into convolutions. The elevated folds of the convolutions are called *gyri,* and the depressed grooves are the *sulci.* Each cerebral hemisphere is subdivided by deep sulci, or *fissures,* into five lobes, four of which are visible from the surface (fig. 9.5). These lobes are the *frontal, parietal, temporal,* and *occipital,* which are visible from the surface, and the deep *insula.*

The **frontal lobe** is the anterior portion of each cerebral hemisphere. A deep fissure, called the *central sulcus,* or *fissure of Rolando,* separates the frontal lobe from the **parietal lobe.** The *precentral gyrus,* (fig. 9.4) involved in motor control, is located in the frontal lobe just in front of the central sulcus. The *postcentral gyrus,* which is located just behind the central sulcus in the parietal lobe, is the primary area of the cortex responsible for the perception of *somatesthetic sensation*—sensation arising from cutaneous, muscle, tendon, and joint receptors.

The precentral (motor) and postcentral (sensory) gyri have been mapped in conscious patients undergoing brain surgery. Electrical stimulation of specific areas of the precentral gyrus was found to cause specific movements, and stimulation of different areas of the postcentral gyrus evoked sensations in specific parts of the body. Typical

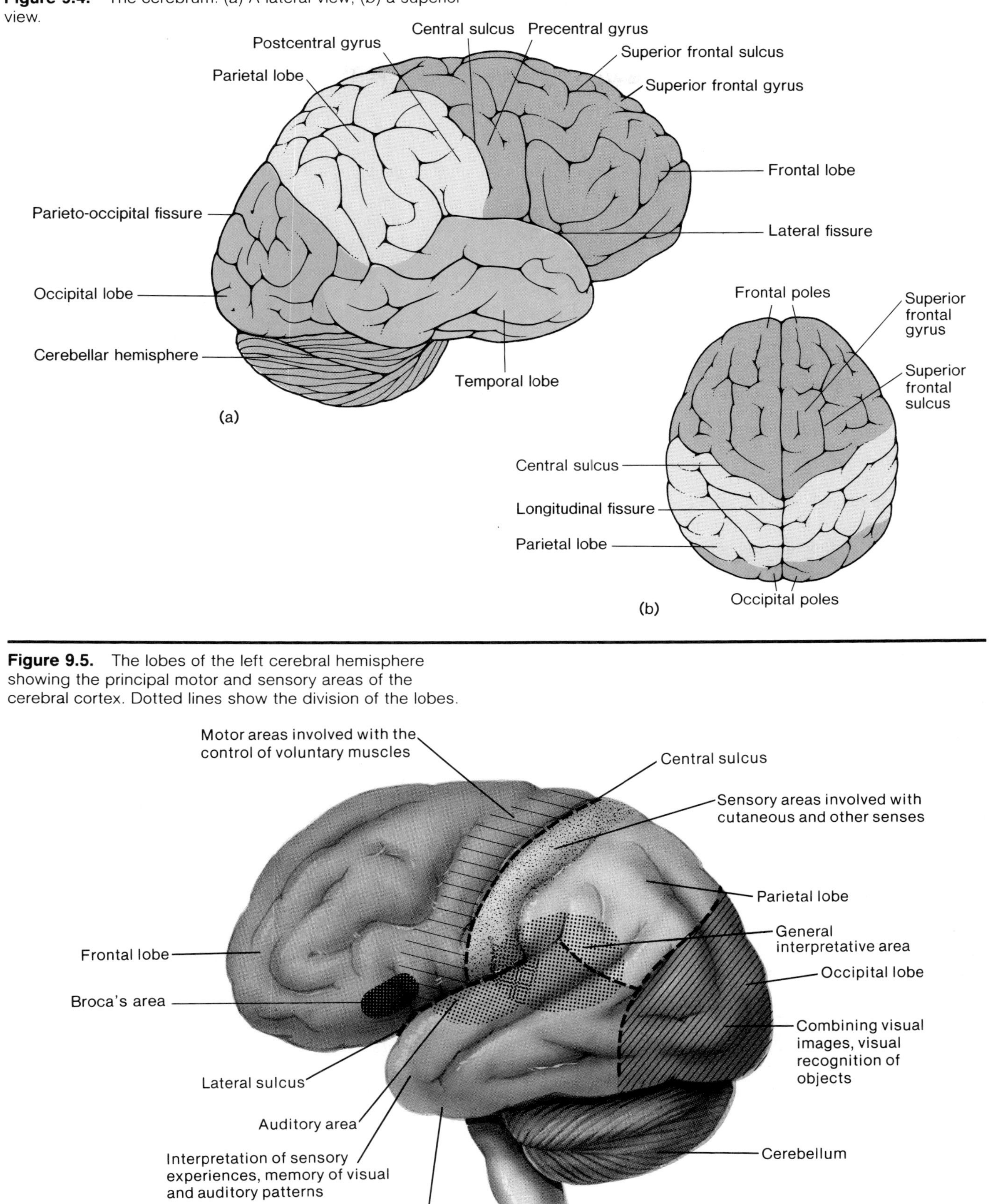

Figure 9.4. The cerebrum. (*a*) A lateral view; (*b*) a superior view.

Figure 9.5. The lobes of the left cerebral hemisphere showing the principal motor and sensory areas of the cerebral cortex. Dotted lines show the division of the lobes.

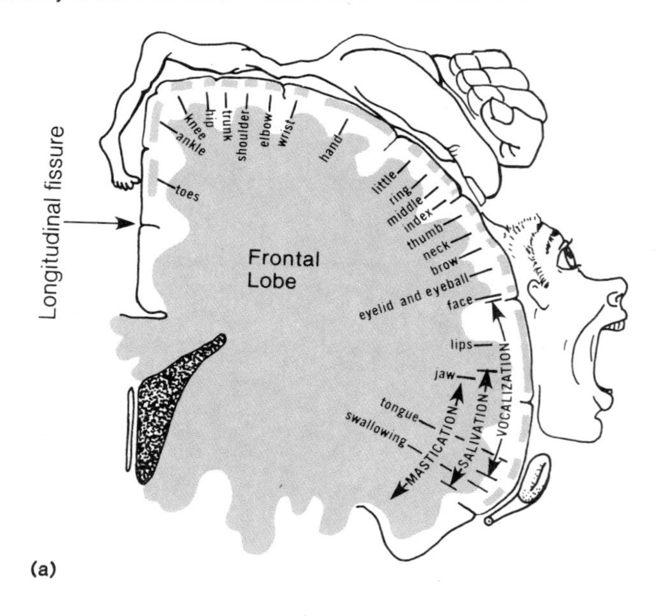

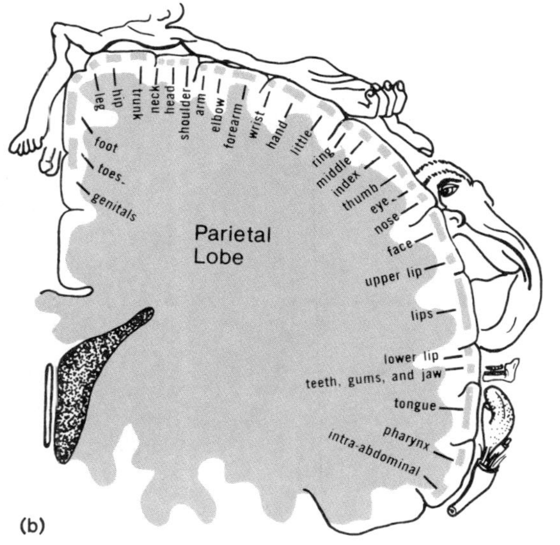

Figure 9.6. Motor and sensory areas of the cerebral cortex. (*a*) Motor areas that control skeletal muscles. (*b*) Sensory areas that receive somatesthetic sensations.

maps of these regions (fig. 9.6) show an upside-down picture of the body, with the superior regions of cortex devoted to the toes and the inferior regions devoted to the head.

A striking feature of these maps is that the area of cortex responsible for different parts of the body do not correspond to the size of the body parts being served. Instead, the body regions with the highest densities of receptors are represented with the largest areas of the sensory cortex, and the body regions with the greatest number of motor innervations are represented with the largest area of motor cortex. The hands and face, therefore, which have a high density of sensory receptors and motor innervation, are served by larger areas of the precentral and postcentral gyri than is the rest of the body.

The **temporal lobe** contains auditory centers that receive sensory fibers from the cochlea of each ear. This lobe is also involved in the interpretation and association of auditory and visual information. The **occipital lobe** is the primary area responsible for vision and for the coordination of eye movements.

Surgical removal of the temporal lobes in monkeys produces a condition called the *Kluver-Bucy syndrome,* and some aspects of this syndrome are observed in humans with temporal lobe damage. One interesting feature of this syndrome is the apparent lack of recognition of objects based on visual cues. Vision is not damaged, but the ability to associate visual information with past experience is impaired.

Electroencephalogram. The synaptic potentials (discussed in chapter 7) produced at the cell bodies and dendrites of the cerebral cortex create electrical currents, which can be measured by electrodes placed on the scalp. A record of these electrical currents is called an *electroencephalogram,* or *EEG.* Deviations from normal EEG patterns can be used clinically to diagnose epilepsy and other abnormal states, and the absence of an EEG can be used to diagnose brain death.

There are normally four types of EEG patterns (fig. 9.7):

1. **Alpha waves** are best recorded from the parietal and occipital regions while a person is awake and relaxed but with the eyes closed. These waves are rhythmic oscillations of about 10–12 cycles/second. The alpha rhythm of a child younger than eight years old occurs at a slightly lower frequency of 4–7 cycles/second.
2. **Beta waves** are strongest from the frontal lobes, especially the area near the precentral gyrus. These waves are produced by visual stimuli and mental activity. Because they respond to stimuli from receptors and are superimposed on the continuous activity patterns, they constitute *evoked activity.* The frequency of beta waves is 13–25 cycles/second.
3. **Theta waves** are emitted from the temporal and occipital lobes. They have a frequency of 5–8 cycles/second and are common in newborn infants. The recording of theta waves in adults generally indicates severe emotional stress and can be a forewarning of a nervous breakdown.

Figure 9.7. Brain waves. (*a*) Technician using an electroencephalograph to take the EEG of a teenaged boy. (*b*) Types of electroencephalograms.

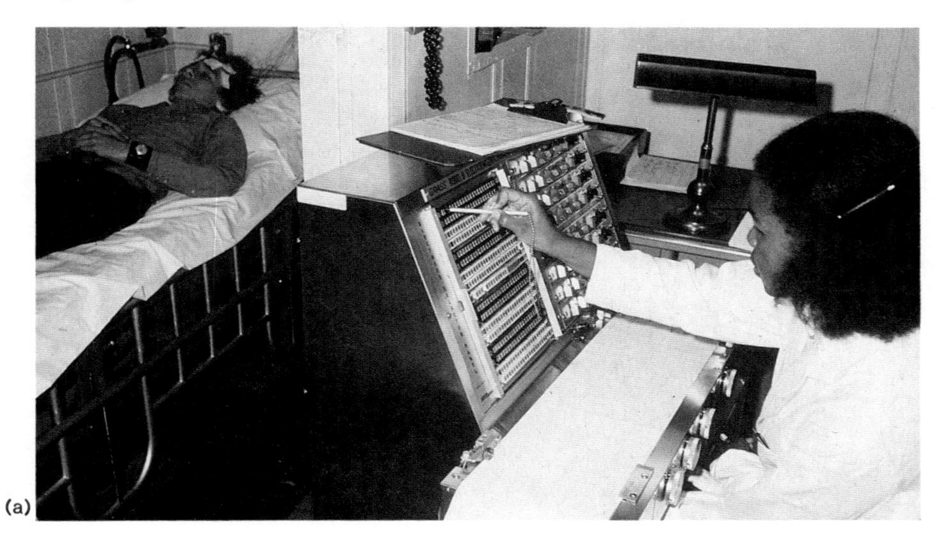

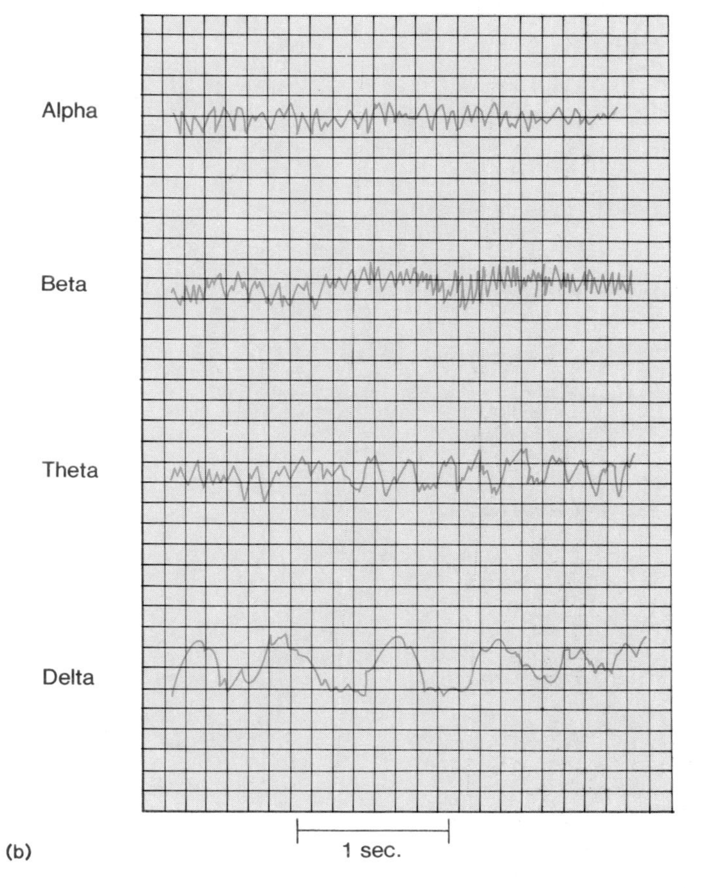

4. **Delta waves** are seemingly emitted in a general pattern from the cerebral cortex. These waves have a frequency of 1–5 cycles/second and are common during sleep and in an awake infant. The presence of delta waves in an awake adult indicates brain damage.

Basal Ganglia. The *basal ganglia* are masses of grey matter, composed of neuron cell bodies, located deep within the white matter of the cerebrum (fig. 9.8). The most prominent of the basal ganglia is the **corpus striatum,** which consists of several masses of nuclei (a *nucleus* is a collection of cell bodies in the CNS). The upper mass is

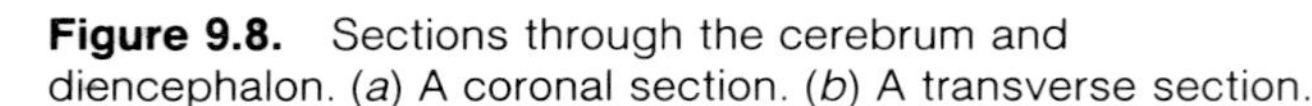

Figure 9.8. Sections through the cerebrum and diencephalon. (*a*) A coronal section. (*b*) A transverse section.

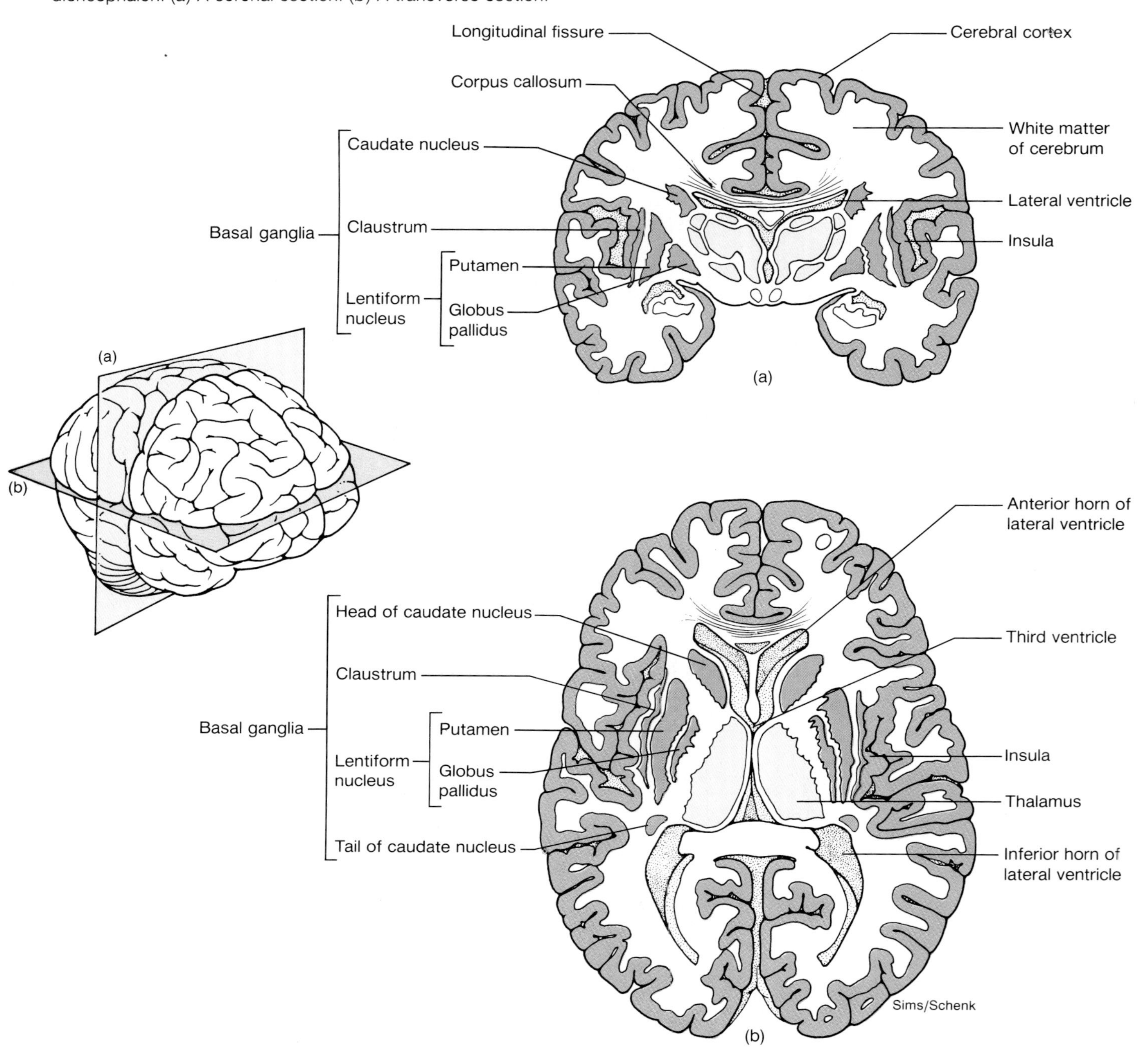

the *caudate nucleus,* which is separated from two lower masses that are collectively called the *lentiform nucleus.* The lentiform nucleus consists of a lateral portion, called the *putamen,* and a medial portion, called the *globus pallidus.* The basal ganglia function in the control of voluntary movements.

Degeneration of the caudate nucleus produces **chorea**—a hyperkinetic disorder characterized by rapid, uncontrolled, jerky movements. Degeneration of dopaminergic neurons from the substantia nigra, a small nucleus considered a part of the basal ganglia, to the caudate nucleus produces **Parkinson's disease.** As discussed in chapter 7, this disease is associated with rigidity, resting tremor, and difficulty in the initiation of voluntary movements.

Figure 9.9. A midsagittal section through the brain. (*a*) A diagram. (*b*) A photograph.

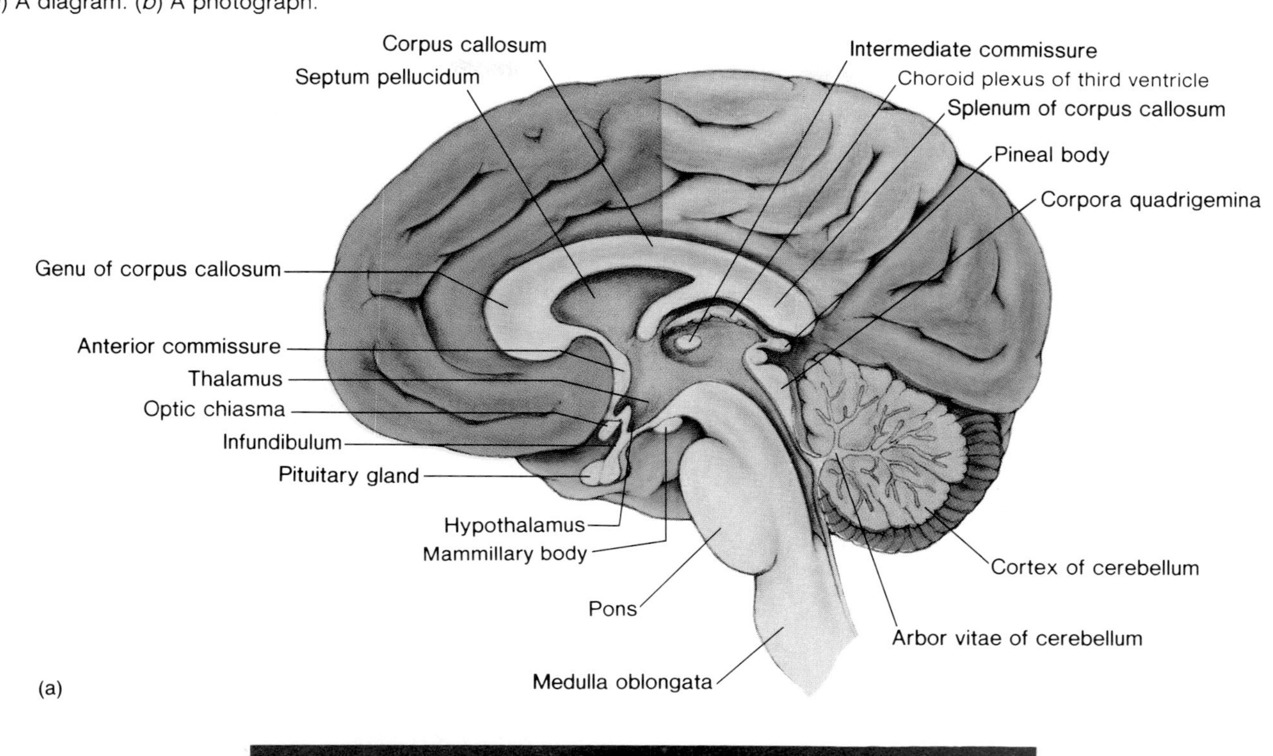

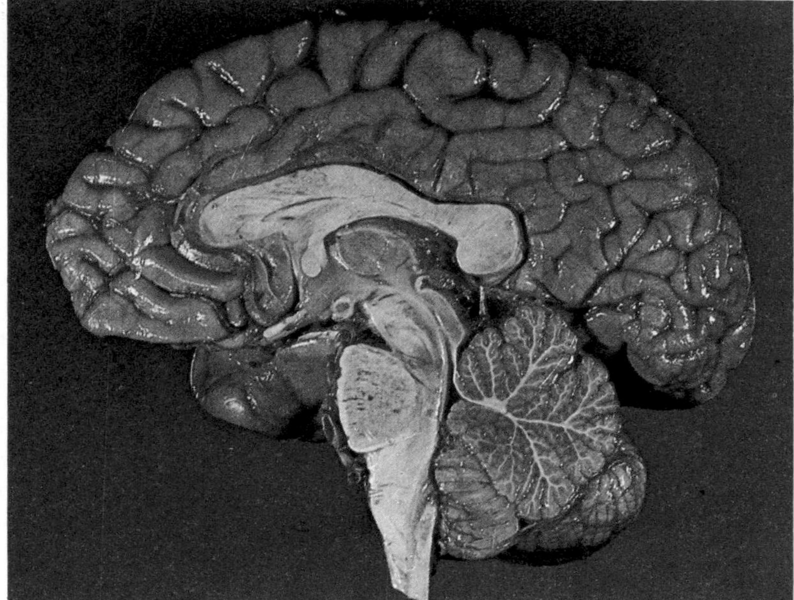

Diencephalon

The *diencephalon,* together with the cerebrum, which almost completely surrounds it, constitutes the part of the brain known as the **forebrain.** The most important structures within the diencephalon are the thalamus, hypothalamus, epithalamus, and neural part of the pituitary gland (neurohypophysis).

The **thalamus,** a mass of grey matter that comprises about four-fifths of the diencephalon, acts as a relay center through which all sensory information (except smell) passes on the way to the cerebrum (fig. 9.9). The **epithalamus** is the dorsal portion of the diencephalon that contains the *choroid plexus* over the third ventricle, where

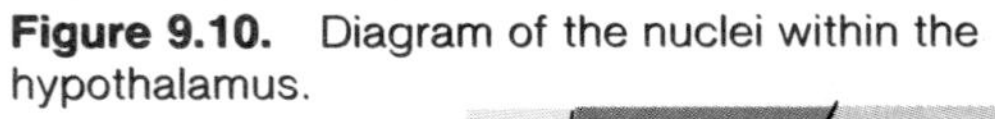

Figure 9.10. Diagram of the nuclei within the hypothalamus.

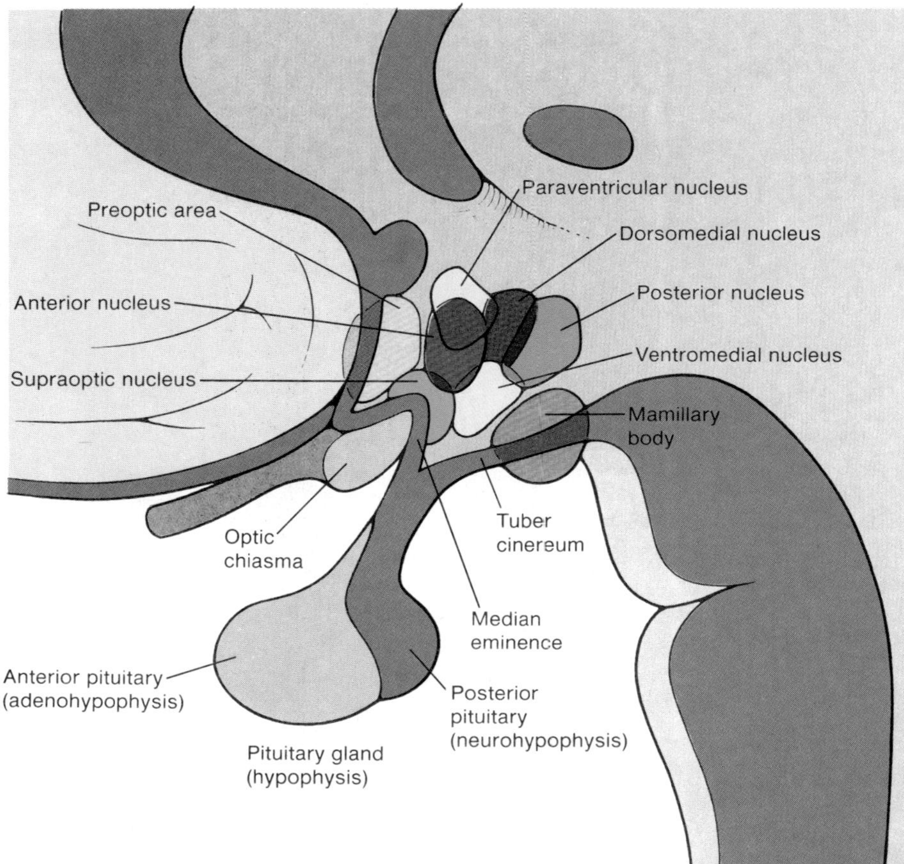

cerebrospinal fluid is formed, and the *pineal gland (epiphysis)*. The pineal gland secretes the hormone *melatonin,* which may play a role in the endocrine control of reproduction (discussed in chapter 20).

The **hypothalamus** is located below the thalamus, where it forms the floor and part of the lateral walls of the third ventricle (fig. 9.9). This extremely important brain region contains neural centers for hunger, thirst, and the regulation of body temperature and pituitary gland secretion. In addition, centers in the hypothalamus contribute to the regulation of sleep, wakefulness, sexual arousal and function, and emotions such as anger, fear, pain, and pleasure. Acting through its connections with the medulla oblongata of the brain stem, the hypothalamus helps to evoke the visceral responses to various emotional states. In its regulation of emotion, the hypothalamus works together with other neural centers collectively called the *limbic system,* which will be discussed in a later section.

Neurons within the *supraoptic* and *paraventricular nuclei* of the hypothalamus (fig. 9.10) produce two hormones—**antidiuretic hormone (ADH),** which is also known as *vasopressin,* and **oxytocin.** These two hormones are transported in axons of the *hypothalamo-hypophyseal tract* to the **neurohypophysis** (posterior pituitary), where they are stored and secreted in response to hypothalamic stimulation. Oxytocin stimulates contractions of the uterus during labor, and ADH stimulates the kidneys to reabsorb water and thus to excrete a smaller volume of urine. Neurons in the hypothalamus also produce hormones known as **releasing hormones** and **inhibiting hormones,** which are transported by the blood of the *hypothalmo-hypophyseal portal system* to the **adenohypophysis** (anterior pituitary). These hypothalamic releasing and inhibiting hormones regulate the secretions of the anterior pituitary and, by this means, regulate the secretions of other endocrine glands (as described in chapter 8).

Mesencephalon

The *mesencephalon,* or **midbrain,** is located between the diencephalon and the pons. The **corpora quadrigemina** are four rounded elevations on the dorsal portion of the midbrain (see fig. 9.9). The upper two of these, called the *superior colliculi,* are involved in visual reflexes. The posterior two, called the *inferior colliculi,* are relay centers for auditory information.

The mesencephalon also contains the cerebral peduncles, red nucleus, and the substantia nigra (not illustrated). The **cerebral peduncles** are a pair of structures composed of ascending and descending fiber tracts. The

Figure 9.11 Nuclei within the pons and medulla oblongata that constitute the respiratory center.

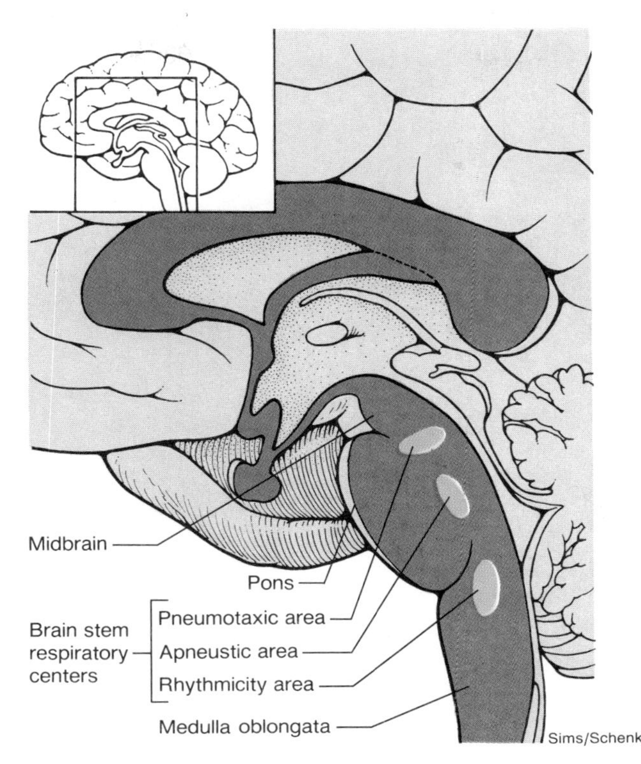

red nucleus is an area of grey matter deep in the midbrain that maintains connections with the cerebrum and the cerebellum and is involved in motor coordination. Another nucleus, the **substantia nigra,** is ventral to the red nucleus and is also involved in motor coordination through its synaptic connections with the basal ganglia, as previously discussed.

Metencephalon

The metencephalon is composed of the pons and the cerebellum. The **pons** can be seen as a rounded bulge on the underside of the brain, between the midbrain and the medulla oblongata (fig. 9.11). Surface fibers in the pons connect to the cerebellum, and deeper fibers are part of motor and sensory tracts that pass from the medulla oblongata, through the pons, and on to the midbrain. Within the pons are several nuclei associated with specific cranial nerves—the trigeminal (V), abducens (VI), facial (VII), and vestibulocochlear (VIII). Other nuclei of the pons cooperate with nuclei in the medulla oblongata to regulate breathing. The two respiratory control centers in the pons are known as the *apneustic* and the *pneumotaxic centers.*

The **cerebellum,** which occupies the inferior and posterior aspect of the cranial cavity, is the second largest structure of the brain, and contains outer grey and inner white matter as does the cerebrum. Fibers from the cerebellum pass through the red nucleus to the thalamus and then to the motor areas of the cerebral cortex. Other fiber tracts connect the cerebellum with the pons, medulla oblongata, and spinal cord. The cerebellum receives input from *proprioceptors* (joint, tendon, and muscle receptors) and, working together with the basal ganglia and motor areas of the cerebral cortex, participates in the coordination of movement.

Damage to the cerebellum produces **ataxia,** which is lack of coordination due to errors in the speed, force, and direction of movement. The movements and speech of those afflicted by ataxia may resemble those of a drunken person. This condition is also characterized by *intention tremor,* which differs from the resting tremor of Parkinson's disease in that it only occurs when intentional movements are made. The person suffering cerebellar damage may reach for an object and miss it, by placing the hand too far to the left or right, and then attempt to compensate by moving the hand in the opposite direction; this can continue and produce oscillations of the limb.

Myelencephalon

The only structure within the myelencephalon is the **medulla oblongata,** often simply called the *medulla.* The medulla is about 3 cm long and is continuous with the pons anteriorly and the spinal cord inferiorly. All the descending and ascending fiber tracts that provide communication between the spinal cord and the brain must

Figure 9.12. The reticular activating system (RAS). The arrows indicate the direction of impulses along nerve pathways that connect with the RAS.

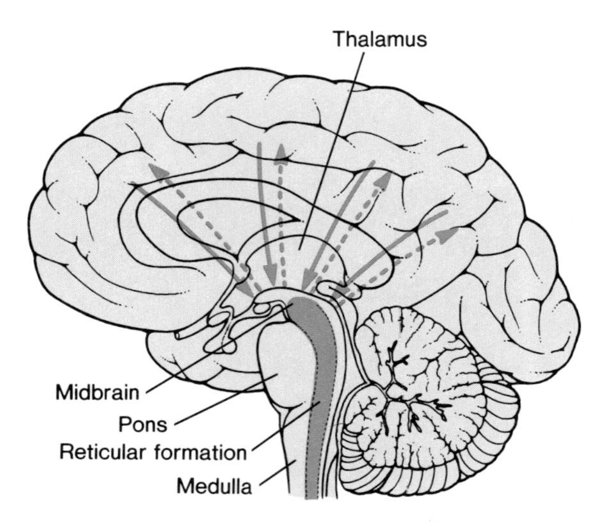

pass through the medulla. These fiber tracts cross to the contralateral side in elevated, triangular structures in the medulla called the **pyramids,** so that the left side of the brain receives sensory information from the right side of the body and vice versa. Similarly, because of crossing, or *decussation,* of fibers, the right side of the brain controls motor activity in the left side of the body and vice versa.

There are several important nuclei within the medulla. The *nucleus ambiguus* and *hypoglossal nucleus* give rise to several cranial nerves (VIII, IX, XI, and XII). The *vagus nuclei* (there is one on each lateral side of the medulla) give rise to the vagus (X) nerves. The *nucleus gracilis* and *nucleus cuneatus* relay sensory information to the thalamus and then to the cerebral cortex. The *inferior olivary nuclei* and the *accessory olivary nuclei* are along the synaptic pathway from the forebrain and midbrain to the cerebellum.

The medulla contains groupings of neurons known as the *vital centers,* because these neurons are required for the regulation of breathing and of cardiovascular responses. The **vasomotor center** controls the autonomic innervation of blood vessels; the **cardioinhibitory center** controls the parasympathetic innervation (via the vagus nerve) of the heart (there does not appear to be a separate cardioaccelerator center); and the **respiratory center** of the medulla acts together with centers in the pons to control breathing.

Reticular Formation. The reticular formation is a complex network of nuclei and nerve fibers within the medulla, pons, midbrain, thalamus, and hypothalamus that functions as the **reticular activating system,** or **RAS** (fig. 9.12). Because of its many interconnections, the RAS is activated in a nonspecific fashion by any modality of sensory information. Nerve fibers from the RAS, in turn, project diffusely to the cerebral cortex; this results in *nonspecific arousal* of the cerebral cortex to incoming sensory information.

The RAS, through its nonspecific arousal of the cortex, helps to maintain a state of alert consciousness. Not surprisingly, there is evidence that general anesthetics may produce unconsciousness by depressing the RAS. Similarly, the ability to fall asleep may be due to the action of specific neurotransmitters that inhibit activity of the RAS.

1. *List the lobes of the cerebrum, and describe their primary functions.*
2. *Identify the location of the sensory and motor cortex, and explain how these areas are organized.*
3. *Identify the location and composition of the basal ganglia, and describe their functions.*
4. *Identify the location and describe the functions of the hypothalamus.*
5. *Identify the structures of the mesencephalon, and describe their functions.*
6. *Describe the functions of the medulla oblongata and pons.*
7. *Describe the composition and functions of the reticular activating system.*

Spinal Cord Tracts

The spinal cord extends from the level of the foramen magnum of the skull to the first lumbar vertebra. The central grey matter of the spinal cord is arranged in the form of an "H," with two *dorsal horns* and two *ventral horns* (also called posterior and anterior horns, respectively).

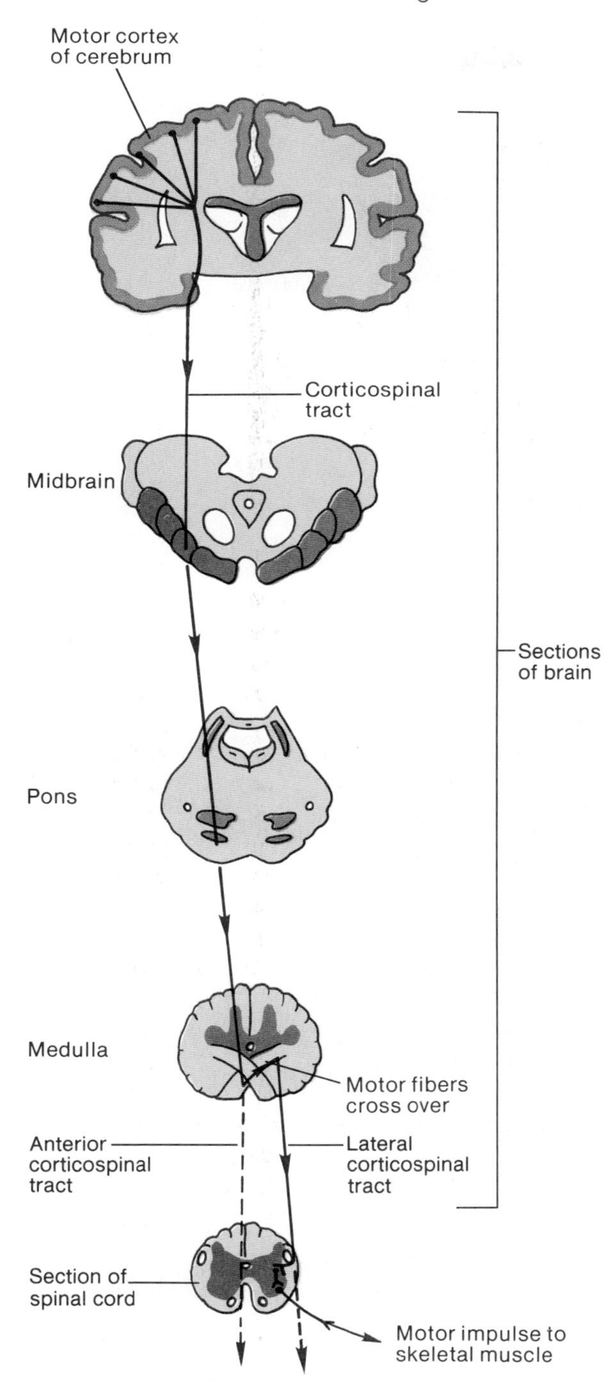

Figure 9.13. Descending tracts composed of motor fibers that cross over within the medulla oblongata.

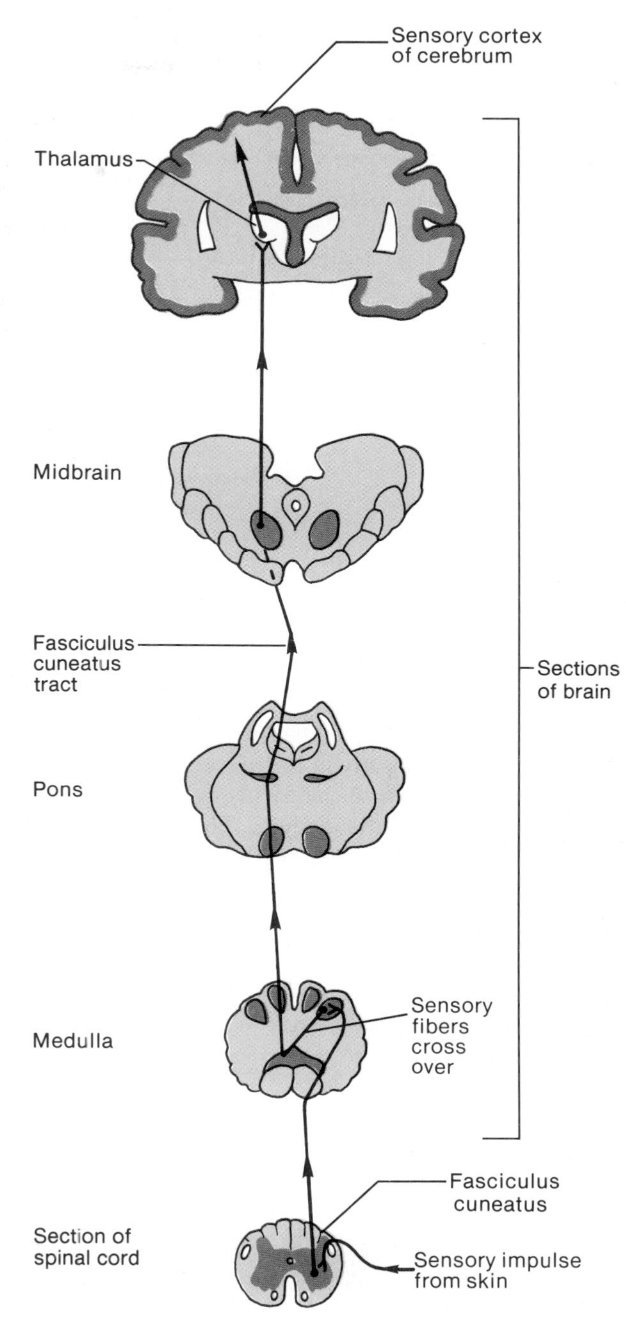

Figure 9.14. Ascending tracts composed of sensory fibers that cross over within the medulla.

The white matter of the spinal cord is composed of ascending and descending fiber tracts. The spinal cord has six columns of white matter called **funiculi,** which are named according to their relative position. These include two *posterior (dorsal) funiculi,* two *anterior (ventral) funiculi,* and two *lateral funiculi.* Each funiculus consists of both ascending and descending tracts. The pathway of information flow within these tracts may cross over, or decussate, in the medulla (as shown in figures 9.13 and 9.14), or the crossing over may occur in the spinal cord. The principal ascending and descending tracts are illustrated in figure 9.15.

Figure 9.15. A cross section showing the principal ascending and descending tracts within the spinal cord.

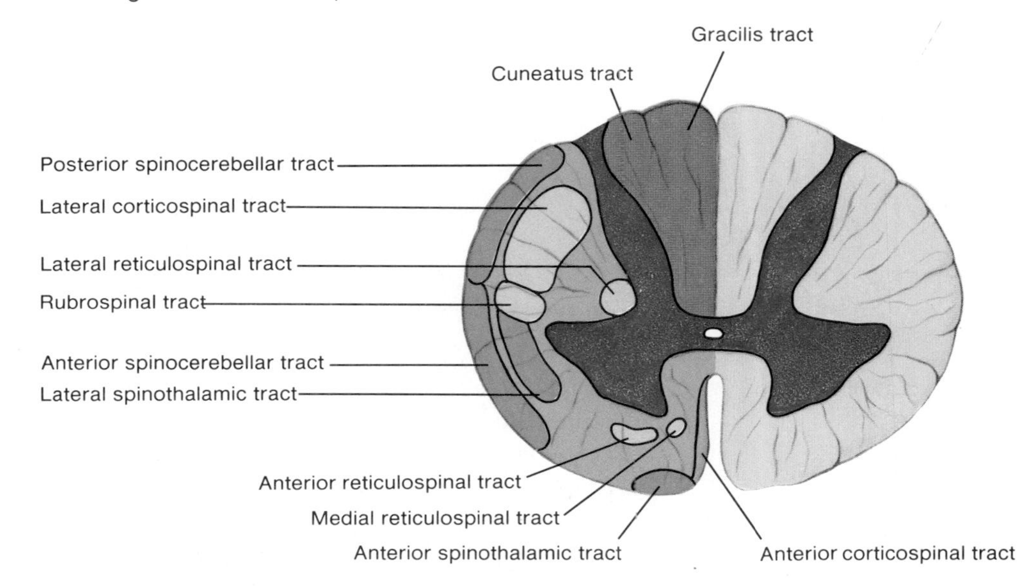

Descending tracts are grouped as either pyramidal or extrapyramidal. **Corticospinal (pyramidal) tracts** descend directly, without synaptic interruption, from the cerebral cortex to the lower motor neurons of the spinal cord. The cell bodies of the neurons that contribute fibers to these tracts are located primarily in the precentral gyrus of the frontal lobe. About 85% of the corticospinal fibers decussate in the pyramids of the medulla. The fibers that cross comprise the **lateral corticospinal tracts** and the remaining uncrossed fibers comprise the **anterior corticospinal tracts.** Because of the decussation of fibers, the right hemisphere primarily controls the musculature on the left side of the body, whereas the left hemisphere controls the musculature on the right side.

The corticospinal tracts appear to be particularly important in voluntary movements that require complex interactions between sensory input and the motor cortex. Speech, for example, is impaired when the corticospinal tracts are damaged in the thoracic region of the spinal cord, whereas involuntary breathing continues. Damage to the pyramidal motor system can be detected clinically by a positive **Babinski reflex,** in which stimulation of the sole of the foot causes extension (upward movement) of the toes. The positive Babinski reflex is normal in infants because neural control is not yet fully developed.

The remaining descending tracts are **extrapyramidal motor tracts,** which originate in the midbrain and brain stem regions (fig. 9.16). Electrical stimulation of the cerebral cortex, cerebellum, and basal ganglia indirectly evokes movement because of synaptic connections within the extrapyramidal system.

The *reticulospinal tracts* are the major descending pathways of the extrapyramidal system. These tracts originate in the reticular formation of the brain stem, which receives either stimulatory or inhibitory input from the cerebrum and the cerebellum. There are no descending tracts from the cerebellum; the cerebellum can only influence motor activity indirectly by way of the vestibular nuclei, red nucleus, and basal ganglia. These structures, in turn, affect lower motor neurons via the *vestibulospinal tracts, rubrospinal tracts,* and reticulospinal tracts. Pathways involving the higher motor neuron control of skeletal movements are illustrated in figure 9.17, and are discussed together with muscle physiology in chapter 11.

The ascending fiber tracts that convey sensory information from cutaneous receptors, proprioceptors, and visceral receptors to the brain are indicated in table 9.1. The physiology of the sense organs and of sensory perception is discussed in more detail in chapter 10.

1. *Explain why each cerebral hemisphere receives sensory input from and directs motor output to the contralateral side of the body.*
2. *List the tracts of the pyramidal motor system, and explain the function of the pyramidal system.*
3. *List the tracts of the extrapyramidal system and explain how this system differs from the pyramidal motor system.*

Figure 9.16. Areas of the brain containing neurons involved in the control of skeletal muscles (higher motor neurons). The thalamus is a relay center between the motor cortex and other brain areas.

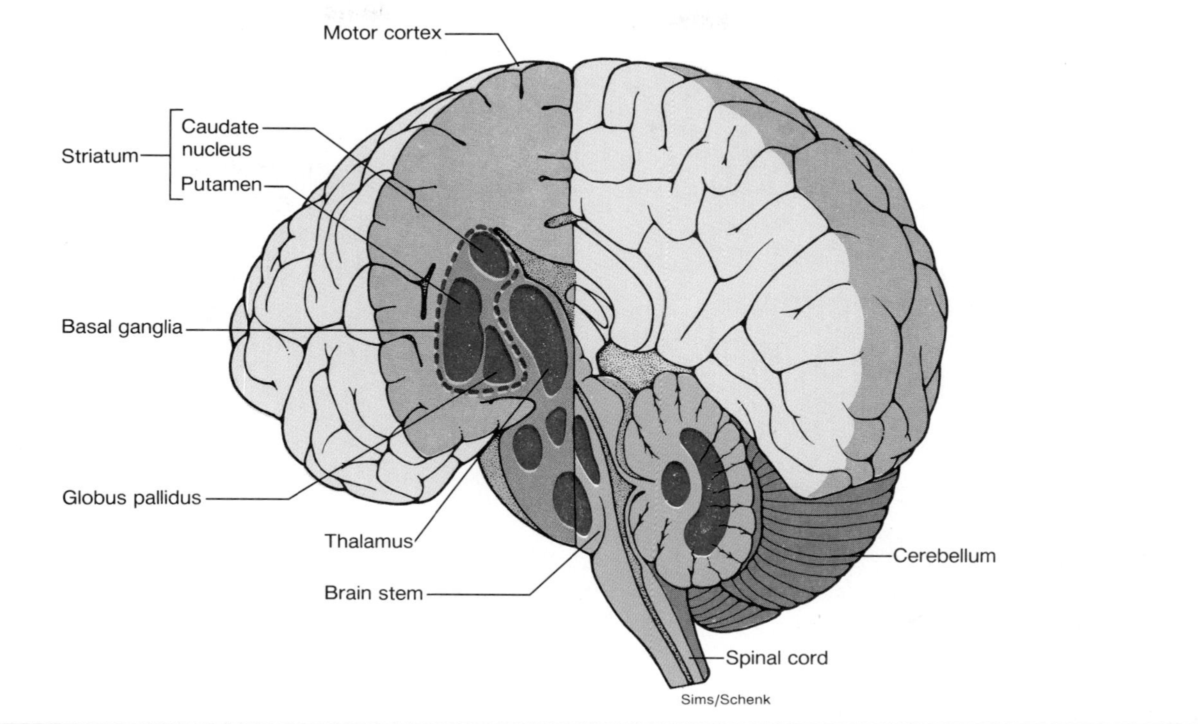

Figure 9.17. Pathways involved in the higher motor neuron control of skeletal muscles.

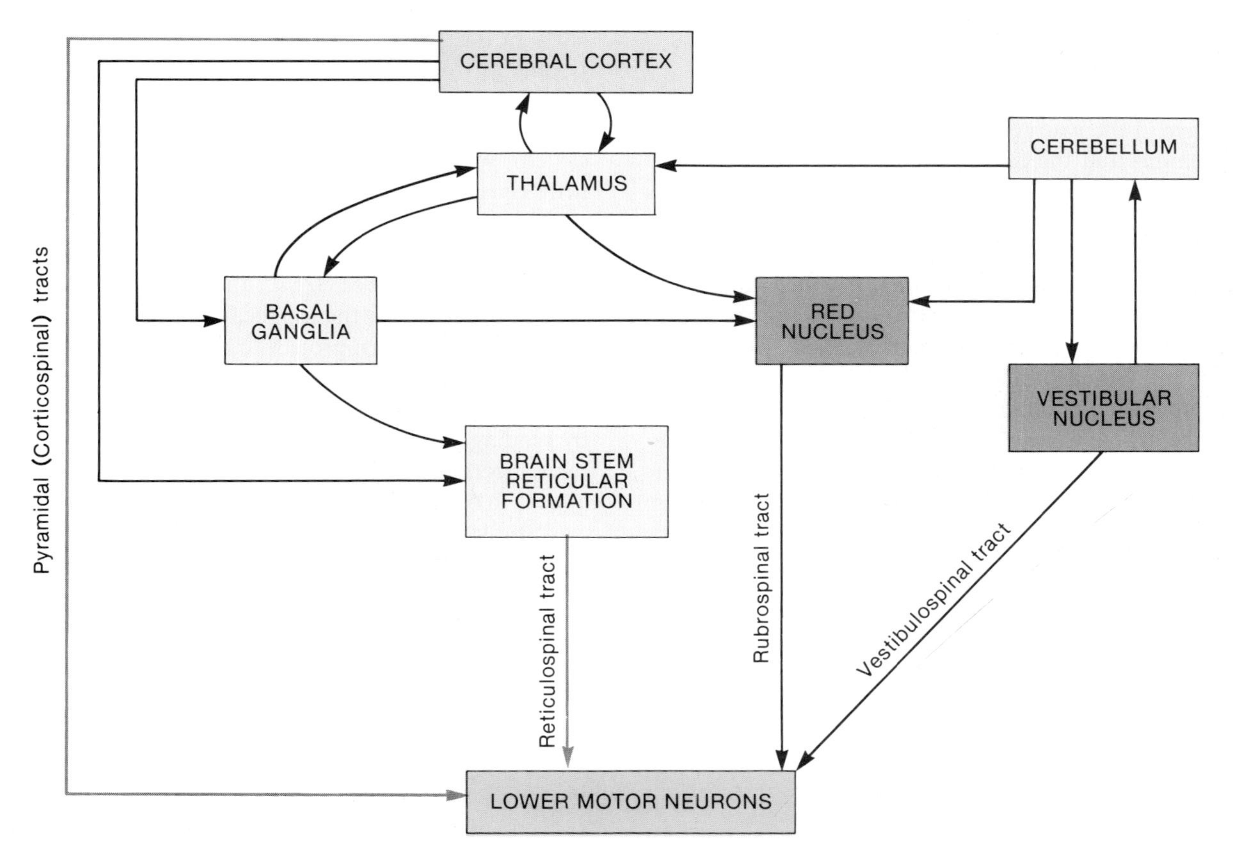

Table 9.1 Principal ascending tracts of spinal cord.

Tract	Funiculus	Origin	Termination	Function
Anterior spinothalamic	Anterior	Posterior horn on one side of cord but crosses to opposite side	Thalamus, then cerebral cortex	Conducts sensory impulses for crude touch and pressure
Lateral spinothalamic	Lateral	Posterior horn on one side of cord but crosses to opposite side	Thalamus, then cerebral cortex	Conducts pain and temperature impulses that are interpreted within cerebral cortex
Fasciculus gracilis and fasciculus cuneatus	Posterior	Peripheral afferent neurons; does not cross over	Nucleus gracilis and nucleus cuneatus of medulla; eventually thalamus, then cerebral cortex	Conducts sensory impulses from skin, muscles, tendons, and joints, which are interpreted as sensations of fine touch, precise pressures, and body movements
Posterior spinocerebellar	Lateral	Posterior horn; does not cross over	Cerebellum	Conducts sensory impulses from one side of body to same side of cerebellum for subconscious proprioception necessary for coordinated muscular contractions
Anterior spinocerebellar	Lateral	Posterior horn; some fibers cross, others do not	Cerebellum	Conducts sensory impulses from both sides of body to cerebellum for subconscious proprioception necessary for coordinated muscular contractions

Source: Van De Graaff, Kent M., *Human Anatomy.*

Emotion and Motivation

The parts of the brain that appear to be of paramount importance in the neural basis of emotional states are the hypothalamus and the **limbic system.** The limbic system is a group of nuclei and fiber tracts that form a ring around the brain stem. The structures of the limbic system include the *cingulate gyrus* (part of the cerebral cortex), *amygdaloid nucleus* (or *amygdala*), *hippocampus,* and the *septal nuclei* (fig. 9.18).

The limbic system was formerly called the *rhinencephalon,* or "smell brain," because it is involved in the central processing of olfactory information. This function may be the primary one of lower vertebrates, in which the limbic system may constitute the entire forebrain, but it is now known that the limbic system of humans is a center for basic emotional drives. The limbic system was derived early in the course of vertebrate evolution, and its tissue is thus phylogenetically older than the *neocortex* of the cerebral hemispheres. There are few synaptic connections between the neocortex and the structures of the limbic system, which perhaps helps to explain why we have so little conscious control over our emotions.

There is a closed circuit of information flow between the limbic system and the thalamus and hypothalamus (fig. 9.18), called the *Papez circuit.* In this circuit, a fiber tract, the *fornix,* connects the hippocampus to the mammillary bodies, which in turn project to the anterior nuclei of the thalamus. The nuclei of the thalamus, in turn, send fibers to the cingulate gyrus, which then completes the circuit by sending fibers to the hippocampus. Through these interconnections, the limbic system and the hypothalamus appear to cooperate in the neural basis of emotional states.

Studies of the functions of these regions include electrical stimulation of specific locations, destruction of tissue (producing *lesions*) in particular sites, and surgical removal, or *ablation,* of specific structures. These studies suggest that the hypothalamus and limbic system are involved in the following processes:

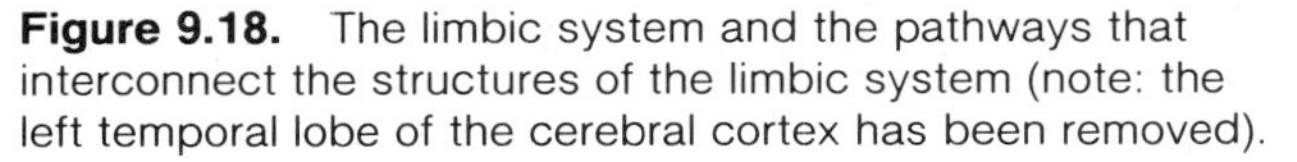

Figure 9.18. The limbic system and the pathways that interconnect the structures of the limbic system (note: the left temporal lobe of the cerebral cortex has been removed).

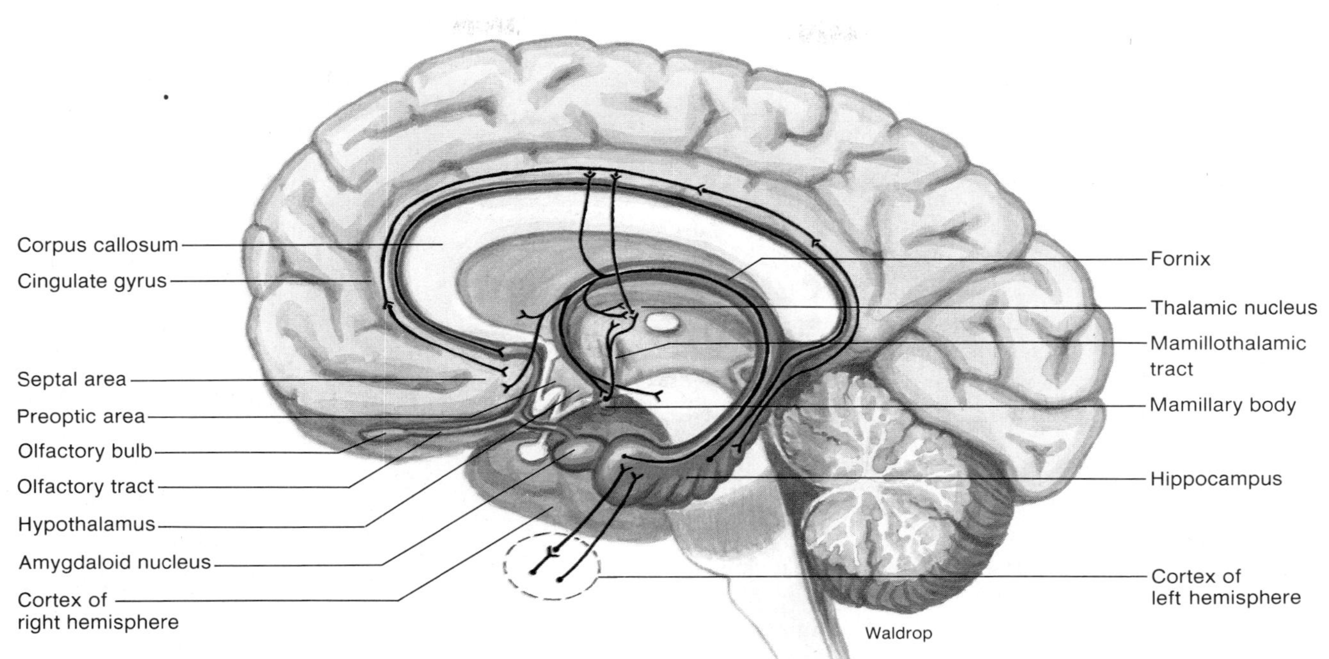

1. **Aggression.** Stimulation of certain areas of the amygdala produce rage and aggression, and lesions of the amygdala can produce docility in experimental animals. Stimulation of particular areas of the hypothalamus can produce similar effects.
2. **Fear.** Fear can be produced by electrical stimulation of the amygdala and hypothalamus, and surgical removal of the limbic system can produce an absence of fear. Monkeys are normally terrified of snakes, for example, but if they have had their limbic system removed they will handle snakes without fear.
3. **Feeding.** The hypothalamus contains both a *feeding center* and a *satiety center.* Electrical stimulation of the former produces overeating, and stimulation of the latter will stop feeding behavior in experimental animals.
4. **Sex.** The hypothalamus and limbic system are involved in the regulation of the sexual drive and sexual behavior, as shown by stimulation and ablation studies in experimental animals. The cerebral cortex, however, is also critically important for the sex drive in lower animals, and the role of the cerebrum is believed to be even more important for the sex drive in humans.
5. **Reward and punishment system.** Electrodes placed in particular sites from the frontal cortex to the hypothalamus can deliver shocks that function as a reward. In rats, this reward is more powerful than food or sex in motivating behavior. Similar studies have been done in some humans, who report a feeling of relaxation and relief from tension, but not ecstasy. Electrodes placed in slightly different positions apparently stimulate a punishment system in experimental animals, who stop their behavior when stimulated in these regions.

---

1. *List the structures of the limbic system, and explain the structural and functional relationships between the limbic system, hypothalamus, and cerebral cortex.*
2. *Describe the techniques used to study the role of the limbic system and hypothalamus in emotions and motivation, and explain some of the results of these studies.*

Higher Brain Functions

The term "higher brain functions" in this section refers to brain functions that we consider to be most developed in humans: language, the ability to learn and remember, and self-awareness or consciousness. At one time these abilities were thought to be the unique domain of the cerebral cortex, which is most developed in humans and also well-developed in gorillas, chimpanzees, and dolphins, which are thought to have some humanlike abilities. More recent research has shown, however, that learning and memory involve subcortical structures as well as the cerebral cortex and that animals which lack a cerebral cortex may have some form of consciousness. Perhaps the most consistent characteristic of higher brain functions is that they are extremely complex; the complexity is so great, in fact, that higher brain function is the area of physiology that is least well understood.

Cerebral Lateralization

The two hemispheres of the cerebral cortex control, via motor fibers originating in the precentral gyrus, skilled movements of the contralateral side of the body. At the same time, somatesthetic sensation from the right side of the body projects to the left postcentral gyrus, and vice versa, because of decussation of fibers. In a similar manner, images falling in the left half of each retina project to the right occipital lobe, and images in the right half of each retina project to the left occipital lobe. Each cerebral hemisphere, however, receives information from both sides of the body because the two hemispheres communicate with each other via the *corpus callosum,* a large tract composed of about 200 million fibers.

The corpus callosum has been surgically cut in some victims of severe epilepsy as a way of alleviating their symptoms. These **split-brain** procedures isolate each hemisphere from the other, but to a casual observer, surprisingly, these split-brain patients do not show evidence of any disability as a result of the surgery. However, in specially designed experiments in which each hemisphere is separately presented with sensory images and asked to perform tasks (speech or writing or drawing with the contralateral hand), it has been learned that each hemisphere is good at certain categories of tasks and poor at others (fig. 9.19).

In a typical experiment, the image of an object may be presented to either the right or left hemisphere (by presenting it to either the left or right visual field only) and the person may be asked to name the object. It was found that, in most people, the task could be performed successfully by the left hemisphere but not by the right. Similar experiments have shown that the left hemisphere is generally the one in which most of the language and analytical abilities reside. These results lead to the concept

Figure 9.19. Different functions of the right and left cerebral hemispheres, as revealed by experiments with people who have had the tract connecting the two hemispheres (the corpus callosum) surgically split.

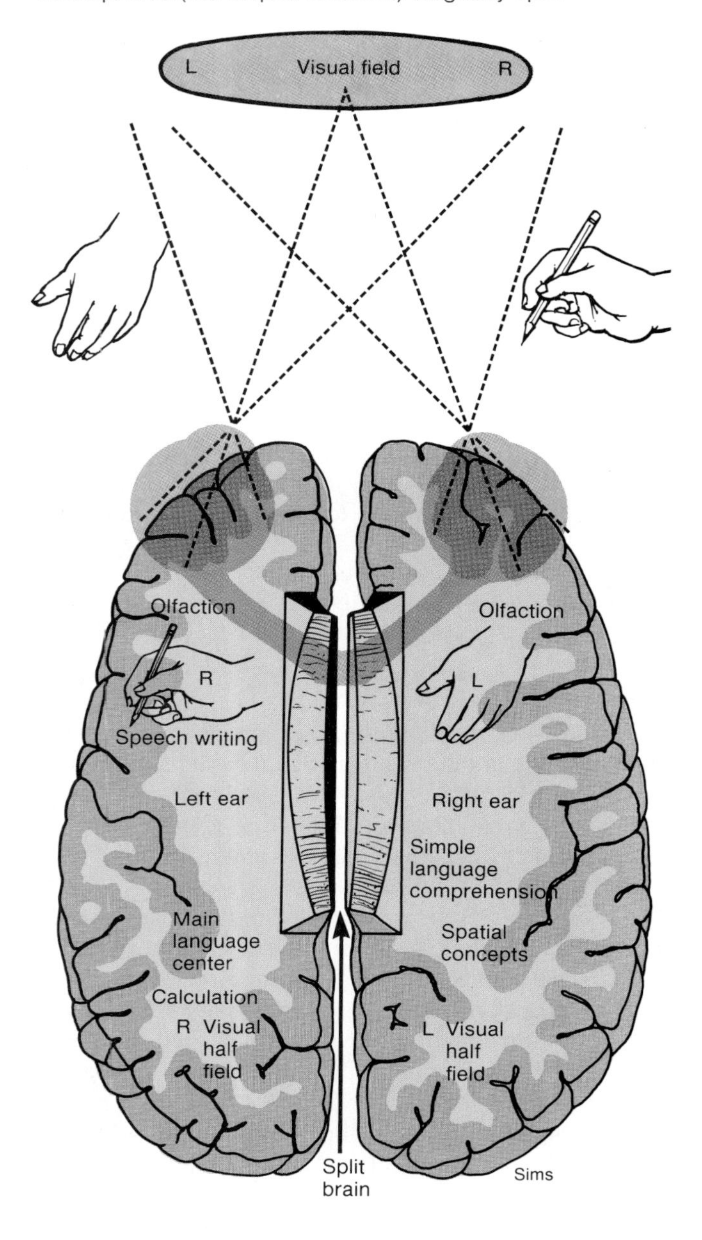

of **cerebral dominance,** which is analogous to the concept of handedness—people generally have better motor competence with one hand than the other. Since most people are right-handed, which is also controlled by the left hemisphere, the left hemisphere was naturally considered to be the dominant hemisphere in most people. Further experiments have shown, however, that the right hemisphere is specialized along different, less obvious lines—rather than one hemisphere being dominant and the other nondominant, the two hemispheres appear to have complementary functions. The term **cerebral lateralization,** or

Figure 9.20. Brain areas involved in the control of speech.

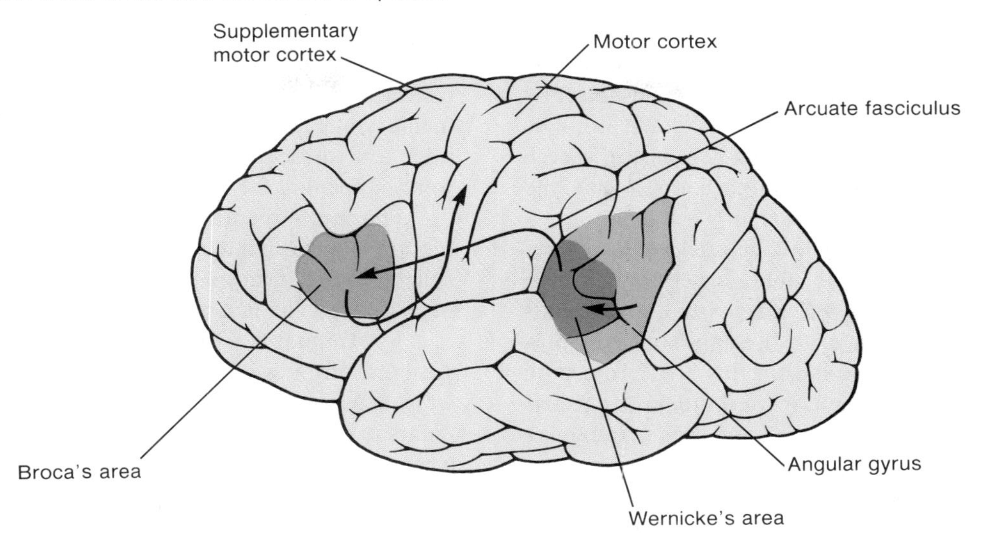

specialization of function to one hemisphere or the other, is thus now preferred to the term *cerebral dominance,* although both terms are currently used.

Experiments have shown that the right hemisphere does have limited verbal ability; more noteworthy is the observation that the right hemisphere is most adept at *visuospatial tasks.* The right hemisphere, for example, can recognize faces better than the left, but it cannot describe facial appearances as well as the left. Acting through its control of the left hand, the right hemisphere is better than the left (controlling the right hand) at arranging blocks or drawing cubes. Patients with damage to the right hemisphere, as might be predicted from the results of split-brain research, have difficulty finding their way around a house and reading maps.

Perhaps as a result of the role of the right hemisphere in the comprehension of patterns and part-whole relationships, the ability to compose music, but not to critically understand it, appears to depend on the right hemisphere. Interestingly, damage to the left hemisphere may cause severe speech problems while leaving the ability to sing unaffected.

The lateralization of functions just described—with the left hemisphere specialized for language and analytical ability, and the right hemisphere specialized for visuospatial ability—is true for 97% of all people. It is true for all right-handers (who comprise 90% of all people) and for 70% of all left-handers. The remaining left-handers are split roughly equally into those who have language-analytical ability in the right hemisphere and those in whom this ability is present in both hemispheres.

It is interesting to speculate that the creative ability of a person may be related to the interaction of information between the right and left hemispheres. This interaction may be greater in left-handed people; a study found that the number of left-handers among college art students was disproportionately higher than in the general population. The observation that Leonardo Da Vinci and Michelangelo were left-handed is interesting in this regard, but clearly is not scientific proof of any hypothesis. Further research on the lateralization of function of the cerebral hemispheres may reveal much more about both brain function and the creative process.

Language

Knowledge of the brain regions involved in language has been gained primarily by the study of *aphasias*—speech and language disorders caused by damage to the brain. The language areas of the brain are primarily located in the left hemisphere of the cerebral cortex in most people, as previously described. As long ago as the nineteenth century, two areas of the cortex—Broca's area and Wernicke's area (fig. 9.20)—were found to be of particular importance in the production of aphasias.

Damage to **Broca's area,** located in the left inferior frontal gyrus, produces an aphasia in which the person is reluctant to speak, and when speech is attempted, it is slow and poorly articulated. People with *Broca's aphasia,* however, have unimpaired comprehension of speech. It should be noted that this speech difficulty is not simply due to a problem in motor control, since the neural control over the musculature of the tongue, lips, larynx, and so on is unaffected. Damage to **Wernicke's area,** located in the superior temporal gyrus, results in speech that is rapid and fluid but does not convey any information. People with *Wernicke's aphasia* produce speech that has been described as a "word salad." The words used may be real words that are chaotically mixed together, or they may be made-up words. Language comprehension has been destroyed; people with Wernicke's aphasia cannot understand either spoken or written language.

It appears that the concept of words to be spoken originates in Wernicke's area and is communicated to Broca's area in a fiber tract called the **arcuate fasciculus.** Broca's area, in turn, sends fibers to the motor cortex (precentral gyrus), which directly controls the musculature of speech. Damage to the arcuate fasciculus produces *conduction aphasia,* which is fluent but nonsensical speech as in Wernicke's aphasia, even though both Broca's and Wernicke's areas are intact.

The **angular gyrus,** located at the junction of the parietal, temporal, and occipital lobes, is believed to be a center for the integration of auditory, visual, and somatesthetic information. Damage to the angular gyrus produces aphasias, which suggests that this area projects to Wernicke's area. Some patients with damage to the left angular gyrus can speak and understand spoken language but cannot read or write. Other patients can write a sentence but cannot read it, presumably due to damage to the projections from the occipital lobe (involved in vision) to the angular gyrus.

Recovery of language ability, by transfer to the right hemisphere after damage to the left hemisphere, is very good in children but decreases after adolescence. Recovery is reported to be faster in left-handed people, possibly because language ability is more evenly divided between the two hemispheres in left-handed people. Some recovery usually occurs after damage to Broca's area, but damage to Wernicke's area produces more severe and permanent aphasias.

Memory

Clinical studies of amnesia suggest that several different brain regions are involved in memory storage and retrieval. Amnesia has been found to result from damage to the temporal lobe of the cerebral cortex, hippocampus, head of the caudate (in Huntington's disease), or the dorsomedial thalamus (in alcoholics suffering from Korsakoff's syndrome with thiamine deficiency). Clinical studies also suggest that there are two major categories of memory: **short-term memory** and **long-term memory.** People with head trauma, for example, and patients with suicidal depression who are treated by *electroconvulsive shock (ECS)* therapy, may lose their memory of recent events but retain their older memories.

The **hippocampus** appears to be required for short-term memory and for the consolidation of that memory into a long-term form. Surgical removal of the left hippocampus impairs the consolidation of short-term verbal memories, and removal of the right hippocampus impairs the consolidation of nonverbal memories. Surgical removal of both the right and left hippocampus was performed in one patient, designated "H.M.," in an effort to treat his epilepsy. After the surgery he was unable to consolidate any short-term memory. He could repeat a phone number and carry out a normal conversation; he could not remember the phone number if momentarily distracted, however, and if the person to whom he was talking left the room and came back a few minutes later, H.M. would have no recollection of seeing that person or of having had a conversation with that person before. Although his memory of events that occurred before the operation was intact, all subsequent events in his life seemed as if they were happening for the first time.

The cerebral cortex is thought to store factual information, with verbal memories lateralized to the left hemisphere and visuospatial information in the right hemisphere. The neurosurgeon Wilder Penfield has electrically stimulated various regions in the brain of awake patients, often evoking visual or auditory memories that were extremely vivid. Electrical stimulation of specific points in the **temporal lobe** evoked specific memories that were so detailed the patient felt that he was reliving the experience. Surgical removal of these regions did not, however, abolish the memory. The amount of memory destroyed by ablation of brain tissue appears to depend more on the amount of brain tissue removed than on the location of the surgery. On the basis of these observations, it appears that the memory may be diffusely located in the brain; stimulation of the correct location of the cortex then retrieves the memory.

Since long-term memory is not abolished by electroconvulsive shock, it seems reasonable to conclude that the consolidation of memory depends on relatively permanent changes in the chemical structure of neurons and their synapses. Experiments suggest that protein synthesis is required for the consolidation of the "memory trace." According to one theory, these proteins may be secreted into the extracellular environment, where they influence synaptic connections. According to another theory, new receptor proteins in the membrane of the postsynaptic neuron are made available as a result of high-frequency stimulation of the presynaptic neuron. This would help to account for the increased sensitivity of postsynaptic neurons to neurotransmitter, as seen in post-tetanic potentiation (discussed in chapter 7). Much more research is obviously needed in this exciting area of physiology before memory can be fully explained at a cellular and molecular level.

---

1. *Define the term* cerebral lateralization, *and explain why this term is preferable to the term* cerebral dominance.
2. *Explain the differences in function of the right and left cerebral hemispheres in right-handed and left-handed people.*
3. *Describe the aphasias produced by damage to Broca's and Wernicke's areas, by damage to the arcuate fasciculus, and by damage to the angular gyrus. Explain how these areas may interact in the production of speech.*
4. *Explain the difference between short-term and long-term memory, and describe the possible roles of different brain regions in memory.*

Summary

Organization of the Central Nervous System p. 214

I. The CNS consists of the brain and spinal cord.
  A. Sensory information arrives at the CNS in sensory neurons of cranial and spinal nerves; fibers of sensory neurons form the dorsal roots of spinal nerves.
  B. The CNS controls effector organs through motor neurons; fibers of motor neurons form the ventral roots of spinal nerves.

II. The cerebrum consists of an outer cortex of grey matter over underlying white matter; it is composed of two cerebral hemispheres that communicate through the corpus callosum.
  A. The cerebral cortex is divided into lobes.
  B. Convolutions in the cortex are called gyri, and the grooves are called sulci.
    1. The precentral gyrus of the frontal lobe is the motor cortex, involved in the control of muscular movements.
    2. The postcentral gyrus of the parietal lobe is the sensory cortex, involved in somatesthetic sensation.
    3. The temporal lobe processes auditory information, and the occipital lobe is involved in vision.
  C. The electroencephalogram records electrical activity of the cerebral cortex.
  D. Nuclei of grey matter located deep within the cerebrum are known as the basal ganglia.
    1. The most prominent of the basal ganglia is the corpus striatum, which consists of the caudate nucleus and the lentiform nucleus.
    2. The basal ganglia are involved in the control of skeletal movements; chorea and Parkinson's diseases result from abnormalities of the basal ganglia.

III. The diencephalon includes the epithalamus, thalamus, hypothalamus, and neural portion of the pituitary gland.
  A. The thalamus acts as a relay center for sensory information.
  B. The hypothalamus controls the activity of the pituitary gland, contains neural centers for hunger, thirst, and regulation of body temperature, and is involved in certain emotional states.
  C. The neurohypophysis (posterior pituitary) secretes ADH and oxytocin, which are produced by neurons in the hypothalamus.

IV. The mesencephalon, or midbrain, contains the corpora quadrigemina, cerebral peduncles, red nucleus, and substantia nigra.
  A. The corpora quadrigemina include the superior colliculi, involved in visual reflexes, and the inferior colliculi, which are auditory relay centers.
  B. The cerebral peduncles are composed of fiber tracts, and the red nucleus and substantia nigra are nuclei involved in motor control.

V. The metencephalon is composed of the pons and cerebellum.
  A. The pons contains nuclei that give rise to some cranial nerves; it also contains two respiratory control centers.
  B. The cerebellum is an important structure involved in the coordination of motor activity.

VI. The myelencephalon consists of the medulla oblongata.
  A. All the ascending and descending fiber tracts pass through the medulla and cross to the contralateral side in the pyramids.
  B. The medulla contains nuclei that give rise to some cranial nerves; it also contains centers involved in regulation of the cardiovascular and respiratory systems.
  C. The reticular formation is an interconnecting network of neurons in the medulla, pons, thalamus, and hypothalamus.
    1. This network functions as the reticular activating system (RAS).
    2. The RAS acts to arouse the cerebral cortex to incoming sensory information.

Spinal Cord Tracts p. 224

I. The white matter of the spinal cord is arranged into columns called funiculi, which contain ascending and descending tracts.
  A. Descending tracts are grouped as either corticospinal (pyramidal) or extrapyramidal.
  B. The corticospinal tracts descend without synaptic interruption from the motor cortex of the cerebrum to the spinal cord.
    1. Most of these fibers cross to the contralateral side, so that the left cerebral hemisphere controls motor activity in the right side of the body, and vice versa.
    2. The corticospinal tracts are involved in voluntary, well-coordinated movements.
  C. The extrapyramidal tracts originate in the brain stem and midbrain.
    1. The cerebellum and basal ganglia influence motor activity indirectly through synaptic connection with the nuclei that give rise to the extrapyramidal tracts.
    2. The major extrapyramidal tract originates in the reticular formation as the reticulospinal tract.

Emotion and Motivation p. 228

I. The hypothalamus and the limbic system appear to be particularly involved in emotions.
  A. The limbic system is a phylogenetically old region of the forebrain that includes the cingulate gyrus, amygdala, hippocampus, and septal nuclei.
  B. Neural circuits connect the limbic system to the hypothalamus, which in turn helps regulate the autonomic nervous system, thus providing visceral responses to emotional states.

II. Stimulation of specific regions of the hypothalamus or of the limbic system or the ablation of specific regions shows that a variety of basic emotions and motivational drives are regulated (at least in part) by these structures.
  A. Rage and aggression can be produced by stimulation of particular areas of the hypothalamus or of the amygdala; fear can be abolished by removal of the limbic system.
  B. There are separate feeding and satiety centers in the hypothalamus governing feelings of hunger and satiety.

C. The sexual drive is complex and involves higher brain areas in the cerebrum, but the limbic system also appears to be important in sexual behavior.
D. Electrical stimulation at specific points in the limbic system and hypothalamus serves as a reward in itself; these observations suggest that there are reward and punishment centers in these regions.

Higher Brain Functions p. 230

I. The two cerebral hemispheres display a degree of specialization of function, which is termed *cerebral lateralization*.
   A. In 97% of all people, the left hemisphere is dominant in terms of its role in language and analytical ability, whereas the right hemisphere is superior in visuospatial ability.
      1. Much of this information has been learned in patients with a "split brain" who have had their corpus callosum cut.
      2. In split-brain patients, sensory input can be presented to one or the other hemisphere, and the response is noted either verbally or by having that hemisphere direct actions of the contralateral hand that it controls.
   B. Although most activities that are casually observed are controlled by the left hemisphere, further observations show that the right hemisphere is important in pattern recognition, musical creation, singing, and the recognition of faces.
   C. About 70% of left-handers have their language ability on the left side of the brain; of the remaining 30%, about half have language function lateralized to the right side, and half show a more even distribution of ability between the two hemispheres.
II. Damage to particular areas of the cerebral cortex produces specific types of aphasias.
   A. Damage to Broca's area produces an aphasia in which language comprehension is unaffected but speech is slow and difficult—this area is believed to project to the motor cortex, which in turn controls the musculature of speech.
   B. Damage to Wernicke's area produces an aphasia in which language comprehension is destroyed, although the person can talk; the speech, however, does not make sense.
      1. Wernicke's area is believed to control Broca's area via the arcuate fasciculus.
      2. Damage to the arcuate fasciculus produces a type of aphasia similar to that produced by damage to Wernicke's area.
   C. The angular gyrus is believed to integrate auditory, visual, and somatesthetic information and to project to Wernicke's area; a person with damage to the angular gyrus may be able to speak and write but not to read.
III. Memory is divided into short-term and long-term memory.
   A. Electroconvulsive shock can cause amnesia of recent events but not of older memories.
   B. The hippocampus is believed to be involved in short-term memory and in the consolidation of short-term memory into long-term memory.
   C. Electrical stimulation of points in the temporal lobe of the cerebral cortex evokes complex, detailed memories; the "memory trace," however, is believed to be more diffusely located in the cortex.
   D. The consolidation of memory requires protein synthesis and probably involves changes in the chemical structure and function of synapses.

Review Activities

Objective Questions

1. The precentral gyrus is
   (a) involved in motor control
   (b) involved in sensory perception
   (c) located in the frontal lobe
   (d) both *a* and *c*
   (e) both *b* and *c*
2. In most people, the right hemisphere controls movement
   (a) of the right side of the body primarily
   (b) of the left side of the body primarily
   (c) of both the right and left sides of the body equally
   (d) of the head and neck only
3. Which of the following statements about the basal ganglia is (are) *true?*
   (a) They are located in the cerebrum.
   (b) They contain the caudate nucleus.
   (c) They are involved in motor control.
   (d) They are part of the extrapyramidal system.
   (e) All of the above are true.
4. Which of the following acts as a relay center for somatesthetic sensation?
   (a) the thalamus
   (b) the hypothalamus
   (c) the red nucleus
   (d) the cerebellum
5. Which of the following statements about the medulla oblongata is *false?*
   (a) It contains nuclei for some cranial nerves.
   (b) It contains the apneustic center.
   (c) It contains the vasomotor center.
   (d) It contains ascending and descending fiber tracts.

6. The reticular activating system
   (a) is composed of neurons that are part of the reticular formation
   (b) is a loose arrangement of neurons with many interconnecting synapses
   (c) is located in the brain stem and midbrain
   (d) functions to arouse the cerebral cortex to incoming sensory information
   (e) all of the above
7. In the control of emotion and motivation, the limbic system works together with the
   (a) pons
   (b) thalamus
   (c) hypothalamus
   (d) cerebellum
   (e) basal ganglia
8. Verbal ability predominates in the
   (a) left hemisphere of right-handed people
   (b) left hemisphere of most left-handed people
   (c) right hemisphere of 97% of all people
   (d) both *a* and *b*
   (e) both *b* and *c*
9. The consolidation of short-term memory into long-term memory appears to be a function of the
   (a) substantia nigra
   (b) hippocampus
   (c) cerebral peduncles
   (d) arcuate fasciculus
   (e) precentral gyrus

Match the nature of the aphasia with its cause:

10. comprehension good, can speak and write, but cannot read (though can see)
11. comprehension good, but speech is slow and difficult (but motor ability not damaged)
12. comprehension poor, speech is fluent but meaningless

   (a) damage to Broca's area
   (b) damage to Wernicke's area
   (c) damage to angular gyrus
   (d) damage to precentral gyrus

Essay Questions

1. Define the term *decussation,* and explain its significance in terms of the pyramidal motor system.
2. Electrical stimulation of the basal ganglia or cerebellum can produce skeletal movements. Describe the pathways by which these brain regions control motor activity.
3. Define the term *ablation,* and give two examples of how this experimental technique has been used to learn about the function of particular brain regions.
4. Explain how "split-brain" patients have been utilized in research on the function of the cerebral hemispheres. Propose experiments that would reveal the lateralization of function in the two hemispheres.
5. What is the evidence that Wernicke's area may control Broca's area? What is the evidence that the angular gyrus has input to Wernicke's area?
6. Provide two reasons why it is believed that there is a difference between short-term and long-term memory, and describe why it is believed that the hippocampus is involved in the consolidation of short-term memory.

Selected Readings

Benson, D. F., and N. Geschwind. 1972. Aphasia and related disturbances. In A. B. Baker, ed. *Clinical neurology,* New York: Harper and Row.

Cote, L. 1981. Basal ganglia, the extrapyramidal motor system, and disease of transmitter metabolism. In E. R. Kandel and J. H. Schwartz, eds. *Principles of neural science.* New York: Elsevier North Holland.

Fine, A. August 1986. Transplantation in the central nervous system. *Scientific American.*

Ganong, W. F. 1985. *Review of medical physiology.* 12th ed. Los Altos, CA: Lange Medical Publishers.

Geschwind, N. April 1972. Language and the brain. *Scientific American.*

Ghez, C. 1981. Cortical control of voluntary movement. In E. R. Kandel and J. H. Schwartz, eds. *Principles of neural science.* New York: Elsevier North Holland.

Hubel, D. H. September 1979. The brain. *Scientific American.*

Kupferman, I. 1981. Learning. In E. R. Kandel and J. H. Schwartz, eds. *Principles of neural science.* New York: Elsevier North Holland.

Lynch, G., and M. Baudry. 1984. The biochemistry of memory: a new and specific hypothesis. *Science* 224:1057.

Lemay, M., and N. Geschwind. 1978. Asymmetries of the human cerebral hemispheres. In A. Caramazza and E. Zurif, eds. *Language acquisition and language breakdown.* Baltimore: Johns' Hopkins University Press.

Routtenberg, A. November 1978. The reward system of the brain. *Scientific American.*

Shashoua, V. E. 1985. The role of extracellular proteins in learning and memory. *American Scientist* 73: 364.

Springer, S. P., and G. Deutch. 1985. *Left brain, right brain.* Rev. ed. New York: W. H. Freeman and Company.

Squire, L. R. 1986. Mechanisms of memory. *Science* 232:1612.

Thompson, R. F. 1985. *The brain.* New York: W. H. Freeman and Company.

Witelson, S. F. 1976. Sex and the single hemisphere: specialization of the right hemisphere for spatial processing. *Science* 193:425.

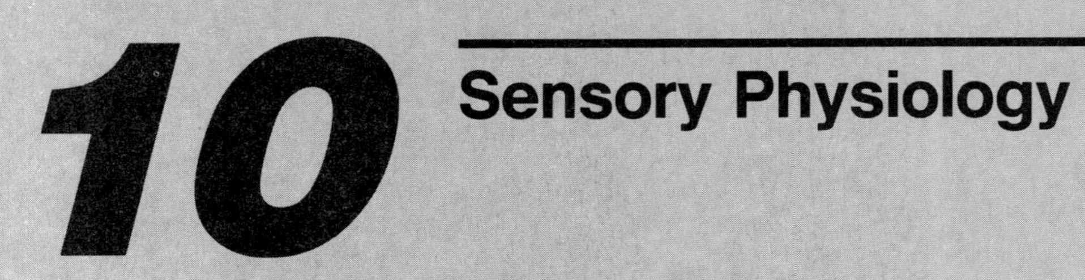

10 Sensory Physiology

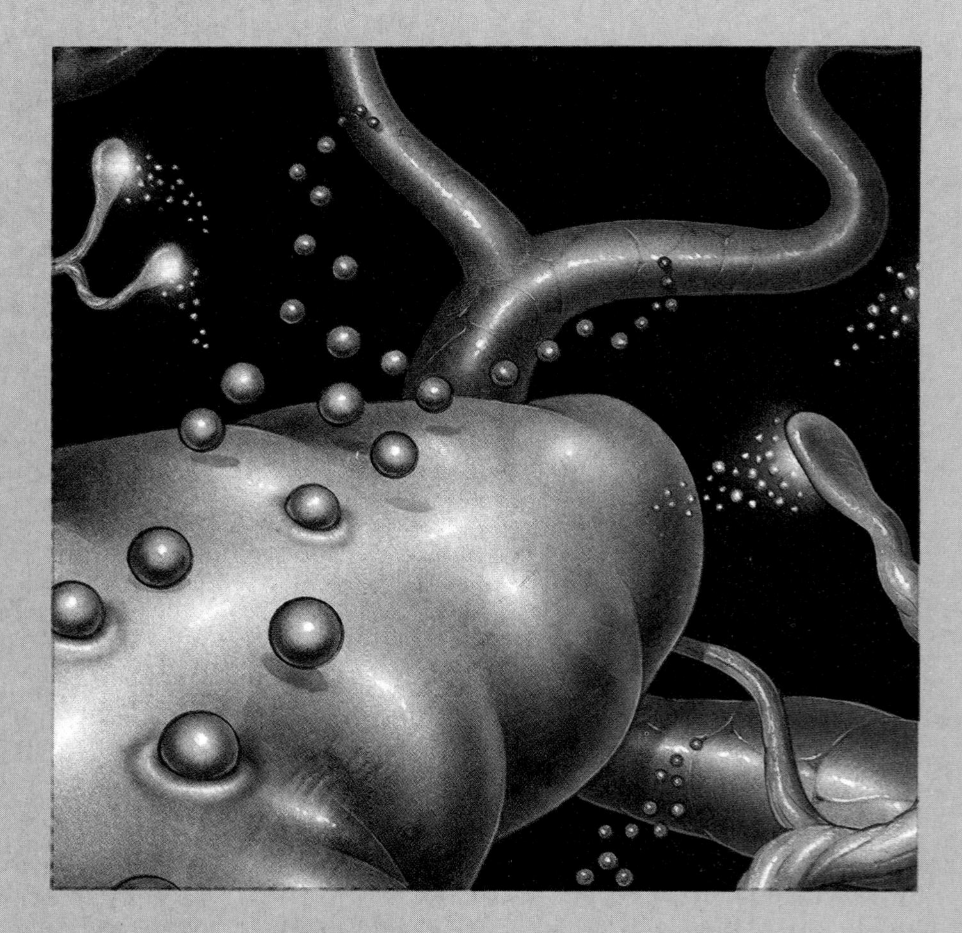

Objectives

By studying this chapter, you should be able to

1. describe the different categories of sensory receptors and explain the differences between tonic and phasic receptors
2. explain the law of specific nerve energies
3. describe the characteristics of the generator potential
4. describe the different types of cutaneous receptors and the neural pathways for the cutaneous senses
5. explain the concepts of receptive fields and lateral inhibition in relation to the cutaneous senses
6. describe the sensory physiology of taste and olfaction
7. describe the structure of the vestibular apparatus and explain how it provides information about acceleration of the body in different directions
8. describe the functions of the outer and middle ear
9. describe the structure of the cochlea and explain how movements of the stapes against the oval window result in vibrations of the basilar membrane
10. explain how the organ of Corti converts mechanical energy into electrical nerve impulses and explain how pitch perception is accomplished
11. describe the structure of the eye and the manner in which images are brought to a focus on the retina
12. explain how visual accommodation is achieved and explain the defects involved in myopia, hyperopia, and astigmatism
13. describe the architecture of the retina and the pathways of light and nerve activity through the retina
14. explain how rhodopsin functions as the visual pigment in rods and explain dark adaptation
15. explain how light affects the electrical activity of rods and their synaptic input to bipolar cells
16. explain the trichromatic theory of color vision
17. describe the differences in synaptic connections and locations of rods and cones and explain why rods provide black-and-white vision under low illumination and cones provide color vision and high acuity under greater light intensities
18. describe the neural pathways from the retina, explaining the differences in pathways taken by input from different regions of the visual field
19. describe the receptive fields of ganglion cells and explain the significance of the arrangement of these receptive fields
20. explain the stimulus requirements of simple, complex, and hypercomplex cortical neurons

Outline

Characteristics of Sensory Receptors

Our perceptions of the world—its textures, colors, and sounds; its warmth, smells, and tastes—are created by the brain from electrochemical nerve impulses delivered to it from sensory receptors. These receptors **transduce** (change) different forms of energy in the "real world" into the energy of nerve impulses, which are conducted into the central nervous system by sensory neurons. Different sensory *modalities*—or qualities of sensation, such as sound, light, pressure, and so on—result from differences in neural pathways and synaptic connections. The brain thus interprets impulses in the auditory nerve as sound and in the optic nerve as sight, even though the impulses themselves are identical in the two nerves.

We know, through the use of scientific instruments, that our senses act as energy filters that allow us to perceive only a narrow range of energy. Vision, for example, is limited to light in the visible spectrum; ultraviolet and infrared light, X rays and radio waves, which are the same type of energy as visible light, cannot normally excite the photoreceptors in the eyes. The perception of cold is entirely a product of the nervous system—there is no such thing as cold in the physical world, only varying degrees of heat. The perception of cold, however, has obvious survival value. Although filtered and distorted by the limitations of sensory function, our perceptions of the world allow us to interact effectively with the environment.

Categories of Sensory Receptors

Sensory receptors can be categorized by their structure and on the basis of different functional criteria. Structurally, the sensory receptors may be the dendritic endings of sensory neurons, which are either free (such as those in the skin, which mediate pain and temperature) or are encapsulated within nonneural structures, such as pressure receptors in the skin (fig. 10.1). The photoreceptors in the retina of the eyes (rods and cones) are highly specialized neurons, which synapse with other neurons in the retina. In the case of taste buds and of hair cells in the inner ears, modified epithelial cells respond to an environmental stimulus and activate sensory neurons.

Functional Categories. Sensory receptors can be grouped according to the type of stimulus energy they transduce. These categories include (1) *chemoreceptors,* such as the taste buds, olfactory epithelium, and the aortic and carotid bodies, which sense chemical changes in the blood; (2) *photoreceptors*—the rods and cones in the retina of the eye; (3) *thermoreceptors,* which respond to heat and cold; and (4) *mechanoreceptors,* which are stimulated by mechanical deformation of the receptor cell membrane—these include touch and pressure receptors in the skin and hair cells within the inner ear. *Nocioreceptors*—or pain

Figure 10.1. Different types of sensory receptors. Free nerve endings (*a*) mediate many cutaneous sensations. Some nerve endings are encapsulated within associated structures: e.g., (*b*) a pacinian corpuscle (*left*) and a Meissner's corpuscle (*right*). Some receptors, such as the taste bud (*c*), are modified epithelial cells that are innervated by sensory neurons.

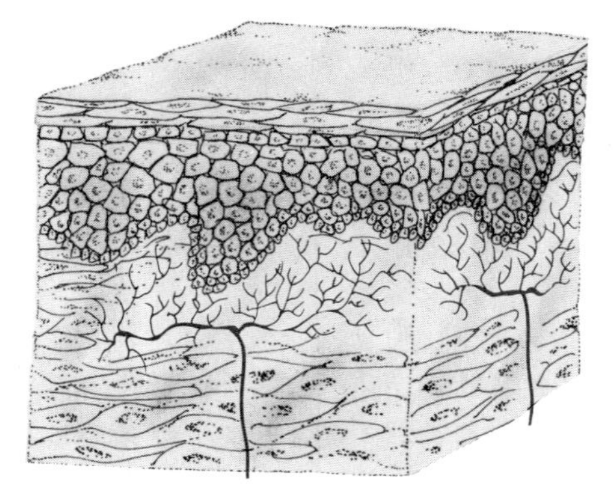

(a)

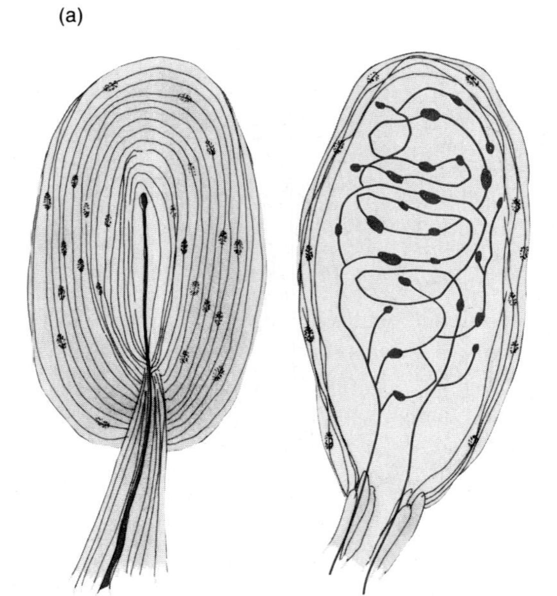

(b)

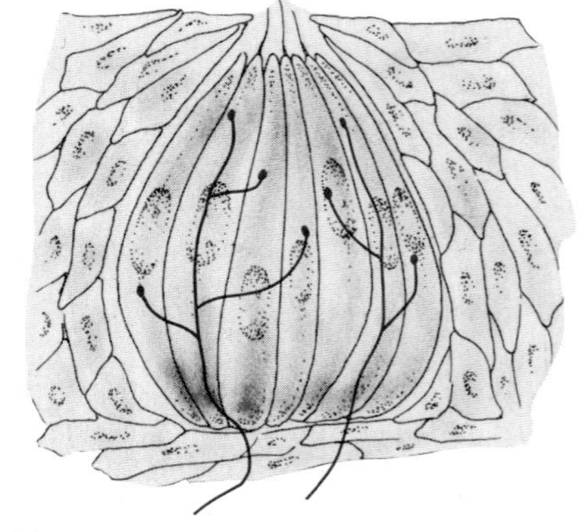

(c)

Figure 10.2. Tonic receptors (*a*) continue to fire at a relatively constant rate as long as the stimulus is maintained. These produce slowly adapting sensations. Phasic receptors (*b*) respond with a burst of action potentials when the stimulus is first applied, but then quickly reduce their rate of firing while the stimulus is maintained. This produces rapidly adapting sensations.

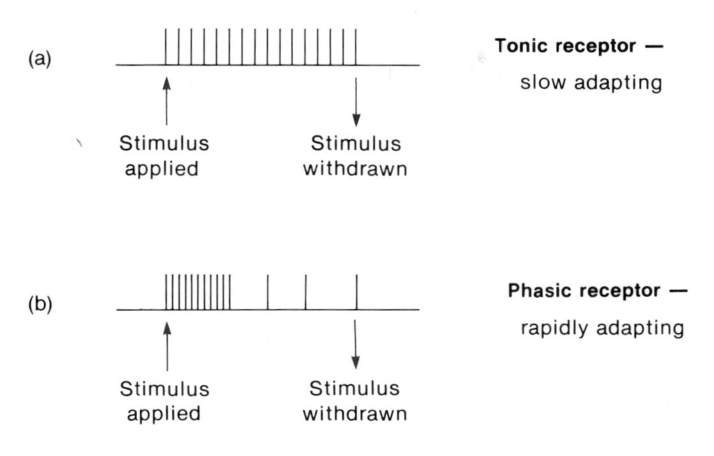

Table 10.1 Classification of receptors based on their normal (or "adequate") stimulus.

Receptor	Normal Stimulus	Mechanisms	Examples
Mechanoreceptors	Mechanical force	Deforms cell membrane of sensory dendrites; or deforms hair cells that activate sensory nerve endings	Cutaneous touch and pressure receptors; vestibular apparatus and cochlea
Pain receptors	Tissue damage	Damaged tissues release chemicals that excite sensory endings	Cutaneous pain receptors
Chemoreceptors	Dissolved chemicals	Chemical interaction affects ionic permeability of sensory cells	Smell and taste (exteroreceptors); osmoreceptors and carotid body chemoreceptors (interoreceptors)
Photoreceptors	Light	Photochemical reaction affects ionic permeability of receptor cell	Rods and cones in retina of eyes

receptors—are stimulated by chemicals released from damaged tissue cells, and thus are a type of chemoreceptor.

Receptors can also be grouped according to the type of sensory information they deliver to the brain. *Proprioceptors* include the muscle spindles, Golgi tendon organs, and joint receptors. These provide a sense of body position and allow fine control of skeletal movements. *Cutaneous receptors* include (1) touch and pressure receptors; (2) warmth and cold receptors; and (3) pain receptors. The receptors that mediate sight, hearing, and equilibrium are grouped together as the *special senses.*

Tonic and Phasic Receptors: Sensory Adaptation. Some receptors respond with a burst of activity when a stimulus is first applied, but then quickly decrease their firing rate—adapt to the stimulus—when the stimulus is maintained. Receptors with this response pattern are called *phasic receptors.* Receptors that produce a relatively constant rate of firing as long as the stimulus is maintained are known as *tonic receptors* (see figure 10.2).

Phasic receptors alert us to changes in sensory stimuli and are in part responsible for the fact that we can cease paying attention to constant stimuli. This ability is called **sensory adaptation.** Odor, touch, and temperature, for example, adapt rapidly; bathwater feels hotter when we first enter it. Sensations of pain, in contrast, adapt little if at all.

Law of Specific Nerve Energies

Stimulation of a sensory nerve fiber produces only one sensation—touch, cold, pain, and so on. According to the **law of specific nerve energies,** the sensation characteristic of each sensory neuron is that produced by its normal, or *adequate stimulus* (table 10.1). The adequate stimulus

Figure 10.3. Sensory stimuli result in the production of local, graded potential changes known as the receptor, or generator, potential (number 1 through 4). If the receptor potential reaches a threshold value of depolarization, it generates action potentials (number 5) in the sensory neuron.

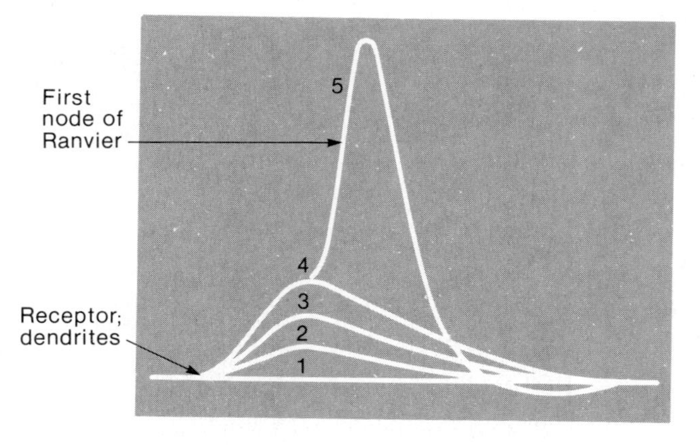

for the photoreceptors of the eye, for example, is light. If these receptors are stimulated by some other means—such as by pressure produced by a punch to the eye—a flash of light (the adequate stimulus) may be perceived.

Paradoxical cold provides another example of the law of specific nerve energies. First, a receptor for cold is located by touching the tip of a cold metal rod to the skin. Sensation then gradually disappears as the rod warms to body temperature. Applying the tip of a rod heated to 45°C to the same spot, however, causes the sensation of cold to reappear. This paradoxical cold is produced because the heat slightly damages receptor endings, and by this means produces an "injury current" that stimulates the receptor.

Regardless of how a sensory neuron is stimulated, therefore, only one sensory modality will be perceived. This specificity is due to the synaptic pathways within the brain that are activated by the sensory neuron. The ability of receptors to function as sensory filters and be stimulated normally by only one type of stimulus (the adequate stimulus) allows the brain to usually perceive the stimulus accurately.

Generator (Receptor) Potential

The electrical behavior of sensory nerve endings is similar to that of the dendrites of other neurons. In response to an environmental stimulus, the sensory endings produce local, graded changes in the membrane potential. In most cases these potential changes are depolarizations, analogous to excitatory postsynaptic potentials (EPSPs, as described in chapter 7). In the sensory endings, however, these potential changes in response to environmental stimulation are called **receptor,** or **generator, potentials,** because they serve to generate action potentials in response to the sensory stimulation.

The pacinian corpuscle, a cutaneous receptor for pressure (see fig. 10.1), can serve as an example. When a light touch is applied to the receptor, a small depolarization (the generator potential) is produced. Increasing the pressure on the pacinian corpuscle increases the magnitude of the generator potential until it reaches the threshold required to produce an action potential (fig. 10.3). The pacinian corpuscle, however, is a phasic receptor; if the pressure is maintained the size of the generator potential produced quickly diminishes. It is interesting to note that this phasic response is a result of the onionlike covering around the dendritic nerve ending; if these layers are peeled off and the nerve ending is stimulated directly it responds in a tonic fashion.

When a tonic receptor is stimulated, the generator potential it produces is proportional to the intensity of the stimulus. After a threshold depolarization is produced, increases in the amplitude of the generator potential result in increases in the *frequency* with which action potentials

Figure 10.4. The response of a tonic receptor to stimuli. As the strength of stimulation is increased, the generator potential increases (*a, b, c*). The amplitude of the generator potential and the length of time it remains above threshold determine the frequency and duration of action potentials (*b, c*).

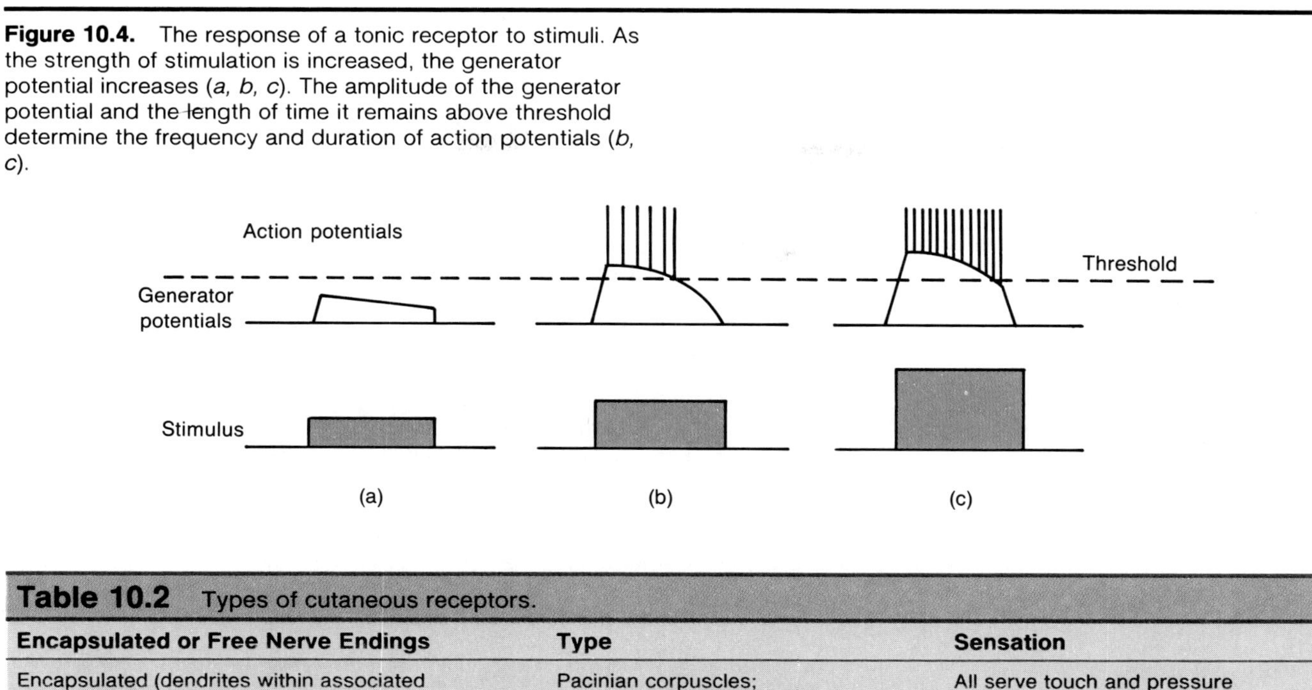

Table 10.2 Types of cutaneous receptors.

Encapsulated or Free Nerve Endings	Type	Sensation
Encapsulated (dendrites within associated structures)	Pacinian corpuscles; Meissner's corpuscles; Krause's end bulbs; Ruffini's end organs	All serve touch and pressure
Free nerve endings	———	Touch, pressure, heat, cold, pain

are produced (fig. 10.4). In this way, the frequency of action potentials that are conducted into the central nervous system serves to code for the strength of the stimulus. As described in chapter 7, this frequency code is needed since the amplitude of action potentials is constant (all-or-none). Acting through changes in action potential frequency, tonic receptors thus provide information about the relative intensity of a stimulus.

1. *Our perceptions are products of our brains; they are incompletely and inconstantly related to physical reality. Explain this statement, using examples of vision and perceptions of cold.*
2. *Define the law of specific nerve energies and the adequate stimulus, and relate these definitions to your answer for question 1.*
3. *Describe sensory adaptation in olfactory and pain receptors. Using a line drawing, relate sensory adaptation to the responses of phasic and tonic receptors.*
4. *Describe how the magnitude of a sensory stimulus is transduced into a receptor potential and how the magnitude of the receptor potential is coded in the sensory nerve fiber.*

Cutaneous Sensations

The cutaneous sensations of touch, pressure, hot and cold, and pain are mediated by the dendritic nerve endings of different sensory neurons. The receptors for hot, cold, and pain are the naked endings of sensory neurons. Sensations of touch and pressure are mediated by both naked dendritic endings and dendrites that are encapsulated within various structures (table 10.2). In *pacinian corpuscles,* for example, the dendritic endings are encased within thirty to fifty onionlike layers of connective tissue (fig. 10.5). These layers absorb some of the pressure when a stimulus is maintained, and thus help to accentuate the phasic response of this receptor.

Figure 10.5. A diagrammatic section of the skin showing the general location and magnified structure of cutaneous receptors.

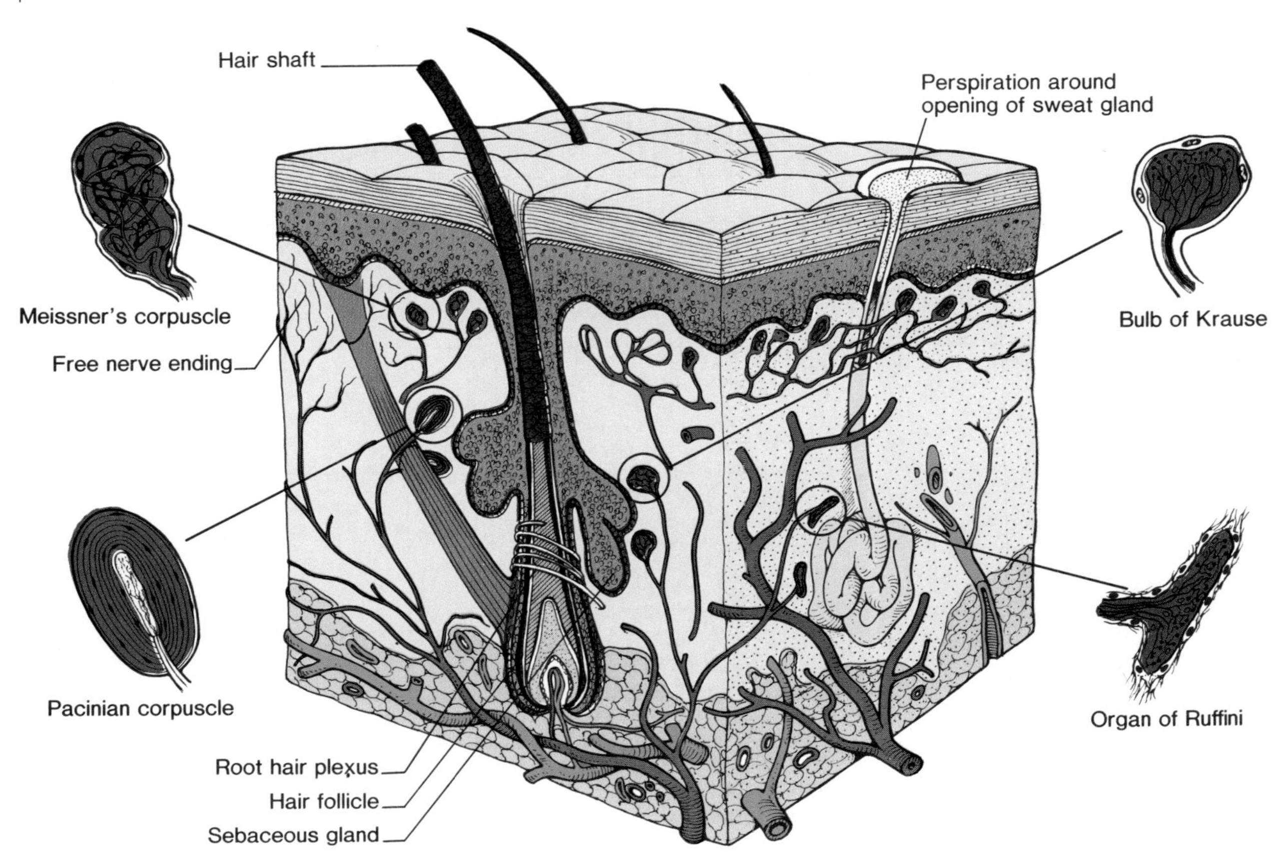

Neural Pathways for Somatesthetic Sensations

The conduction pathways for the *somatesthetic senses*—a term that includes sensations from cutaneous and proprioceptors—are shown in figure 10.6. *Proprioception* and *pressure* are carried by large, myelinated nerve fibers that ascend in the *dorsal columns* of the spinal cord on the same (ipsilateral) side. These fibers do not synapse until they reach the *medulla oblongata* of the brain stem; fibers that carry these sensations from the feet are thus incredibly long. After synapsing in the medulla with other, second-order sensory neurons, information in the latter neurons crosses over to the contralateral side as it ascends via a fiber tract, called the medial lemniscus, to the *thalamus.* Third-order sensory neurons in the thalamus that receive this input in turn project to the postcentral gyrus.

Sensations of *hot, cold,* and *pain* are carried by thin, unmyelinated sensory neurons into the spinal cord. These synapse with second-order interneurons within the spinal cord, which cross over to the contralateral side and ascend to the brain in the *lateral spinothalamic tract.* Fibers that mediate *touch* and *pressure* ascend in the *ventral spinothalamic tract.* Fibers of both spinothalamic tracts synapse with third-order neurons in the thalamus (bypassing the medulla), which in turn project to the postcentral gyrus. Notice that, in all cases, somatesthetic information is carried to the postcentral gyrus in third-order neurons. Also, because of crossing-over, somatesthetic information from each side of the body is projected to the postcentral gyrus of the contralateral cerebral hemisphere.

All somatesthetic information from the same area of the body projects to the same area of the postcentral gyrus, so that a "map" of the body can be drawn on the postcentral gyrus to represent sensory projection points. This map is very distorted, however, because it shows larger areas of cortex devoted to sensation in the face and hands than in other areas in the body. This disproportionately larger area of the cortex devoted to the face and hands reflects the fact that there is a higher density of sensory receptors in these regions.

Receptive Fields and Sensory Acuity

The **receptive field** of a neuron serving cutaneous sensation is the area of skin whose stimulation results in changes in the firing rate of the neuron. Changes in the firing rate of primary sensory neurons affect the firing of second- and third-order neurons, which in turn affects the firing of those

Figure 10.6. Pathways that lead from the cutaneous receptors and proprioreceptors into the postcentral gyrus in the cerebral cortex.

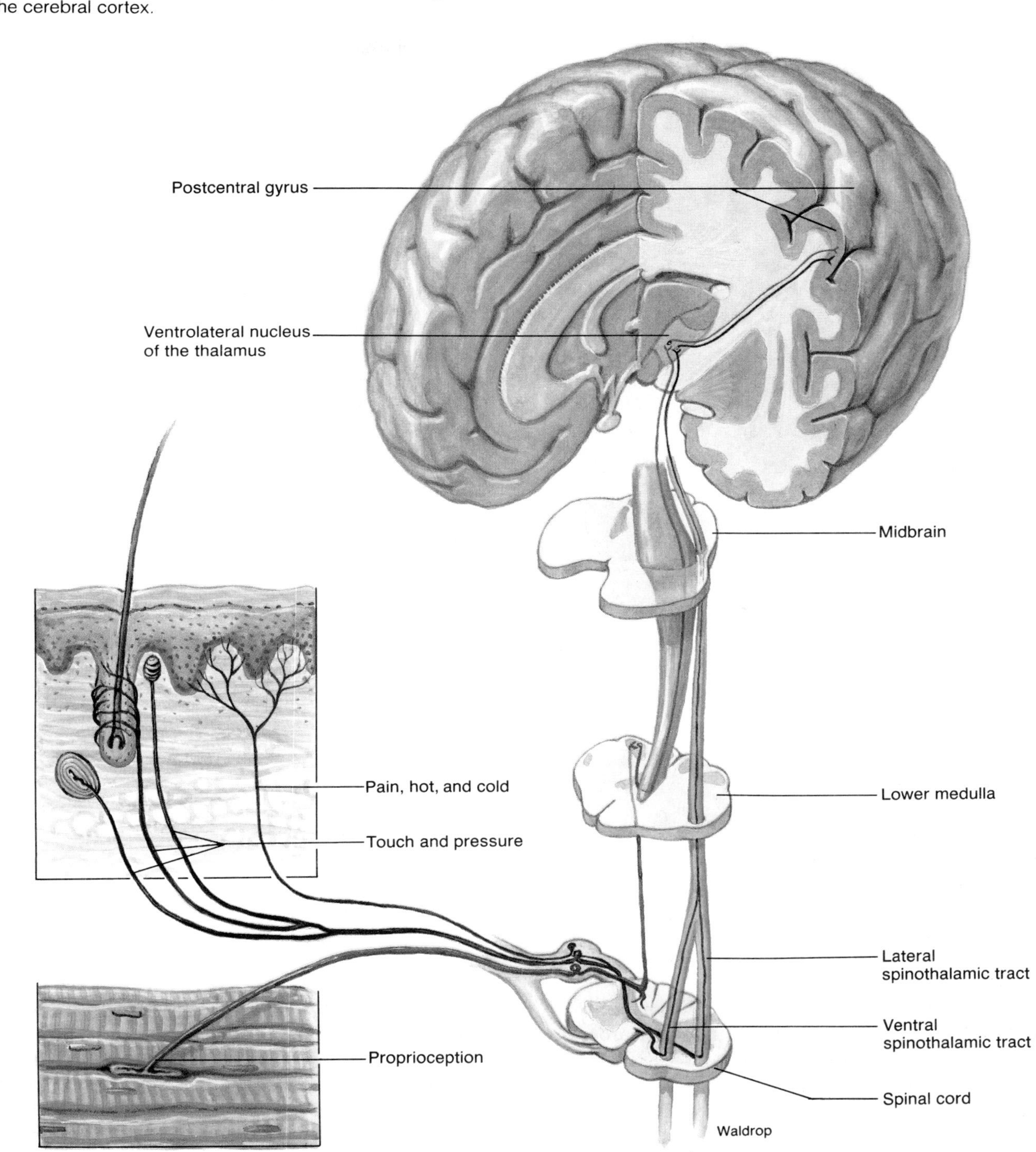

Figure 10.7. The two-point touch threshold test. If each point touches the receptive fields of different sensory neurons, two separate points of touch will be felt. If both caliper points touch the receptive field of one sensory neuron, only one point of touch will be felt.

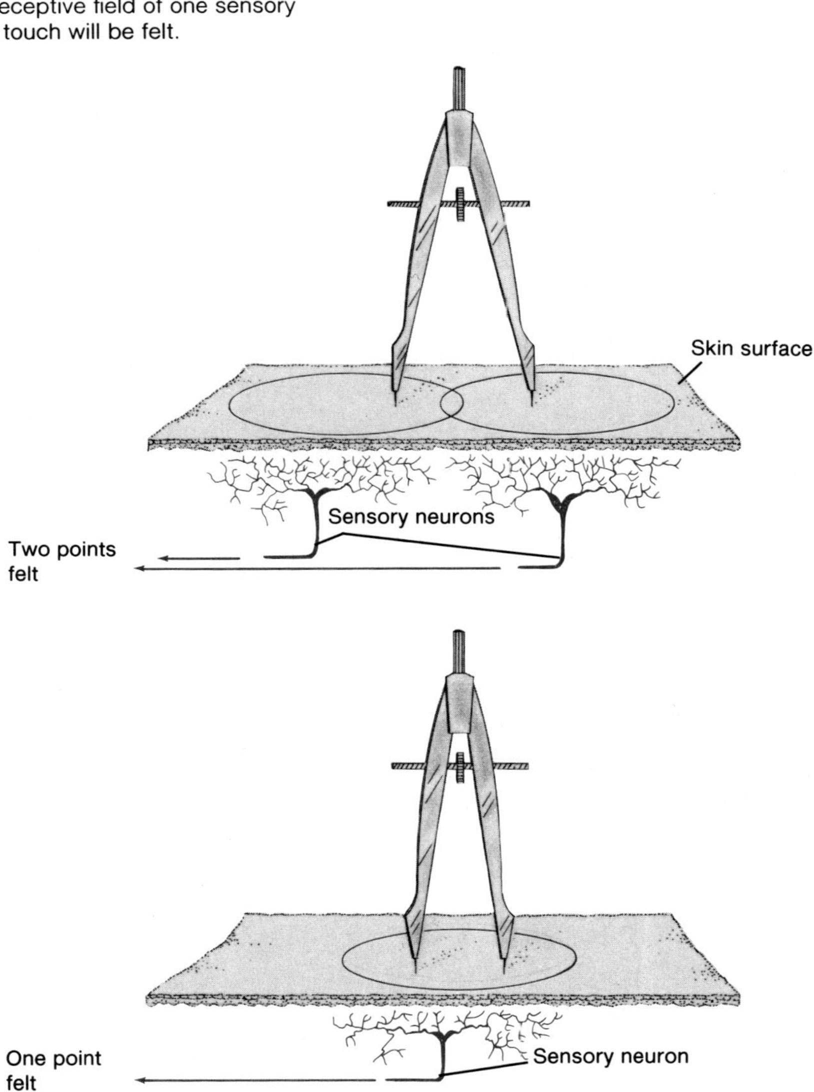

neurons in the postcentral gyrus that receive input from the third-order neurons. Indirectly, therefore, neurons in the postcentral gyrus can be said to have receptive fields in the skin.

The area of each receptive field in the skin varies inversely with the density of receptors in the region. In the back and legs, where a large area of skin is served by relatively few sensory endings, the receptive field of each neuron is correspondingly large. In the fingertips—where a large number of cutaneous receptors serve a small area of skin—the receptive field of each sensory neuron is correspondingly small.

Two-Point Touch Threshold. The approximate size of the receptive fields serving light touch can be measured by the *two-point touch threshold* test. In this procedure, two points of a pair of calipers are lightly touched to the skin at the same time. If the calipers are set sufficiently wide apart, each point will stimulate a different receptive field and a different sensory neuron—two separate points of touch will thus be felt. If the calipers are sufficiently closed, both points will touch the receptive field of only one sensory neuron, and only one point of touch will be felt (fig. 10.7).

Table 10.3 The two-point touch threshold for different regions of the body.

Body Region	Two-Point Touch Threshold (mm)
Big toe	10
Sole of foot	22
Calf	48
Thigh	46
Back	42
Abdomen	36
Upper arm	47
Forehead	18
Palm of hand	13
Thumb	3
First finger	2

Source: From Weinstein, S., and D. R. Kenshalo (editor), The Skin Senses, 1968. Courtesy of Charles C. Thomas, Publisher, Springfield, Illinois.

The two-point touch threshold, which is the minimum distance that can be distinguished between two points of touch, is a measure of the distance between receptive fields. If the two points of the calipers are closer than this distance, only one "blurred" point of touch can be felt. The two-point touch threshold is thus an indication of tactile *acuity,* or the sharpness (acu = needle) of touch perception.

The high tactile acuity of the fingertips is exploited in the reading of *Braille.* Braille symbols consist of dots that are raised 1 mm up from the page and separated from each other by 2.5 mm, which is slightly above the two-point touch threshold in the fingertips (table 10.3). Experienced Braille readers can scan words at about the same speed that a sighted person can read aloud—a rate of about 100 words per minute.

Lateral Inhibition

When a blunt object touches the skin a number of receptive fields may be stimulated. Those receptive fields in the center areas where the touch is strongest will be stimulated more than in neighboring fields where the touch is lighter. We do not usually feel a "halo" of light touch surrounding a center of stronger touch, however. Instead, only a single touch is felt, which is somewhat sharper than the actual shape of the blunt object. This sharpening of sensation is due to a process called *lateral inhibition* (fig. 10.8).

Lateral inhibition and the sharpening of sensation that results occur within the central nervous system. Those sensory neurons whose receptive fields are stimulated most strongly inhibit—via interneurons that pass "laterally" within the CNS—sensory neurons that serve neighboring receptive fields. Lateral inhibition similarly plays a prominent role in pitch discrimination, as described in a later section.

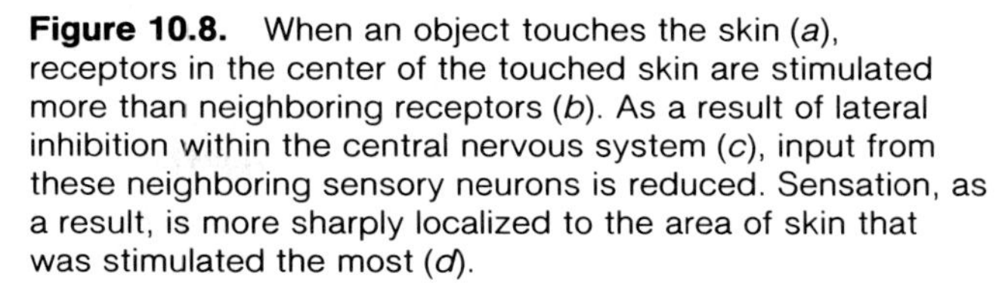

Figure 10.8. When an object touches the skin (*a*), receptors in the center of the touched skin are stimulated more than neighboring receptors (*b*). As a result of lateral inhibition within the central nervous system (*c*), input from these neighboring sensory neurons is reduced. Sensation, as a result, is more sharply localized to the area of skin that was stimulated the most (*d*).

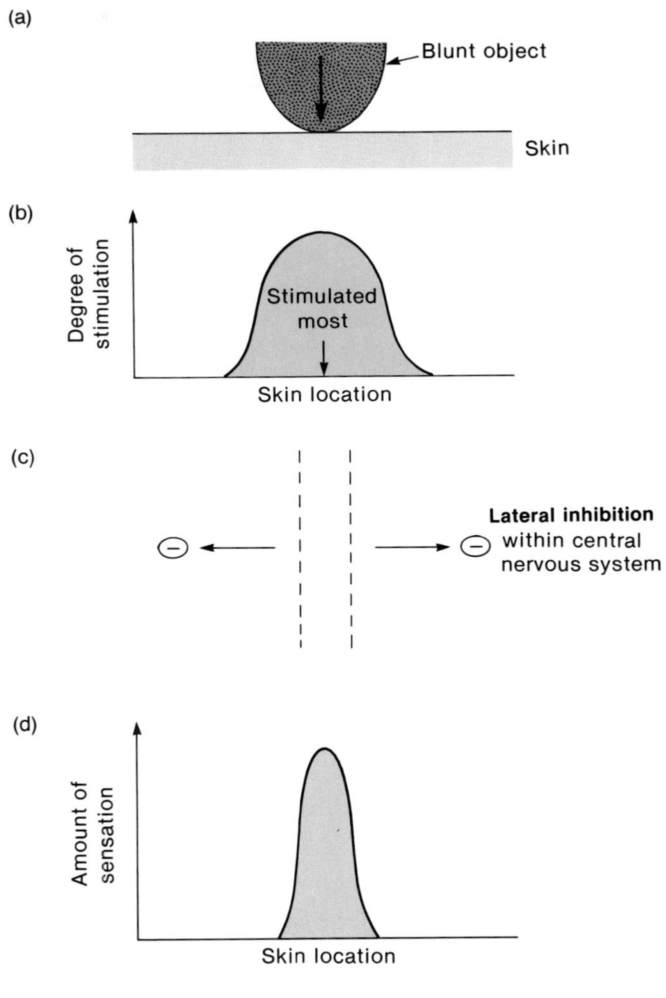

Taste and Olfaction

Chemoreceptors that respond to chemical changes in the internal environment are called **interoceptors;** those that respond to chemical changes in the external environment are **exteroceptors.** Included in the latter category are *taste receptors,* which respond to chemicals dissolved in food or drink, and *olfactory receptors,* which respond to gaseous molecules in the air. This distinction is somewhat arbitrary, however, because odorant molecules in air must first dissolve in fluid within the olfactory mucosa before the sense of smell can be stimulated.

Figure 10.9. A taste bud.

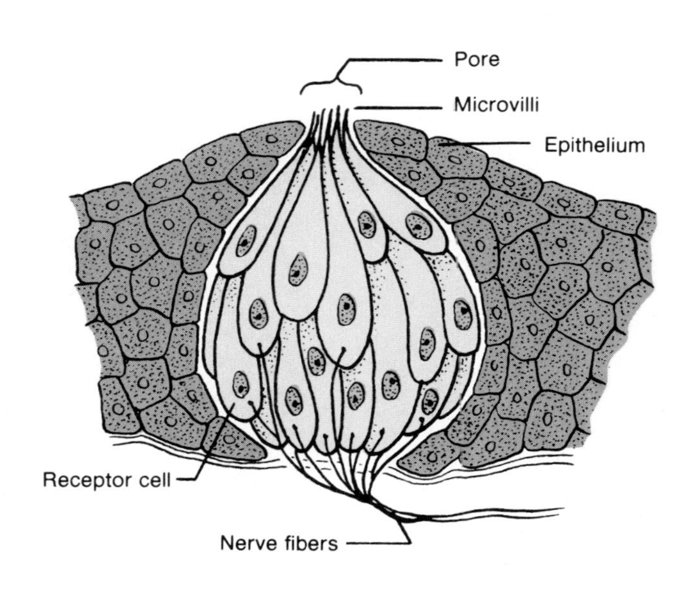

Figure 10.10. Areas of the tongue that are most sensitive to each of the four modalities of taste.

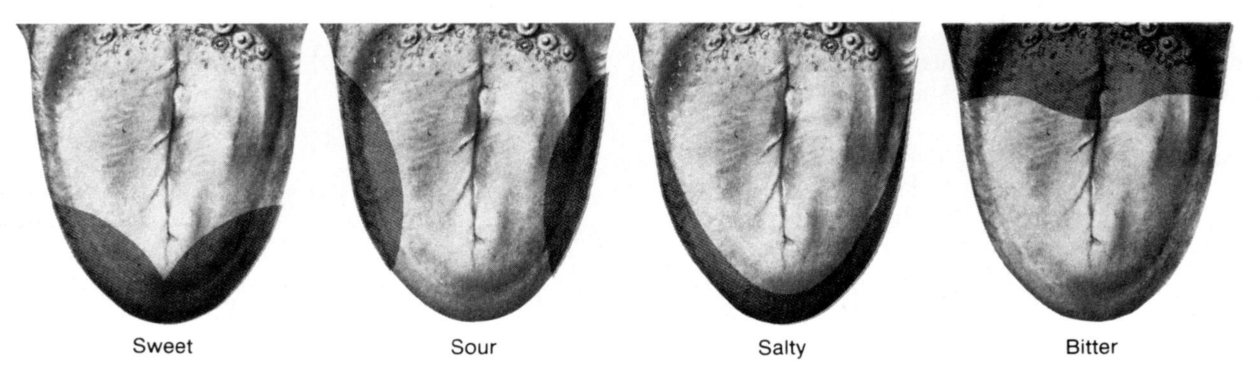

Taste

Taste receptors are specialized epithelial cells that are grouped together into barrel-shaped arrangements called *taste buds,* located in the epithelium of the tongue (fig. 10.9). The cells of the taste buds have microvilli at their apical (top) surface, which is exposed to the external environment through a pore in the surface of the taste bud.

Molecules dissolved in saliva at the surface of the tongue interact with receptor molecules in the microvilli of the taste buds. This interaction stimulates the release of a neurotransmitter chemical from the receptor cells, which in turn stimulates sensory nerve endings that innervate the taste buds. Taste buds in the posterior third of the tongue are innervated by the *glossopharyngeal (ninth cranial) nerve;* those in the anterior two-thirds of the tongue are innervated by the *facial (seventh cranial) nerve.*

There are only four basic modalities of taste, which are sensed most acutely in particular regions of the tongue. These are *sweet* (tip of the tongue), *sour* (sides of the tongue), *bitter* (back of the tongue), and *salty* (over most of the tongue). This distribution is illustrated in figure 10.10.

Figure 10.11. The olfactory epithelium contains receptor neurons that synapse with neurons in the olfactory bulb of the brain.

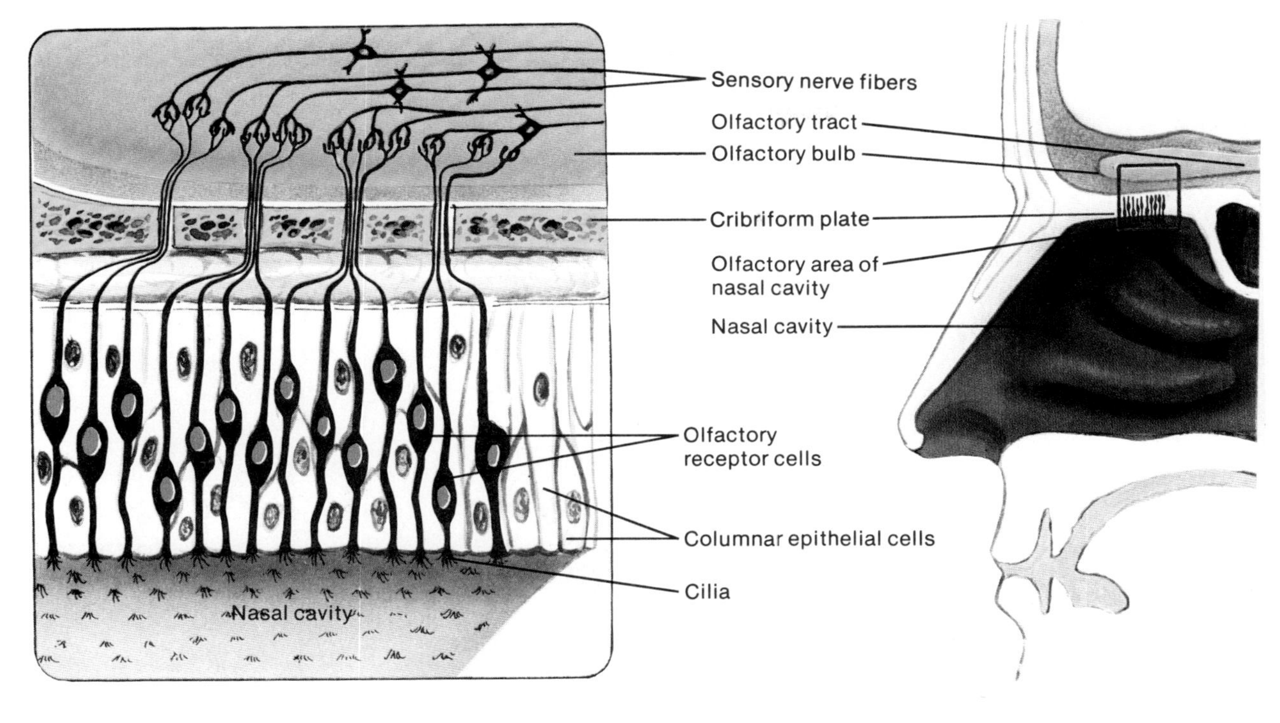

Sour taste is produced by hydrogen ions (H^+); all acids therefore taste sour. Most organic molecules, particularly sugars, taste sweet to varying degrees. Only pure table salt (NaCl) has a pure salty taste—other salts, such as KCl (commonly used in place of NaCl by people with hypertension) taste salty but have bitter overtones. Bitter taste is evoked by quinine and seemingly unrelated molecules.

Olfaction

The olfactory receptors are the dendritic endings of the *olfactory (first cranial) nerve,* in association with epithelial supporting cells. Unlike other sensory modalities, which are relayed to the cerebrum from the thalamus, the sense of olfaction is transmitted directly to the olfactory bulb of the cerebral cortex (fig. 10.11).

Unlike taste, which is divisible into only four modalities, many thousands of different odors can be distinguished by people who are trained in this capacity (as in the perfume and wine industries). The molecular basis of olfaction is not understood; although various theories have attempted to explain families of odors on the basis of similarities in molecular shape and/or charges, such attempts have been only partially successful. The extreme sensitivity of olfaction is possibly as amazing as its diversity—at maximum sensitivity, only one odorant molecule is needed to excite an olfactory receptor.

1. *Using a flow diagram, describe the neural pathways leading from cutaneous pain and pressure receptors to the postcentral gyrus. Indicate where crossing-over occurs.*
2. *Describe the meaning of the term* sensory acuity, *and explain how acuity is affected by the density of receptive fields and by the process of lateral inhibition.*
3. *Describe the distribution of taste receptors in the tongue, and explain what effect damage to the facial nerve might have on taste sensation.*

Figure 10.12. Structures within the inner ear include the cochlea and vestibular apparatus. The vestibular apparatus consists of the utricle and saccule (together called the otolith organs) and the three semicircular canals. Each semicircular canal contains a widened area (the ampulla, which is labeled for only one canal) with sensory hair cells.

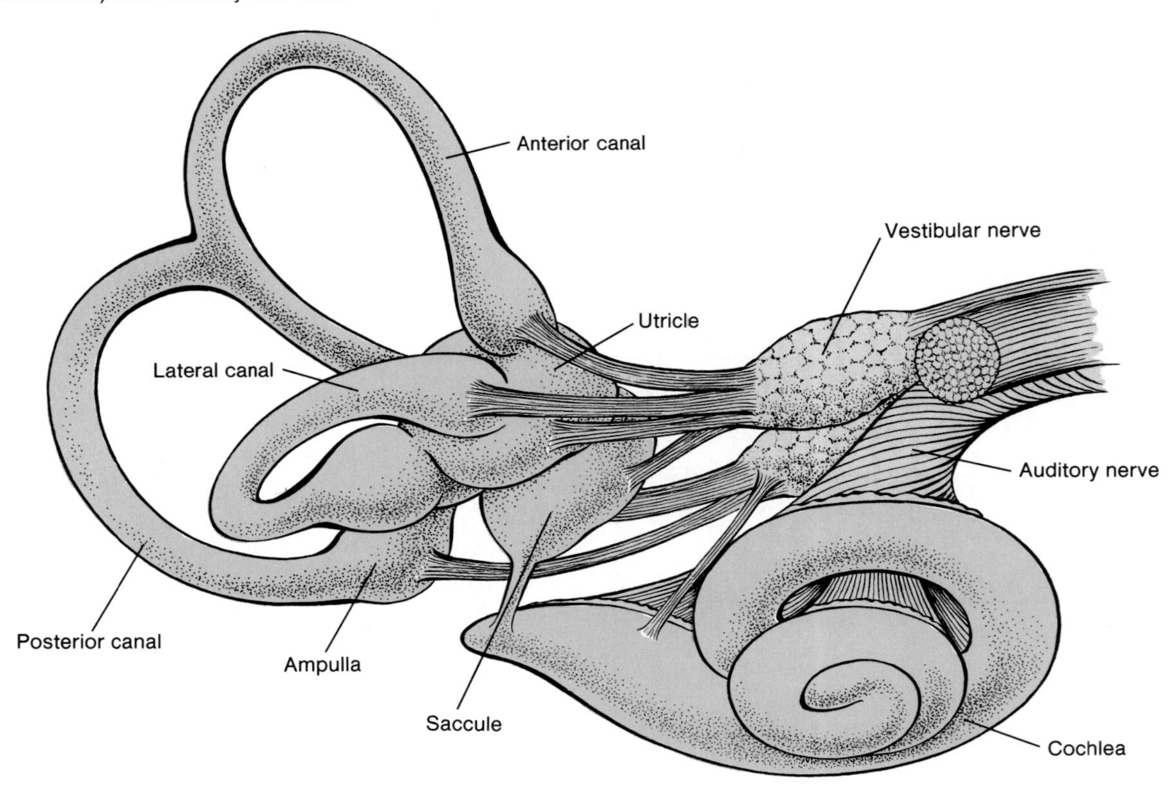

Vestibular Apparatus and Equilibrium

The sense of equilibrium, which provides orientation with respect to gravity, is due to the function of an organ called the **vestibular apparatus.** The vestibular apparatus and a snail-like structure called the *cochlea,* which is involved in hearing, form the *inner ear* within the temporal bones of the skull. The vestibular apparatus consists of two parts: (1) the *otolith organs,* which include the *utricle* and *saccule;* and (2) the *semicircular canals* (fig. 10.12).

The sensory structures of the vestibular apparatus and cochlea are located within a structure called the **membranous labyrinth,** which is filled with a fluid that is similar in composition to intracellular fluid. This fluid is called *endolymph.* The bony structures surrounding the membranous labyrinth in the inner ear contain a fluid called *perilymph,* which is similar in composition to cerebrospinal fluid. The membranous labyrinth (fig. 10.13), in other words, is filled with endolymph and is surrounded by perilymph.

Sensory Hair Cells of the Vestibular Apparatus

The utricle and saccule provide information about *linear acceleration*—changes in velocity when traveling horizontally or vertically. We therefore have a sense of acceleration and deceleration when riding in a car or when skipping rope. A sense of *rotational* or *angular acceleration* is provided by the semicircular canals, which are oriented in three planes like the faces of a cube. This helps us maintain balance when turning the head, spinning, or tumbling.

The receptors for equilibrium are modified epithelial cells called *hair cells,* because they contain twenty to fifty "hairs" (actually microvilli) and one cilium, which is called a *kinocilium* (fig. 10.14). When the hair cells are bent in

Figure 10.13. The labyrinths of the inner ear. The membranous labyrinth is contained within the bony labyrinth.

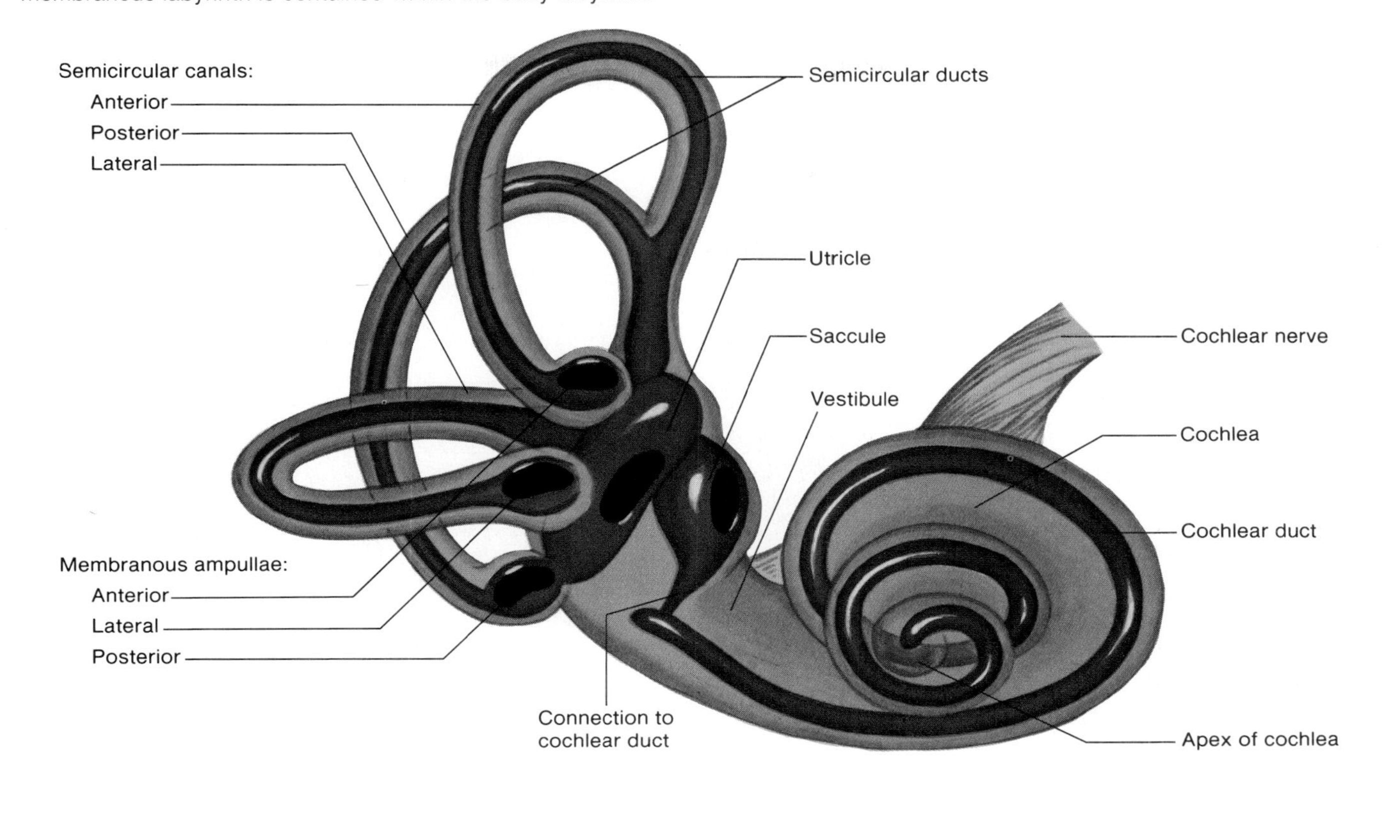

Figure 10.14. Scanning electron micrograph of hairs and kinocilium within the vestibular apparatus.

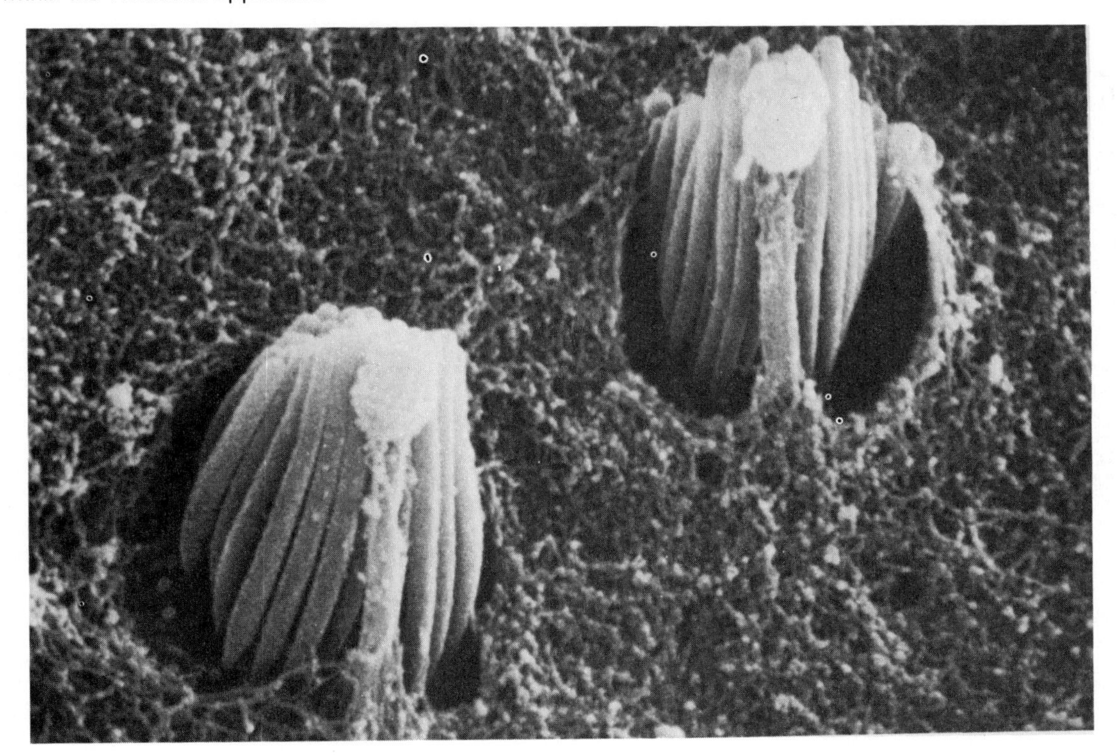

Figure 10.15. (*a*) Sensory hair cells in the vestibular apparatus contain hairs (microvilli) and one kinocilium. (*b*) When hair cells are bent in the direction of the kinocilium, the cell membrane is depressed (see *arrow*) and the sensory neuron innervating the hair cell is stimulated. (*c*) When the hairs are bent in a direction opposite to the kinocilium, the sensory neuron is inhibited.

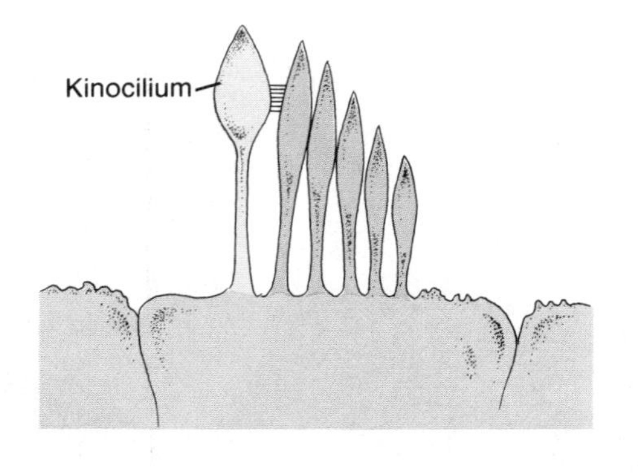

At rest

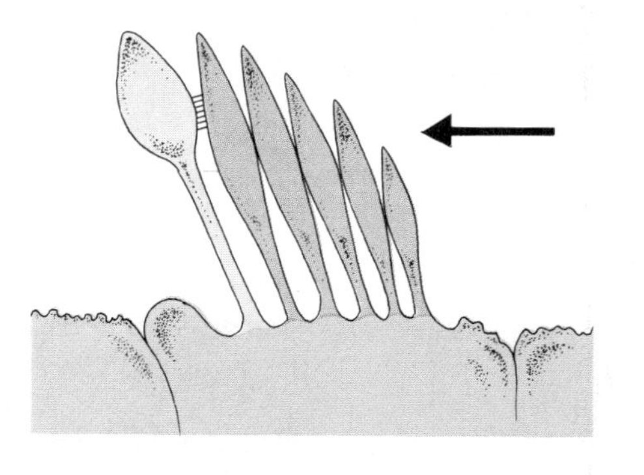

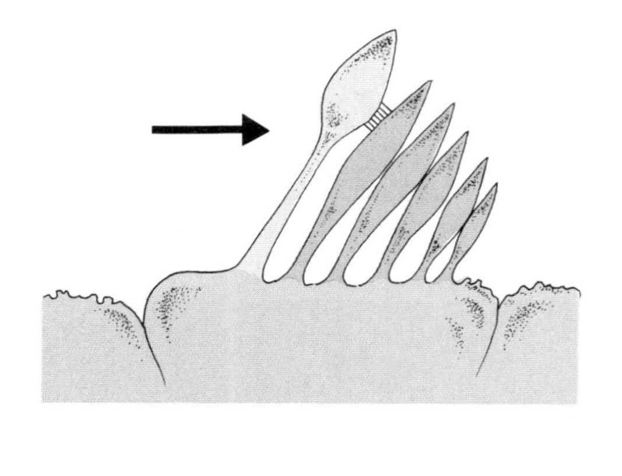

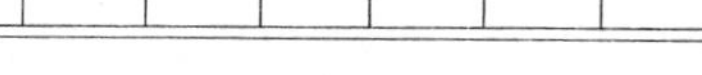

Stimulated

Inhibited

Figure 10.36. A view of the retina as seen with an ophthalmoscope. Optic nerve fibers leave the eyeball at the optic disc to form the optic nerve. Note the blood vessels that can be seen entering the eyeball at the optic disc.

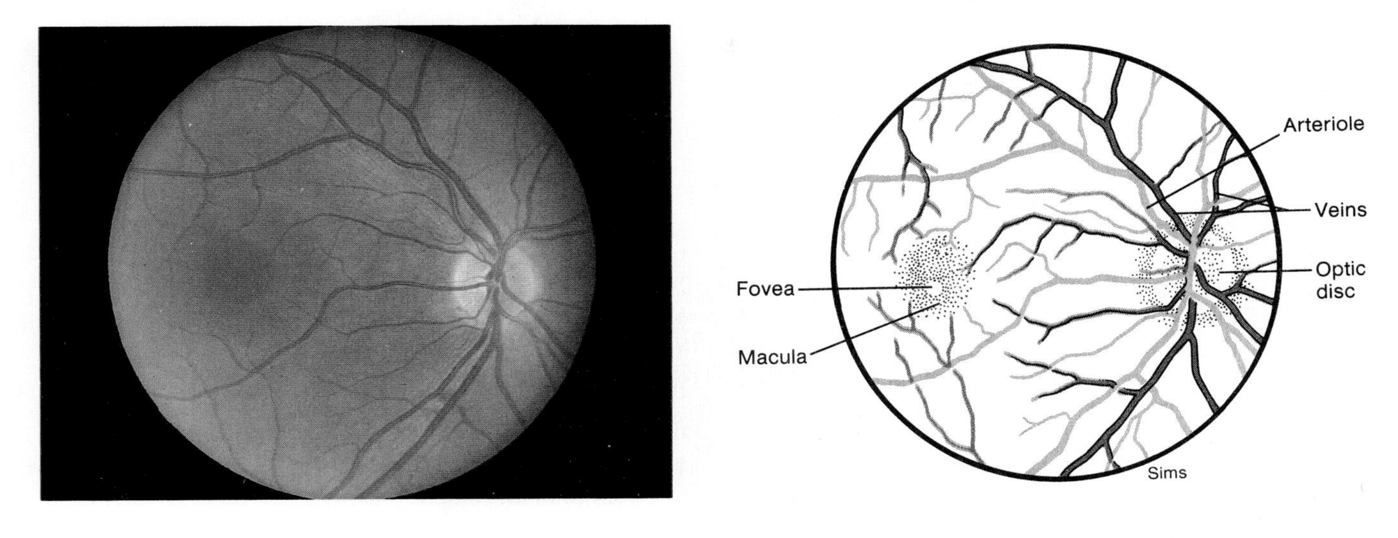

Figure 10.37. The refraction of light waves within the eyeball causes the image of an object to be inverted on the retina.

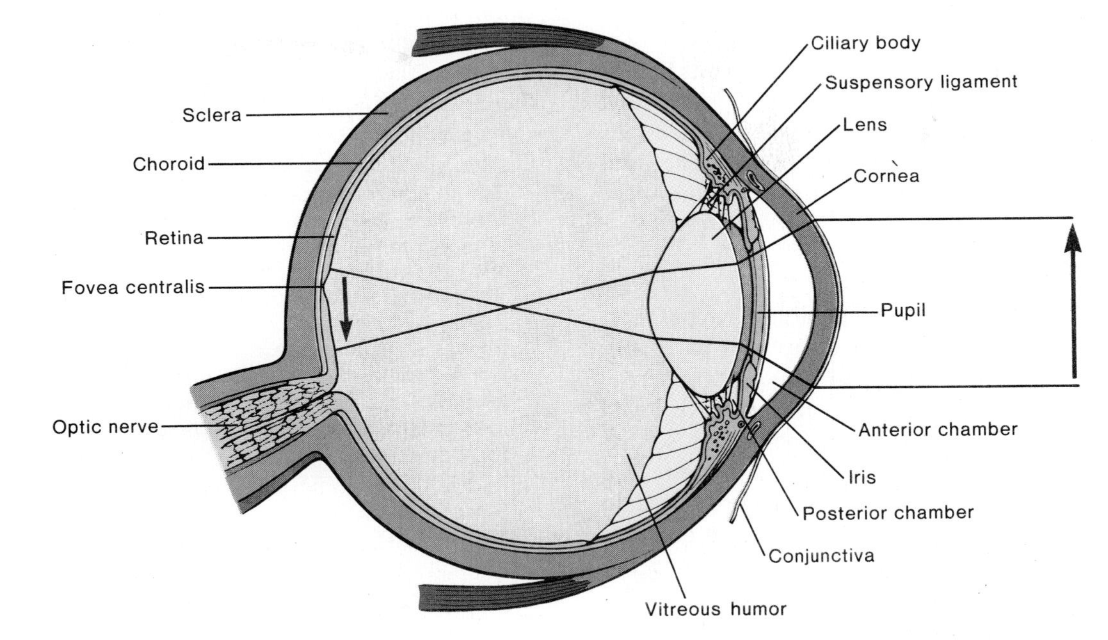

Figure 10.38. Refraction of light in the cornea and lens produces a right-to-left image on the retina. The left side of the visual field is projected to the right half of each retina, while the right side of each visual field is projected to the left half of each retina.

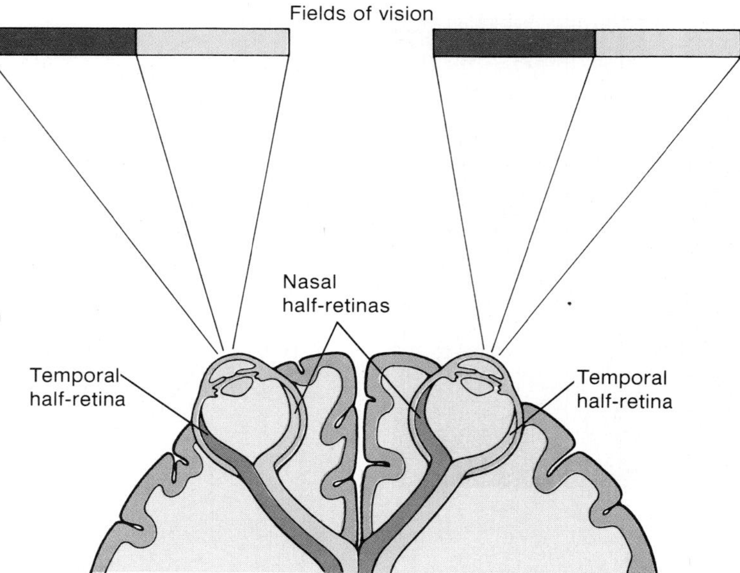

The *visual field*—which is the part of the external world projected onto the retina—is thus reversed in each eye. The cornea and lens focus the right part of the visual field on the left half of the retina of each eye, while the left half of the visual field is focused on the right half of each retina (fig. 10.38). The medial (or nasal) half-retina of the left eye therefore receives the same image as the lateral (or temporal) half-retina of the right eye. The nasal half-retina of the right eye receives the same image as the temporal half-retina of the left eye.

Accommodation

When a normal eye views an object, parallel rays of light are refracted to a point, or *focus,* on the retina (fig. 10.40). If the degree of refraction were to remain constant, movement of the object closer to or farther from the eye would cause corresponding movement of the focal point, so that the focus would either be behind or in front of the retina.

The ability of the eyes to keep the image focused on the retina as the distance between the eye and object is changed, is called **accommodation.** When the object is twenty feet or more from a normal eye, the image is focused on the retina and the lens is in its most flat, least convex form. In this state the muscles of the ciliary body, from which the lens is suspended by zonular fibers, are relaxed. As the object moves closer to the eyes the muscles of the ciliary body contract. This muscular contraction brings the ciliary body closer to the lens, which reduces tension in the zonular fibers suspending the lens. When the tension is reduced, the lens becomes more round and convex as a result of its inherent elasticity (fig. 10.39).

As an object is brought ever closer to the eyes, therefore, the convexity and refractive power of the lens increases and the image remains in focus on the retina. There is, however, a limit to the ability of the eyes to accommodate as an object gets close to the eyes. This *near point of vision* increases with age as a result of the loss of lens elasticity; a printed page must be held farther from the eyes as a result. This loss of accommodating ability due to loss of lens elasticity with age is called **presbyopia** (*presby* = old).

Visual Acuity

Visual acuity refers to the sharpness of vision. The sharpness of an image depends on the *resolving power* of the visual system—that is, on the ability of the visual system to distinguish (resolve) two closely spaced dots. The better the resolving power of the system is, the closer together these dots can be and still be seen as separate; when the resolving power of the system is exceeded, the dots are blurred together as a single image.

Figure 10.39. Changes in the shape of the lens during accommodation. (*a*) The lens is flattened for distant vision when the ciliary muscle fibers are relaxed and the suspensory ligaments are taut. (*b*) The lens is more spherical for closeup vision when the ciliary muscle fibers are contracted and the suspensory ligaments are relaxed.

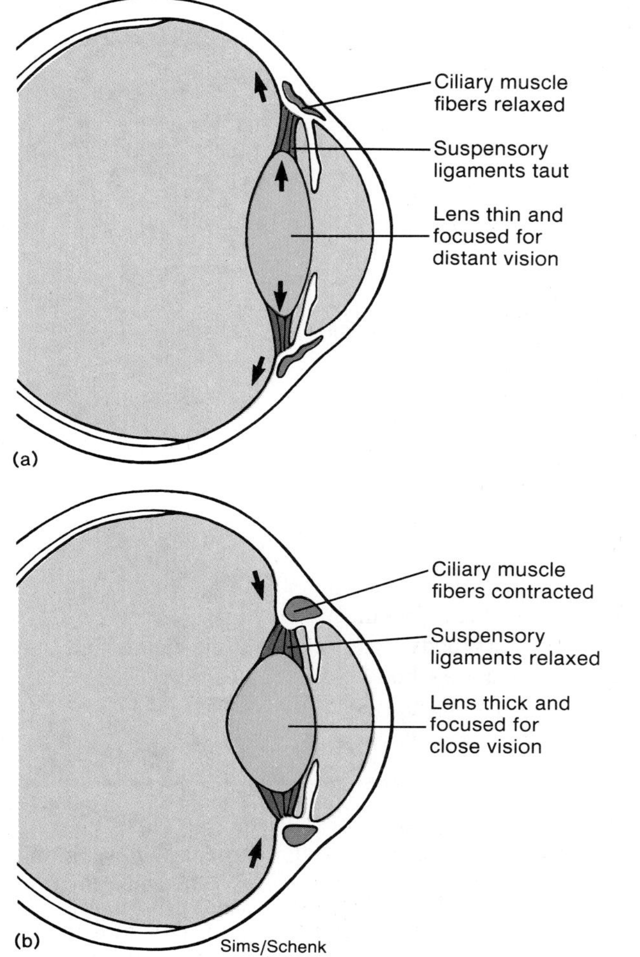

Figure 10.40. In a normal eye (*a*), parallel rays of light are brought to a focus on the retina by refraction in the cornea and lens. If the eye is too long, as in myopia (*b*), the focus is in front of the retina. This can be corrected by a concave lens. If the eye is too short (*c*), as in hyperopia, the focus is behind the retina. This is corrected by a convex lens. In astigmatism (*d*), light refraction is uneven due to an abnormal shape of the cornea or lens.

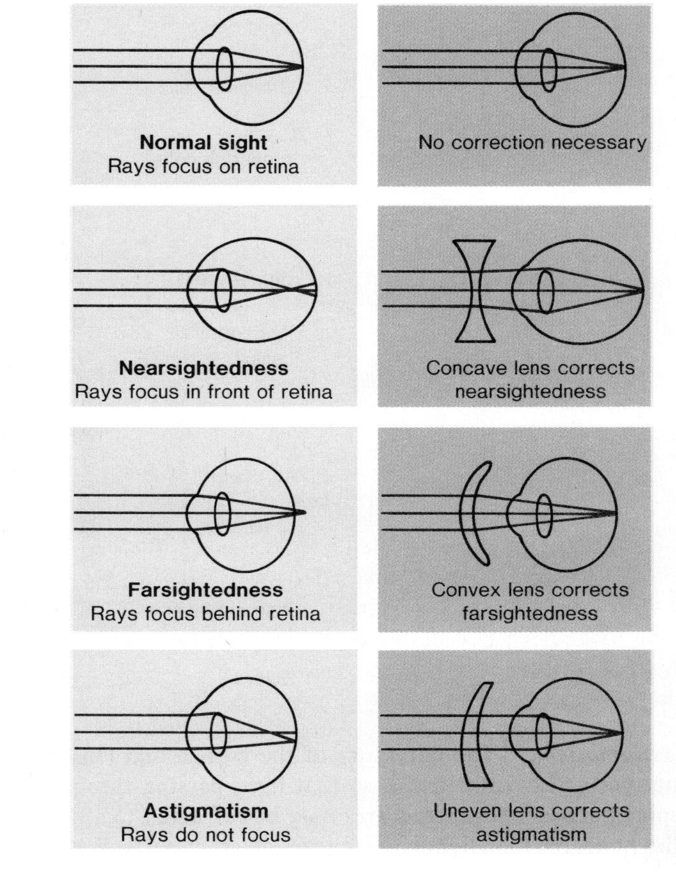

Myopia and Hyperopia. When a person with normal visual acuity stands twenty feet from a *Snellen eye chart* (so that accommodation is not a factor influencing acuity), the line of letters marked "20/20" can be read. If a person has **myopia** (nearsightedness), this line will appear blurred because the focus of this image will be in front of the retina. This is usually caused by the fact that the eyeball is too long. Myopia is corrected by glasses with concave lenses that cause the light rays to diverge; the focus is thus pushed back to the retina (fig. 10.40).

If the eyeballs are too short, the line marked "20/20" will appear blurred because the focus of the image will be behind the retina; the object must thus be placed farther from the eyes to be seen clearly. This condition is called **hyperopia** (farsightedness). Hyperopia is corrected by glasses with convex lenses that increase the convergence of light so that the focus is brought closer to the lens and falls on the retina.

Figure 10.41. The layers of the retina. The retina is inverted, so that light must pass through various layers of nerve cells before reaching the photoreceptors (rods and cones).

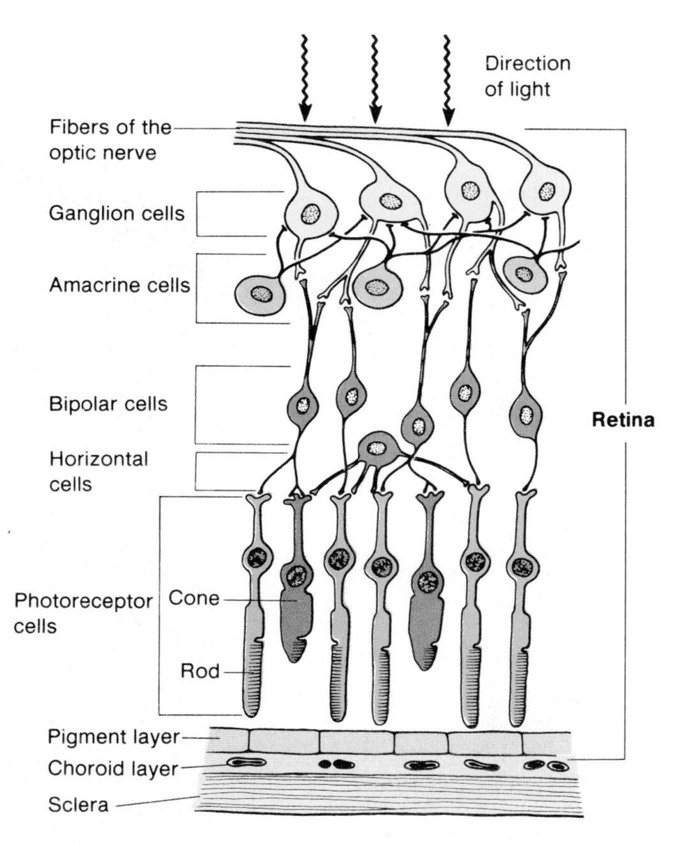

Astigmatism. The curvature of the cornea and lens is not perfectly symmetrical, so that light passing through some parts of these structures may be refracted to a different degree than light passing through other parts. When the asymmetry of the cornea and/or lens is significant, the person is said to have **astigmatism.** If a person with astigmatism views a circle, the image of the circle will not appear clear in all 360 degrees; the parts of the circle that appear blurred can thus be used to map the astigmatism. This condition is corrected by cylindrical lenses that compensate for the asymmetry in the cornea or lens of the eye.

1. *Using a line diagram, explain why an inverse image is produced on the retina. Also explain how the image in one eye corresponds to the image in the other eye.*
2. *Using a line diagram, show how parallel rays of light are brought to a focus on the retina. Explain how this focus is maintained as the distance from the object to the eye is increased or decreased (that is, explain accommodation).*
3. *Use a line diagram to show how a blurred image is produced in presbyopia when an object is brought too close to the eyes. Relate this condition to myopia and hyperopia.*

The Retina

The retina consists of a pigment epithelium, photoreceptor neurons called *rods* and *cones,* and layers of other neurons. The neural layers of the retina are actually a forward extension of the brain. In this sense the optic nerve can be considered a tract, and indeed the myelin sheaths of its fibers are derived from oligodendrocytes (like other CNS nerve fibers) rather than from Schwann cells.

Since the retina is an extension of the brain, the neural layers face outwards, towards the incoming light. Light, therefore, must pass through several neural layers before striking the photoreceptors (fig. 10.41). The photoreceptors then synapse with other neurons, so that nerve impulses are conducted outward in the retina.

The outer layers of neurons that contribute axons to the optic nerve are called *ganglion cells.* This layer receives synaptic input from *bipolar cells* underneath, which in turn receive input from rods and cones. In addition to the flow of information from photoreceptors to bipolar cells to ganglion cells, there are neurons called *horizontal cells,* which synapse with several photoreceptors (and possibly also with bipolar cells), and neurons called *amacrine cells,* which synapse with several ganglion cells.

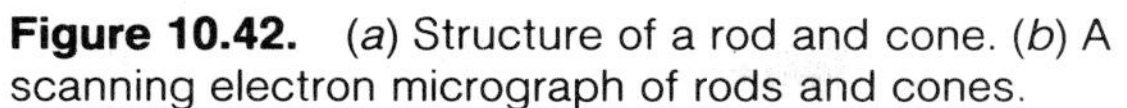

Figure 10.42. (*a*) Structure of a rod and cone. (*b*) A scanning electron micrograph of rods and cones.

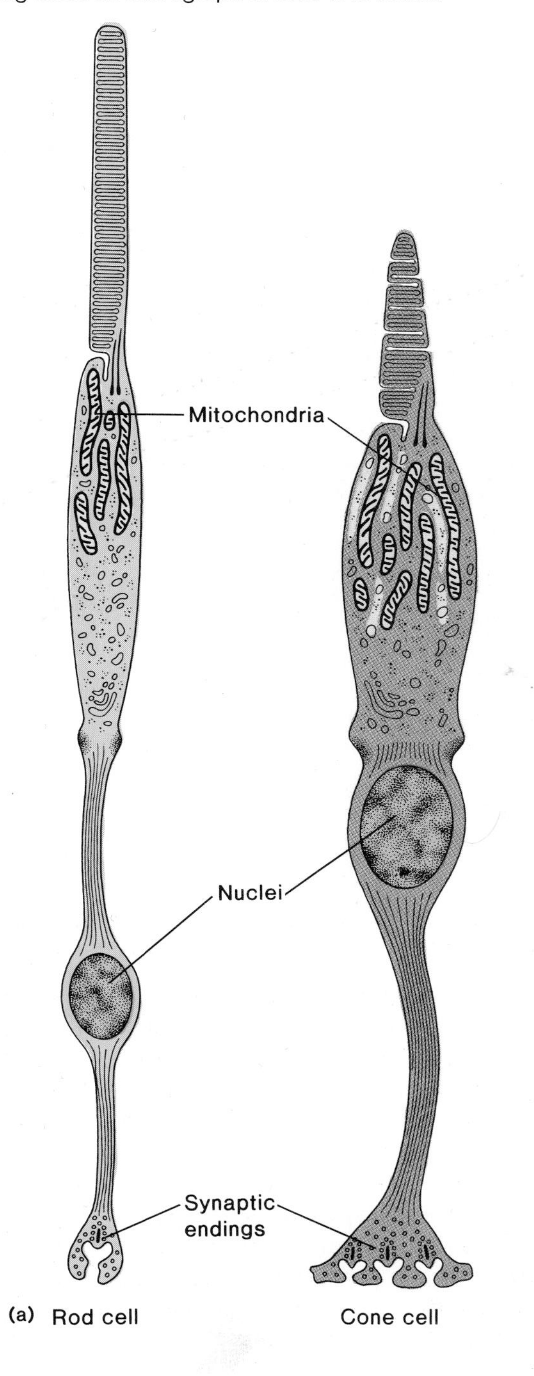

Effect of Light on the Rods

The photoreceptors—rods and cones (fig. 10.42)—are activated when light produces a chemical change in molecules of pigment contained within the membranous lamellae of the outer segments of the receptor cells. Rods contain a purple pigment known as **rhodopsin.** The pigment appears purple (a combination of red and blue), because it transmits light in the red and blue regions of the spectrum, while absorbing light energy in the green region. The wavelength of light that is absorbed best—the *absorption maximum*—is about 500 nm (a green-colored light).

Cars and other objects that are green in color are seen more easily at night (when rods are used for vision) than are red objects. This is because red light is not well absorbed by rhodopsin, and only absorbed light can produce

Figure 10.43. The photopigment rhodopsin consists of the protein opsin combined with 11-*cis* retinaldehyde. In response to light the retinaldehyde is converted to a different form, called all-*trans,* and dissociates from the opsin. This photochemical reaction induces changes in ionic permeability that ultimately results in stimulation of ganglion cells in the retina.

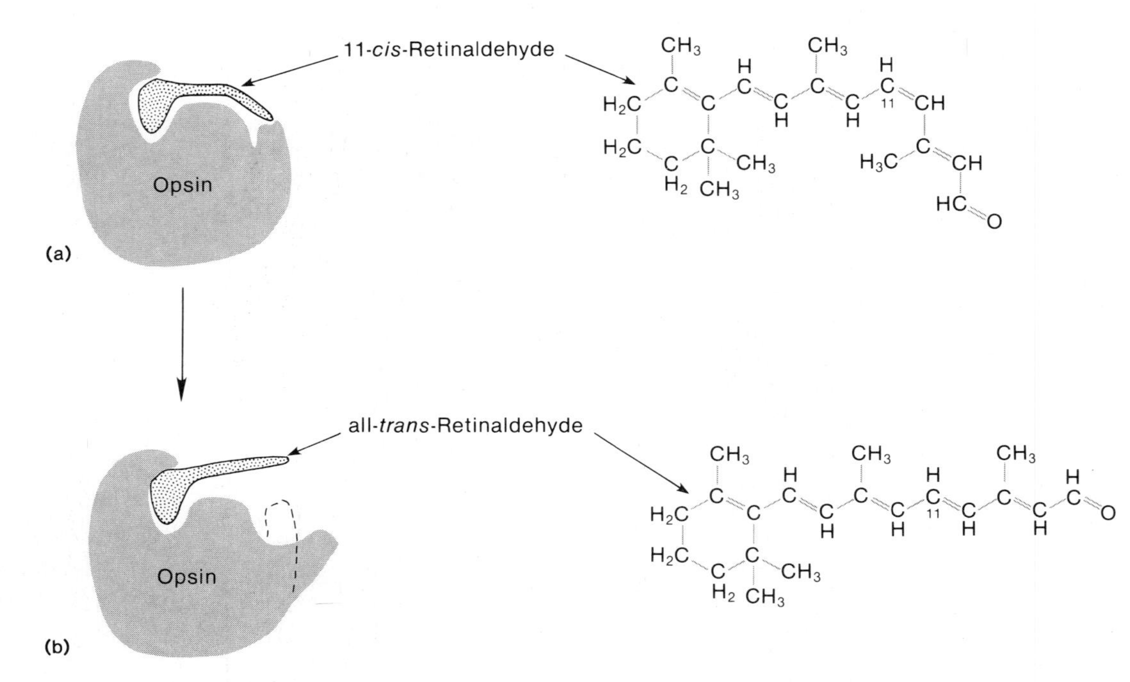

the photochemical reaction that results in vision. In response to absorbed light, rhodopsin dissociates into its two components: a pigment called **retinaldehyde,** derived from vitamin A, and a protein called **opsin.** This reaction is known as the *bleaching reaction.*

Retinaldehyde can exist in two possible configurations (shapes)—one known as the all-*trans* form and one called the 11-*cis* form (fig. 10.43). The all-*trans* form is the most stable, but only the 11-*cis* form is found attached to opsin. In response to absorbed light energy, the 11-*cis* retinaldehyde is converted to the all-*trans* form, causing it to dissociate from the opsin. This dissociation reaction in response to light initiates changes in the ionic permeability of the rod cell membrane and ultimately results in the production of nerve impulses in the ganglion cells. As a result of these effects, rods provide black-and-white vision under conditions of low light intensity (as described in a later section).

Dark Adaptation. The bleaching reaction that occurs in the light results in a lowered amount of rhodopsin in the rods and lowered amounts of visual pigments in the cones. When a light-adapted person first enters a darkened room, therefore, sensitivity to light is low and vision is poor. A gradual increase in photoreceptor sensitivity, known as *dark adaptation,* then occurs, reaching maximal sensitivity at about twenty minutes. The increased sensitivity to low light intensity is due partly to increased amounts of visual pigments produced in the dark. Increased pigments in the cones produce a slight dark adaptation in the first five minutes. Increased rhodopsin in the rods produces a much greater increase in sensitivity to low light levels and is partly responsible for the adaptation that occurs after about five minutes in the dark. In addition to the increased concentration of rhodopsin, other more subtle (and less well-understood) changes occur in the rods that ultimately result in a 100,000-fold increase in light sensitivity in dark-adapted as compared to light-adapted eyes.

Electrical Activity of Retinal Cells

The only neurons in the retina that produce all-or-none action potentials are ganglion cells and amacrine cells. The photoreceptors, bipolar cells, and horizontal cells instead produce only graded depolarizations or hyperpolarizations, analogous to EPSPs and IPSPs.

In the dark, the photoreceptors have a resting membrane potential that is less negative (closer to zero) than that of most other neurons. This is caused by a constant current of Na^+ into the cell, called a *dark current,* through special Na^+ channels. Light causes these Na^+ channels to become blocked; as a result, the photoreceptors become less depolarized than they are in the dark. Light, therefore, causes the photoreceptors to become *hyperpolarized* in comparison to their membrane potential in the dark.

Figure 10.44. There are three types of cones. Each type of cone contains retinaldehyde combined with a different type of protein, producing a pigment that absorbs light maximally at a different wavelength. Color vision, according to the trichromatic theory, is produced by the activity of these blue cones, green cones, and red cones.

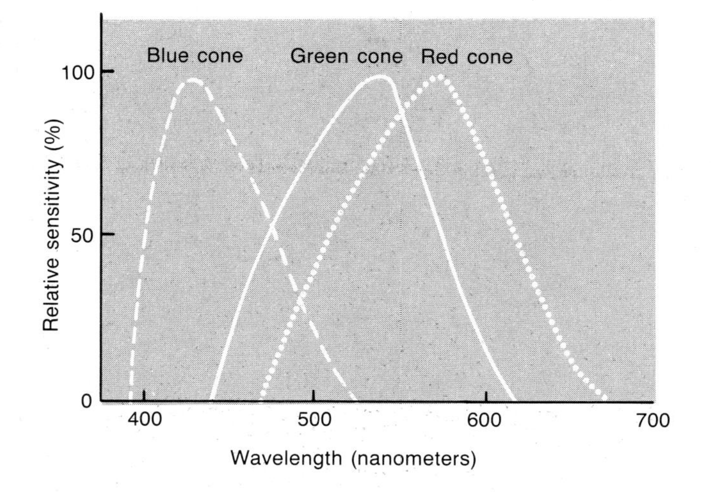

Photoreceptors in the dark release a neurotransmitter chemical at a constant rate at their synapses with bipolar cells. Hyperpolarization of the photoreceptors in response to light is similar to the hyperpolarization of other neurons that occurs during postsynaptic inhibition; the release of neurotransmitter by the photoreceptors is inhibited. The bipolar cells, as a result, may be less stimulated (if the neurotransmitter was excitatory) or less inhibited (if the neurotransmitter was inhibitory). There are both excitatory and inhibitory synapses between photoreceptors and bipolar cells, but it is not known if different transmitter chemicals are involved in this effect or if different bipolar cells respond in opposite ways to the same neurotransmitter. In either case, the bipolar cells may, through the mechanisms described, be either stimulated (by less inhibitory transmitter) or inhibited (by less excitatory transmitter) as a result of the hyperpolarization of photoreceptors in response to light.

Cones and Color Vision

Cones are less sensitive than rods to light, but provide color vision and greater visual acuity, as described in the next section. During the day, therefore, the high light intensity bleaches out the rods, and color vision with high acuity is provided by the cones. According to the **trichromatic theory** of color vision, our perception of a multitude of colors is due to stimulation of only three types of cones. Each type of cone contains retinaldehyde, as in rhodopsin, but this molecule is associated with a different protein than opsin. The protein is different for each of the three cone pigments, and as a result, each of the pigments has a different color. The three colors are blue, green, and red, which correspond to the region of the visible spectrum in which each cone pigment absorbs light maximally (fig. 10.44). Our perception of any given color is produced by the relative degree to which each cone is stimulated by any given wavelength of visible light.

Suppose a person has become dark-adapted in a photographic darkroom over a period of twenty minutes or longer, but needs an increase in light to examine some prints. Since rods do not absorb red light but red cones do, a red light in a photographic darkroom allows vision (because of the red cones), but does not cause bleaching of the rods. When the light is turned off, therefore, the rods will still be dark-adapted and the person will still be able to see.

Color blindness is due to a congenital lack of one or more types of cones. People with normal color vision are *trichromats;* people with only two types of cones are *dichromats.* They may be missing red cones (have *protanopia*), or green cones (have *deuteranopia*), or blue cones (have *tritanopia*). Such a person, for example, may have difficulty distinguishing red from green. People who are *monochromats* have only one cone system and can only see black, white, and shades of gray.

Visual Acuity and Sensitivity

While reading or similarly focusing visual attention on objects in daylight, each eye is oriented so that the image falls within a tiny area of the retina called the **fovea centralis.** The fovea is a pinhead-sized pit (*fovea* = pit) within a yellow area of the retina called the *macula lutea.* The

Figure 10.45. When the eyes "track" an object, the image is cast upon the fovea centralis of the retina. The fovea is literally a "pit" formed by parting of the neural layers, so that light falls directly on the photoreceptors (cones) in this region.

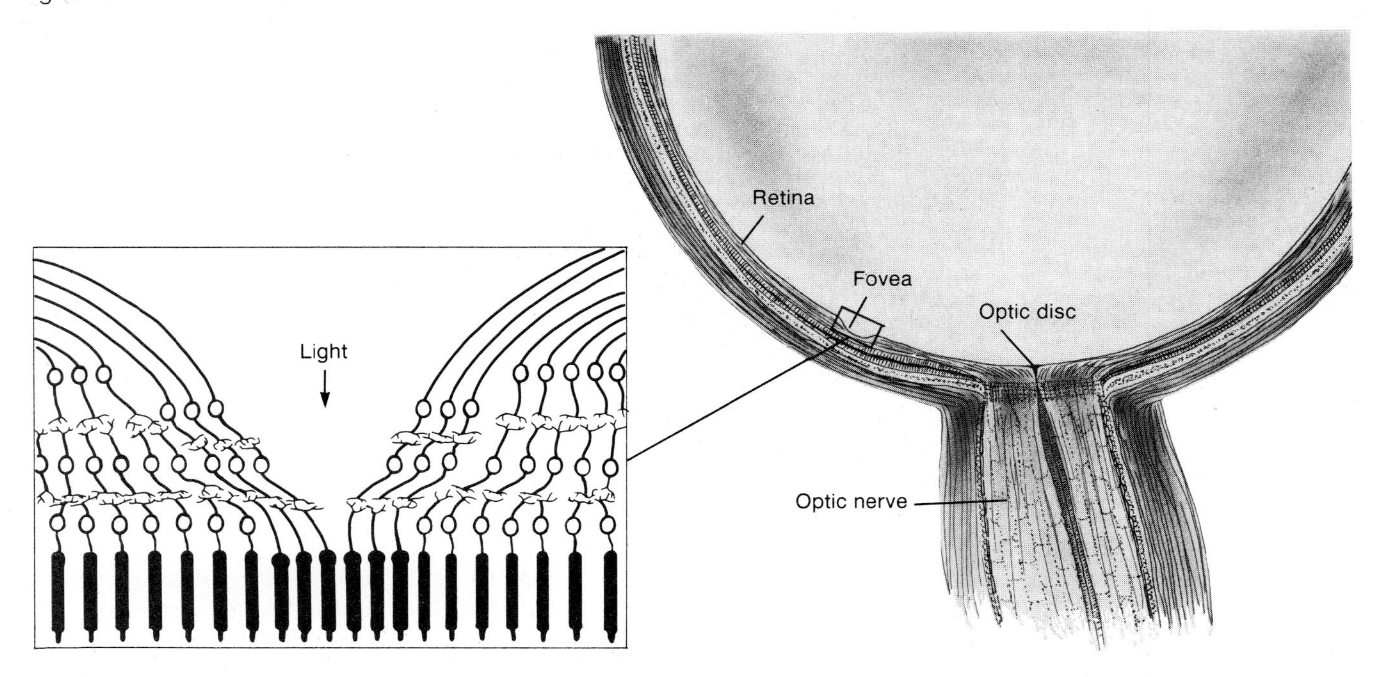

pit is formed as a result of the displacement of neural layers around the fovea, so that light falls directly on photoreceptors in this region (fig. 10.45)—whereas light falling on other areas must pass through several layers of neurons, as previously described.

There are approximately 120 million rods and 6 million cones in each retina, but only about 1.2 million nerve fibers enter the optic nerve of each eye. This gives an overall convergence of photoreceptors on ganglion cells of about 105:1. This number is misleading, however, because the degree of convergence is much lower for cones than for rods, and it is 1:1 in the fovea.

The photoreceptors are distributed in such a way that the fovea contains only cones, whereas more peripheral regions of the retina contain a mixture of rods and cones. Approximately four thousand cones in the fovea provide input to approximately four thousand ganglion cells; each ganglion cell in this region, therefore, has a private line to the visual field. Each ganglion cell thus receives input from an area of retina corresponding to the diameter of one cone (about 2 μm). Peripheral to the fovea, however, many rods synapse with a single bipolar cell, and many bipolar cells synapse with a single ganglion cell. A single ganglion cell outside the fovea thus may receive input from large numbers of rods, corresponding to an area of about 1 mm^2 on the retina (fig. 10.46).

Since each cone in the fovea has a private line to a ganglion cell, and since each ganglion cell receives input from only a tiny region of the retina, visual acuity is greatest and sensitivity to low light is poorest when light falls on the fovea. In dim light only the rods are activated, and vision is best out of the corners of the eye when the image falls away from the fovea. Under these conditions, the convergence of many rods on a single bipolar cell and the convergence of many bipolar cells on a single ganglion cell increase sensitivity to dim light at the expense of visual acuity. Night vision is therefore less distinct than day vision.

Neural Pathways from the Retina

As a result of light refraction by the cornea and lens, the right half of the visual field is projected to the left half of the retina of both eyes (the temporal half of the left retina and the nasal half of the right retina); the left half of the visual field is projected to the right half of the retina of both eyes. The temporal half of the left retina and the nasal half of the right retina therefore see the same image. Axons from ganglion cells in the left (temporal) half of the left retina pass to the left **lateral geniculate body** of the thalamus. Axons from ganglion cells in the nasal half of the right retina cross (decussate) in the *optic chiasm* to synapse also in the left lateral geniculate body. The left lateral geniculate, therefore, receives input from both eyes that relates to the right half of the visual field (fig. 10.47).

Figure 10.46. Since bipolar cells receive input from the convergence of many rods (*a*), and since a number of such bipolar cells converge on a single ganglion cell, rods provide high sensitivity to low levels of light at the expense of visual acuity. The 1 : 1 : 1 ratio of cones to bipolar cells to ganglion cells, (*b*), in contrast, provides high visual acuity, but low sensitivity.

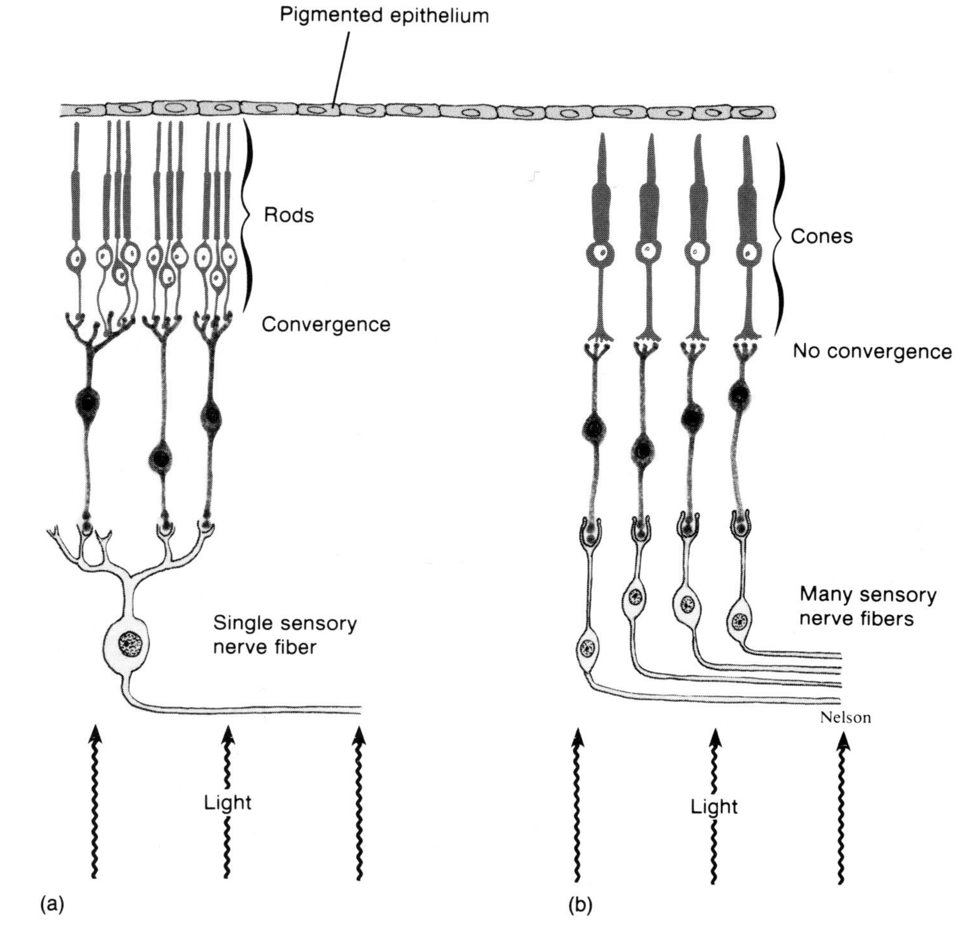

Figure 10.47. The neural pathway leading from the retina to the lateral geniculate body to the visual cortex. As a result of the crossing of optic fibers, the visual cortex of each cerebral hemisphere receives input from the opposite (contralateral) visual field.

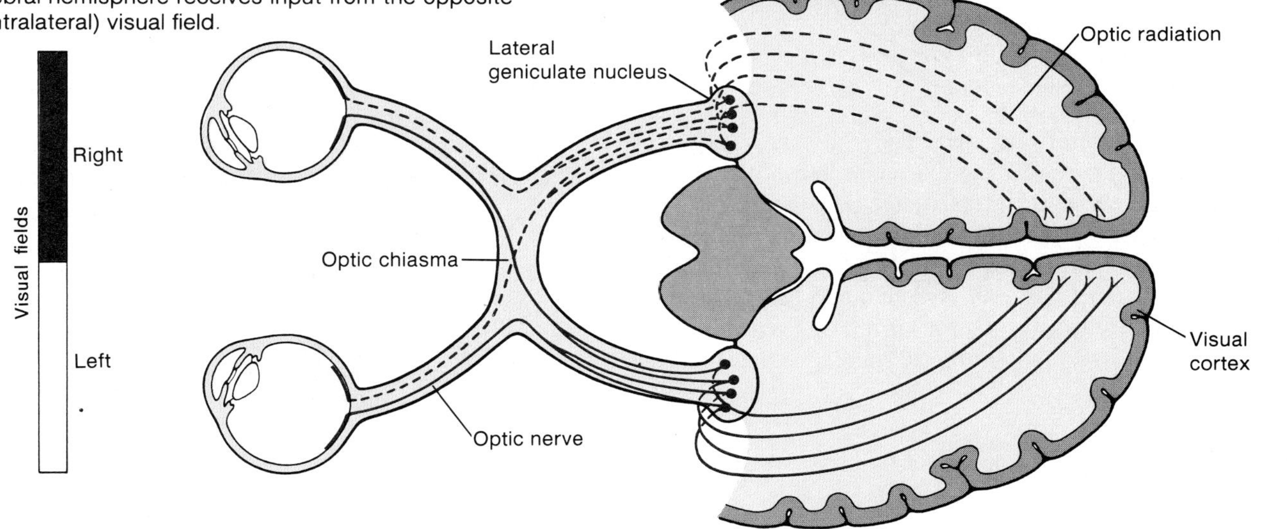

Figure 10.48. Visual fields of the eyes and neural pathways for vision. An overlapping of the visual field of each eye provides binocular vision, which is the ability to perceive depth.

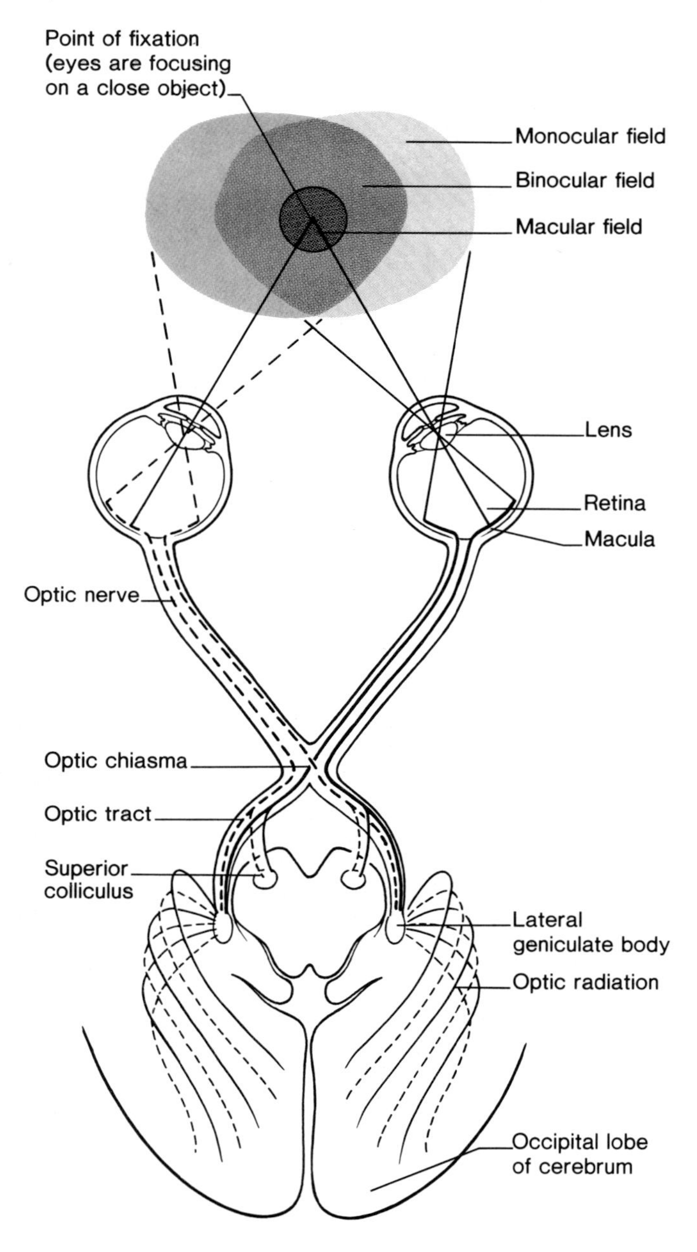

The right lateral geniculate body, similarly, receives input from both eyes relating to the left half of the visual field. Neurons in both lateral geniculate bodies of the thalamus in turn project to the **striate cortex** of the occipital lobe in the cerebral cortex (fig. 10.48). This area is also called area 17, in reference to a numbering system developed by Brodmann. Neurons in area 17 synapse with neurons in areas 18 and 19 of the occipital lobe (fig. 10.49).

Approximately 70% to 80% of the axons from the retina pass to the lateral geniculate bodies and to the striate cortex. This **geniculostriate system** is involved in perception of the visual field. Put another way, the geniculostriate system is needed for answering the question, What is it? Approximately 20% to 30% of the fibers from the retina, however, follow a different path to the *superior colliculus* of the midbrain (also called the *optic tectum*). Axons from the superior colliculus activate motor pathways leading to eye and body movements. The **tectal system,** in other words, is needed for answering the question, Where is it?

Superior Colliculus and Eye Movements. Neural pathways from the superior colliculus to motor neurons in the spinal cord help mediate the "startle" response to the sight of an unexpected intruder. Other nerve fibers from the superior colliculus stimulate the **extrinsic eye muscles** (table 10.5), which are the striated muscles that move the eyes.

There are two types of eye movements coordinated by the superior colliculus. *Smooth pursuit movements* track moving objects and keep the image focused on the fovea centralis. *Saccadic eye movements* are short (lasting 20 to 50 msec), jerky movements that occur while the eyes appear to be still. These saccadic movements are believed to be important in maintaining visual acuity.

The tectal system is also involved in the control of the intrinsic eye muscles—the iris and the muscles of the ciliary body. Shining a light into one eye stimulates the *pupillary reflex* in which both pupils constrict. This is caused by activation of parasympathetic neurons by fibers from the superior colliculus. Postganglionic neurons in the ciliary ganglia behind the eyes, in turn, stimulate constrictor fibers in the iris. Contraction of the ciliary body during *accommodation* also involves stimulation of the superior colliculus.

1. *List the different layers of the retina, and describe the path of light and of nerve activity through these layers.*
2. *Describe the photochemical reaction in the rods, and explain the process of dark adaptation.*
3. *Describe the electrical state of photoreceptors in the dark, and explain how light affects the electrical activity of retinal cells.*
4. *Explain the trichromatic theory of color vision.*
5. *Compare the architecture of the fovea centralis with more peripheral regions of the retina, and explain how this architecture relates to visual acuity and sensitivity.*
6. *Describe how different parts of the visual field are projected on the retinas of both eyes, and describe the neural pathways of this information in the geniculostriate system.*
7. *Describe the neural pathways involved in the tectal system, and explain how these pathways are involved in the control of the extrinsic eye muscles, iris, and ciliary body.*

Figure 10.49. The striate cortex (area 17) and the visual association (areas 18 and 19).

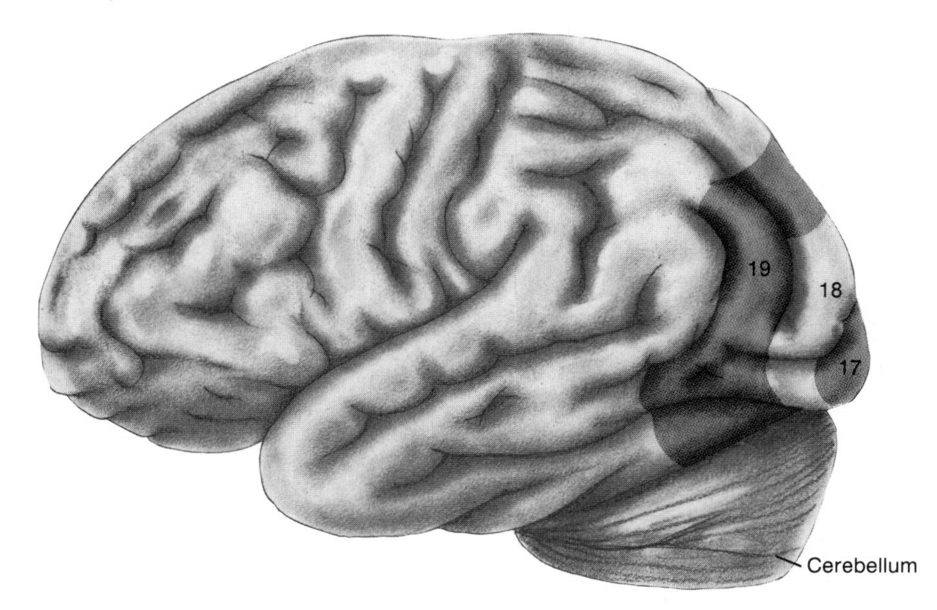

Table 10.5 Muscles of the eye.

Extrinsic Muscles (striated)		
Superior rectus	Oculomotor nerve (III)	Rotates eye upward and toward midline
Inferior rectus	Oculomotor nerve (III)	Rotates eye downward and toward midline
Medial rectus	Oculomotor nerve (III)	Rotates eye toward midline
Lateral rectus	Abducens nerve (VI)	Rotates eye away from midline
Superior oblique	Trochlear nerve (IV)	Rotates eye downward and away from midline
Inferior oblique	Oculomotor nerve (III)	Rotates eye upward and away from midline
Intrinsic Muscles (smooth)		
Ciliary muscles	Oculomotor nerve (III) parasympathetic fibers	Causes suspensory ligaments to relax
Iris, circular muscles	Oculomotor nerve (III) parasympathetic fibers	Causes size of pupil to decrease
Iris, radial muscles	Sympathetic fibers	Causes size of pupil to increase

Neural Processing of Visual Information

Light that is cast on the retina directly affects the activity of photoreceptors and indirectly affects the neural activity in bipolar and ganglion cells. The part of the visual field that affects the activity of a particular ganglion cell can be considered to be its *receptive field*. Since each cone in the fovea has a private line to a ganglion cell, the receptive fields of these ganglion cells are equal to the width of one cone (about 2 μm). Ganglion cells in more peripheral parts of the retina receive input from hundreds of photoreceptors, and thus are influenced by a larger area of the retina (about 1 mm in diameter).

Ganglion Cell Receptive Fields

Studies of the electrical activity of ganglion cells have yielded some surprising results. In the dark, each ganglion cell discharges spontaneously at a slow rate. When the room lights are turned on, the firing rate of many (but not all) ganglion cells increases slightly. A small spot of light that is directed at the center of some ganglion cells' receptive fields, however, stimulates a large increase in firing rate. A small spot of light can thus be a more effective stimulus than larger areas of light.

When the spot of light is moved only a short distance away from the center of the receptive field the ganglion cell responds in an opposite manner. The ganglion cell that was stimulated with light at the center of its receptive field is inhibited by light in the periphery of its field. The response produced by light in the center and by light in the "surround" of the visual field is *antagonistic*. Those ganglion cells that are stimulated by light at the center of their visual fields are said to have **on-center fields.** Ganglion cells that are inhibited by light in the center and stimulated by light in the surround have **off-center fields.**

The reason wide illumination of the retina has less effect than pinpoint illumination is now clear; diffuse illumination gives the ganglion cell conflicting orders—on and off. Because of the antagonism between the center and surround of ganglion cell receptive fields, the activity of each ganglion is a result of the *difference in light intensity* between the center and surround of its visual field. This is a form of *lateral inhibition* that helps to accentuate the contours of images and improve visual acuity.

Lateral Geniculate Bodies

Each of the two lateral geniculate bodies receives input from ganglion cells in both eyes. The right lateral geniculate receives input from the right half of each retina (corresponding to the left half of the visual field); the left lateral geniculate receives input from the left half of each retina (corresponding to the right half of the visual field). Each neuron in the lateral geniculate, however, is activated only by input from one eye. Neurons that are activated by ganglion cells from the left eye are in separate

Figure 10.50. The lateral geniculate nucleus. Each lateral geniculate consists of six layers (numbered 1 through 6 in this figure). Each of these layers receives input from only one eye, with right and left eyes alternating. An arrow through these six layers (see figure) of the left lateral geniculate, for example, encounters corresponding projections from a part of the right visual field in right and left eyes, alternatively, as it passes from the outer to the inner layers.

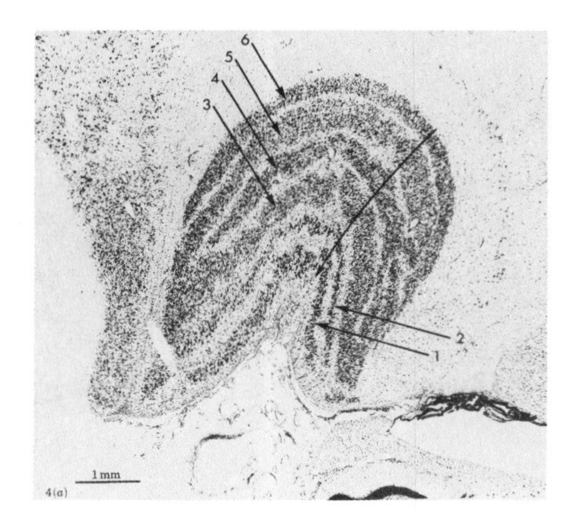

layers within the lateral geniculate from those that are activated by the right eye (fig. 10.50).

The receptive field of each ganglion cell, as previously described, is the part of the retina it "sees" through its photoreceptor input. The receptive field of lateral geniculate neurons, similarly, is the part of the retina it "sees" through its ganglion cell input. Experiments in which the lateral geniculate receptive fields are mapped with a spot of light reveal that they are circular, with an antagonistic center and surround, much like the ganglion cell receptive fields.

The Cerebral Cortex

Projections of nerve fibers from the lateral geniculate bodies to area 17 of the occipital lobe form the *optic radiation* (see fig. 10.48). These fiber projections give area 17 a striped or striated appearance; this area is consequently also known as the striate cortex. Neurons in area 17, in turn, project to areas 18 and 19 of the occipital lobe. Cortical neurons in areas 17, 18, and 19 are thus stimulated indirectly by light on the retina. On the basis of their stimulus requirements, these cortical neurons are classified as simple, complex, and hypercomplex.

Simple Neurons. The receptive fields of simple cortical neurons are rectangular rather than circular. This results from the fact that they receive input from lateral geniculate neurons whose receptive fields are aligned in a particular way (as illustrated in fig. 10.51). Simple cortical neurons are best stimulated by a slit or bar of light that is located in a precise part of the visual field (of either eye) at a precise orientation (fig. 10.52).

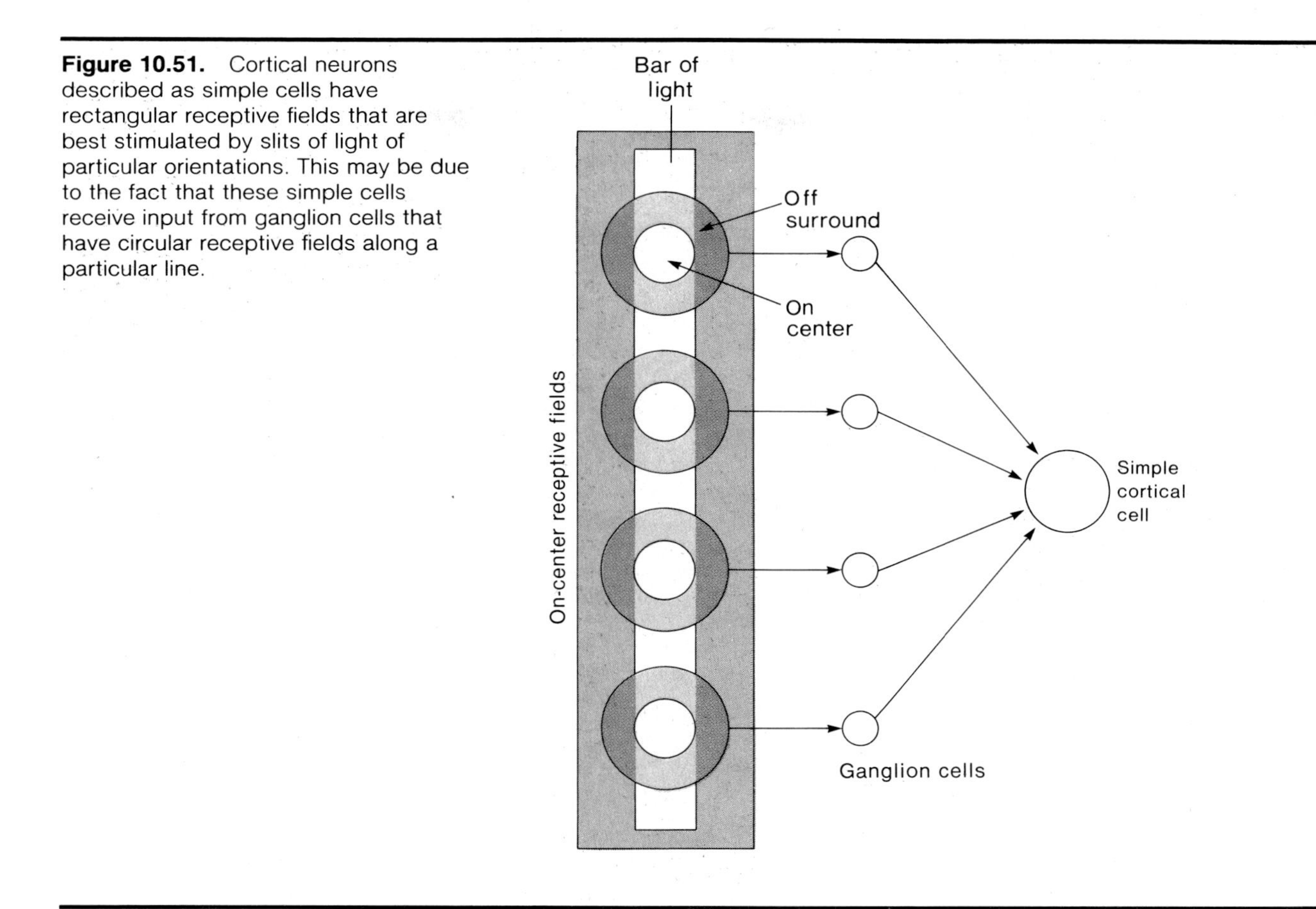

Figure 10.51. Cortical neurons described as simple cells have rectangular receptive fields that are best stimulated by slits of light of particular orientations. This may be due to the fact that these simple cells receive input from ganglion cells that have circular receptive fields along a particular line.

Figure 10.52. Simple cells are best stimulated by a slit or bar of light along a particular orientation within a particular region of the receptive field. The behavior of two different cortical cells (*a*) and (*b*) is illustrated. (Adapted from "Cellular Communication," by Stent, Gunther.)

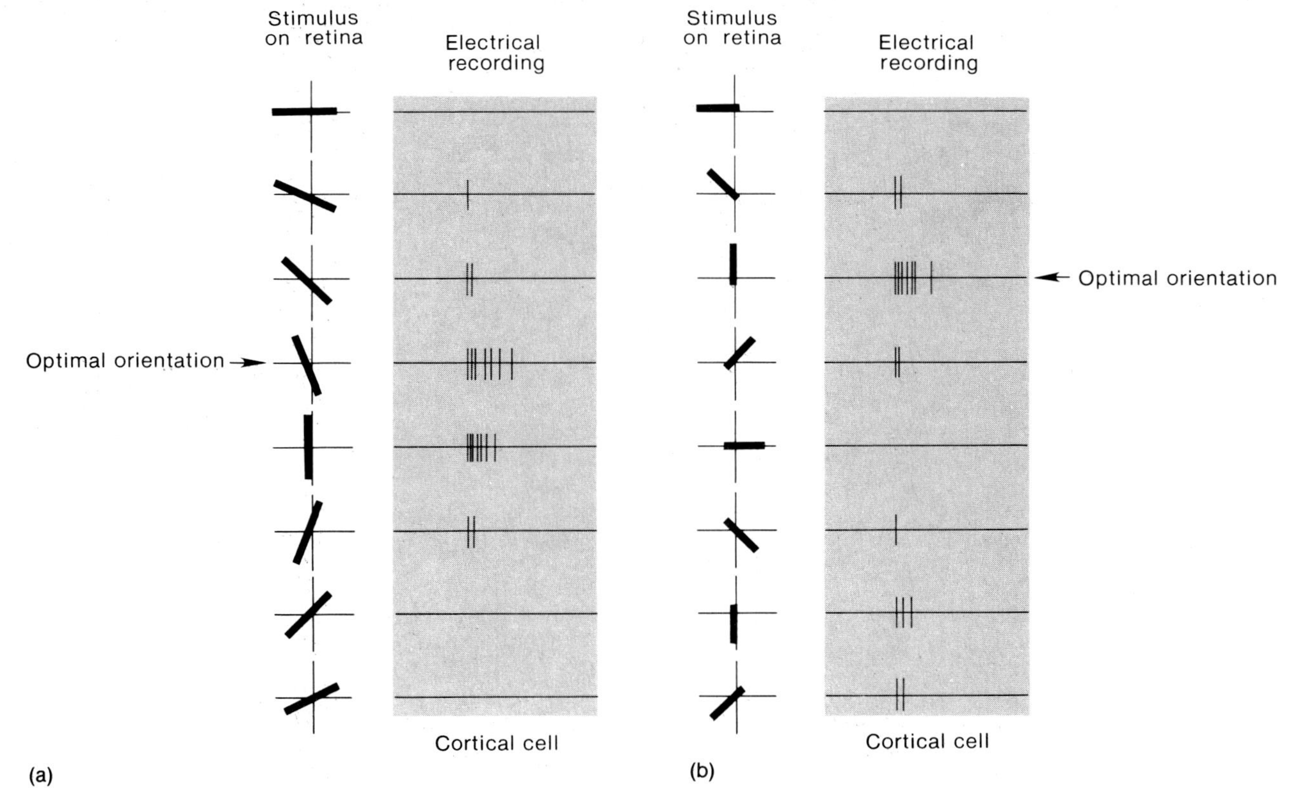

The striate cortex (area 17) contains simple, complex, and hypercomplex neurons. The other visual association areas, designated areas 18 and 19, contain only complex and hypercomplex cells. Complex neurons receive input from simple cells, and hypercomplex neurons receive input from complex cells.

Complex and Hypercomplex Neurons. Complex cells respond best to straight lines with a specific orientation that move in a particular direction through the receptive field. Unlike simple cells, complex cells do not require that the stimulus have a particular position within the receptive field. Hypercomplex cells require that the stimulus be of a particular length or have a particular bend or corner.

The dotlike information from ganglion and lateral geniculate cells is thus transformed in the occipital lobe into information about edges—their position, length, orientation, and movement. Although this represents a high degree of abstraction, the visual association areas of the occipital lobe probably serve as only an early stage in the integration of visual information. Other areas of the brain receive input from the visual association areas and provide meaning to visual perception.

---

1. Describe the response characteristics of ganglion cells to light on the retina, and explain why a small spot of light is a more effective stimulus than general illumination of the retina.
2. Explain how the arrangement of ganglion cells' receptive fields can enhance visual acuity.
3. Describe the stimulus requirements of simple, complex, and hypercomplex cortical cells.

Summary

Characteristics of Sensory Receptors p. 238

I. Sensory receptors may be categorized on the basis of their structure, the stimulus energy they transduce, or the nature of their response.
   A. Receptors may be dendritic nerve endings, specialized neurons, or specialized epithelial cells associated with sensory nerve endings.
   B. Receptors may be chemoreceptors, photoreceptors, thermoreceptors, mechanoreceptors, or nocioreceptors.
      1. Proprioceptors include receptors in the muscles, tendons, and joints.
      2. The sense of sight, hearing, taste, olfaction, and equilibrium are grouped as special senses.
   C. Tonic receptors fire continuously as long as the stimulus is applied; these monitor the presence and intensity of a stimulus.
      1. Phasic receptors respond to stimulus changes; they do not respond to a sustained stimulus.
      2. Phasic receptors therefore help to provide sensory adaptation to sustained stimuli.

II. According to the law of specific nerve energies, each sensory receptor responds with lowest threshold to only one modality of sensation.
   A. That stimulus modality is called the adequate stimulus.
   B. Stimulation of the sensory nerve from a receptor by any means is interpreted in the brain as the adequate stimulus modality of that receptor.

III. Generator potentials are graded changes (usually depolarizations) in the membrane potential of the dendritic endings of sensory neurons.
   A. The magnitude of the potential change of the generator potential is directly proportional to the strength of the stimulus applied to the receptor.
   B. After the generator potential reaches a threshold value, increases in the magnitude of the depolarization results in increased frequency of action potential production in the sensory neuron.

Cutaneous Sensations p. 241

I. Somatesthetic information—from cutaneous and proprioceptors—project by third-order neurons to the postcentral gyrus of the cerebrum.
   A. Proprioception and pressure sensations ascend on the ipsilateral side of the spinal cord, synapse in the medulla and cross to the contralateral side, and ascend to the thalamus in the medial lemniscus; neurons from the thalamus in turn project to the postcentral gyrus.
   B. Sensory neurons from other cutaneous receptors synapse and cross to the contralateral side in the spinal cord and ascend to the thalamus in the lateral and ventral spinothalamic tracts; neurons in the thalamus then project to the postcentral gyrus.

II. The receptive field of a cutaneous sensory neuron is the area of skin that, when stimulated, produces responses in the neuron.
   A. The receptive fields are smaller where the skin has a greater density of cutaneous receptors.
   B. The two-point touch threshold test reveals that the fingertips and mouth areas have a greater density of touch receptors and thus greater sensory acuity than other areas of the body.

III. Lateral inhibition acts to sharpen the sensation by inhibiting the activity of sensory neurons coming from areas of the skin around the area that is most greatly stimulated; the contrast between the most stimulated area and areas of lesser stimulation is thereby increased.

Taste and Olfaction p. 245

I. The sense of taste is mediated by taste buds.
   A. The cells that compose the taste buds are innervated by the glossopharyngeal and facial nerves.
   B. A particular taste bud is most sensitive to one of four taste modalities: sweet, sour, bitter, and salty.

C. Taste buds that are most sensitive to one modality of taste are located in characteristic regions of the tongue.

II. The olfactory receptors are dendritic endings of the olfactory (first cranial) nerve.

Vestibular Apparatus and Equilibrium p. 248

I. The structures for equilibrium and hearing are located in the inner ear in the membranous labyrinth.
   A. The structure involved in equilibrium, known as the vestibular apparatus, consists of the otolith organs (utricle and saccule) and the semicircular canals.
   B. The utricle and saccule provide information about linear acceleration, whereas the semicircular canals provide information about angular acceleration.
   C. The sensory receptors for equilibrium are hair cells that have microvilli and one cilium (called a kinocilium).
      1. When the microvilli are bent in the direction of the kinocilium, the cell membrane becomes depolarized.
      2. When the microvilli are bent in the opposite direction, the membrane becomes hyperpolarized.

II. The hairs of the hair cells in the utricle and saccule protrude into the endolymph of the membranous labyrinth and are embedded within a gelatinous otolith membrane.
   A. When a person is upright, the hairs of the utricle are oriented vertically and those of the saccule are oriented horizontally.
   B. Linear acceleration causes a shearing force between the hairs and the otolith membrane, thus bending the hairs and electrically stimulating the sensory endings.

III. The three semicircular canals are oriented at right angles to each other, like the faces of a cube.
   A. The hair cells are embedded within a gelatinous membrane called the cupula, which projects into the endolymph.
   B. Movement along one of the planes of a semicircular canal causes the endolymph to bend the cupula and stimulate the hair cells.
   C. Stimulation of the hair cells in the vestibular apparatus activates sensory neurons of the vestibulocochlear nerve, which projects to the cerebellum and the vestibular nuclei of the medulla oblongata.
      1. The vestibular nuclei in turn send fibers to the oculomotor center, which controls eye movements.
      2. Spinning and then stopping abruptly can thus cause oscillatory movements of the eyes called nystagmus.

The Ears and Hearing p. 253

I. The outer ear funnels sound waves of a given frequency (measured in hertz) and intensity (measured in decibels) to the tympanic membrane, causing it to vibrate.

II. Vibrations of the tympanic membrane cause movement of the middle ear ossicles—malleus, incus, and stapes—which in turn produces vibrations of the oval window of the cochlea.

III. Vibrations of the oval window set up a traveling wave of perilymph in the scala vestibuli.
   A. This wave can pass around the helicotrema to the scala tympani, or it can reach the scala tympani by passing through the scala media (cochlear duct).
   B. The scala media is filled with endolymph.
      1. The membrane of the cochlear duct that faces the scala vestibuli is called Reissner's membrane.
      2. The membrane that faces the scala tympani is called the basilar membrane.

IV. The sensory structure of the cochlea is called the organ of Corti.
   A. The organ of Corti consists of sensory hair cells on the basilar membrane.
      1. The hairs of the hair cells project upwards into an overhanging tectorial membrane.
      2. The hair cells are innervated by the eighth cranial nerve.
   B. Sounds of high frequency cause maximum displacement of the basilar membrane closer to its base near the stapes; sounds of lower frequency produce maximum displacement of the basilar membrane closer to its apex near the helicotrema.
      1. Displacement of the basilar membrane causes the hairs to bend against the tectorial membrane and stimulate the production of nerve impulses.
      2. Pitch discrimination is thus dependent on the region of the basilar membrane that vibrates maximally to sounds of different frequencies.
      3. Pitch discrimination is enhanced by lateral inhibition.

The Eyes and Vision p. 263

I. Light enters the cornea of the eye, passes through the pupil (the opening of the iris), and then through the lens, from which it is projected to the retina in the back of the eye.
   A. Light rays are bent, or refracted, by the cornea and lens.
   B. Because of refraction, the image on the retina is upside down and right to left.
   C. The right half of the visual field is projected to the left half of the retina in each eye, and vice versa.

II. Accommodation is the ability to maintain a focus on the retina when the distance between the object and the eyes is changed.
   A. Accommodation is produced by changes in the shape and refractive power of the lens.
   B. When the muscles of the ciliary body are relaxed, the suspensory ligament is tight and the lens is pulled to its least convex form.
      1. This gives the lens a low refractive power for distance vision.
      2. As an object is brought closer than 20 feet from the eyes, the ciliary body contracts, the suspensory ligament becomes less tight, and the lens becomes more convex and more powerful.

III. Visual acuity refers to the sharpness of the image and depends in part on the ability of the lens to bring the image to a focus on the retina.
   A. People with myopia have an eyeball that is too long, so that the image is brought to a focus in front of the retina; this is corrected by concave lenses.
   B. People with hyperopia have an eyeball that is too short, so that the image is brought to a focus behind the retina; this is corrected by a convex lens.

C. Astigmatism is the uneven refraction of light around 360 degrees of a circle, caused by uneven curvature of the cornea and/or lens.

The Retina p. 270

I. The retina contains photoreceptor neurons called rods and cones, which synapse with bipolar cells.
  A. When light strikes the rods, it causes the photodissociation of rhodopsin into retinaldehyde and opsin.
    1. This occurs maximally with a light wavelength of 500 nm.
    2. This reaction is called the bleaching reaction.
    3. Photodissociation is caused by the conversion of the 11-*cis* to the all-*trans* form of retinaldehyde, which cannot bond to opsin.
  B. In the dark, more rhodopsin can be produced and contained in the rods, so that the eyes become more sensitive to light; this is responsible for one component of dark adaptation.
  C. The rods provide black-and-white vision under conditions of low light intensity; at higher light intensity the rods are bleached out and the cones provide color vision.

II. In the dark, there is a constant movement of Na^+ into the rods that produces what is known as a "dark current."
  A. When light causes the dissociation of rhodopsin, the Na^+ channels become blocked and the rods become hyperpolarized in comparison to their membrane potential in the dark.
  B. Hyperpolarization of the rods results in their release of less neurotransmitter at their synapses with bipolar cells.
  C. Neurotransmitters from rods cause depolarization of bipolar cells in some cases, and hyperpolarization of bipolar cells in other cases; so when the rods are in light and release less neurotransmitter these effects are inverted.

III. According to the trichromatic theory of color vision, there are three systems of cones each of which responds to one of three colors: red, blue, and green.
  A. Each type of cone contains retinaldehyde attached to a different type of protein.
  B. The names for the cones signify the region of the spectrum in which the cones absorb light maximally.

IV. The fovea centralis contains only cones; more peripheral parts of the retina contain both cones and rods.
  A. Each cone in the fovea synapses with one bipolar cell, which in turn synapses with one ganglion cell.
    1. The ganglion cell that receives input from the fovea thus has a visual field equal to only that part of the retina which activated its cone.
    2. As a result of this 1:1 ratio of cones to bipolar cells, visual acuity is high in the fovea but sensitivity to low light levels is less than in other regions of the retina.
  B. In other regions of the retina, where rods predominate, there are a large number of rods that provide input to each ganglion cell (there is great convergence); visual acuity, as a result, is impaired, but sensitivity to low light levels is improved.

V. The right half of the visual field is projected to the left half of the retina of both eyes.
  A. The left half of the left retina sends fibers to the left lateral geniculate body of the thalamus.
  B. The left half of the right retina also sends fibers to the left lateral geniculate body; this is because these fibers decussate in the optic chiasma.
  C. The left lateral geniculate body thus receives input from the left half of the retina of both eyes, corresponding to the right half of the visual field; the right lateral geniculate receives information about the left half of the visual field.
    1. Neurons in the lateral geniculate bodies send fibers to the striate cortex of the occipital lobes.
    2. The geniculostriate system is involved in providing meaning to the images that form on the retina.
  D. Instead of synapsing in the geniculate bodies, some fibers from the ganglion cells of the retina synapse in the superior colliculus of the midbrain, which controls eye movements.
    1. Since this brain region is also called the optic tectum, this pathway is called the tectal system.
    2. The tectal system enables the eyes to move and track an object; it is also responsible for the pupillary reflex and the changes in lens shape that are needed for accommodation.

Neural Processing of Visual Information p. 278

I. The area of the retina that provides input to a ganglion cell is called the receptive field of the ganglion cell.
  A. The receptive field of a ganglion cell is roughly circular, with an "on" or "off" center and an antagonistic surround.
    1. A spot of light in the center of an "on" receptive field stimulates the ganglion cell; a spot of light in its surround inhibits the ganglion cell.
    2. The opposite is true for ganglion cells with "off" receptive fields.
    3. A larger light that stimulates both the center and the surround of a receptive field has less net effect on a ganglion cell than a smaller spot of light that only illuminates either the center or the surround.
  B. The antagonistic center and surround of the receptive field of ganglion cells provide lateral inhibition, which enhances contours and provides better visual acuity.

II. Each lateral geniculate body receives input from both eyes relating to the same part of the visual field.
   A. The neurons receiving input from each eye are arranged in layers within the lateral geniculate.
   B. The receptive fields of neurons in the lateral geniculate are circular with antagonistic center and surround, much like the receptive field of ganglion cells.

III. Cortical neurons involved in vision are either simple, complex, or hypercomplex.
   A. Simple neurons receive input from neurons in the lateral geniculate, complex neurons receive input from simple cells, and hypercomplex neurons receive input from complex cells.
   B. Simple neurons are best stimulated by a slit or bar of light that is located in a precise part of the visual field and has a precise orientation.
   C. Complex cells respond best to a straight line that has a particular orientation and moves in a particular direction; the position of the line in the visual field is not important.
   D. Hypercomplex cells respond best to lines that have a particular length or have a particular bend or corner.
   E. Information from these cells in the occipital lobe is integrated with information from other areas of the brain to provide meaningful visual perception.

Review Activities

Objective Questions

Match the following:

1. utricle and saccule
2. semicircular canals
3. cochlea

(a) cupula
(b) tectorial membrane
(c) basilar membrane
(d) otolith membrane

4. The dissociation of rhodopsin in the rods in response to light causes
   (a) the Na^+ channels to become blocked
   (b) the rods to secrete less neurotransmitter
   (c) the bipolar cells to become either stimulated or inhibited
   (d) all of the above
5. Tonic receptors
   (a) are fast-adapting
   (b) do not fire continuously to a sustained stimulus
   (c) produce action potentials at a greater frequency as the generator potential is increased
   (d) all of the above
6. Cutaneous receptive fields are smallest in
   (a) the fingertips
   (b) the back
   (c) the thighs
   (d) the arms
7. The process of lateral inhibition
   (a) increases the sensitivity of receptors
   (b) promotes sensory adaptation
   (c) increases sensory acuity
   (d) prevents adjacent receptors from being stimulated
8. The receptors for taste are
   (a) naked sensory nerve endings
   (b) encapsulated sensory nerve endings
   (c) modified epithelial cells
9. The utricle and saccule
   (a) are otolith organs
   (b) are located in the middle ear
   (c) provide a sense of linear acceleration
   (d) both *a* and *c*
   (e) both *b* and *c*
10. Since fibers of the optic nerve that originate in the nasal halves of each retina cross at the optic chiasma, each lateral geniculate receives input from
   (a) both the right and left sides of the visual field of both eyes
   (b) the ipsilateral visual field of both eyes
   (c) the contralateral visual field of both eyes
   (d) the ipsilateral field of one eye and the contralateral field of the other eye
11. When a person with normal vision views an object from a distance of at least 20 feet
   (a) the ciliary muscles are relaxed
   (b) the suspensory ligament is tight
   (c) the lens is in its most flat, least convex shape
   (d) all of the above
12. Glasses with concave lenses help correct
   (a) presbyopia
   (b) myopia
   (c) hyperopia
   (d) astigmatism
13. Parasympathetic nerves that stimulate constriction of the iris (in the pupillary reflex) are activated by neurons in
   (a) the lateral geniculate
   (b) the superior colliculus
   (c) the inferior colliculus
   (d) the striate cortex
14. A bar of light in a specific part of the retina, with a particular length and orientation, is the most effective stimulus for
   (a) ganglion cells
   (b) lateral geniculate cells
   (c) simple cortical cells
   (d) complex cortical cells

Essay Questions

1. Define the term *lateral inhibition,* and give examples of its effects in three sensory systems.
2. Describe the nature of the generator potential, and explain its relationship to stimulus intensity and to frequency of action potential production.
3. Explain how the vestibular apparatus provides information about changes in the position of our body in space.
4. Define accommodation, and explain how it is accomplished. Why is it more of a strain on the eyes to look at a small object close to the eyes than large objects far away?
5. Describe the effects of light on the photoreceptors, and explain how these effects influence the bipolar cells.
6. Explain why images that fall on the fovea centralis are seen more clearly than images that fall on the periphery of the retina. Why are the "corners of the eyes" more sensitive to light than the fovea?
7. Why do rods provide only black-and-white vision? Include a discussion of different types of color blindness in your answer.
8. Why are green-colored objects seen better at night than objects of other colors? What effect does red light in a darkroom have on a dark-adapted eye?
9. Describe the receptive fields of ganglion cells, and explain how the nature of these fields helps to improve visual acuity.

Selected Readings

Boynton, R. M. 1979. *Human color vision.* New York: Holt, Rinehart & Winston.

Goldberg, J. M., and C. Fernandez. 1975. Vestibular mechanisms. *Annual Review of Physiology* 37:129.

Hubel, D. H. 1979. The visual cortex of normal and deprived monkeys. *American Scientist* 67:532.

Hubel, D. H., and T. Wiesel. September 1979. Brain mechanisms of vision. *Scientific American.*

Hudspeth, A. J. February 1983. The hair cells of the inner ear. *Scientific American.*

Loeb, G. E. February 1985. The functional replacement of the ear. *Scientific American.*

MacNichol, E. F., Jr. December 1964.Three-pigment color vision. *Scientific American.*

Nathans, J., D. Thomas, and D. S. Hogness. 1986. Molecular genetics of human vision: The genes encoding blue, green, and red pigments. *Science* 232:193.

O'Brian, D. F. 1982. The chemistry of vision. *Science* 218:961.

Parker, D. E. November 1980. The vestibular apparatus. *Scientific American.*

Pfaffmann, C., Frank, M., and R. Norgren. 1979. Neural mechanisms and behavioral aspects of taste. *Annual Review of Psychology* 30:283.

Rhode, W. S. 1984. Cochlear mechanics. *Annual Review of Physiology* 46:231.

Rushton, W. A. H. March 1975. Visual pigments and color blindness. *Scientific American.*

Van Essen, D. C. 1979. Visual areas of the mammalian cerebral cortex. *Annual Review of Neurosciences* 2:277.

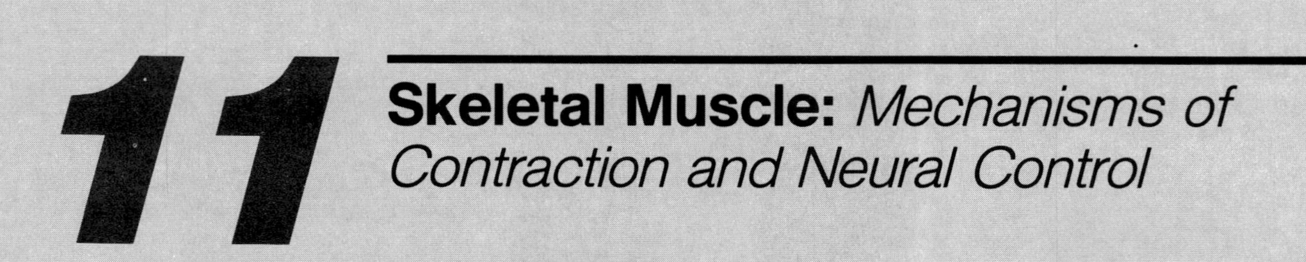

11 Skeletal Muscle: *Mechanisms of Contraction and Neural Control*

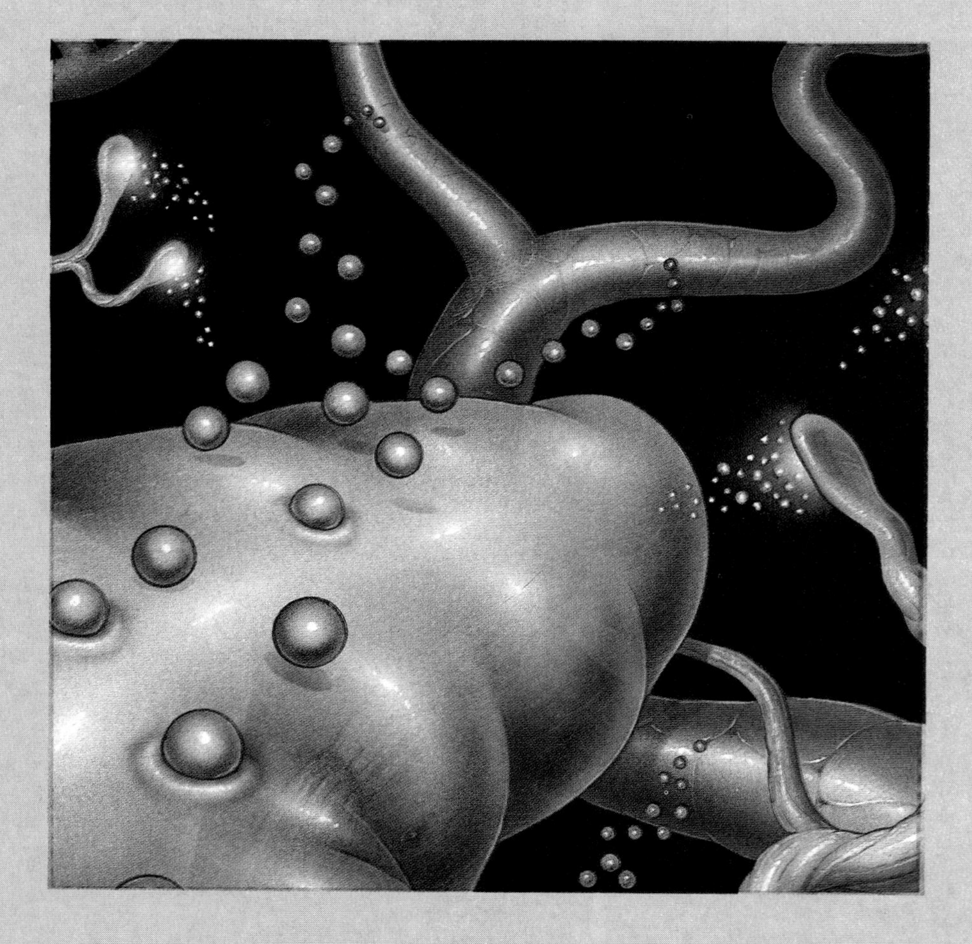

Objectives

By studying this chapter, you should be able to

1. describe the gross and microscopic structure of skeletal muscles
2. describe the all-or-none law of muscle fibers and explain how muscles can produce graded and sustained contractions
3. define a motor unit, explain how motor units vary, and explain how the motor units are used to control muscle contraction
4. distinguish between isometric and isotonic contraction and describe how the series elastic component affects muscle contraction
5. describe the structure and functions of muscle spindles and explain the mechanisms involved in a stretch reflex
6. explain the function of Golgi tendon organs and explain why a slow, gradual muscle stretch might avoid the spasm that could be caused by a rapid stretch
7. explain reciprocal innervation and describe the neural pathways involved in a crossed extensor reflex
8. describe the significance of gamma motoneurons in the neural control of muscle contraction and in the maintenance of muscle tone
9. describe the neural pathways involved in the pyramidal and extrapyramidal systems
10. describe the structure of sarcomeres and of myofibrils and explain how this structure accounts for the striated appearance of muscle fibers
11. explain the sliding filament theory of contraction
12. explain the events that occur during cross-bridge cycles and the role of ATP in muscle contraction
13. explain the physiological roles of tropomyosin and troponin and the role of Ca^{++} in excitation-contraction coupling
14. describe how action potentials in a nerve and muscle fiber affect the contractile mechanism and explain the role of the sarcoplasmic reticulum in muscle contraction and relaxation
15. explain the differences in structure and function between slow-twitch, fast-twitch, and intermediate fibers
16. explain how muscles fatigue and what changes occur in muscle fibers in response to physical training

Outline

Figure 11.1. Actions of antagonistic muscles in the leg.

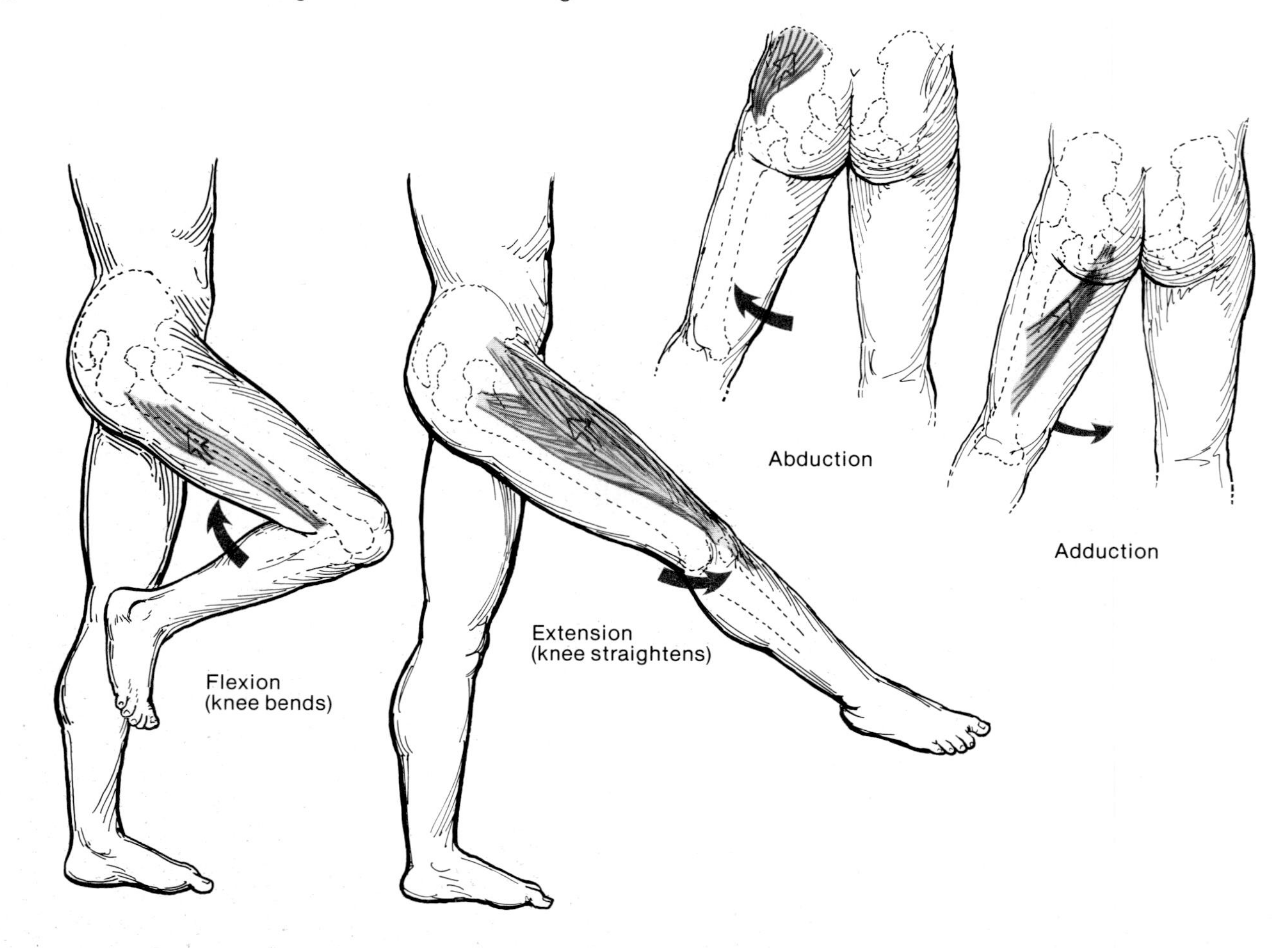

Table 11.1 Categories of skeletal muscle actions.

Category	Action
Extensor	Increases the angle at a joint
Flexor	Decreases the angle at a joint
Abductor	Moves limb away from the midline of the body
Adductor	Moves limb towards the midline of the body
Levator	Moves insertion upward
Depressor	Moves insertion downward
Rotator	Rotates a bone along its axis
Sphincter	Constricts an opening

Structure and Actions of Skeletal Muscles

Skeletal muscles are usually attached to bone on each end by tough connective-tissue *tendons.* When a muscle shortens during contraction its more movable attachment, known as its *insertion,* is pulled towards its less movable attachment (the *origin*). A variety of different skeletal movements are possible, depending on the type of joint involved and the attachments of the muscles (table 11.1 and fig. 11.1). When *flexor* muscles contract, for example, they decrease the angle of a joint. Contraction of *extensor* muscles increases the angle of their attached bones at the joint. Flexors and extensors that attach to the same bones are thus *antagonistic muscles.*

Figure 11.2. The relationship between muscle fibers and the connective tissues of the tendon, epimysium, perimysium, and endomysium.

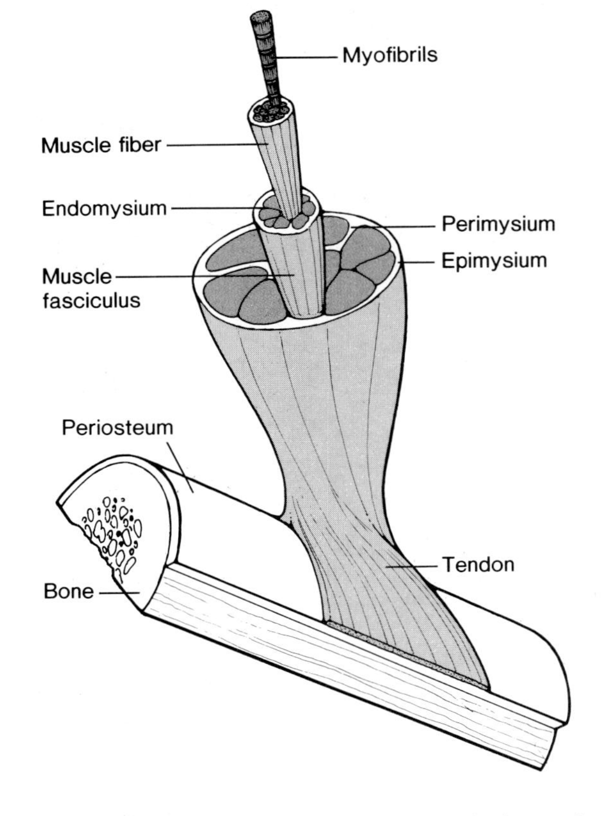

The position of the limbs, for example, is determined by the actions of a variety of antagonistic muscles. In addition to the positions of flexion and extension, a limb can be moved away from the midline of the body by *abductor* muscles, and brought inward towards the midline by contraction of *adductor* muscles. In all cases, these skeletal movements are produced by the shortening of the appropriate muscle groups—the *agonists*—while the antagonist muscles remain relaxed.

Structure of Skeletal Muscles

The fibrous connective tissue proteins within the tendons continue in an irregular arrangement around the muscle to form a sheath known as the *epimysium* (*epi* = above; *my* = muscle). Connective tissue from this outer sheath extends into the body of the muscle, subdividing it into columns, or *fascicles* (e.g., the "strings" in stringy meat). Each of these fascicles is thus surrounded by its own connective tissue sheath, known as the *perimysium* (*peri* = around).

Figure 11.3. Appearance of skeletal muscle fibers in the light microscope (note the striated appearance produced by alternating dark A bands and light I bands).

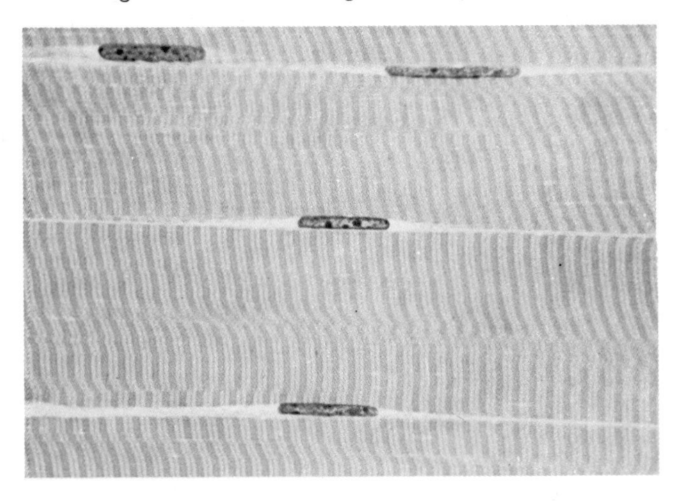

Dissection of a muscle fascicle under a microscope reveals that it, in turn, is composed of many **muscle fibers** (or *myofibers*) surrounded by wisps of connective tissue called *endomysium* (fig. 11.2). These fibers are the muscle cells. Since the connective tissue of the tendons, epimysium, perimysium, and endomysium is continuous, muscle fibers do not normally pull out of the tendons when they contract.

Despite their unusual fiber shape, muscle cells have the same organelles that are present in other cells: mitochondria, intracellular membranes, glycogen granules, and others. Unlike most other cells in the body, skeletal muscle fibers are multinucleate—that is, they contain many nuclei. The most distinctive feature of the microscopic appearance of skeletal muscle fibers, however, is their **striated** appearance (fig. 11.3). The striations (stripes) are produced by alternating dark and light bands that appear to cross the width of the fiber.

The dark bands are able to polarize visible light (are *anisotropic*) and have therefore been named *A bands*. The light bands do not polarize light (are *isotropic*) and are thus called *I bands*. At high magnification, thin dark lines can be seen in the middle of the I bands. These were once thought to be membranes and were accordingly labeled *Z lines* (for *Zwischenscheibe*, a German word for "membrane").

Figure 11.4. The setup for observing the contractile behavior of the isolated gastrocnemius muscle of a frog.

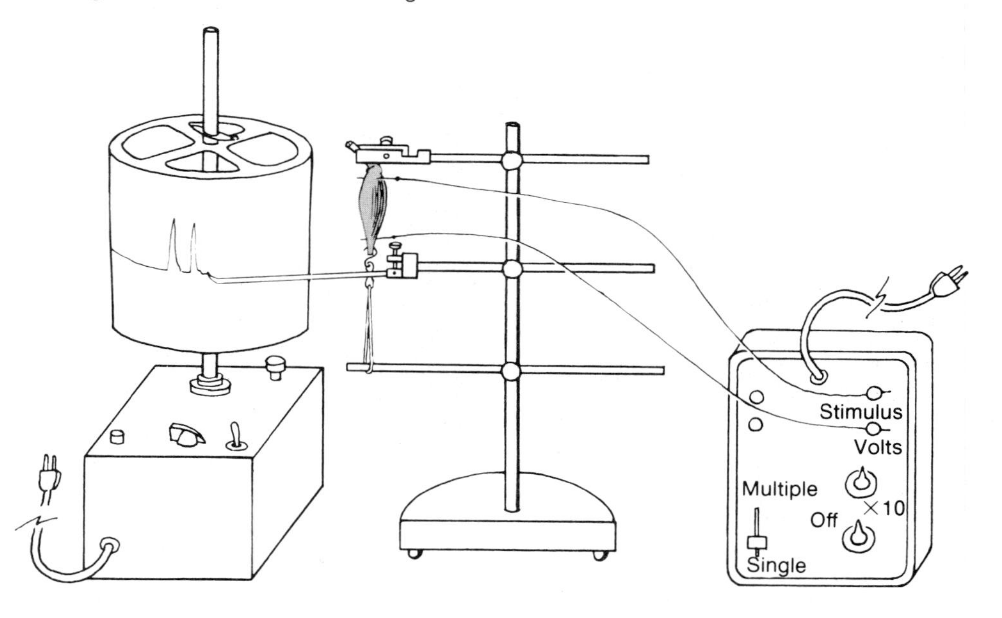

Figure 11.5. The physiograph Mark III recorder. This is one of many types of electronic recording devices that receive electrical signals produced by transducers and record these signals on moving paper.

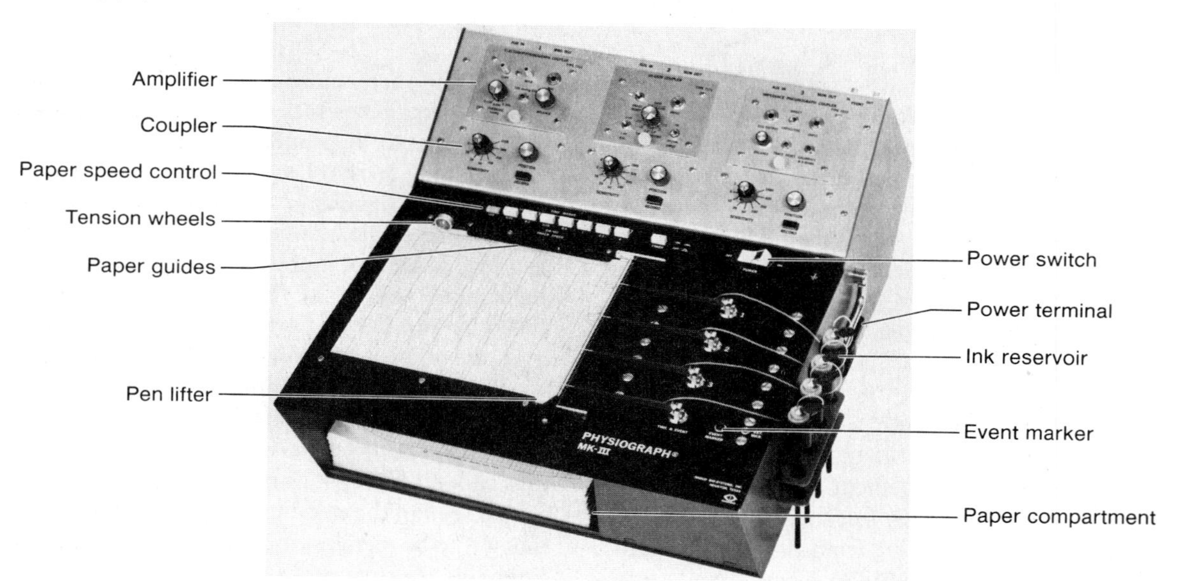

Twitch, Summation, and Tetanus

The contractile behavior of skeletal muscles is more easily studied *in vitro* (outside the body) than *in vivo* (within the body). In these studies a muscle, such as the gastrocnemius (calf muscle) of a frog, is usually mounted so that one end is fixed and the other is movable. In the classic studies of this kind, the movable end of the muscle directly produces deflections of a pen, which writes on a rotating drum recorder (fig. 11.4). When more modern equipment is used, the mechanical force of the muscle contraction is transduced into an electric current, which can be amplified and displayed as pen deflections in a multi-channel recorder (fig. 11.5). In this way, the contractile behavior of the whole muscle in response to experimentally administered electric shocks can be studied.

When the muscle is stimulated with a single electric shock of a sufficient voltage, it quickly contracts and relaxes. This response is called a **twitch.** Increasing the stimulus voltage increases the strength of the twitch up to a maximum. The strength of a muscle contraction can thus be *graded,* or varied—an obvious requirement for the proper control of skeletal movements. If a second electric shock is delivered immediately after the first, it will produce a second twitch that may partially "ride piggyback" on the first. This response is called **summation.**

If the stimulator is set to deliver an increasing frequency of electric shocks automatically, the relaxation time between successive twitches will get shorter and shorter, as the strength of contraction increases in amplitude. This is **incomplete tetanus.** Finally, at a particular "fusion frequency" of stimulation, there is no visible relaxation between successive twitches (fig. 11.6). Contraction is smooth and sustained, very much like normal muscle contraction *in vivo.* This smooth, sustained contraction is called **complete tetanus.** The term *tetanus* should not be confused with the disease of the same name, which is accompanied by a painful state of muscle contracture, or *tetany.*

Research laboratories, using an experimental setup analogous to the one previously described, can study the contractile behavior of *an isolated muscle fiber.* The isolated fiber produces a twitch of submaximal strength in response to a single action potential. Since the fiber recovers from its refractory period before it has finished its twitch, it can be stimulated a number of times to produce a summation of twitches. Continued stimulation of the isolated fiber can then result in tetanus, analogous to the behavior of the whole muscle previously described.

Stimulation of the whole muscle *in vitro* with an electric stimulator or stimulation of muscle fibers *in vivo* by motor axons, however, usually results in the production of many action potentials in the stimulated muscle fibers. Within the whole muscle, therefore, muscle fibers are normally stimulated maximally and thus produce *all-or-none* contractions. In this case, gradations in the strength of skeletal muscle contractions result from summation of the contractions of different *numbers* of muscle fibers and not from variations in the strength of the contractions of individual muscle fibers. Stronger muscle contractions, in other words, are produced by the stimulation of greater numbers of muscle fibers.

Series Elastic Component. In order for a muscle to shorten when it contracts, and thus to move its insertion toward its origin, the noncontractile parts of the muscle and the connective tissue of its tendons must first be pulled tight. These structures, particularly the tendons, have elasticity—they resist distension, and when the distending force is released, they tend to spring back to their resting lengths. Since the tendons are in series with the force of muscle contraction, they provide the **series elastic component** of muscle contraction.

Figure 11.6. (*a*) A recording of summation of muscle twitches using a physiograph. (*b*) An illustration of summation, tetanus, and fatigue in an isolated muscle.

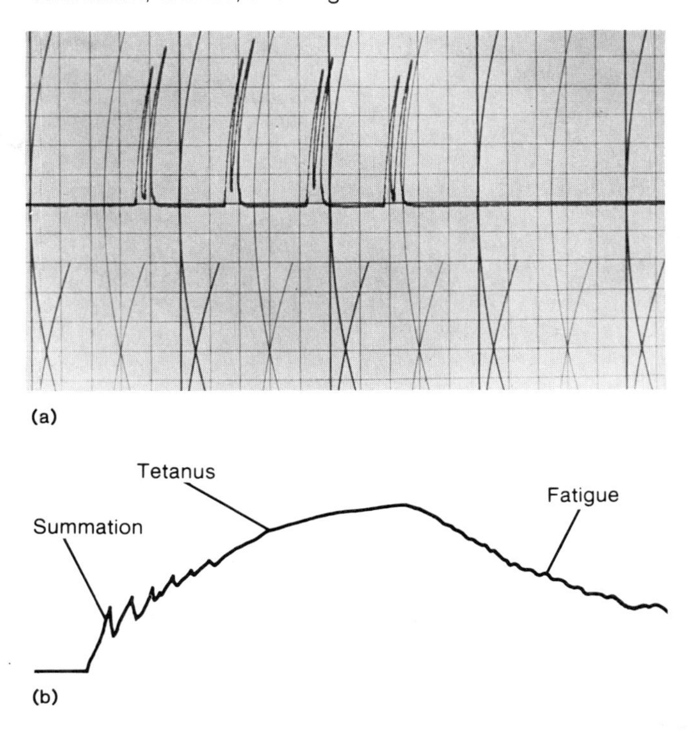

When the gastrocnemius muscle was stimulated with a single electric shock in the previously described experiment, the amplitude of the twitch was reduced because some of the force of contraction was used to stretch the series elastic component. Delivery of a second shock quickly after the first thus produced a greater degree of muscle shortening than the first shock, culminating at the fusion frequency of stimulation with complete tetanus, in which the strength of contraction was much greater than that of individual twitches.

Some of the energy used to stretch the series elastic component during muscle contraction is released when the muscle relaxes. This elastic recoil helps the muscles to return to their resting length, and is of particular importance for the muscles involved in breathing. As we will see in chapter 15, inspiration is produced by muscle contraction and expiration is produced by the elastic recoil of the thoracic structures that were stretched during inspiration.

Figure 11.7. (*a*) A photomicrograph of neuromuscular junctions formed by somatic motor neurons. Note that a single axon branches to innervate a number of muscle fibers. (*b*) An illustration of a neuromuscular junction.

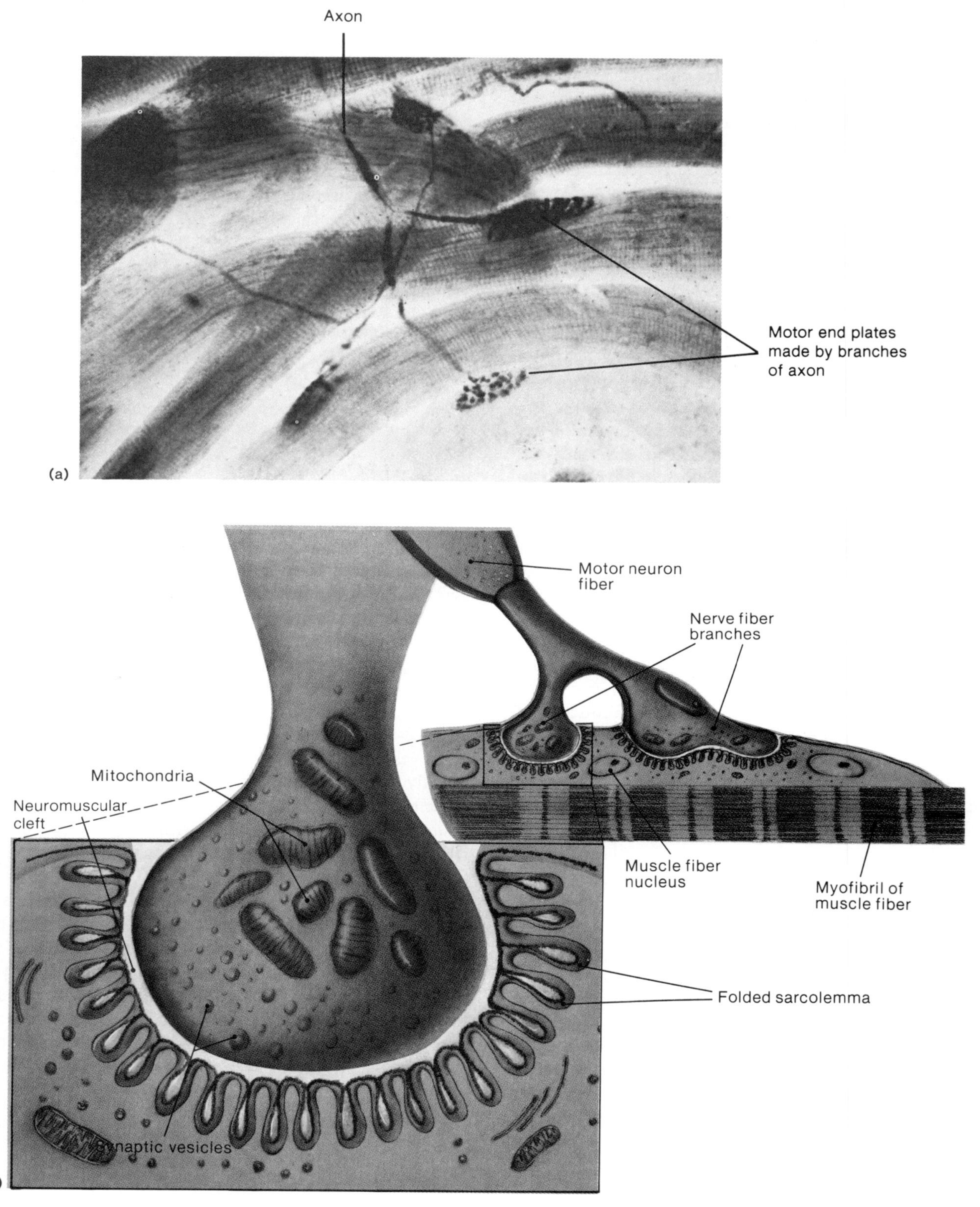

Figure 11.8. A diagram illustrating the innervation of muscle fibers by different motor units. (Actually, many more muscle fibers would be included in a single motor unit than is shown in this drawing.)

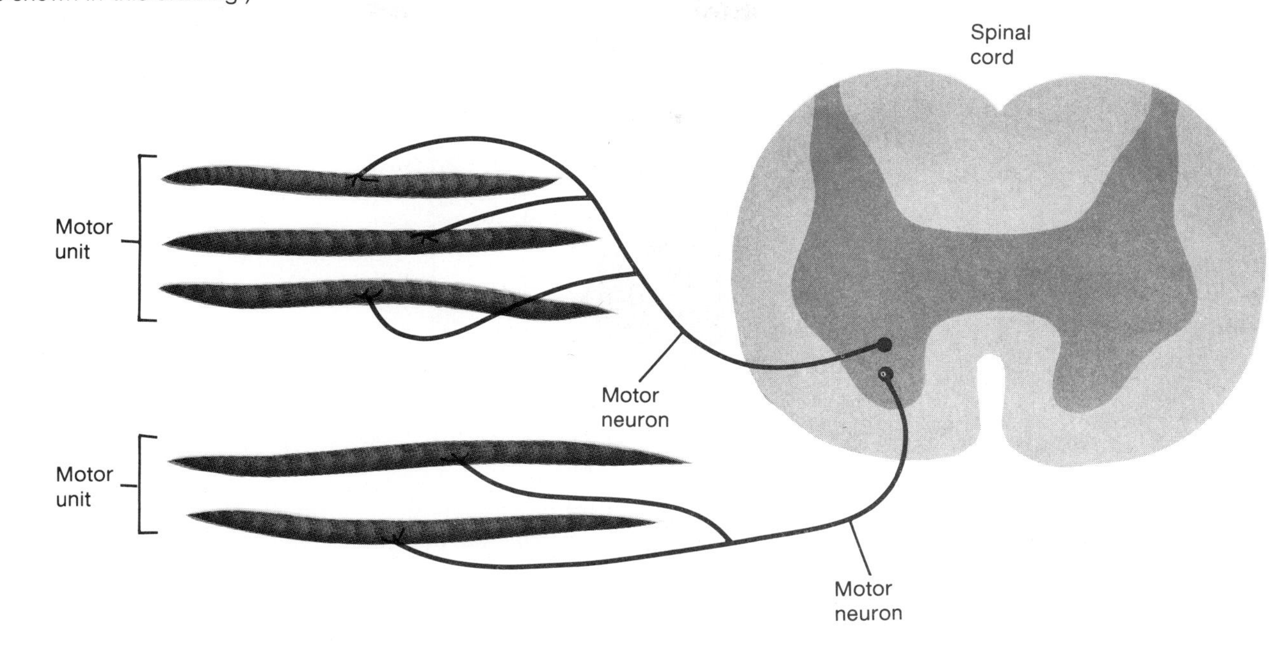

Motor Units

In vivo, each muscle fiber receives a single axon terminal from a somatic motor neuron (fig. 11.7), which stimulates it to contract by liberating acetylcholine at the neuromuscular junction (described in chapter 7). The cell body of a somatic motor neuron is located in the ventral horn of the grey matter of the spinal cord and gives rise to a single axon that emerges in the ventral root of a spinal nerve. Each axon, however, can produce a number of collateral branches to innervate an equal number of muscle fibers. Each somatic motor neuron together with all of the muscle fibers that it innervates is known as a **motor unit** (fig. 11.8).

Whenever a somatic motor neuron is activated, all of the muscle fibers that it innervates are stimulated to contract with all-or-none twitches. Graded contractions of whole muscles *in vivo* are produced by variations in the number of motor units that are activated, and smooth tetanic contractions are produced by rapid, asynchronous stimulation of different motor units.

Fine neural control over the strength of muscle contraction is optimal when there are many small motor units involved. In the extraocular muscles that position the eyes, for example, the *innervation ratio* of an average motor unit is one neuron per twenty-three muscle fibers, which affords a fine degree of control. The innervation ratio of the gastrocnemius, in contrast, averages one neuron per one thousand muscle fibers. Stimulation of these motor units results in more powerful contractions at the expense of finer gradations in contraction strength.

All of the motor units controlling the gastrocnemius, however, are not the same size. Innervation ratios vary from 1:100 to 1:2000. Neurons that innervate fewer muscle fibers have smaller cell bodies and are stimulated by lower levels of excitatory input (EPSPs) than the larger neurons that have larger innervation ratios. The smaller motor units, as a result, are the ones that are used most often. When contractions of greater strength are required, larger and larger motor units are activated; this is known as **recruitment** of motor units.

Figure 11.9. (*a*) Isometric and (*b*) isotonic contraction.

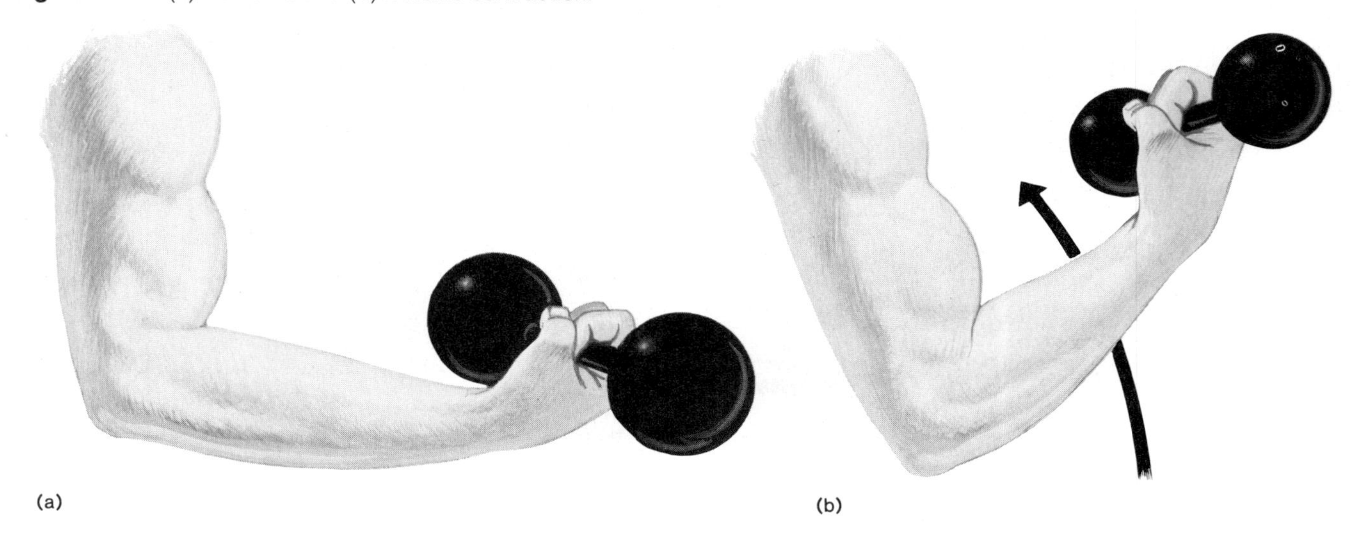

Isotonic and Isometric Contractions

In order for muscle fibers to shorten when they contract, they must generate a force that is greater than the opposing forces that act to prevent movement of the muscle's insertion. Flexion of the forearm, for example, occurs against the force of gravity and the weight of the objects being lifted (fig. 11.9). The tension produced by the contraction of each muscle fiber separately is insufficient to overcome these opposing forces, but the combined contractions of many muscle fibers may be sufficient to overcome the opposing force and flex the forearm. In this case the muscle and all its fibers shorten in length.

Contraction that results in muscle shortening is called *isotonic contraction,* so-called because the force of contraction remains relatively constant throughout the shortening process (iso = same; tonic = strength). If the opposing forces are too great or the number of motor units activated is too few to shorten the muscle, however, the contraction is called *isometric* (literally, "same length").

Isometric contraction can be voluntarily produced, for example, by lifting a weight and maintaining the forearm in a partially flexed position. One can then increase the amount of muscle tension produced by recruiting more motor units until the muscle begins to shorten; at this point, isometric contraction is converted to isotonic contraction.

1. *Illustrate and verbally describe muscle twitch, summation, and tetanus.*
2. *Describe how graded muscle contractions result from fiber twitches that are all-or-none and how a smooth muscle contraction (tetanus) is produced* in vivo.
3. *Illustrate one motor unit with an innervation ratio of 1:5. Describe the functional significance of the innervation ratio.*
4. *Try to "make a muscle" with your arm and while doing so, feel your* biceps brachii *and your* triceps brachii. *Are these muscles contracting isotonically or isometrically? How is this type of contraction produced? (Include the antagonism of these two muscle groups in your answer.)*

Lower Motor Neuron Control of Skeletal Muscles

The **lower motor neurons** (often shortened to *motoneurons*) are those previously described in the spinal cord that directly stimulate muscle contraction (table 11.2). The activity of these neurons is influenced by (1) sensory feedback from the muscles and tendons and (2) facilitory and inhibitory effects from **higher motor neurons** in the brain. Lower motor neurons are thus said to be the *final common pathway* by which sensory stimuli and higher brain centers exert control over skeletal movements.

Table 11.2 Some terms used to describe the neural control of skeletal muscles.

Term	Description
1. Lower motoneurons	Neurons whose axons innervate skeletal muscles—also called the "final common pathway" in the control of skeletal muscles
2. Higher motoneurons	Neurons in the brain that are involved in the control of skeletal movements and that act by facilitating or inhibiting (usually by way of interneurons) the activity of the lower motoneurons
3. Alpha motoneurons	Lower motoneurons whose fibers innervate ordinary (extrafusal) muscle fibers
4. Gamma motoneurons	Lower motoneurons whose fibers innervate the muscle spindle fibers (intrafusal fibers)
5. Agonist/antagonist	Pair of muscles or muscle groups that insert on the same bone, the agonist being the muscle of reference
6. Synergist	A muscle whose action facilitates the action of the agonist
7. Ipsilateral/contralateral	Ipsilateral—to the same side, or the side of reference; contralateral—the opposite side
8. Afferent/efferent	Afferent neurons—sensory; efferent neurons—motor

Figure 11.10. A sensory neuron, an association neuron (interneuron), and a somatic motor neuron at the spinal cord level.

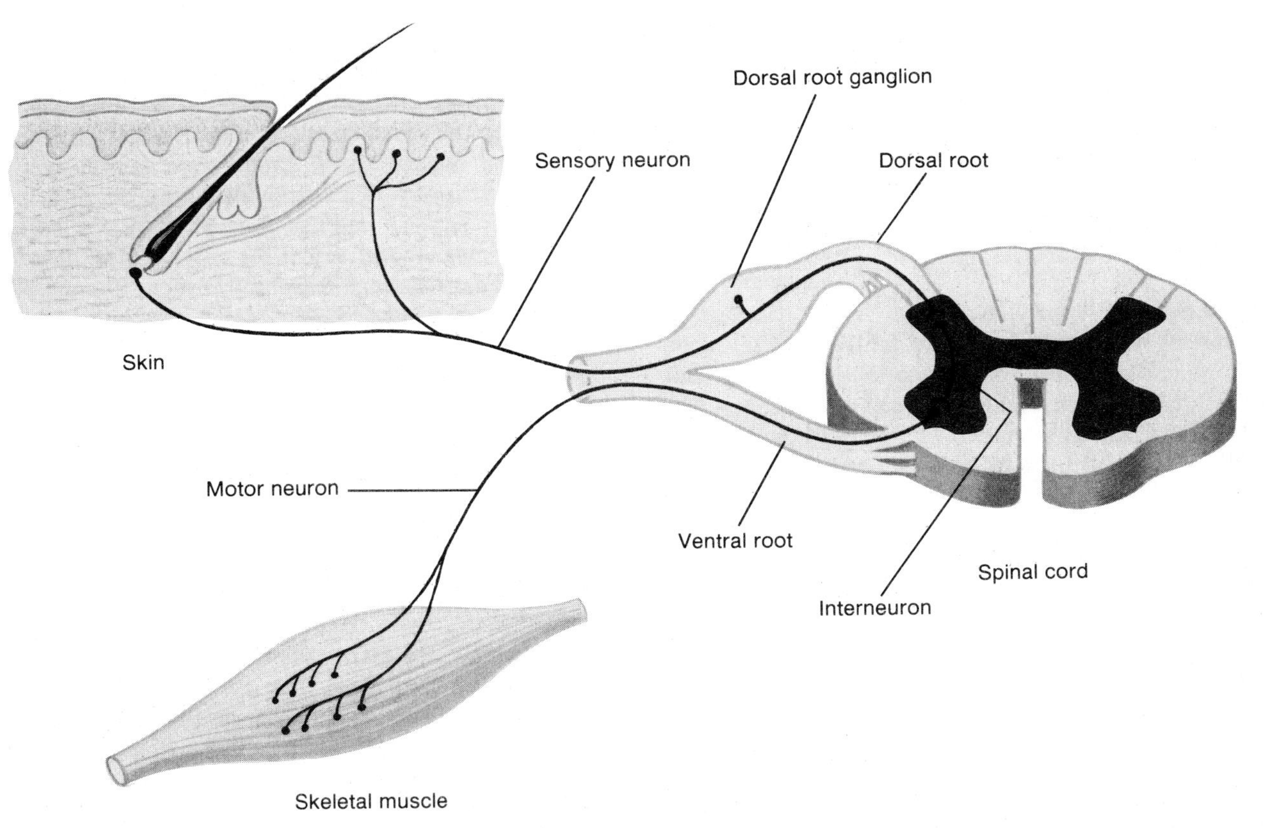

The cell bodies of lower motor neurons are located in the ventral horn of the grey matter of the spinal cord. Axons from these cell bodies leave the ventral side of the spinal cord to form the *ventral roots* of spinal nerves (fig. 11.10). The *dorsal roots* of spinal nerves contain sensory fibers whose cell bodies are located in the *dorsal root ganglia.* Both sensory (*afferent*) and motor (*efferent*) fibers join in a common connective tissue sheath to form the spinal nerves at each segment of the spinal cord. In the lumbar region (the small of the back), there are about twelve thousand sensory and six thousand motor fibers per spinal nerve.

Table 11.3 Spindle apparatus content of selected skeletal muscles.

Muscle		Muscle Weight (g)	Average Number of Spindles	Number of Spindles per Gram Muscle
Gastrocnemius	leg	7.6	35	5
Rectus femoris	leg	8.36	104	12
Tibialis anterior	leg	4.57	71	15
Semitendinosis	leg	6.41	114	18
Soleus	leg	2.49	56	23
Fifth interossei—foot		0.33	29	88
Fifth interossei—hand		0.21	25	119

About 375,000 cell bodies have been counted in a lumbar segment—a number far larger than can be accounted for by the number of motor neurons. Most of these neurons do not contribute fibers to the spinal nerve, but rather serve as *interneurons* whose fibers conduct impulses up, down, and across the central nervous system. Those fibers that conduct impulses to higher spinal cord segments and the brain form *ascending tracts.* Those fibers that conduct to lower spinal segments contribute to *descending tracts.* And those that cross the midline of the CNS to synapse on the opposite side are part of *commissural tracts.* Interneurons can thus conduct impulses up and down on the same, or *ipsilateral,* side, and can affect neurons on the opposite, or *contralateral,* side of the central nervous system.

Muscle Spindle Apparatus

In order for the nervous system to control skeletal movements properly, it must receive continuous sensory feedback information concerning the effects of its actions. This sensory information includes (1) the tension that the muscle exerts on its tendons, provided by the **Golgi tendon organs** and (2) muscle length, provided by the **muscle spindle apparatus.** The spindle apparatus, so-called because it is wider in the center and tapers towards the ends, functions as a length detector. Muscles that require the finest degree of control, such as the muscles of the hand, have the highest density of spindles (table 11.3).

Each spindle apparatus contains several thin muscle cells, called *intrafusal fibers* (*fusus* = spindle), packaged within a connective tissue sheath. Like the stronger and more numerous "ordinary" muscle fibers—the *extrafusal fibers*—the spindles insert into tendons on each end of the muscle. Spindles are therefore said to be in parallel with the extrafusal fibers.

There are two types of intrafusal fibers. One type, the *nuclear bag* fibers, have their nuclei arranged in a loose aggregate in the central regions of the fibers. The other type of intrafusal fibers have their nuclei arranged in rows and are called *nuclear chain* fibers. There are likewise two types of sensory nerve endings that serve these intrafusal fibers. **Primary,** or **annulospiral, endings** wrap around the central regions of the nuclear bag and chain fibers (fig. 11.11), and **secondary,** or **flower-spray, endings** are located near the ends of the nuclear chain fibers.

Since the spindles are arranged in parallel with the extrafusal muscle fibers, stretching a muscle causes its spindles to stretch. This stimulates both the primary and secondary sensory endings. The primary endings, however, are most stimulated at the onset of stretch, whereas the secondary endings respond in a more tonic fashion as stretch is maintained. Sudden, rapid stretching of a muscle activates both types of sensory endings and is thus a more powerful stimulus for the muscle spindles than slower, more gradual stretching. Since the activation of the sensory endings in muscle spindles produces a reflex contraction, the force of this reflex contraction is greater in response to rapid stretch than to gradual stretch.

Alpha and Gamma Motor Neurons

There are two types of lower motor neurons in the spinal cord that innervate skeletal muscles. The motor neurons that innervate the extrafusal muscle fibers are called **alpha motoneurons,** and those that innervate the intrafusal fibers are called **gamma motoneurons** (fig. 11.11). The alpha motoneurons are larger and more rapidly conducting (60–90 meters per second) than the thinner, more slowly conducting (10–40 meters per second) gamma motoneurons. Since only the extrafusal muscle fibers are sufficiently strong and numerous to cause a muscle to shorten, only stimulation of the alpha motoneurons can cause muscle contraction that results in skeletal movements.

Figure 11.27. The production and utilization of phosphocreatine in muscles.

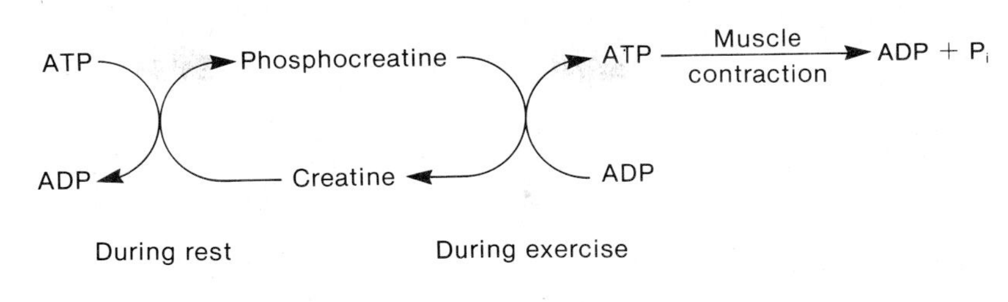

Figure 11.28. Creatine kinase in plasma is separated from other proteins and identified by the rate of its movement in an electric field. The height of each peak indicates concentration; and the position of each peak indicates the isoenzyme form. As can be seen in the *lower figure,* the plasma concentration of creatine kinase is elevated in both heart disease (myocardial infarction) and skeletal muscle disease, but the two types of disease may be distinguished in this test by the appearance of a different isoenzyme in myocardial infarction than in muscle disease.

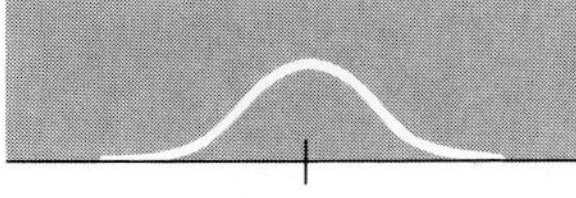

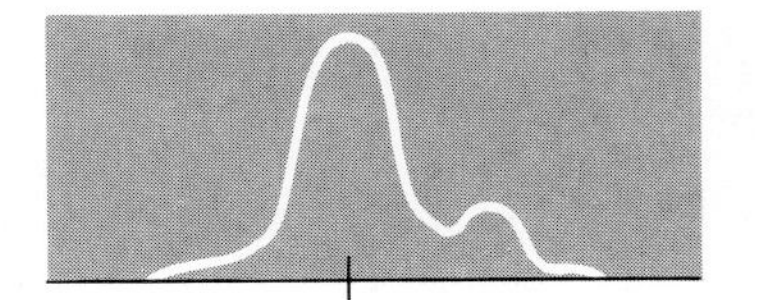

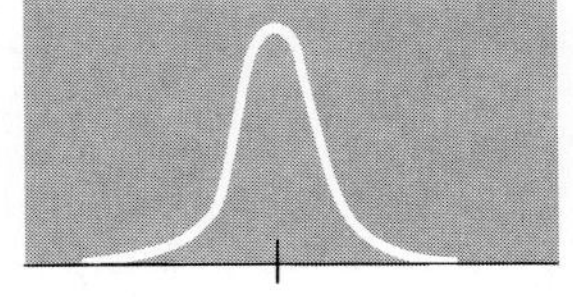

Figure 11.29. Skeletal muscle (of a cat) stained to indicate the activity of myosin ATPase. Different types of fibers can be distinguished by the intensity of the stain.

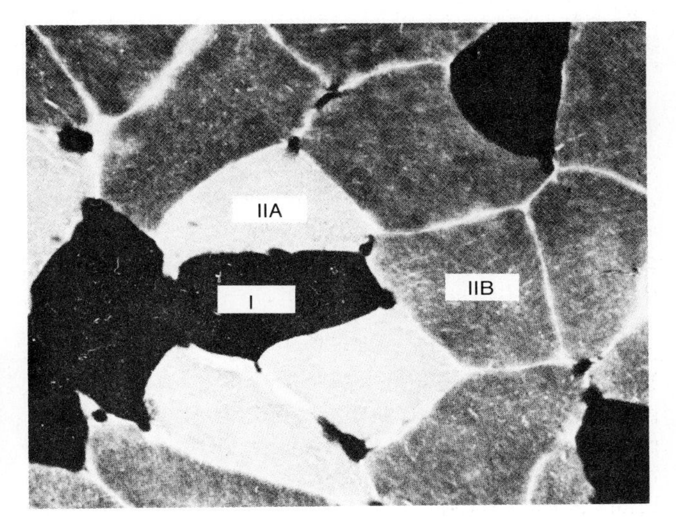

The phosphocreatine concentration within muscle cells is greater than three times the concentration of ATP and represents a ready reserve of high-energy phosphate that can be donated directly to ADP (fig. 11.27). During times of rest, the depleted reserve of phosphocreatine can be restored by the reverse reaction—phosphorylation of creatine with phosphate derived from ATP.

The enzyme that transfers phosphate between creatine and ATP is called **creatine kinase.** Skeletal muscle and heart muscle have two different forms of this enzyme (they have different isoenzymes, as described in chapter 4). The skeletal muscle isoenzyme is found to be elevated in the blood of people with *muscular dystrophy* (degenerative disease of skeletal muscles). The plasma concentration of the isoenzyme characteristic of heart muscle is elevated as a result of *myocardial infarction* (damage to heart muscle), and measurements of this enzyme are thus used as a means of diagnosing this condition (fig. 11.28).

Slow- and Fast-Twitch Fibers

Skeletal muscle fibers can be divided on the basis of their contraction speed (time required to reach maximum tension) into **slow-twitch,** or **type I fibers,** and **fast-twitch,** or **type II fibers.** These differences are associated with different myosin ATPase isoenzymes, designated "slow" and "fast," by which the two fiber types can be distinguished when they are appropriately stained (fig. 11.29). The extraocular muscles that position the eyes, for example, have a high proportion of fast-twitch fibers and reach maximum tension in about 7.3 msec (milliseconds—thousandths of a second); the soleus muscle in the leg, in contrast, has a high proportion of slow-twitch fibers and requires about 100 msec to reach maximum tension (fig. 11.30).

Figure 11.30. A comparison of the rates with which maximum tension is developed in three muscles. These include the relatively fast-twitch extraocular (*a*) and gastrocnemius (*b*) muscles and the slow-twitch soleus muscle (*c*).

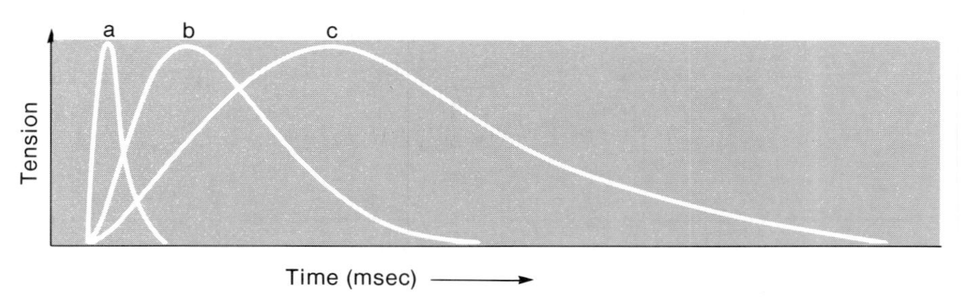

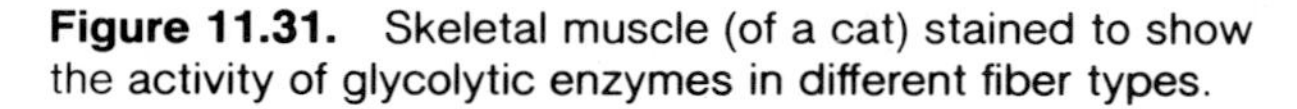
Figure 11.31. Skeletal muscle (of a cat) stained to show the activity of glycolytic enzymes in different fiber types.

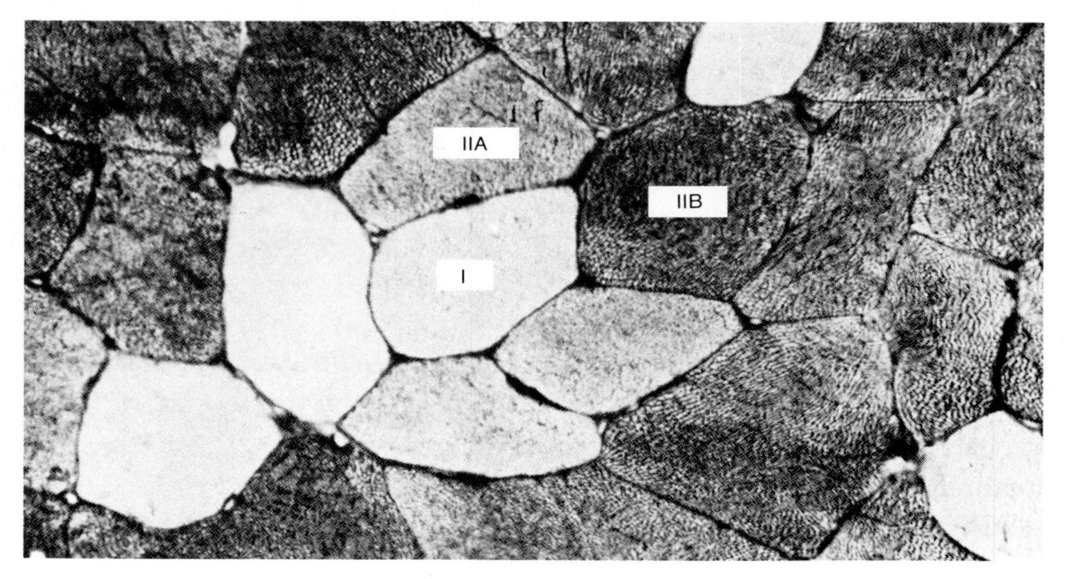

Muscles like the soleus are *postural muscles* that must be able to sustain a contraction for a long period of time without fatigue. This is aided by other characteristics of slow-twitch (type I) fibers that endow them with a high capacity for aerobic respiration. Slow-twitch fibers are served by large numbers of capillaries, have numerous mitochondria and aerobic respiratory enzymes, and have a high concentration of *myoglobin* pigment. Myoglobin is a red pigment—hence the alternate name of *red fibers* for these muscle cells—which is related to the hemoglobin pigment of blood and serves to improve the delivery of oxygen to the slow-twitch fibers.

The thicker, fast-twitch (type II) fibers have a lower capillary supply, fewer mitochondria, and less myoglobin; hence, these fibers are also called *white fibers.* Fast-twitch fibers are adapted to respire anaerobically by a large store of glycogen and a high concentration of glycolytic enzymes, which enable these fibers to be distinguished when appropriately stained (fig. 11.31). In addition to the type I (slow-twitch) and type II (fast-twitch) fibers, human muscles may also have an intermediate form of fibers, which are fast-twitch but also have a high aerobic capability. These are sometimes called type IIA, to distinguish them from the anaerobically adapted fast-twitch fibers (which are then labeled IIB). A comparison of these fiber types is summarized in table 11.10.

Table 11.10 Characteristics of red, intermediate, and white muscle fibers.

	Red (Type I)	Intermediate (Type IIA)	White (Type IIB)
Diameter	Small	Intermediate	Large
Z-line thickness	Wide	Intermediate	Narrow
Glycogen content	Low	Intermediate	High
Resistance to fatigue	High	Intermediate	Low
Capillaries	Many	Many	Few
Myoglobin content	High	High	Low
Respiration type	Aerobic	Aerobic	Anaerobic
Twitch rate	Slow	Fast	Fast
Myosin ATPase content	Low	High	High

Interestingly, the conduction rate of motor neurons that innervate fast-twitch fibers is faster (80–90 meters per second) than the conduction rate to slow-twitch fibers (60–70 meters per second). The fiber type indeed seems to be determined by the motor neuron. When the motor neurons to different fiber types are switched in experimental animals, the previously fast-twitch fibers become slow and the slow-twitch fibers become fast. As expected from these observations, all of the muscle fibers innervated by the same motor neuron (that are part of the same motor unit) are of the same type.

A muscle such as the gastrocnemius (calf muscle) contains both fast- and slow-twitch fibers, although fast-twitch fibers predominate. A given somatic motor axon, however, only innervates muscle fibers of one type, and the motor units composed of slow-twitch fibers tend to be smaller (have fewer fibers) than the motor units of fast-twitch fibers. Since motor units are recruited from smaller to larger when increasing effort is required, as previously described, the smaller motor units with slow-twitch fibers would be used most often in routine activities. Larger motor units with fast-twitch fibers, which can exert a great deal of force but which respire anaerobically and thus fatigue quickly, would be used relatively infrequently and for only short periods of time.

Muscle Fatigue

Muscle fatigue may be defined as the inability to maintain a particular muscle tension when the contraction is sustained or to reproduce a particular tension during rhythmic contraction over time. Fatigue during a sustained maximal contraction, when all the motor units are used and the rate of neural firing is maximal, appears to be due to an accumulation of extracellular K^+. This reduces the membrane potential of muscle fibers and interferes with their ability to produce action potentials. Fatigue under these circumstances lasts only a short time, and maximal tension can again be produced after less than a minute rest.

Fatigue during moderate exercise involving rhythmical contractions occurs as the slow-twitch fibers deplete their reserve glycogen and fast-twitch fibers are increasingly recruited. Fast-twitch fibers obtain their energy through anaerobic respiration, converting glucose to lactic acid, and this results in a rise in intracellular H^+ and a fall in pH. The decrease in muscle pH, in turn, inhibits the activity of key glycolytic enzymes so that the rate of ATP production is reduced. Since ATP is needed for all active transport, the decrease in ATP is believed to result in a loss of intracellular Ca^{++} to the extracellular environment, so that excitation-contraction coupling is hindered in the muscle cells. The ability of the muscle to maintain a particular tension is thus decreased, first due to loss of the contribution of smaller, slow-twitch motor units and then due to the inability of the fast-twitch muscle fibers to contract as lactic acid accumulates.

Note that the decrease in ATP within the muscle cell causes it to fatigue due to interference with excitation-contraction coupling, not due to direct interference with the cross-bridge cycle. This is supported by the observation that, even at maximal exhaustion, the amount of ATP in muscle fibers is only reduced by about 25% (although the amount of phosphocreatine may be completely depleted). Actually, the interference with excitation-contraction coupling can be thought of as a protective mechanism, because if the ATP were significantly depleted the muscle would enter a state analogous to rigor mortis. This does not occur in a living muscle, even during maximal exercise, because ATP can be quickly reformed, using the phosphate group from creatine phosphate, and if ATP does become depleted by a certain amount, the muscle fatigues due to interference with excitation-contraction coupling.

Adaptations to Exercise

When exercise is performed at low levels of effort, such that the body does not need to consume more oxygen than 50% of its maximum oxygen consumption rate, the energy for muscle contraction is obtained almost entirely from aerobic cell respiration. Anaerobic cell respiration, with its consequent production of lactic acid, contributes to the energy requirements as the exercise level rises and more than 60% of the maximal oxygen uptake is required. The maximal oxygen uptake is about ten times higher than the resting oxygen uptake rate in average adults, but it can reach up to twenty times higher than the resting rate in top endurance-trained athletes. These athletes can produce less lactic acid at a given level of exercise than the average person and, therefore, are less subject to fatigue than the average person.

Since the fiber types are determined by their innervations, endurance-training cannot change fast-twitch (type II) fibers to slow-twitch (type I) fibers. All fiber types, however, adapt to endurance training by an increase in myoglobin and aerobic respiratory enzymes, so that the maximal oxygen uptake can be increased by up to 20% through this training. In addition to changes in aerobic capacity, fibers show an increase in their content of triglycerides, which serves as an alternate energy source helping to spare their stores of glycogen. A summary of the changes that occur as a result of endurance training is presented in table 11.11.

Endurance training does not increase the size of muscles. Muscle enlargement is produced only by frequent bouts of muscle contraction against a high resistance—as in weight lifting. As a result of this latter type of training, muscle fibers become thicker, and the muscle therefore grows by hypertrophy (increase in cell size, rather than number). This happens as the myofibrils within a muscle fiber get thicker, due to the addition of new sarcomeres and myofibrils. After a myofibril attains a certain thickness it may split into two myofibrils, which may then each become thicker due to the addition of sarcomeres. Muscle hypertrophy, in summary, is associated with an increase in the number and size of myofibrils within the muscle fibers.

Table 11.11 Summary of the effects of endurance training (long-distance running, swimming, bicycling, etc.) on skeletal muscles.

1. Improved ability to obtain ATP from oxidative phosphorylation
2. Increased size and number of mitochondria
3. Less lactic acid produced per given amount of exercise
4. Increased myoglobin content
5. Increased intramuscular triglyceride content
6. Increased lipoprotein lipase (enzyme needed to utilize lipids from blood)
7. Increased proportion of energy derived from fat, less from carbohydrates
8. Lower rate of glycogen depletion during exercise
9. Improved efficiency in extracting oxygen from blood

---

1. *Draw a figure illustrating the relationship between ATP and creatine phosphate, and explain the physiological significance of this relationship.*
2. *Describe the characteristics of slow- and fast-twitch fibers (including intermediate fibers). Explain how the fiber types are determined and the functions of different fiber types.*
3. *Explain the different causes of muscle fatigue, and relate the causes of fatigue to the fiber types.*
4. *Describe the effects of endurance training and strength training on the fiber characteristics of the muscles.*

Summary

Structure and Actions of Skeletal Muscles p. 288

I. Skeletal muscles are attached by tendons to bones.
   A. Skeletal muscles are composed of separate cells or fibers that are attached in parallel to the tendons.
   B. Skeletal muscle fibers are striated.
      1. The dark striations are called A bands and the light regions are called I bands.
      2. Z lines are located in the middle of each I band.

II. Muscles *in vitro* can exhibit twitch, summation, and tetanus.
   A. The rapid contraction and relaxation of muscle fibers is called a twitch.
   B. A whole muscle also produces a twitch in response to a single electrical pulse *in vitro*.
      1. The stronger the electric shock, the stronger the muscle twitch—whole muscle can give a graded contraction.
      2. The graded contraction of whole muscles is due to different numbers of fibers participating in the contraction.
   C. The summation of fiber twitches can occur so rapidly that the muscle produces a smooth, sustained contraction known as tetanus.

III. The contraction of muscle fibers *in vivo* is stimulated by somatic motor neurons.
   A. Each somatic motor nerve fiber branches to innervate a number of muscle fibers.
   B. The motor neuron and the muscle fibers it innervates is called a motor unit.
      1. When a muscle is composed of many motor units (such as in the hand), there is fine control of muscle contraction.

2. The large muscles of the leg have relatively few motor units, which are correspondingly large.
3. Sustained contractions are produced by the asynchronous stimulation of different motor units.

IV. When a muscle exerts tension without shortening, the contraction is termed isometric; when shortening does occur, the contraction is isotonic.

Lower Motor Neuron Control of Skeletal Muscles p. 294

I. The somatic motor neurons that innervate the muscles are called the lower motor neurons.
   A. Alpha motoneurons innervate the ordinary, or extrafusal, muscle fibers—these are the fibers that produce muscle shortening during contraction.
   B. Gamma motoneurons innervate the intrafusal fibers of the muscle spindles.

II. Muscle spindles are length detectors in the muscle.
   A. Spindles consist of several intrafusal fibers wrapped together and in parallel with the extrafusal fibers.
   B. Stretching of the muscle stretches the spindles; this excites sensory endings in the spindle apparatus.
      1. Impulses in the sensory neurons travel into the spinal cord in the dorsal roots of spinal nerves.
      2. The sensory neuron makes a synapse directly with an alpha motoneuron within the spinal cord—this produces a monosynaptic reflex.
      3. The alpha motoneuron stimulates the extrafusal muscle fibers to contract, thus relieving the stretch—this is called the stretch reflex.
   C. The activity of gamma motoneurons tighten the spindles, thus making them more sensitive to stretch and better able to monitor the length of the muscle, even during muscle shortening.

III. The Golgi tendon organs monitor the tension that the muscle exerts on its tendons.
   A. As the tension increases, sensory neurons from Golgi tendon organs inhibit the activity of alpha motoneurons.
   B. This is a disynaptic reflex, because the sensory neurons synapse with interneurons, which in turn make inhibitory synapses with motoneurons.

IV. A crossed-extensor reflex occurs when a foot steps on a tack.
   A. Sensory input from the injured foot causes stimulation of flexor muscles and inhibition of the antagonistic extensor muscles.
   B. The sensory input also crosses the spinal cord to cause stimulation of extensor and inhibition of flexor muscles in the contralateral leg.

Higher Motor Neuron Control of Skeletal Muscles p. 302

I. Neurons in the brain that affect the lower motor neurons are called higher motor neurons.
   A. The fibers of neurons in the precentral gyrus, or motor cortex, descend to the lower motor neurons as the lateral and ventral corticospinal tracts.
      1. Most of these fibers cross to the contralateral side in the brain stem, forming structures called the pyramids; this system is therefore called the pyramidal system.
      2. The left side of the brain thus controls the musculature on the right side, and vice versa.
      3. The pyramidal system is responsible for fine control of voluntary movements.
   B. Other descending motor tracts are part of the extrapyramidal system.
      1. The neurons of the extrapyramidal system make numerous synapses in different areas of the brain, including the midbrain, brain stem, basal ganglia, and cerebellum.
      2. Damage to the cerebellum produces intention tremor.
      3. The degeneration of fibers in the basal ganglia that use dopamine as a transmitter produces Parkinson's disease.

II. Most of the fibers of descending tracts synapse with spinal interneurons, which in turn synapse with the lower motor neurons.
   A. Alpha and gamma motoneurons are usually stimulated at the same time, or co-activated.
   B. The stimulation of gamma motoneurons keeps the muscle spindles under tension and sensitive to stretch.
   C. Higher motor neurons, primarily in the basal ganglia, also exert inhibitory effects on gamma motoneurons.
      1. Damage to the spinal cord at first causes muscles to become flaccid as stimulation by higher motor neurons is blocked.
      2. After a period of time, the gamma motoneurons become hyperexcitable, since they are freed from higher motor neuron inhibition and cause muscle spasticity and clonus.

Mechanisms of Contraction p. 305

I. Skeletal muscle cells, or fibers, contain structures called myofibrils.
   A. Each myofibril is striated with dark (A) and light (I) bands; there are Z lines in the middle of each I band.
   B. The A bands contain thick filaments composed of myosin.
      1. The edges of each A band also contain thin filaments overlapped with the thick filaments.
      2. The central regions of the A bands contain only thick filaments—these regions are the H bands.
   C. The I bands contain only thin filaments, which are composed primarily of the protein called actin.
   D. Thin filaments are formed of globular actin subunits known as G-actin. A protein known as tropomyosin is also located periodically in the thin filaments; another protein—troponin—is attached to the tropomyosin.

II. Myosin cross-bridges extend out from the thick filaments to the thin filaments.
   A. At rest, the cross-bridges are not attached to actin.
      1. The cross-bridge heads function as ATPase enzymes.
      2. ATP is split into ADP and P_i, activating the cross-bridge.
   B. When the activated cross-bridges attach to actin, they undergo a power stroke and in the process release ADP and P_i.

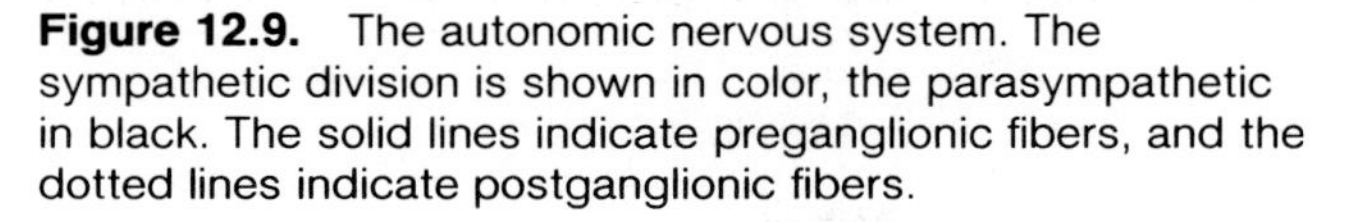

Figure 12.9. The autonomic nervous system. The sympathetic division is shown in color, the parasympathetic in black. The solid lines indicate preganglionic fibers, and the dotted lines indicate postganglionic fibers.

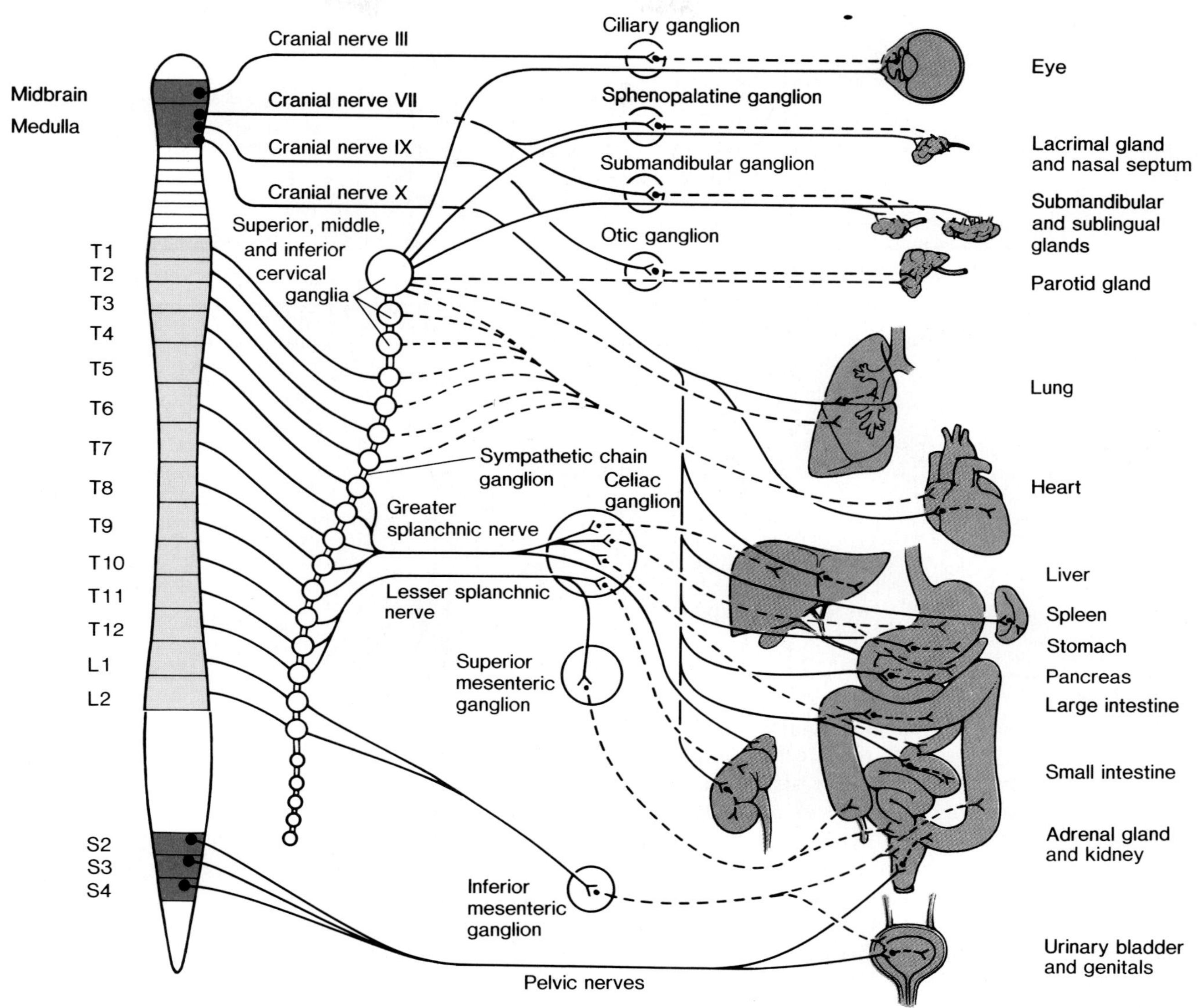

innervated by preganglionic sympathetic fibers originating in the thoracic level of the spinal cord; they secrete epinephrine into the blood in response to this neural stimulation. The effects of epinephrine are complementary to those of the neurotransmitter norepinephrine, which is released from postganglionic sympathetic nerve endings. For this reason and because the adrenal medulla is stimulated as part of the mass activation of the sympathetic system, the two are often grouped together as the **sympathoadrenal system.**

Parasympathetic (Craniosacral) Division

The parasympathetic system is also known as the *craniosacral division* of the autonomic system. This is because its preganglionic fibers originate in the brain (specifically, the midbrain and the medulla oblongata of the brain stem) and in the second through fourth sacral levels of the spinal cord. These preganglionic parasympathetic fibers synapse in ganglia that are located next to—or actually within—the organs innervated. These parasympathetic ganglia,

Table 12.4 The sympathetic (thoracolumbar) system.

Parts of Body Innervated	Spinal Origin of Preganglionic Fibers	Origin of Postganglionic Fibers
Eye	C–8 and T–1	Cervical ganglia
Head and neck	T–1 to T–4	Cervical ganglia
Heart and lungs	T–1 to T–5	Upper thoracic (paravertebral) ganglia
Upper extremity	T–2 to T–9	Lower cervical and upper thoracic (paravertebral) ganglia
Upper abdominal viscera	T–4 to T–9	Celiac and superior mesenteric (collateral) ganglia
Adrenal	T–10 and T–11	—
Urinary and reproductive systems	T–12 to L–2	Celiac and inferior mesenteric (collateral) ganglia
Lower extremities	T–9 to L–2	Lumbar and upper sacral (paravertebral) ganglia

Table 12.5 The parasympathetic (craniosacral) system.

Effector Organs	Origin of Preganglionic Fibers	Nerve	Location of Terminal Ganglia
Eye (ciliary and iris muscles)	Midbrain (cranial)	Oculomotor (third cranial) nerve	Ciliary ganglion
Lacrimal, mucous, and salivary glands in head	Medulla oblongata (cranial)	Facial (seventh cranial) nerve	Sphenopalatine and submandibular ganglia
Parotid (salivary) gland	Medulla oblongata (cranial)	Glossopharyngeal (ninth cranial) nerve	Otic ganglion
Heart, lungs, gastrointestinal tract, liver, pancreas	Medulla oblongata (cranial)	Vagus (tenth cranial) nerve	Terminal ganglia in or near organ
Lower half of large intestine, rectum, urinary bladder, and reproductive organs	S–2 to S–4 (sacral)	Through pelvic spinal nerves	Terminal ganglia near organs

which are called **terminal ganglia,** supply the postganglionic fibers that synapse with the effector cells. Tables 12.4 and 12.5 show the comparative structures of the sympathetic and parasympathetic divisions. It should be noted that, unlike sympathetic fibers, most parasympathetic fibers do not travel within spinal nerves. Cutaneous effectors (blood vessels, sweat glands, and arrector pili muscles) and blood vessels in skeletal muscles thus receive sympathetic but not parasympathetic innervation.

Four of the twelve pairs of cranial nerves contain preganglionic parasympathetic fibers. These are the oculomotor (III), facial (VII), glossopharyngeal (IX), and vagus (X) nerves. Parasympathetic fibers within the first three of these cranial nerves synapse in ganglia located in the head; fibers in the vagus nerve synapse in terminal ganglia located in many regions of the body.

The oculomotor nerve contains somatic motor and parasympathetic fibers that originate in the oculomotor nuclei of the midbrain. These parasympathetic fibers synapse in the *ciliary ganglion,* whose postganglionic fibers innervate the ciliary muscle and constrictor fibers in the iris of the eye. Preganglionic fibers that originate in the medulla oblongata travel in the facial nerve to the *pterygopalatine ganglion,* also called the sphenopalatine ganglion, which sends postganglionic fibers to the nasal mucosa, pharynx, palate, and lacrimal glands. Another group of fibers in the facial nerve terminate in the *submandibular ganglion,* which sends postganglionic fibers to the submandibular and sublingual salivary glands. Preganglionic fibers of the glossopharyngeal nerve synapse in the *otic ganglion,* which sends postganglionic fibers to innervate the parotid salivary gland.

Table 12.6 Some comparisons between the structure of the sympathetic and parasympathetic systems.

	Sympathetic	Parasympathetic
Origin of Preganglionic Outflow	Thoracolumbar levels of spinal cord	Midbrain, hindbrain, and sacral levels of spinal cord
Location of Ganglia	Chain of paravertebral ganglia and prevertebral (collateral) ganglia	Terminal ganglia in or near effector organs
Distribution of Postganglionic Fibers	Throughout the body	Mainly limited to the head and the viscera of the chest, abdomen, and pelvis
Divergence of Impulses from Pre- to Postganglionic Fibers	Great divergence (one preganglionic may activate twenty postganglionic fibers)	Little divergence (one preganglionic only activates a few postganglionic fibers)
Mass Discharge of System as a Whole	Yes	Not normally

Other nuclei in the medulla oblongata contribute preganglionic fibers to the very long *tenth cranial,* or *vagus nerve* (the "vagrant" or "wandering" nerve). The preganglionic fibers in the vagus synapse with postganglionic neurons that are actually located *within* the innervated organs. The preganglionic vagus fibers are thus quite long, and provide parasympathetic innervation to the heart, lungs, esophagus, stomach, pancreas, liver, small intestine, and the upper half of the large intestine. Postganglionic parasympathetic fibers arise from terminal ganglia within these organs and synapse with effector cells (smooth muscles and glands).

Preganglionic fibers from the sacral levels of the spinal cord provide parasympathetic innervation to the lower half of the large intestine, rectum, and to the urinary and reproductive systems. These fibers, like those of the vagus, synapse with terminal ganglia located within the effector organs.

Parasympathetic nerves to the visceral organs thus consist of preganglionic fibers, whereas sympathetic nerves to these organs contain postganglionic fibers. A composite view of the sympathetic and parasympathetic systems is provided in figure 12.9, and these comparisons are summarized in table 12.6.

1. *Using a simple line diagram, illustrate the sympathetic pathway (a) from the spinal cord to the heart, and (b) from the spinal cord to the adrenal gland. Label the preganglionic and postganglionic fibers and the ganglion.*
2. *Describe the mass activation of the sympathetic system, and explain the significance of the term* sympathoadrenal system.
3. *Using a simple line diagram, illustrate the parasympathetic pathway from the brain to the heart. Compare the location of the pre- and postganglionic fibers and the location of their ganglia with those of the sympathetic division.*

Functions of the Autonomic Nervous System

The sympathetic and parasympathetic divisions of the autonomic system affect the visceral organs in different ways. Mass activation of the *sympathetic system* prepares the body for intense physical activity in emergencies; the heart rate increases, blood glucose rises, and blood is diverted to the skeletal muscles (away from the visceral organs and skin). These and other effects are listed in table 12.7. The "theme" of the sympathetic system has been aptly summarized in a phrase: **fight or flight.**

The effects of *parasympathetic nerve* stimulation are in many ways opposite to the effects of sympathetic stimulation. The parasympathetic system, however, is not normally activated as a whole. Stimulation of separate parasympathetic nerves can result in slowing of the heart, dilation of visceral blood vessels, and an increased activity of the digestive tract (table 12.7). The different responses of a visceral organ to sympathetic and parasympathetic nerve activity are due to the fact that the postganglionic fibers of these two divisions release different neurotransmitters.

Neurotransmitters of the Autonomic System

Acetylcholine (*ACh*) is the neurotransmitter of all preganglionic fibers (both sympathetic and parasympathetic). Acetylcholine is also the transmitter released by all parasympathetic postganglionic fibers at their synapses with effector cells (fig. 12.10). Transmission at the autonomic ganglia and at synapses of postganglionic nerve fibers is thus said to be **cholinergic.**

Table 12.7 Effects of autonomic nerve stimulation on various visceral effector organs.

Effector Organ	Sympathetic Effect	Parasympathetic Effect
Eye		
Iris (radial muscle)	Dilates pupil	———
Iris (sphincter muscle)	———	Constricts pupil
Ciliary muscle	Relaxes (for far vision)	Contracts (for near vision)
Glands		
Lacrimal (tear)	———	Stimulates secretion
Sweat	Stimulates secretion	———
Salivary	Decreases secretion; saliva becomes thick	Increases secretion; saliva becomes thin
Stomach	———	Stimulates secretion
Intestine	———	Stimulates secretion
Adrenal medulla	Stimulates secretion of hormones	———
Heart		
Rate	Increases	Decreases
Conduction	Increases rate	Decreases rate
Strength	Increases	———
Blood Vessels	Mostly constricts; affects all organs	Dilates in a few organs (e.g., penis)
Lungs		
Bronchioles (tubes)	Dilates	Constricts
Mucous glands	Inhibits secretion	Stimulates secretion
Gastrointestinal Tract		
Motility	Inhibits movement	Stimulates movement
Sphincters	Stimulates closing	Inhibits closing
Liver	Stimulates hydrolysis of glycogen	———
Adipose (fat) cells	Stimulates hydrolysis of fat	———
Pancreas	Inhibits exocrine secretions	Stimulates exocrine secretions
Spleen	Stimulates contraction	———
Urinary Bladder	Helps set muscle tone	Stimulates contraction
Piloerector Muscles	Stimulates erection of hair and goose bumps	———
Uterus	If pregnant: contraction If not pregnant: relaxation	———
Penis	Erection; ejaculation	Erection (due to vasodilation)

The neurotransmitter released by most postganglionic sympathetic nerve fibers is **norepinephrine** (*noradrenalin*). Transmission at these synapses is thus said to be **adrenergic.** There are a few exceptions to this rule: some sympathetic fibers that innervate blood vessels in skeletal muscles, as well as sympathetic fibers to sweat glands, release ACh (are cholinergic).

In view of the fact that the cells of the adrenal medulla are derived from postganglionic sympathetic neurons, it is not surprising that the hormones they secrete (normally about 85% epinephrine and 15% norepinephrine) are similar to the transmitter of postganglionic sympathetic neurons. Epinephrine differs from norepinephrine only in the presence of an additional methyl (CH_3) group, as shown in figure 12.11. Epinephrine, norepinephrine, and dopamine (a transmitter within the CNS) are collectively termed **catecholamines.** The catecholamines, together with serotonin (another transmitter within the CNS), are often referred to as *monoamines,* or *biogenic amines.*

Responses to Adrenergic Stimulation

Adrenergic stimulation—by epinephrine in the blood and by norepinephrine released from sympathetic nerve endings—has both excitatory and inhibitory effects. The heart, dilatory muscles of the iris, and the smooth muscles of many blood vessels are stimulated to contract. The smooth muscles of the bronchioles and of some blood vessels, however, are inhibited from contracting; adrenergic chemicals, therefore, cause these structures to dilate.

Since excitatory and inhibitory effects can be produced in different tissues by the same chemical, the responses clearly depend on the biochemistry of the tissue cells rather than on the intrinsic properties of the chemical. Included in the biochemical differences among the target tissues for catecholamines are differences in the *membrane receptor proteins* for these chemical agents. There are two major classes of these receptor proteins, designated **alpha** (α) and **beta** (β) **receptors.**

Further experiments have revealed that there are two subtypes of each category of adrenergic receptor. These are designated by subscripts: α_1 and α_2; β_1 and β_2. Scientists have developed compounds that selectively bind

Figure 12.10. Neurotransmitters of the autonomic motor system. ACh = acetylcholine; NE = norepinephrine; E = epinephrine. Those nerves that release ACh are called cholinergic; those nerves that release NE are called adrenergic. The adrenal medulla secretes both epinephrine (85%) and norepinephrine (15%) as hormones into the blood.

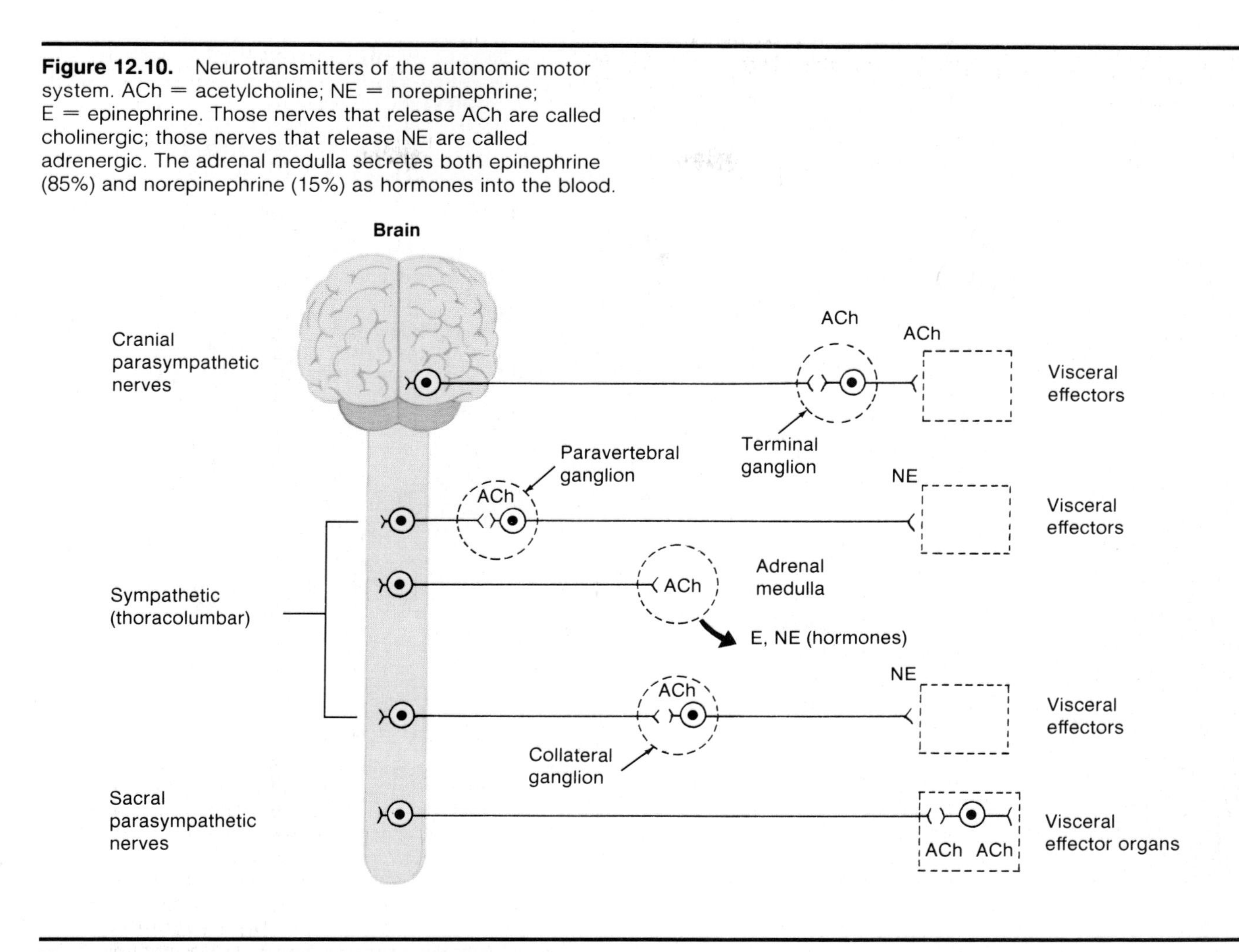

Figure 12.11. The structure of the catecholamines norepinephrine and epinephrine.

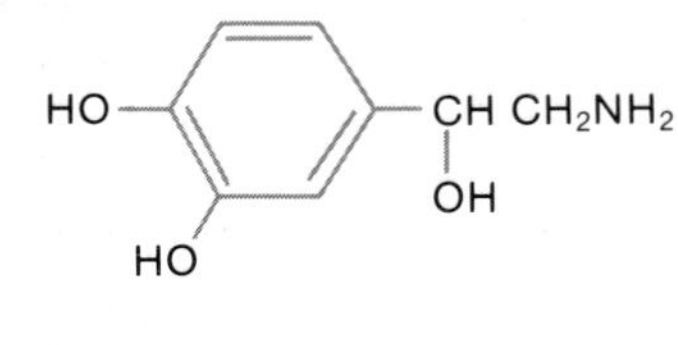

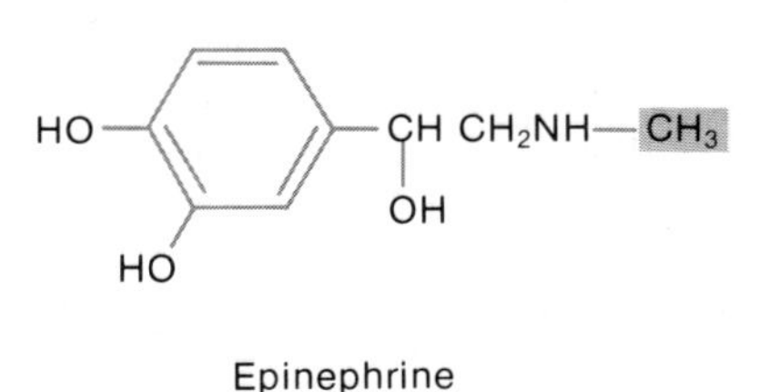

more to one or the other type of adrenergic receptor and, by this means, either promote or inhibit the normal action produced when epinephrine binds to the receptor. By using these selective compounds, it has been possible to determine the adrenergic receptor subtype in each organ and to correlate that receptor with the physiological effect of autonomic neurotransmitters in the organs (table 12.8). Both types of beta receptors appear to produce these effects by stimulating the production of cyclic AMP (discussed in chapter 7) within the target cells. The activation of the α_2 receptors has the opposite effect—cyclic AMP production is blocked. The response to α_1 receptors may be mediated by a rise in the cytoplasmic concentration of Ca^{++}, which serves as a "second messenger" in the target cell instead of cyclic AMP. Each of these changes, it should be remembered, ultimately results in the characteristic response of the tissue to the autonomic neurotransmitter.

Review of table 12.8 reveals certain generalities about the actions of adrenergic receptors. The stimulation of alpha-adrenergic receptors consistently causes constriction of smooth muscles. One can thus state that the vasoconstrictor effect of sympathetic nerves always results

Table 12.8 Adrenergic and cholinergic effects of sympathetic and parasympathetc nerves.

	Effect of			
	Sympathetic		Parasympathetic	
Organ	*Action*	*Receptor*	*Action*	*Receptor*
Eye				
Iris				
Radial muscle	Contracts	α	. . .	. . .
Circular muscle	. . .	. . .	Contracts	M
Heart				
Sinoatrial node	Accelerates	β_1	Decelerates	M
Contractility	Increases	β_1	Decreases (atria)	M
Vascular smooth muscle				
Skin, splanchnic vessels	Contracts	α	. . .	. . .
Skeletal muscle vessels	Relaxes	β_2	. . .	. . .
	Relaxes	M*	. . .	. . .
Bronchiolar smooth muscle	Relaxes	β_2	Contracts	M
Gastrointestinal tract				
Smooth muscle				
Walls	Relaxes	β_2	Contracts	M
Sphincters	Contracts	α	Relaxes	M
Secretion	. . .	. . .	Increases	M
Myenteric plexus	Inhibits	α	. . .	. . .
Genitourinary smooth muscle				
Bladder wall	Relaxes	β_2	Contracts	M
Sphincter	Contracts	α	Relaxes	M
Uterus, pregnant	Relaxes	β_2	. . .	. . .
	Contracts	α	. . .	. . .
Penis, seminal vesicles	Ejaculation	α	Erection	M
Skin				
Pilomotor smooth muscle	Contracts	α	. . .	. . .
Sweat glands				
Thermoregulatory	Increases	M	. . .	. . .
Apocrine (stress)	Increases	α	. . .	. . .

*Vascular smooth muscle in skeletal muscle has sympathetic cholinergic dilator fibers.

Adrenergic receptors are indicated as alpha (α) or beta (β); cholinergic receptors are indicated as muscarinic (M).

Modified from Katzung, B. G. 1984. *Basic and Clinical Pharmacology,* 2d ed., p. 60. Los Altos, Calif.: Lange Medical Publications.

from the activation of alpha-adrenergic receptors. The vasodilation effect of sympathetic nerves—which is of relatively less importance—may be produced by the activation of beta-adrenergic receptors or by means of cholinergic receptors (for vessels in skeletal muscles). The effects of beta-adrenergic activation are more complex; these receptors stimulate the relaxation of smooth muscles (in the digestive tract, brochioles, and uterus, for example), but stimulate contraction of cardiac muscle and promote an increase in cardiac rate.

The use of drugs that selectively stimulate or block (act as *agonists* or *antagonists*) adrenergic receptors is of great clinical benefit. People with hypertension, for example, may receive a beta-blocking drug, such as *propranolol,* which reduces the cardiac rate. People with asthma may be treated with epinephrine, which stimulates bronchodilation (a β_2 effect), but since epinephrine also stimulates the heart (a β_1 effect), more selective β_2 agonist drugs are available for the treatment of asthma. Compounds that stimulate α_1 adrenergic receptors, such as *phenylephrine,* are often part of medications to relieve a stuffy nose (by causing vasoconstriction in the nasal mucosa).

Although it may at first seem paradoxical, some people with hypertension are treated with an α_2 agonist drug—*clonidine.* Stimulation of the α_2 receptors, it may be recalled, results in the inhibition of cyclic AMP production and thus has just the opposite effect to that of stimulation of beta-adrenergic receptors. It is currently believed that presynaptic axon terminals in the brain contain α_2 receptors and that stimulation of these receptors inhibits the release of neurotransmitter from the nerve terminals; this may represent a negative feedback control of the amount of neurotransmitter released. It is known that clonidine, by stimulating α_2 receptors in the CNS, inhibits the central activation of the sympathetic nervous system and thus serves to decrease the heart rate and lower the blood pressure of hypertensive patients.

Responses to Cholinergic Stimulation

Somatic motor neurons, all preganglionic autonomic neurons, and all postganglionic parasympathetic neurons are cholinergic—they release acetylcholine as a neurotransmitter. The cholinergic effects of somatic motor neurons and preganglionic autonomic neurons are always excitatory. The cholinergic effects of postganglionic parasympathetic fibers are usually excitatory, but there are notable exceptions—the parasympathetic fibers innervating the heart, for example, cause slowing of the heart rate. It is useful to remember that the effects of parasympathetic stimulation are, in general, opposite to the effects of sympathetic stimulation.

There are two known subtypes of cholinergic receptors. The drug *muscarine,* derived from some poisonous mushrooms, stimulates the cholinergic receptor proteins in the target organs of postganglionic parasympathetic nerve fibers (such as in the heart, eye, and digestive system). Muscarine does not stimulate ACh receptor proteins in autonomic ganglia or at the neuromuscular junction of skeletal muscle fibers. The ACh receptors stimulated by muscarine are therefore called **muscarinic receptors,** and the effects produced by parasympathetic nerves in their target organs are called *muscarinic effects* (indicated in table 12.8).

The drug *nicotine,* derived from the tobacco plant, specifically stimulates cholinergic transmission of preganglionic fibers at the autonomic ganglia and activation of the neuromuscular junction of skeletal muscles. These ACh receptors are thus called **nicotinic receptors** to distinguish them from the muscarinic receptors. The drug *curare* specifically blocks nicotinic receptors but has little effect on muscarinic receptors.

The muscarinic effects of ACh are specifically inhibited by the drug **atropine,** derived from the deadly nightshade plant (*Atropa belladonna*). Indeed, extracts of this plant were used by women during the middle ages to dilate their pupils (atropine inhibits parasympathetic stimulation of the iris). This was done to enhance their beauty (belladonna—beautiful woman). Atropine is used clinically today to dilate pupils during eye examinations, to dry mucous membranes of the respiratory tract prior to general anesthesia, and to inhibit spasmodic contractions of the lower digestive tract.

Organs with Dual Innervation

Many organs receive a dual innervation—they are innervated by both sympathetic and parasympathetic fibers. When this occurs, the effects of these two divisions may be antagonistic, complementary, or cooperative.

Antagonistic Effects. The effects of sympathetic and parasympathetic innervation of the pacemaker region of the heart is the best example of the antagonism of these two systems. In this case sympathetic and parasympathetic fibers innervate the same cells. Adrenergic stimulation from sympathetic fibers increases the heart rate, and cholinergic stimulation from parasympathetic fibers inhibits the pacemaker cells and, thus, decreases the heart rate. Antagonism is also seen in the digestive tract, where sympathetic nerves inhibit and parasympathetic nerves stimulate intestinal movements and secretions.

The effects of sympathetic and parasympathetic stimulation on the diameter of the pupil of the eye are analogous to the reciprocal innervation of flexor and extensor skeletal muscles by somatic motor neurons. This is because the iris contains antagonistic muscle layers. Contraction of the radial muscles, which is stimulated by

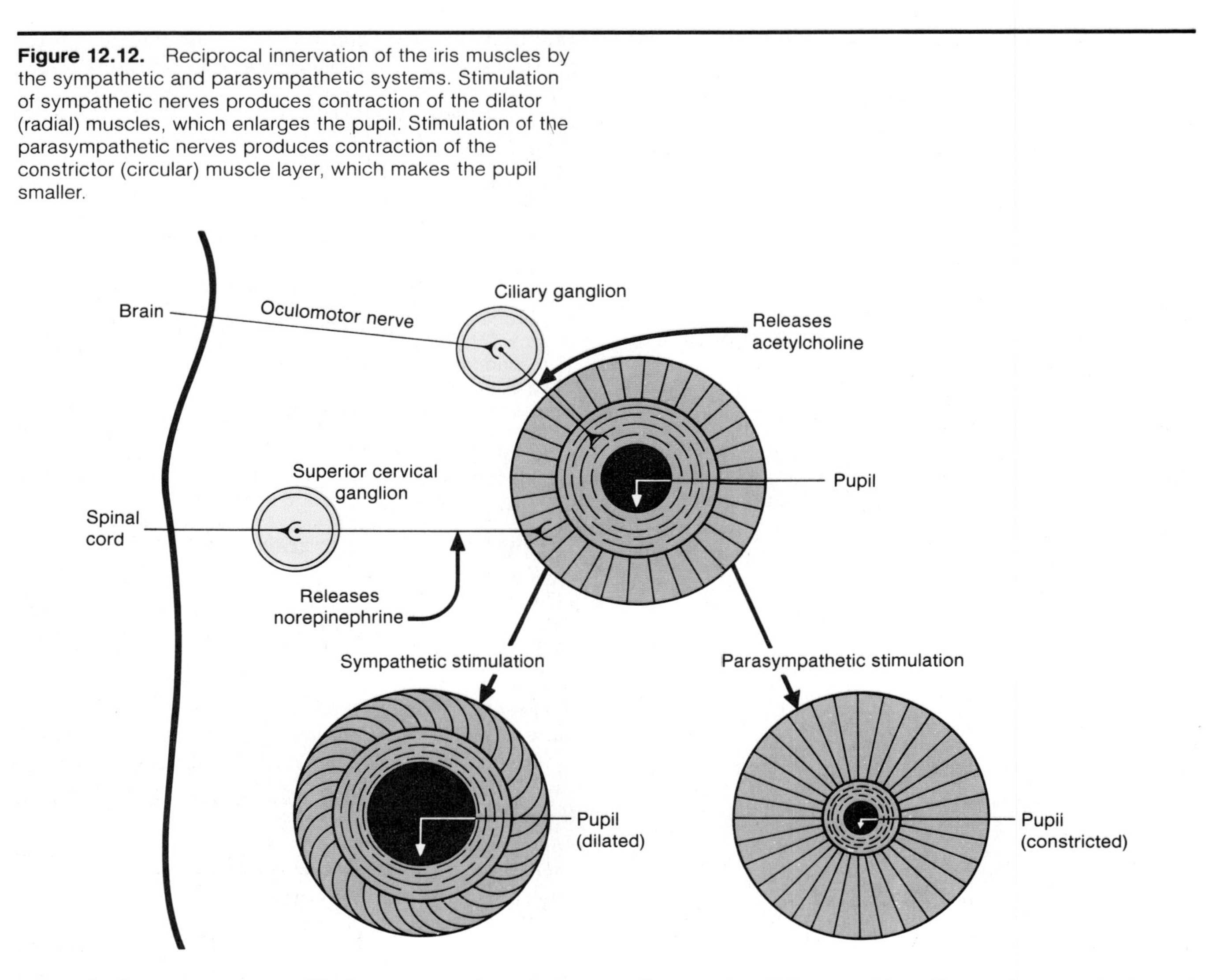

Figure 12.12. Reciprocal innervation of the iris muscles by the sympathetic and parasympathetic systems. Stimulation of sympathetic nerves produces contraction of the dilator (radial) muscles, which enlarges the pupil. Stimulation of the parasympathetic nerves produces contraction of the constrictor (circular) muscle layer, which makes the pupil smaller.

sympathetic nerves, causes dilation; contraction of the circular muscles, which are innervated by parasympathetic nerve endings, causes constriction of the pupils (fig. 12.12).

Complementary Effects. The effects of sympathetic and parasympathetic stimulation on salivary gland secretion are complementary. The secretion of watery saliva is stimulated through parasympathetic nerves, which also stimulate the secretion of other exocrine glands in the digestive tract. Impulses through sympathetic nerves stimulate the constriction of blood vessels throughout the digestive tract. The resultant decrease in blood flow to the salivary glands causes the production of a thicker, more viscous saliva.

Cooperative Effects. The effects of sympathetic and parasympathetic stimulation on the urinary and reproductive systems are cooperative. Erection of the penis, for example, is due to vasodilation resulting from parasympathetic nerve stimulation; ejaculation is due to stimulation through sympathetic nerves. Although the contraction of the urinary bladder is myogenic (independent of nerve stimulation), it is promoted in part by the action of parasympathetic nerves. This *micturition,* or urination, urge and reflex is also enhanced by sympathetic nerve activity, which increases the tone of the bladder muscles. Emotional states that are accompanied by high sympathetic nerve activity may thus result in reflex urination at bladder volumes that are normally too low to trigger this reflex.

Organs without Dual Innervation

Although most organs are innervated by both sympathetic and parasympathetic nerves, some—including the adrenal medulla, arrector pili muscles, sweat glands, and most blood vessels—receive only sympathetic innervation. In these cases regulation is achieved by increases or decreases in the "tone" (firing rate) of the sympathetic fibers. Constriction of the blood vessels, for example, is produced by increased sympathetic activity, which stimulates alpha-adrenergic receptors, and vasodilation results from decreased sympathetic nerve stimulation.

The sympathoadrenal system is required for *nonshivering thermogenesis:* animals deprived of their sympathetic system and adrenals cannot tolerate cold stress. The sympathetic system itself is required for proper thermoregulatory responses to heat. In a hot room, for example, decreased sympathetic stimulation produces dilation of the blood vessels in the surface of the skin, which increases cutaneous blood flow and provides better heat radiation. During exercise, in contrast, there is increased sympathetic activity, which causes constriction of the blood vessels in the skin of the limbs and stimulation of sweat glands in the trunk.

The eccrine sweat glands in the trunk secrete a watery fluid in response to sympathetic stimulation. Evaporation of this dilute sweat helps to cool the body. The eccrine sweat glands also secrete a chemical called **bradykinin** in response to sympathetic stimulation. Bradykinin stimulates dilation of the surface blood vessels near the sweat glands, helping to radiate heat. At the conclusion of exercise, sympathetic stimulation is reduced and blood flow to the surface of the limbs is increased, which aids in the elimination of metabolic heat. Notice that all of these thermoregulatory responses are achieved without the direct involvement of the parasympathetic system.

---

1. *Define the meaning of the terms* adrenergic *and* cholinergic, *and use these terms to describe the neurotransmitters of different autonomic nerve fibers.*
2. *List the effects of sympathoadrenal stimulation on different effector organs, and indicate which effects are due to alpha or beta receptor stimulation.*
3. *Describe the effects of the drug* atropine, *and explain these effects in terms of the actions of the parasympathetic system.*
4. *Explain how the sympathetic and parasympathetic systems can have antagonistic, cooperative, and complementary effects, and give examples.*

Table 12.9 Some reflexes stimulated by sensory input from afferent fibers in the vagus that are transmitted to centers in the medulla oblongata.

Organs	Type of Receptors	Reflex Effects
Lungs	Stretch receptors	Inhibit further inhalation; stimulate an increase in cardiac rate and vasodilation
	Type J receptors	Stimulated by pulmonary congestion—produce feelings of breathlessness and cause a reflex fall in cardiac rate and blood pressure
Aorta	Chemoreceptors	Stimulated by rise in CO_2 and fall in O_2—produce increased rate of breathing, fall in heart rate, and vasoconstriction
	Baroreceptors	Stimulated by increased blood pressure—produce a reflex decrease in heart rate
Heart	Atrial stretch receptors	Inhibit antidiuretic hormone secretion, thus increasing the volume of urine excreted
	Stretch receptors in ventricles	Produce a reflex decrease in heart rate and vasodilation
Gastrointestinal tract	Stretch receptors	Feelings of satiety, discomfort, and pain

Control of the Autonomic System by Higher Brain Centers

Visceral functions are regulated, to a large degree, by autonomic reflexes. In most autonomic reflexes sensory input is directed to brain centers, which in turn regulate the activity of descending pathways to preganglionic autonomic neurons. The neural centers that directly control the activity of autonomic nerves are influenced by higher brain areas as well as by sensory input.

The **medulla oblongata** of the brain stem is the area that most directly controls the activity of the autonomic system. Almost all autonomic responses can be elicited by experimental stimulation of the medulla, which contains centers for the control of the cardiovascular, pulmonary, urinary, reproductive, and digestive systems. Much of the sensory input to these centers travels in the afferent fibers of the vagus nerve, which is a mixed nerve containing both sensory and motor fibers. These reflexes are listed in table 12.9.

Autonomic dysreflexia, a serious condition producing rapid elevations in blood pressure that can lead to stroke (cerebrovascular accident), occurs in 85% of people with quadriplegia and others with spinal cord lesions above the sixth thoracic level. Lesions to the spinal cord first produce the symptoms of spinal shock, characterized by the loss of both skeletal muscle and autonomic reflexes. After a period of time both types of reflexes return in an exaggerated state; the skeletal muscles may become spastic, due to absence of higher inhibitory influences, and the visceral organs experience denervation hypersensitivity. Patients in this state have difficulty emptying their urinary bladders and must often be catheterized.

Noxious stimuli, such as overdistension of the urinary bladder, can result in reflex activation of the sympathetic nerves below the spinal cord lesion. This produces goose bumps, cold skin, and vasoconstriction in the regions served by the spinal cord below the level of the lesion. The rise in blood pressure resulting from this vasoconstriction activates pressure receptors that transmit impulses along sensory nerve fibers to the medulla. In response to this sensory input, the medulla directs a reflex slowing of the heart and vasodilation. Since descending impulses are blocked by the spinal lesion, however, the skin is warm and moist (due to vasodilation and sweat gland secretion) above the lesion, but cold below the level of spinal cord damage.

Hypothalamus

The hypothalamus (fig. 12.13) is an extremely important brain region, located just above (superior to) the pituitary gland, which in turn is located above the posterior portion of the roof of the mouth. By means of efferent fibers to the brain stem and posterior pituitary and by means of hormones that regulate the anterior pituitary, the hypothalamus serves to orchestrate somatic, autonomic, and endocrine responses during various behavioral states.

Experimental stimulation of different areas of the hypothalamus can evoke the autonomic responses characteristic of aggression, sexual behavior, eating, or satiety. Chronic stimulation of the lateral hypothalamus, for example, can make an animal eat and become obese, whereas stimulation of the medial hypothalamus inhibits eating. Other areas contain osmoreceptors that stimulate thirst and the secretion of antidiuretic hormone (ADH) from the posterior pituitary.

The hypothalamus is also where the body's "thermostat" is located. Experimental cooling of the preoptic-anterior hypothalamus causes shivering (a somatic response) and nonshivering thermogenesis (sympathetic responses). Experimental heating of this hypothalamic area results in hyperventilation (stimulated by somatic motor nerves), vasodilation, salivation, and sweat gland secretion (stimulated by sympathetic nerves).

The coordination of sympathetic and parasympathetic reflexes by the medulla oblongata is thus integrated with the control of somatic and endocrine responses by the hypothalamus. The activities of the hypothalamus are in turn influenced by higher brain centers.

Limbic System, Cerebellum, and Cerebrum

The limbic system is a group of fiber tracts and nuclei that form a ring (a limbus) around the brain stem. It includes the cingulate gyrus of the cerebral cortex, the hypothalamus, the fornix (a fiber tract), the hippocampus, and the amygdaloid nucleus. These structures, which were derived early in the course of vertebrate evolution, were once called the *rhinencephalon,* or "smell brain," because of their importance in the central processing of olfactory information.

In higher vertebrates, these structures are now recognized as centers involved in basic emotional drives—such as anger, fear, sex, and hunger—and in short-term memory (described in chapter 9). Complex circuits between the hypothalamus and other parts of the limbic system contribute visceral responses to emotions.

The autonomic correlates of motion sickness—nausea, sweating, and cardiovascular changes—are abolished by cutting the efferent tracts of the cerebellum. This demonstrates that impulses from the cerebellum to the medulla influence activity of the autonomic system. Experimental and clinical observations have also demonstrated that the frontal and temporal lobes of the cerebral cortex influence lower brain areas as part of their involvement in emotion and personality.

One of the most dramatic examples of the role of higher brain areas in personality and emotion is the famous crowbar accident, which occurred in 1848. A twenty-five-year-old railroad foreman, Phineas P. Gage, was tamping gunpowder into a hole in a rock when it exploded. The rod—three feet, seven inches long and one and one-fourth inches thick—was driven through his left eye, passed through his brain, and emerged in the back of his skull.

After a few minutes of convulsions, Gage got up, rode a horse three-quarters of a mile into town, and walked up a long flight of stairs to see a doctor. He recovered well, with no noticeable sensory or motor deficits. His associates, however, noted striking personality changes. Before the accident Gage was a responsible, capable, and financially prudent man. Afterwards, he appeared to lose social inhibitions, engaged in gross profanity (which he did not do before the accident), and seemed to be tossed about by chance whims. He was eventually fired from his job, and his previous friends remarked that he was "no longer Gage."

Figure 13.18. The structure of a medium-sized artery and vein showing the relative thickness and composition of the tunicas.

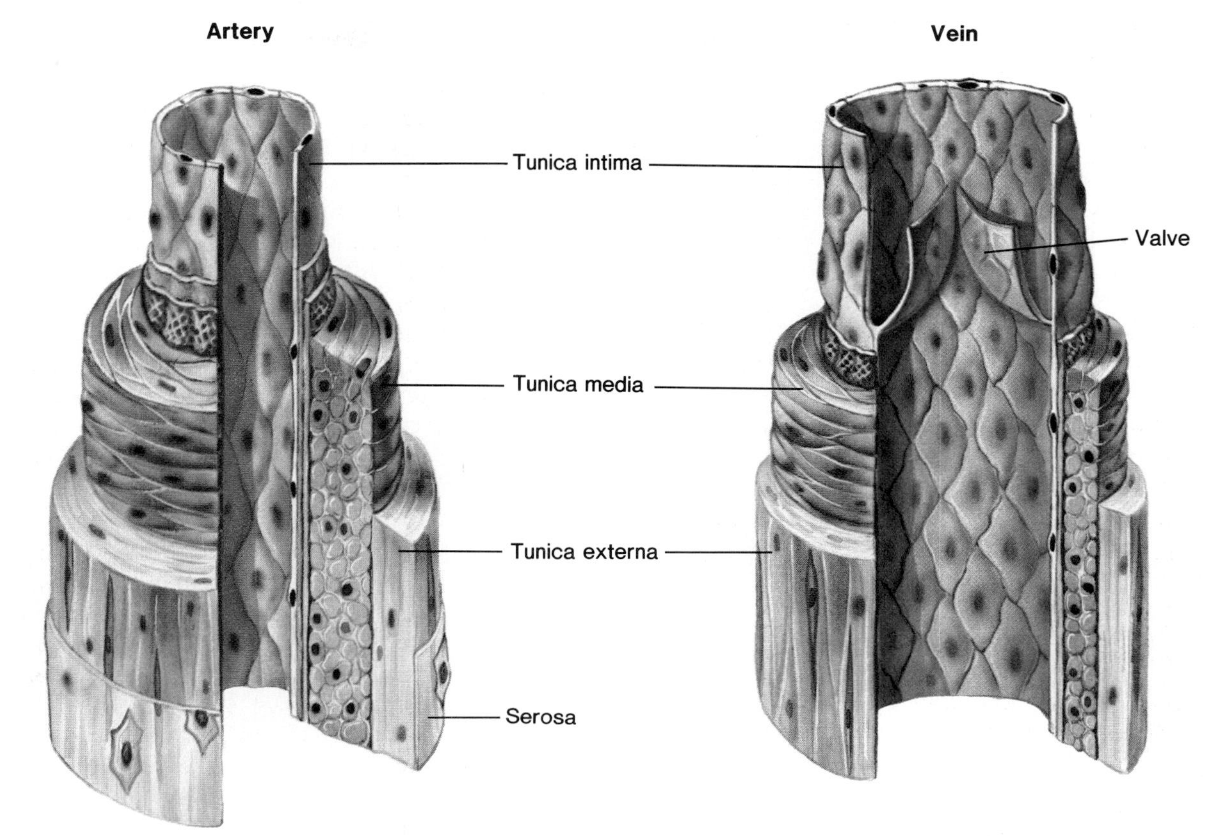

The small arteries and arterioles are less elastic than the larger arteries and have a thicker layer of smooth muscle for their diameters. Unlike the larger **elastic arteries,** therefore, the smaller **muscular arteries** retain almost the same diameter as the pressure of the blood rises and falls during the heart's pumping activity. Since arterioles and small muscular arteries have narrow lumina, they provide the greatest resistance to blood flow through the arterial system.

Small muscular arteries that are 100 μm or less in diameter branch to form smaller arterioles (20–30 μm in diameter). In some tissues, blood from the arterioles can enter the venules directly through *arteriovenous anastomoses.* In most cases, however, blood from arterioles passes into capillaries (fig. 13.19). Capillaries are the narrowest of blood vessels (7–10 μm in diameter), and serve as the "business end" of the circulatory system in which exchanges of gases and nutrients between the blood and the tissues occur.

Capillaries

The arterial system branches extensively (table 13.3) to deliver blood to over forty billion capillaries in the body. The extensiveness of these branchings is indicated by the fact that all tissue cells are located within a distance of only 60–80 μm of a capillary and by the fact that capillaries provide a total surface area of 1,000 square miles for exchanges between blood and tissue fluid.

Figure 13.19. The microcirculation. Metarterioles (arteriovenous anastomoses) provide a path of least resistance between arterioles and venules. Precapillary sphincter muscles regulate the flow of blood through the capillaries.

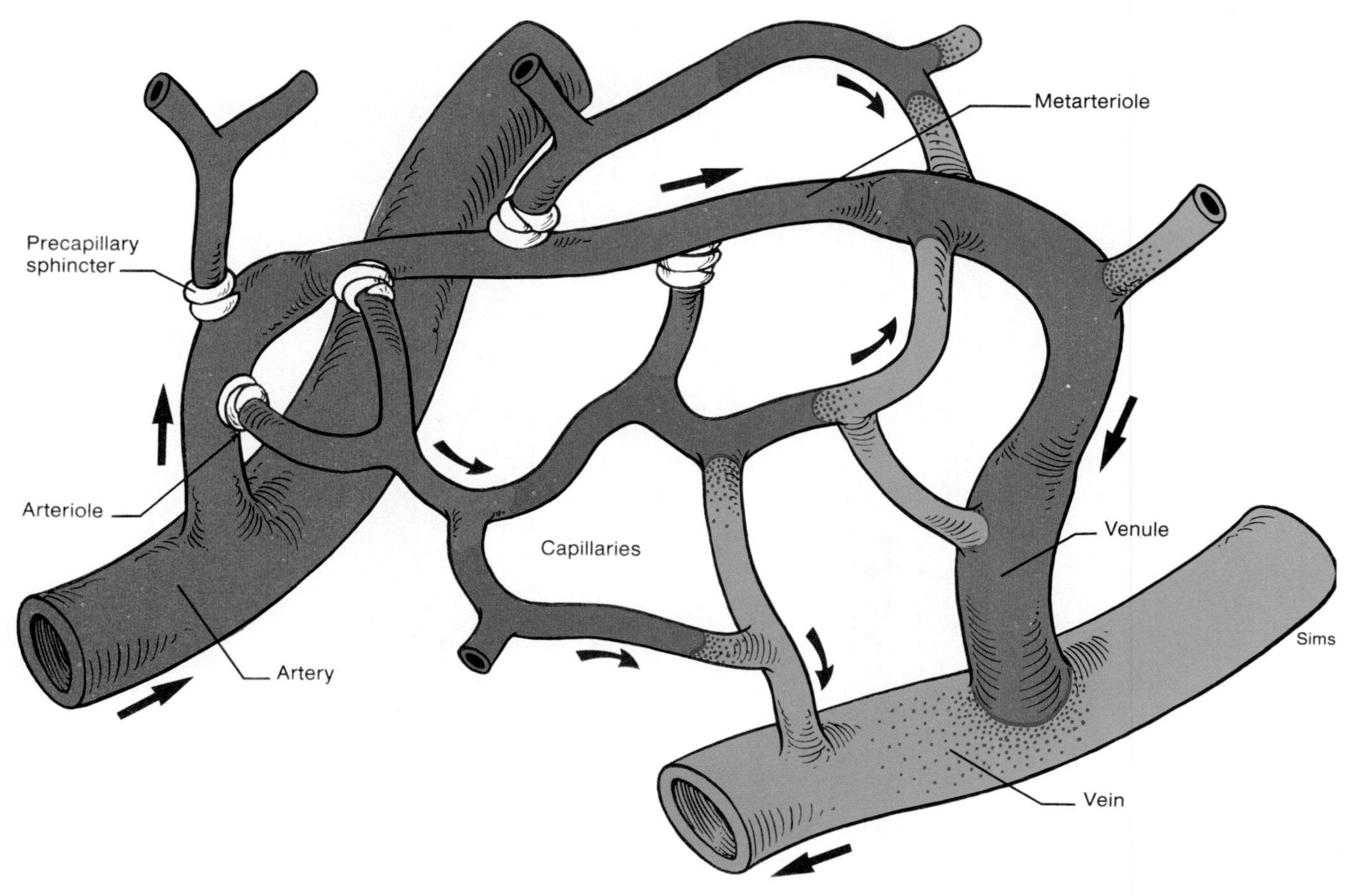

Table 13.3 Characteristics of the vascular supply to the mesenteries in a dog.

Kind of Vessel	Diameter (mm)	Number	Total Cross-Sectional Area (cm^2)	Length (cm)	Total Volume (cm^3)
Aorta	10	1	0.8	40	30
Large arteries	3	40	3.0	20	60
Main artery branches	1	600	5.0	10	50
Terminal branches	0.6	1,800	5.0	1	25
Arterioles	0.02	40,000,000	125	0.2	25
Capillaries	0.008	1,200,000,000	600	0.1	60
Venules	0.03	80,000,000	570	0.2	110
Terminal veins	1.5	1,800	30	1	30
Main venous branches	2.4	600	27	10	270
Large veins	6.0	40	11	20	220
Vena cava	12.5	1	1.2	40	50
					930

Source: *Animal Physiology: Principles and Adaptations*, 3d ed., by Malcolm S. Gordon. Macmillan Publishing Company. Copyright © 1977 by Malcolm S. Gordon.

Figure 13.20. An electron micrograph of a capillary in a coronary muscle. Notice the thin intercellular channel and the fact that the capillary wall is only one cell thick. Arrows show some of the many pinocytotic vesicles.

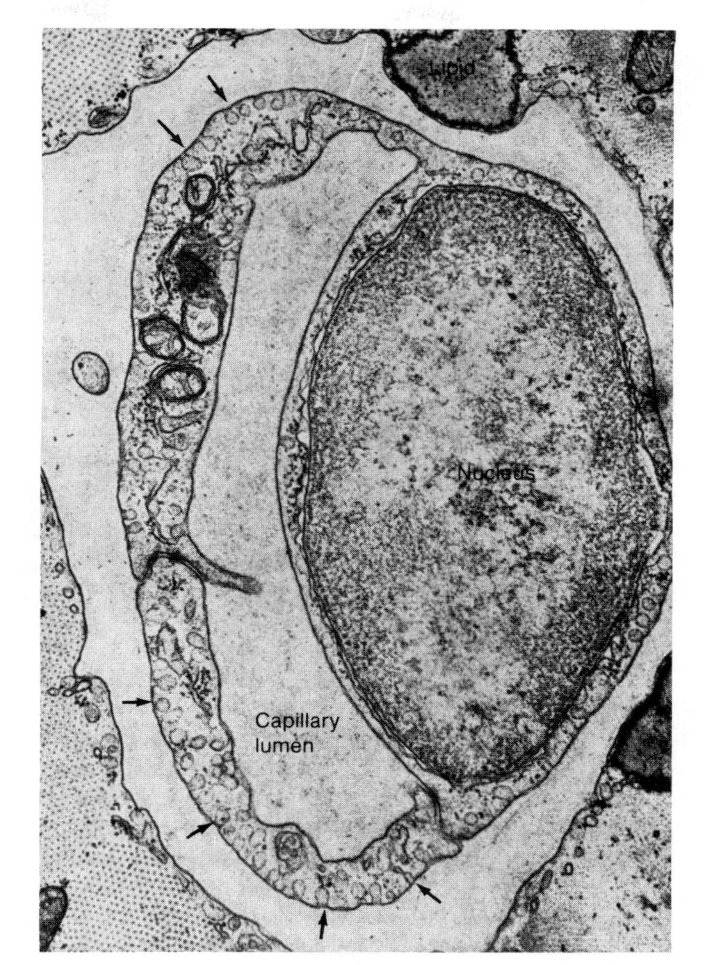

Despite their large number, capillaries contain only about 250 ml of blood at any time, out of a total blood volume of about 5,000 ml (most is contained in the venous system). The amount of blood flowing through a particular capillary bed is determined in part by the action of the **precapillary sphincter muscles.** These muscles allow only 5%–10% of the capillary beds in skeletal muscles, for example, to be open at rest. Blood flow to an organ is regulated by the action of these precapillary sphincters and by the degree of resistance to blood flow (due to constriction or dilation) provided by the small arteries and arterioles in the organ.

Unlike the vessels of the arterial and venous systems, the walls of capillaries are composed of only one cell layer—a simple squamous epithelium, or endothelium (fig. 13.20). The absence of smooth muscle and connective-tissue layers permits a more rapid rate of transport of materials between the blood and the tissues.

Types of Capillaries. Different organs have different types of capillaries, which are distinguished by significant differences in structure. In terms of their endothelial lining, these capillary types include those that are *continuous,* those that are *discontinuous,* and those that are *fenestrated.*

Continuous capillaries are those in which adjacent endothelial cells are closely joined together. These are found in muscles, lungs, adipose tissue, and in the central nervous system. Continuous capillaries in the CNS lack intercellular channels; this fact contributes to the blood-brain barrier. Continuous capillaries in other organs have narrow intercellular channels (about 40–45 Å in width), which allow the passage of molecules other than protein between the capillary blood and tissue fluid (fig. 13.21).

Figure 13.21. Diagrams of continuous, fenestrated, and discontinuous capillaries as they appear in the electron microscope. This classification is derived from the continuity of the endothelial layer.

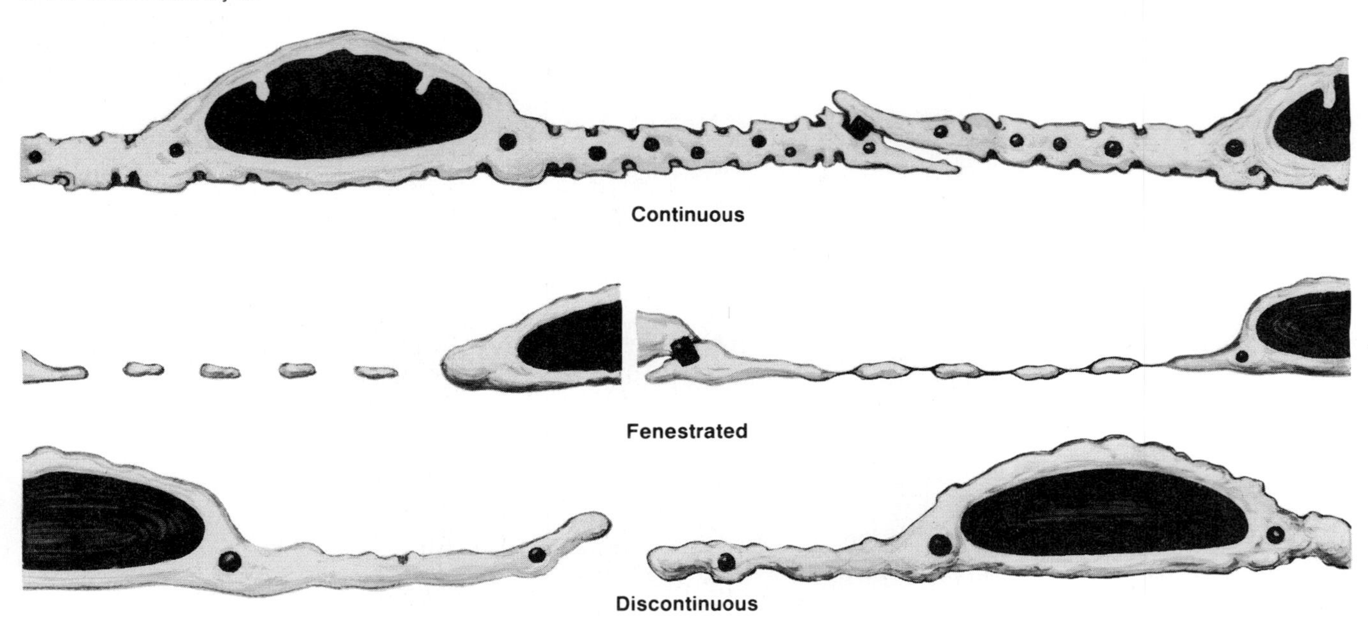

Examination of endothelial cells with an electron microscope has revealed the presence of pinocytotic vesicles (fig. 13.20), which suggests that the intracellular transport of material may occur across the capillary walls. This type of transport appears to be the only mechanism of capillary exchange available within the central nervous system and may account, in part, for the selective nature of the blood-brain barrier.

The kidneys, endocrine glands, and intestines have *fenestrated capillaries,* characterized by wide intercellular pores (800–1,000 Å) that are covered by a layer of mucoprotein, which may serve as a diaphragm. In the bone marrow, liver, and spleen, the distance between endothelial cells is so great that these *discontinuous* capillaries appear as little cavities (*sinusoids*) in the organ.

Veins

The average pressure in the veins is only 2 mm Hg, compared to a much higher average arterial pressure of about 100 mm Hg. These pressures represent the hydrostatic pressure that the blood exerts on the walls of the vessels, and the numbers indicate the differences from atmospheric pressure.

The low venous pressure is insufficient to return blood to the heart, particularly from the lower limbs. Veins, however, pass between skeletal muscle groups that produce a massaging action as they contract (fig. 13.22). As the veins are squeezed by contracting skeletal muscles, a one-way flow of blood to the heart is insured by the presence of **venous valves.** The ability of these valves to prevent the flow of blood away from the heart was demonstrated in the seventeenth century by William Harvey (fig. 13.23). After applying a tourniquet to a subject's arm, Harvey found that he could push the blood in a bulging vein toward the heart but not in the reverse direction.

The effect of the massaging action of skeletal muscles on venous blood flow is often described as the **skeletal muscle pump.** The rate of venous return to the heart is dependent, in large part, on the action of skeletal muscle pumps. When these pumps are less active, as when a person stands still or is bedridden, blood accumulates in the veins and causes them to bulge. When a person is more active, blood returns to the heart at a faster rate and less is left in the venous system.

The accumulation of blood in the veins of the legs over a long period of time, as may occur in people with occupations that require standing still all day, can cause the veins to stretch to the point where the venous valves are no longer efficient. This can produce **varicose veins.** During walking the movements of the foot activate the soleus muscle pump. This effect can be produced in bedridden people by upward and downward manipulations of the feet.

Figure 13.22. The action of the one-way venous valves. Contraction of skeletal muscles helps to pump blood toward the heart, but is prevented from pushing blood away from the heart by closure of the venous valves.

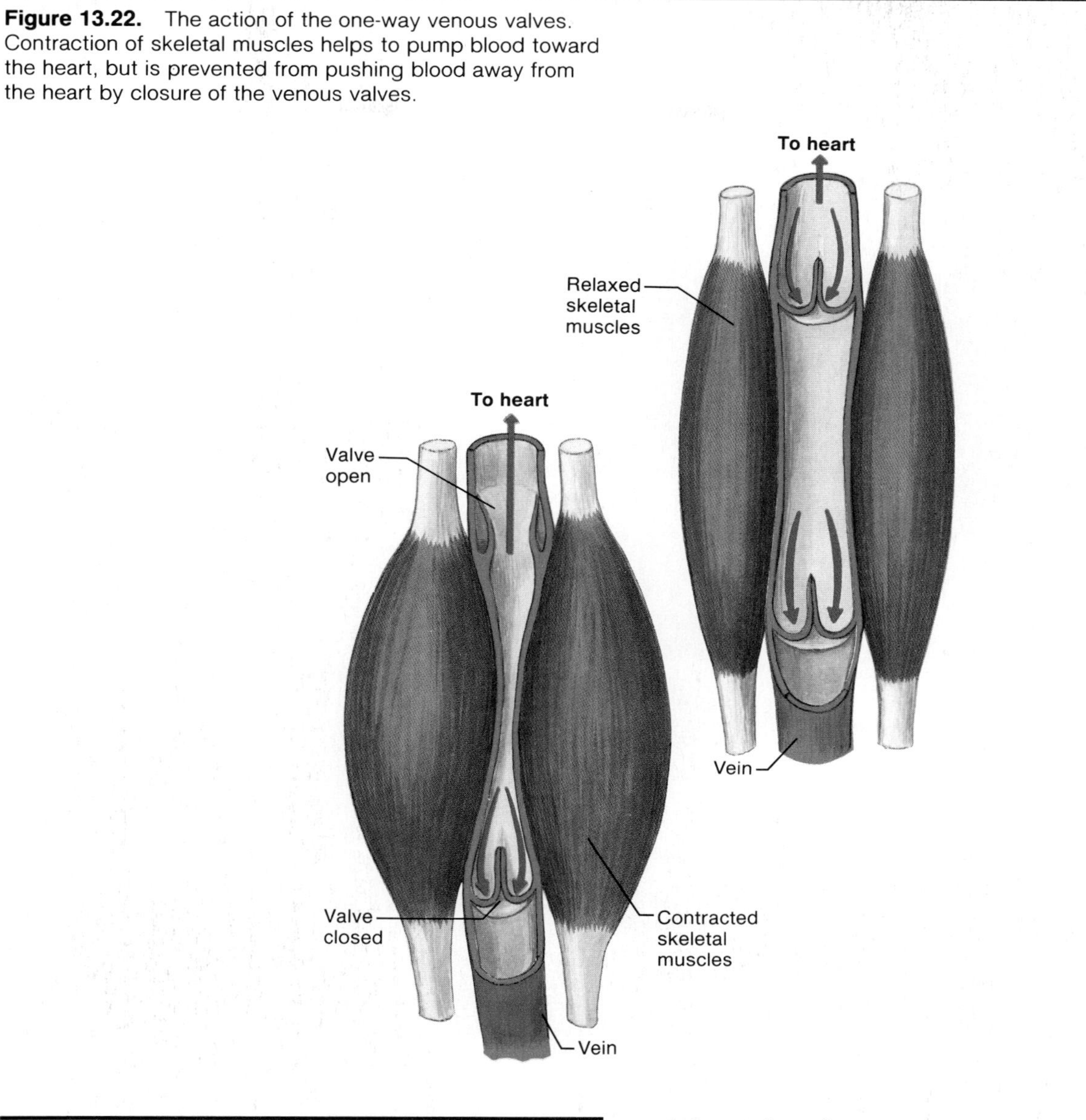

Figure 13.23. A classical demonstration by William Harvey of the existence of venous valves that prevent the flow of blood away from the heart.

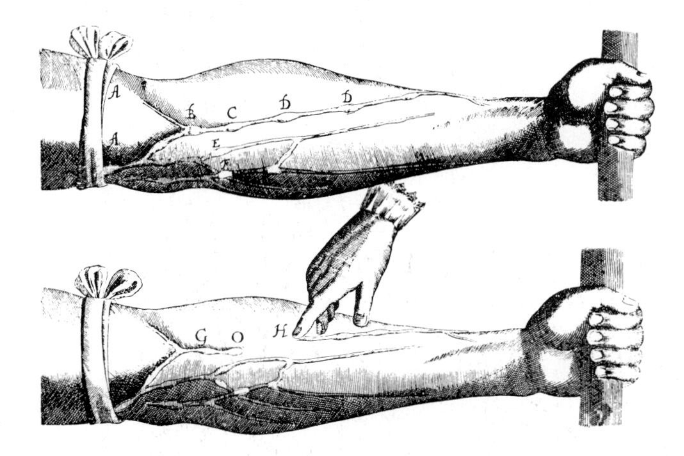

Atherosclerosis

Atherosclerosis is the most common form of arteriosclerosis (hardening of the arteries) and, through its contribution to heart disease and stroke, is responsible for about 50% of the deaths in the United States. In atherosclerosis, localized *plaques,* or **atheromas,** protrude into the lumen of the artery and thus reduce blood flow. The atheromas additionally serve as sites for *thrombus* (blood clot) formation, which can further occlude the blood supply to an organ (fig. 13.24).

It is currently believed that atheromas begin as benign tumors of smooth muscle cells that have migrated from the tunica media to the intima and proliferated by

Figure 13.24. (*a*) The lumen (cavity) of a human coronary artery is partially occluded by an atheroma and (*b*) almost completely occluded by a thrombus. (*c*) The structure of an atheroma is diagrammed.

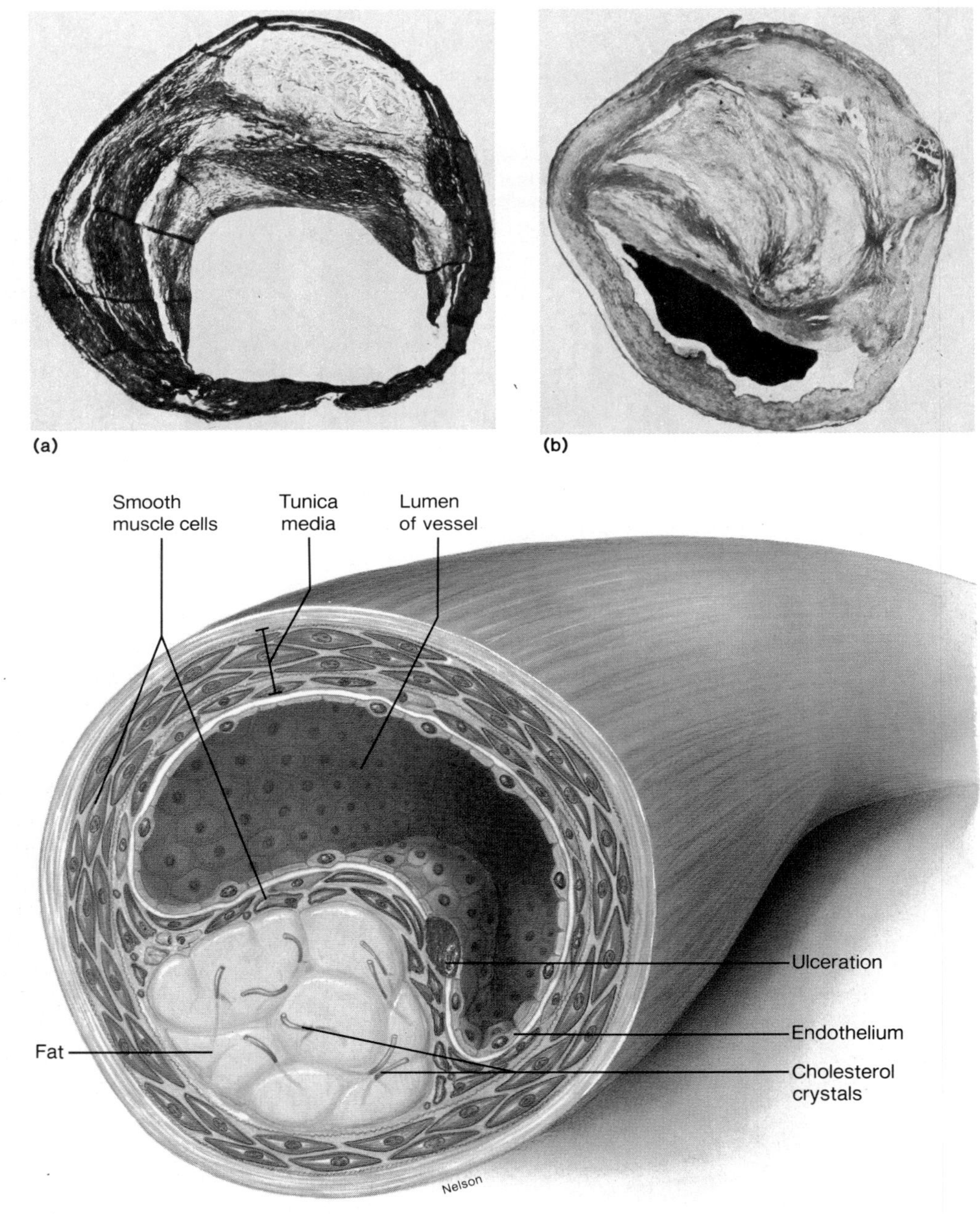

cell division. In later stages, these cells accumulate cholesterol and other lipids, giving them a "foamy cell" appearance. In fully developed plaques, cholesterol and other lipids accumulate outside the smooth muscle cells, and these accumulations can become calcified. Damage to the endothelium that covers the atheromas results in the exposure of subendothelial connective tissue to the blood. This may contribute to further growth of the atheroma and to the formation of blood clots.

Risk factors in the development of atherosclerosis include advanced age, smoking, hypertension, and high blood cholesterol concentrations. It is currently believed that an intact and properly functioning endothelium protects against atherosclerosis. If the endothelium in a particular region of an artery is removed, or if it is damaged in some way, growth factors, or *mitogens* (chemicals that stimulate mitosis), may stimulate proliferation of the smooth muscle cells and the growth of an atheroma. These growth factors may be derived from blood platelets (cell fragments in the blood involved in clotting), endothelial cells, and/or monocytes (a type of white blood cell).

Cholesterol and Plasma Lipoproteins. There is good evidence that high blood cholesterol is associated with an increased risk of atherosclerosis. This high blood cholesterol can be produced by a diet rich in cholesterol and saturated fat, or it may be the result of an inherited condition known as *familial hypercholesteremia.* This condition is inherited as a single dominant gene; individuals who inherit two of these genes have extremely high cholesterol concentrations (regardless of diet) and usually suffer heart attacks during childhood.

Cholesterol is carried to the arteries by plasma proteins called **low-density lipoproteins (LDL).** These particles, produced by the liver, consist of a core of cholesterol surrounded by a layer of phospholipids (to make the particle water-soluble) and a protein. Cells in various organs contain receptors for the protein in LDL; when LDL attaches to its receptors, the cell engulfs the LDL by receptor-mediated endocytosis (described in chapter 6) and utilizes the cholesterol for different purposes. Most of the LDL in the blood is removed in this way by the liver.

Once LDL has passed through the endothelium of an artery, it may stimulate monocytes to enter the area and engulf the cholesterol (thereby becoming "foam cells"). The monocytes may then be stimulated by LDL to secrete a growth factor that either begins or contributes to the development of an atheroma. A high blood concentration of LDL favors these events. Recent evidence shows that people who eat a diet high in cholesterol and saturated fat and people with familial hypercholesteremia, have a high blood LDL concentration because their tissues (principally the liver) have a low number of LDL receptors. With fewer LDL receptors the liver is less able to remove the LDL from the blood, the blood LDL concentration is raised, and the risk of atherosclerosis is greatly increased.

Excessive cholesterol may be released from cells and travel in the blood as **high-density lipoproteins (HDL),** which are removed by the liver. Since the cholesterol in HDL does not travel to the blood vessels, it does not contribute to atherosclerosis. Indeed, a high proportion of cholesterol in HDL as compared to LDL is beneficial, since it indicates that cholesterol may be traveling away from the blood vessels to the liver. The concentration of HDL-cholesterol appears to be higher and the risk of atherosclerosis lower in people who exercise regularly. The HDL-cholesterol concentration, for example, is higher in marathon runners than in joggers and is higher in joggers than in inactive men. Women in general have higher HDL-cholesterol concentrations and a lower risk of atherosclerosis than men.

Ischemic Heart Disease. A tissue is said to be **ischemic** when it receives an inadequate supply of oxygen because of an inadequate blood flow. The most common cause of myocardial ischemia is atherosclerosis of the coronary arteries. The adequacy of blood flow is relative—it depends on the metabolic requirements of the tissue for oxygen. An obstruction in a coronary artery, for example, may allow sufficient blood flow at rest but may produce ischemia when the heart is stressed by exercise or emotional conditions.

Myocardial ischemia is associated with increased concentrations of blood lactic acid produced by anaerobic respiration of the ischemic tissue. This condition often causes substernal pain, which may also be referred to the left shoulder and arm, as well as other areas. This referred pain is called **angina pectoris.** People with angina frequently take nitroglycerin or related drugs that help relieve the ischemia and pain. These drugs are effective because they stimulate vasodilation, which improves circulation to the heart and decreases the work that the heart must perform to eject blood into the arteries.

Myocardial cells are adapted to respire aerobically and cannot respire anaerobically for more than a few minutes. If ischemia and anaerobic respiration continue for more than a few minutes, *necrosis* (cellular death) may occur in the areas most deprived of oxygen. A sudden, irreversible injury of this kind is called a **myocardial infarction,** or **MI.**

Figure 13.25. Depression of the S-T segment of the electrocardiogram as a result of myocardial ischemia.

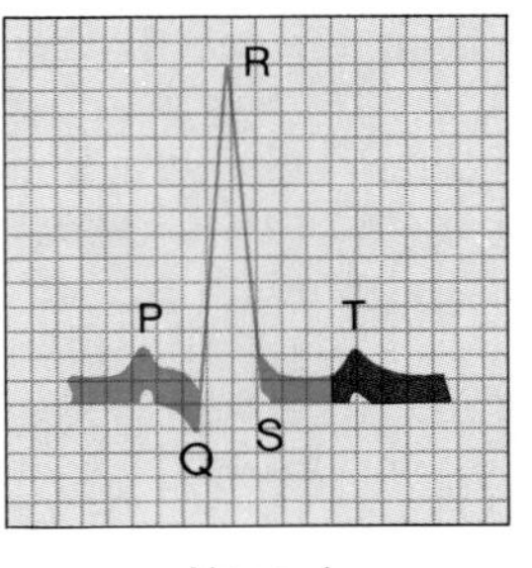

Normal

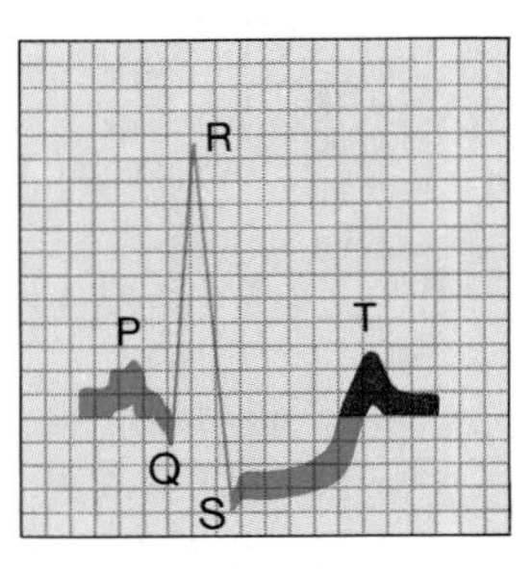

Ischemia

Table 13.4 Changes in the enzyme activity in plasma following a myocardial infarction.

Serum Enzyme	Earliest Increase (hr)	Maximum Concentration (hr)	Return to Normal (days)	Amplitude of Increase (× normal)
Creatine kinase	3 to 6	24 to 36	3	7
Malate dehydrogenase	4 to 6	24 to 48	5	4
AST	6 to 8	24 to 48	4 to 6	5
Lactate dehydrogenase	10 to 12	48 to 72	11	3
α-Hydroxybutyrate dehydrogenase	10 to 12	48 to 72	13	3 to 4
Aldolase	6 to 8	24 to 48	4	4
ALT	Usually normal, unless there are other complications			
Isocitrate dehydrogenase	Usually normal			

Source: From Montgomery, Rex, Dryer, Robert L., Conway, Thomas W., and Spector, Arthur A.: Biochemistry: a case-oriented approach, ed. 4, St. Louis, 1983, The C. V. Mosby Co.

Myocardial ischemia may be detected by changes in the S-T segment of the electrocardiogram (fig. 13.25). The diagnosis of myocardial infarction is aided by measurement of the concentration of enzymes in the blood that are released by the infarcted tissue. Plasma concentrations of *creatine phosphokinase* (*CPK*), for example, increase within three to six hours after the onset of symptoms and return to normal after three days. Plasma levels of *lactate dehydrogenase* (*LDH*) reach a peak within forty-eight to seventy-two hours after the onset of symptoms and remain elevated for about eleven days (table 13.4).

1. *Describe the basic structural pattern of arteries and veins. Describe how arteries and veins differ in structure and how these differences contribute to the resistance function of arteries and the capacitance function of veins.*
2. *Describe the functional significance of the "skeletal muscle pump," and illustrate the action of venous valves.*
3. *Explain the functions of capillaries, and describe the structural differences between capillaries in different organs.*
4. *Explain how cholesterol is carried in the plasma and how the concentrations of cholesterol carriers are related to the risk of developing atherosclerosis.*
5. *Explain how angina pectoris is produced and the relationship of this symptom to conditions in the heart.*

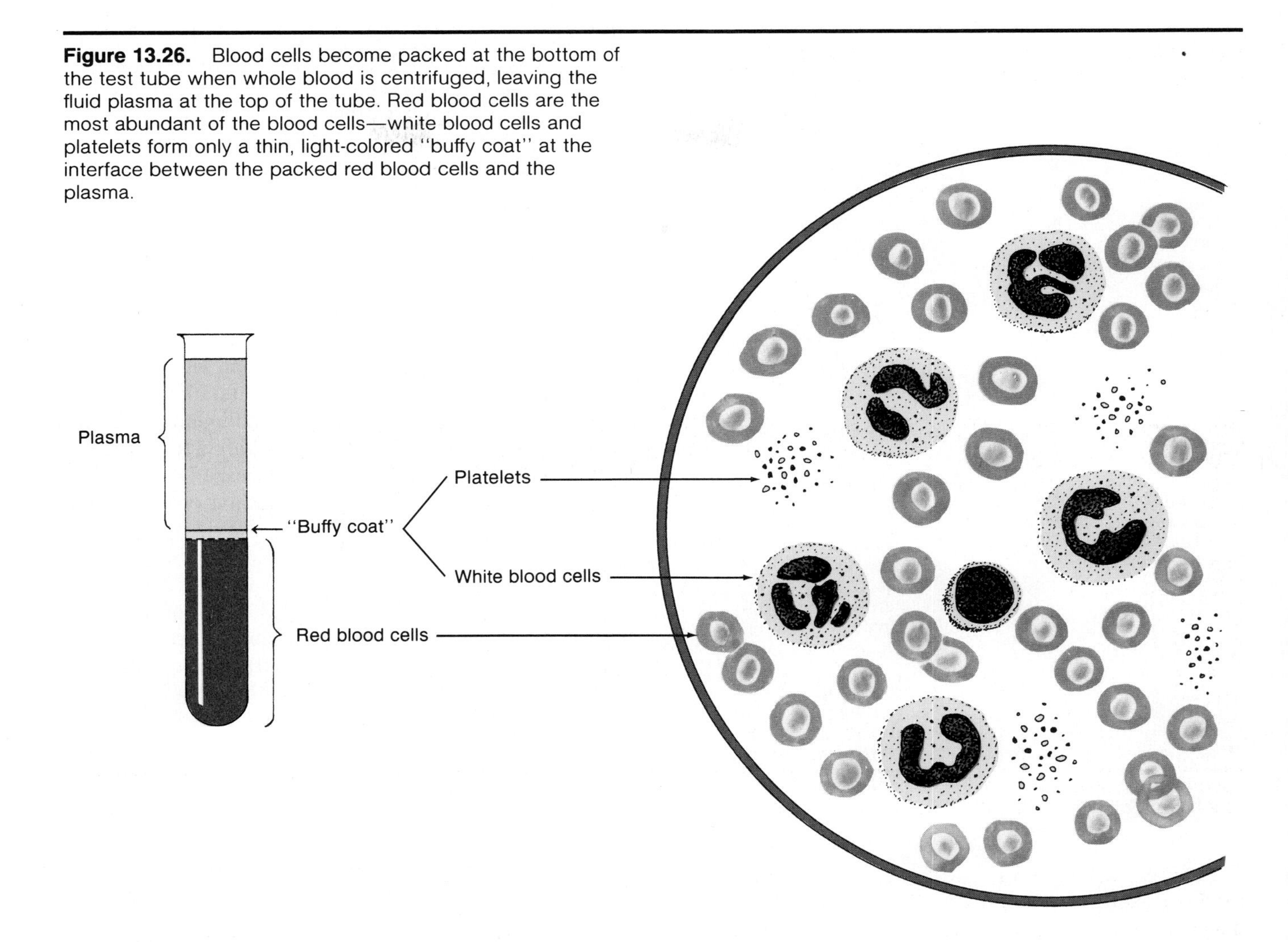

Figure 13.26. Blood cells become packed at the bottom of the test tube when whole blood is centrifuged, leaving the fluid plasma at the top of the tube. Red blood cells are the most abundant of the blood cells—white blood cells and platelets form only a thin, light-colored "buffy coat" at the interface between the packed red blood cells and the plasma.

Composition of the Blood

The average-sized adult has about 5 liters of blood, which constitute about 8% of the total body weight. Blood leaving the heart is referred to as *arterial blood.* Arterial blood, with the exception of that going to the lungs, is bright red in color due to a high concentration of oxyhemoglobin (the combination of oxygen with hemoglobin) in the erythrocytes. *Venous blood* is blood returning to the heart and, except for the venous blood from the lungs, has a darker color (due to hemoglobin that is no longer combined with oxygen). Blood has a viscosity that ranges between 4.5 and 5.5. This means that it flows thicker than water, which has a viscosity of 1. Blood has a pH range of 7.35 to 7.45 and a temperature within the thorax of the body of about 38° C (100.4° F).

Blood is composed of a cellular portion, called **formed elements,** and a fluid portion, called **plasma.** When a blood sample is centrifuged, the heavier formed elements are packed into the bottom of the tube, leaving plasma at the top (fig. 13.26). The formed elements constitute approximately 45% of the total blood volume (the *hematocrit*), and the plasma accounts for the remaining 55%.

Plasma

Plasma is a straw-colored liquid consisting of water and dissolved solutes. Sodium ion is the major solute of plasma in terms of its concentration (Na^+ contributes most to the total osmolality of plasma). In addition to Na^+, plasma

Table 13.5 Representative normal plasma values.

Measurement	Normal Range
Blood volume	80–85 ml/kg body weight
Blood osmolality	285–295 mOsm
Blood pH	7.35–7.45
Enzymes	
Creatine phosphokinase (CPK)	Female: 5–35 mU/ml Male: 5–55 mU/ml
Lactic dehydrogenase (LDH)	60–120 U/ml
Phosphatase (acid)	Female: 0.01–0.56 Sigma U/ml Male: 0.13–0.63 Sigma U/ml
Hematology values	
Hematocrit	Female: 37%–48% Male: 45%–52%
Hemoglobin	Female: 12–16 g/100 ml Male: 13–18 g/100 ml
Red blood cell count	4.2–5.9 million/mm³
White blood cell count	4,300–10,880/mm³
Hormones	
Testosterone	Male: 300–1,100 ng/100 ml Female: 25–90 ng/100 ml
Adrenocorticotrophic Hormone (ACTH)	15–70 pg/ml
Growth hormone	Children: over 10 ng/ml Adult male: below 5 ng/ml
Insulin	6–26 μU/ml (fasting)
Ions	
Bicarbonate	24–30 mmol/l
Calcium	2.1–2.6 mmol/l
Chloride	100–106 mmol/l
Potassium	3.5–5.0 mmol/l
Sodium	135–145 mmol/l
Organic molecules (other)	
Cholesterol	120–220 mg/100 ml
Glucose	70–110 mg/100 ml (fasting)
Lactic acid	0.6–1.8 mmol/l
Protein (total)	6.0–8.4 g/100 ml
Triglyceride	40–150 mg/100 ml
Urea nitrogen	8–25 mg/100 ml
Uric acid	3–7 mg/100 ml

Source: *New England Journal of Medicine,* 1980.

contains many other salts and ions, as well as organic molecules such as metabolites, hormones, enzymes, antibodies, and other proteins. The values of some of these constituents of plasma are shown in table 13.5.

Plasma Proteins. Plasma proteins constitute 7%–9% of the plasma. The three types of proteins are albumins, globulins, and fibrinogen. **Albumins** account for most (60%–80%) of the plasma proteins and are the smallest in size. They are produced by the liver and serve to provide the osmotic pressure needed to draw water from the surrounding tissue fluid into the capillaries. This action is needed to maintain blood volume and pressure. **Globulins** are divided into three subtypes: **alpha globulins, beta globulins,** and **gamma globulins.** The alpha and beta globulins are produced by the liver and function to transport lipids and fat-soluble vitamins in the blood. Gamma globulins are antibodies produced by lymphocytes (one of the formed elements found in blood and lymphoid tissues) and function in immunity.

The Formed Elements of Blood

The formed elements of blood include two types of blood cells: **erythrocytes,** or red blood cells, and **leukocytes,** or white blood cells. Erythrocytes are by far the most numerous of these two types: a cubic millimeter of blood contains 5.1 million to 5.8 million erythrocytes in males and 4.3 million to 5.2 million erythrocytes in females. The same volume of blood, in contrast, contains only 5,000 to 9,000 leukocytes.

Erythrocytes. Erythrocytes are flattened, biconcave discs, about 7 μm in diameter and 2.2 μm thick. Their unique shape relates to their function of transporting oxygen and provides an increased surface area through which gas can diffuse (fig. 13.27). Erythrocytes lack a nucleus and mitochondria (they get energy from anaerobic respiration). Because of these deficiencies, erythrocytes have a circulating life-span of only about 120 days before they are destroyed by phagocytic cells in the liver, spleen, and bone marrow.

Each erythrocyte contains approximately 280 million *hemoglobin* molecules, which give blood its red color. Each hemoglobin molecule consists of a protein, called globin, and an iron-containing pigment, called heme. The iron group of heme is able to combine with oxygen in the lungs and release oxygen in the tissues.

Anemia is present when there is an abnormally low hemoglobin concentration and/or red blood cell count. The most common cause of this condition is a deficiency in iron **(iron-deficiency anemia),** which is an essential component of the hemoglobin molecule. In **pernicious anemia,** there is inadequate availability of vitamin B_{12}, which is needed for red blood cell production. This results from atrophy of the glandular mucosa of the stomach, which normally secretes a substance, called *intrinsic factor,* that is needed for absorption of vitamin B_{12} obtained in the diet. **Aplastic anemia** is anemia due to destruction of the bone marrow, which may be caused by chemicals (including benzene and arsenic) and by X rays.

Figure 13.27. The structure of an erythrocyte.

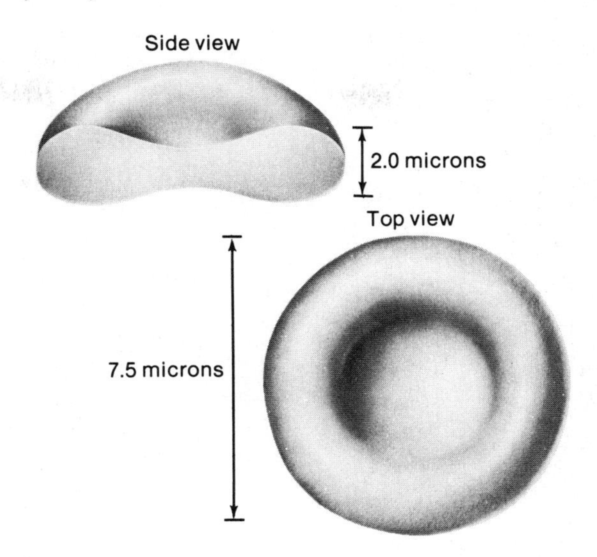

Leukocytes. Leukocytes, or white blood cells, differ from erythrocytes in several ways. Leukocytes contain nuclei and mitochondria and can move in an amoeboid fashion (erythrocytes are not able to move independently). Because of their ameboid ability, leukocytes can squeeze through pores in capillary walls and get to a site of infection, whereas erythrocytes usually remain confined within blood vessels. The movement of leukocytes through capillary walls is called *diapedesis.*

Leukocytes, which are almost invisible under the microscope unless they are stained, are classified according to their stained appearance. Those leukocytes that have granules in their cytoplasm are called **granular leukocytes,** and those that do not are called **agranular** (or nongranular) **leukocytes.** The granular leukocytes are also identified by their oddly shaped nuclei, which in some cases are contorted into lobes separated by thin strands. The granular leukocytes are therefore known as *polymorphonuclear leukocytes* (abbreviated PMN).

The stain used to identify leukocytes is usually a mixture of a pink-to-red stain called eosin and a blue-to-purple stain called a "basic stain." Granular leukocytes with pink-staining granules are therefore called *eosinophils,* and those with blue-staining granules are called *basophils.* Those with granules that have little affinity for either stain are *neutrophils.* Neutrophils are the most abundant type of leukocyte, comprising 50%–70% of the leukocytes in the blood.

There are two types of agranular leukocytes: lymphocytes and monocytes. *Lymphocytes* are usually the second most numerous type of leukocyte; they are small cells with round nuclei and little cytoplasm. *Monocytes,* in contrast, are the largest of the leukocytes and generally have kidney- or horseshoe-shaped nuclei. In addition to these two cell types, there are smaller numbers of large lymphocytes, or *plasma cells,* that may be difficult to distinguish from monocytes. Plasma cells produce and secrete large amounts of antibodies.

Thrombocytes. Thrombocytes, or **platelets,** are the smallest of the formed elements and are actually fragments of large cells called *megakaryocytes,* found in bone marrow. (This is why the term *formed elements* is used rather than *blood cells* to describe erythrocytes, leukocytes, and thrombocytes.) The fragments that enter the circulation as platelets lack nuclei but, like leukocytes, are capable of ameboid movement. The platelet count per cubic millimeter of blood is 130,000 to 360,000. Platelets survive about five to nine days and then are destroyed by the spleen and liver.

Platelets play an important role in blood clotting. They constitute the major portion of the mass of the clot, and phospholipids in their cell membranes serve to activate the clotting factors in plasma that result in threads of fibrin,

Figure 13.28. Types of formed elements.

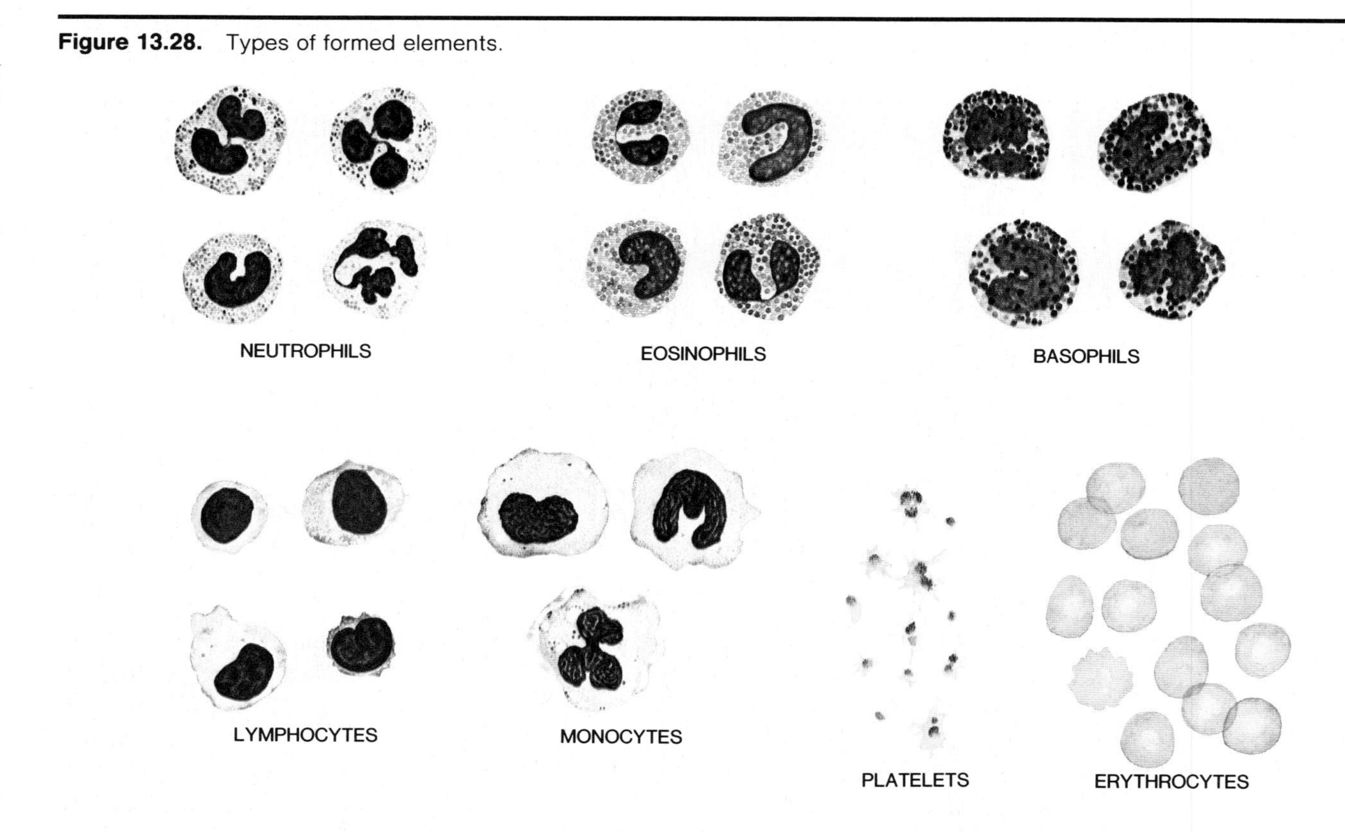

which reinforce the platelet plug. Platelets that attach together in a blood clot also release a chemical called *serotonin,* which stimulates constriction of the blood vessels and thus reduces the flow of blood to the injured area.

The appearance of the formed elements of the blood is shown in figure 13.28, and a summary of the characteristics of these formed elements is presented in table 13.6.

Blood cell counts are an important source of information in determining the health of a person. An abnormal increase in erythrocytes, for example, is termed **polycythemia** and is indicative of several dysfunctions. An abnormally low red blood cell count is **anemia.** An elevated leukocyte count, called **leukocytosis,** is often associated with localized infection. A large number of immature leukocytes within a blood sample is diagnostic of the disease **leukemia.**

1. *List the different classes of plasma proteins, and describe their functions.*
2. *Distinguish between the different types of leukocytes, and explain why leukocytes may be found in connective tissues as well as in blood.*
3. *Describe the origin of platelets and their function.*

Blood Clotting

Trauma to a blood vessel initiates a sequence of events that ultimately results in the formation of a clot. Another sequence of events follows the healing of the blood vessel that results in the dissolution of the clot.

When a blood vessel is injured, a number of physiological mechanisms are activated that promote **hemostasis,** or the cessation of bleeding. Breakage of the endothelial lining of a vessel exposes collagen proteins from the subendothelial connective tissue to the blood. This initiates three separate, but overlapping, hemostatic mechanisms: (1) vasoconstriction; (2) the formation of a platelet plug; and (3) the production of a web of fibrin proteins around the platelet plug.

Functions of Platelets

In the absence of vessel damage, platelets are repelled from each other and from the endothelial lining of vessels. The repulsion of platelets from an intact endothelium is believed to be due to *prostacyclin,* a derivative of prostaglandins, produced within the endothelium. Mechanisms that prevent platelets from sticking to the blood vessels and to each other are obviously needed to prevent inappropriate blood clotting.

Table 13.6 Formed elements of the blood.

Component	Description	Number Present	Function
Erythrocyte (red blood cell)	Biconcave disc without nucleus; contains hemoglobin; survives 100–120 days	4,000,000 to 6,000,000/mm^3	Transports oxygen and carbon dioxide
Leukocytes (white blood cells)		5,000 to 10,000/mm^3	Aid in defense against infections by microorganisms
Granulocytes	About twice the size of red blood cells; cytoplasmic granules present; survive 12 hours to 3 days		
1. Neutrophil	Nucleus with 2–5 lobes; cytoplasmic granules stain slightly pink	54%–62% of white cells present	Phagocytic
2. Eosinophil	Nucleus bilobed; cytoplasmic granules stain red in eosin stain	1%–3% of white cells present	Helps to detoxify foreign substances; secretes enzymes that break down clots
3. Basophil	Nucleus lobed; cytoplasmic granules stain blue in hematoxylin stain	Less than 1% of white cells present	Releases anticoagulant heparin
Agranulocytes	Cytoplasmic granules absent; survive 100–300 days		
1. Monocyte	2–3 times larger than red blood cell; nuclear shape varies from round to lobed	3%–9% of white cells present	Phagocytic
2. Lymphocyte	Only slightly larger than red blood cell; nucleus nearly fills cell	25%–33% of white cells present	Provides specific immune response (including antibodies)
Thrombocyte (platelet)	Cytoplasmic fragment; survives 5–9 days	130,000 to 360,000/mm^3	Clotting

Source: Van De Graaff, Kent M., *Human Anatomy.* © 1984 Wm. C. Brown Publishers, Dubuque, Iowa. All Rights Reserved. Reprinted by permission.

Damage to the endothelium of vessels exposes subendothelial tissue to the blood. Platelets are able to stick to exposed collagen proteins that have become coated with a protein (*von Willebrand factor*) secreted by endothelial cells. Platelets contain secretory granules; when platelets stick to collagen, they *degranulate* as the secretory granules release their products. These products include *ADP* (adenosine diphosphate), *serotonin,* and a prostaglandin called *thromboxane* A_2. This event is known as the **platelet release reaction.**

Serotonin and thromboxane A_2 stimulate vasoconstriction, which helps to decrease blood flow to the injured vessel. Phospholipids that are exposed on the platelet membrane participate in the activation of clotting factors.

The release of ADP and thromboxane A_2 from platelets that are stuck to exposed collagen makes other platelets in the vicinity "sticky," so that they adhere to those stuck to the collagen. The second layer of platelets, in turn, undergoes a platelet release reaction, and the ADP and thromboxane A_2 that are secreted cause additional platelets to aggregate at the site of injury. This produces a **platelet plug** in the damaged vessel, which is strengthened by the activation of plasma-clotting factors.

In order to undergo a release reaction, the production of prostaglandins by the platelets is required. **Aspirin** inhibits the conversion of arachidonic acid (a cyclic fatty acid) into prostaglandins and thus inhibits the release reaction and consequent formation of a platelet plug. The ingestion of excessive amounts of aspirin can thus significantly prolong bleeding time, which is why blood donors and women in the last trimester of pregnancy are advised to avoid aspirin.

Clotting Factors: Formation of Fibrin

The platelet plug is strengthened by a meshwork of insoluble protein fibers known as **fibrin** (fig. 13.29). Blood clots therefore contain platelets, fibrin, and usually trapped red blood cells that give the clot a red color (clots formed in arteries generally lack red blood cells and are gray in color). Finally, contraction of the platelets in the process of *clot retraction* forms a more compact and effective plug (fig. 13.30). Fluid squeezed from the clot as it retracts is called *serum,* which is plasma without fibrinogen (the soluble precursor of fibrin).

There are two pathways that result in the conversion of fibrinogen into fibrin. Blood left in a test tube will clot without the addition of any external chemicals; the

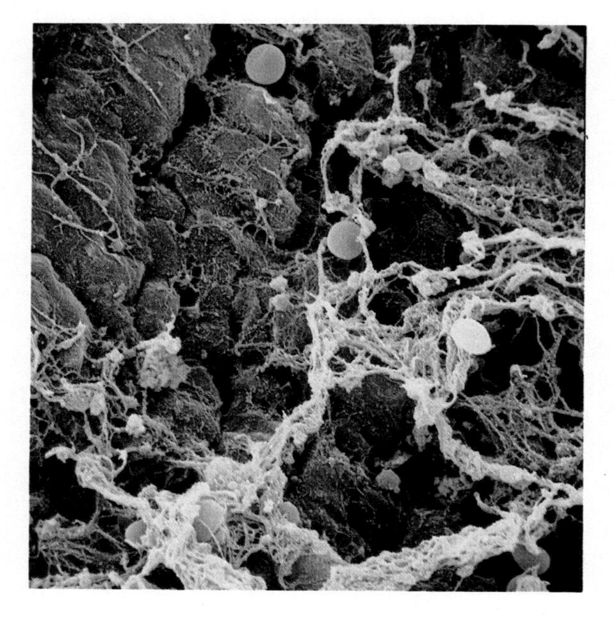

Figure 13.29. A scanning electron micrograph showing threads of fibrin.

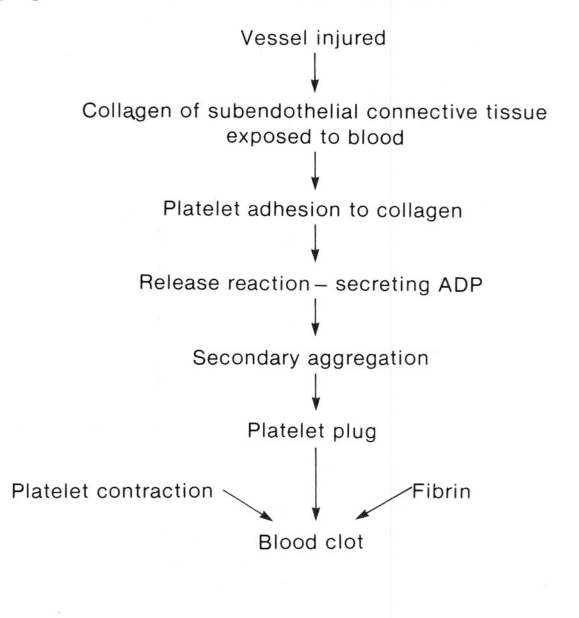

Figure 13.30. The sequence of events leading to platelet aggregation and the formation of a blood clot.

Table 13.7 The plasma clotting factors.

Factor	Name	Function	Pathway
I	Fibrinogen	converted to fibrin	common
II	Prothrombin	enzyme	common
III	Tissue thromboplastin	cofactor	extrinsic
IV	Calcium ions (Ca^{++})	cofactor	intrinsic, extrinsic, and common
V	Proaccelerin	cofactor	common
VII	Proconvertin	enzyme	extrinsic
VIII	Antihemophilic factor	cofactor	intrinsic
IX	Plasma thromboplastin component; Christmas factor	enzyme	intrinsic
X	Stuart-Prower factor	enzyme	common
XI	Plasma thromboplastin antecedent	enzyme	intrinsic
XII	Hageman factor	enzyme	intrinsic
XIII	Fibrin stabilizing factor	enzyme	common

pathway that produces this clot is thus called the **intrinsic pathway.** The intrinsic pathway also produces clots in damaged blood vessels when collagen is exposed to plasma. Damaged tissues, however, release a chemical that initiates a "shortcut" to the formation of fibrin. Since this chemical is not part of blood, the shorter pathway is called the **extrinsic pathway.**

Intrinsic Pathway. The intrinsic pathway is initiated by the exposure of plasma to negatively charged surfaces, such as that provided by collagen or glass. This activates a plasma protein called **factor XII** (table 13.7), which is a protein-digesting enzyme (protease). Active factor XII in turn activates another plasma protein—**factor XI**—by cleaving part of its inactive precursor. Activated factor XI, in turn, activates **factor IX.**

The next steps in the sequence require the presence of phospholipids, which are provided by platelets, and Ca^{++}. The combination of these with active factor IX and **factor VIII** forms a complex that activates **factor X.** A complex is then formed between active factor X, **factor V,** platelet phospholipids, and Ca^{++} that converts **prothrombin** (inactive **factor II**) to **thrombin.**

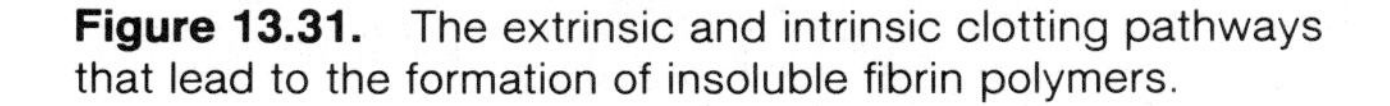

Figure 13.31. The extrinsic and intrinsic clotting pathways that lead to the formation of insoluble fibrin polymers.

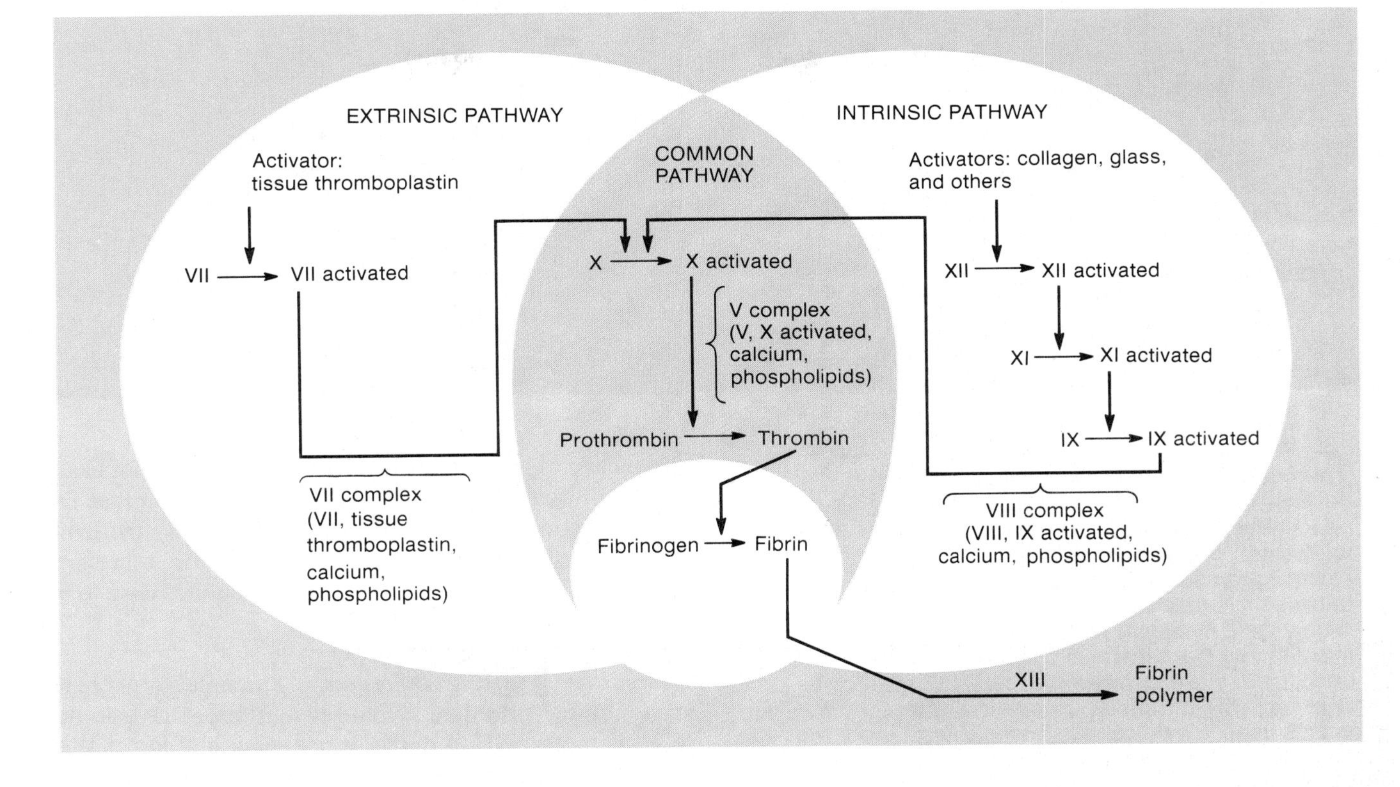

Thrombin is a protease that converts the soluble protein **fibrinogen** (**factor I**) into **fibrin** monomers. These monomers are joined together to form insoluble fibrin polymer by the action of **factor XIII.** The intrinsic clotting sequence is shown on the right side of figure 13.31.

Extrinsic Pathway. The formation of fibrin can occur more rapidly as a result of the release of **tissue thromboplastin,** or **factor III,** from damaged tissue cells. Tissue thromboplastin activates and combines with **factor VII;** factor VII, together with phospholipids and Ca^{++}, forms a complex (the VII complex) that activates factor X. The extrinsic clotting sequence is shown on the left side of figure 13.31.

The extrinsic and intrinsic pathways overlap at this point (see fig. 13.31) into a common pathway. The V complex, formed by active factor X, factor V, phospholipids, and Ca^{++} converts prothrombin to thrombin, which in turn converts fibrinogen to fibrin.

Dissolution of Clots

As the damaged blood vessel wall is repaired, activated factor XII promotes the conversion of another inactive molecule in plasma, *prekallikrein,* to the active form called **kallikrein.** Kallikrein, in turn, catalyzes the conversion of *plasminogen* into the active molecule called **plasmin.**

Figure 13.32. Events that produce dissolution of the blood clot and vasodilation.

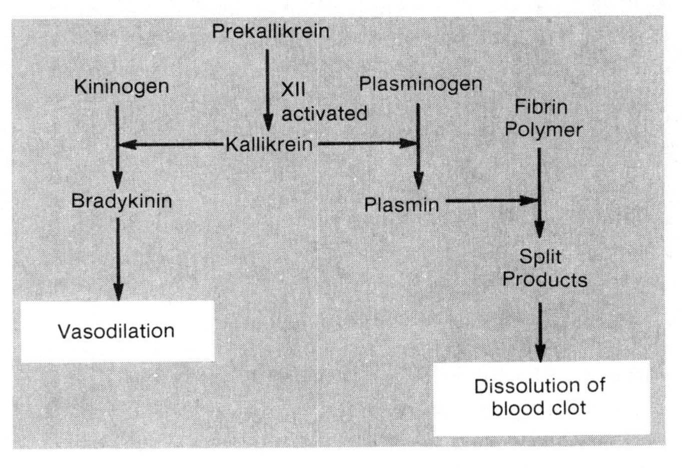

Plasmin is an enzyme that digests fibrin into "split products," thus promoting dissolution of the clot (fig. 13.32). Since a clotting factor (factor XII) initiates the pathway leading to dissolution of the clot, the action of plasmin represents the completion of a delayed negative feedback loop.

Table 13.8 Some acquired and inherited defects in the clotting mechanism.

Category	Cause of Disorder	Comments
Acquired clotting disorders	Vitamin K deficiency	Inadequate formation of prothrombin and other clotting factors in the liver
	Aspirin	Inhibits prostaglandin production, resulting in defective platelet release reaction
Anticoagulants	Coumarin	Competes with the action of vitamin K
	Heparin	Inhibits activity of thrombin
	Citrate	Combines with Ca^{++} and thus inhibits the activity of many clotting factors
Inherited clotting disorders	Hemophilia A (defective factor VIII_{AHF})	Recessive trait carried on X chromosome; results in delayed formation of fibrin
	Von Willebrand's disease (defective factor VIII_{VWF})	Dominant trait carried on autosomal chromosome; impaired ability of platelets to adhere to collagen in subendothelial connective tissue
	Hemophilia B (defective factor IX), also called Christmas disease	Recessive trait carried on X chromosome; results in delayed formation of fibrin

There are a number of plasminogen activators besides kallikrein. Some of these are endogenous—activators produced by tissues and leukocytes. **Streptokinase** is a potent activator of plasminogen but is classified as exogenous because it is made by bacteria. Streptokinase is widely used medically to dissolve clots. It may be injected into the general circulation or injected specifically into a coronary vessel that has become occluded by a thrombus (blood clot). Streptokinase is also used in medical laboratories to measure the plasminogen levels in blood samples.

Kallikrein, in addition to its function in the formation of plasmin, also stimulates the formation of a molecule called **bradykinin** from its precursor *(kininogen)*. Bradykinin stimulates vasodilation, which helps to counter the vasoconstrictor effects of prostaglandins and serotonin released by platelets during the formation of a clot.

Anticoagulants. Clotting of the blood in test tubes can be prevented by the addition of *citrate* or *EDTA,* which chelates (binds to) calcium. By this means Ca^{++} levels in the blood that can participate in the clotting sequence are lowered, and clotting is inhibited. A mucoprotein called *heparin* can also be added to the tube to prevent clotting. Heparin activates a plasma protein called antithrombin III, which combines with and inactivates thrombin. Heparin is also given intravenously during certain medical procedures to prevent clotting. Patients may also be given *coumarins* as anticoagulants. The coumarins prevent blood clotting by competing with vitamin K, which is needed for the formation of factors II, VII, IX, and X by the liver. In contrast to the immediate effects of heparin, therefore, coumarin must be given to a patient for several days to be effective.

Hereditary Clotting Disorders. Examples of hereditary clotting disorders include two different genetic defects in factor VIII. A defect in one subunit of factor VIII prevents this factor from participating in the intrinsic clotting pathway. This genetic disease, called **hemophilia A,** is an X-linked recessive trait that is prevalent in the royal families of Europe. A defect in another subunit of factor VIII results in **von Willebrand's disease.** In this disease, rapidly circulating platelets are unable to stick to collagen and a platelet plug cannot be formed. Some acquired and inherited defects in the clotting system are summarized in table 13.8.

1. *Describe how a platelet plug is formed when a vessel is cut.*
2. *List the steps shared in common by the intrinsic and extrinsic clotting pathways.*
3. *Explain the meaning of the terms* intrinsic *and* extrinsic *in terms of the clotting pathways, and describe how these pathways differ from each other.*
4. *Describe the steps that lead to the formation of plasmin, and explain how this might be regarded as a negative feedback mechanism.*

Table 14.4 Extrinsic control of vascular resistance and blood flow.

Extrinsic Agent	Effect	Comments
Sympathetic nerves		
Alpha-adrenergic	Vasoconstriction	It occurs throughout the body. This is the dominant effect of sympathetic nerve stimulation on the vascular system.
Beta-adrenergic	Vasodilation	There is some activity in arterioles in skeletal muscles and in coronary vessels, but effects are masked by dominant alpha-receptor-mediated constriction.
Cholinergic	Vasodilation	Effects are localized to arterioles in skeletal muscles and are only activated during defense (fight-or-flight) reaction.
Parasympathetic nerves	Vasodilation	Effects are primarily restricted to the gastrointestinal tract, external genitals, and salivary glands and have little effect on total peripheral resistance.
Angiotensin II	Vasoconstriction	A powerful vasoconstrictor produced as a result of secretion of renin from the kidneys, it may function to help maintain adequate filtration pressure in kidneys when systemic blood flow and pressure are reduced.
ADH (vasopressin)	Vasoconstriction	Although the effects of this hormone on vascular resistance and blood pressure in anesthetized animals are well documented, the importance of these effects in conscious humans is controversial.
Histamine	Vasodilation	It promotes localized vasodilation during inflammation and allergic reactions.
Bradykinins	Vasodilation	Bradykinins are polypeptides secreted by sweat glands that promote local vasodilation.
Prostaglandins	Vasodilation or vasoconstriction	Prostaglandins are cyclic fatty acids that can be produced by most tissues, including blood vessel walls. Prostaglandin I_2 is a vasodilator, whereas thromboxane A_2 is a vasoconstrictor. The physiological significance of these effects is presently controversial.

The vasodilation that occurs in response to tissue metabolism can be demonstrated by constricting the blood supply to an area for a short time and then removing the constriction. The constriction allows metabolic products to accumulate by preventing venous drainage of the area. When the constriction is removed and blood flow resumes, the metabolic products that have accumulated cause vasodilation. The tissue thus appears red. This response is called **reactive hyperemia.** A similar increase in blood flow occurs in skeletal muscles and other organs as a result of increased metabolism. This is called **active hyperemia.** Intrinsic control mechanisms are summarized in table 14.5.

1. *Describe the relationship between blood flow rate, arterial blood pressure, and vascular resistance.*
2. *Describe the relationship between vascular resistance and the radius of a vessel. Explain how blood flow can be diverted from one organ to another.*
3. *Explain how vascular resistance and blood flow are regulated by (1) sympathetic adrenergic fibers, (2) sympathetic cholinergic fibers, and (3) parasympathetic fibers.*
4. *Define* autoregulation, *and explain how this is accomplished through myogenic and metabolic mechanisms.*

Table 14.5 The intrinsic control of vascular resistance and blood flow.

Category	Agent (↑ = increase; ↓ = decrease)	Comments
Myogenic	↑ Blood pressure	Stretching of the arterial wall as the blood pressure rises directly stimulates increased smooth muscle tone (vasoconstriction).
Metabolic	↓ Oxygen ↑ Carbon dioxide ↓ pH ↑ Adenosine ↑ K^+	Local changes in gas and metabolite concentrations act directly on vascular smooth muscle walls to produce vasodilation in the systemic circulation. The importance of different agents varies in different organs.
Myogenic	—	It helps to maintain relatively constant rates of blood flow and pressure within an organ despite changes in systemic arterial pressure (autoregulation).
Metabolic	—	It aids in autoregulation of blood flow and also helps to shunt increased amounts of blood to organs with higher metabolic rates (active hyperemia).

Blood Flow to the Heart and Skeletal Muscles

Blood flow to the heart and skeletal muscles is regulated by both extrinsic and intrinsic mechanisms. These mechanisms provide increased rates of blood flow when the metabolic requirements of these tissues are increased during exercise.

Survival requires that the heart and brain receive adequate rates of blood flow at all times. The ability of skeletal muscles to respond quickly in emergencies and to maintain continued high levels of activity may also be critically important for survival. During such times, high rates of blood flow to the skeletal muscles must be maintained without compromising blood flow to the heart and brain. This is accomplished by mechanisms that increase the cardiac output and that divert a higher proportion of the cardiac output to the heart, skeletal muscles, and brain and away from the viscera and skin.

Aerobic Requirements of the Heart

The coronary arteries supply an enormous number of capillaries, which are packed within the myocardium at a density of about 2,500–4,000 per cubic millimeter of tissue. Fast-twitch skeletal muscles, in contrast, have a capillary density of 300–400 per cubic millimeter of tissue. Each myocardial cell, as a consequence, is within 10 μm of a capillary (compared to an average distance in other organs of 60–80 μm). The exchange of gases by diffusion between myocardial cells and capillary blood thus occurs very quickly.

Contraction of the myocardium squeezes the coronary arteries. Unlike blood flow in all other organs, flow in the coronary vessels thus decreases in systole and increases during diastole. The myocardium, however, contains large amounts of *myoglobin,* a pigment related to hemoglobin, which stores oxygen during diastole and releases its oxygen during systole. In this way, the myocardial cells can receive a continuous supply of oxygen even though coronary blood flow is temporarily reduced during systole.

Table 14.6 Some vascular and metabolic comparisons between heart and skeletal muscle.

	Cardiac Muscle	Skeletal Muscle
Number of capillaries	4×	1×
Mean blood flow	10–20×	1×
Myolemma (sarcolemma)	Thin, low resistance	Thicker, higher resistance
Sarcomere length	1×	1.7×
Glycogen concentration	Maintained	Depleted by fasting, diabetes
Glycolytic enzyme systems	1×	2×, strongly developed
Creatine phosphate concentration	1×	6×
Anaerobic energy production	Beating: 2 min Arrested: 30–90 min	Up to 40% total energy
Lactic acid production	Terminal mechanism	Frequent when incurring O_2 debt
Ability to incur O_2 debt	1×	4×
Increased O_2 requirement met primarily by	Increased flow	Increased extraction
Oxygen consumption	3×	1×
Oxygen extraction at rest	Near maximal	Significant reserve
Increased O_2 consumption with increased work	2×	30×
Myoglobin	Present	Present in red skeletal muscle
Mitochondria	Abundant, giant	Fewer, smaller
Krebs cycle enzymes	2–3×	1×
Cytochrome-C	6×	1×
Myosin ATPase activity	1×	3×

Source: Brachfeld, N. The Physiology of Muscular Exercise, *Primary Cardiology,* June, 1979, p. 112. Reproduced with permission.

In addition to containing large amounts of myoglobin, heart muscle contains numerous mitochondria and aerobic respiratory enzymes. This indicates that—even more than slow-twitch skeletal muscles—the heart is extremely specialized for aerobic respiration (table 14.6). The normal heart always respires aerobically, even during heavy exercise when the metabolic demand for oxygen can rise to five times resting levels. This increased oxygen requirement is met by a corresponding increase in coronary blood flow, from about 80 ml at rest to about 400 ml per minute per 100 g tissue during heavy exercise.

Figure 14.16. An angiogram of the left coronary artery in a patient (*a*) when the ECG was normal and (*b*) when the ECG showed evidence of myocardial ischemia. Notice that a coronary artery spasm—see arrow in (*b*)—appears to accompany the ischemia.

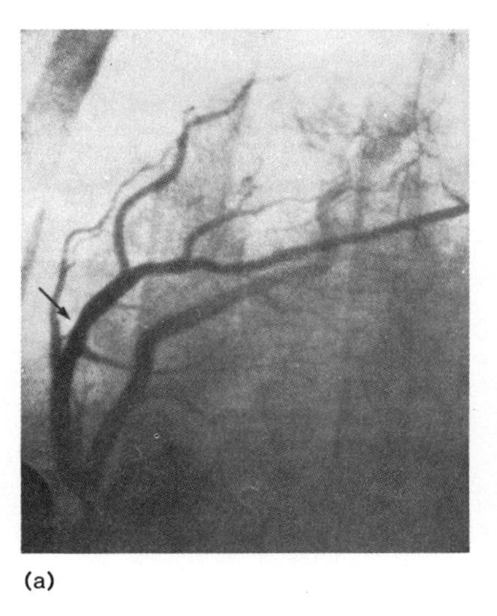

(a)

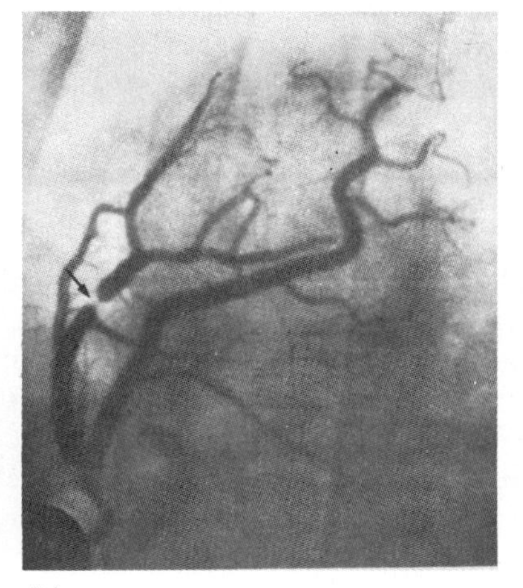

(b)

Regulation of Coronary Blood Flow

Sympathetic nerve fibers, through stimulation of alpha-adrenergic receptors in the coronary arterioles, produce a relatively high vascular resistance in the coronary circulation at rest. Vasodilation of coronary vessels may be produced in part by sympathetic nerve activation of beta-adrenergic receptors. Most of the vasodilation that occurs during exercise, however, is due to intrinsic metabolic control mechanisms. As the metabolism of the myocardium increases, local accumulation of carbon dioxide and adenosine acts directly on the vascular smooth muscles to cause relaxation and vasodilation.

Under abnormal conditions, the blood flow to the myocardium may be inadequate, resulting in myocardial ischemia. This can result from blockage by atheromas and/or blood clots or from muscular spasm of a coronary artery (fig. 14.16). Occlusion of a coronary artery can be visualized by a technique called *selective coronary arteriography.* In this procedure, a catheter (plastic tube) is inserted into a brachial or femoral artery all the way to the opening of the coronary arteries in the aorta, and radiographic contrast material is injected. The picture thus obtained is called an **angiogram.**

If the occlusion is sufficiently great a coronary bypass operation may be performed. In this procedure a length of blood vessel, usually taken from the saphenous vein in the leg, is sutured to the aorta and to the coronary artery at a location beyond the site of the occlusion (fig. 14.17).

Figure 14.17. A diagram of coronary artery bypass surgery.

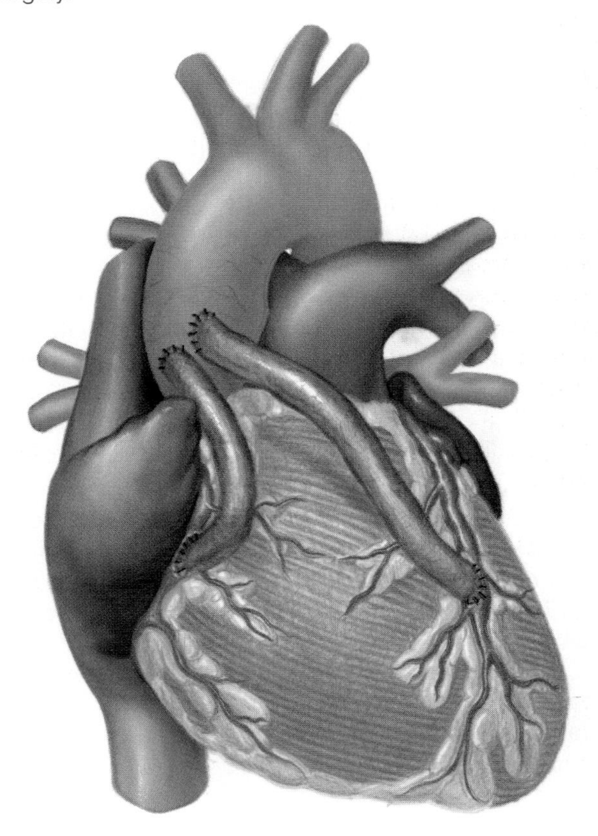

Table 14.7 Changes in skeletal muscle blood flow under conditions of rest and exercise.

Condition	Blood Flow (ml/min)	Mechanism
Rest	1,000	High adrenergic sympathetic stimulation of vascular alpha-receptors causing vasoconstriction
Beginning exercise	Increased	Dilation of arterioles in skeletal muscles due to cholinergic sympathetic nerve activity
Heavy exercise	20,000	Fall in alpha-adrenergic activity Increased sympathetic cholinergic activity Increased metabolic rate of exercising muscles producing intrinsic vasodilation

Regulation of Blood Flow through Skeletal Muscles

The arterioles in skeletal muscles, like those of the coronary circulation, have a high vascular resistance at rest as a result of alpha-adrenergic sympathetic stimulation. This produces a relatively low rate of blood flow per tissue weight (4–6 ml per minute per 100 g tissue), but because muscles have such a large mass, this still accounts for 20–25% of the total blood flow in the body at rest.

In addition to adrenergic fibers (those that release norepinephrine), there are also sympathetic cholinergic fibers in skeletal muscles. The release of acetylcholine (ACh) from these fibers stimulates vasodilation as part of the "fight-or-flight" response to any stressful state, including that existing just prior to exercise (table 14.7). The increased blood flow that results from this vasodilation may provide an extra edge that improves skeletal muscle performance once exercise begins.

As exercise progresses, the vasodilation and increased skeletal muscle blood flow that occur are almost entirely due to intrinsic metabolic control. The high metabolic rate of skeletal muscles during exercise causes local changes such as increased carbon dioxide concentrations, decreased pH (due to carbonic acid and lactic acid), decreased oxygen, increased extracellular K^+, and the secretion of adenosine. Like the intrinsic control of the coronary circulation, these changes cause vasodilation of arterioles in skeletal muscles. This decreases the vascular resistance and increases the rate of blood flow. This effect is combined with the recruitment of capillaries by the opening of precapillary sphincter muscles (only 5–10% of the skeletal muscle capillaries are open at rest). As a result of these changes, skeletal muscles can receive as much as 85% of the total blood flow in the body during maximal exercise.

While the vascular resistance in skeletal muscles decreases during exercise, the resistance to flow through visceral organs and skin increases. This increased resistance occurs because of vasoconstriction stimulated by adrenergic sympathetic fibers and results in decreased rates of blood flow through these organs. During exercise, therefore, the blood flow to skeletal muscles increases because of increased total blood flow (cardiac output) and because blood is diverted away from the viscera and skin. Blood flow to the heart also increases during exercise, whereas blood flow to the brain does not appear to change significantly (fig. 14.18).

Table 14.8 Relationship between age and maximum cardiac rate.

Age	Maximum Cardiac Rate
20–29	190 beats/min
30–39	160 beats/min
40–49	150 beats/min
50–59	140 beats/min
60 and above	130 beats/min

Although the skeletal muscles consume large amounts of oxygen during exercise, the oxygen concentration of arterial blood is usually not reduced. This is due to the fact that the rate and depth of breathing is also increased during exercise and to the fact that the pulmonary blood flow increases to match the increased systemic blood flow. The oxygen concentration of venous blood, however, can be decreased to one-half or one-third the resting levels by exercising muscles, which indicates that the muscles are extracting a much higher proportion of the oxygen delivered to them in the arterial blood. The ability of exercising muscles to extract oxygen from arterial blood is improved by endurance training, which results in increased capillary density and aerobic enzymes in the trained muscles.

Changes in Cardiac Output during Exercise

During exercise the cardiac output can increase fivefold, from about 5 L per minute to about 25 L per minute. This is primarily due to an increase in cardiac rate. The cardiac rate, however, can only increase up to a maximum value (table 14.8), which is determined mainly by the person's age. In very well-trained athletes the stroke volume can also increase significantly, allowing these individuals to achieve a cardiac output during strenuous exercise as much as six or seven times greater than their resting values.

Figure 14.18. The distribution of blood flow (cardiac output) during rest and heavy work. At rest, the cardiac output is 5 L per minute (*bottom of figure*); during heavy work the cardiac output increases to 25 L per minute (*top of figure*). During rest, for example, the brain receives 15% of 5 L per minute (= 750 ml/min.), whereas during exercise it receives 3% to 4% of 25 L per minute (0.03 × 25 = 750 ml/min.). Flow to the skeletal muscles increases more than twentyfold, because the total cardiac output increases (from 5 L/min. to 25 L/min.) and because the percent of the total received by the muscles increases from 15% to 80%.

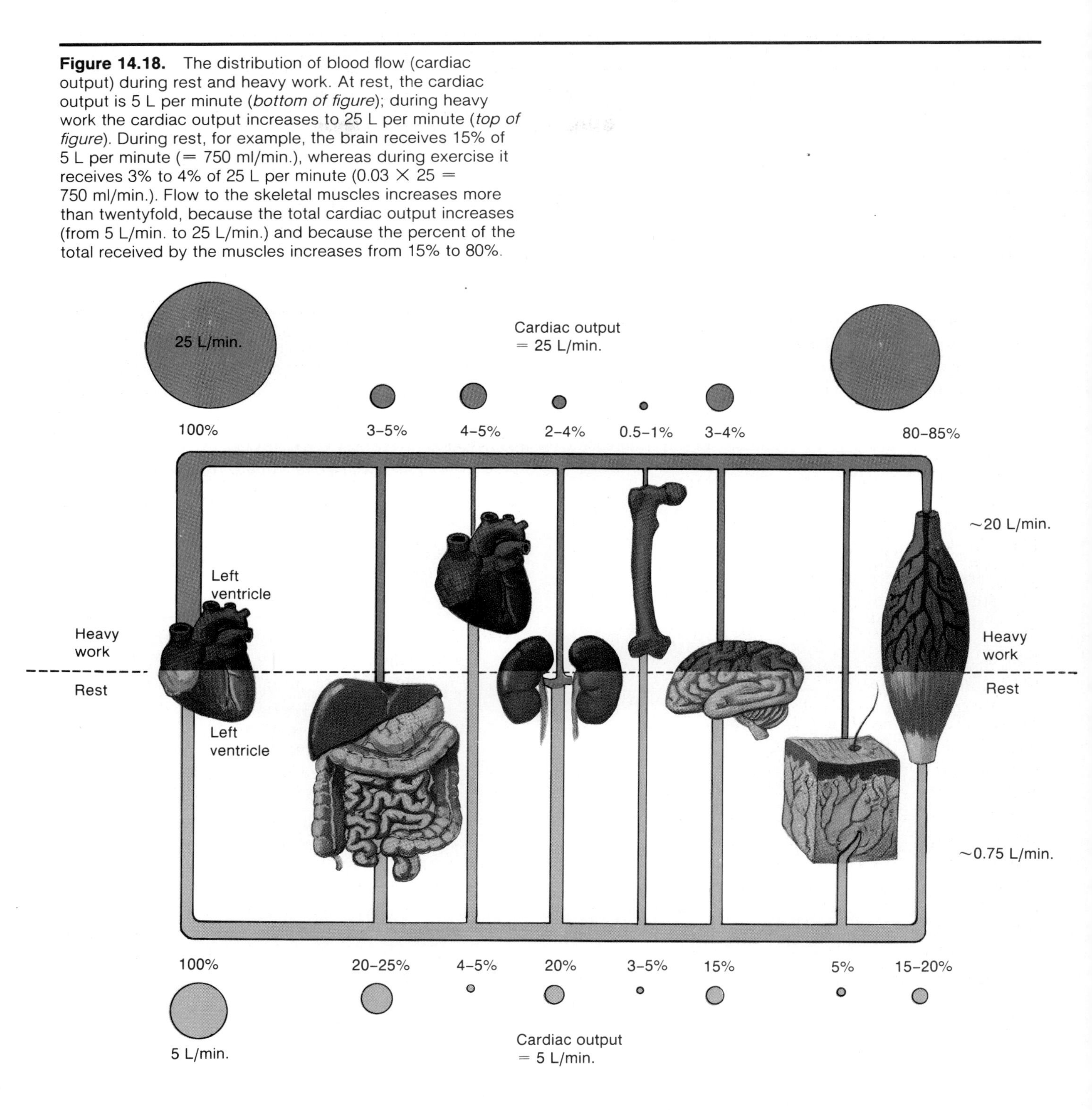

Figure 14.19. Cardiovascular adaptations to exercise.

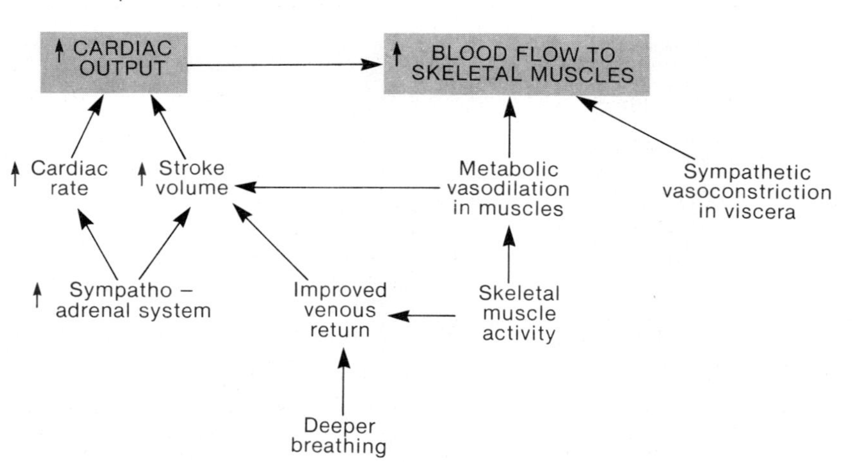

In most people the stroke volume can only increase from 10% to 35% during exercise. The fact that the stroke volume can increase at all during exercise may at first be surprising, in view of the fact that the heart has less time to fill with blood between beats when it is pumping faster. Despite the faster beat, however, the end-diastolic volume during exercise is not decreased. This is because the venous return is aided by the improved action of the skeletal muscle pumps and by increased respiratory movements during exercise (fig. 14.19). Since the end-diastolic volume is not significantly changed during exercise, any increase in stroke volume that occurs must be due to an increase in the proportion of blood ejected per stroke.

The proportion of the end-diastolic volume ejected per stroke can increase from 67% at rest to as much as 90% during heavy exercise. This increased *ejection fraction* is produced by the increased contractility that results from sympathoadrenal stimulation. There may also be a decrease in total peripheral resistance as a result of vasodilation in the exercising skeletal muscles, which decreases the afterload and thus further augments the increase in stroke volume. The cardiovascular changes that occur during exercise are summarized in table 14.9.

Endurance training often results in a lowering of the resting cardiac rate and an increase in the resting stroke volume. The lowering of the resting cardiac rate results from a greater degree of inhibition of the SA node by the vagus nerve. The increased resting stroke volume is believed to be due to an increase in blood volume; indeed, studies have shown that the blood volume can increase by about 500 ml after only eight days of training. These adaptations enable the trained athlete to produce a larger proportionate increase in cardiac output and achieve a higher absolute cardiac output during exercise. This large cardiac output is the major factor in the improved oxygen delivery to skeletal muscles that occurs as a result of endurance training.

1. *Describe blood flow and oxygen delivery to the myocardium during systole and diastole.*
2. *Describe how blood flow to the heart is affected by exercise, and explain how blood flow to the heart is regulated at rest and during exercise.*
3. *Describe the mechanisms that produce vasodilation of the arterioles in skeletal muscles during exercise, and describe three other mechanisms that increase skeletal muscle blood flow during exercise.*
4. *Explain how the stroke volume can increase during exercise despite the fact that the filling times are reduced at high cardiac rates.*

Blood Flow to the Brain and Skin

The examination of cerebral and cutaneous blood flow is a study in contrasts. Cerebral blood flow is regulated primarily by intrinsic mechanisms; cutaneous blood flow is regulated by extrinsic mechanisms. Cerebral blood flow is relatively constant; cutaneous blood flow exhibits more variation than blood flow in any other organ. The brain is the organ that can least tolerate and the skin is the organ that can most tolerate low rates of blood flow.

Table 14.9 Cardiovascular changes that occur during moderate exercise.

Variable	Change	Mechanisms
Cardiac output	Increased	Cardiac rate and stroke volume increased
Cardiac rate	Increased	Increased sympathetic nerve activity; decreased activity of the vagus nerve
Stroke volume	Increased	Increased myocardial contractility due to stimulation by sympathoadrenal system; decreased total peripheral resistance
Total peripheral resistance	Decreased	Vasodilation of arterioles in skeletal muscles (and in skin when thermoregulatory adjustments are needed)
Arterial blood pressure	Increased	Increased systolic and pulse pressure due primarily to increased cardiac output; diastolic pressure rises less due to decreased total peripheral resistance
End-diastolic volume	Unchanged	Decreased filling time at high cardiac rates is compensated by increased venous pressure, increased activity of the skeletal muscle pump, and decreased intrathoracic pressure aiding the venous return
Blood flow to heart and muscles	Increased	Increased muscle metabolism produces intrinsic vasodilation; aided by increased cardiac output and increased vascular resistance in visceral organs
Blood flow to visceral organs	Decreased	Vasoconstriction in digestive tract, liver, and kidneys due to sympathetic nerve stimulation
Blood flow to skin	Increased	Metabolic heat produced by exercising muscles produces reflex (involving hypothalamus) that reduces sympathetic constriction of arteriovenous shunts and arterioles
Blood flow to brain	Unchanged	Autoregulation of cerebral vessels, which maintains constant cerebral blood flow despite increased arterial blood presure

Cerebral Circulation

When the brain is deprived of oxygen for a few seconds, the person loses consciousness; irreversible brain injury may occur after a few minutes. For these reasons, the cerebral blood flow is remarkably constant at about 750 ml per minute. This amounts to about 15% of the total cardiac output at rest.

Unlike the coronary and skeletal muscle blood flow, cerebral blood flow is not normally influenced by sympathetic nerve activity. Only when the mean arterial pressure rises to about 200 mm Hg do sympathetic nerves cause a significant degree of vasoconstriction in the cerebral circulation. This vasoconstriction helps to protect small, thin-walled arterioles from bursting under the pressure and thus helps to prevent cerebrovascular accident (stroke).

In the normal range of arterial pressures, cerebral blood flow is regulated almost exclusively by intrinsic mechanisms. These mechanisms help insure a constant rate of blood flow despite changes in systemic arterial pressure—a process called *autoregulation.* The autoregulation of cerebral blood flow is achieved by both myogenic and metabolic mechanisms.

Myogenic regulation occurs when there is variation in systemic arterial pressure. The cerebral arteries automatically dilate when the blood pressure falls and constrict when the pressure rises. This helps to maintain a constant flow rate during the normal pressure variations that occur during rest, exercise, and emotional states.

The cerebral vessels are also sensitive to the carbon dioxide concentration of arterial blood. When the carbon dioxide concentration rises, as a result of inadequate ventilation (hypoventilation), the cerebral arterioles dilate. This is believed to be due to decreases in the pH of cerebrospinal fluid rather than to a direct effect of CO_2 on the cerebral vessels. Conversely, when the arterial CO_2 falls below normal during hyperventilation, the cerebral vessels constrict. The resulting decrease in cerebral blood flow is responsible for the dizziness that occurs during hyperventilation.

Figure 14.20. Computerized picture of blood flow distribution in the brain after injecting the carotid artery with a radioactive isotope. In (*a*), on the left, the subject followed a moving object with his eyes. High activity is seen over the occipital lobe of the brain. In (*a*), on the right, the subject listened to spoken words. Notice that the high activity is seen over the temporal lobe (the auditory cortex). In (*b*), on the left, the subject moved his fingers on the side of the body opposite to the cerebral hemisphere being studied. In (*b*), on the right, the subject counted to twenty. High activity is shown over the mouth area of the motor cortex, the supplementary motor area, and the auditory cortex.

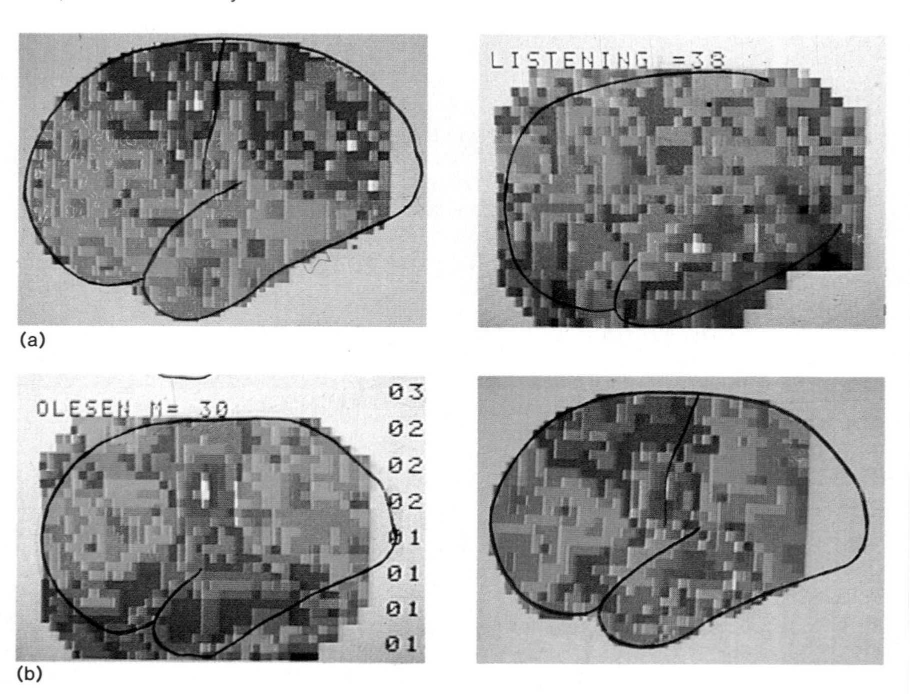

The cerebral arterioles are exquisitely sensitive to local changes in metabolic activity, so that those brain regions with the highest metabolic activity get the most blood. Indeed, areas of the brain that control specific processes have been mapped by the changing patterns of blood flow that result when these areas are activated. Vision and hearing, for example, increase blood flow to the appropriate sensory areas of the cerebral cortex, whereas motor activities such as movements of the eyes, arms, and organs of speech result in different patterns of blood flow (fig. 14.20).

Cutaneous Blood Flow

The skin is the outer covering of the body and as such serves as the first line of defense against invasion by disease-causing organisms. The skin, as the interface between the internal and external environments, also serves to help maintain a constant deep-body temperature despite changes in the ambient (external) temperature, a process called *thermoregulation.* The small thickness and large size of the skin (1.0–1.5 mm thick; 1.7–1.8 square meters in surface area) make it an effective radiator of heat when the body temperature is greater than the ambient temperature. The transfer of heat from the body to the external environment is aided by the flow of warm blood through capillary loops near the surface of the skin.

Blood flow through the skin is adjusted to maintain deep-body temperature at about 37° C (98.6° F). These adjustments are made by variations in the degree of constriction or dilation of ordinary arterioles and of unique

Figure 14.21. Circulation in the skin showing arteriovenous anastomoses, which function as shunts allowing blood to be diverted directly from the arteriole to the venule, thus bypassing superficial capillary loops.

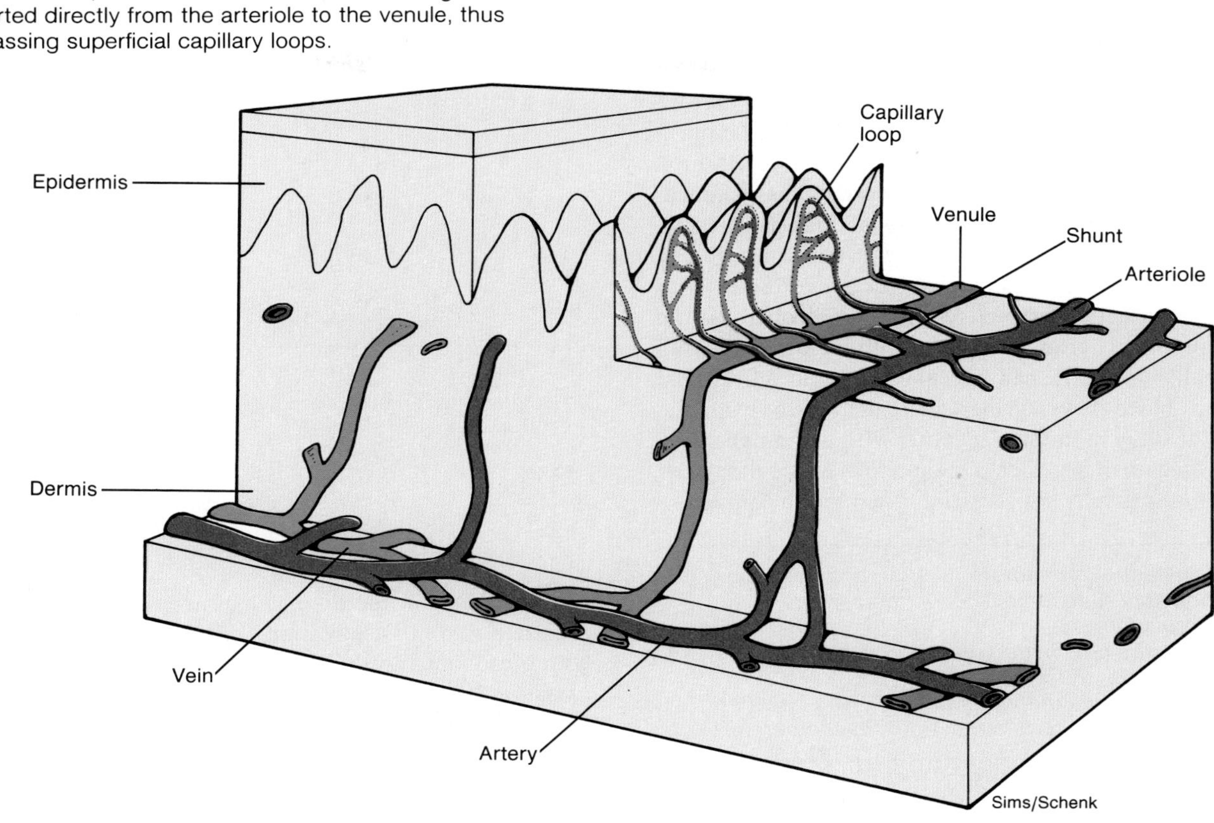

arteriovenous anastomoses (fig. 14.21). These latter vessels, found predominantly in the fingertips, palms of the hands, toes, soles of the feet, ears, nose, and lips, shunt (divert) blood directly from arterioles to deep venules, thus bypassing superficial capillary loops. Both the ordinary arterioles and the arteriovenous anastomoses are innervated by sympathetic nerve fibers. When the ambient temperature is cold, sympathetic nerves stimulate cutaneous vasoconstriction; cutaneous blood flow is thus decreased, so that less heat will be lost from the body. Since the arteriovenous anastomoses also constrict, the skin may appear rosy as a result of the fact that blood is diverted to the superficial capillary loops. In spite of this rosy appearance, however, the total cutaneous blood flow and rate of heat loss is lower than under usual conditions.

Skin can tolerate an extremely low blood flow rate in cold weather because its metabolic rate decreases when the ambient temperature decreases. In cold weather, therefore, the skin requires less blood. In very cold weather, however, blood flow to the skin can be so severely restricted that the tissue does die: this is *frostbite.* Blood flow to the skin can vary from less than 20 ml per minute at maximal vasoconstriction to as much as 3–4 L per minute at maximal vasodilation.

As the temperature warms, cutaneous arterioles in the hands and feet dilate as a result of decreased sympathetic nerve activity. Continued warming causes dilation of arterioles in other areas of the skin. If the resulting increase in cutaneous blood flow is not sufficient to cool the body, secretion of the sweat glands may be stimulated. Sweat helps to cool the body as it evaporates from the surface of the skin. Also, the sweat glands secrete **bradykinin,** a polypeptide that stimulates vasodilation. This increases blood flow to the skin and to the sweat glands, so that larger volumes of more dilute sweat are produced.

Under the usual conditions of ambient temperature, the cutaneous vascular resistance is high and the blood flow is low when a person is not exercising. In the preexercise state of "fight or flight," sympathetic nerve activity reduces cutaneous blood flow still further. During exercise, however, the need to maintain a deep-body temperature takes precedence over the need to maintain an adequate systemic blood pressure. As the body temperature rises during exercise, vasodilation in cutaneous vessels occurs together with vasodilation in the exercising muscles. This can produce an even greater lowering of total peripheral resistance. If exercise is performed in hot and humid weather and if restrictive clothing is worn that increases skin temperature and cutaneous vasodilation, a dangerously low blood pressure may be produced after exercise has ceased and the cardiac output has declined. People have lost consciousness and even died as a result.

Changes in cutaneous blood flow occur as a result of changes in sympathetic nerve activity. Since the activity of the sympathetic nervous system is controlled by the brain, emotional states, acting through control centers in the medulla oblongata, can affect sympathetic activity and cutaneous blood flow. During fear reactions, for example, vasoconstriction in the skin together with activation of the sweat glands can produce a pallor and a "cold sweat." Other emotions may cause vasodilation and blushing.

1. *Define the term* autoregulation, *and describe how this is accomplished in the cerebral circulation.*
2. *Describe how hyperventilation can cause dizziness.*
3. *Explain how cutaneous blood flow is adjusted to maintain a constant deep-body temperature.*

Blood Pressure

Resistance to flow in the arterial system is greatest in the arterioles because these vessels have the smallest diameters. Although the total blood flow through a system of arterioles must be equal to the flow in the larger vessel that gave rise to those arterioles, the narrow diameter of each arteriole reduces the flow rate in each according to Poiseuille's law. Blood flow rate and pressure is thus reduced in the capillaries, which are located downstream of the high resistance imposed by the arterioles. The blood pressure upstream of the arterioles—in the medium and large arteries—is correspondingly increased (fig. 14.22).

The blood pressure and flow rate within the capillaries are further reduced by the fact that their total cross-sectional area is much greater, due to their large number, than the cross-sectional areas of arteries and arterioles (fig. 14.23). Thus, although each capillary is much narrower than each arteriole, the capillary beds served by arterioles do not provide as great a resistance to blood flow as do the arterioles.

Figure 14.22. A constriction increases blood pressure upstream (analogous to the arterial pressure) and decreases pressure downstream (analogous to capillary and venous pressure).

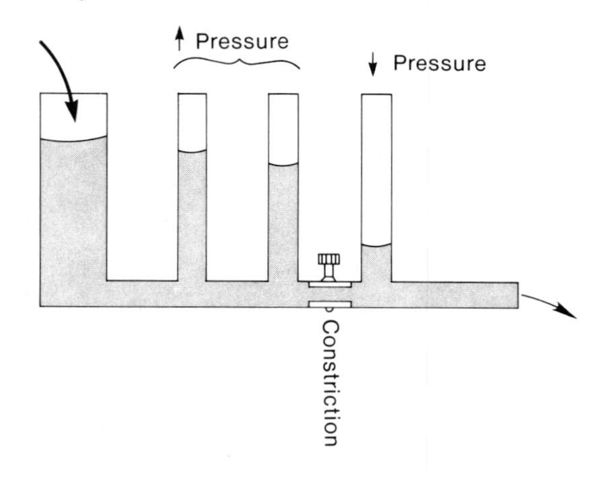

Variations in the diameter of arterioles due to vasoconstriction and vasodilation thus simultaneously affect both blood flow through capillaries and the *arterial blood pressure* "upstream" from the capillaries. An increase in total peripheral resistance due to vasoconstriction of arterioles can raise arterial blood pressure. Blood pressure can also be raised by an increase in the cardiac output. This may be due to elevations in cardiac rate or stroke volume, which in turn is affected by other factors. The three most important variables affecting blood pressure are the **cardiac rate, blood volume,** and **total peripheral resistance.** An increase in any of these three, if not compensated by a decrease in another variable, will result in an increased blood pressure.

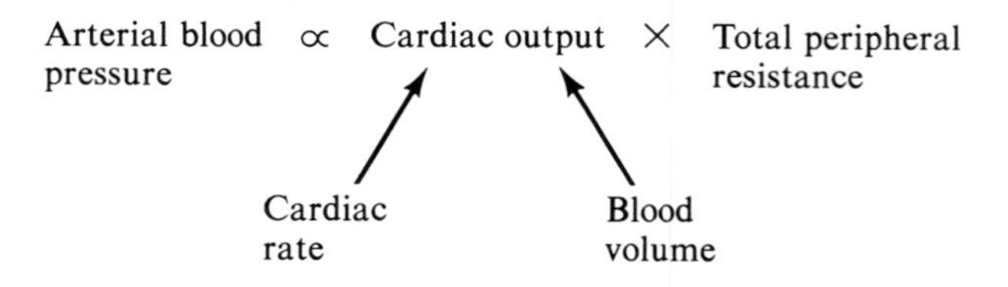

Blood pressure can thus be regulated by the kidneys, which control blood volume, and by the sympathoadrenal system. Increased activity of the sympathoadrenal system can raise blood pressure by stimulating vasoconstriction of arterioles (thus raising total peripheral resistance) and by promoting an increased cardiac output. Sympathetic stimulation can also affect blood volume indirectly, by stimulating constriction of renal blood vessels and thus reducing urine output.

Figure 14.23. As blood passes from the aorta to the smaller arteries, arterioles, and capillaries, the cross-sectional area increases as the pressure decreases.

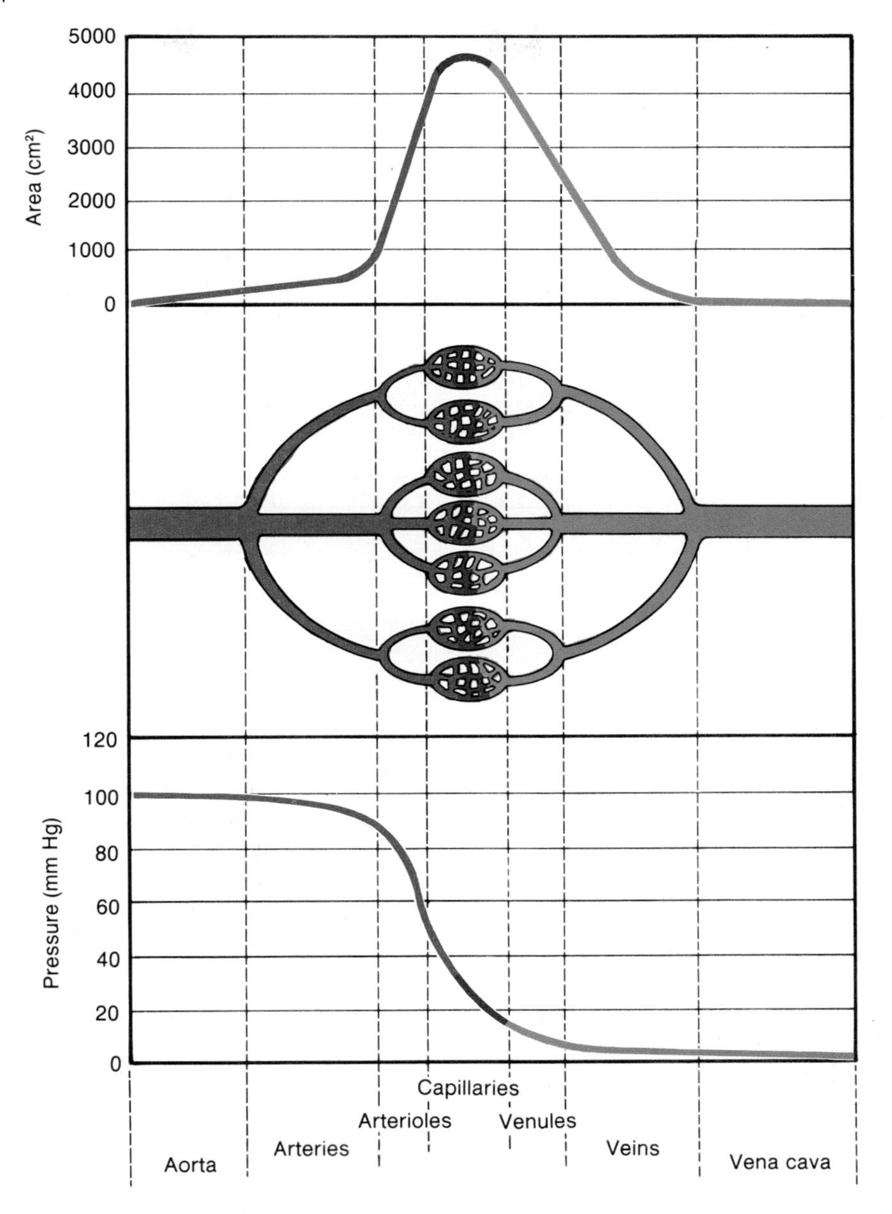

Regulation of Blood Pressure

In order for blood pressure to be maintained within limits, specialized receptors for pressure are needed. These **baroreceptors** are stretch receptors located in the *aortic arch* and in the *carotid sinuses.* An increase in pressure causes the walls of these arterial regions to stretch and stimulate the activity of sensory nerve endings (fig. 14.24). A fall in pressure below the normal range, in contrast, causes a decrease in the frequency of action potentials produced by these sensory nerve fibers.

Sensory nerve activity from the baroreceptors ascends via the vagus and glossopharyngeal nerves to the medulla oblongata, which directs the autonomic system to respond appropriately. **Vasomotor control centers** in the medulla control vasoconstriction/vasodilation and hence help regulate total peripheral resistance. **Cardiac control centers** in the medulla regulate the cardiac rate (fig. 14.25).

Figure 14.24. Action potential frequency in sensory nerve fibers from baroreceptors in the carotid sinus and aortic arch. As the blood pressure increases, the baroreceptors become increasingly stretched. This results in an increase in the frequency of action potentials that are transmitted to the cardiac and vasomotor control centers in the medulla oblongata.

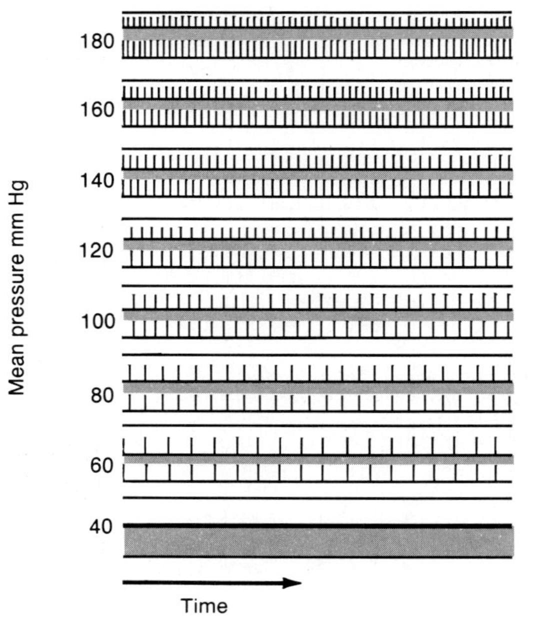

Figure 14.25. The baroreceptor reflex. Sensory stimuli from baroreceptors in the carotid sinus and the aortic arch, acting via control centers in the medulla oblongata, affect the activity of sympathetic and parasympathetic nerve fibers in the heart.

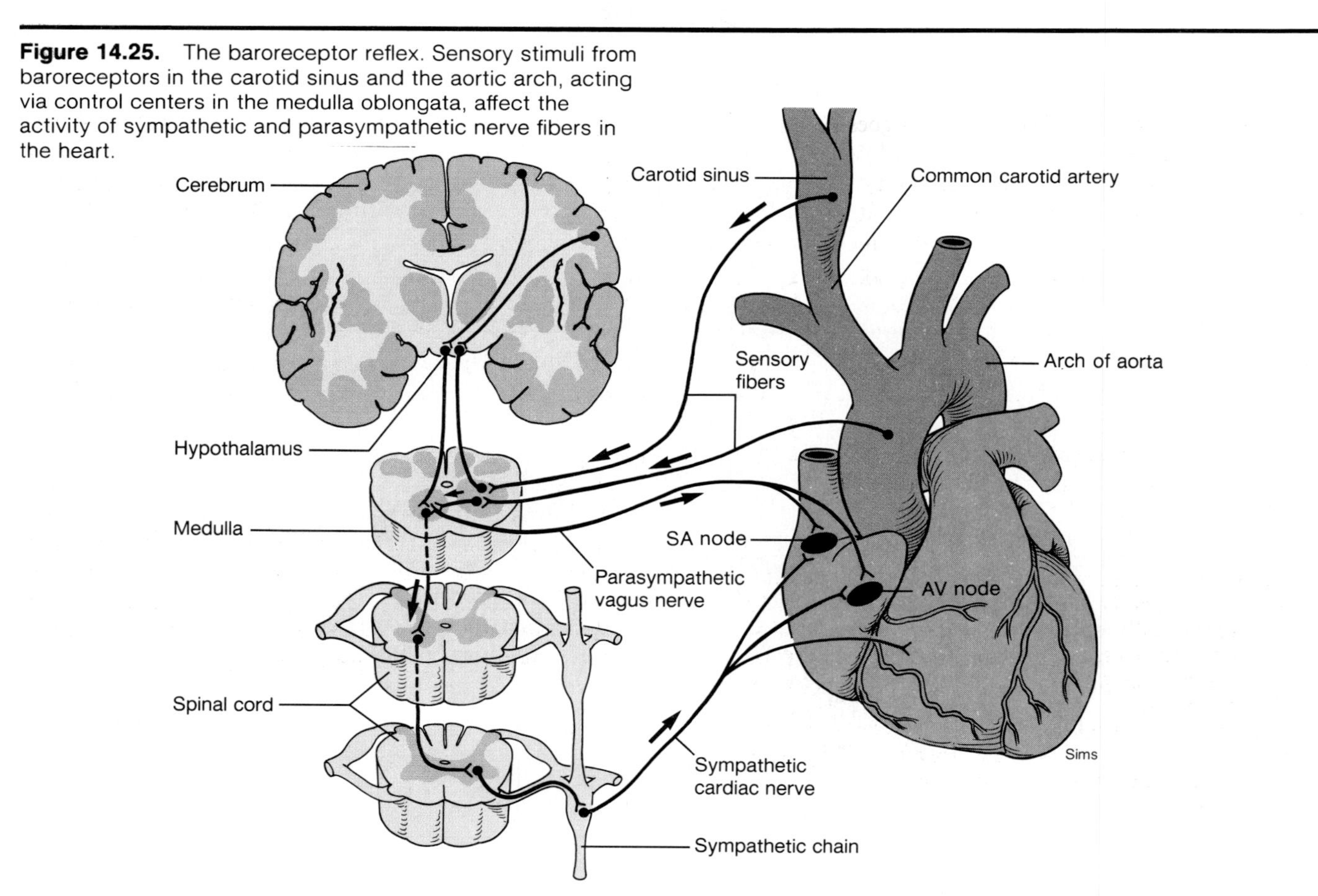

Figure 14.26. Compensations for the upright posture induced by the baroreceptor reflex. Numbers indicate the sequence of cause-and-effect steps.

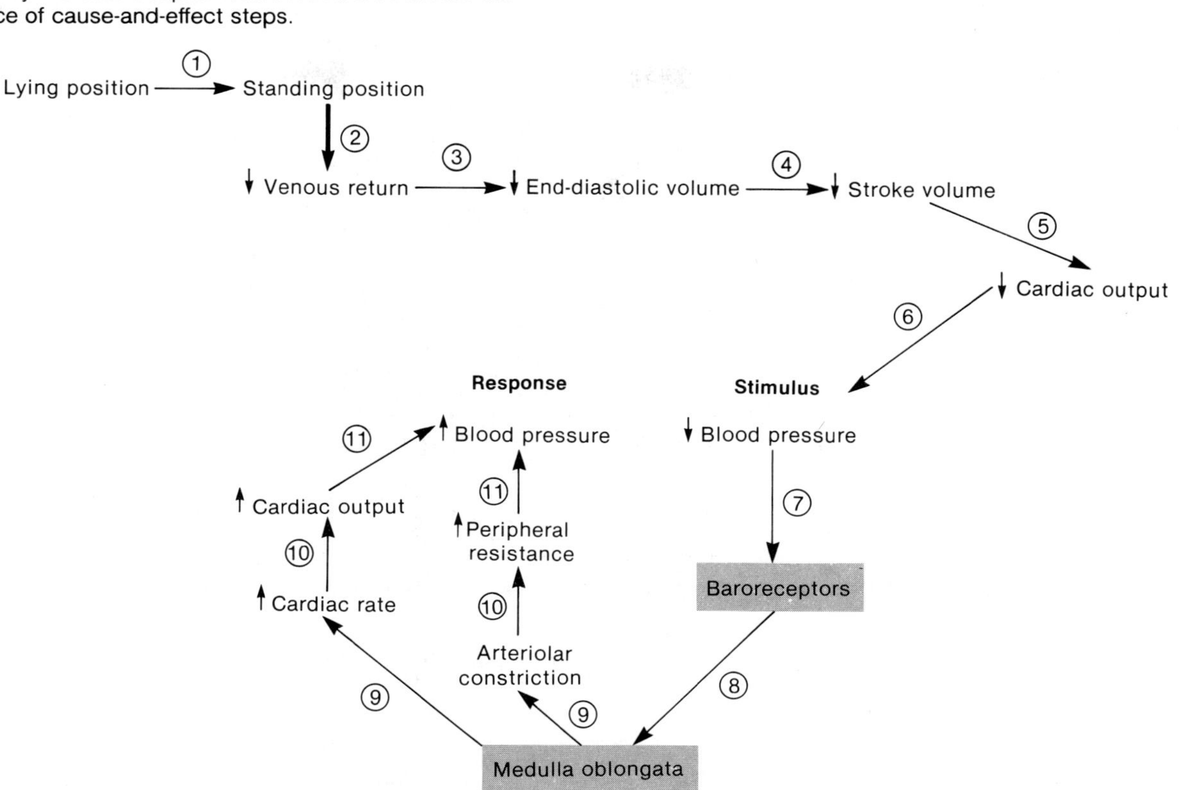

Baroreceptor Reflex. When a person goes from a lying to a standing position, there is a shift of 500–700 ml of blood from the veins of the thoracic cavity to veins in the lower extremities, which expand to contain the extra volume of blood. This pooling of blood reduces the venous return and cardiac output. The resulting fall in blood pressure is almost immediately compensated by the baroreceptor reflex. A decrease in baroreceptor sensory information, traveling in the ninth and tenth cranial nerves, to the medulla oblongata inhibits parasympathetic activity and promotes sympathetic nerve activity, resulting in increased cardiac rate and vasoconstriction. These responses help maintain an adequate blood pressure upon standing (fig. 14.26).

An effective baroreceptor reflex helps to maintain adequate blood flow to the brain upon standing. If the baroreceptor sensitivity is reduced, perhaps by atherosclerosis, an uncompensated fall in pressure may occur when a person stands up too rapidly. This condition—called **postural,** or **orthostatic, hypotension** (hypotension = low blood pressure)—can make a person feel dizzy or even faint because of inadequate perfusion of the brain.

The baroreceptor reflex can also mediate the opposite response. When the blood pressure rises above an individual's normal range, the baroreceptor reflex causes a slowing of the cardiac rate and vasodilation. Manual massage of the carotid sinus, a procedure sometimes employed by physicians to reduce tachycardia and lower blood pressure, also evokes this reflex. Such carotid massage should be used cautiously, however, because the intense vagus-nerve-induced slowing of the cardiac rate could cause loss of consciousness (as occurs in emotional fainting).

Other Reflexes Controlling Blood Pressure. The reflex control of ADH secretion by osmoreceptors in the hypothalamus and the control of angiotensin II production and aldosterone secretion by the juxtaglomerular apparatus of the kidneys have been previously discussed. Antidiuretic hormone and aldosterone increase blood volume, and angiotensin II stimulates vasoconstriction.

Another reflex that may be important to blood pressure regulation is initiated by stretch receptors located in the left atrium of the heart. These receptors are activated by increased venous return to the heart and produce (1) reflex tachycardia, as a result of increased sympathetic nerve activity; and (2) inhibition of ADH secretion, resulting in larger volumes of urine excretion and a lowering of blood volume.

Measurement of Blood Pressure

Stephen Hales (1677–1761) accomplished the first documented measurement of blood pressure by inserting a cannula into the artery of a horse and measuring the height to which blood would rise in the vertical tube. Modern clinical blood pressure measurements, fortunately, are less direct. The indirect, or **auscultatory, method** of blood pressure measurement is based on the correlation of blood pressure and arterial sounds.

In the auscultatory method, an inflatable rubber bladder within a cloth cuff is wrapped around the upper arm and a stethoscope is applied over the brachial artery (fig. 14.27). The artery is normally silent before inflation of the cuff because blood normally travels in a smooth *laminar flow* through the arteries. The term laminar means layered—blood in the central axial stream moves the fastest, and blood flowing closer to the artery wall moves more slowly. There is little transverse movement between these layers that would produce mixing.

The laminar flow that normally occurs in arteries produces little vibration and is thus silent. When the artery is pinched, however, blood flow through the constriction becomes turbulent, which causes the artery to vibrate and produce sounds (much like the sounds produced by water through a kink in a garden hose). The tendency of the cuff pressure to constrict the artery is opposed by the blood pressure. Thus, in order to constrict the artery, the cuff pressure must be greater than the diastolic blood pressure. If the cuff pressure is also greater than the systolic blood pressure the artery would be pinched off and silent. *Turbulent flow* and sounds produced by vibrations of the artery as a result of this flow, therefore, only occur when the cuff pressure is greater than the diastolic and less than the systolic blood pressure.

Figure 14.27. The use of a pressure cuff and sphygmomanometer to measure blood pressure.

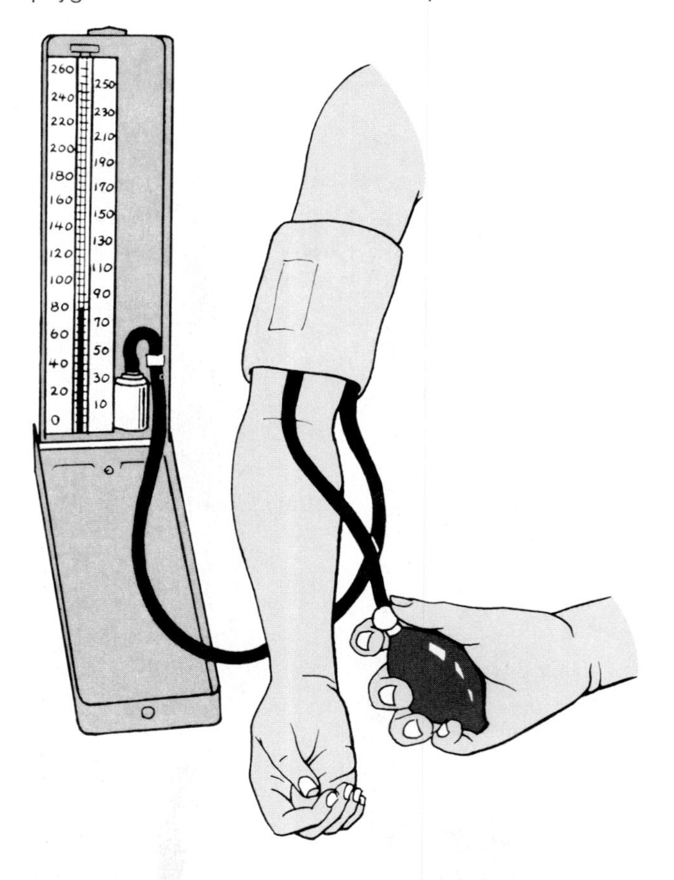

Suppose that a person has a systolic pressure of 120 mm Hg and a diastolic pressure of 80 mm Hg (the average normal values). When the cuff pressure is between 80 and 120 mm Hg, the artery will be closed during diastole and opened during systole. As the artery begins to open with every systole, turbulent flow of blood through the constriction will create vibrations that are heard as the **sounds of Korotkoff,** as shown in figure 14.28. These are usually "tapping" sounds because the artery closes and silence resumes with every diastole. It should be understood that the sounds of Korotkoff are *not* "lub-dub" sounds produced by closing of the heart valves (those sounds can only be heard on the chest, not on the brachial artery).

Initially, the cuff is usually inflated to produce a pressure greater than the systolic pressure so that the artery is pinched off and silent. The pressure in the cuff is read from an attached meter called a *sphygmomanometer.* A valve is then turned to allow the release of air from the

Figure 14.28. Korotkoff sounds are produced by the turbulent flow of blood through the partially constricted brachial artery. This occurs when the cuff pressure is greater than the diastolic pressure but less than the systolic pressure.

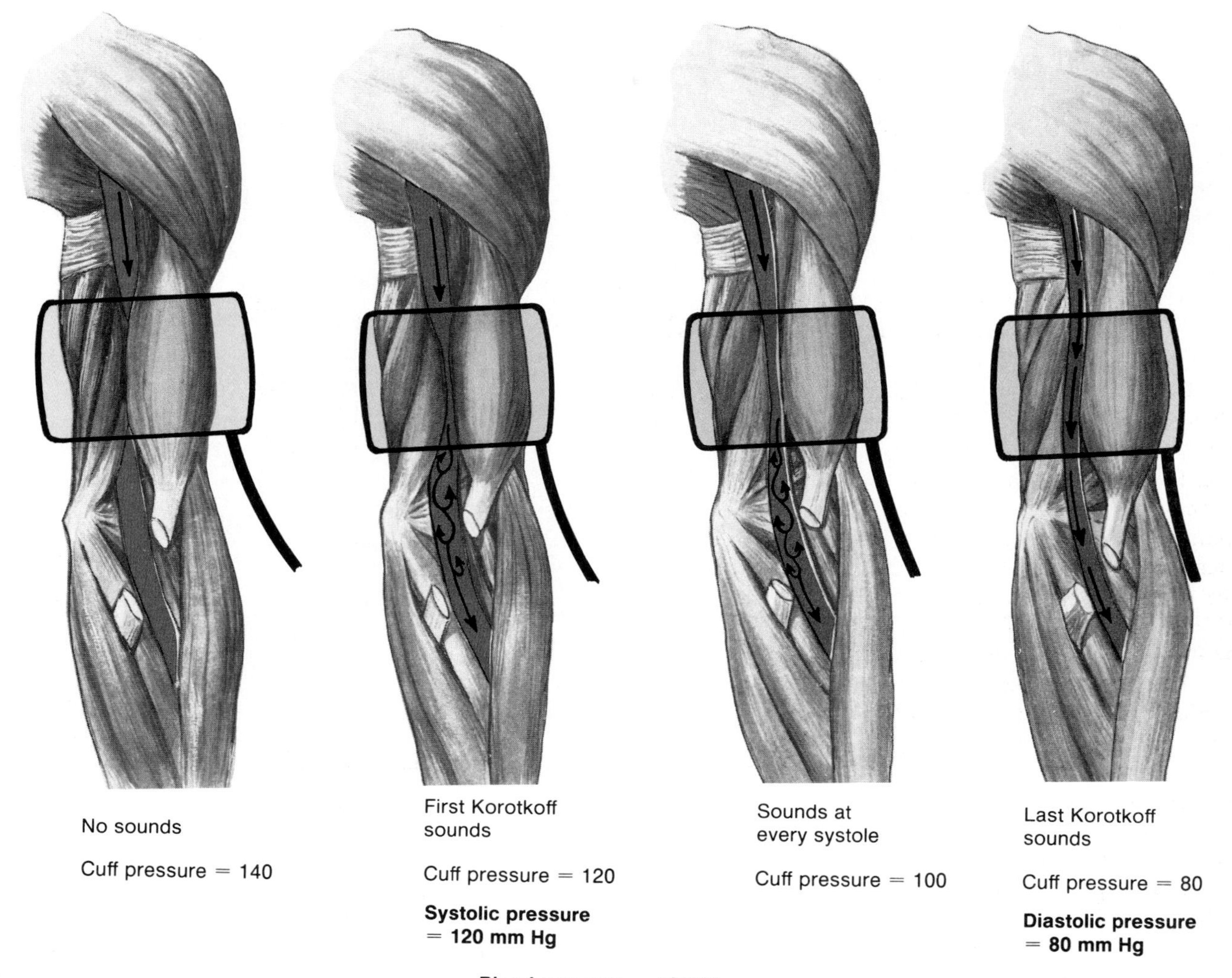

cuff, causing a gradual decrease in cuff pressure. When the cuff pressure is equal to the systolic pressure, the **first sound** of Korotkoff is heard as blood passes in a turbulent flow through the constricted opening of the artery.

Korotkoff sounds will continue to be heard at every systole as long as the cuff pressure remains greater than the diastolic pressure. When the cuff pressure becomes equal to or less than the diastolic pressure, the sounds disappear because the artery remains open (fig. 14.29). The **last sound** of Korotkoff thus occurs when the cuff pressure is equal to the diastolic pressure.

Different phases in the measurement of blood pressure are identified on the basis of the quality of the Korotkoff sounds (fig. 14.30). In some people, the Korotkoff sounds do not disappear even when the cuff pressure is reduced to zero (zero pressure means that it is equal to atmospheric pressure). In these cases—and often routinely—the onset of muffling of the sounds (phase 4 in fig. 14.30) is used rather than the onset of silence (phase 5) as an indication of diastolic pressure. Normal blood pressure values are shown in table 14.10.

Figure 14.29. The indirect, or auscultatory, method of blood-pressure measurement. Korotkoff sounds, produced by turbulent blood flow through a constricted artery, occur whenever the cuff pressure is less than the systolic blood pressure and greater than the diastolic blood pressure. As a result, the first Korotkoff sound is heard when the cuff pressure is equal to the systolic blood pressure, and the last sound is heard when the cuff pressure and diastolic blood pressures are equal.

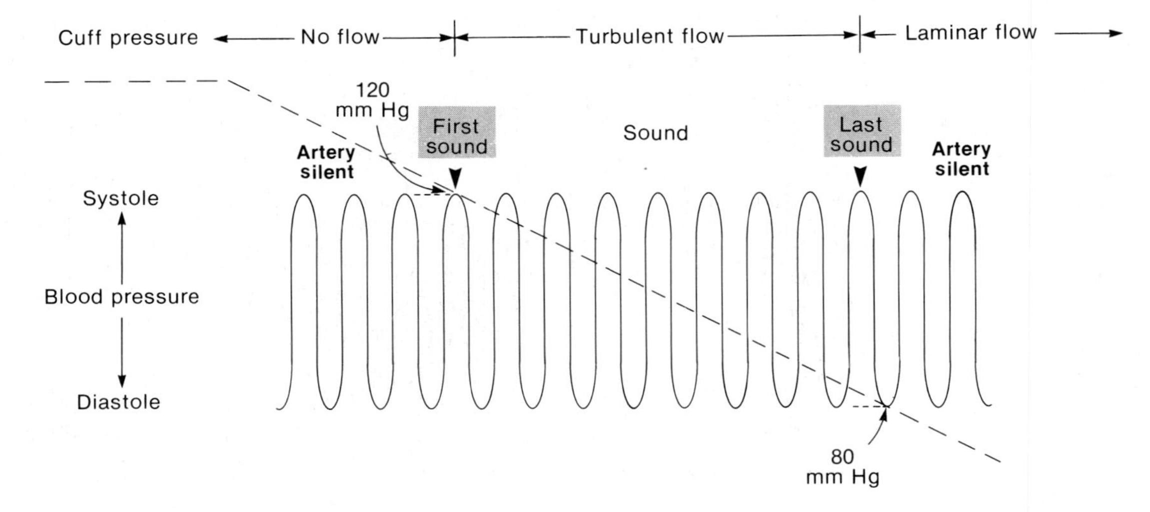

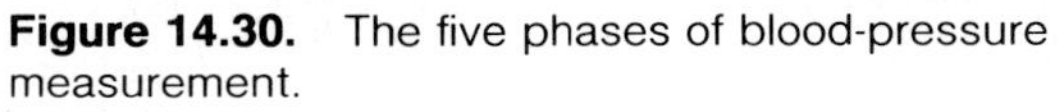
Figure 14.30. The five phases of blood-pressure measurement.

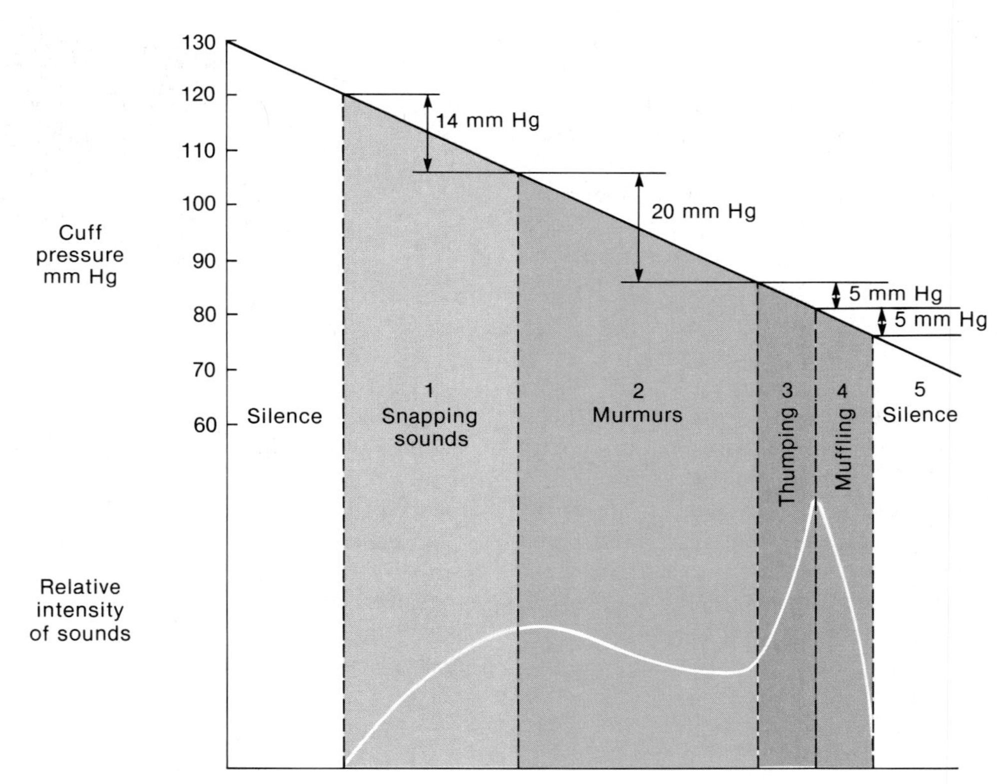

Table 14.10 Normal arterial blood pressure at different ages.

Age	Systolic		Diastolic		Age	Systolic		Diastolic	
	Men	Women	Men	Women		Men	Women	Men	Women
1 day	70				16 years	118	116	73	72
3 days	72				17 years	121	116	74	72
9 days	73				18 years	120	116	74	72
3 weeks	77				19 years	122	115	75	71
3 months	86				20–24 years	123	116	76	72
6–12 months	89	93	60	62	25–29 years	125	117	78	74
1 year	96	95	66	65	30–34 years	126	120	79	75
2 years	99	92	64	60	35–39 years	127	124	80	78
3 years	100	100	67	64	40–44 years	129	127	81	80
4 years	99	99	65	66	45–49 years	130	131	82	82
5 years	92	92	62	62	50–54 years	135	137	83	84
6 years	94	94	64	64	55–59 years	138	139	84	84
7 years	97	97	65	66	60–64 years	142	144	85	85
8 years	100	100	67	68	65–69 years	143	154	83	85
9 years	101	101	68	69	70–74 years	145	159	82	85
10 years	103	103	69	70	75–79 years	146	158	81	84
11 years	104	104	70	71	80–84 years	145	157	82	83
12 years	106	106	71	72	85–89 years	145	154	79	82
13 years	108	108	72	73	90–94 years	145	150	78	79
14 years	110	110	73	74	95–106 years	145	149	78	81
15 years	112	112	75	76					

From Diem, K., and Lentner, C., Eds., *Documenta Geigy Scientific Tables,* 7th ed., J. R. Geigy S. A., Basle, Switzerland, 1970. With permission.

Pulse Pressure and Mean Arterial Pressure

The **pulse pressure** is equal to the difference between the systolic and diastolic pressures. If a person has a blood pressure of 120/80 (systolic/diastolic), therefore, the pulse pressure would equal 40 mm Hg. This value is significant because it represents the pressure difference between the "heart side" of an artery and the "capillary side" of the artery when the heart contracts. The pulse pressure is thus the driving pressure or ΔP that moves the arterial blood towards the capillaries (fig. 14.31).

$$\text{Pulse pressure} = \text{Systolic pressure} - \text{Diastolic pressure}$$

The **mean arterial pressure** represents an average of the systolic and diastolic blood pressures. This is usually calculated by adding one-third of the pulse pressure to the diastolic pressure. If a person has a blood pressure of 120/80, for example, the mean arterial pressure will be $80 + \frac{1}{3}(40) = 93$ mm Hg.

$$\text{Mean arterial pressure} = \text{Diastolic pressure} + \frac{1}{3} \times \text{Pulse pressure}$$

Figure 14.31. The pulse pressure provides the driving force for blood flow in the arteries.

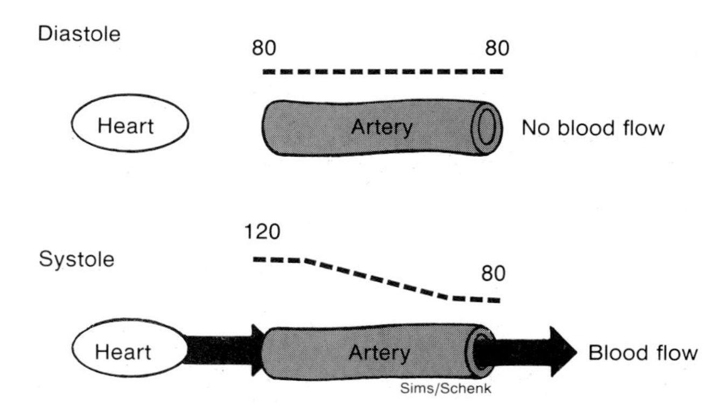

Table 14.11 Effects of different variables on blood pressure measurements.

Variable	Diastolic Pressure	Systolic Pressure	Pulse Pressure
$\uparrow$Cardiac rate	Increases (more)	Increases (less)	Decreases
$\uparrow$Peripheral resistance	Increases (more)	Increases (less)	Decreases
$\uparrow$Blood volume*	Increases (less)	Increases (more)	Increases
$\uparrow$Stroke volume	Increases (less)	Increases (more)	Increases

*Conversely, a fall in blood volume (as in dehydration) produces a decrease in pulse pressure.

A rise in total peripheral resistance and cardiac rate increases the diastolic pressure more than it increases the systolic pressure. When the baroreceptor reflex is activated by going from a lying to a standing position, for example, the diastolic pressure usually increases by 5–10 mm Hg, whereas the systolic pressure is either unchanged or slightly reduced (as a result of decreased venous return). People with hypertension (high blood pressure) who usually have elevated total peripheral resistance and cardiac rates, likewise have a greater increase in diastolic than in systolic pressure. Dehydration or blood loss results in decreased cardiac output and thus also produces a decrease in pulse pressure.

An increase in cardiac output, in contrast, raises the systolic pressure more than it raises the diastolic pressure (though both pressures do rise). This occurs during exercise, for example, when the blood pressure may rise to values as high as 200/100 (yielding a pulse pressure of 100 mm Hg). The effects of different variables on blood pressure are summarized in table 14.11.

1. *Describe the relationship between blood pressure and the total cross-sectional area of arteries, arterioles, and capillaries. Describe how arterioles influence blood flow through capillaries and arterial blood pressure.*
2. *Describe how the baroreceptor reflex helps to compensate for a fall in blood pressure. Explain how a person who is severely dehydrated would have a rapid pulse.*
3. *Describe how the sounds of Korotkoff are produced and how these sounds are used to measure blood pressure.*
4. *Define pulse pressure and explain its physiological significance.*

Hypertension, Shock, and Congestive Heart Failure

An understanding of the normal function, or physiology, of the cardiovascular system is prerequisite to the study of its *pathophysiology,* or mechanisms of abnormal function. Since the mechanisms that regulate cardiac output, blood flow, and blood pressure are highlighted in particular disease states, a study of pathophysiology at this time can strengthen the students' understanding of the mechanisms involved in normal function.

Hypertension

Approximately 20% of all adults in the United States have *hypertension*—blood pressure in excess of the normal range for the person's age and sex. Hypertension that is a result of ("secondary to") known disease processes is logically called **secondary hypertension.** Secondary hypertension comprises only about 10% of the hypertensive population. Hypertension that is the result of complex and poorly understood processes is not-so-logically called **primary,** or **essential, hypertension.**

Diseases of the kidneys and arteriosclerosis of the renal arteries can cause secondary hypertension because of high blood volume. More commonly, the reduction of renal blood flow can raise blood pressure by stimulating the secretion of vasoactive chemicals from the kidneys. Experiments in which the renal artery is pinched, for example, produce hypertension that is associated (at least initially) with elevated renin secretion. These and other causes of secondary hypertension are summarized in table 14.12. 14.12.

Essential Hypertension. The vast majority of people with hypertension have essential hypertension. An increased total peripheral resistance is a universal characteristic of this condition. Cardiac rate and the cardiac output are elevated in many, but not all, of these cases.

The secretion of renin, which is correlated with angiotensin II production and aldosterone secretion, is likewise variable. Although some people with essential hypertension have low renin secretion, most have either normal or elevated levels of renin secretion. Renin secretion in the normal range is inappropriate for people with

Table 14.12 Some possible causes of secondary hypertension.

System Involved	Examples	Mechanisms
Kidneys	Kidney disease Renal artery disease	Decreased urine formation Secretion of vasoactive chemicals
Endocrine	Excess catecholamines (tumor of adrenal medulla) Excess aldosterone (Conn's syndrome)	Increased cardiac output and total peripheral resistance Excess salt and water retention by the kidneys
Nervous	Increased intracranial pressure Damage to vasomotor center	Activation of sympathoadrenal system Activation of sympathoadrenal system
Cardiovascular	Complete heart block; patent ductus arteriosus Arteriosclerosis of aorta; coarctation of aorta	Increased stroke volume Decreased distensibility of aorta

Figure 14.32. The sequence of events that has been proposed as a possible cause of essential hypertension.

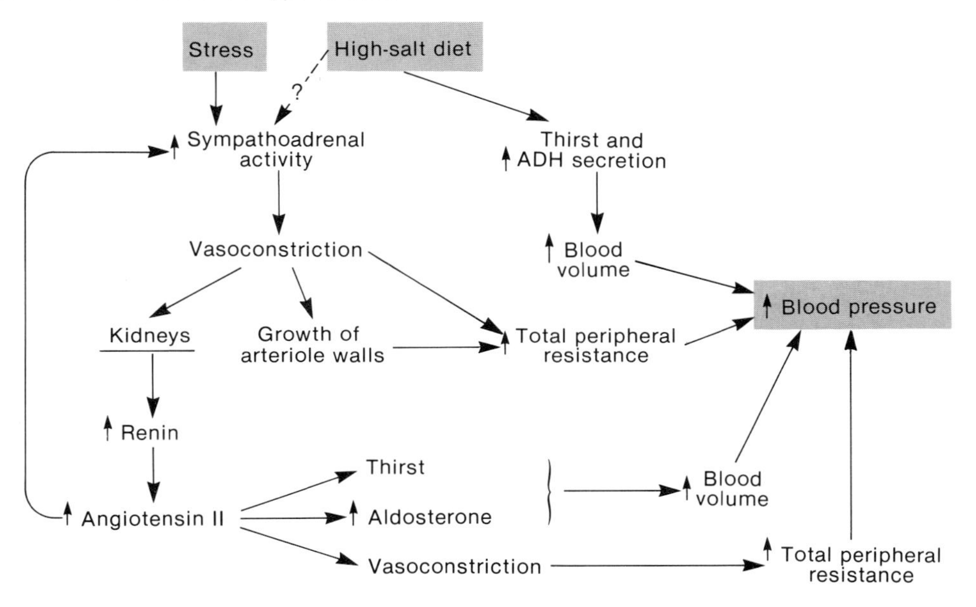

hypertension, since high blood pressure should inhibit renin secretion and, through a lowering of aldosterone, result in greater excretion of salt and water. Inappropriately high levels of renin secretion could thus contribute to hypertension by promoting (via stimulation of aldosterone secretion) salt and water retention and high blood volume.

Sustained high stress (acting via the sympathetic nervous system) and high salt intake appear to act synergistically in the development of hypertension. There is some evidence that Na^+ enhances the vascular response to sympathetic stimulation. Further, sympathetic nerves can cause constriction of the renal blood vessels and thus decrease the excretion of salt and water. In one study, hypertensive students given an IQ test (which was apparently stressful) had decreased renal blood flow, increased renin secretion, decreased excretion of salt and water, and a greater increase in blood pressure than did students with normal blood pressure. There also appears to be a genetic component to these responses; students who were normotensive but who had a family history of hypertension had responses that were intermediate between hypertensive students and students who were normotensive with no family history of hypertension.

The interactions between salt intake, sympathetic nerve activity, cardiovascular responses to sympathetic nerve activity, kidney function, and genetics make it difficult to sort out the cause-and-effect sequence that leads to essential hypertension. Many people have suggested that there is no single cause-and-effect sequence but rather a web of causes and effects (fig. 14.32). This is currently controversial.

Table 14.13 Examples and mechanisms of action of some antihypertensive drugs.

Category of Drugs	Examples	Mechanisms
Extracellular fluid volume depletors	Thiazide diuretics	Increase volume of urine excreted, thus lowering blood volume
Sympathoadrenal system inhibitors	Clonidine; alpha-methyldopa	Acts on brain to decrease sympathoadrenal stimulation
	Guanethidine; reserpine	Depletes norepinephrine from sympathetic nerve endings
	Propranolol	Blocks beta-adrenergic receptors, decreasing cardiac output and/or renin secretion
	Phentolamine	Blocks alpha-adrenergic receptors, decreasing sympathetic vasoconstriction
Vasodilators	Hydralazine; sodium nitroprusside	Causes vasodilation by acting directly on vascular smooth muscle

Dangers of Hypertension. If other factors remain constant, blood flow increases as arterial blood pressure increases. People with hypertension thus have adequate perfusion of their organs with blood until the hypertension causes vascular damage. Hypertension, as a result, is usually without symptoms until a dangerous amount of vascular damage is produced.

Hypertension is dangerous because (1) high arterial pressure increases the afterload, making it more difficult for the ventricles to eject blood; this increases the amount of work that the heart must perform and may result in pathological changes in heart structure and function, leading to congestive heart failure; (2) high pressure may damage cerebral blood vessels, leading to cerebrovascular accident (stroke); and (3) it contributes to the development of atherosclerosis, which can itself lead to heart disease and stroke as previously described.

Treatment of Hypertension. Hypertension is usually treated by restricting salt intake and by taking drugs that act in a variety of ways. Most commonly, these drugs are *diuretics* that increase urine volume, thus decreasing blood volume and pressure. Sympathetic-blocking drugs are also often used; drugs that block beta-adrenergic receptors (such as propranolol) decrease cardiac rate. Various vasodilators (table 14.13) may also be used to decrease total peripheral resistance.

Circulatory Shock

Circulatory shock occurs when there is inadequate blood flow and/or oxygen utilization by the tissues. Some of the signs of shock (table 14.14) are a result of inadequate tissue perfusion; other signs of shock are produced by cardiovascular responses that help to compensate for the poor tissue perfusion (table 14.15). When these compensations are effective, they (together with emergency medical care) are able to reestablish adequate tissue perfusion. In some cases, however, and for reasons that are not clearly understood, the shock may progress to an irreversible stage and death may result.

Hypovolemic Shock. The term *hypovolemic shock* refers to circulatory shock due to low blood volume, as might be caused by hemorrhage (bleeding), dehydration, or burns. This is accompanied by decreased blood pressure and decreased cardiac output. In response to these changes, the sympathoadrenal system is activated by means of the baroreceptor reflex. As a result, tachycardia is produced and vasoconstriction occurs in the skin, digestive tract, kidneys, and muscles. Decreased blood flow through the kidneys stimulates renin secretion and activation of the renin-angiotensin-aldosterone system. A person in hypovolemic shock thus has low blood pressure, rapid pulse, cold, clammy skin, and little urine excretion.

Since the resistance in the coronary and cerebral circulations is not increased, blood is diverted to the heart and brain at the expense of other organs. Interestingly, a similar response occurs in diving mammals and, to a lesser degree, in Japanese pearl divers during prolonged submersion. These responses help to deliver blood to the two organs that have the highest requirements for aerobic metabolism.

Vasoconstriction in organs other than the brain and heart raises total peripheral resistance, which helps (along with the reflex increase in cardiac rate) to compensate for the drop in blood pressure due to low blood volume. Constriction of arterioles also decreases capillary blood flow and capillary filtration pressure. Less filtrate is formed, as a result, while the osmotic return of fluid to the capillary is increased because of increased plasma colloid osmotic pressure. The blood volume is thus raised at the expense of tissue fluid volume. Blood volume is also conserved by decreased urine production, which occurs as a result of vasoconstriction in the kidneys and the water-conserving effects of ADH and aldosterone, which are secreted in increased amounts during shock.

Table 14.14 Signs of shock.

	Early Sign	Late Sign
Blood pressure	Decreased pulse pressure Increased diastolic pressure	Decreased systolic pressure
Urine	Decreased Na^+ concentration Increased osmolality	Decreased volume
Blood pH	Increased pH (alkalosis) due to hyperventilation	Decreased pH (acidosis) due to "metabolic" acids
Effects of poor tissue perfusion	Slight restlessness; occasionally warm, dry skin	Cold, clammy skin "Cloudy" senses

Source: Wilson, R. F., ed., *Principles and Techniques of Critical Care,* vol. 1. © 1977 Upjohn Company.

Table 14.15 Cardiovascular reflexes that help to compensate for circulatory shock.

Organ(s)	Compensatory Mechanisms
Heart	Sympathoadrenal stimulation produces increased cardiac rate and increased stroke volume, due to "positive inotropic effect" on myocardial contractility
Digestive tract and skin	Decreased blood flow due to vasoconstriction as a result of sympathetic nerve stimulation (alpha-adrenergic effect)
Kidneys	Decreased urine production as a result of sympathetic-nerve-induced constriction of renal arterioles; increased salt and water retention due to increased aldosterone and antidiuretic hormone (ADH) secretion

Other Causes of Circulatory Shock. A rapid fall in blood pressure occurs in **anaphylactic shock** as a result of severe allergic reaction (usually to bee stings or penicillin). This results from the widespread release of histamine, which causes vasodilation and thus decreases total peripheral resistance. A rapid fall in blood pressure also occurs in **neurogenic shock,** in which sympathetic tone is decreased, usually because of upper spinal cord damage or spinal anesthesia. **Cardiogenic shock** results from cardiac failure, as defined by a cardiac output that is inadequate to maintain tissue perfusion.

Congestive Heart Failure

Cardiac failure occurs when the cardiac output is insufficient to maintain the blood flow required by the body. This may be due to heart disease—resulting from myocardial infarction or congenital defects—or to hypertension, which increases the afterload of the heart. The most common causes of left pump failure are myocardial infarction, aortic valve stenosis, and incompetence of the aortic and bicuspid (mitral) valves. Failure of the right pump is usually caused by prior failure of the left pump.

Heart failure can also result from disturbance in the electrolyte concentrations of the blood. Excessive plasma K^+ concentration decreases the resting membrane potential of myocardial cells; low blood Ca^{++} reduces excitation-contraction coupling. High blood K^+ and low blood Ca^{++} can thus cause the heart to stop in diastole. Conversely, low blood K^+ and high blood Ca^{++} can arrest the heart in systole.

The term *congestive* is often used in describing heart failure because of the increased venous volume and pressure that results. Failure of the left pump, for example, raises the left atrial pressure and produces pulmonary congestion and edema, which causes shortness of breath and fatigue. Failure of the right pump results in increased right atrial pressure, which produces congestion and edema in the systemic circulation.

People with congestive heart failure are often treated with the drug *digitalis.* Digitalis appears to inhibit the action of Na^+/K^+ pumps in the cell membranes, causing a rise in the intracellular concentrations of Na^+. This in turn apparently causes the intracellular concentrations of Ca^{++} to rise, which strengthens the contractions of the heart.

The compensatory responses that occur during congestive heart failure are similar to those that occur during hypovolemic shock. Activation of the sympathoadrenal system stimulates cardiac rate, contractility of the ventricles (fig. 14.33), and constriction of arterioles. As in hypovolemic shock, renin secretion is increased and urine output is reduced.

As a result of these compensations, chronically low cardiac output is associated with elevated blood volume and dilation and hypertrophy of the ventricles. These changes can themselves be dangerous. Elevated blood volume places a work overload on the heart, and the enlarged ventricles have a higher metabolic requirement for oxygen. These problems are often treated with drugs that increase myocardial contractility (such as digitalis), drugs that are vasodilators (such as nitroglycerin), and diuretic drugs that lower blood volume by increasing the volume of urine excreted.

1. *Explain how stress and a high-salt diet can contribute to hypertension.*
2. *Using a flow chart to show cause and effect, explain why a person in hypovolemic shock may have a fast pulse and cold, clammy skin.*
3. *Describe the compensatory mechanisms that act to raise blood volume during cardiovascular shock.*
4. *Describe congestive heart failure, and explain the compensatory responses that occur during this condition.*

Figure 14.33. The relationship between end-diastolic volume and ventricular pressure (a measure of contraction strength). At a given end-diastolic volume the ventricular pressure is decreased during heart failure and increased by sympathetic stimulation. Heart failure may be compensated by sympathetic stimulation.

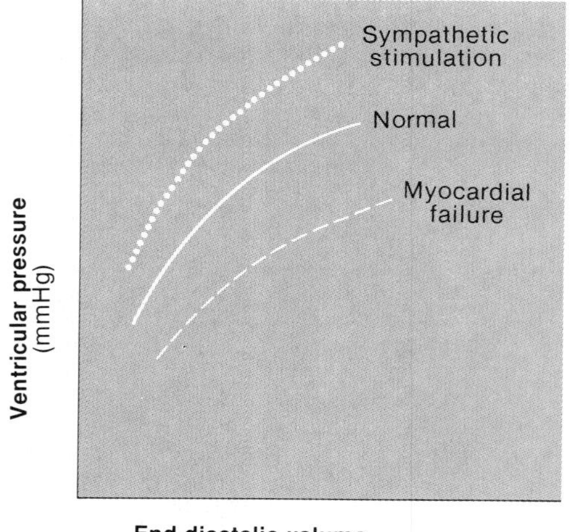

Summary

Cardiac Output p. 386

I. Cardiac rate is increased by sympathoadrenal stimulation and decreased by the effects of parasympathetic fibers that innervate the SA node.

II. Stroke volume is regulated both extrinsically and intrinsically.
  - A. The Frank-Starling law of the heart describes the way the end-diastolic volume, through various degrees of myocardial stretching, influences the contraction strength of the myocardium and thus the stroke volume.
  - B. The end-diastolic volume is called the preload; the total peripheral resistance, through its effect on arterial blood pressure, provides an afterload that acts to reduce the stroke volume.
  - C. At a given end-diastolic volume, the amount of blood ejected depends on contractility; strength of contraction is increased by sympathoadrenal stimulation.

III. The venous return of blood to the heart is dependent largely on the total blood volume and mechanisms that improve the flow of blood in veins.
  - A. The total blood volume is regulated by the kidneys.
  - B. The venous flow of blood to the heart is aided by the action of skeletal muscle pumps and the effects of breathing.

Blood Volume p. 390

I. Tissue fluid is formed from and returns to the blood.
  - A. The hydrostatic pressure of the blood forces fluid from the arteriolar ends of capillaries into the interstitial spaces of the tissues.
  - B. Since the colloid osmotic pressure of plasma is greater than tissue fluid, water returns by osmosis to the venular ends of capillaries.
  - C. Excess tissue fluid is returned to the venous system by lymphatic vessels.
  - D. Edema occurs when there is an accumulation of tissue fluid.

II. The kidneys control the blood volume by regulating the amount of filtered fluid that will be reabsorbed.
  - A. Antidiuretic hormone stimulates reabsorption of water from the kidney filtrate and thus acts to maintain the blood volume.
  - B. A decrease in blood flow through the kidneys activates the renin-angiotensin system.
  - C. Angiotensin II stimulates vasoconstriction and the secretion of aldosterone by the adrenal cortex.
  - D. Aldosterone acts on the kidneys to promote the retention of salt and water.

Vascular Resistance and Blood Flow p. 397

I. Poiseuille's law describes the fact that blood flow is directly related to the pressure difference between the two ends of a vessel and is inversely related to the resistance to blood flow through the vessel.
II. Extrinsic regulation of vascular resistance is provided mainly by the sympathetic nervous system, which stimulates vasoconstriction of arterioles in the viscera and skin.
III. Intrinsic control of vascular resistance allows organs to autoregulate their own blood flow rates.
   A. Myogenic regulation occurs when vessels constrict or dilate as a direct response to a rise or fall in blood pressure.
   B. Metabolic regulation occurs when vessels dilate in response to the local chemical environment within the organ.

Blood Flow to the Heart and Skeletal Muscles p. 402

I. The heart normally always respires aerobically because of its high capillary supply, myoglobin content, and enzyme content.
II. During exercise, when the heart's metabolism increases, intrinsic metabolic mechanisms stimulate vasodilation of the coronary vessels and thus increase coronary blood flow.
III. Just prior to exercise and at the start of exercise, blood flow through skeletal muscles increases due to vasodilation caused by cholinergic sympathetic nerve fibers; during exercise, intrinsic metabolic vasodilation occurs.
IV. Since cardiac output can increase by a factor of five or more during exercise, the heart and skeletal muscles receive an increased proportion of a higher total blood flow.
   A. The cardiac rate increases due to lower activity of the vagus nerve and higher activity of the sympathetic nerve.
   B. The venous return is faster because of higher activity of the skeletal muscle pumps and increased breathing.
   C. Increased contractility of the heart, combined with a decrease in total peripheral resistance, can result in a higher stroke volume.

Blood Flow to the Brain and Skin p. 406

I. Cerebral blood flow is regulated both myogenically and metabolically.
   A. Cerebral vessels automatically constrict if the systemic blood pressure rises too high.
   B. Metabolic products cause local vessels to dilate and supply more active areas with more blood.
II. The skin has unique arteriovenous anastomoses, which can shunt the blood away from surface capillary loops.
   A. Sympathetic nerve fibers cause constriction of cutaneous arterioles.
   B. As a thermoregulatory response, there is increased cutaneous blood flow and increased flow through surface capillary loops when body temperature rises.

Blood Pressure p. 410

I. Baroreceptors in the aortic arch and carotid sinuses affect, via the sympathetic nervous system, the cardiac rate and the total peripheral resistance.
   A. The baroreceptor reflex causes pressure to be maintained when an upright posture is assumed and can cause a lowered pressure when the carotid sinuses are massaged.
   B. Other mechanisms that affect blood volume help to regulate blood pressure.
II. Blood pressure is commonly measured indirectly by auscultation of the brachial artery when a pressure cuff is inflated and deflated.
   A. The first sound of Korotkoff, caused by turbulent flow of blood through a constriction in the artery, occurs when the cuff pressure equals the systolic pressure.
   B. The last sound of Korotkoff is heard when the cuff pressure equals the diastolic blood pressure.
III. The pulse pressure, which is the difference between systolic and diastolic pressures, represents the driving force for blood flow through the arterial system.

Hypertension, Shock, and Congestive Heart Failure p. 418

I. Hypertension, or high blood pressure, is classified as either primary or secondary.
   A. Primary hypertension, which is also called essential hypertension when it is benign, may be the result of the interaction of many mechanisms that raise the blood volume, cardiac output, and/or peripheral resistance.
   B. Secondary hypertension is the direct result of known, specific diseases.
II. Circulatory shock occurs when there is inadequate delivery of oxygen to the organs of the body.
   A. In hypovolemic shock, low blood volume causes low blood pressure; this may progress to an irreversible state that cannot be stopped by compensations.
   B. The fall in blood volume and pressure stimulates various reflexes that produce a rise in cardiac rate, shift of fluid from the tissues into the vascular system, decrease in urine volume, and vasoconstriction.
III. Congestive heart failure occurs when the cardiac output is insufficient to supply the blood flow required by the body; the term "congestive" is used to describe the increased venous volume and pressure that results.

Review Activities

Objective Questions

1. According to the Frank-Starling law, the strength of ventricular contraction is
   (a) directly proportional to the end-diastolic volume
   (b) inversely proportional to the end-diastolic volume
   (c) independent of the end-diastolic volume
2. In the absence of compensations, the stroke volume will decrease when
   (a) blood volume increases
   (b) venous return increases
   (c) contractility increases
   (d) arterial blood pressure increases
3. Which of the following statements about tissue fluid is *false?*
   (a) It contains the same glucose and salt concentration as plasma.
   (b) It contains a lower protein concentration than plasma.
   (c) Its colloid osmotic pressure is greater than that of plasma.
   (d) Its hydrostatic pressure is less than that of plasma.
4. Edema may be caused by
   (a) high blood pressure
   (b) decreased plasma protein concentration
   (c) leakage of plasma protein into tissue fluid
   (d) blockage of lymphatic vessels
   (e) all of the above
5. Both ADH and aldosterone act to
   (a) increase urine volume
   (b) increase blood volume
   (c) increase total peripheral resistance
   (d) all of the above
6. The greatest resistance to blood flow occurs in the
   (a) large arteries
   (b) medium-sized arteries
   (c) arterioles
   (d) capillaries
7. If a vessel were to dilate to twice its previous radius and if pressure remained constant, blood flow through this vessel would
   (a) increase by a factor of 16
   (b) increase by a factor of 4
   (c) increase by a factor of 2
   (d) decrease by a factor of 2
8. The sounds of Korotkoff are produced by
   (a) closing of the semilunar valves
   (b) closing of the AV valves
   (c) the turbulent flow of blood through an artery
   (d) elastic recoil of the aorta
9. Vasodilation in the heart and skeletal muscles during exercise is primarily due to the effects of
   (a) alpha-adrenergic stimulation
   (b) beta-adrenergic stimulation
   (c) cholinergic stimulation
   (d) products released by the exercising muscle cells
10. Blood flow in the coronary circulation is
    (a) increased during systole
    (b) increased during diastole
    (c) constant throughout the cardiac cycle
11. Blood flow in the cerebral circulation
    (a) varies with systemic arterial pressure
    (b) is regulated primarily by the sympathetic system
    (c) is maintained constant within physiological limits
    (d) is increased during exercise
12. Which of the following organs is able to tolerate the greatest restriction in blood flow? The
    (a) brain
    (b) heart
    (c) skeletal muscles
    (d) skin
13. Arteriovenous shunts in the skin
    (a) divert blood to superficial capillary loops
    (b) are closed when the ambient temperature is very cold
    (c) are closed when the deep-body temperature rises much above 37° C.
    (d) all of the above

Essay Questions

1. Define the terms *contractility, preload,* and *afterload,* and explain how these factors affect the cardiac output.
2. Explain, using the Frank-Starling law, how the stroke volume is affected by (a) bradycardia and (b) a "missed beat."
3. Which part of the cardiovascular system contains the most blood? Which part provides the greatest resistance to blood flow? Which part provides the greatest cross-sectional area? Explain.
4. Explain how the kidneys regulate blood volume.
5. A person who is dehydrated drinks more and urinates less. Explain the mechanisms involved.
6. Explain, using Poiseuille's law, how arterial blood flow can be diverted from one organ system to another.
7. Describe the mechanisms that increase the cardiac output during exercise and that increase the rate of blood flow to the heart and skeletal muscles.
8. Explain how an anxious person may have a cold, clammy skin and how the skin becomes hot and flushed on a hot, humid day.

Selected Readings

Atlas, S. A. 1986. Atrial natriuretic factor: Renal and systemic effects. *Hospital Practice* 21:67.

Berne, R. M., and M. N. Levy. 1981. *Cardiovascular physiology.* 4th ed. St. Louis: The C. V. Mosby Company.

Braunwald, E. 1974. Regulation of the circulation. *New England Journal of Medicine* 290: first part, p. 1124; second part, p. 1420.

Brody, M. J., J. R. Haywood, and K. B. Toun. 1980. Neural mechanisms in hypertension. *Annual Review of Physiology* 42:441.

Cantin, M., and J. Genest. February 1986. The heart as an endocrine gland. *Scientific American.*

Carafol, E., and J. T. Penniston. November 1985. The calcium signal. *Scientific American.*

Del Zoppo, G. J., and L. A. Harker. 1984. Blood vessel interaction in coronary disease. *Hospital Practice* 19:163.

Donald, D. E., and J. T. Shepard. 1980. Autonomic regulation of the peripheral circulation. *Annual Review of Physiology* 42:429.

Folkow, B., and E. Neil. 1971. *Circulation.* London: Oxford University Press.

Franciosa, J. A. 1981. Hypertensive left heart failure: Pathogenesis and therapy. *Hospital Practice* 16:165.

Granger, D. E., and P. R. Kvietys. 1981. The splanchnic circulation: Intrinsic regulation. *Annual Review of Physiology* 43:409.

Herd, J. A. 1984. Cardiovascular response to stress in man. *Annual Review of Physiology* 46:177.

Hilton, P. J. 1986. Cellular sodium transport in essential hypertension. *New England Journal of Medicine* 314:222.

Hilton, S. M., and K. M. Spyer. 1980. Central nervous regulation of vascular resistance. *Annual Review of Physiology* 42:399.

Kaplan, N. M. 1980. The control of hypertension: A therapeutic breakthrough. *American Scientist* 68:537.

Katz, A. M. 1975. Congestive heart failure. *New England Journal of Medicine* 293:1184.

Kontos, H. A. 1981. Regulation of the cerebral circulation. *Annual Review of Physiology* 43:397.

Laragh, J. H. 1985. Atrial natriuretic hormone, the renin-aldosterone axis, and blood pressure-electrolyte homeostasis. *New England Journal of Medicine* 313:1330.

Light, K. C., J. P. Loepke, P. A. Obrist, and P. W. Willis, IV. 1983. Psychological stress induces sodium and fluid retention in men at high risk for hypertension. *Science* 220:249.

McCarron, D. A. et al. 1984. Blood pressure and nutrient intake in the United States. *Science* 224:1392.

Nadel, E. R. 1985. Physiological adaptations to aerobic training. *American Scientist* 73:334.

Olsson, R. A. 1981. Local factors regulating cardiac and skeletal muscle blood flow. *Annual Review of Physiology* 43:385.

Ross, J., Jr. 1983. The failing heart and circulation. *Hospital Practice* 18:151.

Schatz, I. J. 1983. Orthostatic hypotension: Diagnosis and treatment. *Hospital Practice* 18:59.

Stephensen, R. B. 1984. Modification of reflex regulation of blood pressure by behavior. *Annual Review of Physiology* 46:133.

Vatner, S. F., and E. Braunwald. 1975. Cardiovascular control mechanisms in the conscious state. *New England Journal of Medicine* 293:970.

Weber, K. T., J. S. Janicki, and W. Laskey. 1983. The mechanics of ventricular function. *Hospital Practice* 18:113.

Weinberger, M. H. 1986. Dietary sodium and blood pressure. *Hospital Practice* 21:55.

Zelis, R., S. F. Flaim, A. J. Liedke, and S. H. Nellis. 1981. Cardiovascular dynamics in the normal and failing heart. *Annual Review of Physiology* 43:455.

15 Respiratory Physiology

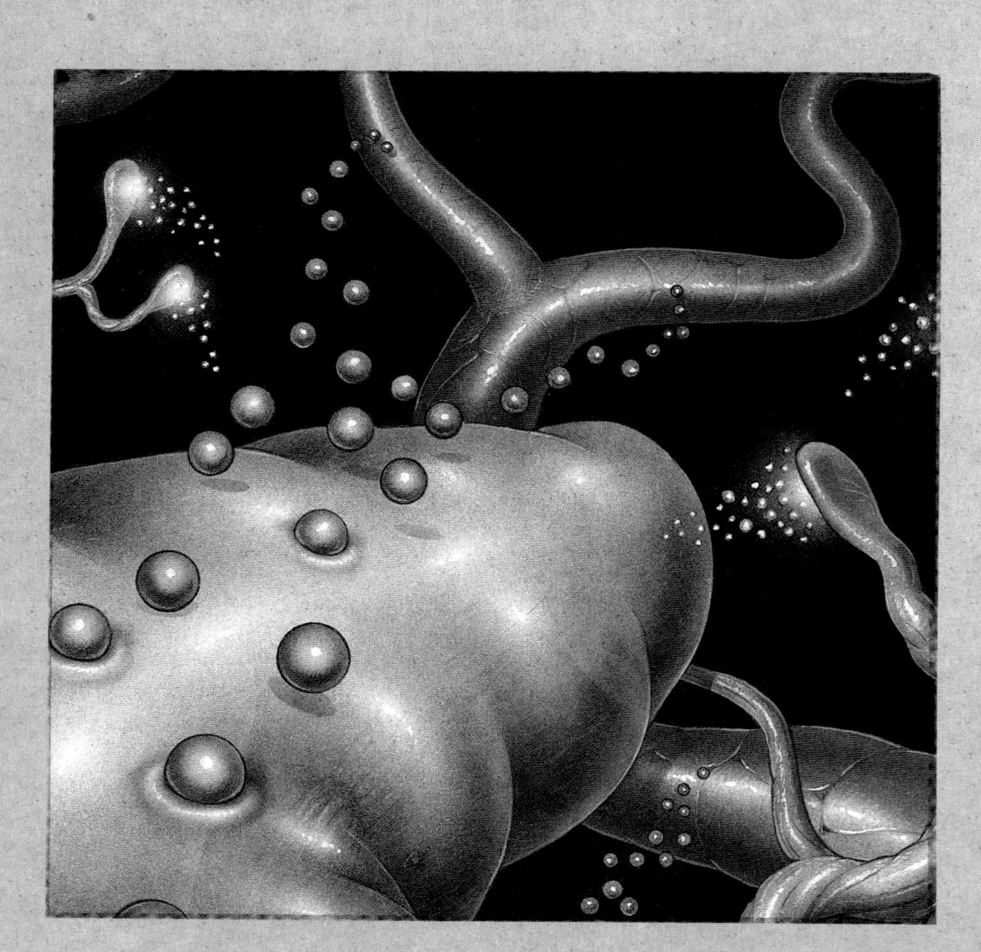

Objectives

By studying this chapter, you should be able to

1. describe the structures that comprise the lungs and explain their functions; describe the compartmentation of the thoracic cavity
2. explain how the intrapulmonary and intrapleural pressures vary during ventilation and how Boyle's law applies to these pressure changes
3. define the terms *compliance* and *elasticity* and explain how these lung properties affect ventilation
4. explain the significance of surface tension in lung mechanics, how the law of LaPlace applies to lung function, and the role of pulmonary surfactant
5. explain how inspiration and expiration are accomplished in unforced breathing and describe the accessory respiratory muscles that are used in forced breathing
6. define the various lung volumes and capacities that can be measured by spirometry and explain how obstructive diseases may be detected by the FEV test
7. describe the nature of some pulmonary disorders, including asthma, bronchitis, emphysema, and fibrosis
8. explain Dalton's law and describe how the partial pressure of a gas in a mixture of gases is calculated
9. explain Henry's law, describe how the partial pressure of oxygen and carbon dioxide in a fluid (such as blood) is measured, and explain the clinical significance of these measurements
10. describe the roles of the medulla oblongata, pons, and cerebral cortex in the regulation of breathing
11. explain why the P_{CO_2} and pH of blood, rather than its P_{O_2}, serve as the primary chemical stimuli for breathing
12. explain how the chemoreceptors in the medulla oblongata and the peripheral chemoreceptors in the aortic and carotid bodies respond to changes in P_{CO_2}, pH, and P_{O_2}
13. describe the Hering-Breuer reflex and its significance
14. describe the different forms of hemoglobin and their significances
15. describe the loading and unloading reactions and explain how the extent of these reactions is influenced by the P_{O_2} and affinity of hemoglobin for oxygen
16. describe the oxyhemoglobin dissociation curve, explain the reason for and the significance of its shape, and demonstrate how the curve is used to derive the percent unloading of oxygen
17. explain how oxygen transport is influenced by changes in blood pH and temperature and explain the effect and physiological significance of 2,3-DPG on oxygen transport
18. list the different forms of carbon dioxide transport in the blood and explain the chloride shift in the tissues and the reverse chloride shift in the lungs
19. explain how carbon dioxide affects blood pH and describe how hypoventilation and hyperventilation affect acid-base balance
20. describe the hyperpnea of exercise and explain how the anaerobic threshold is affected by endurance training
21. explain the compensations of the respiratory system to life at a high altitude

Outline

Figure 15.1. A diagram showing the relationship between lung alveoli and pulmonary capillaries.

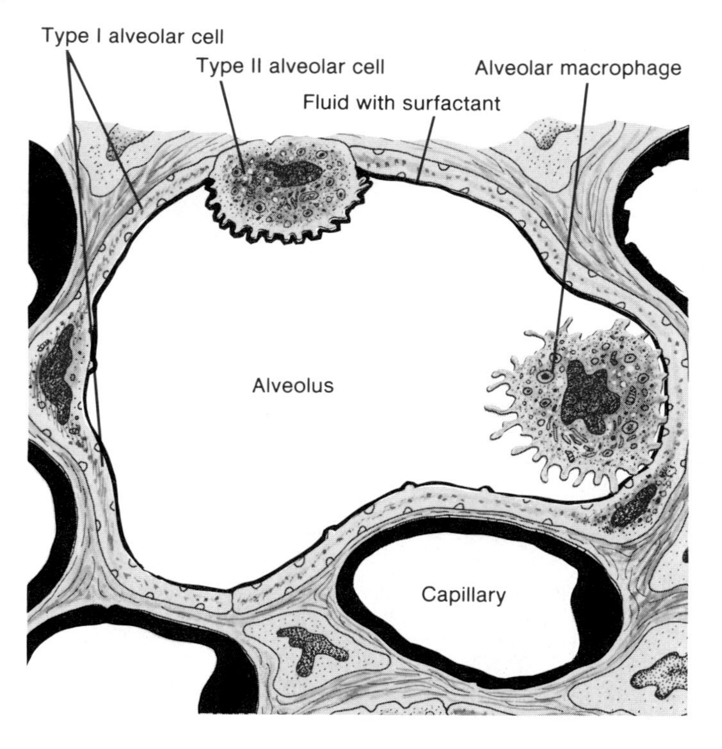

The Respiratory System

The term *respiration* includes three separate but related functions: (1) **ventilation** (breathing); (2) **gas exchange,** between the air and blood in the lungs and between the blood and tissues; and (3) **oxygen utilization** by the tissues in the energy-liberating reactions of cell respiration. Ventilation and the exchange of gases (oxygen and carbon dioxide) between the air and blood are together called *external respiration.* Gas exchange between the blood and tissues and oxygen utilization by the tissues are together known as *internal respiration.*

Ventilation is the mechanical process that moves air into and out of the lungs. Since air in the lungs has a higher oxygen concentration than the blood, oxygen diffuses from air to blood. Carbon dioxide, similarly, moves from the blood to the air within the lungs by diffusion. As a result of this gas exchange, the inspired air contains more oxygen and less carbon dioxide than the expired air. More importantly, blood leaving the lungs (in the pulmonary veins) contains a higher oxygen and a lower carbon dioxide concentration than the blood delivered to the lungs in the pulmonary artery. This results from the fact that the lungs function to bring the blood into gaseous equilibrium with the air.

Gas exchange between the air and blood occurs entirely by diffusion through lung tissue. This diffusion occurs very rapidly because there is a high surface area within the lungs and a very short diffusion distance between blood and air. The fact that blood in the pulmonary veins and systemic arteries is almost in complete equilibrium with the inspired air is testimony to the high efficiency of normal lung function.

Structure of the Respiratory System

Gas exchange in the lungs occurs across about 300 million tiny (0.25–0.50 mm in diameter) "air sacs," known as **alveoli.** The enormous number of these structures provides a high surface area—60–80 square meters, or about 760 square feet—for diffusion of gases. The diffusion rate is further increased by the fact that each alveolus is only one cell-layer thick, so that the total "air-blood barrier" is only two cells across (an alveolar cell and a capillary endothelial cell), or about 2 μm. This is an average distance because the type II alveolar cells are thicker than the type I cells (fig. 15.1). Where the basement membranes of capillary endothelial cells fuse with those of type I alveolar cells, the diffusion distance is less than 1 μm (fig. 15.2).

Figure 15.2. An electron micrograph of a capillary within the thin interalveolar septum that separates two adjacent alveoli. Note the short distance separating the alveolar space on one side (left, in this figure) from the capillary. RBC = red blood cell; BM = basement membrane; IS = interstitial connective tissue.

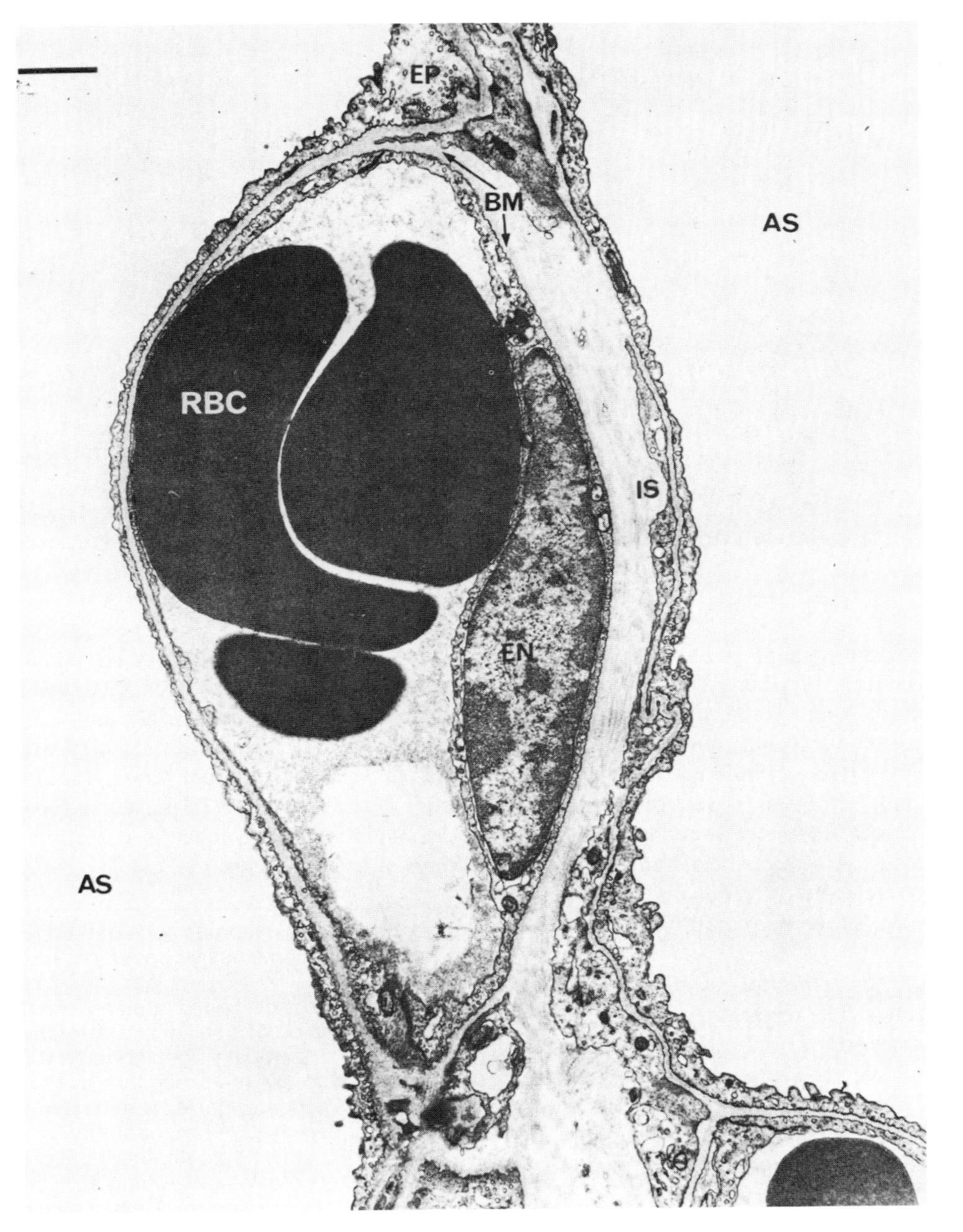

Figure 15.3. (*a*) A scanning electron micrograph showing lung alveoli and a small bronchiole. (*b*) The alveoli under higher power (the arrow points to an alveolar pore through which air can pass from one alveolus to another).

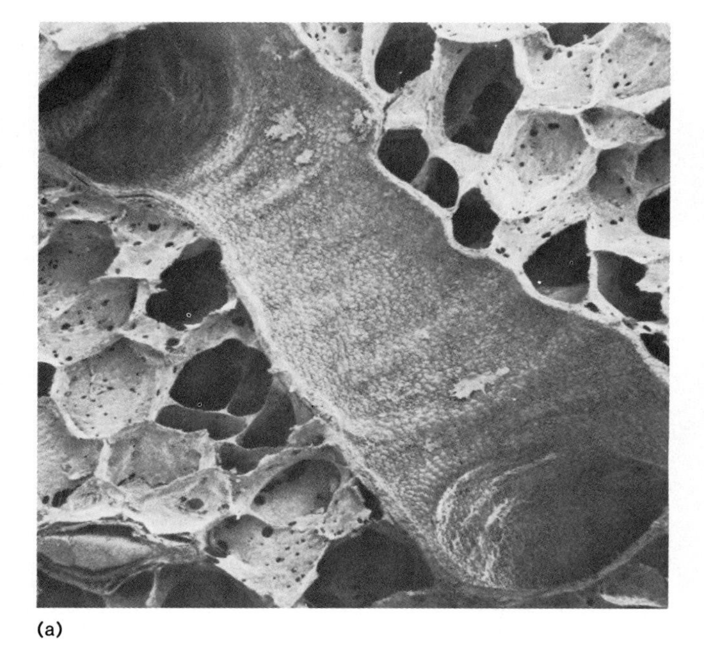

(a)

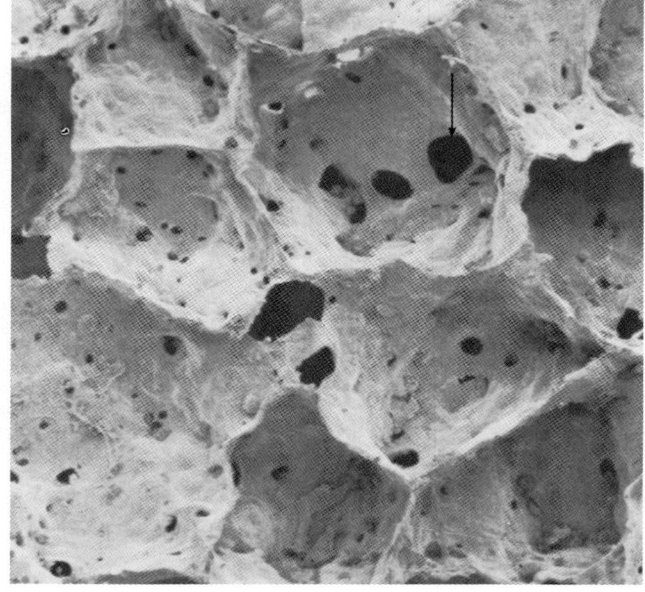

(b)

Alveoli are polyhedral in shape and are usually clustered together, like the units of a honeycomb. Air within one member of a cluster can enter other members through tiny pores. These clusters of alveoli usually occur at the ends of *respiratory bronchioles,* which are the very thin air tubes that end blindly in alveolar sacs. Individual alveoli also occur as separate outpouchings along the length of respiratory bronchioles. Although the distance between each respiratory bronchiole and its terminal alveoli is only about 0.5 mm, these units together comprise most of the mass of the lungs.

The air passages of the respiratory system are divided into two functional zones. The **respiratory zone** is the region where gas exchange occurs, and it therefore includes the respiratory bronchioles (because they contain separate outpouchings of alveoli) and the terminal clusters of alveolar sacs (fig. 15.3). The **conducting zone** includes all of the anatomical structures through which air passes before reaching the respiratory zone (fig. 15.4).

Air enters the respiratory bronchioles from *terminal bronchioles,* which are narrow airways formed from many successive divisions of the right and left *primary bronchi.* These two large air passages, in turn, are continuous with the *trachea,* or windpipe, which is located in the neck in front of the esophagus (a muscular tube carrying food to the stomach). The trachea is a sturdy tube supported by rings of cartilage (fig. 15.5).

If the trachea becomes occluded through inflammation, excessive secretion, trauma, or aspiration of a foreign object, it may be necessary to create an emergency opening into this tube so that ventilation can still occur. A **tracheotomy** is the process of surgically opening the trachea, and a **tracheostomy** is the procedure of inserting a tube into the trachea to permit breathing and to keep the passageway open. A tracheotomy should be performed only by a competent physician, however, as there is great risk of cutting a recurrent laryngeal nerve or the carotid artery, which is also in this area.

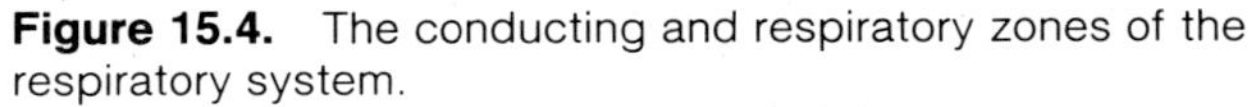

Figure 15.4. The conducting and respiratory zones of the respiratory system.

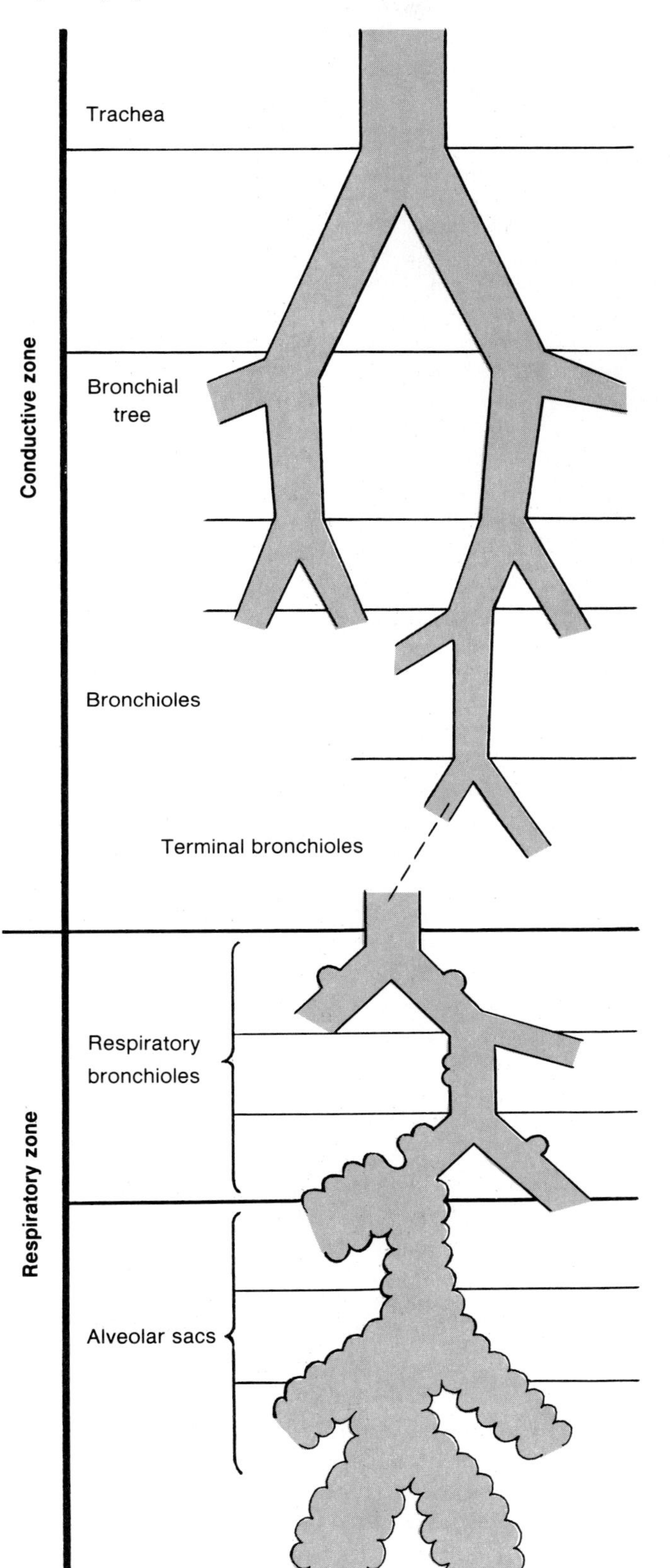

Figure 15.5. (*a*) A plastic cast of the conducting airways from the trachea to the terminal bronchioles. The relationship between these airways and the structure of the lungs is illustrated in (*b*).

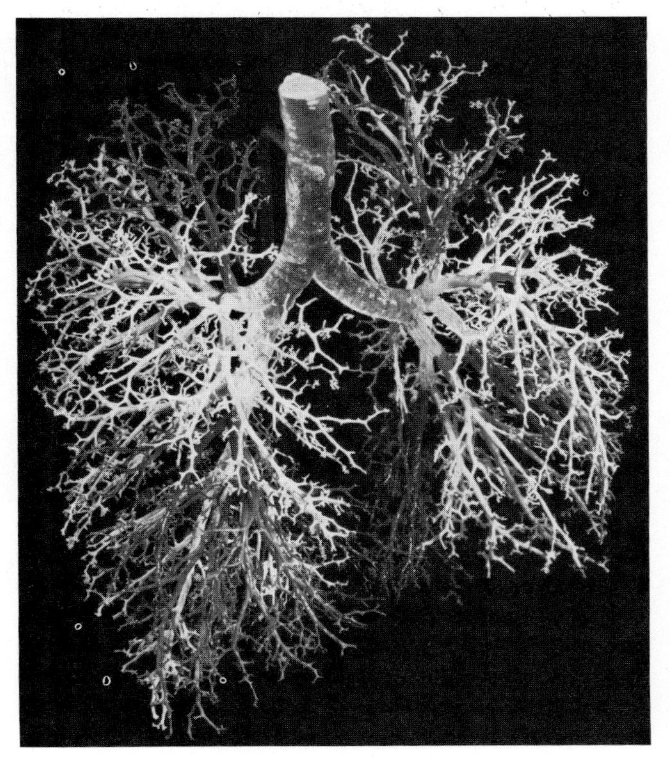

(a)

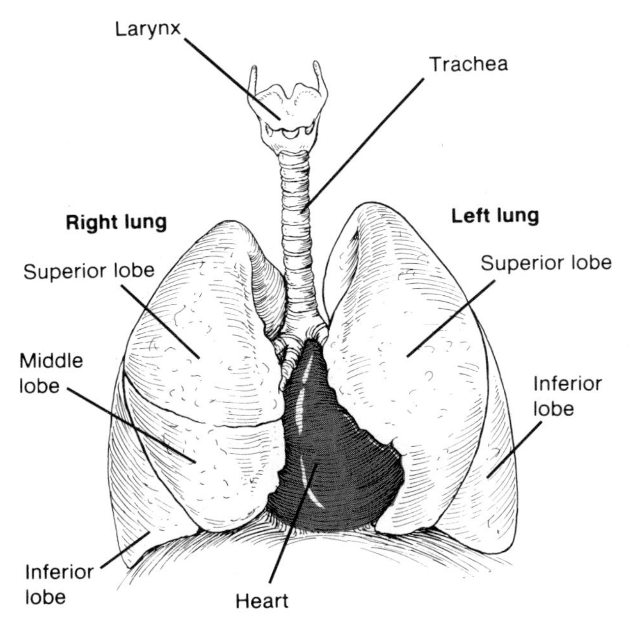

(b)

Figure 15.6. A photograph of the larynx, showing the true and false vocal cords and the glottis.

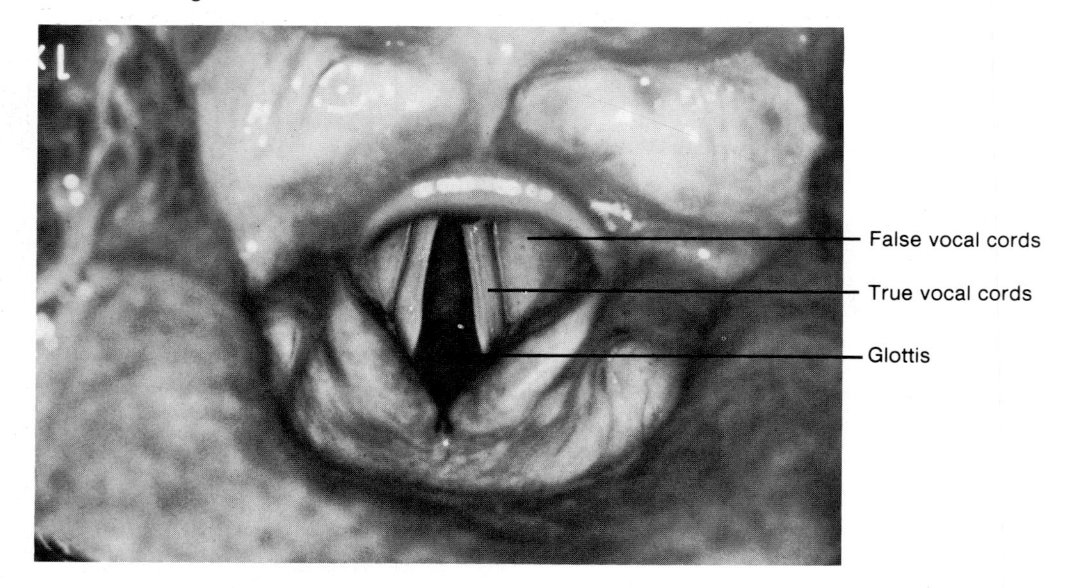

Air enters the trachea from the *pharynx,* which is the cavity located behind the palate that receives the contents of both the oral and nasal passages. In order for air to enter or leave the trachea and lungs, however, it must pass through a valvelike opening called the *glottis* between the vocal cords. The vocal cords are part of the *larynx,* or voice box, which guards the entrance to the trachea (fig. 15.6). The Adam's apple in the neck is produced by a protruding part of the larynx.

The conducting zone of the respiratory system, in summary, consists of the mouth, nose, pharynx, larynx, trachea, primary bronchi, and all successive branchings of the bronchioles up to and including the terminal bronchioles. In addition to conducting air into the respiratory zone, these structures serve two additional functions: *humidification* of the inspired air and *filtration* and *cleaning.*

Regardless of the temperature and humidity of the atmosphere, when the inspired air reaches the respiratory zone it is at a temperature of 37° C (body temperature), and it is saturated with water vapor. The first function is needed to maintain a constant internal body temperature, and the latter function is needed to protect delicate lung tissue from desiccation.

Mucus secreted by cells of the conducting zone serves to trap small particles in the inspired air and thereby performs a filtration function. This mucus is moved along at a rate of 1–2 centimeters per minute by cilia projecting from the tops of epithelial cells that line the conducting zone (fig. 15.7). There are about three hundred cilia per cell that beat in a coordinated fashion to move mucus towards the pharynx, where it can either be swallowed or expectorated.

Figure 15.7. A scanning electron micrograph of a bronchial wall showing cilia, which help to cleanse the lung by moving trapped particles.

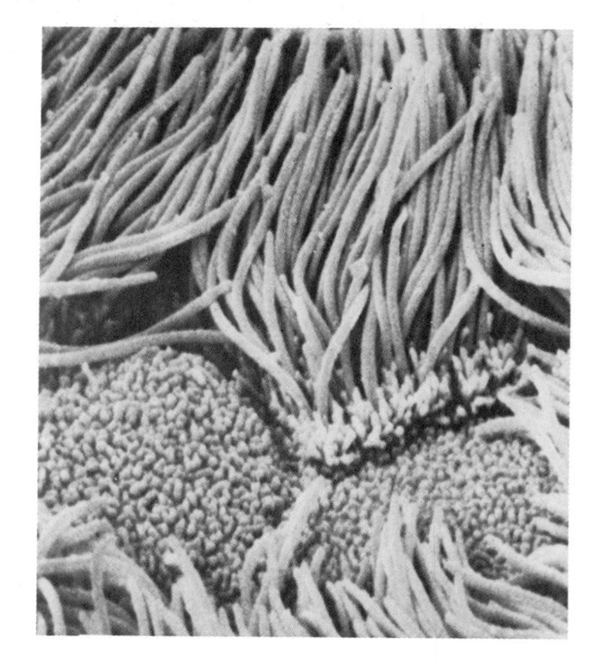

As a result of this filtration function, particles larger than about 6 μm do not normally enter the respiratory zone of the lungs. The importance of this function is evidenced by the disease called *black lung,* which occurs in miners who inhale too much carbon dust and therefore develop pulmonary fibrosis (as described in a later section). The alveoli themselves are normally kept clean by

Figure 15.8. A cross section of the thoracic cavity showing the mediastinum and pleural membranes.

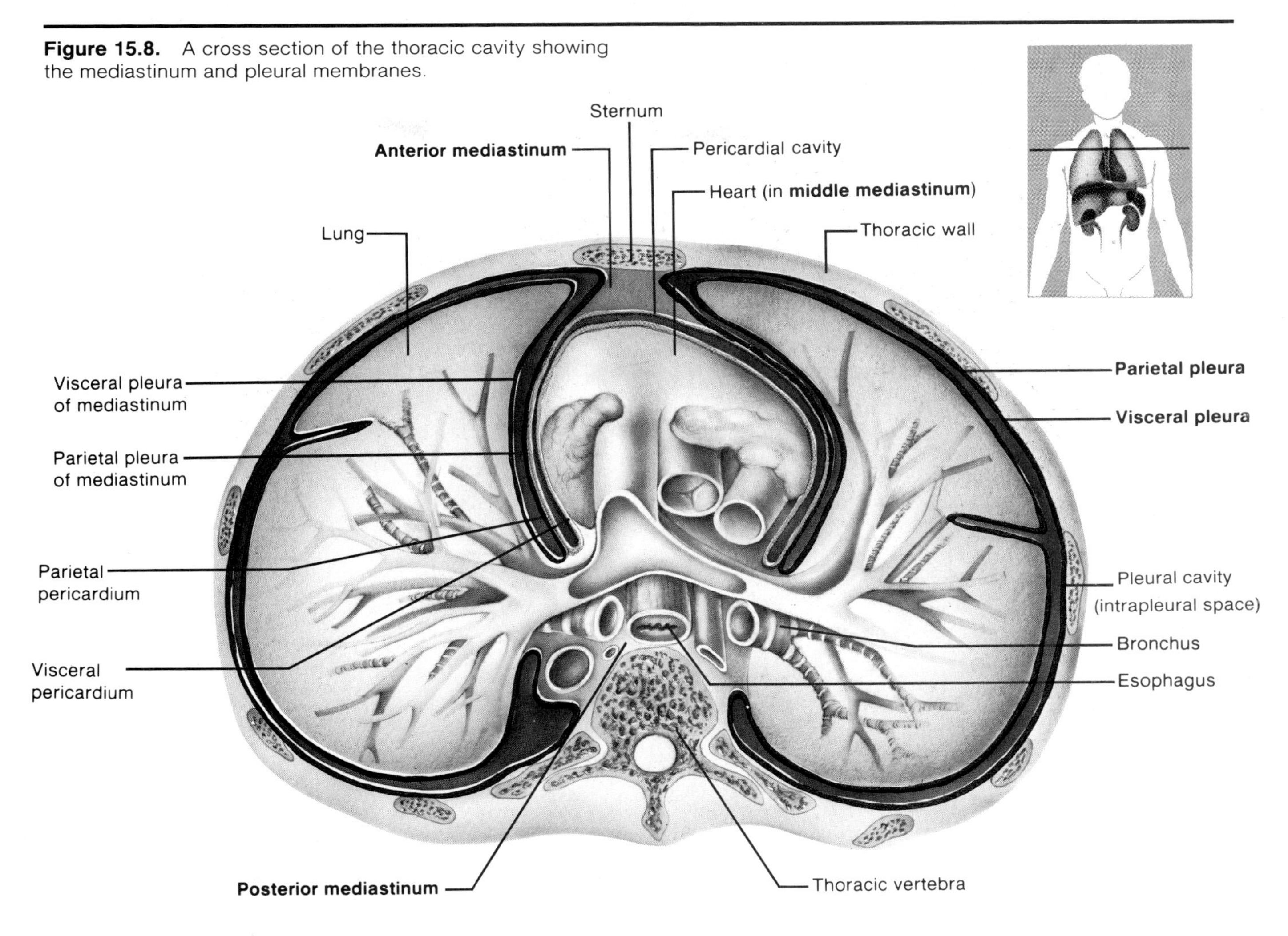

the action of *macrophages* (literally, "big eaters") that reside within them. The cleansing action of cilia and macrophages in the lungs has been shown to be diminished by cigarette smoke.

Thoracic Cavity

The *diaphragm,* a dome-shaped sheet of striated muscle, divides the body cavity into two parts. The area below the diaphragm, or the *abdominal cavity,* contains the liver, pancreas, gastrointestinal tract, spleen, genitourinary tract, and other organs. Above the diaphragm, the chest, or *thoracic cavity,* contains the heart, large blood vessels, trachea, esophagus, and thymus gland in the central region and is filled elsewhere by the right and left lungs.

The structures in the central region—or *mediastinum*—are enveloped by a double layer of wet epithelial membranes, called the *pleural membranes.* One membrane of this double layer is continuous with the *parietal pleural membrane,* a wet epithelial membrane that lines the inside of the thoracic wall. The other layer is continuous with the *visceral pleural membranes* that cover the surface of the lungs (fig. 15.8).

The lungs normally fill the thoracic cavity so that the visceral pleural membranes covering the lungs are pushed against the parietal pleural membrane lining the thoracic wall. There is normally little or no air between the visceral and parietal pleural membranes. There is, however, a "potential space"—called the *intrapleural space*—that can become a real space if a lung collapses. The normal position of the lungs in the thoracic cavity is shown in the radiograph in figure 15.9.

---

1. *Describe the structures involved in gas exchange in the lungs and how this process occurs.*
2. *Describe the structures and functions of the conducting zone of the respiratory system.*
3. *Describe how each lung is packaged separately in pleural membranes, and describe the relationship between the visceral and parietal pleural membranes.*

Figure 15.9. Radiographic (X ray) views of the chest of a normal female (*a*) and a normal male (*b*).

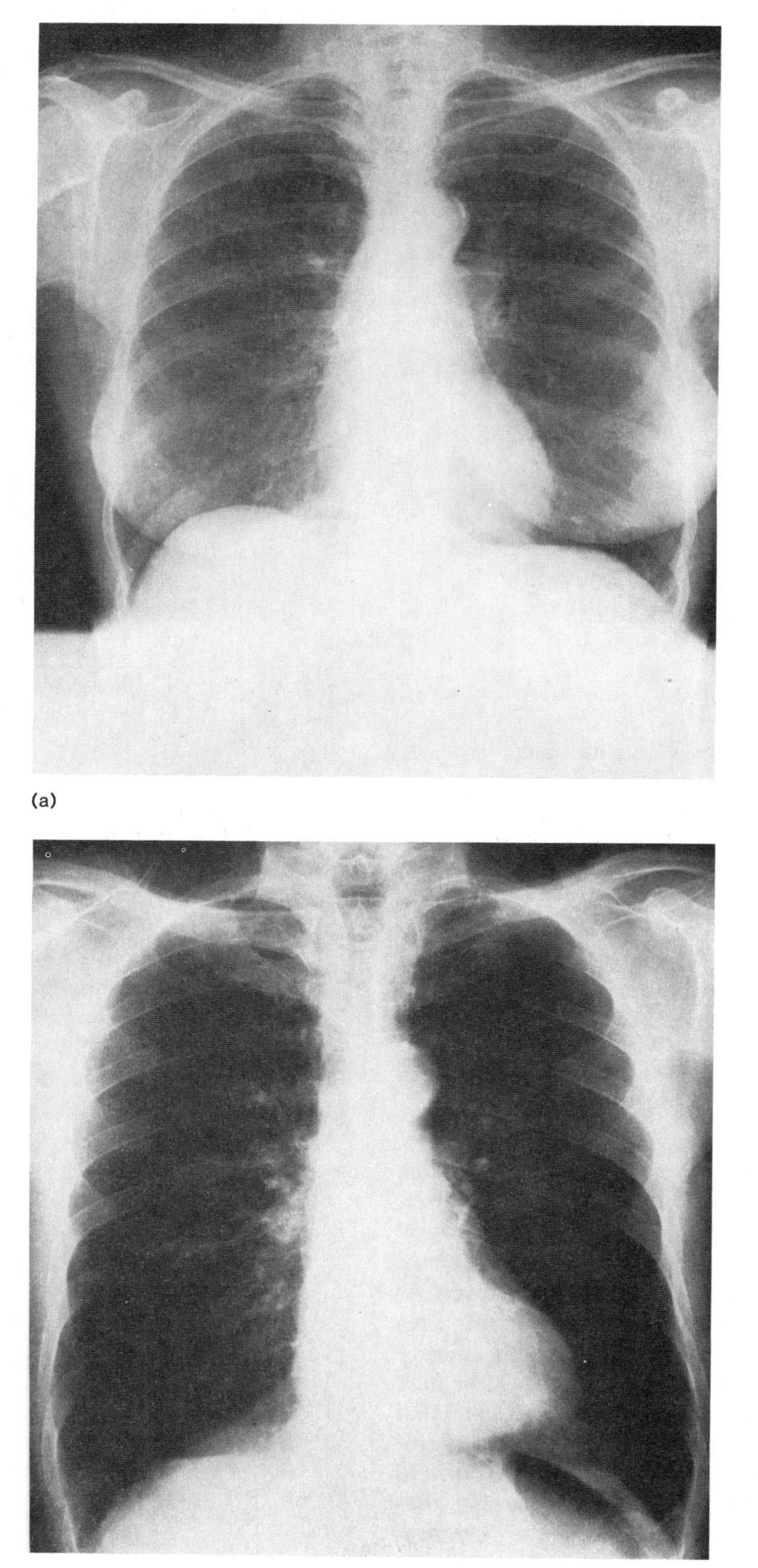

Physical Aspects of Ventilation

Movement of air from the conducting zone to the terminal bronchioles occurs as a result of the pressure difference between the two ends of the airways. Air flow through bronchioles, like blood flow through blood vessels, is directly proportional to the pressure difference and inversely proportional to the frictional resistance to flow. The pressure differences in the pulmonary system are induced by changes in lung volumes. Ventilation is thus influenced by the physical properties of the lungs, including their compliance, elasticity, and surface tension. The wet, serous membranes of the visceral and parietal pleurae are normally against each other, so that the lungs are stuck to the chest wall in the same manner that two wet pieces of glass stick to each other. The *intrapleural space* between the two wet membranes contains only a thin layer of fluid secreted by the pleural membranes. The pleural cavity in a healthy, living organism is thus potential rather than real; it can become real only in abnormal situations when air enters the intrapleural space. Since the lungs normally remain against the chest wall, they get larger and smaller together with the thoracic cavity during respiratory movements.

Intrapulmonary and Intrapleural Pressures

Air enters the lungs during inspiration because the atmospheric pressure is greater than the **intrapulmonary,** or **intra-alveolar, pressure.** Since the atmospheric pressure does not usually change, the intrapulmonary pressure must fall below atmospheric pressure to cause inspiration. A pressure below that of the atmosphere is called a *subatmospheric pressure,* or *negative pressure.* During quiet inspiration, for example, the intrapulmonary pressure may become 3 mm Hg less than the pressure of the atmosphere. This subatmospheric pressure is commonly shown as −3 mm Hg. Expiration, conversely, occurs when the intrapulmonary pressure is greater than the atmospheric pressure. During quiet expiration, for example, the intrapulmonary pressure may rise to at least +3 mm Hg over the atmospheric pressure.

The lack of air in the intrapleural space produces a subatmospheric **intrapleural pressure,** which is lower than the intrapulmonary pressure (table 15.1). There is thus a pressure difference across the wall of the lung—called the **transpulmonary pressure**—which is the difference between the intrapulmonary pressure and the intrapleural pressure. Since the pressure within the lungs (intrapulmonary pressure) is greater than outside the lungs (intrapleural pressure), the difference in pressure (transpulmonary pressure) acts to expand the lungs as the thoracic volume expands during inspiration.

Table 15.1 Intrapulmonary and intrapleural pressures in normal, quiet breathing, and the transpulmonary pressure acting to expand the lungs.

	Inspiration	Expiration
Intrapulmonary pressure (mm Hg)	−3 (to zero)	+3 (to zero)
Intrapleural pressure (mm Hg)	−6	−3
Transpulmonary pressure (mm Hg)	+3	+6

Note: Pressures indicate mm Hg below or above atmospheric pressure. Intrapleural pressure is normally always negative (subatmospheric).

Boyle's Law. Changes in intrapulmonary pressure occur as a result of changes in lung volume. This follows from **Boyle's law,** which states that *the pressure of a gas is inversely proportional to its volume.* An increase in lung volume during inspiration decreases intrapulmonary pressure to subatmospheric levels; air therefore goes in. A decrease in lung volume raises the intrapulmonary pressure above that of the atmosphere, thus pushing air out. These changes in lung volume occur as a consequence of changes in thoracic volume produced by muscle contraction.

Physical Properties of the Lungs

In order for inspiration to occur, the lungs must be able to expand when stretched; they must have high *compliance.* In order for expiration to occur, the lungs must get smaller when this stretching force is released; they must have *elasticity.* The tendency to get smaller is also aided by *surface tension* forces within the alveoli.

Compliance. The lungs are very distensible—they are, in fact, about one hundred times more distensible than a toy balloon. Another term for distensibility is **compliance,** which is defined as *the change in lung volume per change in transpulmonary pressure.* A given transpulmonary pressure, in other words, will cause greater or lesser expansion, depending on the compliance of the lungs.

The compliance of the lungs is reduced whenever there is more resistance to distension. If the lungs were filled with concrete (as an extreme example), a given transpulmonary pressure would produce no increase in lung volume and no air would enter; the compliance would be zero. The infiltration of lung tissue with connective tissue proteins, a condition called *pulmonary fibrosis,* similarly decreases lung compliance (fig. 15.10). In *emphysema,* in which alveolar tissue is destroyed, the lungs are less resistant to distension and have a greater compliance (if no fibrosis is present).

Figure 15.10. The volume of air inspired as a function of the transpulmonary pressure. The slope of each line (change in volume per unit change in pressure) in the linear regions is called the *compliance* and is a measure of the distensibility of the lungs.

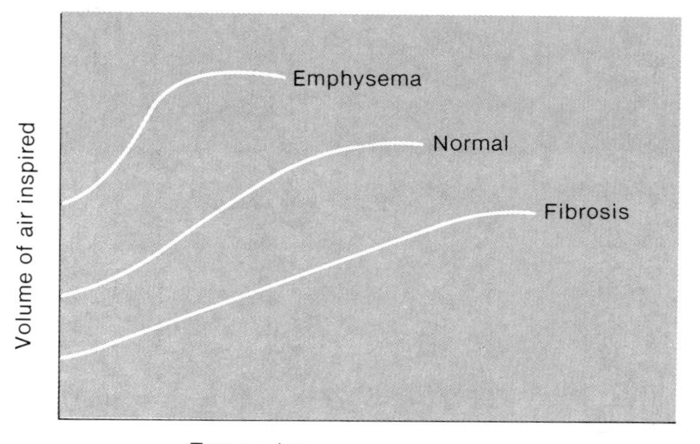

Elasticity. The term **elasticity** refers to the tendency of a structure to return to its initial size after being distended. The lungs are very elastic, due to a high content of elastin proteins, and resist distension. Since the lungs are normally stuck to the chest wall, they are always in a state of elastic tension. This tension increases during inspiration when the lungs are stretched and is reduced by elastic recoil during expiration. The elasticity of the lungs and of other thoracic structures thus aids in pushing the air out during expiration.

The elastic nature of lung tissue is revealed when air enters the intrapleural space (as a result of an open chest wound, for example). This condition is called a **pneumothorax,** which is shown in figure 15.11. As air enters the intrapleural space, the intrapleural pressure rises until it is equal to the atmospheric pressure. When the intrapleural pressure is the same as the intrapulmonary pressure the lung can no longer expand. Not only does the lung not expand during inspiration, it actually collapses away from the chest wall as a result of elastic recoil. Fortunately, a pneumothorax usually occurs only in one lung since each lung is in a separate pleural compartment.

Figure 15.11. A pneumothorax of the right lung. The right side of the thorax appears uniformly dark because it is filled with air; the space between the ribs is also greater than on the left due to release from the elastic tension of the lungs. The left lung appears denser (less dark) because of shunting of blood from the right to the left lung.

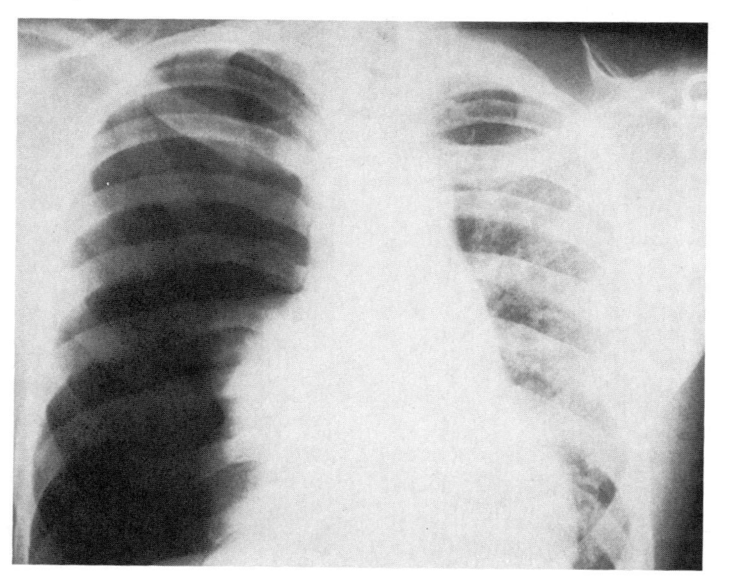

Figure 15.12. Water molecules at the surface have a greater attraction for other water molecules than for air. The surface molecules are thus attracted to each other and pulled tightly together by the attractive forces of water underneath. This produces surface tension.

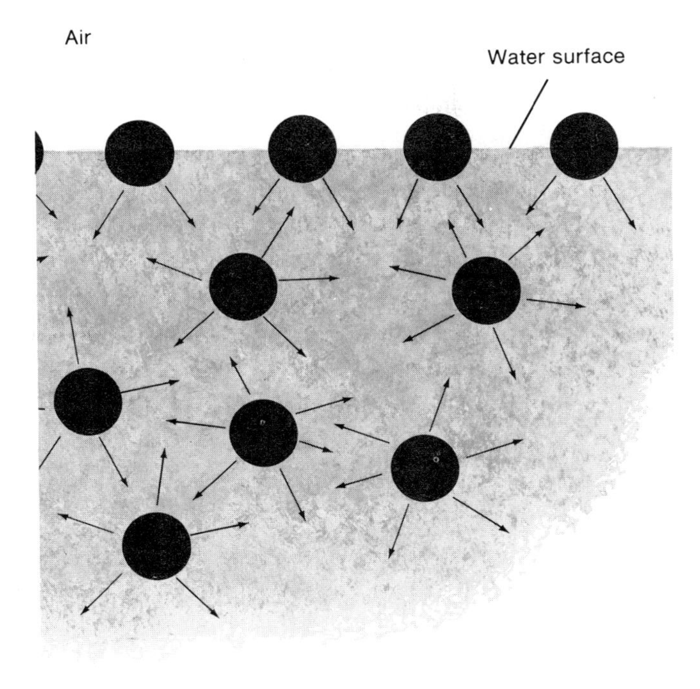

Figure 15.13. According to the Law of LaPlace, the pressure created by surface tension should be greater in the smaller alveolus (*right*) than in the larger alveolus (*left*). This implies that (without surfactant) smaller alveoli would collapse and empty their air into larger alveoli.

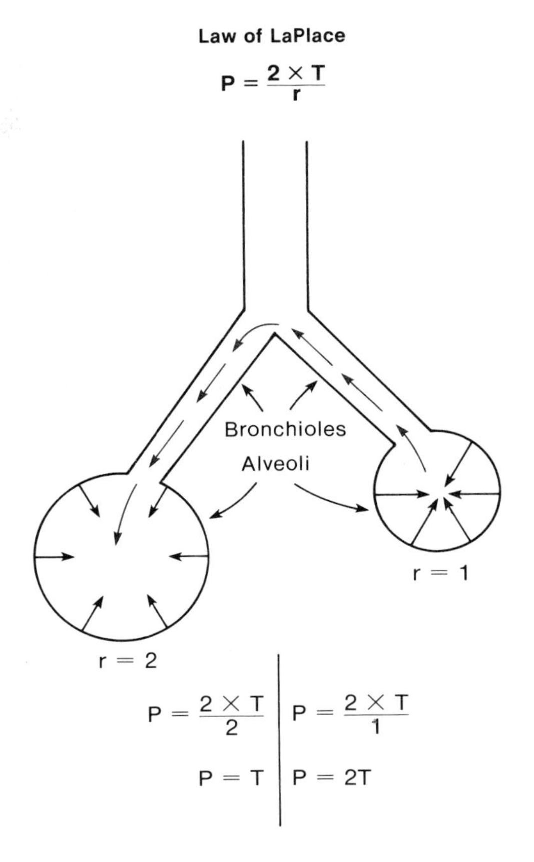

Surface Tension. The forces that act to resist distension include elastic resistance and **surface tension** that is exerted by fluid in the alveoli. Although the alveoli are relatively dry, they do contain a very thin film of fluid, much like soap bubbles. Surface tension is created by the fact that water molecules at the surface are attracted more to other water molecules than to air. As a result, the surface water molecules are pulled tightly together by attractive forces from underneath (fig. 15.12).

The surface tension of an alveolus produces a force that is directed inward and, as a result, creates pressure within the alveolus. As described by the **Law of LaPlace,** the pressure thus created is directly proportional to the surface tension and *inversely proportional to the radius* of the alveolus. According to this law, the pressure in a smaller alveolus would be greater than in a larger alveolus if the surface tension is the same in both. The greater pressure of the smaller alveolus would then cause it to empty its air into the larger one (fig. 15.13). This does not normally occur because as an alveolus decreases in size, its surface tension as well as its radius is reduced. The cause of the reduced surface tension is described in the next section.

Surfactant and the Respiratory Distress Syndrome

Alveolar fluid contains a phospholipid known as dipalmitoyl lecithin, which is probably attached to a protein and functions to lower surface tension. This compound is called **lung surfactant,** which is a contraction of the term *surface active agent.* Because of the presence of surfactant, the surface tension in the alveoli is lower at any given lung volume than would be predicted if surfactant were absent. Further, the ability of surfactant to lower surface tension improves as the alveoli get smaller during expiration. Surfactant thus helps to prevent the alveoli from collapsing as a result of surface tension. Even after a forceful expiration, the alveoli remain open and a *residual volume* of air remains in the lungs. Since the alveoli do not collapse, less surface tension has to be overcome to inflate them at the next inspiration.

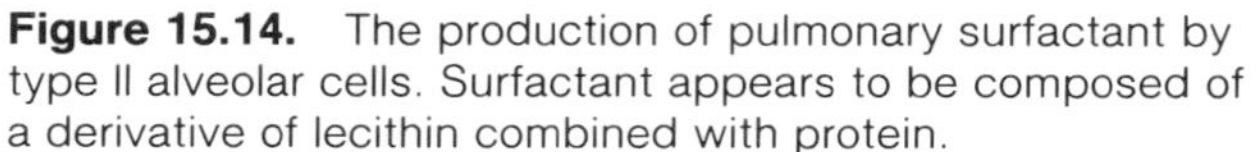

Figure 15.14. The production of pulmonary surfactant by type II alveolar cells. Surfactant appears to be composed of a derivative of lecithin combined with protein.

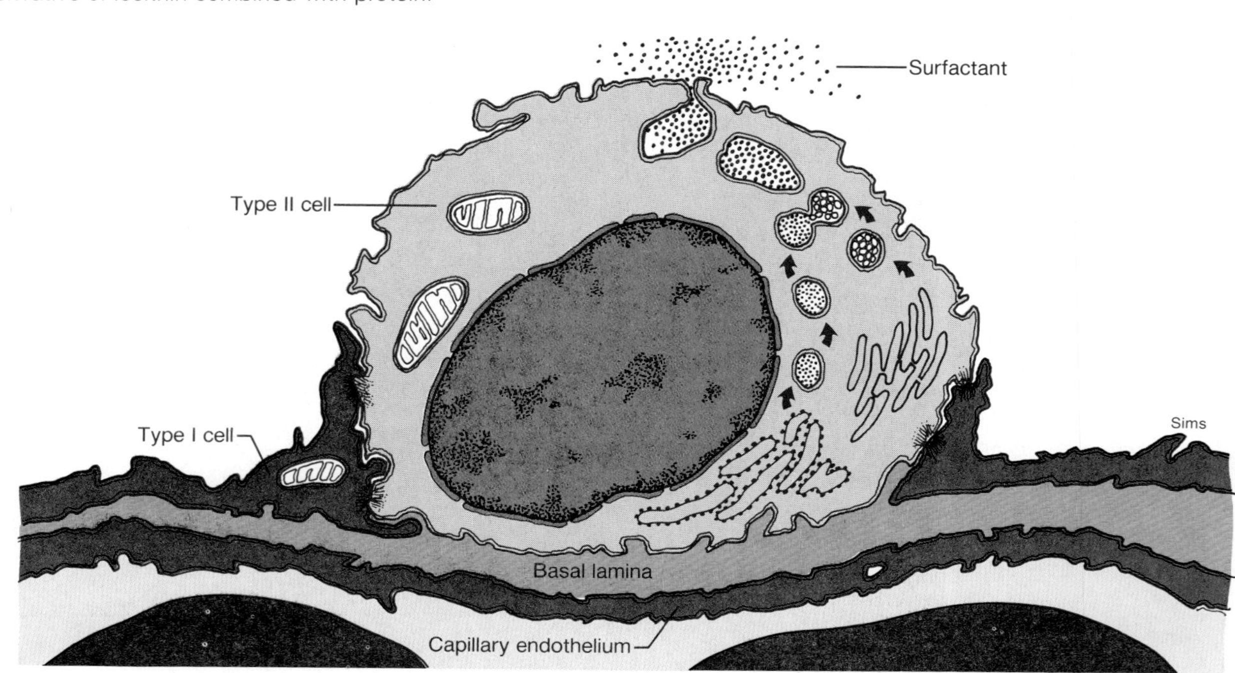

Surfactant is produced by type II alveolar cells (fig. 15.14) in late fetal life. Since surfactant does not start to be produced until about the eighth month, premature babies are sometimes born with lungs that lack sufficient surfactant, and their alveoli are collapsed as a result. This condition is called **respiratory distress syndrome.** It is also called **hyaline membrane disease,** because the high surface tension causes plasma fluid to leak into the alveoli, producing a glistening "membrane" appearance (and pulmonary edema). This condition does not occur in all premature babies; the rate of lung development depends on hormonal conditions (thyroxine and hydrocortisone primarily) and on genetic factors.

Even under normal conditions, the first breath of life is a difficult one because the newborn must overcome great surface tension forces in order to inflate its partially collapsed alveoli. The transpulmonary pressure required for the first breath is fifteen to twenty times that required for subsequent breaths, and an infant with respiratory distress syndrome must duplicate this effort with every breath. Fortunately, many babies with this condition can be saved by mechanical ventilators that keep them alive long enough for their lungs to mature and manufacture sufficient surfactant.

1. *Describe how the intrapulmonary and intrapleural pressures change during inspiration and expiration, and explain the reasons for these changes in terms of Boyle's law.*
2. *Define the terms* compliance *and* elasticity, *and explain how these lung properties affect inspiration and expiration.*
3. *Describe lung surfactant, and explain why the alveoli would collapse in the absence of surfactant.*

Mechanics of Breathing

Breathing, or **pulmonary ventilation,** refers to the movement of air into and out of the respiratory system. This movement of air occurs as a result of differences between the atmospheric and the intrapulmonary pressures; air goes in when the intrapulmonary pressure is subatmospheric, and air goes out when the intrapulmonary pressure rises above that of the atmosphere. As previously discussed, these pressure changes within the lungs occur as a consequence of changes in thoracic volume.

The thorax must be semirigid yet flexible. It must be sufficiently rigid so that it can protect vital organs and provide attachments for many short, powerful muscles. It must be flexible to function as a bellows during the ventilation cycle. The rigidity and the surfaces for muscle attachment are provided by the bony composition of the rib cage. The rib cage is pliable, however, because the ribs

Figure 15.15. A change in lung volume, as shown by radiographs, during expiration (*a*) and inspiration (*b*). The increase in lung volume during full inspiration is shown by comparison with the lung volume in full expiration (*dashed lines*).

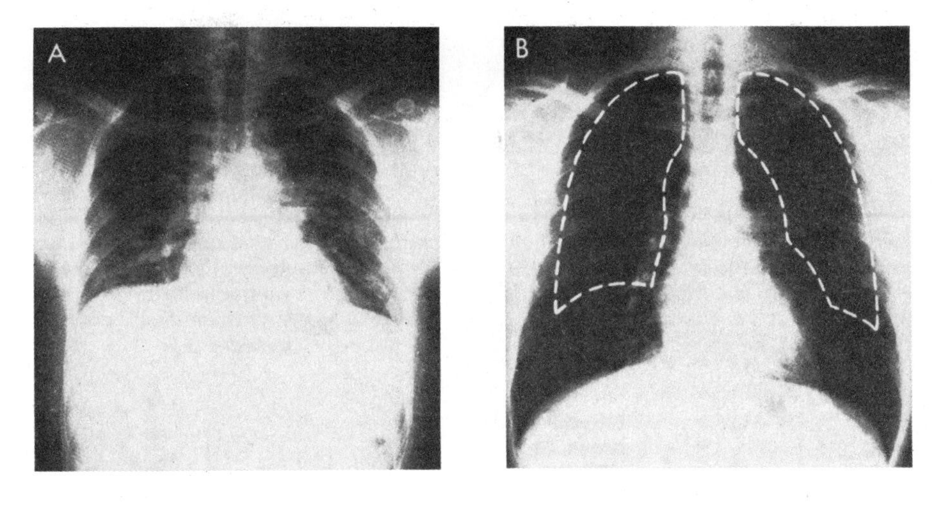

Figure 15.16. The position of the principal muscles of inspiration.

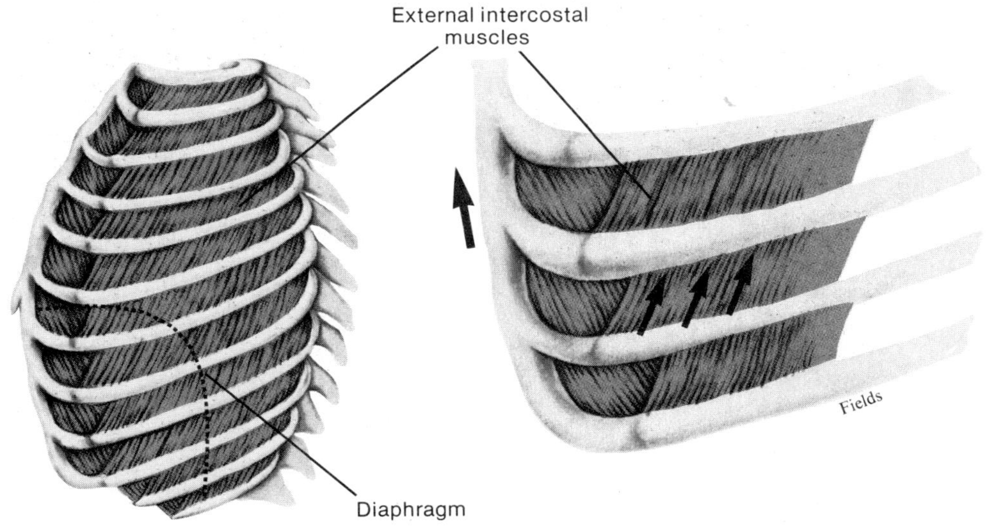

are separate from one another and because most ribs (the upper ten of the twelve pairs) are attached to the sternum by resilient costal cartilages. The vertebral attachments likewise provide considerable mobility. The structure of the rib cage and associated cartilages provide continuous elastic tension, so that when stretched by muscle contraction during inspiration, the rib cage can return passively to its resting dimensions when the muscles relax. This elastic recoil is greatly aided by the elasticity of the lungs.

Pulmonary ventilation consists of two phases, inspiration and expiration. Inspiration (inhalation) and expiration (exhalation) are accomplished by alternately increasing and decreasing the volumes of the thorax and lungs (fig. 15.15).

Inspiration and Expiration

The thoracic cavity increases in size during inspiration in three directions: anteroposteriorly, laterally, and vertically. This is accomplished by contractions of the **diaphragm** and the **external intercostal** muscles (fig. 15.16). A contraction of the dome-shaped diaphragm downward increases the thoracic volume vertically. A simultaneous contraction of the external intercostals increases the lateral and anteroposterior dimensions of the thorax.

Other thoracic muscles become involved in forced (deep) inspiration. The most important of these is the scalenus, followed by the pectoralis minor, and in extreme

Table 15.2 Summary of the mechanisms involved in normal, quiet ventilation and forced ventilation.

	Inspiration	Expiration
Normal, quiet breathing	Contraction of the diaphragm and external intercostal muscles increases the thoracic and lung volume, decreasing intrapulmonary pressure to about −3 mm Hg.	Relaxation of the diaphragm and external intercostals, plus elastic recoil of lungs, decreases lung volume and increases intrapulmonary pressure to about +3 mm Hg.
Forced ventilation	Inspiration, aided by contraction of accessory muscles such as the scalenes and sternocleidomastoid, decreases intrapulmonary pressure to −20 mm Hg or less.	Expiration, aided by contraction of abdominal muscles and internal intercostal muscles, increases intrapulmonary pressure to +30 mm Hg or more.

Figure 15.17. The mechanics of pulmonary ventilation. (*a*) The lungs and pleural cavities prior to inspiration. (*b*) During inspiration, the diaphragm is contracted and the rib cage elevated; intrapleural pressure is reduced and air inflates the lungs. (*c*) Inspiration is completed as intrapulmonary pressure is equal to atmospheric pressure. (*d*) As the muscles of inspiration are relaxed, the rib cage recoils and the intrapleural and intrapulmonary pressures are raised until the intrapulmonary pressure equals the atmospheric pressure (*a*).

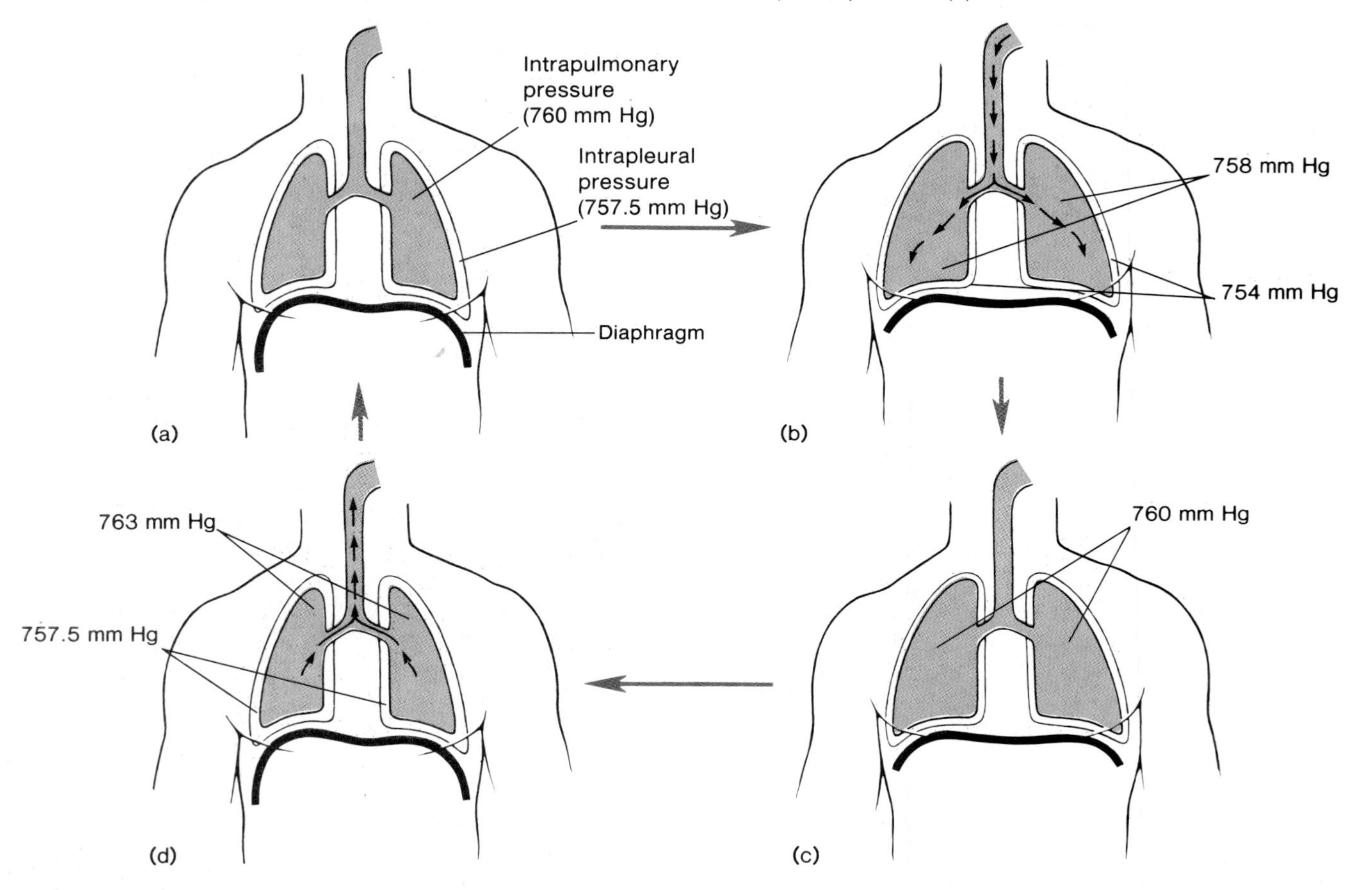

cases the sternocleidomastoid. Contraction of these muscles elevates the ribs in an anteroposterior direction, while at the same time the upper rib cage is stabilized so that the intercostals become more effective.

Quiet expiration is a passive process. After becoming stretched by contractions of the diaphragm and thoracic muscles, the thorax and lungs recoil as a result of their elastic tension when the respiratory muscles relax. The decrease in lung volume raises the pressure within the alveoli above the atmospheric pressure and pushes the air out.

During forced expiration the *internal intercostal* muscles contract and depress the rib cage. The abdominal muscles may also aid expiration, because when they contract they force abdominal organs up against the diaphragm and further decrease the volume of the thorax. By this means the intrapulmonary pressure can rise 20 or 30 mm Hg above the atmospheric pressure. The events that occur during inspiration and expiration are summarized in table 15.2 and shown in figure 15.17.

Figure 15.18. A spirometer (Collins 9L respirometer) used to measure lung volumes and capacities.

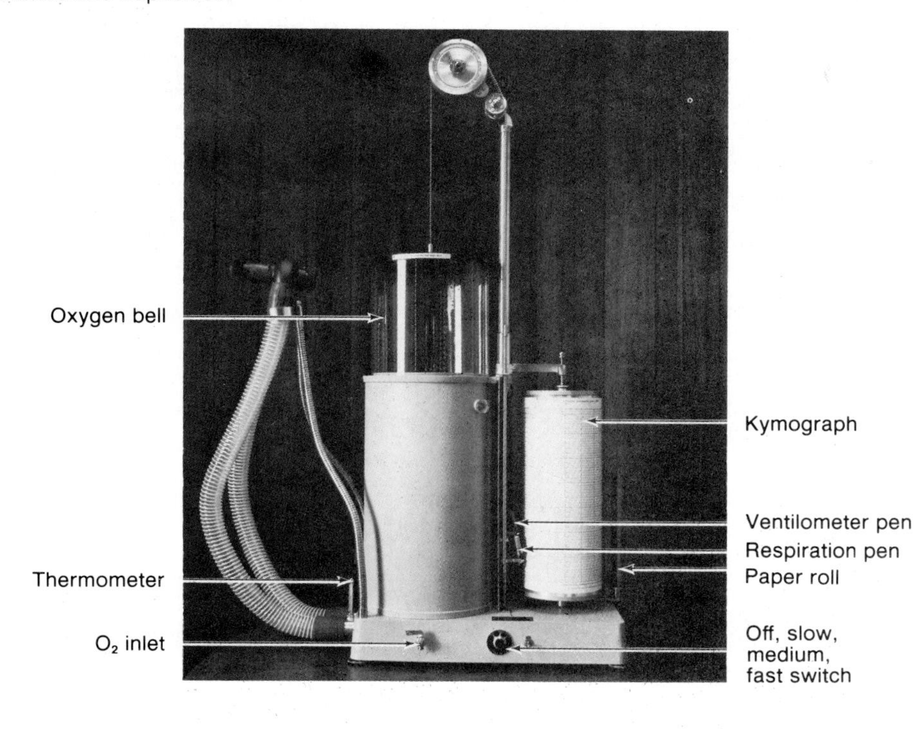

Figure 15.19. A spirogram showing lung volumes and capacities at rest.

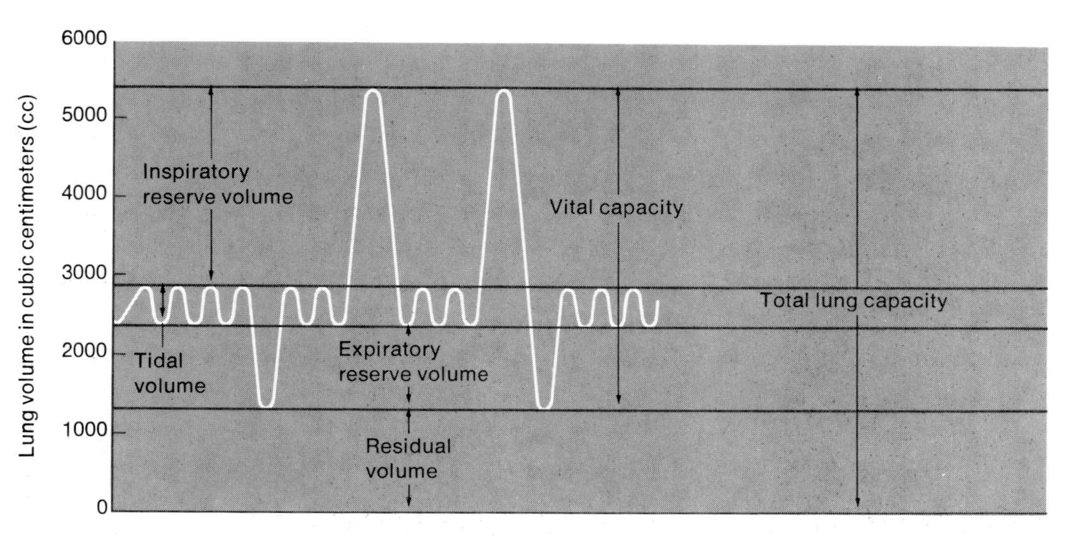

Pulmonary Function Tests

Pulmonary function may be assessed clinically by means of a technique known as *spirometry.* In this procedure, a subject breathes in a closed system in which air is trapped within a light plastic bell floating in water. The bell moves up when the subject exhales and down when the subject inhales. The movements of the bell cause corresponding movements of a pen, which traces a record of the breathing on a rotating drum recorder (fig. 15.18).

The appearance of a spirogram taken during quiet, resting breathing is shown in figure 15.19, and the various lung volumes and capacities are defined in table 15.3. During quiet breathing, the amount of air expired in each breath (the **tidal volume**) is equal to about 500 ml. This is about 12% of the **vital capacity** (the maximum amount of air that can be expired after a maximum inspiration). Multiplying the tidal volume at rest times the number of breaths per minute yields a **total minute volume** of about

Table 15.3 Definitions of terms used to describe lung volumes and capacities.

Term	Definition
Lung volumes	The four nonoverlapping components of the total lung capacity
Tidal volume	The volume of gas inspired or expired in an unforced respiratory cycle
Inspiratory reserve volume	The maximum volume of gas that can be inspired from the end of a tidal inspiration
Expiratory reserve volume	The maximum volume of gas that can be expired from the end of a tidal expiration
Residual volume	The volume of gas remaining in the lungs after a maximum expiration
Lung capacities	Measurements that are the sum of two or more lung volumes
Total lung capacity	The total amount of gas in the lungs at the end of a maximum inspiration
Vital capacity	The maximum amount of gas that can be expired after a maximum inspiration
Inspiratory capacity	The maximum amount of gas that can be inspired at the end of a tidal expiration
Functional residual capacity	The amount of gas remaining in the lungs at the end of a tidal expiration

Figure 15.20. An illustration of the one-second forced expiratory volume ($FEV_{1.0}$) spirometry test for detecting obstructive pulmonary disorders.

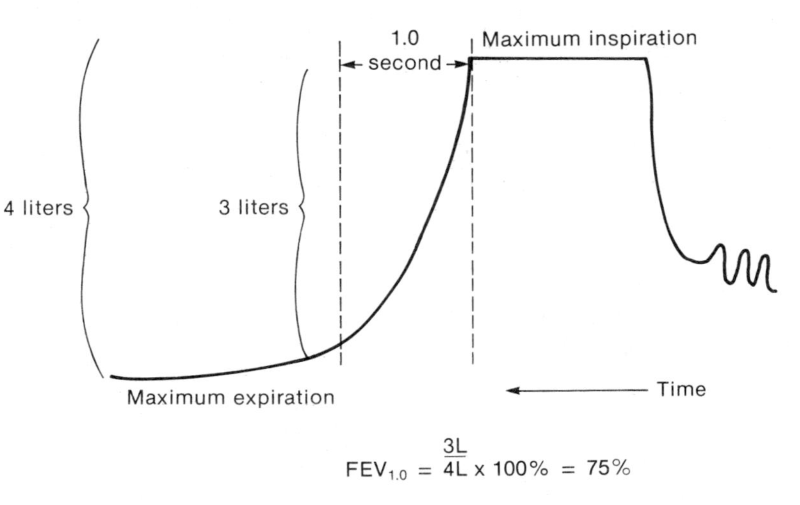

$$FEV_{1.0} = \frac{3L}{4L} \times 100\% = 75\%$$

6 L per minute. During exercise, the tidal volume can increase to as much as 50% of the vital capacity, and the number of breaths per minute likewise increases to produce a total minute volume as high as 100–200 L per minute.

Spirometry is useful in the diagnosis of lung diseases. In purely *restrictive disorders,* such as pulmonary fibrosis, the vital capacity is reduced below normal. In disorders that are only obstructive, such as asthma, the vital capacity is normal because lung tissue is not damaged. Bronchoconstriction and the increased airway resistance that results, however, makes expiration more difficult and takes a longer time. *Obstructive disorders* are thus diagnosed by tests that measure the rate of expiration. One such test is the **forced expiratory volume (FEV),** in which the percent of the vital capacity that can be exhaled in the first second ($FEV_{1.0}$), second ($FEV_{2.0}$), and third ($FEV_{3.0}$) seconds is measured (fig. 15.20). An $FEV_{1.0}$ that is significantly less than 75% to 85% suggests the presence of obstructive pulmonary disease.

Bronchoconstriction often occurs in response to the inhalation of noxious agents in the air, such as from smoke or smog. The $FEV_{1.0}$ has, therefore, been used by researchers to determine the effects of various components of smog and of passive cigarette smoke inhalation on pulmonary function. These studies have shown that it is unhealthy to exercise on very smoggy days and that inhalation of smoke from other people's cigarettes in a closed environment can measurably affect pulmonary function.

Pulmonary Disorders

People with pulmonary disorders frequently complain of **dyspnea,** which is a subjective feeling of "shortness of breath." Dyspnea may occur even when ventilation is normal, however, and may not occur even when total minute volume is very high as in exercise. Some of the terms used to define ventilation are defined in table 15.4.

Table 15.4 Definitions of some terms used to describe ventilation.

Term	Definition
Air spaces	Alveolar ducts, alveolar sacs, and alveoli
Airways	Structures that conduct air from the mouth and nose to the respiratory bronchioles
Alveolar ventilation	Removal and replacement of gas in alveoli; equal to the tidal volume minus the volume of dead space
Anatomical dead space	Volume of the conducting airways to the zone where gas exchange occurs
Apnea	Cessation of breathing
Dyspnea	Unpleasant, subjective feeling of difficult or labored breathing
Eupnea	Normal, comfortable breathing at rest
Hyperventilation	Alveolar ventilation that is excessive in relation to metabolic rate; results in abnormally low alveolar CO_2
Hypoventilation	An alveolar ventilation that is low in relation to metabolic rate; results in abnormally high alveolar CO_2
Physiological dead space	Combination of anatomical dead space and underventilated or underperfused alveoli that do not contribute normally to blood-gas exchange
Pneumothorax	Presence of gas in the pleural space (the space between the visceral and parietal pleural membranes) causing lung collapse
Torr	Synonymous with millimeters of mercury (760 mm Hg = 760 torr)

Asthma. The obstruction of air flow through the bronchioles may occur as a result of excessive mucus secretion, inflammation, and/or contraction of the smooth muscles in the bronchioles. **Asthma** results from bronchiolar constriction, which increases airway resistance and makes breathing difficult. Constriction of the bronchiolar smooth muscles is stimulated by leukotrienes (molecules related to prostaglandins) and to a lesser degree by histamine, released by leukocytes and mast cells. This can be provoked by an allergic reaction or by the release of acetylcholine from parasympathetic nerve endings. There is evidence that some people with asthma may have a decreased sensitivity to the beta-adrenergic effects of epinephrine, which stimulate bronchodilation; the effects of chemicals that promote bronchoconstriction may therefore be enhanced.

Epinephrine and related compounds stimulate beta-adrenergic receptors in the bronchioles and by this means promote bronchodilation. Asthmatics can, therefore, take epinephrine as an inhaled spray to relieve the symptoms of an asthma attack. It has been learned that there are two subtypes of beta receptors for epinephrine and that the subtype in the heart (called beta-1) is different from the one in the bronchioles (beta-2). By utilizing these differences, compounds have been developed that can more selectively stimulate the beta-2 adrenergic receptors and cause bronchodilation, without having as great an effect on the heart as epinephrine does.

Emphysema. Alveolar tissue is destroyed in emphysema, resulting in fewer but larger alveoli (fig. 15.21). This produces a decreased surface area for gas exchange and a decreased ability of the bronchioles to remain open during expiration. Collapse of the bronchioles as a result

Figure 15.21. Photomicrographs of tissue from a normal lung and from the lung of a person with emphysema. In emphysema, lung tissue is destroyed, resulting in the presence of fewer and larger alveoli.

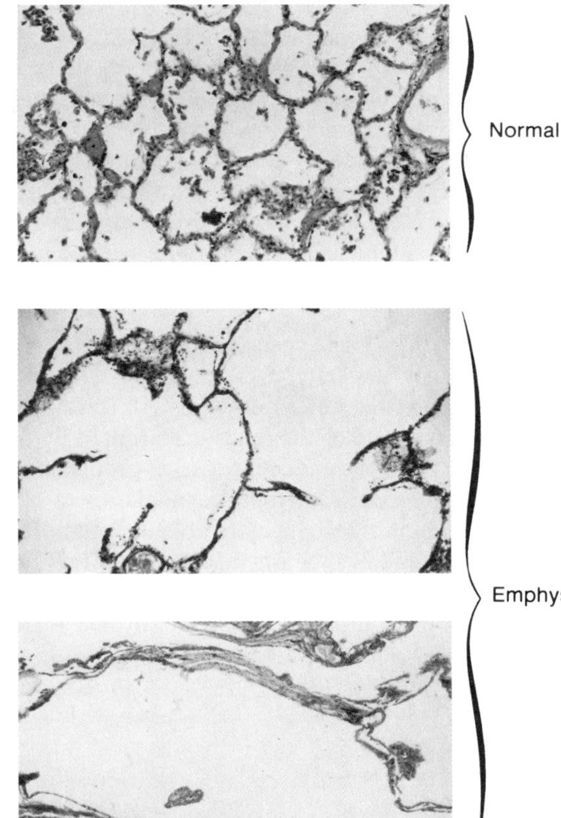

Table 15.5 Pulmonary disorders caused by inhalation of noxious agents containing particles less than 6 μm in size.

Condition	Agent	Pulmonary Disorders
Silicosis	Silica	Fibrosis, obstructive emphysema, tuberculosis, cor pulmonale
Anthracosis (Black lung)	Coal dust	Similar to silicosis (most coal dust contains silica)
Asbestosis	Asbestos fibers	Diffuse fibrosis, cor pulmonale, pulmonary malignancy

of the compression of the lungs during expiration produces *air trapping,* which further decreases the efficiency of gas exchange in the alveoli.

There are different types of emphysema. The most common type occurs almost exclusively in people who have smoked cigarettes heavily over a period of years. A component of cigarette smoke apparently stimulates the macrophages and leukocytes to secrete proteolytic (protein-digesting) enzymes that destroy lung tissues. A less common type of emphysema results from the genetic inability to produce a plasma protein called alpha-1-antitrypsin. This protein normally inhibits proteolytic enzymes such as trypsin and thus normally protects the lungs against the effects of enzymes that are released from alveolar macrophages.

Chronic bronchitis and emphysema, the most common causes of respiratory failure, together are called **chronic obstructive pulmonary disease (COPD).** In addition to the more direct obstructive and restrictive aspects of these conditions, other pathological changes may occur. These include edema, inflammation, hyperplasia (increased cell number), zones of pulmonary fibrosis, pneumonia, pulmonary emboli (traveling blood clots), and heart failure. Patients with severe emphysema may eventually develop *cor pulmonale*—pulmonary hypertension with hypertrophy and the eventual failure of the right ventricle.

Pulmonary Fibrosis. Under certain conditions, for reasons that are poorly understood, lung damage leads to pulmonary fibrosis instead of emphysema. In this condition the normal structure of the lungs is disrupted by the accumulation of fibrous connective tissue proteins. Fibrosis can result, for example, from the inhalation of particles less than 6 μm in size, which are able to accumulate in the respiratory zone of the lungs. Included in this category is *anthracosis,* or black lung, which is produced by the inhalation of carbon particles from coal dust (table 15.5).

1. *Describe the actions of the diaphragm and external intercostal muscles during inspiration, and explain how quiet expiration is produced.*
2. *Describe how forced inspiration and forced expiration are produced.*
3. *Define the terms* tidal volume *and* vital capacity. *Describe how the total minute volume is calculated, and explain how this value changes during exercise.*
4. *Describe how the vital capacity and the forced expiratory volume measurements are affected by asthma and pulmonary fibrosis, and explain why these measurements are affected.*

Gas Exchange in the Lungs

The atmosphere is an ocean of gas that exerts pressure on all objects within it. The amount of this pressure can be measured with a glass U-tube filled with fluid. One end of the U-tube is exposed to the atmosphere, while the other side is continuous with a sealed vacuum tube. Since the atmosphere presses on the exposed side, but not on the side connected to the vacuum tube, atmospheric pressure pushes fluid in the U-tube up on the vacuum side to a height determined by the atmospheric pressure and the density of the fluid. Water, for example, would be pushed up to a height of 33.9 feet (10,332 mm) at sea level, whereas mercury (Hg)—which is more dense—is raised to a height of 760 mm. As a matter of convenience, therefore, devices used to measure atmospheric pressure (barometers) use mercury rather than water. The atmospheric pressure at sea level is thus said to be equal to *760 mm Hg* (or *760 torr*), which is also described as a pressure of *one atmosphere* (fig. 15.22).

According to **Dalton's law,** the total pressure of a gas mixture (such as air) is equal to the sum of the pressures that each gas in the mixture would exert independently. The pressure that each gas would exert independently is equal to the product of the total pressure times the fractional composition of the gas in the mixture. The total pressure of the gas mixture is thus equal to the sum of the partial pressures of the constituent gases. Since oxygen

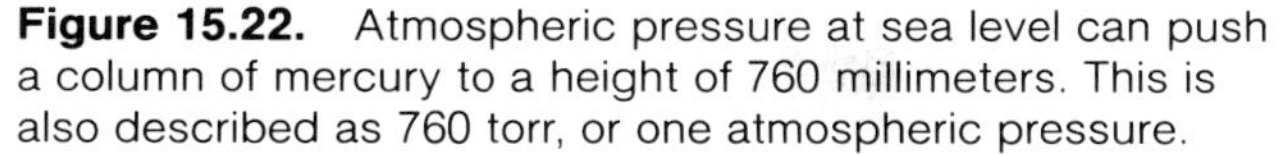

Figure 15.22. Atmospheric pressure at sea level can push a column of mercury to a height of 760 millimeters. This is also described as 760 torr, or one atmospheric pressure.

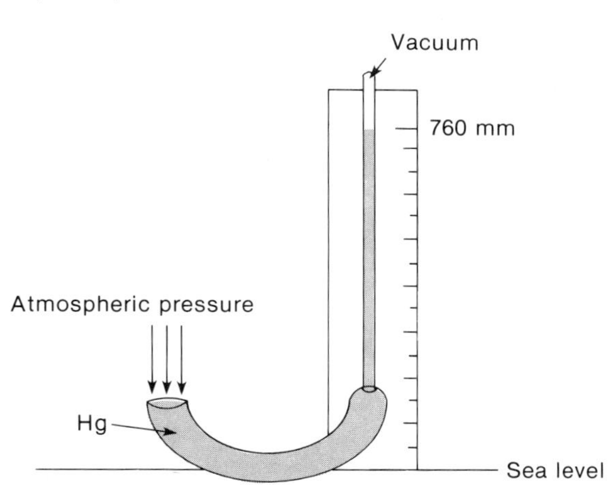

Table 15.6 The effect of altitude on P_{O_2}.

Changes in P_{O_2} at Various Altitudes

Altitude (Feet above Sea Level)	Atmospheric Pressure (mm Hg)	P_{O_2} in Air (mm Hg)	P_{O_2} in Alveoli (mm Hg)	P_{O_2} in Arterial Blood (mm Hg)
0	760	159	105	100
2,000	707	148	97	92
4,000	656	137	90	85
6,000	609	127	84	79
8,000	564	118	79	74
10,000	523	109	74	69
20,000	349	73	40	35
30,000	226	47	21	19

comprises about 21 percent of the atmosphere, for example, its partial pressure (abbreviated P_{O_2}) is 21% of 760, or about 159 mm Hg. Since nitrogen comprises about 79% of the atmosphere, its partial pressure is equal to $0.79 \times 760 = 601$ mm Hg. These two gases thus contribute about 99% of the total pressure of 760 mm Hg:

$$P_{\text{dry atmosphere}} = P_{N_2} + P_{O_2} + P_{CO_2} = 760 \text{ mm Hg}$$

Calculation of P_{O_2}

As one goes to a higher altitude, the total atmospheric pressure and the partial pressure of the constituent gases decrease (table 15.6). At Denver, for example (5,000 feet above sea level), the atmospheric pressure is decreased to 619 mm Hg, and the P_{O_2} is therefore reduced to $619 \times 0.21 = 130$ mm Hg. At the peak of Mount Everest (at 29,000 feet), the P_{O_2} is only 42 mm Hg. As one descends below sea level, as in ocean diving, the total pressure increases by one atmosphere for every 33 feet. At 33 feet therefore, the pressure equals $2 \times 760 = 1{,}520$ mm Hg. At 66 feet, the pressure equals three atmospheres.

Inspired air contains variable amounts of moisture. By the time the air has passed into the respiratory zone of the lungs, however, it is normally saturated with water vapor (has a relative humidity of 100%). The capacity of air to contain water vapor depends on its temperature; since the temperature of the respiratory zone is constant at 37° C, its water vapor pressure is also constant (47 mm Hg).

Figure 15.23. Partial pressures of gases in the inspired air and the alveolar air.

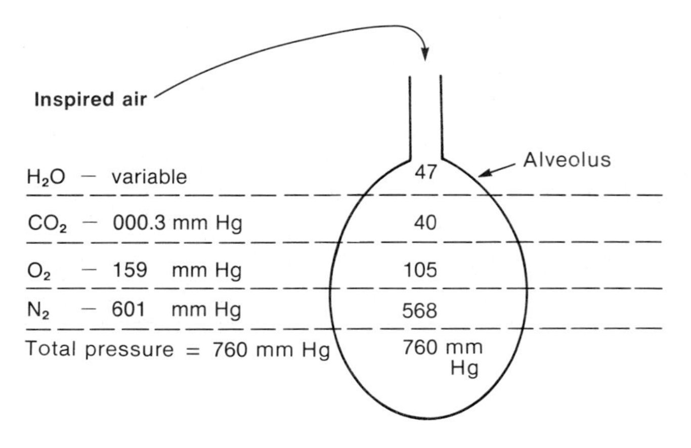

Figure 15.24. Scanning electron micrographs of plastic casts of alveolar capillaries. In the upper photograph only the capillaries are visible. In the lower photograph, plastic also entered the alveoli so that the relationship between alveoli and capillaries can be seen. (From: *Tissues and Organs: A Text Atlas of Scanning Electron Microscopy* by R. G. Kessel and R. Kardon. W. H. Freeman and Company. © 1979.)

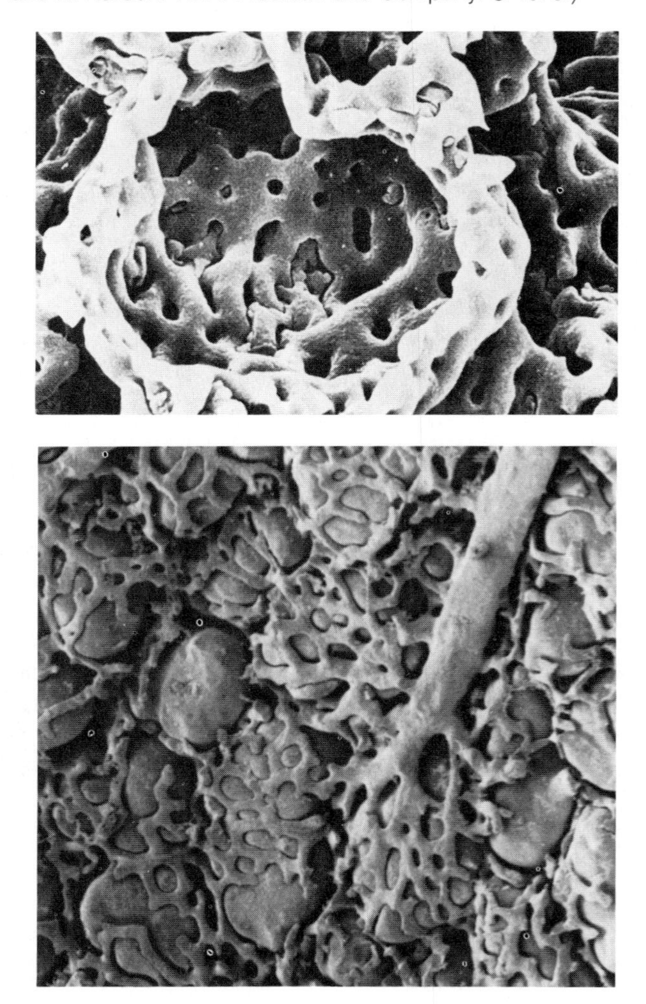

Water vapor, like the other constituent gases, contributes a partial pressure to the total atmospheric pressure. Since the total atmospheric pressure is constant (depending only on the height of the air mass), the water vapor "dilutes" the contribution of other gases to the total pressure:

$$P_{\text{wet atmosphere}} = P_{N_2} + P_{O_2} + P_{CO_2} + P_{H_2O}$$

When the effect of water vapor pressure is considered, the partial pressure of oxygen in the inspired air is decreased at sea level to

$$P_{O_2} \text{ (sea level)} = 0.21 \times (760 - 47) = 150 \text{ mm Hg}$$

As a result of gas exchange in the alveoli, the P_{O_2} of *alveolar air* is further reduced to about 105 mm Hg. A comparison of the partial pressures of the inspired air with the partial pressures of alveolar air is shown in figure 15.23.

Partial Pressures of Gases in Blood

The enormous surface area of alveoli and the short diffusion distance between alveolar air and the capillary blood quickly help to bring the blood into gaseous equilibrium with the alveolar air. This function is further aided by the fact that each alveolus is surrounded by so many capillaries that they form an almost continuous sheet of blood around the alveoli (fig. 15.24).

When a fluid and a gas, such as blood and alveolar air, are at equilibrium, the amount of gas dissolved in the fluid reaches a maximum value. This value depends, according to **Henry's law** (table 15.7), on (1) the solubility of the gas in the fluid, which is a physical constant; (2) the temperature of the fluid—more gas can be dissolved in cold water than warm water; and (3) the partial pressure of the gas. Since the temperature of the blood does not vary significantly, the amount of dissolved gas depends directly on its partial pressure. When water—or plasma—is brought into equilibrium with air at a P_{O_2} of 100 mm Hg, for example, the fluid will contain 0.3 ml O_2 per 100 ml of fluid at 37° C. If the P_{O_2} of the gas were reduced by half, the amount of dissolved oxygen would also be reduced by half.

Table 15.7 Some of the physical laws that are important in ventilation and gas exchange.

Physical Law	Description
Boyle's law	The volume of a gas is inversely proportional to its pressure when the temperature and mass are constant (PV = constant).
Dalton's law	The total pressure of a gas mixture is equal to the sum of the partial pressures of its constituent gases. The partial pressure of each gas is the pressure it would exert if it alone occupied the total volume of the mixture.
Graham's law	The rate of diffusion of a gas through a liquid is directly proportional to its solubility and inversely proportional to its density (or gram molecular weight).
Henry's law	The weight of a gas dissolved in a liquid at a given temperature is proportional to the partial pressure of the gas.
LaPlace's law	The inward pressure tending to collapse a bubble is equal to two times its surface tension divided by the radius of the bubble.

Figure 15.25. Blood-gas measurements using a P_{O_2} electrode. (*a*) The electrical current generated by the oxygen electrode is calibrated so that the needle of the blood-gas machine points to the P_{O_2} of the gas with which the fluid is in equilibrium. (*b*) Once standardized in this way, the electrode can be inserted in a fluid, such as blood, and the P_{O_2} of this solution can be measured.

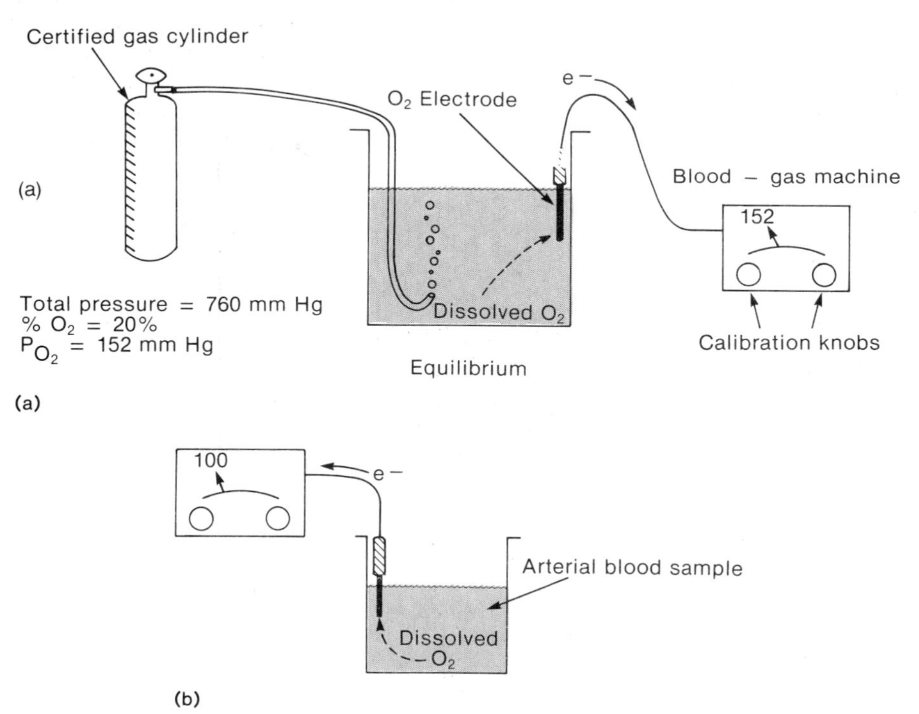

Blood-Gas Measurements. Measurement of the oxygen content of blood (in ml of O_2 per 100 ml blood) is a laborious procedure. Fortunately, an **oxygen electrode** that produces an electric current in proportion to the amount of *dissolved oxygen* has been developed. If this electrode is placed in a fluid while oxygen is artificially bubbled into it, the current produced by the oxygen electrode will increase up to a maximum value. At this maximum value the fluid is *saturated* with oxygen—that is, all of the oxygen that can be dissolved at that temperature and P_{O_2} is dissolved. At a constant temperature the amount dissolved, and thus the electric current, depend only on the P_{O_2} of the gas.

As a matter of convenience, it can now be said that the *fluid has the same P_{O_2} as the gas.* If it is known that the gas has a P_{O_2} of 152 mm Hg, for example, the deflection of the needle produced on a scale by the oxygen electrode can be calibrated at 152 mm Hg (fig. 15.25). The actual amount of dissolved oxygen under these circumstances is not known or very important (it can be looked up in solubility tables, if desired), since this is simply a linear function of the P_{O_2}. A lower P_{O_2} indicates that less oxygen is dissolved, and a higher P_{O_2} indicates that more oxygen is dissolved.

If the oxygen electrode is next inserted into an unknown sample of blood, the P_{O_2} of that sample can be read directly from the previously calibrated scale. Suppose, as illustrated in figure 15.25, the blood sample has a P_{O_2} of 100 mm Hg. Since alveolar air has a P_{O_2} of about 105 mm Hg, this reading indicates that the blood is almost in complete equilibrium with the alveolar air.

The oxygen electrode responds only to the oxygen dissolved in water or plasma; it cannot respond to the oxygen that is hidden in red blood cells. Most of the oxygen in blood, however, is located in the red blood cells attached to hemoglobin. The oxygen content of whole blood thus depends on both its P_{O_2} and its red blood cell and hemoglobin content. At a P_{O_2} of about 100 mm Hg, whole blood normally contains 20 ml O_2 per 100 ml blood; of this amount, only 0.3 ml O_2 is dissolved in the plasma and 19.7 ml O_2 is within the red blood cells. Since only the 0.3 ml O_2 per 100 ml blood affects the P_{O_2} measurement, this measurement would be unchanged if the red blood cells were removed from the sample.

Significance of Blood P_{O_2} and P_{CO_2} Measurements

Since blood P_{O_2} measurements are not directly affected by the oxygen in red blood cells, the P_{O_2} does not provide a measurement of the oxygen content of whole blood. It does, however, provide a good index of *lung function.* If the inspired air has a normal P_{O_2} but the arterial P_{O_2} is below normal, for example, gas exchange in the lungs must be impaired. Measurements of arterial P_{O_2} thus provide valuable information in the treatment of people with pulmonary diseases, during surgery (when breathing may be depressed by anesthesia), and in the care of premature babies with respiratory distress syndrome.

When the lungs are functioning properly, the P_{O_2} of systemic arterial blood is only 5 mm Hg less than the P_{O_2} of alveolar air. Hyperventilation, therefore, cannot significantly increase the blood P_{O_2}. At a normal P_{O_2} of about 100 mm Hg, hemoglobin is almost completely loaded with oxygen. An increase in blood P_{O_2}—produced, for example, by breathing 100% oxygen from a gas tank—thus cannot significantly increase the amount of oxygen contained in the blood. This can, however, significantly increase the amount of oxygen dissolved in the plasma (because the amount dissolved is directly determined by the P_{O_2}). If the P_{O_2} doubles, the amount of oxygen dissolved in the plasma also doubles, but the total oxygen content of whole blood increases only slightly since most of the oxygen by far is not in plasma but in the red blood cells.

Since the oxygen carried by red blood cells must first dissolve in plasma before it can diffuse to the tissue cells, however, a doubling of the blood P_{O_2} means that the *rate of oxygen delivery* to the tissues would double under these conditions. For this reason, breathing from a tank of 100% oxygen (with a P_{O_2} of 760 mm Hg) would significantly increase oxygen delivery to the tissues, although it would have little effect on the total oxygen content of blood.

An electrode that produces a current in response to dissolved carbon dioxide has also been developed, so that the P_{CO_2} of blood can be measured together with its P_{O_2}. Blood in the systemic veins, which is delivered to the lungs by the pulmonary arteries, usually has a P_{O_2} of 40 mm Hg and a P_{CO_2} of 46 mm Hg. After gas exchange in the alveoli of the lungs, blood in the pulmonary veins and systemic arteries has a P_{O_2} of about 100 mm Hg and a P_{CO_2} of 40 mm Hg (fig. 15.26). The values in arterial blood are relatively constant and clinically significant, because they reflect lung function. Blood-gas measurements of venous blood are not performed clinically because these values are far more variable; venous P_{O_2} is much lower and P_{CO_2} much higher after exercise, for example, than at rest, whereas arterial values are not significantly affected by usual changes in physical activity.

Pulmonary Circulation and Ventilation/Perfusion Ratios

In a fetus, the pulmonary circulation has a high vascular resistance as a result of the fact that the lungs are partially collapsed. This high vascular resistance helps to shunt blood from the right to the left atrium through the foramen ovale, and from the pulmonary artery to the aorta through the ductus arteriosus (described in chapter 13). After birth, the foramen ovale and ductus arteriosus close, and the vascular resistance of the pulmonary circulation falls sharply. This fall in vascular resistance at birth is due to (1) opening of the vessels as a result of the subatmospheric intrapulmonary pressure and physical stretching of the lungs during inspiration and (2) to dilation of the pulmonary arterioles in response to increased alveolar P_{O_2}.

In the adult, the right ventricle (like the left) has an output of about 5.5 L per minute. The rate of blood flow through the pulmonary circulation is thus equal to the flow rate through the systemic circulation. Blood flow rate, as described in chapter 14, is directly proportional to the pressure difference between the two ends of the vessel and inversely proportional to the vascular resistance. In the systemic circulation, the mean arterial pressure is 90–100 mm Hg, and the pressure of the right atrium is 0 mm Hg; the pressure difference is thus about 100 mm Hg. The mean pressure of the pulmonary artery, in contrast, is only 15 mm Hg and the pressure of the left atrium is 5 mm Hg. The driving pressure in the pulmonary circulation is thus 15 − 5, or 10 mm Hg.

Figure 15.26. The P_{O_2} and P_{CO_2} of blood as a result of gas exchange in the lung alveoli and gas exchange between systemic capillaries and body cells.

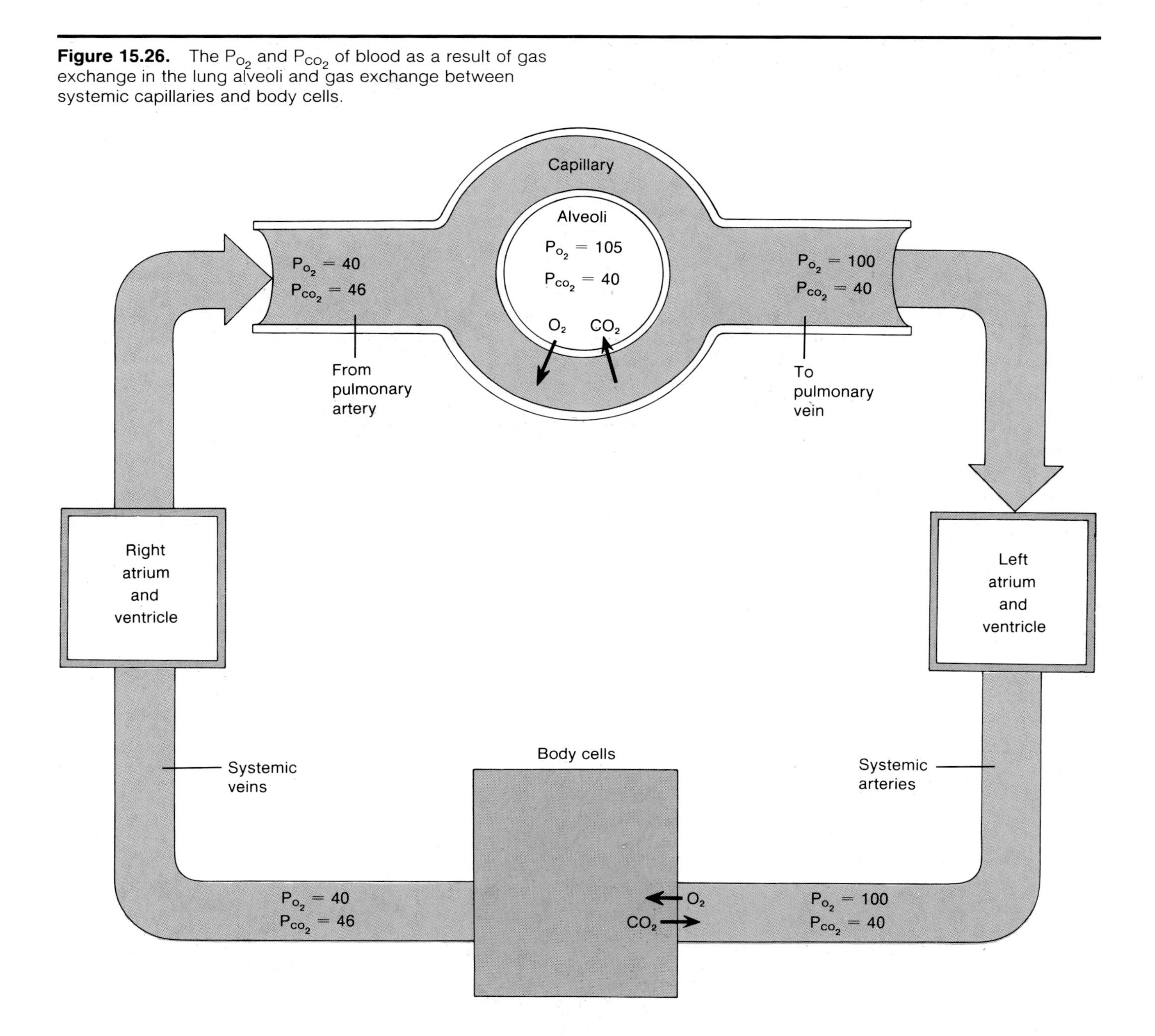

Since the driving pressure in the pulmonary circulation is only one-tenth that of the systemic circulation, and since the flow rates are equal, it follows that the pulmonary vascular resistance must be one-tenth that of the systemic vascular resistance. The pulmonary circulation, in other words, is a low resistance, low pressure pathway. The low pulmonary blood pressure helps to protect against *pulmonary edema* (there is less filtration pressure in pulmonary capillaries than in systemic capillaries), which is a dangerous condition that can impede ventilation and gas exchange. Pulmonary edema can occur when there is pulmonary hypertension produced, for example, by left heart failure.

Pulmonary arterioles constrict when the alveolar P_{O_2} is low and dilate as the alveolar P_{O_2} is raised. This response is opposite to that of systemic arterioles, which dilate in response to low tissue P_{O_2} (as described in chapter 14). Dilation of the systemic arterioles when the P_{O_2} is low helps to supply more blood and oxygen to the tissues; constriction of the pulmonary arterioles when the alveolar P_{O_2} is low helps to decrease blood flow to alveoli that are inadequately ventilated.

Figure 15.27. Ventilation, blood flow, and ventilation/perfusion ratios at the apex and base of the lungs. The ratios indicate that the apex is relatively overventilated and the base underventilated in relation to their blood flows. Such uneven matching of ventilation to perfusion results in the fact that the blood leaving the lungs has a P_{O_2} that is slightly less (by 4–5 mm Hg) than the P_{O_2} of alveolar air.

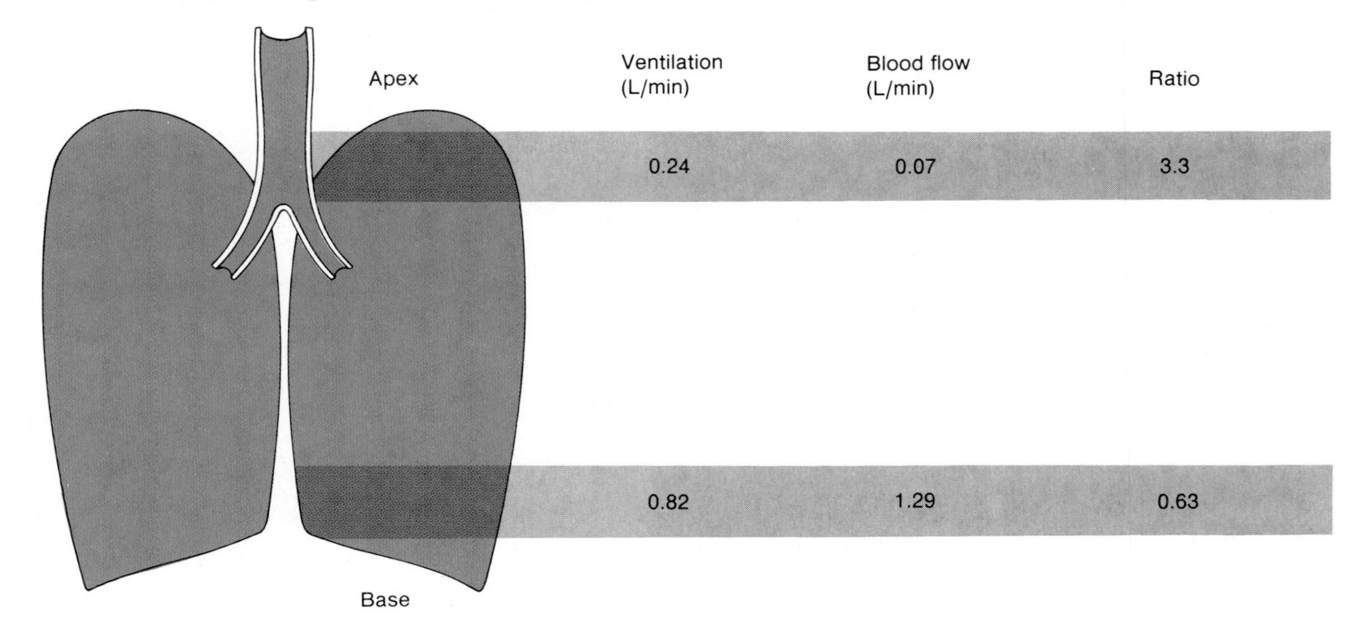

Constriction of the pulmonary arterioles where the alveolar P_{O_2} is low and dilation of arterioles where the alveolar P_{O_2} is high helps to *match ventilation to perfusion* (the term *perfusion* refers to blood flow). If this autoregulation of blood flow did not occur, blood from less-well-ventilated alveoli (with a low P_{O_2}) would mix with blood from well-ventilated alveoli and the blood leaving the lungs would have a lowered P_{O_2} as a result of this dilution effect.

Dilution of the P_{O_2} of pulmonary vein blood actually does occur to some degree despite autoregulation of pulmonary blood flow. When a person stands upright, gravity causes a greater blood flow to the base of the lungs than to the apex (top). Ventilation likewise decreases from base to apex, but this decrease is not proportionate to the decrease in blood flow. The *ventilation/perfusion ratio* at the apex is thus high (0.24 L air per minute divided by 0.07 L blood per minute equals a ratio of 3.4/1.0), while at the base of the lungs it is low (0.82 L air per minute divided by 1.29 L blood per minute equals a ratio of 0.6/1.0). This is illustrated in figure 15.27.

Functionally, the alveoli at the apex of the lungs are thus overventilated and are actually larger, compared to alveoli at the base. The apices of the lungs, in other words, are underperfused. This mismatch of ventilation/perfusion ratios is responsible for the 5 mm Hg difference in P_{O_2} between alveolar air and arterial blood. This difference is not significant because, as previously described, the hemoglobin of normal arterial blood is almost completely saturated with oxygen.

Disorders Caused by High Partial Pressures of Gases

The total atmospheric pressure increases by one atmosphere (760 mm Hg) for every 10 m (33 feet) below sea level. If one dives 10 m below sea level, therefore, the partial pressures and amounts of dissolved gases in the plasma are twice that at sea level. At 66 feet they are three times, and at 100 feet they are four times the values at sea level. The increased amounts of nitrogen and oxygen dissolved in the blood plasma under these conditions can have serious effects on the body.

Oxygen Toxicity. Although breathing 100% oxygen at one or two atmospheres pressure can be safely tolerated for a few hours, higher partial oxygen pressures can be very dangerous. *Oxygen toxicity* develops rapidly when the P_{O_2} rises above about 2.5 atmospheres. This is apparently caused by the oxidation of enzymes and other destructive changes that can damage the nervous system and lead to coma and death. For these reasons, deep sea divers commonly use gas mixtures in which oxygen is diluted with inert gases such as nitrogen (as in ordinary air) or helium.

Hyperbaric oxygen—oxygen at greater than one atmosphere pressure—is often used to treat conditions such as carbon monoxide poisoning, circulatory shock, and gas gangrene. Before the dangers of oxygen toxicity were realized these hyperbaric oxygen treatments sometimes resulted in tragedy. Particularly tragic were the cases of **retrolental fibroplasia,** in which damage to the retina and blindness resulted from hyperbaric oxygen treatment of premature babies with hyaline membrane disease.

Nitrogen Narcosis. Although at sea level nitrogen is physiologically inert, larger amounts of dissolved nitrogen under hyperbaric conditions have deleterious effects. Since it takes time for the nitrogen to dissolve, these effects usually don't appear until the person has remained submerged over an hour. *Nitrogen narcosis* resembles alcohol intoxication; depending on the depth of the dive, the diver may experience "rapture of the deep" or may become so drowsy that he or she is totally incapacitated.

Decompression Sickness. The amount of nitrogen dissolved in the plasma decreases as the diver ascends to sea level, due to a progressive decrease in the P_{N_2}. If the diver surfaces slowly, a large amount of nitrogen can diffuse through the alveoli and be eliminated in the expired breath. If decompression occurs too rapidly, however, bubbles of nitrogen gas (N_2) can form in the blood and block small blood channels, producing muscle and joint pain as well as more serious damage. These effects are known as *decompression sickness,* or the bends.

The cabins of airplanes that fly long distances at high altitudes (30,000 to 40,000 feet) are pressurized so that the passengers and crew are not exposed to the very low atmospheric pressures of these altitudes. If a cabin were to become rapidly depressurized at high altitude, much less nitrogen could remain dissolved at the greatly lowered pressure. People in this situation, like the divers that ascend too rapidly, would thus experience decompression sickness.

1. *Describe how the P_{O_2} of air is calculated and how this value is affected by altitude, diving, and water vapor pressure.*
2. *Describe how blood P_{O_2} measurements are taken, and explain the physiological and clinical significance of these measurements.*
3. *Explain how the arterial P_{O_2} and the oxygen content of whole blood is affected by (1) hyperventilation; (2) breathing from a tank containing 100% oxygen; (3) anemia (low red blood cell count and hemoglobin concentration); and (4) high altitude.*
4. *Describe the ventilation/perfusion ratios of the lungs, and explain why systemic arterial blood has a slightly lower P_{O_2} than alveolar air.*

Regulation of Breathing

Inspiration and expiration are produced by the contractions and relaxations of skeletal muscles in response to activity in somatic motor neurons. The activity of these motor neurons, in turn, is controlled by neurons in the respiratory control centers of the brain stem and by neurons in the cerebral cortex.

The automatic control of breathing is regulated by nerve fibers that descend in the lateral and ventral white matter of the spinal cord from the medulla oblongata. The voluntary control of breathing is a function of the cerebral cortex and involves nerve fibers that descend in the corticospinal tracts. The separation of the voluntary and involuntary pathways is dramatically illustrated in the condition called **Ondine's curse,** in which neurological damage abolishes the automatic but not the voluntary control of breathing. People with this condition must consciously force themselves to breathe and be put on artificial respirators when they sleep.

Brain Stem Respiratory Centers

A loose aggregation of neurons in the reticular formation of the *medulla oblongata* forms the **rhythmicity center** that controls automatic breathing. The rhythmicity center consists of interacting pools of neurons that fire either during inspiration (*I neurons*) or expiration (*E neurons*). The I neurons project to and stimulate spinal motoneurons that innervate the respiratory muscles. Expiration is a passive process that occurs when the I neurons are inhibited by the activity of the E neurons. The activity of I and E neurons varies in a reciprocal way, so that a rhythmic pattern of breathing is produced. The cycle of inspiration and expiration is thus intrinsic to the neural activity of the medulla. The rhythmicity center in the medulla is divided into a dorsal group of neurons, which regulates the activity of the phrenic nerves to the diaphragm, and a ventral group, which controls the motor neurons to the intercostal muscles.

The activity of the medullary rhythmicity center is influenced by centers in the *pons*. As a result of research in which the brain stem is destroyed at different levels, two respiratory control centers have been identified in the pons. One area—the **apneustic center**—appears to promote inspiration by stimulating the I neurons in the medulla. The other pontine area—called the **pneumotaxic**

Figure 15.28. Approximate locations of the brain stem respiratory centers.

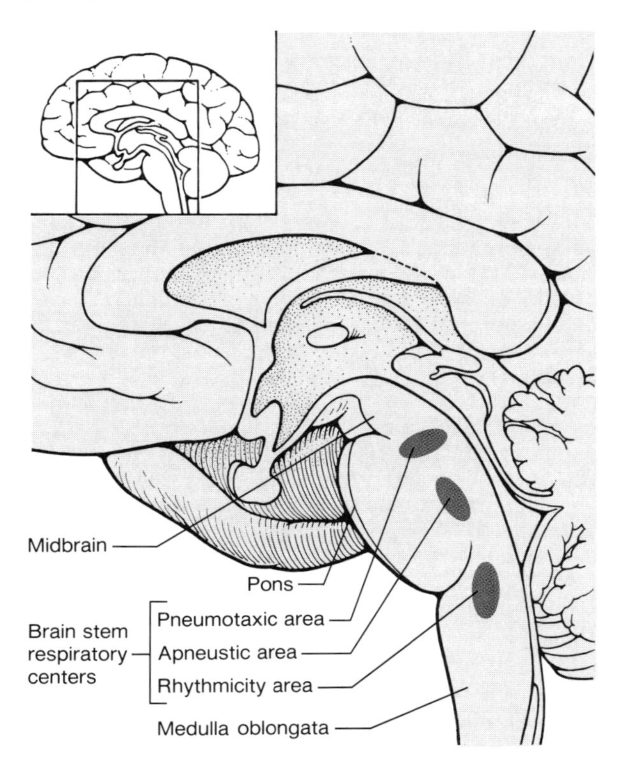

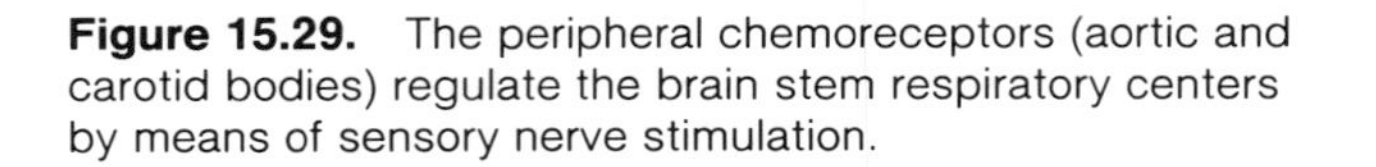

Figure 15.29. The peripheral chemoreceptors (aortic and carotid bodies) regulate the brain stem respiratory centers by means of sensory nerve stimulation.

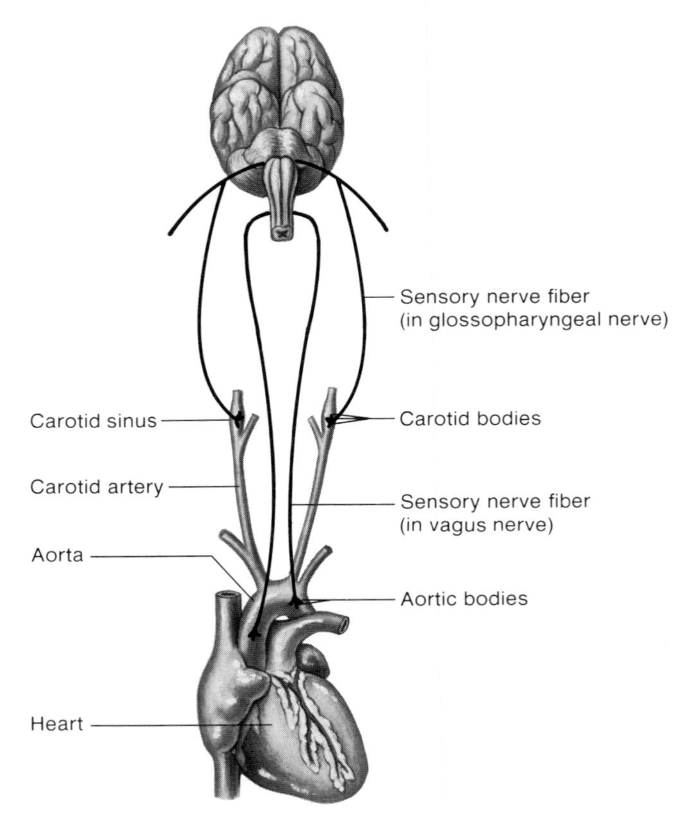

center—seems to antagonize the apneustic center and inhibit inspiration (fig. 15.28). The apneustic center is believed to provide a tonic, or constant, stimulus for inspiration which is cyclically inhibited by the activity of the pneumotaxic center.

The automatic control of breathing is also influenced by input from receptors sensitive to the chemical composition of the blood. There are two groups of *chemoreceptors* that respond to changes in blood P_{CO_2}, pH, and P_{O_2}. These are the **central chemoreceptors** in the medulla oblongata and the **peripheral chemoreceptors.** The peripheral chemoreceptors include a couple of *aortic bodies,* located in the aortic arch, and a pair of *carotid bodies,* located in the carotid artery on each side near the division into the external and internal carotids (fig. 15.29).

The peripheral chemoreceptors control breathing indirectly via sensory nerve fibers to the medulla. The aortic bodies send sensory information to the medulla in the vagus (tenth) cranial nerve; the carotid bodies stimulate sensory fibers in the glossopharyngeal (ninth) cranial nerve. The neural and sensory control of ventilation is summarized in figure 15.30.

Effects of Blood P_{CO_2} and pH on Ventilation

Chemoreceptor input to the brain stem modifies the rate and depth of breathing so that, under normal conditions, arterial P_{CO_2}, pH, and P_{O_2} remain relatively constant. If hypoventilation (inadequate ventilation) occurs, P_{CO_2} quickly rises and pH falls. The fall in pH is due to the fact that carbon dioxide can combine with water to form carbonic acid. The oxygen content of the blood decreases much more slowly, because there is a large "reservoir" of oxygen attached to hemoglobin. During hyperventilation,

Figure 15.30. The control of ventilation by the central nervous system. The feedback effects of pulmonary stretch receptors and "irritant" receptors are not shown.

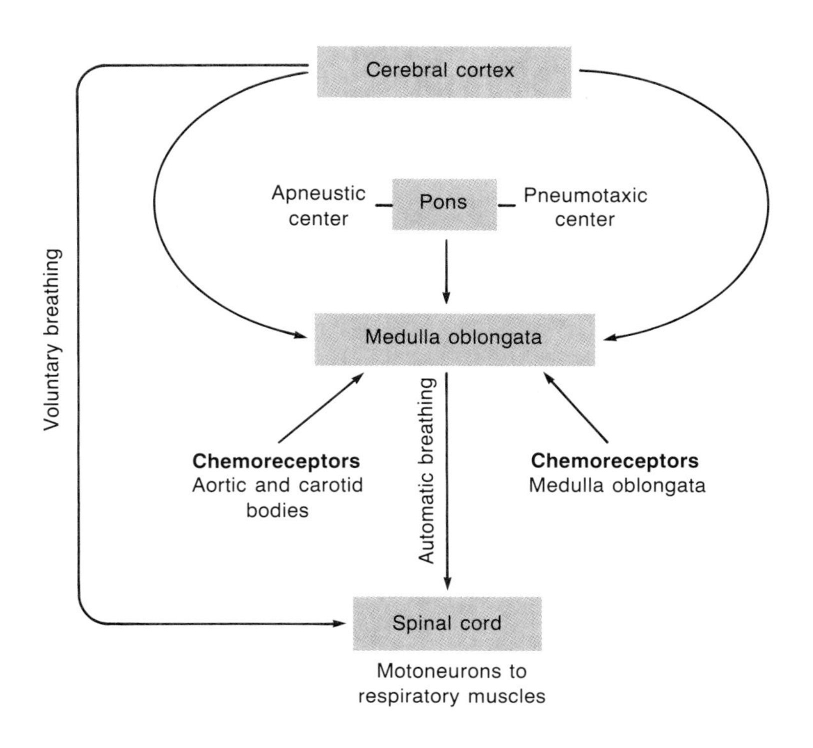

conversely, blood P_{CO_2} quickly falls and pH rises due to the excessive elimination of carbonic acid. The oxygen content of blood, on the other hand, is not significantly increased by hyperventilation (hemoglobin in arterial blood is 97% saturated during normal ventilation).

The blood P_{CO_2} and pH are, therefore, more immediately affected by changes in ventilation than is the oxygen content. Indeed, changes in P_{CO_2} provide a sensitive index of ventilation, as shown in table 15.8. In view of these facts, it is not surprising that changes in P_{CO_2} provide the most potent stimulus for the reflex control of ventilation. Ventilation, in other words, is adjusted to maintain a constant P_{CO_2}; proper oxygenation of the blood occurs naturally as a side product of this reflex control.

Table 15.8 The effect of ventilation, as measured by total minute volume (breathing rate × tidal volume), on the P_{CO_2} of arterial blood.

Total Minute Volume	Arterial P_{CO_2}	Type of Ventilation
2 L/min	80 mm Hg	Hypoventilation
4–5 L/min	40 mm Hg	Normal ventilation
8 L/min	20–25 mm Hg	Hyperventilation

Figure 15.31. Negative feedback control of ventilation through changes in blood P_{CO_2} and pH.

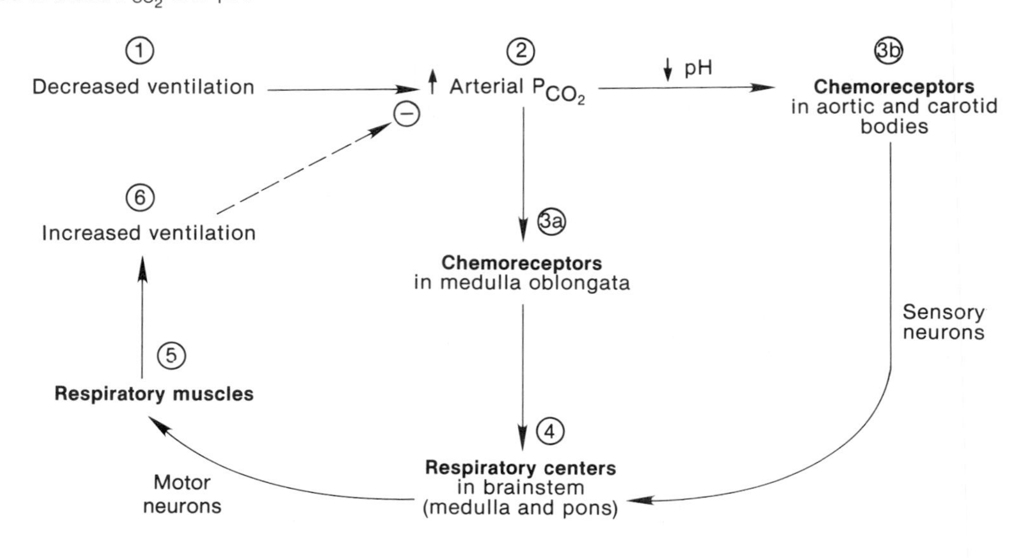

The rate and depth of ventilation are normally adjusted to maintain an arterial P_{CO_2} of 40 mm Hg. Hypoventilation causes a rise in P_{CO_2}—a condition called *hypercapnia.* Hyperventilation, conversely, results in *hypocapnia.* Chemoreceptor regulation of breathing in response to changes in P_{CO_2} is illustrated in figure 15.31.

Chemoreceptors in the Medulla. The chemoreceptors most sensitive to changes in the arterial P_{CO_2} are located in the ventral area of the medulla, near the exit of the ninth and tenth cranial nerves. These chemoreceptor neurons are anatomically separate from, but synaptically communicate with, the neurons of the respiratory control center in the medulla.

An increase in arterial P_{CO_2} causes a rise in the H^+ concentration of the blood as a result of increased carbonic acid concentrations. The H^+ in the blood, however, cannot cross the blood-brain barrier and, therefore, cannot influence the medullary chemoreceptors. Carbon dioxide in the arterial blood can cross the blood-brain barrier and, through the formation of carbonic acid, can lower the pH of cerebrospinal fluid. This fall in cerebrospinal fluid pH directly stimulates the medullary chemoreceptors when there is a rise in arterial P_{CO_2}.

The chemoreceptors in the medulla are ultimately responsible for 70%–80% of the increased ventilation that occurs in response to a sustained rise in arterial P_{CO_2}. This response, however, takes several minutes. The immediate increase in ventilation that occurs when P_{CO_2} rises is produced by stimulation of the peripheral chemoreceptors.

Peripheral Chemoreceptors. The aortic and carotid bodies are stimulated directly by a rise in the H^+ concentration (fall in pH) of arterial blood. This is normally produced by an increase in P_{CO_2} and carbonic acid. The retention of CO_2 during hypoventilation thus stimulates the medullary chemoreceptors through a lowering of cerebrospinal fluid pH and stimulates peripheral chemoreceptors through a lowering of blood pH.

People who hyperventilate during psychological stress are sometimes told to breathe into a paper bag so that they re-breathe their expired air that is enriched in CO_2. This procedure helps to raise their blood P_{CO_2} back up to the normal range. This is needed because hypocapnia causes cerebral vasoconstriction. In addition to producing dizziness, the cerebral ischemia that results can lead to acidotic conditions in the brain that, through stimulation of the medullary chemoreceptors, can cause further hyperventilation. Breathing into a paper bag can thus relieve the hypocapnia and stop the hyperventilation.

Table 15.9 Definitions of some terms used to describe blood oxygen and carbon dioxide levels.

Term	Definition	Term	Definition
Hypoxemia	The oxygen content or P_{O_2} is lower than normal in arterial blood.	Hypercapnia, or hypercarbia	The P_{CO_2} of systemic arteries is higher than 40 mm Hg. Usually this occurs when the ventilation is inadequate for a given metabolic rate (hypoventilation). Antonyms are *hypocapnia* and *hypocarbia* (usually produced by hyperventilation).
Hypoxia	The oxygen content or P_{O_2} is lower than normal in the lungs, blood, or tissues. This is a more general term than hypoxemia. Tissues can be hypoxic, for example, even though there is no hypoxemia (as when the blood flow is occluded).		

Table 15.10 The sensitivity of chemoreceptors to changes in blood gases and pH.

Stimulus	Chemoreceptor(s)	Comments
↑ P_{CO_2}	Medulla oblongata; aortic and carotid bodies	Medullary chemoreceptors are sensitive to the pH of cerebrospinal fluid (CSF). Diffusion of CO_2 from the blood into the CSF lowers the pH of CSF by forming carbonic acid. Similarly, the aortic and carotid bodies are stimulated by a fall in blood pH induced by increases in blood CO_2.
↓ pH	Aortic and carotid bodies	Peripheral chemoreceptors are stimulated by decreased blood pH independent of the effect of blood CO_2. Chemoreceptors in the medulla are not affected by changes in blood pH because H^+ cannot cross the blood-brain barrier.
↓ P_{O_2}	Carotid bodies	Low blood P_{O_2} (hypoxemia) augments the chemoreceptor response to blood P_{CO_2} and can stimulate ventilation directly when the P_{O_2} falls below 50 mm Hg.

Effects of Blood P_{O_2} on Ventilation

Under normal conditions, blood P_{O_2} affects breathing only indirectly, by influencing the chemoreceptor sensitivity to changes in P_{CO_2}. Chemoreceptor sensitivity to P_{CO_2} is augmented by a low P_{O_2} (so ventilation is increased at a high altitude, for example) and is decreased by a high P_{O_2}. If the blood P_{O_2} is raised by breathing 100% oxygen, therefore, the breath can be held longer because the response to increased P_{CO_2} is blunted.

When the blood P_{CO_2} is held constant by experimental techniques, the P_{O_2} of arterial blood must fall from 100 mm Hg to below 50 mm Hg before ventilation is significantly stimulated. This stimulation is apparently due to a direct effect of P_{O_2} on the carotid bodies. Since this degree of *hypoxemia,* or low blood oxygen (table 15.9), does not normally occur even in breath-holding, P_{O_2} does not normally exert this direct effect on breathing.

In emphysema, when there is a chronic retention of carbon dioxide, the chemoreceptors eventually lose their ability to respond to increases in the arterial P_{CO_2}. The abnormally high P_{CO_2}, however, enhances the sensitivity of the carotid bodies to a fall in P_{O_2}. For people with emphysema, breathing may thus be stimulated by a *hypoxic drive* rather than by increases in blood P_{CO_2}.

The effect of changes in the blood P_{CO_2}, pH, and P_{O_2} on chemoreceptors and the regulation of ventilation are summarized in table 15.10.

There are a variety of disease processes that can produce cessation of breathing during sleep, or *sleep apnea.* **Sudden infant death syndrome (SIDS)** is an especially tragic form of sleep apnea that claims the lives of about ten thousand babies annually in the United States. Victims of this condition are apparently healthy two-to-five-month-old babies who die in their sleep without apparent reason—hence, the layman's term of *crib death.* These deaths seem to be caused by failure of the respiratory control mechanisms in the brain stem and/or by failure of the carotid bodies to be stimulated by reduced arterial oxygen.

Abnormal breathing patterns often appear prior to death by brain damage or heart disease. The most common of these abnormal patterns is **Cheyne-Stokes breathing,** in which the depth of breathing progressively increases and then progressively decreases. These cycles of increasing and decreasing tidal volumes may be followed by periods of apnea of varying durations. Cheyne-Stokes breathing may be caused by neurological damage or by insufficient oxygen delivery to the brain. The latter may result from heart disease or from a brain tumor that diverts a large part of the vascular supply from the respiratory centers.

Table 15.11 Pulmonary receptors and reflexes served by sensory nerve fibers in the vagus (tenth cranial nerve).

Stimulus	Receptors	Comments
Stretch of lungs at inspiration	Stretch receptors	The Hering-Breuer reflex. This reflex is believed to be important in the control of breathing in the newborn but not in the adult at normal tidal volumes.
Stretch of lungs at expiration	Stretch receptors	Like the Hering-Breuer reflex, this is not believed to be significant in adults at normal tidal volumes.
Pulmonary congestion	Type J (juxta-capillary) receptors	This reflex may produce feelings of dyspnea at high altitudes and during severe exercise.
Irritation	"Irritant receptors"	This reflex causes reflex constriction of bronchioles in response to irritation from smoke, smog, and other noxious agents.

Pulmonary Stretch and Irritant Reflexes

The lungs contain various types of receptors that influence the brain stem respiratory control centers via sensory fibers in the vagus (table 15.11). Irritant receptors in the lungs, for example, stimulate reflex constriction of the bronchioles in response to smoke and smog. Similarly, sneezing, sniffing, and coughing may be stimulated by irritant receptors in the nose, larynx, and trachea.

The **Hering-Breuer reflex** is stimulated by pulmonary stretch receptors. The activation of these receptors during inspiration inhibits the respiratory control centers, making further inspiration increasingly difficult. This helps to prevent undue distension of the lungs and may contribute to the smoothness of the ventilation cycles. A similar inhibitory reflex may occur during expiration. The Hering-Breuer reflex appears to be important in the control of normal ventilation in the newborn. Pulmonary stretch receptors in adults, however, are probably not active at normal resting tidal volumes (500 ml per breath) but may contribute to respiratory control at high tidal volumes.

1. *Describe the effects of voluntary hyperventilation and breath-holding on arterial P_{CO_2}, pH, and oxygen content. Indicate the relative degree of changes in these values.*
2. *Using a flow chart to show a negative feedback loop, explain the relationship between ventilation and arterial P_{CO_2}.*
3. *Explain the effect of increased arterial P_{CO_2} on (1) chemoreceptors in the medulla oblongata and (2) chemoreceptors in the aortic and carotid bodies.*
4. *Explain the role of arterial P_{O_2} in the regulation of breathing. Explain why ventilation increases when a person goes to a high altitude.*

Hemoglobin and Oxygen Transport

If the lungs are functioning properly, blood leaving in the pulmonary veins and traveling in the systemic arteries has a P_{O_2} of about 100 mm Hg, indicating a plasma oxygen concentration of about 0.3 ml O_2 per 100 ml blood. The total oxygen content of the blood, however, is not known; this depends not only on the P_{O_2} but also on the hemoglobin concentration. If the P_{O_2} and hemoglobin concentration are normal, arterial blood contains about 20 ml of O_2 per 100 ml of blood (fig. 15.32).

Hemoglobin

Most of the oxygen in the blood is contained within the red blood cells, where it is chemically bonded to **hemoglobin.** Each hemoglobin molecule consists of (1) a protein globin part, composed of four polypeptide chains; and (2) four nitrogen-containing, disc-shaped organic pigment molecules, called *hemes* (fig. 15.33).

The protein part of hemoglobin is composed of two identical *alpha chains,* which are each 141 amino acids long, and two identical *beta chains,* which are each 146 amino acids long. Each of the four polypeptide chains is combined with one heme group. In the center of each heme group is a single atom of iron, which can combine with one molecule of oxygen (O_2). One hemoglobin molecule can thus combine with four molecules of oxygen; since there are about 280 million hemoglobin molecules per red blood cell, each red blood cell can carry over a billion molecules of oxygen.

Normal heme contains iron in the reduced form (Fe^{++}, or ferrous iron). In this form, the iron can share electrons and bond to oxygen to form **oxyhemoglobin.** When oxyhemoglobin dissociates to release oxygen to the tissues, the heme iron is still in the reduced (Fe^{++}) form and the hemoglobin is called **deoxyhemoglobin,** or **reduced hemoglobin.** The term *oxyhemoglobin* is thus not equivalent to *oxidized* hemoglobin; hemoglobin does not lose an electron (and become oxidized) when it combines with oxygen.

Figure 15.32. Plasma and whole blood that are brought into equilibrium with the same gas mixture have the same P_{O_2} and thus the same amount of dissolved oxygen molecules (shown as black dots). The oxygen content of whole blood, however, is much higher than that of plasma because of the binding of oxygen to hemoglobin.

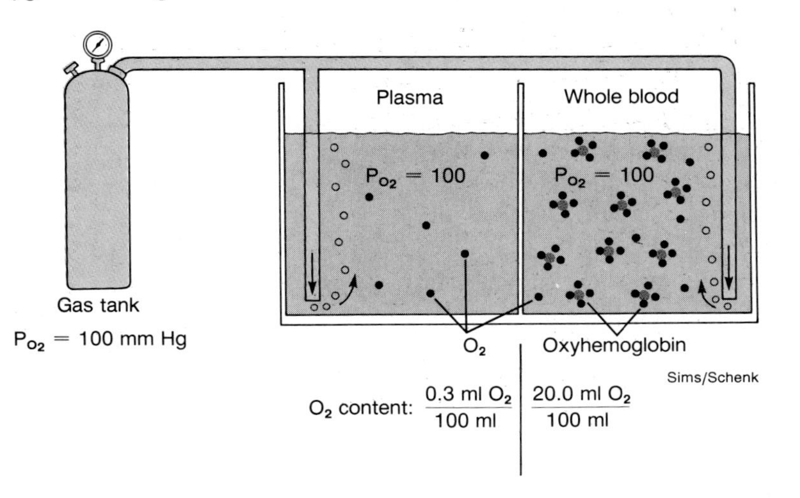

Figure 15.33. (*a*) An illustration of the three-dimensional structure of hemoglobin, in which the two alpha and two beta polypeptide chains are shown; the four heme groups are represented as flat structures with iron (*dark spheres*) in the centers. (*b*) The chemical structure of heme.

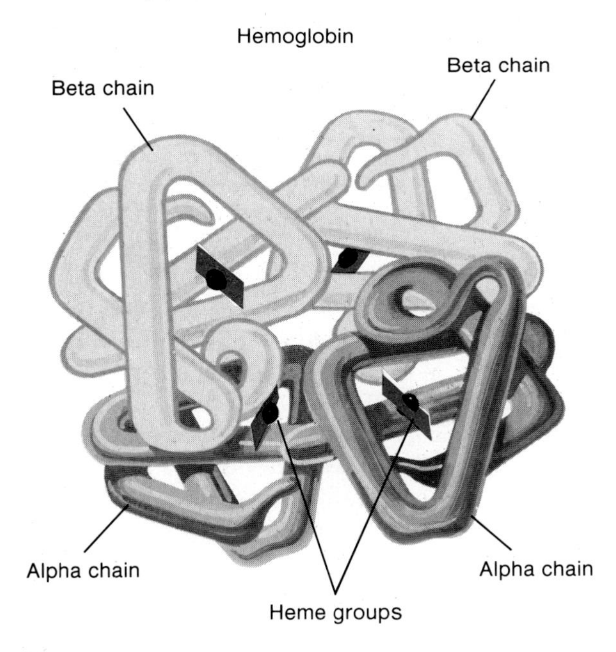

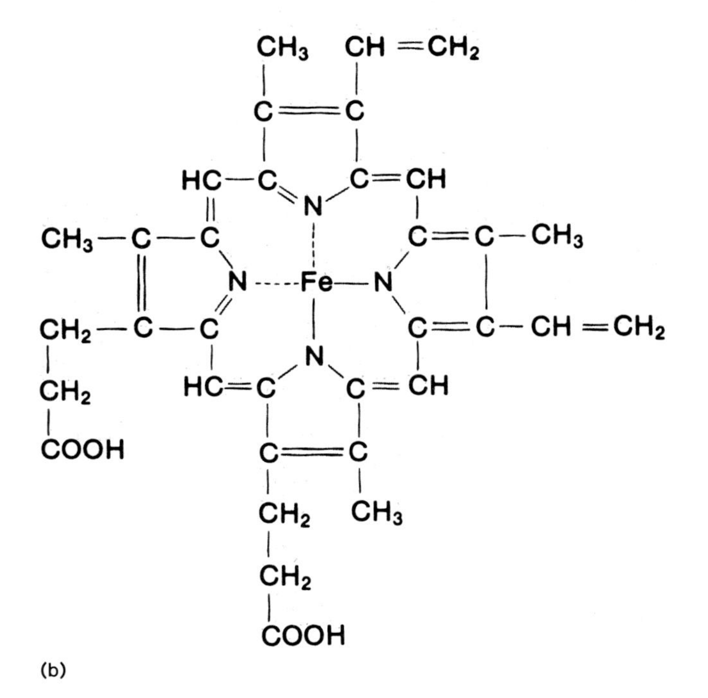

Oxidized hemoglobin, or **methemoglobin,** has iron in the oxidized (Fe^{+++}, or ferric) state. Methemoglobin thus lacks the electron it needs to form a bond with oxygen and cannot participate in oxygen transport. Blood normally contains only a small amount of methemoglobin, but certain drugs can increase this amount.

Carboxyhemoglobin is another abnormal form of hemoglobin, in which the reduced heme is combined with *carbon monoxide* instead of oxygen. Since the bond with carbon monoxide is about 210 times stronger than the bond with oxygen, carbon monoxide tends to displace oxygen in hemoglobin and remains attached to hemoglobin as the blood passes through systemic capillaries. The transport of oxygen to the tissues is thus reduced.

According to federal standards, the percent carboxyhemoglobin in the blood of active nonsmokers should not be higher than 1.5%. Studies have shown, however, that these values can range as high as 3% or more in nonsmokers and as high as 10% or more in smokers in some cities. Although these high levels may not cause immediate problems in healthy people, long-term effects on health might occur. People with respiratory or cardiovascular diseases would be particularly vulnerable to the negative effects of carboxyhemoglobin on oxygen transport.

Hemoglobin Concentration. The *oxygen-carrying capacity* of whole blood is determined by its concentration of normal hemoglobin. If the hemoglobin concentration is below normal—a condition called **anemia**—the oxygen concentration of the blood is reduced below normal. Conversely, when the hemoglobin concentration is increased above the normal range—as occurs in **polycythemia** (high red blood cell count)—the oxygen-carrying capacity of blood is increased accordingly. This can occur as an adaptation to life at a high altitude.

The production of hemoglobin and red blood cells in bone marrow is controlled by a hormone called **erythropoietin,** produced primarily by the kidneys. The production of erythropoietin—and thus the production of red blood cells—is stimulated when the delivery of oxygen to the kidneys and other organs is lower than normal. Red blood cell production is also promoted by androgens, which explains why the hemoglobin concentration in men averages 1–2 g per 100 ml higher than in women.

The Loading and Unloading Reactions. Reduced hemoglobin and oxygen combine to form oxyhemoglobin; this is called the **loading reaction.** Oxyhemoglobin, in turn, dissociates to yield deoxyhemoglobin and free oxygen molecules; this is the **unloading reaction.** The loading reaction occurs in the lungs and the unloading reaction occurs in the systemic capillaries.

Loading and unloading can thus be shown as a reversible reaction:

$$\text{Deoxyhemoglobin} + O_2 \underset{\text{(tissues)}}{\overset{\text{(lungs)}}{\rightleftharpoons}} \text{Oxyhemoglobin}$$

The extent that the reaction will go in each direction depends on two factors: (1) the P_{O_2} of the environment and (2) the *affinity,* or bond strength, between hemoglobin and oxygen. High P_{O_2} drives the loading reaction; at the high P_{O_2} of the pulmonary capillaries almost all the deoxyhemoglobin molecules combine with oxygen. Low P_{O_2} in the systemic capillaries drives the reaction in the opposite direction to promote unloading. The extent of this unloading depends on how low the P_{O_2} values are.

The affinity (bond strength) between hemoglobin and oxygen also influences the loading and unloading reactions. A very strong bond would favor loading but inhibit unloading; a weak bond would hinder loading but improve unloading. The bond strength between hemoglobin and oxygen is normally strong enough so that 97% of the hemoglobin leaving the lungs is in the form of oxyhemoglobin, yet the bond is sufficiently weak so that adequate amounts of oxygen are unloaded to sustain aerobic respiration in the tissues. Under normal resting conditions only about 22% of the oxygen is unloaded; this satisfies the tissue needs for oxygen yet maintains an oxygen reserve in the blood for emergency conditions.

The Oxyhemoglobin Dissociation Curve

Blood in the systemic arteries, at a P_{O_2} of 100 mm Hg, has a *percent oxyhemoglobin saturation* of 97% (which means that 97% of the hemoglobin is in the form of oxyhemoglobin). This blood is delivered to the systemic capillaries, where oxygen diffuses into the tissue cells and is consumed in aerobic respiration. Blood leaving in the systemic veins is thus reduced in oxygen; it has a P_{O_2} of about 40 mm Hg and a percent saturation of about 75% (table 15.12). In other words, blood entering the tissues contains 20 ml O_2 per 100 ml blood, and blood leaving the tissues contains 15.5 ml O_2 per 100 ml blood (fig. 15.34). Thus, 22%, or 4.5 ml of O_2 out of 20 ml O_2 per 100 ml blood, is unloaded to the tissues.

A graphic illustration of the percent oxyhemoglobin saturation at different values of P_{O_2} is called an **oxyhemoglobin dissociation curve** (fig. 15.34). The values in this graph are obtained by subjecting samples of blood *in vitro* (outside the body) to different partial oxygen pressures.

Table 15.12 The relationship between percent oxyhemoglobin saturation and P_{O_2} (at pH = 7.40 and temperature = 37°C).

P_{O_2}(mm Hg)	100	80	61	45	40	36	30	26	23	21	19
Percent oxyhemoglobin	97	95	90	80	75	70	60	50	40	35	30
	Arterial blood				Venous blood						

Figure 15.34. The percent of oxyhemoglobin saturation and the blood oxygen content are shown at different values of P_{O_2}. Notice that there is about a 25% decrease in percent oxyhemoglobin as the blood passes through the tissue from arteries to veins, resulting in the unloading of approximately 5 ml O_2 per 100 ml to the tissues.

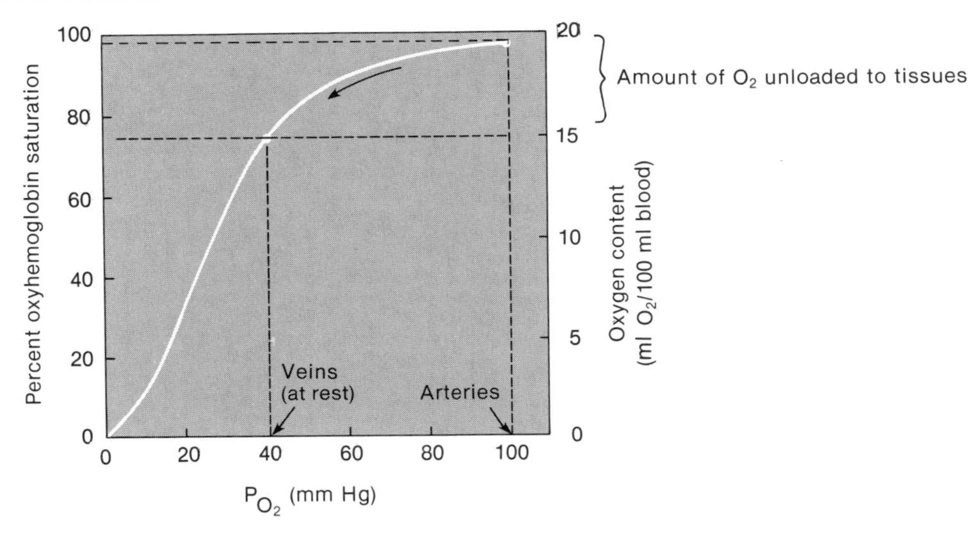

The percent oxyhemoglobin saturation obtained by this procedure, however, can be used to predict what the percent unloading would be *in vivo* (within the body) with a given difference in arterial and venous P_{O_2} values.

Figure 15.34 shows the difference between the arterial and venous P_{O_2} and the percent oxyhemoglobin saturation at rest. The relatively large amount of oxyhemoglobin remaining in the venous blood at rest functions as an oxygen reservoir. If a person stops breathing, there is a sufficient reserve of oxygen in the blood to keep the brain and heart alive for approximately four to five minutes in the absence of cardiopulmonary resuscitation (CPR) techniques. This reserve supply of oxygen can also be tapped when the tissue's requirements for oxygen are raised.

The oxyhemoglobin dissociation curve is S-shaped, or *sigmoidal.* The fact that it is relatively flat at high P_{O_2} values indicates that changes in P_{O_2} within this range have little effect on the loading reaction. One would have to ascend as high as 10,000 feet, for example, before the oxyhemoglobin saturation of arterial blood would decrease from 97% to 93%. At more common elevations the percent oxyhemoglobin saturation would not be significantly different from the 97% value at sea level.

At the steep part of the sigmoidal curve, however, small changes in P_{O_2} values produce large differences in percent saturation. A decrease in *venous* P_{O_2} from 40 mm Hg to 30 mm Hg, as might occur during mild exercise, corresponds to a change in percent saturation from 75% to 58%. Since the *arterial* percent saturation is usually still 97% during exercise, this change in venous percent saturation indicates that more oxygen has been unloaded to the tissues. The difference between the arterial and venous percent saturations indicates the percent unloading: in these examples, 97% minus 75% = 22% unloading at rest, and 97% minus 58% = 39% unloading during mild exercise. During heavier exercise, the venous P_{O_2} can drop to 20 mm Hg or less, indicating a percent unloading in excess of 70%.

Figure 15.35. A decrease in blood pH (an increase in H^+ concentration) decreases the affinity of hemoglobin for oxygen at each P_{O_2} value, resulting in a "shift to the right" of the oxyhemoglobin dissociation curve. A curve that is shifted to the right has a lower percent oxyhemoglobin saturation at each P_{O_2}, but the effect is more marked at lower P_{O_2} values. This is called the *Bohr effect.*

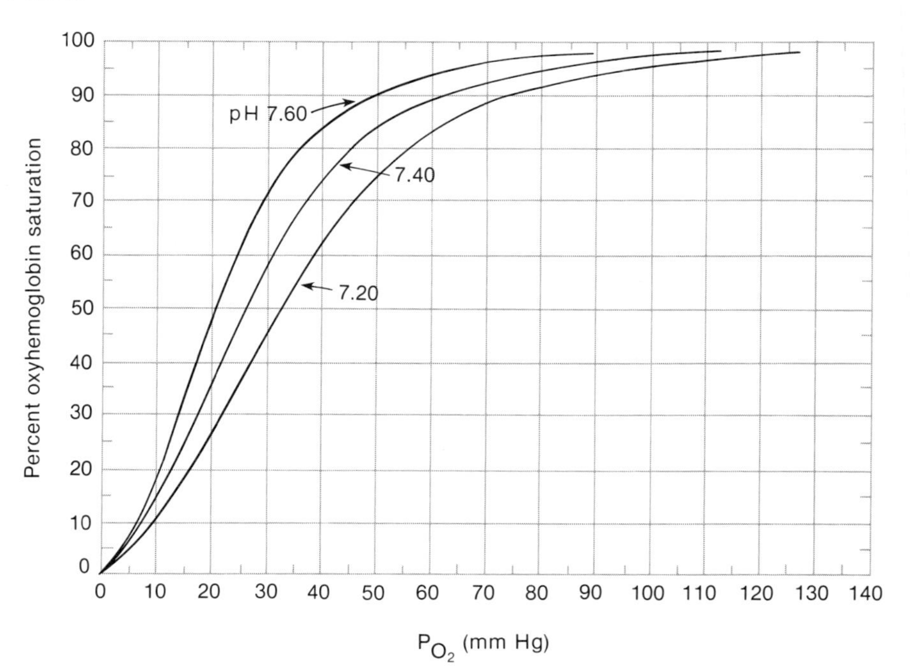

Table 15.13 Effect of pH on hemoglobin affinity for oxygen and oxyhemoglobin "unloading."

pH	Affinity	Arterial O_2 Content per 100 ml	Venous O_2 Content per 100 ml	O_2 Unloaded to Tissues per 100 ml
7.40	Normal	19.8 ml O_2	14.8 ml O_2	5.0 ml O_2
7.60	Increased	20.0 ml O_2	17.0 ml O_2	3.0 ml O_2
7.20	Decreased	19.2 ml O_2	12.6 ml O_2	6.6 ml O_2

Effect of pH and Temperature on Oxygen Transport

In addition to changes in P_{O_2}, the loading and unloading reactions are influenced by changes in the bond strength, or affinity, of hemoglobin for oxygen. The affinity is decreased when the pH is lowered and increased when the pH is raised; this is called the **Bohr effect.** When the affinity of hemoglobin for oxygen is reduced, there is less loading of the blood with oxygen in the lungs but greater unloading of oxygen in the tissues. A weakening of the bond between hemoglobin and oxygen, however, has less effect at the higher P_{O_2} values in the lungs than at the lower P_{O_2} values of the tissues. A lowering of pH thus results in slightly less oxygen loading in the lungs and significantly more oxygen unloading in the tissues. The net effect is that the tissues receive more oxygen when the blood pH is lowered (table 15.13). Since the pH can be decreased by carbon dioxide (through the formation of carbonic acid), the Bohr effect helps to provide a little more oxygen to the tissues when their carbon dioxide output (and metabolism) is increased.

When the percent oxyhemoglobin saturation at different pH values is graphed as a function of P_{O_2}, the dissociation curve is shown to be shifted to the right by a lowering of pH and shifted to the left by a rise in pH (fig. 15.35). If the percent unloading is calculated by subtracting the percent oxyhemoglobin saturation at given P_{O_2} values for arterial and venous blood, it will be clear that a *shift to the right* of the curve indicates a greater oxygen unloading, whereas a *shift to the left* indicates less unloading but slightly more oxygen loading in the lungs.

Figure 15.36. The oxyhemoglobin dissociation curve is shifted to the right as the temperature increases, indicating a lowered affinity of hemoglobin for oxygen at each P_{O_2}. This effect, like the Bohr effect (see figure 15.35), is more marked at lower P_{O_2} values.

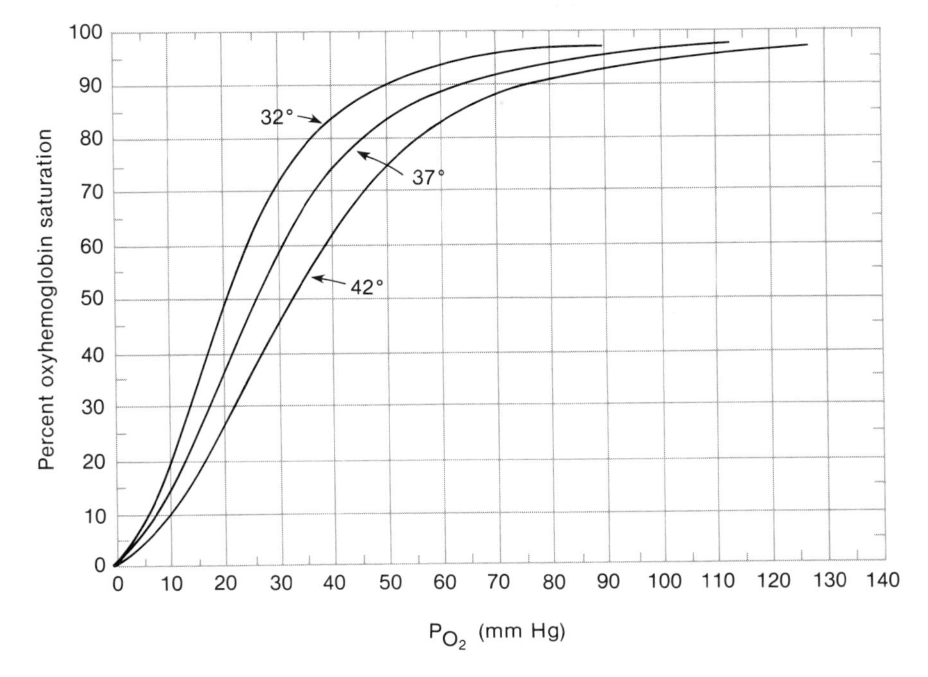

Table 15.14 Factors that affect the affinity of hemoglobin for oxygen and the position of the oxyhemoglobin dissociation curve.

Factor	Affinity	Position of Curve	Comments
↓ pH	Decreased	Shift to the right	Called the Bohr effect; increases oxygen delivery during hypercapnia
↑ Temperature	Decreased	Shift to the right	Increases oxygen unloading during exercise and fever
↑ 2,3-DPG	Decreased	Shift to the right	Increases oxygen unloading when there is a decrease in total hemoglobin or total oxygen content; an adaptation to anemia and high-altitude living

When oxyhemoglobin dissociation curves are constructed at constant pH values but at different temperatures, it can be seen that the affinity of hemoglobin for oxygen is decreased by a rise in temperature. An increase in temperature weakens the bond between hemoglobin and oxygen and thus has the same effect as a fall in pH; the oxyhemoglobin dissociation curve is shifted to the right (fig. 15.36). At higher temperatures, therefore, more oxygen is unloaded to the tissues than would be the case if the bond strength were constant. This effect can significantly increase the delivery of oxygen to muscles that are warmed during exercise.

Effect of 2,3-DPG on Oxygen Transport

Mature red blood cells lack both nuclei and mitochondria. Without mitochondria they cannot respire aerobically and thus cannot use the oxygen they carry. Red blood cells, therefore, obtain energy through the anaerobic respiration of glucose. At a certain point in the glycolytic pathway there is a "side reaction" in red blood cells that results in a unique product—**2,3-diphosphoglyceric acid (2,3-DPG).**

The enzyme that produces 2,3-DPG is inhibited by oxyhemoglobin. When the oxyhemoglobin concentration is decreased, therefore, the production of 2,3-DPG is increased. This increase in 2,3-DPG production can occur when the total hemoglobin concentration is low (in anemia) or when the P_{O_2} is low (at a high altitude, for example). 2,3-DPG combines with deoxyhemoglobin and makes it more stable. At the P_{O_2} values in the tissue capillaries, therefore, a higher proportion of the oxyhemoglobin will be converted to deoxyhemoglobin by unloading its oxygen. Increased concentrations of 2,3-DPG in red blood cells thus increase oxygen unloading (table 15.14) and shift the oxyhemoglobin dissociation curve to the right.

Figure 15.37. A comparison of the dissociation curves for hemoglobin and for myoglobin. At the P_{O_2} of venous blood, the myoglobin retains almost all of its oxygen, indicating a higher affinity than hemoglobin for oxygen. The myoglobin does, however, release its oxygen at the very low P_{O_2} values found inside the mitochondria.

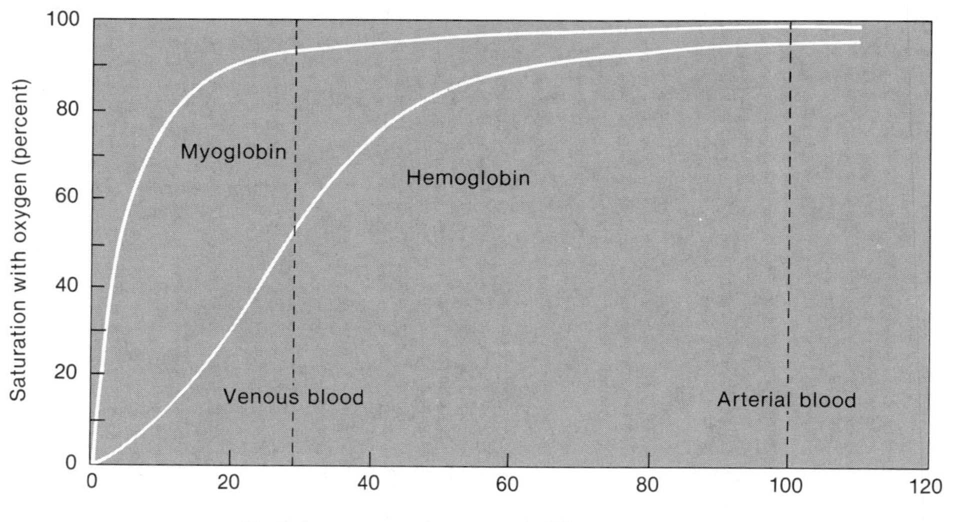

The importance of 2,3-DPG within red blood cells is now recognized in blood banking. Old, stored red blood cells can lose their ability to produce 2,3-DPG as they lose their ability to metabolize glucose. Modern techniques for blood storage, therefore, include the addition of energy substrates for respiration and phosphate sources needed for the production of 2,3-DPG.

Anemia. When the total blood hemoglobin concentration is reduced below normal in anemia, each red blood cell produces increased amounts of 2,3-DPG. A normal hemoglobin concentration of 15 g per 100 ml unloads about 4.5 ml O_2 per 100 ml at rest. If the hemoglobin concentration were reduced by half, you might expect that the tissues would receive only half the normal amount of oxygen (2.25 ml O_2 per 100 ml). It has been shown, however, that as much as 3.3 ml O_2 per 100 ml are unloaded to the tissues under these conditions. This occurs as a result of a rise in 2,3-DPG production that produces a decrease in hemoglobin affinity for oxygen.

Hemoglobin F. The effects of 2,3-DPG are also important in the transfer of oxygen from maternal to fetal blood. The mother has hemoglobin molecules composed of two alpha and two beta chains, as previously described, whereas the fetal hemoglobin contains two alpha and two *delta chains* in place of beta chains (delta chains differ from beta chains in 37 of their amino acids). Normal adult hemoglobin in the mother (*hemoglobin A*) is able to bond to 2,3-DPG. Fetal hemoglobin (*hemoglobin F*), in contrast, cannot bond to 2,3-DPG, and thus has a higher affinity for oxygen at a given P_{O_2} than does hemoglobin A. Since hemoglobin F can have a higher percent oxyhemoglobin saturation than hemoglobin A at a given P_{O_2}, oxygen is transferred from the maternal to the fetal blood as these two come into close proximity in the placenta.

Muscle Myoglobin

Myoglobin is a red pigment found exclusively in striated muscle fibers. In particular, slow-twitch, aerobically respiring skeletal fibers and cardiac muscle fibers are rich in myoglobin. Myoglobin is similar to hemoglobin, but it has one rather than four hemes and, therefore, can combine with only one molecule of oxygen.

Myoglobin has a higher affinity for oxygen than does hemoglobin, and its dissociation curve is therefore to the left of the oxyhemoglobin dissociation curve (fig. 15.37). The shape of the myoglobin curve is also different from the oxyhemoglobin dissociation curve; it is rectangular, indicating that oxygen will be released only when the P_{O_2} gets very low. The differences in behavior between myoglobin and hemoglobin highlight the importance of the tetrameric structure of hemoglobin. When one heme group of hemoglobin combines with oxygen a shape change occurs that allows the other heme groups to combine with oxygen more easily. The same interaction beween the four subunits of hemoglobin allows the hemoglobin to unload its oxygen at higher values of P_{O_2} than can myoglobin.

Table 15.15 Some differences between normal hemoglobin A and two mutant hemoglobin forms.

	Hemoglobin		
	A	S	C
DNA Base Triplet	CTT	CAT	TTT
mRNA Codon	GAA	GUA	AAA
Amino Acid in Position 6	Glutamic acid	Valine	Lysine

Since the P_{O_2} in mitochondria is very low (because oxygen is incorporated into water here), myoglobin may act as a "middleman" in the transfer of oxygen from blood to the mitochondria within muscle cells. Myoglobin may also have an oxygen storage function, which is of particular importance in the heart. During diastole, when the coronary blood flow is greatest, the myoglobin can load up with oxygen. This stored oxygen can then be released during systole, when the coronary arteries are squeezed closed by the contracting myocardium.

Inherited Defects in Hemoglobin Structure and Function

There are a number of hemoglobin diseases that are produced by inherited (congenital) defects in the protein part of hemoglobin. **Sickle-cell anemia**—a disease affecting 8%–11% of the black population of the United States—for example, is caused by an abnormal form of hemoglobin called *hemoglobin S*. Hemoglobin S differs from normal hemoglobin A in only one amino acid: valine is substituted for glutamic acid in position 6 on the beta chains. This amino acid substitution is caused by a single base change in the region of DNA that codes for the beta chains (table 15.15).

Under conditions of low blood P_{O_2}, hemoglobin S comes out of solution and cross links to form a "paracrystalline gel" within the red blood cells. This causes the characteristic sickle shape of red blood cells (fig. 15.38) and makes them less flexible. Since red blood cells must be able to bend in the middle to pass through many narrow capillaries, a decrease in their flexibility may cause them to block small blood channels and produce organ ischemia. The decreased solubility of hemoglobin S in solutions of low P_{O_2} is used in the diagnosis of sickle-cell anemia and sickle-cell trait (the carrier state, in which a person has the genes for both hemoglobin A and hemoglobin S).

Figure 15.38. (*a*) A sickled red blood cell as seen in the light microscope. (*b*) Normal cells. (*c*). Sickled red blood cells as seen in the scanning electron microscope.

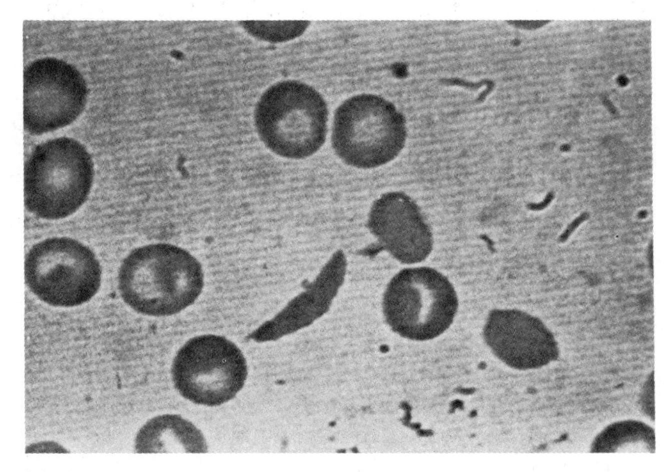

(a)

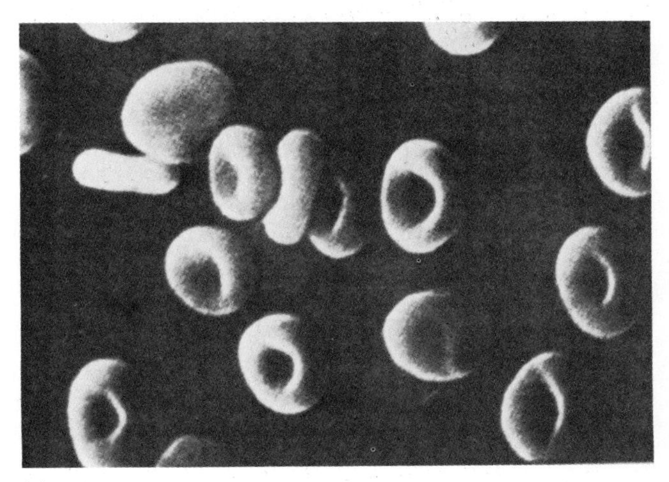

(b)

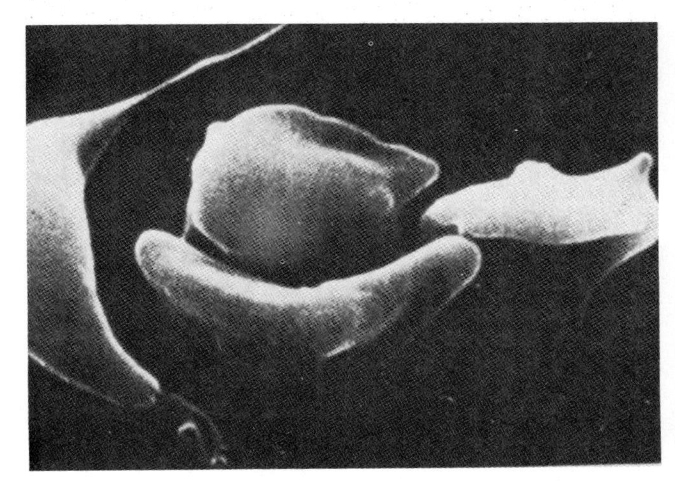

(c)

Thalassemia is a family of hemoglobin diseases found predominantly among people of Mediterranean ancestry. In *alpha thalassemia,* there is decreased synthesis of the alpha chains of hemoglobin, whereas in *beta thalassemia* the synthesis of the beta chains is impaired. One of the compensations for thalassemia is increased synthesis of delta chains, resulting in the retention of large amounts of hemoglobin F (fetal hemoglobin) into adulthood. Although this may partially compensate for the anemia, it has a major drawback; hemoglobin F has a higher affinity for oxygen than hemoglobin A and thus cannot unload as much oxygen to the tissues.

Some types of abnormal hemoglobins have been shown to be advantageous in the environments in which they evolved. A person who is a carrier for sickle-cell anemia, for example (and who therefore has both hemoglobin A and hemoglobin S), has a high resistance to malaria. This is because the parasite that causes this disease cannot live in red blood cells that contain hemoglobin S.

---

1. *Describe the effect of P_{O_2} on the loading and unloading reactions. Illustrate this with a graph, and label the regions of the graph that show when loading and unloading occur.*
2. *Draw an oxyhemoglobin dissociation curve, and label the values found in the arterial blood and venous blood under resting conditions. Use this graph to show an increase in the percent unloading that can occur during exercise.*
3. *Describe how changes in pH and temperature affect the hemoglobin affinity for oxygen and the position of the oxyhemoglobin dissociation curve. Explain the effect of these changes on oxygen transport.*
4. *Explain how a person who is anemic or a person at high altitude could have an increase in the percent unloading of oxygen by hemoglobin.*

Carbon Dioxide Transport and Acid-Base Balance

Carbon dioxide is carried by the blood in three forms: (1) as *dissolved CO_2*—since carbon dioxide is about twenty-one times more soluble than oxygen in water, a substantial portion (about one-tenth) of the total blood CO_2 is dissolved in plasma; (2) as *carbaminohemoglobin*—about one-fifth of the total blood CO_2 is carried attached to an amino acid in hemoglobin (carbaminohemoglobin should not be confused with carboxyhemoglobin, which is a combination of hemoglobin with carbon monoxide); and (3) as *carbonic acid* and *bicarbonate,* which account for most of the CO_2 carried by the blood.

Carbon dioxide is able to combine with water to form carbonic acid. This reaction occurs spontaneously in the plasma at a slow rate but occurs much more rapidly within the red blood cells due to the catalytic action of the enzyme **carbonic anhydrase.** Since this enzyme is confined to the red blood cells, most of the carbonic acid is produced there rather than in the plasma. The formation of carbonic acid from CO_2 and water is favored by the high P_{CO_2} found in tissue capillaries (this is an example of the *law of mass action,* described in chapter 4).

$$CO_2 + H_2O \xrightarrow[\text{high } P_{CO_2}]{\text{carbonic anhydrase}} H_2CO_3$$

The Chloride Shift

As a result of catalysis by carbonic anhydrase within the red blood cells, large amounts of carbonic acid are produced as blood passes through the systemic capillaries. The buildup of carbonic acid concentrations within the red blood cells favors the dissociation of these molecules into *H^+* (protons, which contribute to the acidity of a solution) and *HCO_3^-* (bicarbonate):

$$H_2CO_3 \longrightarrow H^+ + HCO_3^-$$

The H^+ released by the dissociation of carbonic acid are largely buffered by their combination with deoxyhemoglobin within the red blood cells. Although the unbuffered H^+ are free to diffuse out of the red blood cells, more bicarbonate diffuses outward into the plasma than does H^+. As a result of the "trapping" of H^+ within the red blood cells by their attachment to hemoglobin and the outward diffusion of bicarbonate, the inside of the red blood cell gains a net positive charge. This attracts chloride ions (Cl^-), which move into the red blood cells as HCO_3^- moves out. This exchange of anions as blood moves through the tissue capillaries is called the **chloride shift** (fig. 15.39).

Figure 15.39. An illustration of carbon dioxide transport by the blood and the "chloride shift." Carbon dioxide is transported in three forms: as dissolved CO_2 gas, attached to hemoglobin as carbaminohemoglobin, and as carbonic acid and bicarbonate.

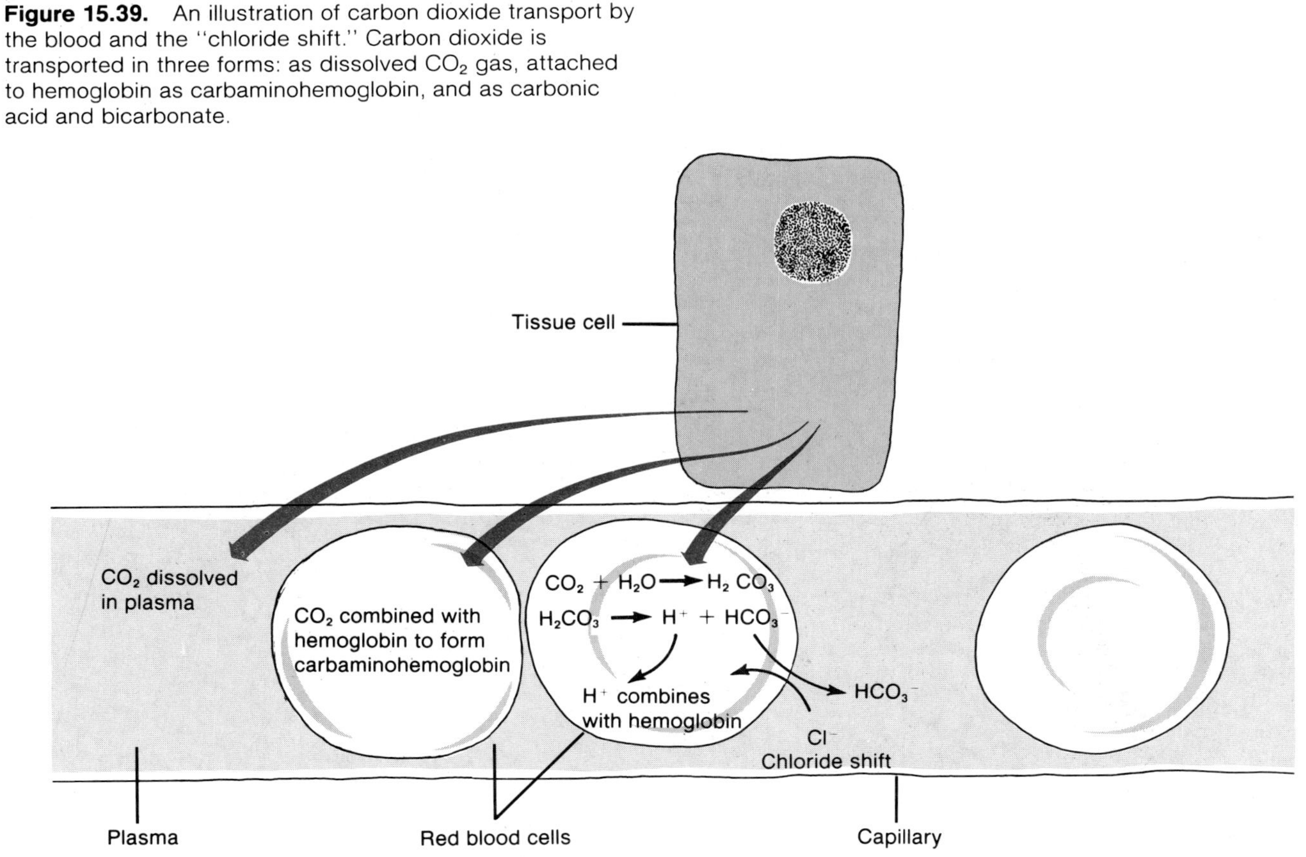

There is an interesting interaction between the transport of oxygen and carbon dioxide. The bonding of H^+ by oxyhemoglobin promotes the unloading of oxygen (the Bohr effect) and thus promotes the conversion of oxyhemoglobin to deoxyhemoglobin. Since deoxyhemoglobin bonds H^+ more than does oxyhemoglobin, the act of unloading its oxygen improves the ability of hemoglobin to buffer the H^+ released by carbonic acid, and this (by the law of mass action) favors the continued production of carbonic acid. The formation of carbonic acid, in summary, enhances oxygen unloading (the Bohr effect), and oxygen unloading improves the ability of the blood to form carbonic acid and transport carbon dioxide.

When blood reaches the pulmonary capillaries, deoxyhemoglobin is converted to oxyhemoglobin. Since oxyhemoglobin has a lower affinity for H^+ than does deoxyhemoglobin, H^+ are released within the red blood cells. This attracts HCO_3^- from the plasma, which combines with H^+ to form carbonic acid:

$$H^+ + HCO_3^- \longrightarrow H_2CO_3$$

Under conditions of low P_{CO_2}, as occurs in the pulmonary capillaries, carbonic anhydrase catalyzes the conversion of carbonic acid to carbon dioxide and water:

$$H_2CO_3 \xrightarrow[\text{low } P_{CO_2}]{\text{carbonic anhydrase}} CO_2 + H_2O$$

In summary, the carbon dioxide produced by the tissue cells is converted within the systemic capillaries, mostly through the action of carbonic anhydrase in the red blood cells, to carbonic acid. With the buildup of carbonic acid concentrations in the RBCs, the carbonic acid dissociates into bicarbonate and H^+, which results in the chloride shift.

Figure 15.40. Carbon dioxide is released from the blood as it travels through the pulmonary capillaries. During this time a "reverse chloride shift" occurs and carbonic acid is transformed into CO_2 and H_2O.

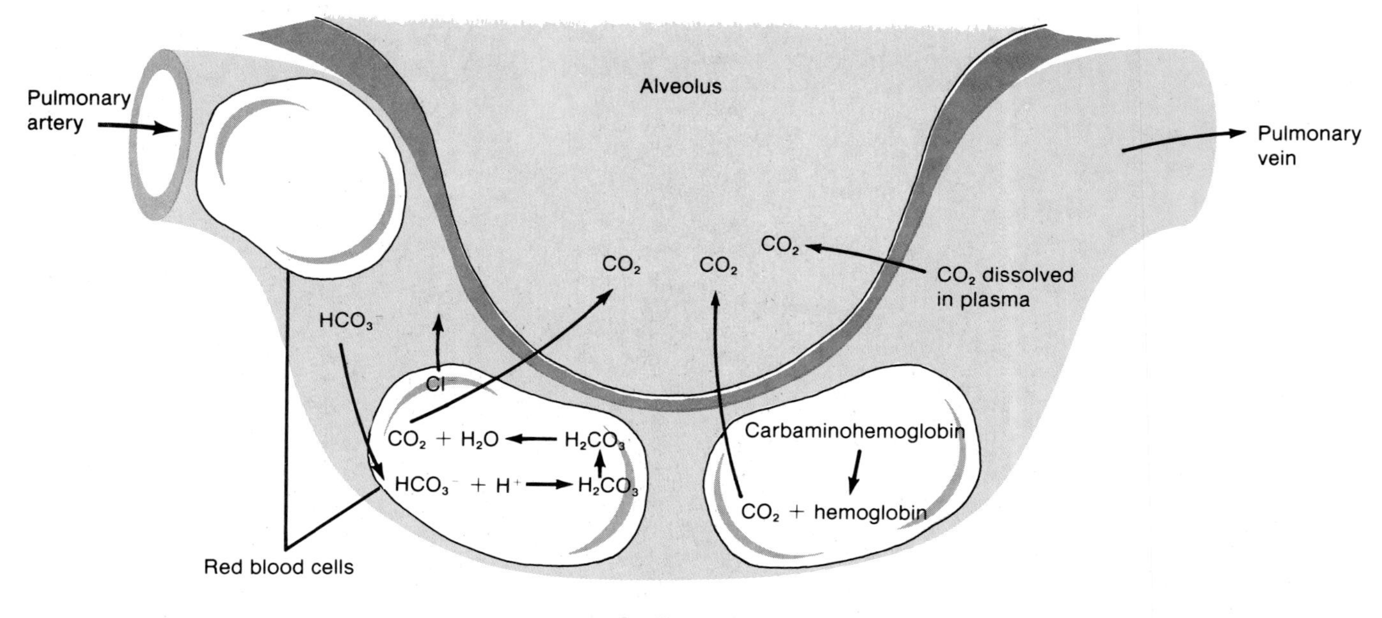

A *reverse chloride shift* operates in the pulmonary capillaries to convert carbonic acid and bicarbonate to CO_2 gas, which is eliminated in the expired breath (fig. 15.40). The P_{CO_2}, carbonic acid, H^+, and bicarbonate concentrations in the systemic arteries are thus maintained relatively constant by normal ventilation.

Ventilation and Acid-Base Balance

Normal systemic arterial blood has a pH of 7.35 to 7.45. In other words, arterial blood has a H^+ concentration of about $10^{-7.4}$ molar. Some of these H^+ are derived from carbonic acid, and some are derived from nonvolatile *metabolic acids* (fatty acids, ketone bodies, lactic acid, and others) that cannot be eliminated in the expired breath.

Under normal conditions the H^+ released by metabolic acids do not affect blood pH because these H^+ combine with HCO_3^- and are thereby removed from solution. *Bicarbonate is the major buffer in the plasma* and acts to maintain a blood pH of 7.4 despite the constant production of nonvolatile metabolic acids by the tissues. In this buffering process, some of the HCO_3^- released from the red blood cells during the chloride shift is converted into H_2CO_3 in the plasma. Normally, however, there is still a buffer reserve of free bicarbonate that can help protect against unusually large additions of metabolic acids to the blood. These processes are illustrated in figure 15.41.

Normal plasma, therefore, contains free bicarbonate, carbonic acid, and H^+ concentrations indicated by a pH of 7.4. If the H^+ concentration of the blood should fall, the carbonic acid produced by the buffering reaction can dissociate and serve as a source of additional H^+. If the H^+ concentration should rise, bicarbonate can remove this excess H^+ from solution. Carbonic acid and bicarbonate are thus said to function as a *buffer pair.*

Acidosis and Alkalosis. A fall in blood pH below 7.35 is called **acidosis,** because the pH is to the acid-side of normal. Acidosis does not mean acidic (pH less than 7); a blood pH of 7.2, for example, represents serious acidosis. Similarly, a rise in blood pH above 7.45 is known as **alkalosis.** There are two components to acid-base balance: *respiratory* and *metabolic.* Respiratory acidosis and alkalosis are due to abnormal concentrations of carbonic acid, as a result of abnormal ventilation; metabolic acidosis and alkalosis result from abnormal amounts of H^+ derived from nonvolatile metabolic acids (table 15.16).

Ventilation is normally adjusted to keep pace with the metabolic rate. *Hypoventilation* occurs when the rate of CO_2 production exceeds the rate at which it is "blown off" in ventilation. Under these conditions, carbonic acid production is excessively high compared to ventilation, and **respiratory acidosis** occurs. In *hyperventilation,* the depletion of carbonic acid raises the pH, and **respiratory alkalosis** occurs.

Figure 15.41. Bicarbonate released into the plasma from red blood cells functions to buffer H^+ produced by the ionization of metabolic acids (lactic acid, fatty acids, ketone bodies, and others).

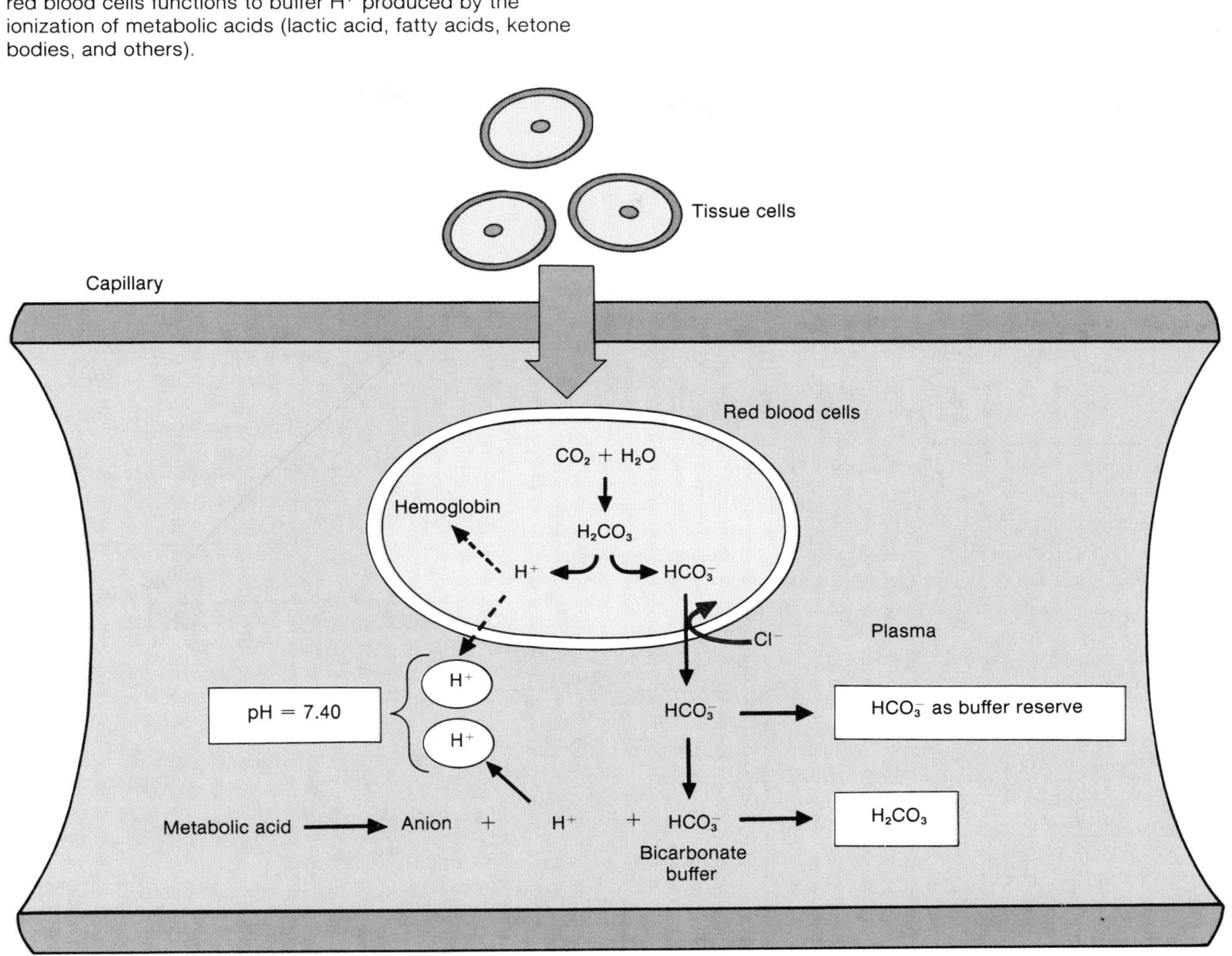

Table 15.16 Definition of terms used to describe acid-base balance.

Terms	Definitions
Acidosis, respiratory	Increased carbon dioxide retention (due to hypoventilation), which can result in the accumulation of carbonic acid and thus a fall in blood pH below normal
Acidosis, metabolic	Increased production of "nonvolatile" acids such as lactic acid, fatty acids, and ketone bodies, or loss of blood bicarbonate (such as by diarrhea) resulting in a fall in blood pH below normal
Alkalosis, respiratory	A rise in blood pH due to loss of CO_2 and carbonic acid (through hyperventilation)
Alkalosis, metabolic	A rise in blood pH produced by loss of nonvolatile acids (as in excessive vomiting) or by excessive accumulation of bicarbonate base
Compensated acidosis or alkalosis	Metabolic acidosis or alkalosis are partially compensated by opposite changes in blood carbonic acid levels (through changes in ventilation). Respiratory acidosis or alkalosis are partially compensated by increased retention or excretion of bicarbonate in the urine.

Table 15.17 The effect of lung function on blood acid-base balance.

Condition	pH	P_{CO_2}	Ventilation	Cause or Compensation
Normal	7.35–7.45	39–41 mm Hg	Normal	Not applicable
Respiratory acidosis	Low	High	Hypoventilation	Cause of the acidosis
Respiratory alkalosis	High	Low	Hyperventilation	Cause of the alkalosis
Metabolic acidosis	Low	Low	Hyperventilation	Compensation for acidosis
Metabolic alkalosis	High	High	Hypoventilation	Compensation for alkalosis

Metabolic acidosis can occur when the production of nonvolatile acids is abnormally increased. In uncontrolled diabetes mellitus, for example, ketone bodies (derived from fatty acids) may accumulate and produce *ketoacidosis.* In order for metabolic acidosis to occur, however, the buffer reserve of bicarbonate must first be depleted (this is why metabolic acidosis can also be produced by the excessive loss of bicarbonate, as in diarrhea). Until the buffer reserve is depleted the pH remains normal—ketosis can occur, for example, without ketoacidosis. **Metabolic alkalosis,** less common than metabolic acidosis, can result when there is a loss of acidic gastric juice by vomiting or when there is an excessive intake of bicarbonate (from stomach antacids or from an intravenous solution).

Compensations for Acidosis and Alkalosis. A change in blood pH, produced by alterations in either the respiratory or metabolic component of acid-base balance, can be compensated by a change in the other component. Metabolic acidosis, for example, is partially compensated by hyperventilation (the aortic and carotid bodies are directly stimulated by blood H^+), which causes a decrease in carbonic acid. People with metabolic acidosis would thus have a low pH accompanied by a low blood P_{CO_2} as a result of the hyperventilation. Metabolic alkalosis, similarly, is partially compensated by the retention of carbonic acid due to hypoventilation (table 15.17).

A person with respiratory acidosis has a low pH and a high blood P_{CO_2} due to hypoventilation. This condition can be partially compensated by the kidneys, which help to regulate the blood bicarbonate concentration. Two organs thus regulate blood acid-base balance: the lungs (regulating the respiratory component) and the kidneys (regulating the metabolic component). The role of the kidneys in acid-base balance is discussed in more detail in chapter 16.

1. *List the ways that carbon dioxide is carried by the blood. Using equations, show how carbonic acid and bicarbonate are formed.*
2. *Describe the events that occur in the chloride shift in the systemic capillaries, and describe the reverse chloride shift that occurs in the pulmonary capillaries.*
3. *Describe the functions of bicarbonate and carbonic acid in blood.*
4. *Describe how hyperventilation and hypoventilation affect the blood pH, and explain the mechanisms involved.*

Effect of Exercise and High Altitude on Respiratory Function

Changes in ventilation and oxygen delivery occur during exercise and during acclimatization to a high altitude. These changes help to compensate for the increased metabolic rate during exercise and for the decreased arterial P_{O_2} at a high altitude.

Ventilation during Exercise

Immediately upon exercise, the rate and depth of breathing increase to produce a total minute volume that is many times the resting value. This increased ventilation, particularly in well-trained athletes, is exquisitely matched to the simultaneous increase in oxygen consumption and carbon dioxide production by the exercising muscles. The arterial blood P_{O_2}, P_{CO_2}, and pH, thus, remain surprisingly constant during exercise (fig. 15.42).

It is tempting to suppose that ventilation increases during exercise as a result of the increased CO_2 production by the exercising muscles. Ventilation increases together with increased CO_2 production, however, so that blood measurements of P_{CO_2} during exercise are not significantly higher than at rest. The mechanisms responsible for the increased ventilation during exercise must therefore be more complex.

Figure 15.42. The effect of moderate and heavy exercise on arterial blood gases and pH. Notice that there are no consistent and significant changes in these measurements during the first several minutes of moderate and heavy exercise and that only the P_{CO_2} changes (actually decreases) during more prolonged exercise.

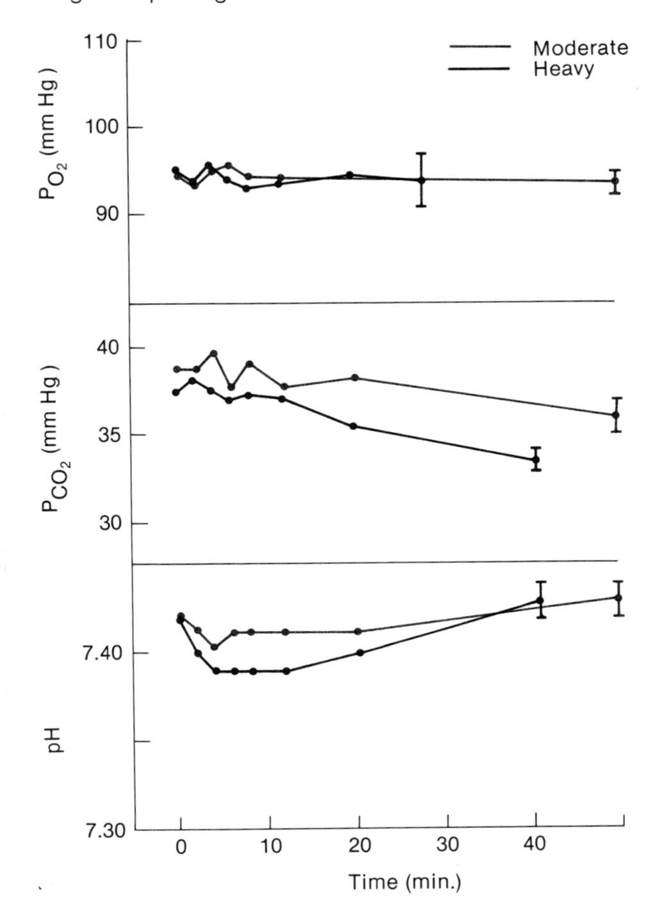

Two groups of mechanisms—*neurogenic* and *humoral*—have been proposed to explain the increased ventilation that occurs during exercise. Possible neurogenic mechanisms include the following: (1) sensory nerve activity from the exercising limbs may stimulate the respiratory muscles, either directly or via the brain stem respiratory centers; and/or (2) input from the cerebral cortex may stimulate the brain stem centers to modify ventilation. These neurogenic theories help to explain the immediate increase in ventilation that occurs at the beginning of exercise.

Rapid and deep ventilation continues after exercise has stopped, suggesting that humoral (chemical) factors in the blood may also stimulate ventilation during exercise. Since the P_{O_2}, P_{CO_2}, and pH of the blood samples from exercising subjects are within the resting range, these humoral theories propose that (1) the P_{CO_2} and pH in the region of the chemoreceptors may be different from these values "downstream" where blood samples are taken; and/or (2) there may be cyclic variations in these values that stimulate the chemoreceptors but cannot be detected by blood samples. The evidence suggests that both neurogenic and humoral mechanisms are involved in the *hyperpnea,* or increased ventilation, of exercise.

Anaerobic Threshold and Endurance Training. The ability of the cardiopulmonary system to deliver adequate amounts of oxygen to the exercising muscles at the beginning of exercise may be insufficient, as a result of the time lag required to make proper cardiovascular adjustments. During this time, therefore, the muscles respire anaerobically and a "stitch in the side"—probably due to hypoxia of the diaphragm—may develop. After the cardiovascular adjustments have been made, a person may experience a "second wind" when the muscles receive sufficient oxygen for their needs.

Continued heavy exercise can cause a person to reach the **anaerobic threshold,** which is the maximum rate of oxygen consumption that can be attained before blood lactic acid levels rise as a result of anaerobic respiration. This occurs when 50%–60% of the maximal oxygen uptake of the person has been reached. The rise in lactic acid levels is not due to a malfunction of the cardiopulmonary system; indeed, the arterial oxygen hemoglobin saturation remains at 97% and venous blood draining the muscles contains unused oxygen. The anaerobic threshold, however, is higher in endurance-trained athletes than it is in other people.

The rise in blood lactic acid that occurs when the anaerobic threshold is exceeded is due to the inability of the exercising muscles to increase their oxygen consumption rate sufficiently to prevent anaerobic respiration. Endurance training increases the skeletal muscle content of myoglobin, mitochondria, and Krebs cycle enzymes. These muscles, therefore, are able to utilize more of the oxygen delivered to them by the arterial blood. At a given level of exercise, in other words, the venous blood that drains from muscles in endurance-trained people contains a lower

Table 15.18 Changes in respiratory function during exercise.

Variable	Change	Comments
Ventilation	Increased	This is not hyperventilation because ventilation is matched to increased metabolic rate. Mechanisms responsible for increased ventilation are not well understood.
Blood gases	No change	Blood-gas measurements during light, moderate, and heavy exercise show little change because ventilation is increased to match increased muscle oxygen consumption and carbon dioxide production.
Oxygen delivery to muscles	Increased	Although the total oxygen content and P_{O_2} do not increase during exercise, there is an increased rate of blood flow to the exercising muscles.
Oxygen extraction by muscles	Increased	Increased oxygen consumption lowers the tissue P_{O_2} and lowers the affinity of hemoglobin for oxygen (due to the effect of increased temperature). More oxygen, as a result, is unloaded so that venous blood contains a lower oxyhemoglobin saturation than at rest. This effect is enhanced by endurance training.

percent oxyhemoglobin than in other people. The effects of exercise and endurance training on respiratory function are summarized in table 15.18.

Acclimatization to High Altitude

When a person who is from a region near sea level moves to a significantly higher altitude, several adjustments in the respiratory system must be made to compensate for the decreased atmospheric pressure and P_{O_2} at the higher elevation. These adjustments include changes in ventilation, in the hemoglobin affinity for oxygen, and in the total hemoglobin concentration.

Changes in Ventilation. A decrease in arterial P_{O_2} makes the chemoreceptors more sensitive to increases in P_{CO_2}; a decrease in P_{O_2} below 50 mm Hg directly stimulates the carotid bodies to increase ventilation. These changes cause *hyperventilation,* which becomes stabilized after a few days at about 2–3 L per minute more than the total minute volume at sea level. As a result of this hyperventilation, the arterial P_{CO_2} decreases from 40 mm Hg (its value at sea level) to about 29 mm Hg.

Hyperventilation cannot, of course, increase blood P_{O_2} above that of the inspired air. The P_{O_2} of arterial blood is therefore low at high altitudes. In the Peruvian Andes, for example, the normal arterial P_{O_2} is reduced from 100 mm Hg (at sea level) to 45 mm Hg. The loading of hemoglobin with oxygen is therefore incomplete, producing an oxyhemoglobin saturation that is decreased from 97% (at sea level) to 81%.

Hemoglobin Affinity for Oxygen. Normal arterial blood at sea level only unloads about 22% of its oxygen to the tissues at rest; the percent saturation is reduced from 97% in arterial blood to 75% in venous blood. As a partial compensation for the decrease in oxygen content at high altitude, the affinity of hemoglobin for oxygen is reduced, so that a higher proportion of oxygen is unloaded. This occurs because the low oxyhemoglobin content of red blood cells stimulates the production of 2,3-DPG, which in turn decreases the hemoglobin affinity for oxygen.

At very high altitudes, however, the story becomes more complex. At the summit of Mount Everest, in one study, the very low arterial P_{O_2} (28 mm Hg) stimulated intense hyperventilation, so that the arterial P_{CO_2} was decreased to 7.5 mm Hg. The resultant respiratory alkalosis (arterial pH greater than 7.7) caused a shift to the left of the oxyhemoglobin dissociation curve despite the antagonistic effects of increased 2, 3-DPG concentrations. It was suggested that the increased affinity of hemoglobin for oxygen caused by the respiratory alkalosis may have been beneficial at such a high altitude because it would have increased the loading of hemoglobin with oxygen in the lungs.

Increased Hemoglobin and Red Blood Cell Production. In response to tissue hypoxia, the kidneys secrete the hormone **erythropoietin.** Erythropoietin stimulates the bone marrow to increase its production of hemoglobin and red blood cells. In the Peruvian Andes, for example, people have a total hemoglobin concentration that is increased from 15 g per 100 ml (at sea level) to 19.8 g per 100 ml. Although the percent oxyhemoglobin saturation is still lower than at sea level, the total oxygen content of the blood is actually greater—22.4 ml O_2 per 100 ml compared to a sea level value of about 20 ml O_2 per 100 ml. These compensations of the respiratory system to high altitude are summarized in table 15.19.

1. *Describe the effect of exercise on the blood values of P_{O_2}, P_{CO_2}, and pH, and explain how ventilation might be increased during exercise.*
2. *Explain why endurance-trained athletes have a higher anaerobic threshold than other people.*
3. *Describe the changes that occur in the respiratory system during acclimatization to life at a high altitude.*

Table 15.19 Changes in respiratory function during acclimatization to high altitude.

Variable	Change	Comments
Partial oxygen pressure	Decreased	Due to decreased total atmospheric pressure
Percent oxyhemoglobin saturation	Decreased	Due to lower P_{O_2} in pulmonary capillaries
Ventilation	Increased	Due at first to low arterial P_{O_2}; later to changes in blood P_{CO_2} as chemoreceptor sensitivity increases
Total hemoglobin	Increased	Due to stimulation by erythropoietin; raises oxygen capacity of blood to partially or completely compensate for the reduced partial pressure
Oxyhemoglobin affinity	Decreased	Due to increased DPG within the red blood cells; results in a higher percent unloading of oxygen to the tissues, which may partially or completely compensate for the reduced arterial oxyhemoglobin saturation

Summary

The Respiratory System p. 428

I. Alveoli are small, numerous, thin-walled air sacs that provide an enormous surface area for gas diffusion.
   A. The region of the lungs where gas exchange with the blood occurs is known as the respiratory zone.
   B. The trachea, bronchi, and bronchioles that deliver air to the respiratory zone comprise the conducting zone.

II. The thoracic cavity is limited by the chest wall and diaphragm.
   A. The structures of the thoracic cavity are covered by thin, wet pleural membranes.
   B. The lungs are covered by a visceral pleura, which is normally against the parietal pleura, which lines the chest wall.
   C. The potential space between the two pleural membranes is called the intrapleural space.

Physical Aspects of Ventilation p. 435

I. The intrapleural and intrapulmonary pressures vary during ventilation.
   A. The intrapleural pressure is always less than the intrapulmonary pressure.
   B. The intrapulmonary pressure is subatmospheric during inspiration and greater than the atmospheric pressure during expiration.
   C. Pressure changes in the lungs are produced by variations in lung volume, in accordance with the inverse relationship between the volume and pressure of a gas described by Boyle's law.

II. The mechanics of ventilation are influenced by the physical properties of the lungs.
   A. The compliance of the lungs is the change in lung volume per change in transpulmonary pressure (the difference between intrapulmonary and intrapleural pressure).
   B. The elasticity of the lungs refers to their tendency to recoil after distension.
   C. The surface tension of the fluid in the alveoli exerts a force directed inwards, which acts to resist distension.

III. On first consideration, it would seem that the surface tension in the alveoli would create a pressure that would cause smaller alveoli to collapse and empty their air into large alveoli.
   A. This would occur because the pressure caused by a given amount of surface tension would be greater in smaller alveoli than in larger alveoli, as described by the law of LaPlace.
   B. Collapse of alveoli due to surface tension does not normally occur, however, because the presence of pulmonary surfactant (a combination of phospholipid and protein) acts to lower surface tension.
   C. In hyaline membrane disease, the lungs of premature infants collapse due to the absence of surfactant.

Mechanics of Breathing p. 438

I. Inspiration and expiration are accomplished by contraction and relaxation of striated muscles.
   A. During quiet inspiration, the diaphragm and external intercostal muscles contract and thus increase the volume of the thorax.
   B. During quiet expiration, these muscles relax, and the elastic recoil of the lungs and thorax causes a decrease in thoracic volume.
   C. Forced inspiration and expiration are aided by contraction of the accessory respiratory muscles.

II. Spirometry aids the diagnosis of a number of pulmonary disorders.
   A. In restrictive disease, such as pulmonary fibrosis, the vital capacity measurement is decreased below normal.
   B. In obstructive disease, such as asthma and bronchitis, the forced expiratory volume is reduced below normal because of increased airway resistance to air flow.

III. Asthma results from bronchoconstriction; emphysema and chronic bronchitis are frequently referred to as chronic obstructive pulmonary disease.

Gas Exchange in the Lungs p. 444

I. According to Dalton's law, the total pressure of a gas mixture is equal to the sum of the pressures that each gas in the mixture would exert independently.

A. The partial pressure of a gas in a dry gas mixture is thus equal to the total pressure times the percent composition of that gas in the mixture.
B. Since the total pressure of a gas mixture decreases with altitude above sea level, the partial pressures of the constituent gases likewise decrease with altitude.
C. When the partial pressure of a gas in a wet gas mixture is calculated, the water vapor pressure must be taken into account.

II. According to Henry's law, the amount of gas that can be dissolved in a fluid is directly proportional to the partial pressure of that gas in contact with the fluid.
A. The concentrations of oxygen and carbon dioxide that are dissolved in plasma are proportional to an electric current generated by special electrodes that react with these gases.
B. Normal arterial blood has a P_{O_2} of 100 mm Hg, indicating a concentration of dissolved oxygen of 0.3 ml per 100 ml of blood; the oxygen contained in red blood cells (about 19.7 ml per 100 ml blood) does not affect the P_{O_2} measurement.

III. The P_{O_2} and P_{CO_2} measurements of arterial blood provide information about lung function.

IV. In addition to proper ventilation of the lungs, blood flow (perfusion) in the lungs must be adequate and matched to air flow (ventilation) in order for adequate gas exchange to occur.

V. Abnormally high partial pressures of gases in blood can cause a variety of disorders, including oxygen toxicity, nitrogen narcosis, and decompression sickness.

Regulation of Breathing p. 451

I. The rhythmicity center in the medulla oblongata directly controls the muscles of respiration.
A. Activity of the inspiratory and expiratory neurons varies in a reciprocal way to produce an automatic breathing cycle.
B. Activity in the medulla is influenced by the apneustic and pneumotaxic centers in the pons, as well as by sensory feedback information.
C. Conscious breathing involves direct control by the cerebral cortex via corticospinal tracts.

II. Breathing is affected by chemoreceptors sensitive to the P_{CO_2}, pH, and P_{O_2} of the blood.
A. The P_{CO_2} of the blood and consequent changes in pH are usually of greater importance than the blood P_{O_2} in the regulation of breathing.
B. Central chemoreceptors in the medulla oblongata are sensitive to changes in blood P_{CO_2}, because these changes cause the pH of cerebrospinal fluid to change.
C. The peripheral chemoreceptors in the aortic and carotid bodies are sensitive to changes in blood P_{CO_2} indirectly, because of consequent changes in blood pH.

III. Decreases in blood P_{O_2} directly stimulate breathing only when the blood P_{O_2} is less than 50 mm Hg; a drop in P_{O_2} also stimulates breathing indirectly, by making the chemoreceptors more sensitive to changes in P_{CO_2} and pH.

IV. At tidal volumes of one liter or more, inspiration is inhibited by stretch receptors in the lungs (the Hering-Breuer reflex); there is a similar deflation reflex.

Hemoglobin and Oxygen Transport p. 456

I. Hemoglobin is composed of two alpha and two beta polypeptide chains and four heme groups that contain a central atom of iron.
A. When the iron is in the reduced form and not attached to oxygen, the hemoglobin is called deoxyhemoglobin, or reduced hemoglobin; when it is attached to oxygen, it is called oxyhemoglobin.
B. If the iron is attached to carbon monoxide, the hemoglobin is called carboxyhemoglobin; when the iron is in an oxidized state and unable to transport any gas, the hemoglobin is called methemoglobin.
C. Deoxyhemoglobin combines with oxygen in the lungs (the loading reaction) and breaks its bonds with oxygen in the tissue capillaries (the unloading reaction); the extent of each reaction is determined by the P_{O_2} and the affinity of hemoglobin for oxygen.

II. A graph of percent oxyhemoglobin saturation at different values of P_{O_2} is called an oxyhemoglobin dissociation curve.
A. At rest, the difference between arterial and venous oxyhemoglobin saturations indicates that about 22% of the oxyhemoglobin unloads its oxygen to the tissues.
B. During exercise, the venous P_{O_2} and percent oxyhemoglobin saturation are decreased, indicating that a higher percent of the oxyhemoglobin unloaded its oxygen to the tissues.

III. The pH and temperature of the blood influence the affinity of hemoglobin for oxygen and thus the extent of loading and unloading.
A. A fall in pH decreases the affinity, and a rise in pH increases the affinity of hemoglobin for oxygen; this is called the Bohr effect.
B. A rise in temperature decreases the affinity of hemoglobin for oxygen.
C. When the affinity is decreased, the oxyhemoglobin dissociation curve is shifted to the right; this indicates a greater percentage unloading of oxygen to the tissues.

IV. The affinity of hemoglobin for oxygen is also decreased by an organic molecule in the red blood cells called 2,3-diphosphoglyceric acid (2,3-DPG).
A. Since oxyhemoglobin inhibits 2,3-DPG production, there will be higher 2,3-DPG concentrations when the oxyhemoglobin is decreased due to anemia or low P_{O_2} (as in high altitude).

B. If a person is anemic, the lowered hemoglobin concentration is partially compensated by the fact that a higher percent of the oxyhemoglobin will unload its oxygen due to the effect of 2,3-DPG.

C. Since fetal hemoglobin cannot bond 2,3-DPG, it has a higher affinity for oxygen than the mother's hemoglobin; this facilitates the transfer of oxygen to the fetus.

V. Striated muscles have myoglobin, a pigment related to hemoglobin, which can combine with oxygen and deliver it to the muscle cell mitochondria at low values of P_{O_2}.

VI. Inherited defects in the amino acid composition of hemoglobin are responsible for such diseases as sickle-cell anemia and thalassemia.

Carbon Dioxide Transport and Acid-Base Balance p. 464

I. Red blood cells contain an enzyme called carbonic anhydrase, which catalyzes the reversible reaction whereby carbon dioxide and water are used to form carbonic acid.

A. This reaction is favored by the high P_{CO_2} in the tissue capillaries, and as a result, carbon dioxide produced by the tissues is converted into carbonic acid in the red blood cells.

B. Carbonic acid then ionizes to form H^+ and HCO_3^- (bicarbonate).

C. Since much of the H^+ is buffered by hemoglobin, but more bicarbonate is free to diffuse outward, an electrical gradient is established which draws Cl^- into the red blood cells; this is called the chloride shift.

D. A reverse chloride shift occurs in the lungs; in this process, the low P_{CO_2} favors the conversion of carbonic acid to carbon dioxide, which can be exhaled.

II. By adjusting the blood concentration of carbon dioxide and thus of carbonic acid, the process of ventilation helps to maintain proper acid-base balance of the blood.

A. Normal arterial blood pH is 7.40; a pH less than 7.35 is termed acidosis, and a pH greater than 7.45 is termed alkalosis.

B. Hyperventilation causes respiratory alkalosis, and hypoventilation causes respiratory acidosis.

C. Metabolic acidosis stimulates hyperventilation, which can cause a respiratory alkalosis as a partial compensation.

Effect of Exercise and High Altitude on Respiratory Function p. 468

I. During exercise there is increased ventilation, or hyperpnea, which can be matched to the increased metabolic rate so that the arterial blood P_{CO_2} remains normal.

A. This hyperpnea may be caused by proprioceptor information, cerebral input, and/or changes in arterial P_{CO_2} and pH.

B. During heavy exercise the anaerobic threshold may be reached at about 50%–60% of the maximal oxygen uptake; at this point, lactic acid is released into the blood by the muscles.

C. Endurance training enables the muscles to utilize oxygen more effectively, so that greater levels of exercise can be performed before the anaerobic threshold is reached.

II. Acclimatization to a high altitude involves changes that help to deliver oxygen more effectively to the tissues despite reduced arterial P_{O_2}.

A. Hyperventilation occurs in response to the low P_{O_2}.

B. The red blood cells produce more 2,3-DPG, which lowers the affinity of hemoglobin for oxygen and improves the unloading reaction.

C. The kidneys produce the hormone erythropoietin, which stimulates the bone marrow to increase its production of red blood cells, so that more oxygen can be carried by the blood at given values of P_{O_2}.

Review Activities

Objective Questions

1. Which of the following statements about intrapulmonary and intrapleural pressure is true?
   (a) The intrapulmonary pressure is always subatmospheric.
   (b) The intrapleural pressure is always greater than the intrapulmonary pressure.
   (c) The intrapulmonary pressure is greater than the intrapleural pressure.
   (d) The intrapleural pressure equals the atmospheric pressure.

2. If the transpulmonary pressure equals zero,
   (a) a pneumothorax has probably occurred
   (b) the lungs cannot inflate
   (c) elastic recoil causes the lungs to collapse
   (d) all of the above

3. The maximum amount of air that can be expired after a maximum inspiration is the
   (a) tidal volume
   (b) forced expiratory volume
   (c) vital capacity
   (d) maximum expiratory flow rate

4. If the blood lacked red blood cells but the lungs were functioning normally,
   (a) the arterial P_{O_2} would be normal
   (b) the oxygen content of arterial blood would be normal
   (c) both a and b
   (d) neither a nor b

5. If a person were to dive with scuba equipment to a depth of sixty-six feet, which of the following statements would be *false?*
   (a) The arterial P_{O_2} would be three times normal.
   (b) The oxygen content of plasma would be three times normal.
   (c) The oxygen content of whole blood would be three times normal.
6. Which of the following would be most affected by a decrease in the affinity of hemoglobin for oxygen? The
   (a) arterial P_{O_2}
   (b) arterial percent oxyhemoglobin saturation
   (c) venous oxyhemoglobin saturation
   (d) arterial P_{CO_2}
7. If a person with normal lung function were to hyperventilate for several seconds, there would be a significant
   (a) increase in the arterial P_{O_2}
   (b) decrease in the arterial P_{CO_2}
   (c) increase in the arterial percent oxyhemoglobin saturation
   (d) decrease in the arterial pH
8. Erythropoietin is produced by the
   (a) kidneys
   (b) liver
   (c) lungs
   (d) bone marrow
9. The affinity of hemoglobin for oxygen is decreased under conditions of
   (a) acidosis
   (b) fever
   (c) anemia
   (d) acclimatization to a high altitude
   (e) all of the above
10. Most of the carbon dioxide in the blood is carried in the form of
    (a) dissolved CO_2
    (b) carbaminohemoglobin
    (c) carbonic acid and bicarbonate
    (d) carboxyhemoglobin
11. The bicarbonate concentration of the blood would be decreased during
    (a) metabolic acidosis
    (b) respiratory acidosis
    (c) metabolic alkalosis
    (d) respiratory alkalosis
12. The chemoreceptors in the medulla are directly stimulated by
    (a) CO_2 from the blood
    (b) H^+ from the blood
    (c) H^+ in cerebrospinal fluid that is derived from blood CO_2
    (d) decreased arterial P_{O_2}
13. The rhythmic control of breathing is produced by the activity of inspiratory and expiratory neurons in the
    (a) medulla oblongata
    (b) apneustic center of the pons
    (c) pneumotaxic center of the pons
    (d) cerebral cortex
14. Which of the following occurs during hypoxemia?
    (a) increased ventilation
    (b) increased production of 2,3-DPG
    (c) increased production of erythropoietin
    (d) all of the above
15. During exercise, which of the following statements is true?
    (a) The arterial percent oxyhemoglobin saturation is decreased.
    (b) The venous percent oxyhemoglobin saturation is decreased.
    (c) The arterial P_{CO_2} is measurably increased.
    (d) The arterial pH is measurably decreased.

Essay Questions

1. Using a flow diagram to show cause and effect, explain how contraction of the diaphragm produces inspiration.
2. Radiographic (X-ray) pictures show that the ribs of a person with a pneumothorax are expanded and farther apart. Explain why this occurs.
3. Explain, using a flow chart, how a rise in blood P_{CO_2} stimulates breathing. Include both the central and peripheral chemoreceptors in your answer.
4. A person with ketoacidosis may hyperventilate. Explain why this occurs, and explain why this hyperventilation can be stopped by an intravenous fluid containing bicarbonate.
5. What blood measurements can be performed to detect (*a*) anemia, (*b*) carbon monoxide poisoning, and (*c*) poor lung function?
6. Explain how measurements of blood P_{CO_2}, bicarbonate, and pH are affected by hypoventilation and hyperventilation.
7. Explain how blood pH and bicarbonate concentrations are affected by respiratory and metabolic acidosis.
8. How would an increase in the red blood cell content of 2,3-DPG affect the P_{O_2} of venous blood? Explain your answer.

Selected Readings

Avery, M. E., N. S. Wang, and H. W. Taeusch, Jr. March 1975. The lung of the newborn infant. *Scientific American.*

Berger, A. J., R. A. Mitchel, and J. W. Severinghaus. 1977. Regulation of respiration. *New England Journal of Medicine* 297: first part, p. 92; second part, p. 138; third part, p. 194.

Bramble, D. M., and D. R. Carrier. 1983. Running and breathing in mammals. *Science* 219:251.

Browning, R. J. 1982. Pulmonary disease: Back to basics (part 1); Putting blood gases to work (part 2). *Diagnostic Medicine.* First part: Jan/Feb, p. 39; second part: March/April, p. 59.

Cherniak, N. S. 1986. Breathing disorders during sleep. *Hospital Practice* 21:81.

Dantzker, D. R. 1986. Physiology and pathophysiology of pulmonary gas exchange. *Hospital Practice 21*:135.

Finch, C. A., and C. Lenfant. 1972. Oxygen transport in man. *New England Journal of Medicine* 286:407.

Flenley, D. C., and P. M. Warren. 1983. Ventilatory response to O_2 and CO_2 during exercise. *Annual Review of Physiology* 45:415.

Fraser, R. G., and J. A. P. Pare. 1977. *Structure and function of the lung.* 2d ed. Philadelphia: W. B. Saunders Co.

Guz, A. 1975. Regulation of respiration in man. *Annual Review of Physiology* 37:303.

Haddad, G. G., and R. B. Mellins. 1984. Hypoxia and respiratory control in early life. *Annual Review of Physiology* 46:629.

Irsigler, G. B., and J. W. Severinghaus. 1980. Clinical problems of ventilatory control. *Annual Review of Medicine* 31:109.

Kassirer, J. P., and N. E. Madias. 1980. Respiratory acid-base disorders. *Hospital Practice* 15:57.

Macklem, P. T. 1986. Respiratory muscle dysfunction. *Hospital Practice* 21:83.

Massaro, D. 1986. Oxygen: Toxicity and tolerance. *Hospital Practice* 21:95.

Murray, J. F. 1985. The lungs and heart failure. *Hospital Practice 20*:55.

Nadel, E. R. 1985. Physiological adaptations to aerobic training. *American Scientist 73*:334.

Naeye, R. L. April 1980. Sudden infant death. *Scientific American.*

Perutz, M. F. December 1978. Hemoglobin structure and respiratory transport. *Scientific American.*

Rigatto, H. 1984. Control of ventilation in the newborn. *Annual Review of Physiology* 46:661.

Roussos, C., and P. T. Macklem. 1982. The respiratory muscles. *New England Journal of Medicine* 307:786.

Shannon, D. C., and D. H. Kelly. 1982. SIDS and near-SIDS. *New England Journal of Medicine* 306: first part, p. 959; second part, p. 1022.

Snapper, J. R., and K. L. Bingham. 1986. Pulmonary edema. *Hospital Practice* 21:87.

Tobin, M. J. 1986. Update on strategies in mechanical ventilation. *Hospital Practice* 21:69.

Walker, D. W. 1984. Peripheral and central chemoreceptors in the fetus and newborn. *Annual Review of Physiology* 46:687.

West, J. B. 1984. Human physiology at extreme altitudes on Mount Everest. *Science* 223:784.

Whipp, B. J. 1983. Ventilatory control during exercise in humans. *Annual Review of Physiology* 45:393.

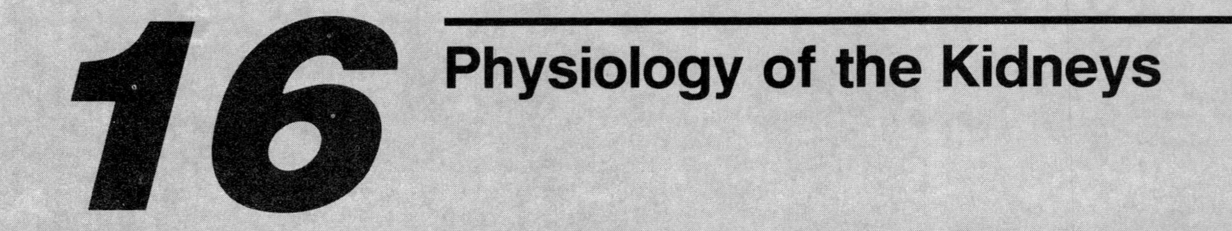

16 Physiology of the Kidneys

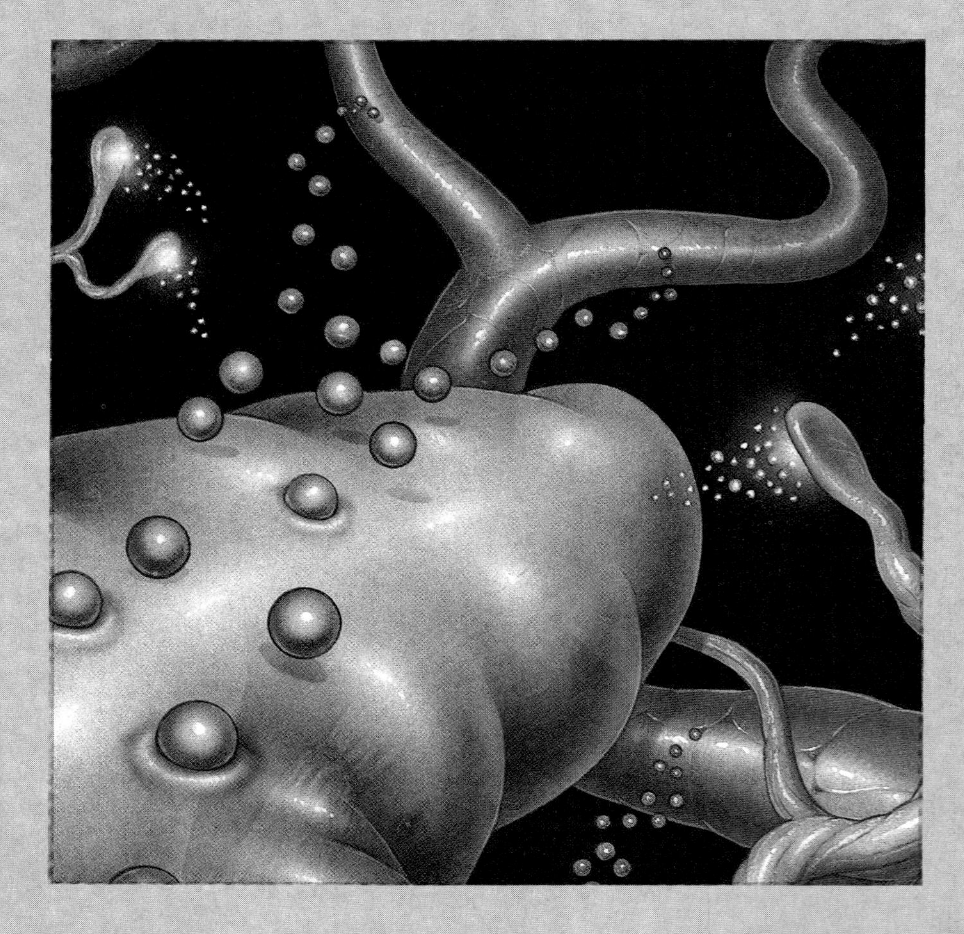

Objectives

By studying this chapter, you should be able to

1. describe the different regions of the nephron tubules and explain the anatomic relationship between the tubules and the gross structure of the kidney
2. describe the structural and functional relationships between the nephron tubules and their associated blood vessels
3. describe the composition of glomerular ultrafiltrate and explain how it is produced, how it enters Bowman's capsule, and how the glomerular filtration rate is regulated
4. explain how the proximal convoluted tubule reabsorbs salt and water and relate the mechanisms of reabsorption to the structure of the tubular epithelial cells
5. describe the processes of active transport and osmosis in the ascending and descending limbs of the loop of Henle and explain how these processes function to produce a countercurrent multiplier system
6. explain how the vasa recta function in countercurrent exchange to maintain a high osmolality in the tissue fluid of the renal medulla, and state the significance of this high osmolality
7. explain how antidiuretic hormone (ADH) functions to regulate the final urine volume
8. describe the mechanisms of glucose and amino acid reabsorption and explain the meanings of the terms *transport maximum* and *renal plasma threshold*
9. describe the meaning of the term *renal plasma clearance rate*, explain why the clearance of rate of inulin is equal to the glomerular filtration rate (GFR), and explain how the GFR is calculated
10. explain how the clearance rate of different molecules is determined and how the processes of reabsorption and secretion affect the clearance rate measurement
11. explain how the renal plasma clearance rate of para-aminohippuric acid (PAH) can be used to measure the total renal blood flow
12. explain the mechanism of Na^+ reabsorption in the distal tubule and why this reabsorption occurs together with the secretion of K^+
13. describe the effects of aldosterone on the distal convoluted tubule and how aldosterone secretion is affected by the blood Na^+ and K^+ concentration
14. describe the structure of the juxtaglomerular apparatus and explain how activation of the renin-angiotensin system results in the stimulation of aldosterone secretion
15. describe the interaction between plasma K^+ and H^+ concentrations and explain how this affects the tubular secretion of these ions
16. distinguish between the respiratory and metabolic components of acid-base balance and describe the role of the kidneys in the regulation of acid-base balance
17. describe the different mechanisms by which substances can act as diuretics and explain why some cause excessive loss of K^+ whereas others are classified as potassium-sparing diuretics

Outline

Figure 16.1. The anatomical locations of the kidneys, ureters, and urinary bladder.

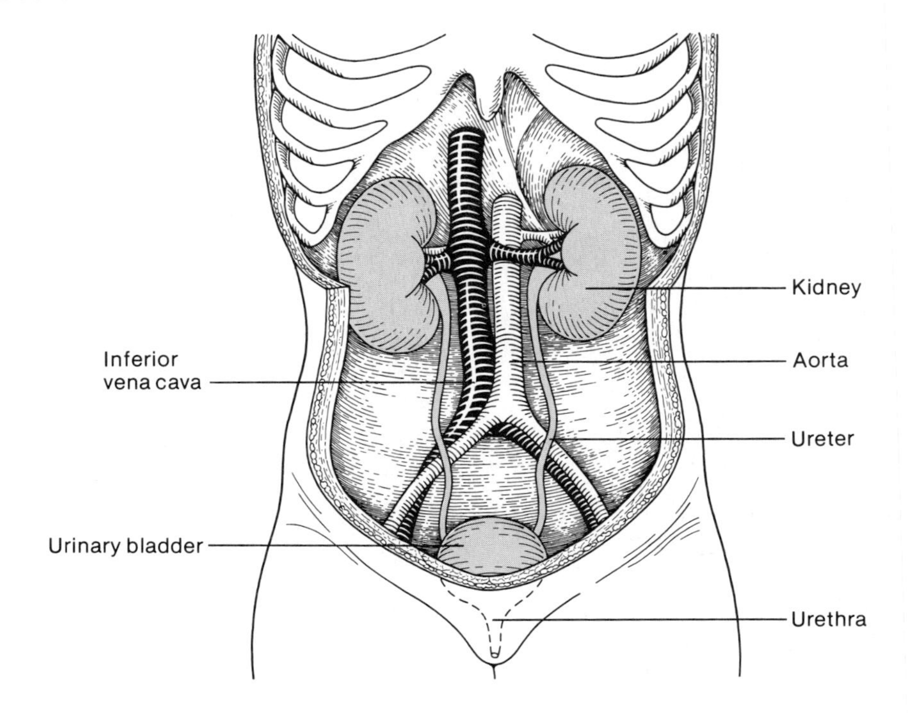

Structure and Function of the Kidneys

The primary function of the kidneys is regulation of the extracellular fluid (plasma and tissue fluid) environment in the body. This function is accomplished through the formation of urine, which is a modified filtrate of plasma. In the process of urine formation, the kidneys regulate (1) the volume of blood plasma and thus contribute significantly to the regulation of blood pressure; (2) the concentration of waste products in the blood; (3) the concentration of electrolytes (Na^+, K^+, HCO_3^-, and other ions) in the plasma; and (4) the pH of plasma. In order to understand how these functions are performed by the kidneys, a knowledge of kidney structure is required.

Gross Structure of the Kidney

The paired right and left kidneys are located in the abdominal cavity below the diaphragm and the liver. Each kidney in an adult weighs approximately 160–175 g, and is about 10–12 cm long and 5–6 cm wide—about the size of a fist. Urine produced in the kidneys is drained into the *renal pelvis* (= basin), and from there it is channeled via two long ducts—the *ureters*—to the single *urinary bladder* (fig. 16.1).

A coronal section of the kidney shows two distinct regions (fig. 16.2). The outer **cortex,** in contact with the capsule, is reddish brown and granular in appearance because of its many capillaries. The deeper region, or **medulla,** is lighter in color and striped in appearance because of the presence of microscopic tubules and blood vessels. The medulla is composed of eight to fifteen conical *renal pyramids,* separated by *renal columns.*

The cavity of the kidney collects and transports urine from the kidney to the ureter. It is divided into several portions. Each pyramid projects into a small depression called a *minor calyx* (the plural form is *calyces*). Several minor calyces unite to form a *major calyx.* In turn, the major calyces join to form the funnel-shaped *renal pelvis.* The renal pelvis is actually an expanded portion of the ureter in the kidney and serves to collect urine from the calyces and transport it to the ureter.

Microscopic Structure of the Kidney

The **nephron** is the functional unit of the kidney that is responsible for the formation of urine. Each kidney contains more than a million nephrons. A nephron consists of **tubules** and associated small blood vessels. Fluid formed by capillary filtration enters the tubules and is subsequently modified by transport processes; the resulting fluid that leaves the tubules is urine.

Figure 16.2. The internal structures of a kidney. (*a*) A coronal section showing the structure of the cortex, medulla, and renal calyces. (*b*) A diagrammatic magnification of a renal pyramid and cortex to depict the tubules. (*c*) A diagrammatic view of a single nephron.

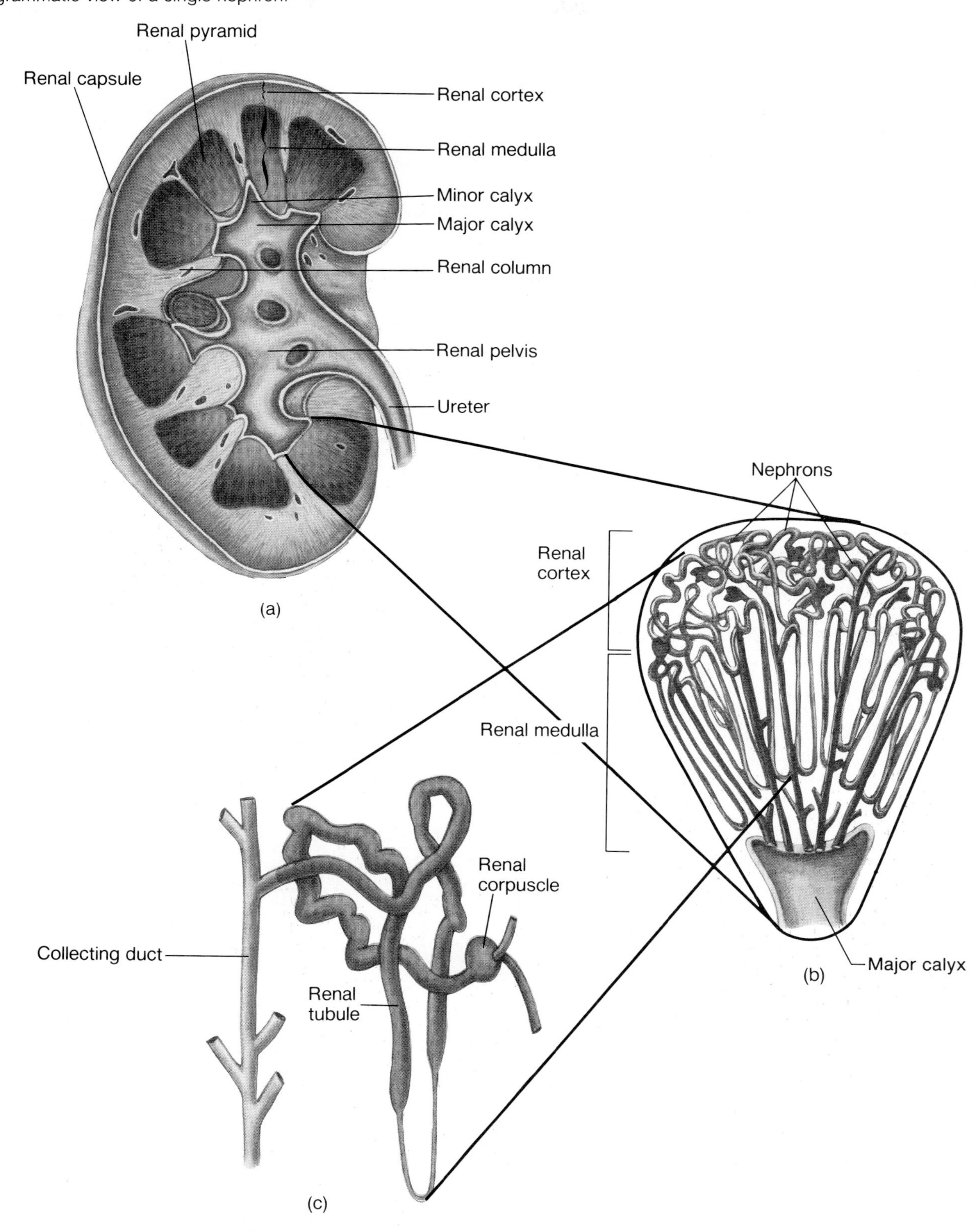

Figure 16.3. The vascular structure of the kidneys. (*a*) An illustration of the major arterial supply and (*b*) a scanning electron micrograph of the glomeruli. (From *Tissues and Organs: A Text-Atlas of Scanning Electron Microscopy* by R. G. Kessel and R. Kardon. W. H. Freeman and Company © 1979.)

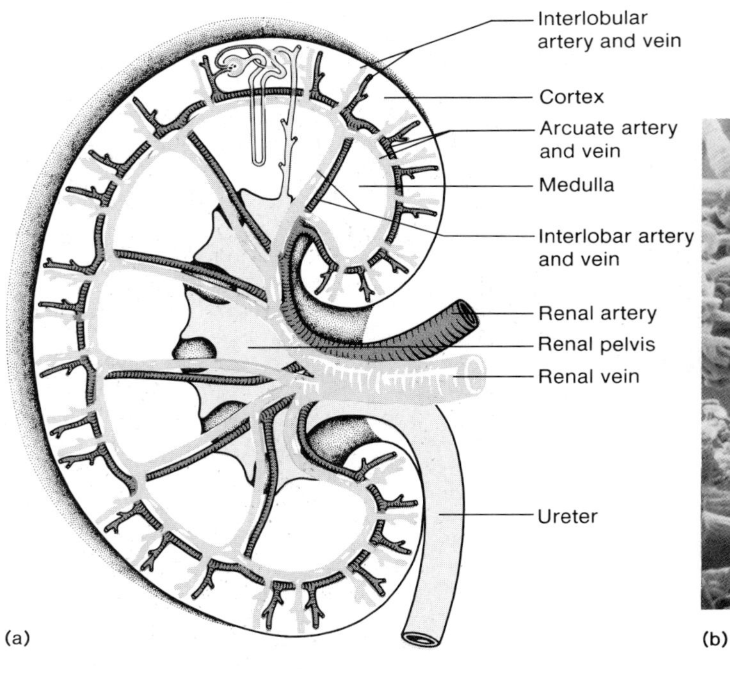

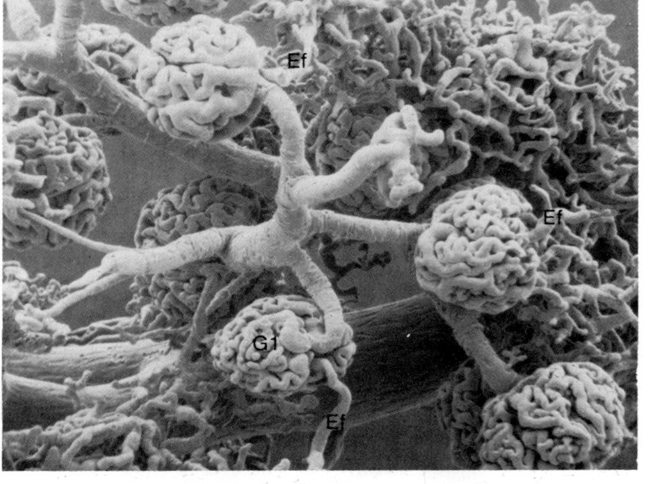

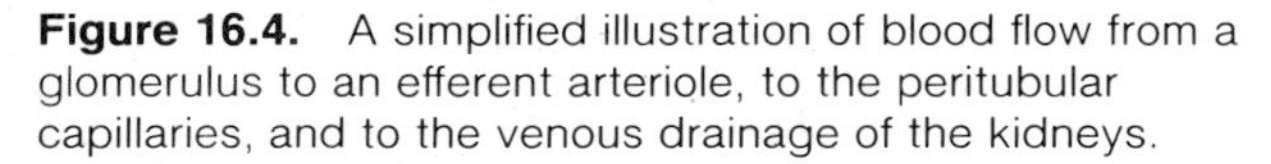

Figure 16.4. A simplified illustration of blood flow from a glomerulus to an efferent arteriole, to the peritubular capillaries, and to the venous drainage of the kidneys.

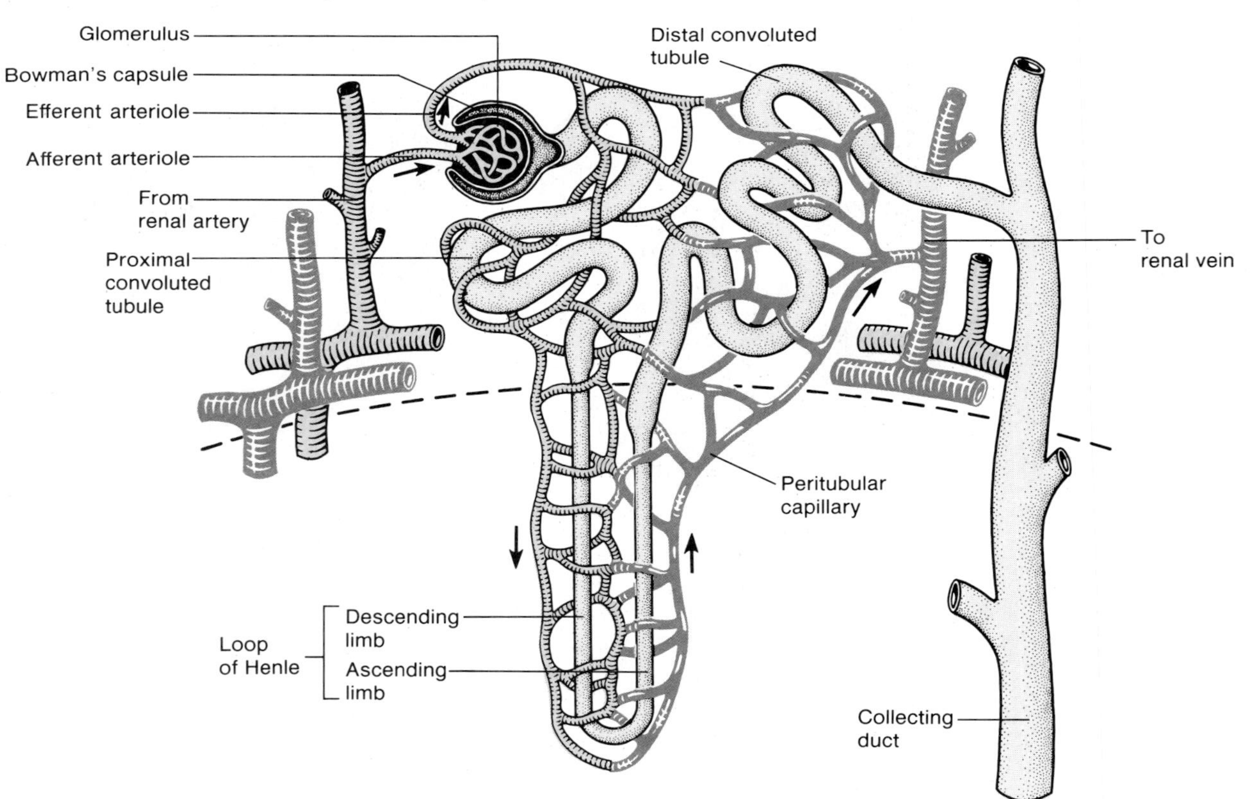

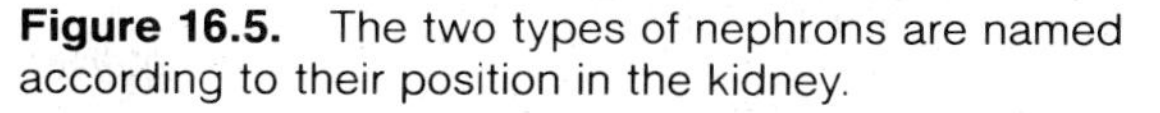
Figure 16.5. The two types of nephrons are named according to their position in the kidney.

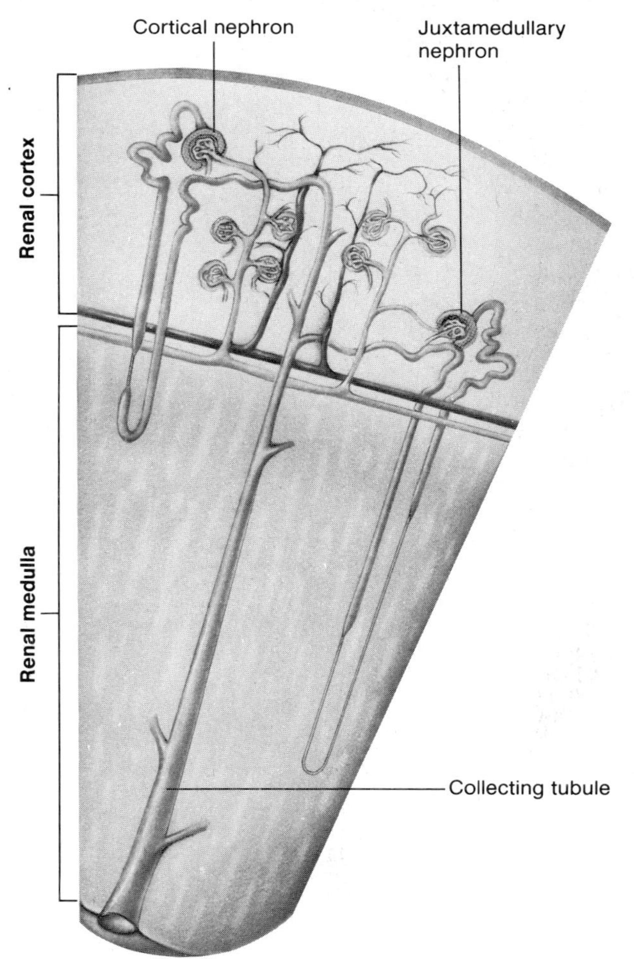

Renal Blood Vessels. Arterial blood enters the kidney through the *renal artery,* which divides into *interlobar arteries* (fig. 16.3), that pass between the pyramids through the renal columns. *Arcuate arteries* branch from the interlobar arteries at the boundary of the cortex and medulla. Many *interlobular arteries* radiate from the arcuate arteries and subdivide into numerous *afferent arterioles* (fig. 16.4), which are microscopic in size. The afferent arterioles deliver blood into capillary networks, called **glomeruli,** which produce a blood filtrate that enters the urinary tubules. The blood remaining in the glomerulus leaves through an *efferent arteriole,* which delivers the blood into another capillary network, the *peritubular capillaries,* that surrounds the tubules. This blood is drained into veins that parallel the course of the arteries in the kidney and are named the *interlobular veins, arcuate veins,* and *interlobar veins.* The interlobar veins descend between the pyramids, converge, and leave the kidney as a single *renal vein* that empties into the inferior vena cava.

Nephron Tubules. The tubular portion of a nephron consists of a glomerular capsule, a proximal convoluted tubule, a descending limb of the loop of Henle, an ascending limb of the loop of Henle, and a distal convoluted tubule (fig. 16.4).

The **glomerular (Bowman's) capsule** surrounds the glomerulus. The glomerular capsule and its associated glomerulus are located in the cortex of the kidney and together constitute the *renal corpuscle.* The glomerular capsule contains an inner visceral layer of epithelium around the glomerular capillaries and an outer parietal layer. The space between these two layers receives the glomerular filtrate.

Filtrate in the glomerular capsule passes into the lumen of the **proximal convoluted tubule.** The wall of the proximal convoluted tubule consists of a single layer of cuboidal cells, containing millions of microvilli; these microvilli serve to increase the surface area for reabsorption. In the process of reabsorption, salt, water, and other molecules needed by the body are transported from the lumen, through the tubular cells, and into the surrounding peritubular capillaries.

The glomerulus, glomerular capsule, and proximal convoluted tubule are located in the renal cortex. Fluid passes from the proximal convoluted tubule to the **loop of Henle.** This fluid is carried into the medulla in the **descending limb** of the loop and returns to the cortex in the **ascending limb** of the loop. Back in the cortex, the tubule becomes coiled again and is called the **distal convoluted tubule.** In contrast to the proximal tubule, the distal convoluted tubule is shorter and has relatively few microvilli. The distal convoluted tubule is the last segment of the nephron and terminates as it empties into a collecting duct.

There are two types of nephrons, which are classified according to their position in the kidney and the lengths of their loops of Henle. Nephrons that originate in the inner one-third of the cortex—called *juxtamedullary nephrons*—have longer loops of Henle than the more numerous *cortical nephrons,* which originate in the outer two-thirds of the cortex (fig. 16.5).

The distal convoluted tubules of several nephrons drain into a **collecting duct.** Fluid is then drained by the collecting duct from the cortex to the medulla as the collecting duct passes through a renal pyramid. This fluid, now called urine, passes into a minor calyx. Urine is then funneled through the renal pelvis and out of the kidney in the ureter.

1. *Describe the "theme" of kidney function in a single sentence, and list the components of this functional theme.*
2. *Trace the course of blood flow through the kidney from the renal artery to the renal vein.*
3. *Trace the course of tubular fluid from the glomerular capsules to the ureter.*
4. *Draw a diagram of the tubular component of a nephron. Label the segments, and indicate which parts are in the cortex and which are in the medulla.*

Figure 16.6. (*a*) The inner (visceral) layer of Bowman's capsule is composed of podocytes, as shown in this scanning electron micrograph. Very fine extensions of these podocytes form foot processes, or pedicels, that interdigitate around the glomerular capillaries. Spaces between adjacent pedicels form the "filtration slits." (*b*) An illustration of the relationship between glomerular capillaries and the inner layer of Bowman's capsule.

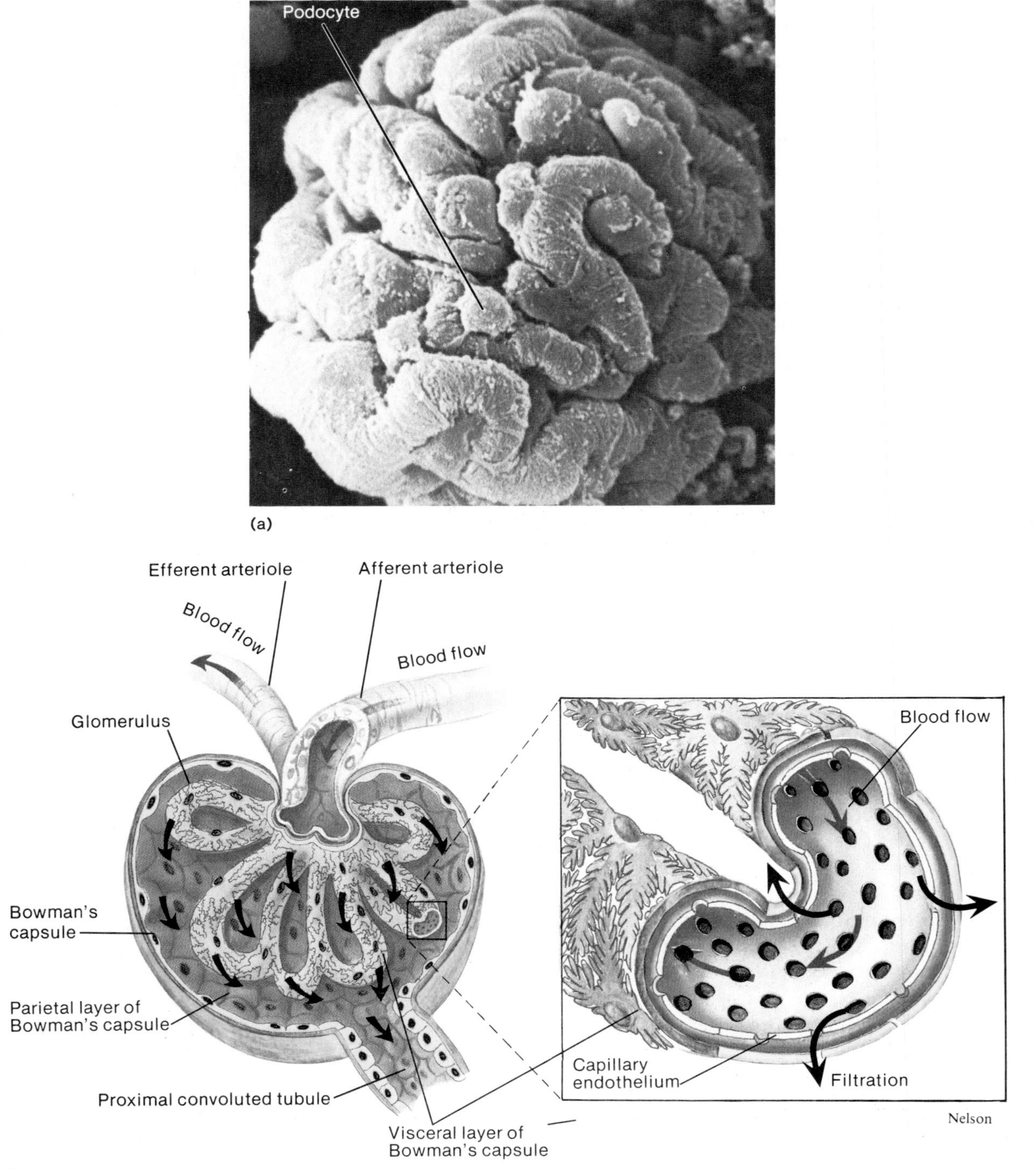

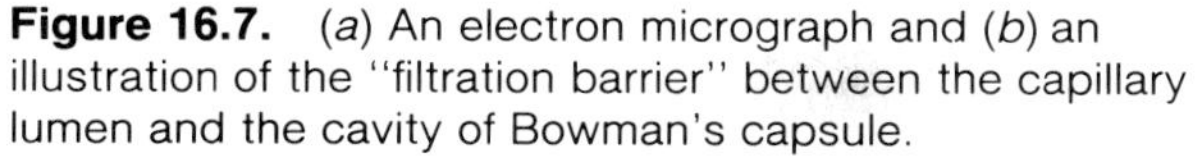

Figure 16.7. (*a*) An electron micrograph and (*b*) an illustration of the "filtration barrier" between the capillary lumen and the cavity of Bowman's capsule.

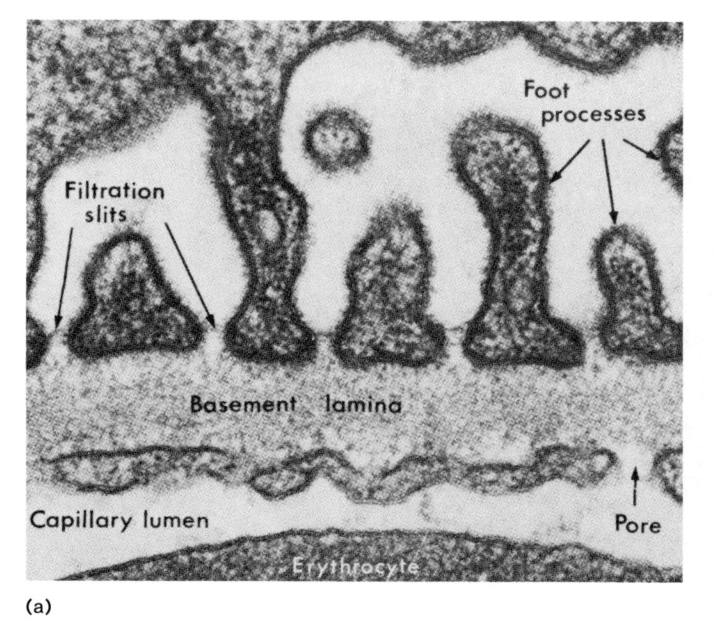

(a)

(b)

Glomerular Filtration

The glomerular capillaries have extremely large pores (200–500 Å in diameter) and are thus said to be *fenestrated.* As a result of these large pores (fenestra), glomerular capillaries are one hundred to four hundred times more permeable to plasma water and dissolved solutes than are the capillaries of skeletal muscles. Although the pores of glomerular capillaries are large, they are still small enough to prevent the passage of red blood cells, white blood cells, and platelets into the filtrate.

Before the filtrate can enter the interior of the glomerular capsule it must pass through the capillary pores, the basement membrane (a thin layer of glycoproteins immediately outside the endothelial cells), and the inner, visceral layer of the glomerular capsule. The inner layer of the glomerular capsule is composed of unique cells, called *podocytes,* with numerous cytoplasmic extensions, known as *pedicels* (fig. 16.6). These pedicels interdigitate, like fingers wrapped around the glomerular capillaries. The narrow slits between adjacent pedicels provide the passageways through which filtered molecules must pass to enter the interior of the glomerular capsule (fig. 16.7).

Although the glomerular capillary pores are apparently large enough to permit the passage of proteins, the fluid that enters the capsular space is almost completely free of plasma proteins. This exclusion of plasma proteins from the filtrate is partially a result of their negative charges, which hinder their passage through the negatively charged glycoproteins in the basement membrane of the capillaries. The large size and negative charges of plasma proteins may also restrict their movement through the filtration slits between pedicels.

Glomerular Ultrafiltrate

The fluid that enters the glomerular capsule is called *ultrafiltrate* (fig. 16.8), because it is formed under pressure (the hydrostatic pressure of the blood). This is similar to the formation of tissue fluid by other capillary beds in the body. The force favoring filtration is opposed by a counter force developed by the hydrostatic pressure of fluid in the glomerular capsule. Also, since the protein concentration of the tubular fluid is low (less than 2–5 mg per 100 ml) compared to that of plasma (6–8 g per 100 ml), the greater colloid osmotic pressure of plasma promotes the osmotic return of filtered water. When these opposing forces are subtracted from the hydrostatic pressure of the glomerular capillaries, a *net filtration pressure* of approximately 10 mm Hg is obtained.

Figure 16.8. The formation of glomerular ultrafiltrate. Proteins (*large circles*) are not filtered, but smaller plasma solutes (*dots*) easily enter the glomerular ultrafiltrate.

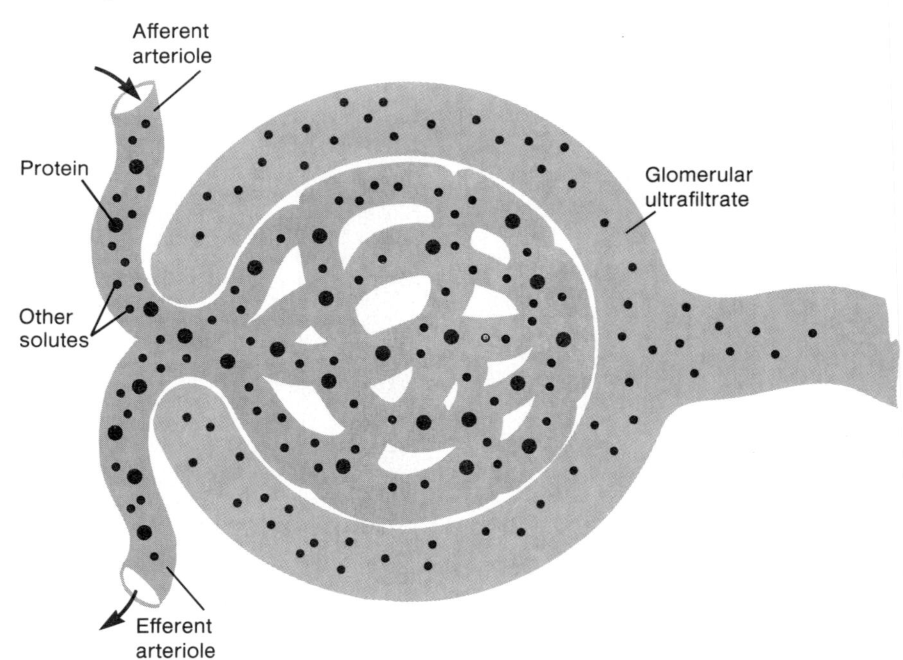

Because glomerular capillaries are extremely permeable and have a high surface area, this modest net filtration pressure produces an extraordinarily large volume of filtrate. The **glomerular filtration rate (GFR)** averages 115 ml per minute in women and 125 ml per minute in men. This is equivalent to 7.5 L per hour or 180 L per day (about 45 gallons)! Since the total blood volume averages about 5 L, this means that the total blood volume is filtered into the urinary tubules every forty minutes. Most of the filtered water must obviously be returned immediately to the vascular system, or people would literally urinate to death within several minutes.

Regulation of Glomerular Filtration Rate

Vasoconstriction or dilation of afferent arterioles affects the rate of blood flow to the glomerulus and thus affects the glomerular filtration rate. Changes in the diameter of the afferent arterioles result from both extrinsic (sympathetic innervation) and intrinsic regulatory mechanisms.

Sympathetic Nerve Effects. An increase in sympathetic nerve activity, as occurs during the fight-or-flight reaction and exercise, stimulates constriction of afferent arterioles. This is an alpha-adrenergic effect, which helps to preserve blood volume and to divert blood to the muscles and heart. A similar effect occurs during cardiovascular shock, in which sympathetic nerve activity stimulates vasoconstriction. The decreased GFR and the resulting decreased rate of urine formation help to compensate for the rapid drop of blood pressure under these circumstances (fig. 16.9).

Renal Autoregulation. When the direct effect of sympathetic stimulation is experimentally removed, the effect of systemic blood pressure on GFR can be observed. Under these conditions, surprisingly, the GFR remains relatively constant despite changes in mean arterial pressure within a range of 70–180 mm Hg (normal mean arterial pressure is 100 mm Hg). The ability of the kidneys to maintain a relatively constant GFR in the face of fluctuating blood pressures is called **renal autoregulation.**

Renal autoregulation results from the effects of locally produced chemicals on the afferent arterioles (effects on the efferent arterioles are believed to be of secondary importance). When systemic arterial pressure falls toward a mean of 70 mm Hg, the afferent arterioles dilate, and when the pressure rises, the afferent arterioles constrict. Blood flow to the glomeruli and GFR can thus remain within the autoregulatory range of blood pressure values. The effects of different regulatory mechanisms on GFR are summarized in table 16.1.

Figure 16.9. The effect of increased sympathetic nerve activity on kidney function and other physiological processes.

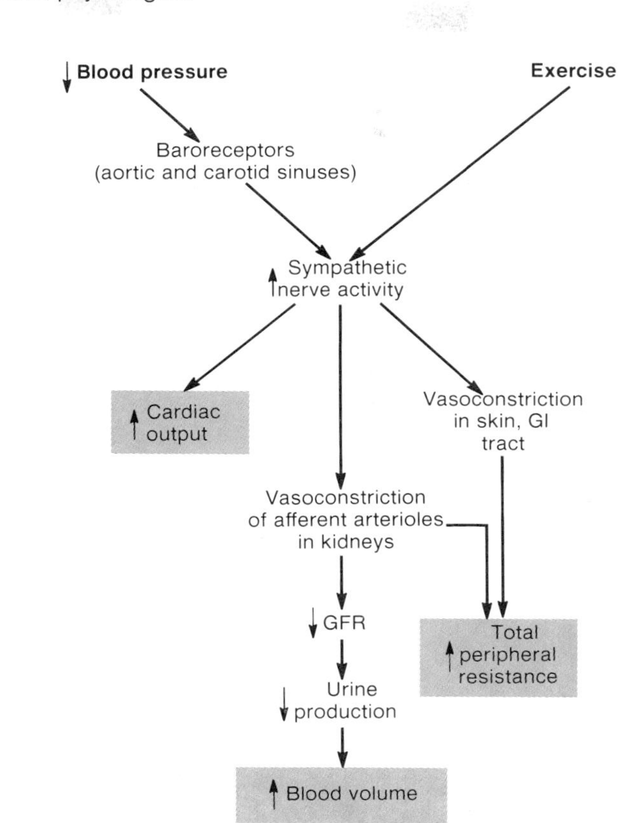

Table 16.1 Regulation of glomerular filtration rate (GFR).

Regulation	Stimulus	Afferent Arteriole	GFR
Sympathetic nerves	Activation by aortic and carotid baroreceptors or by higher brain centers	Constricts	Decreases
Autoregulation	Decreased blood pressure	Dilates	No change
Autoregulation	Increased blood pressure	Constricts	No change

1. *Describe the structures that plasma fluid must pass through to enter the glomerular capsule. Explain how proteins are excluded from the filtrate.*
2. *Describe the forces that affect the formation of glomerular ultrafiltrate.*
3. *Describe the effect of sympathetic innervation on the glomerular filtration rate, and explain the meaning of the term "renal autoregulation."*

Reabsorption of Salt and Water

Although about 180 L per day of glomerular ultrafiltrate are produced, the kidneys normally excrete only 1–2 L per day of urine. Approximately 99% of the filtrate must thus be returned to the vascular system, while 1% is excreted in the urine. The urine volume, however, varies according to the needs of the body. When a well-hydrated person drinks a liter or more of water, urine volume increases to 16 ml per minute (the equivalent of 23 L per day if this were to continue for twenty-four hours). In severe dehydration, when the body needs to conserve water, only 0.3 ml per minute, or 400 ml per day, of urine are produced. A volume of 400 ml per day of urine is needed to excrete the amount of metabolic wastes produced by the body; this is called the *obligatory water loss.* When water in excess of this amount is excreted, the urine volume is increased and its concentration is decreased.

Regardless of the body's state of hydration, it is clear that most of the filtered water must be returned to the vascular system to maintain blood volume and pressure.

Figure 16.10. Plasma water and its dissolved solutes (except proteins) enter the glomerular ultrafiltrate, but most of these filtered molecules are reabsorbed. The term *reabsorption* refers to the transport of molecules out of the tubular filtrate back into the blood.

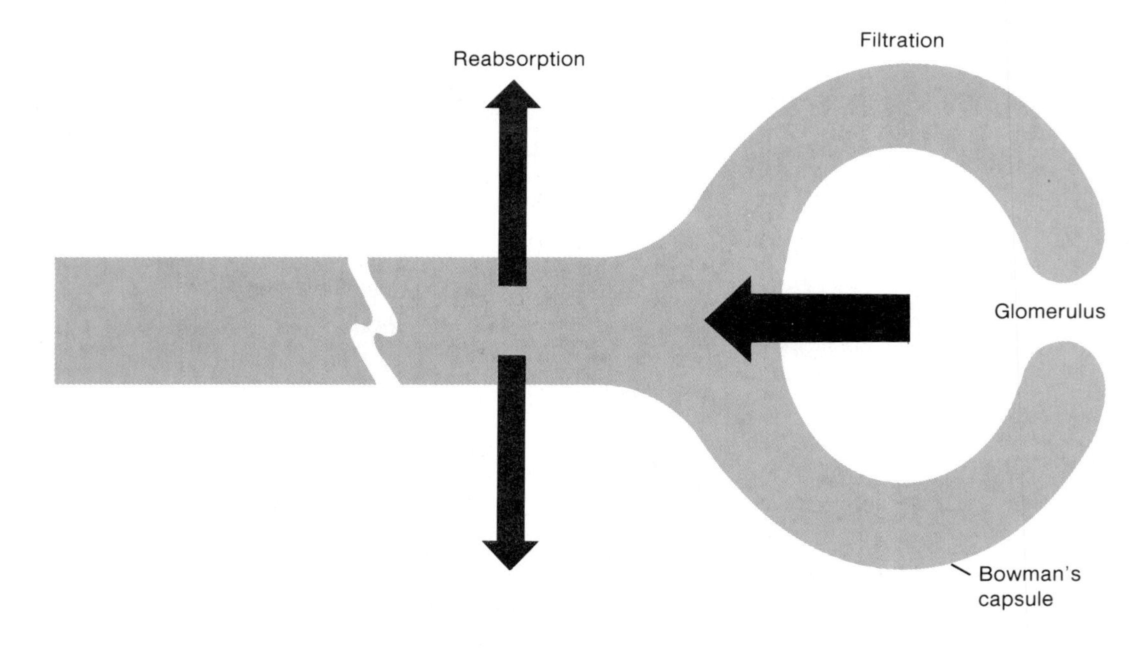

The return of filtered molecules from the tubules to the blood is called **reabsorption** (fig. 16.10). It is important to realize that the transport of water always occurs passively by *osmosis;* there is no such thing as active transport of water. A concentration gradient must thus be created between tubular fluid and blood that favors the osmotic return of water to the vascular system.

Reabsorption in the Proximal Tubule

Since all plasma solutes, with the exception of proteins, are able to enter the glomerular ultrafiltrate freely, the osmolality of the filtrate is essentially the same as that of plasma (300 milliosmoles per liter, or 300 mOsm). The filtrate is thus said to be isosmotic with the plasma. Osmosis thus cannot occur unless the concentration of plasma in the peritubular capillaries and the concentration of filtrate are altered by active transport processes. This is achieved by the active transport of Na^+ from the filtrate to the peritubular blood.

Active and Passive Transport. The epithelial cells that compose the wall of the proximal tubule are joined together by gap junctions only on their apical sides—that is, the sides of each cell that are closest to the lumen of the tubule (fig. 16.11). Each cell therefore has four exposed surfaces: the apical side facing the lumen, which contains microvilli, the opposite, basal side facing the peritubular capillaries, and the lateral sides, facing the narrow clefts between adjacent epithelial cells.

The concentration of Na^+ in the glomerular ultrafiltrate—and thus in the fluid entering the proximal tubule—is the same as in plasma. The epithelial cells of the tubule, however, have a much lower Na^+ concentration. This lower Na^+ concentration is partially due to the low permeability of the cell membrane to Na^+ and partially due to the active transport of Na^+ out of the cell by Na^+/K^+ pumps, as described in chapter 6. In the cells of the proximal tubule, the Na^+/K^+ pumps are located in the basal and lateral sides of the cell membrane but not in the apical membrane. As a result of the action of these active

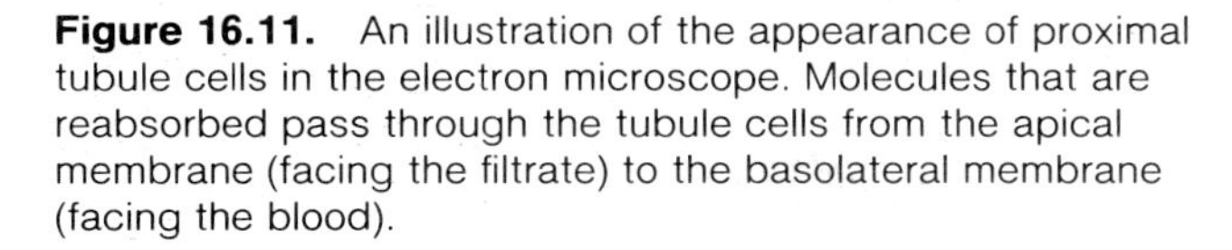

Figure 16.11. An illustration of the appearance of proximal tubule cells in the electron microscope. Molecules that are reabsorbed pass through the tubule cells from the apical membrane (facing the filtrate) to the basolateral membrane (facing the blood).

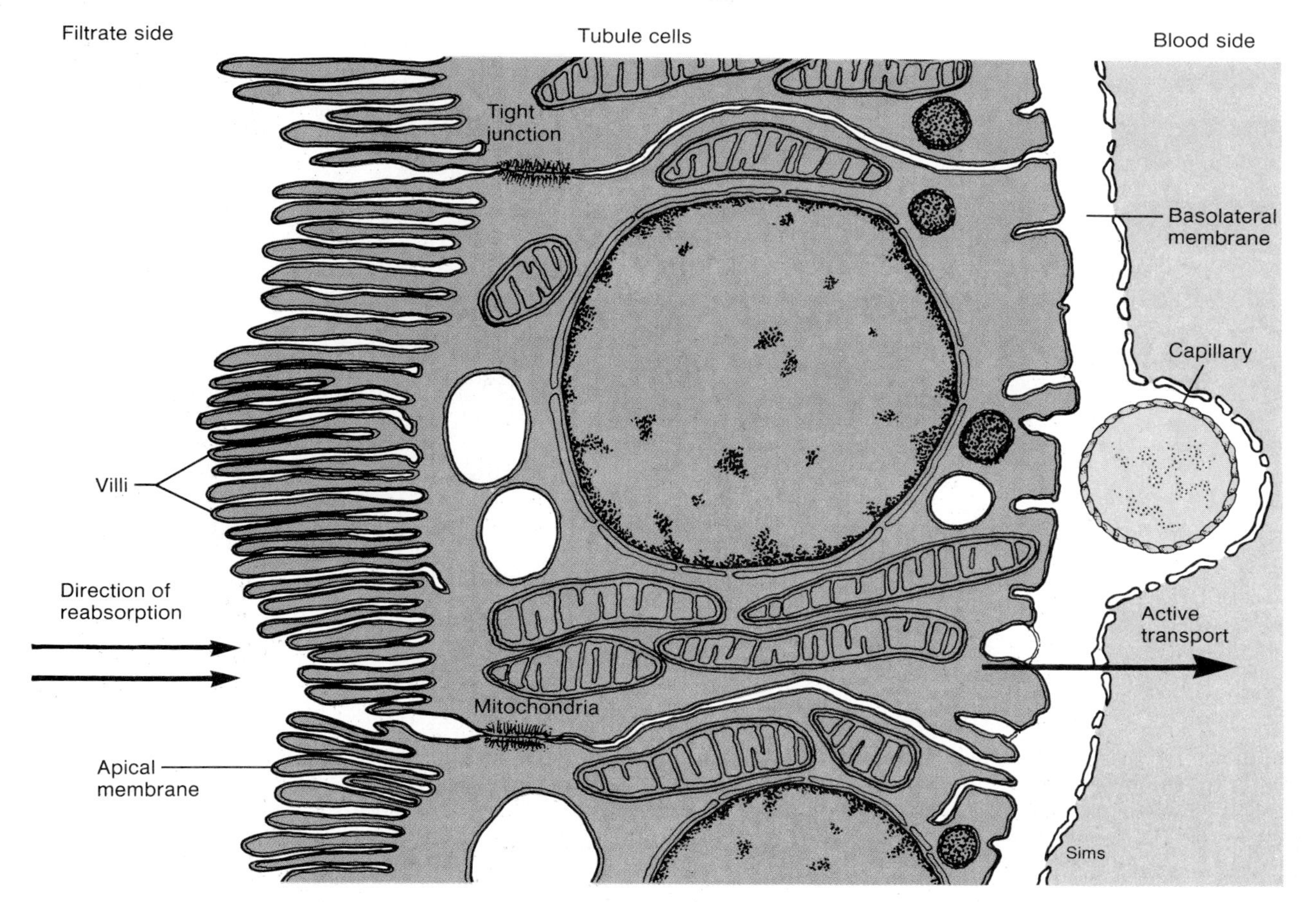

transport pumps, a concentration gradient is created that favors the diffusion of Na^+ from the tubular fluid, across the apical cell membranes, and into the epithelial cells of the proximal tubule. The Na^+ is then extruded into the surrounding tissue fluid by the Na^+/K^+ pumps.

The removal of Na^+ from the tubular fluid and its subsequent appearance in the tissue fluid surrounding the epithelial cells of the proximal tubule create an electrical gradient that favors the passive transport of Cl^- towards the higher Na^+ concentration in the tissue fluid. As a result of the accumulation of NaCl, the osmolality and osmotic pressure of the tissue fluid surrounding the epithelial cells is increased above that of the tubular fluid. This is particularly true of the tissue fluid between the lateral membranes of adjacent epithelial cells, where the narrow spaces permit the accumulated NaCl to become less diluted.

An osmotic gradient is thus created beween the tubular fluid and the tissue fluid surrounding the proximal tubule. Since the cells of the proximal tubule are permeable to water, water moves by osmosis from the tubular fluid into the epithelial cells and then across the basal and lateral sides of the epithelial cells into the tissue fluid. The salt and water which were reabsorbed from the tubular fluid can then move passively into the surrounding peritubular capillaries and in this way be returned to the blood (fig. 16.12).

Significance of Proximal Tubule Reabsorption. Approximately 65% of the salt and water in the original glomerular ultrafiltrate is reabsorbed across the proximal tubule and returned to the vascular system. The volume

Figure 16.12. Mechanisms of salt and water reabsorption in the proximal tubule. Sodium is actively transported out of the filtrate, and chloride follows passively by electrical attraction. Water follows the salt out of the tubular filtrate into the peritubular capillaries by osmosis.

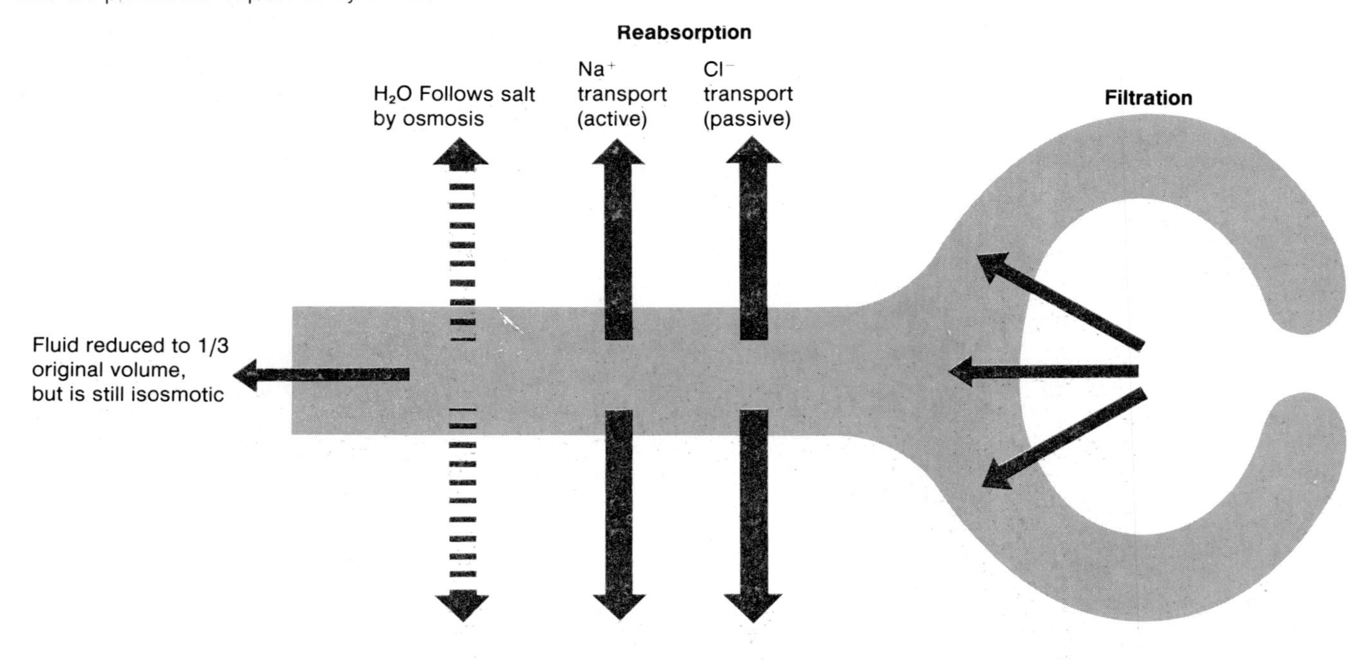

of tubular fluid remaining is reduced accordingly, but this fluid is still isosmotic with the blood (has a concentration of 300 mOsm). This results from the fact that the cell membranes in the proximal tubule are freely permeable to water so that water and salt are removed in proportionate amounts.

An additional smaller amount of salt and water is returned to the vascular system by reabsorption in the loop of Henle. This reabsorption, like that in the proximal tubule, occurs constantly regardless of the person's state of hydration. Unlike reabsorption in later regions of the nephron, it is not subject to hormonal regulation. Approximately 85% of the filtered salt and water is, therefore, reabsorbed in a constant, unregulated fashion in the early regions of the nephron (proximal tubule and loop of Henle). This reabsorption is very costly in terms of energy expenditures, accounting for as much as 6% of the calories consumed by the body at rest.

Since 85% of the original glomerular ultrafiltrate is immediately reabsorbed in the early region of the nephron, only 15% of the initial filtrate remains to enter the distal convoluted tubule and collecting duct. This is still a large volume of fluid—15% × GFR (180 L per day) = 27 L per day—that must be reabsorbed to varying degrees in accordance with the body's state of hydration. This "fine tuning" of the percent reabsorption and urine volume is accomplished by the action of hormones on the later regions of the nephron.

The Countercurrent Multiplier System

Water cannot be actively transported across the tubule wall, and osmosis of water cannot occur if the tubular fluid and surrounding tissue fluid are isotonic to each other. In order for water to be reabsorbed by osmosis, the surrounding tissue fluid must be hypertonic. The osmotic pressure of the tissue fluid in the renal medulla is, in fact, raised to over four times that of plasma. This results partly from the fact that the tubule bends; the geometry of the loop of Henle allows interaction to occur between the descending and ascending limbs. Since the ascending limb is the active partner in this interaction, its properties will be described before those of the descending limb.

Ascending Limb of the Loop of Henle. Salt (NaCl) is actively extruded from the ascending limb into the surrounding tissue fluid. Until recently, it was thought that the cells of the ascending limb accomplish this by actively pumping out Cl^- and that Na^+ follows the Cl^- passively by electrical attraction (the opposite of the situation in the proximal tubule). Newer evidence, however, suggests that Na^+, K^+, and Cl^- move from the filtrate into the ascending limb cells, in a ratio of 1 Na^+ to 1 K^+ to 2 Cl^-, and that the Na^+ is then actively transported across the basolateral membrane by the Na^+/K^+ pump. The ion that is actively transported is Na^+; Cl^- follows the Na^+ passively because of electrical attraction, and K^+ probably diffuses back into the filtrate.

The ascending limb is structurally divisible into two regions: a *thin segment,* nearest to the tip of the loop, and a *thick segment* of varying lengths, which carries the filtrate outwards into the cortex and into the distal convoluted tubule. Some scientists believe that only the cells of the thick segments of the ascending limb are capable of actively transporting NaCl from the filtrate into the surrounding tissue fluid.

Regardless of the mechanism of active transport or the region of the ascending limb in which this transport occurs, the net effect is the same as in the proximal tubule: salt (NaCl) is extruded into the surrounding tissue fluid. Unlike the epithelial walls of the proximal tubule, however, the walls of the ascending limb of the loop of Henle are *not permeable to water.* The tubular fluid thus becomes increasingly dilute as it ascends toward the cortex, whereas the tissue fluid around the loops of Henle in the medulla becomes increasingly more concentrated. By means of these processes, the tubular fluid that enters the distal tubule in the cortex is made hypotonic (with a concentration of about 100 mOsm), whereas the tissue fluid in the medulla is made hypertonic.

Descending Limb of the Loop of Henle. The deeper regions of the medulla, around the tips of the loops of juxtamedullary nephrons, reach a concentration of 1200–1400 mOsm. In order to reach this high a concentration, the salt pumped out of the ascending limb must accumulate in the tissue fluid. This occurs as a result of the properties of the descending limb and as a result of the fact that blood vessels around the loop do not carry back all of the extruded salt to the general circulation.

The descending limb does not actively transport salt. Instead, it is *passively permeable* to water and perhaps also to salt (this is currently controversial). The wall of the descending limb is like a porous plastic membrane; water and perhaps salt are free to diffuse according to their concentration gradients. Since the renal medulla is hypertonic to the fluid entering the descending limb, water moves by osmosis out of the descending limb and is removed by peritubular capillaries. At the same time, salt may diffuse into the descending tubule since it is present at a higher concentration in the surrounding tissue fluid. The concentration of tubular fluid is thus increased, and its volume is decreased, as it descends toward the tips of the loops.

As a result of these passive transport processes in the descending limb, the fluid that "rounds the bend" at the tip of the loop has the same osmolality as the surrounding tissue fluid (1200–1400 mOsm). There is, therefore, a higher salt concentration arriving in the ascending limb than there would be if the descending limb delivered isotonic fluid. Salt transport by the ascending limb is increased accordingly, so that the "saltiness" of the tissue fluid is multiplied (fig. 16.13).

Countercurrent Multiplication. Countercurrent flow (flow in opposite directions) in the ascending and descending limbs and the close proximity of the two limbs allow interaction to occur. Since the concentration of the tubular fluid in the descending limb reflects the concentration of surrounding tissue fluid, and since the concentration of this tissue fluid is raised by the active extrusion of salt from the ascending limb, a *positive feedback* mechanism is created. The more salt the ascending limb extrudes, the more concentrated will be the fluid that returns to it from the descending limb. This positive feedback mechanism multiplies the concentration of tissue fluid and descending limb fluid and is thus called the **countercurrent multiplier system.**

The countercurrent multiplier system recirculates salt and thus traps some of the salt that enters the loop of Henle in the tissue fluid of the renal medulla. This system results in a gradually increasing concentration of renal tissue fluid from the cortex to the inner medulla; the osmolality of tissue fluid increases from 300 mOsm (isotonic) in the cortex to 1200–1400 mOsm in the deepest part of the medulla.

Figure 16.13. The countercurrent multiplier system. The extrusion of sodium chloride from the ascending limb makes the surrounding tissue fluid more concentrated. This concentration is multiplied by the fact that the descending limb is passively permeable so that its fluid increases in concentration as the surrounding tissue fluid becomes more concentrated. The transport properties of the loop and their effect on tubular fluid concentration is shown in (*a*). The values of these changes in osmolality, together with the effect on surrounding tissue fluid concentration, are shown in (*b*).

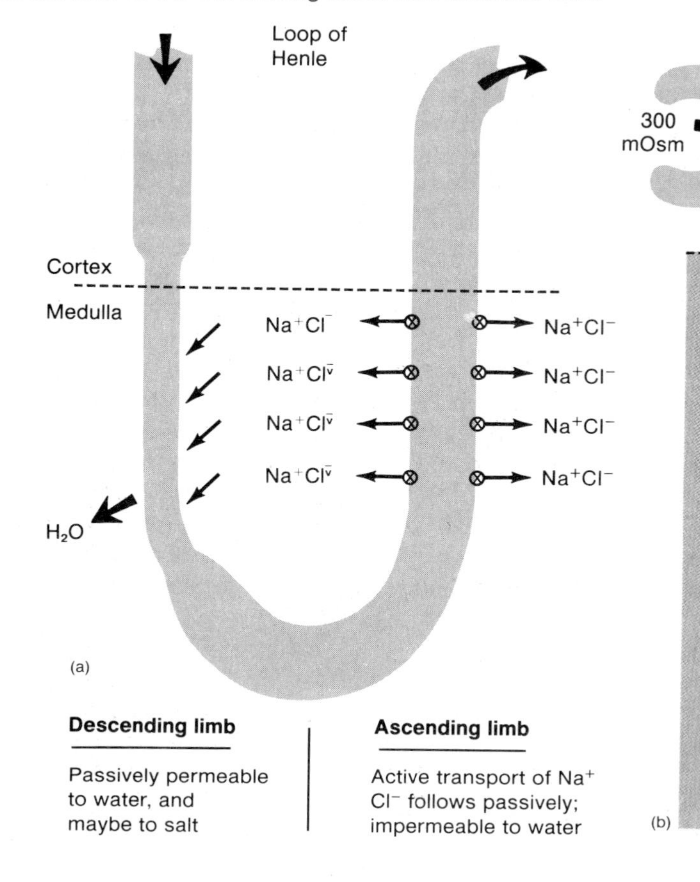

Vasa Recta. In order for the countercurrent multiplier system to be effective, most of the salt that is extruded from the ascending limbs must remain in the tissue fluid of the medulla, while most of the water that leaves the descending limbs must be removed by the blood. This is accomplished by vessels known as the **vasa recta,** which form long capillary loops that parallel the long loops of Henle of the juxtamedullary nephrons (see fig. 16.16).

The vasa recta maintain the hypertonicity of the renal medulla by means of a mechanism known as **countercurrent exchange.** Salt and other solutes (such as urea) that are present at high concentrations in the medullary tissue fluid diffuse into the blood as the blood descends into the capillary loops of the vasa recta, but then passively diffuse out of the ascending vessels and back into the descending vessels (where the concentration is lower). Solutes are thus recirculated and trapped within the medulla. Water, in contrast, diffuses out of the descending vessels and into the ascending vessels (where the osmotic pressure is higher) and is thus transported out of the medulla (fig. 16.14).

Possible Effects of Urea. Experimental evidence suggests that active transport of Na^+ may only occur in the thick segments of the ascending limbs. The thin segments of the ascending limbs, which are located in the deeper regions of the medulla, may not be able to extrude salt actively. Since salt does leave the thin segments, it must leave by diffusion even though the surrounding tissue fluid is at least as concentrated as the tubular fluid. Some investigators therefore conclude that molecules other than salt—specifically urea—contribute to the hypertonicity of the tissue fluid. Urea may accumulate in the medullary tissue fluid as a result of the recycling of urea between the collecting duct and the loop of Henle (fig. 16.15). The transport properties of different tubule segments are summarized in table 16.2.

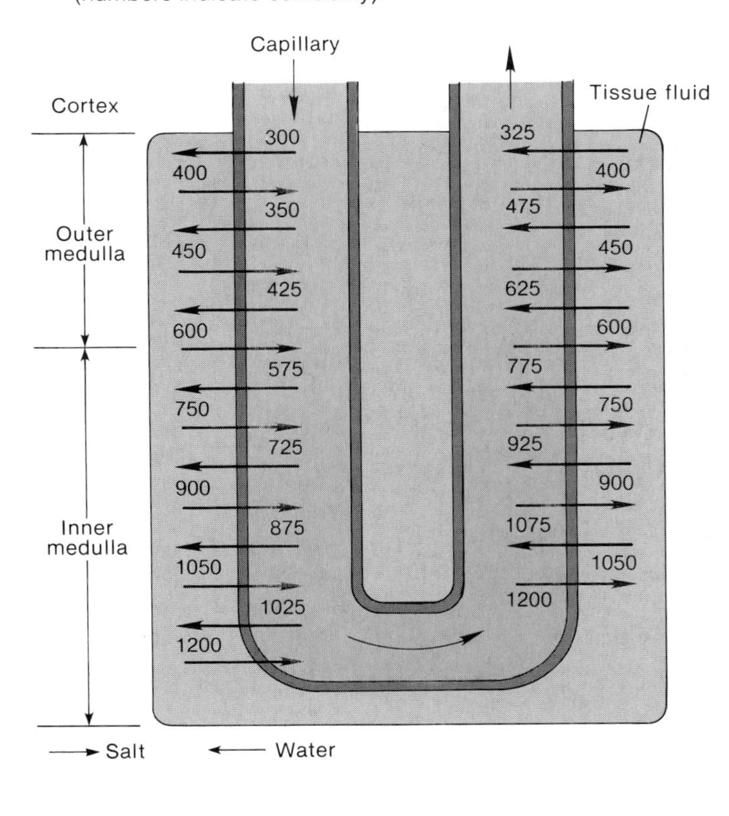

Figure 16.14. Countercurrent exchange in the vasa recta. The diffusion of salt and water first into and then out of these blood vessels helps to maintain the "saltiness" (hypertonicity) of the interstitial fluid in the renal medulla (numbers indicate osmolality).

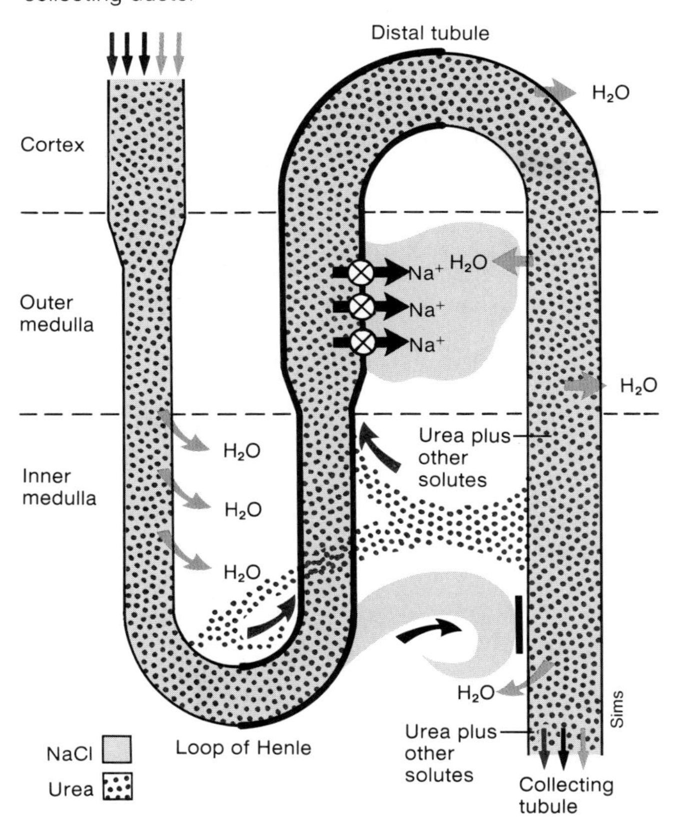

Figure 16.15. According to some authorities, urea diffuses out of the collecting duct and contributes significantly to the concentration of the interstitial fluid in the renal medulla. The active transport of Na^+ out of the thick segments of the ascending limbs also contributes to the hypertonicity of the medulla so that water is reabsorbed by osmosis from the collecting ducts.

Table 16.2 Properties of different nephron segments in regard to the concentrating-diluting mechanisms of the kidney.

Nephron Segment	Active Transport	Passive Transport		
		Salt	Water	Urea
Proximal tubule	Na^+	Cl^-	Yes	Yes
Descending limb of Henle's loop	None	Maybe	Yes	No
Thin segment of ascending limb*	Na^+ or none	Cl^- or NaCl	No	Yes
Thick segment of ascending limb	Na^+	Cl^-	No	No
Distal tubule	Na^+	No	No	No
Collecting duct**	Slight Na^+	No	Yes (ADH) or Slight (no ADH)	Yes

*The thin segment may actively transport Na^+ in the same manner as the thick segment, or it may only permit the passive diffusion of NaCl. Both possibilities are presented in this table.
**The permeability of the collecting duct to water depends on the presence of ADH.

Figure 16.16. The countercurrent multiplier system in the loop of Henle and countercurrent exchange in the vasa recta help create a hypertonic renal medulla. Under the influence of antidiuretic hormone (ADH), the collecting duct is more permeable to water so that water is drawn by osmosis out into the hypertonic renal medulla and into the peritubular capillaries.

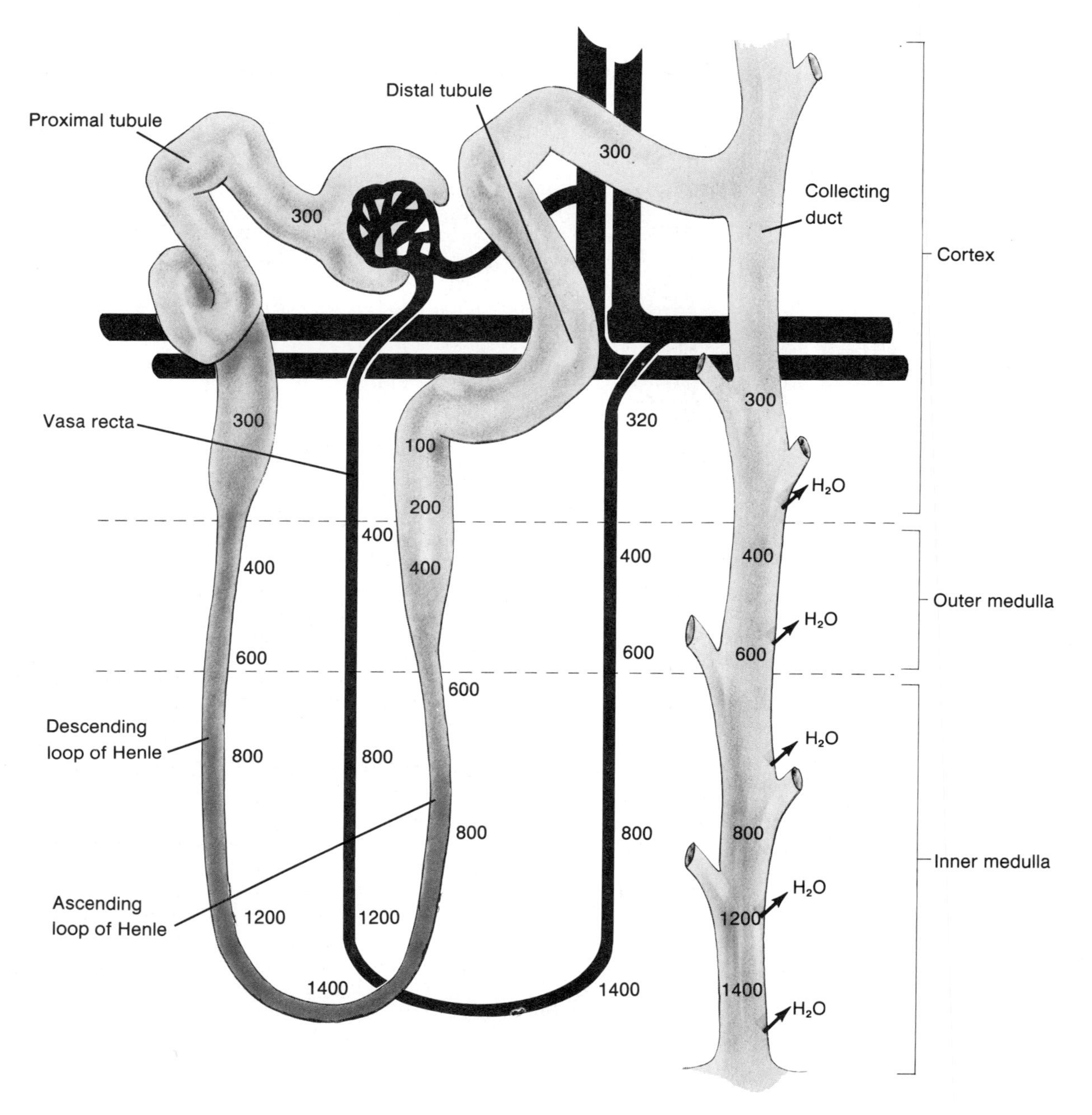

Table 16.3 Antidiuretic hormone secretion and action.

Stimulus	Receptors	Secretion of ADH	Effects on	
			Urine Volume	Blood
↑Osmolality (dehydration)	Osmoreceptors in hypothalamus	Increased	Decreased	Increased water retention; decreased blood osmolality
↓Osmolality	Osmoreceptors in hypothalamus	Decreased	Increased	Water loss increases blood osmolality
↑Blood volume	Stretch receptors in left atrium	Decreased	Increased	Decreased blood volume
↓Blood volume	Stretch receptors in left atrium	Increased	Decreased	Increased blood volume

Collecting Duct: Effect of Antidiuretic Hormone (ADH)

As a result of the recycling of salt between the ascending and descending limbs and possibly of the recycling of urea between the collecting duct and the loop of Henle, the medullary tissue fluid is made very hypertonic. The collecting ducts must transport their fluid through this hypertonic environment in order to empty their contents of urine into the calyces. Since the distal convoluted tubules are impermeable to water, the fluid that enters the collecting ducts in the cortex is hypotonic as a result of the active extrusion of salt by the ascending limbs of the loops.

The walls of the collecting ducts are *permeable to water but not to salt.* Since the surrounding tissue fluid in the renal medulla is very hypertonic, as a result of the countercurrent multiplier system, water is drawn out of the collecting ducts by osmosis. This water does not dilute the surrounding tissue fluid because it is transported by capillaries to the general circulation. In this way, most of the water remaining in the filtrate after reabsorption in the proximal tubules is returned to the vascular system (fig. 16.16).

The osmotic gradient created by the countercurrent multiplier system provides the force for water reabsorption through the collecting ducts. The rate of this reabsorption, however, is determined by the permeability of the collecting duct cell membranes to water. The permeability of the collecting duct to water, in turn, is determined by the concentration of **antidiuretic hormone (ADH)** in the blood. When the concentration of ADH is increased, the collecting ducts become more permeable to water, and more water is reabsorbed. A decrease in ADH, conversely, results in less reabsorption of water and thus in the excretion of a larger volume of more dilute urine.

ADH is produced by neurons in the hypothalamus and is secreted from the posterior pituitary gland. The secretion of ADH is stimulated when osmoreceptors in the hypothalamus respond to an increase in blood osmotic pressure. During dehydration, therefore, when the plasma becomes more concentrated, increased secretion of ADH promotes increased permeability of the collecting ducts to water. In severe dehydration only the minimal amount of water needed to eliminate the body's wastes is excreted. This minimum, about 400 ml per day, is limited by the fact that urine cannot become more concentrated than the medullary tissue fluid surrounding the collecting ducts. Under these conditions about 99.8% of the initial glomerular ultrafiltrate is reabsorbed.

A person in a state of normal hydration excretes about 1.5 L per day of urine, indicating that 99.2% of the glomerular ultrafiltrate volume is reabsorbed. Notice that small changes in percent reabsorption translate into large changes in urine volume. Increasing water ingestion—and thus decreasing ADH secretion (table 16.3)—results in correspondingly larger volumes of urine excretion. It should be noted that even in the complete absence of ADH some water is still reabsorbed through the collecting ducts.

Diabetes insipidus is a disease associated with the inadequate secretion or action of ADH. The collecting ducts are thus not very permeable to water and, therefore, a large volume (5–10 L per day) of dilute urine is produced. The dehydration that results causes intense thirst, but a person with this condition has difficulty drinking enough to compensate for the large volumes of water lost in the urine.

1. *Describe the mechanisms for salt and water reabsorption in the proximal tubule.*
2. *Compare the transport of Na^+, Cl^-, and water across the walls of the proximal tubule, ascending and descending limbs of the loop of Henle, and collecting duct.*
3. *Explain the interaction of the ascending and descending limbs of the loop and how this interaction results in a hypertonic renal medulla.*
4. *Explain how ADH helps the body to conserve water, and describe how variations in ADH secretion affect the volume and concentration of urine.*

Figure 16.17. Secretion refers to the active transport of substances from the peritubular capillaries into the tubular fluid. This transport is in a direction opposite to that of reabsorption.

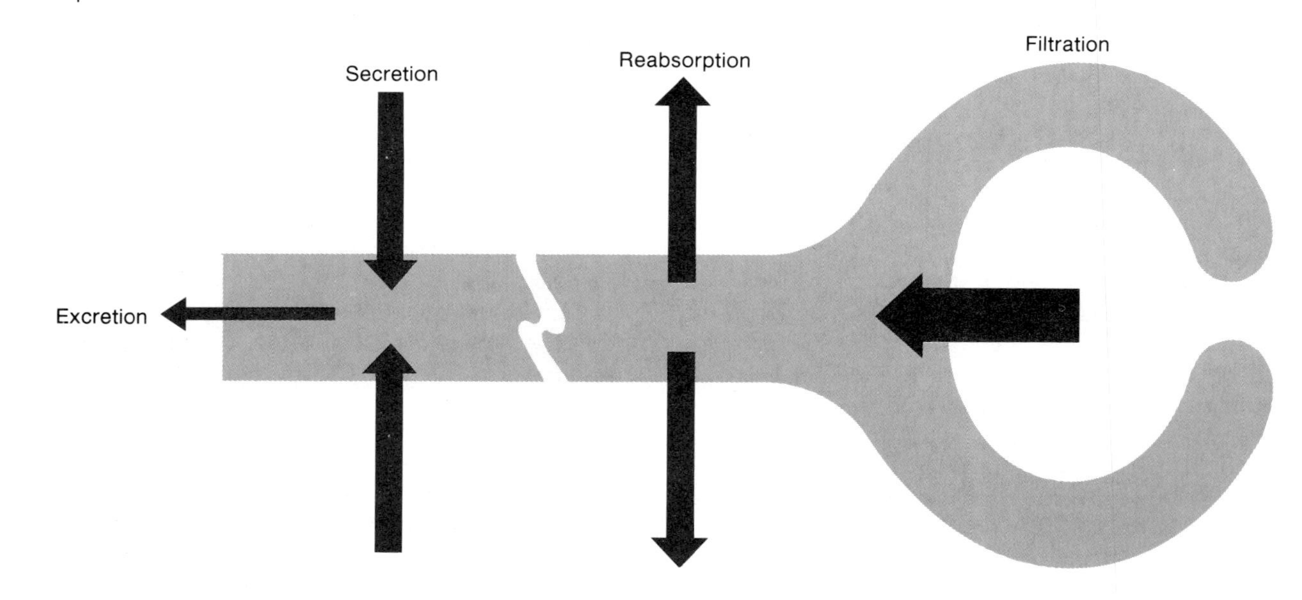

Table 16.4 The effects of filtration, reabsorption, and secretion on renal clearance rates.

Term	Means	Effect on Renal Clearance
Filtered	A substance enters the glomerular ultrafiltrate	Some or all of a filtered substance may enter the urine and be "cleared" from the blood.
Reabsorbed	The transport of a substance from the filtrate, through tubular cells, and into the blood	Reabsorption decreases the rate at which a substance is cleared; clearance rate is less than the glomerular filtration rate (GFR).
Secreted	The transport of a substance from peritubular blood, through tubular cells, and into the filtrate	When a substance is secreted by the nephrons, its clearance rate is greater than the GFR.

Renal Clearance Rates

One of the major functions of the kidneys is the excretion of waste products such as urea, creatinine, and other molecules. These molecules are filtered through the glomerulus into the glomerular capsule along with water, salt, and other plasma solutes. In addition, some waste products can gain access to the urine by a process called **secretion** (fig. 16.17). Secretion is the opposite of reabsorption by active transport. Molecules that are secreted move out of the peritubular capillaries and into the tubular cells, from which they are actively transported into the tubular lumen. In this way, molecules that were not filtered from the blood in the glomerulus can still be excreted in the urine.

Although most (about 99%) of the filtered water is returned to the vascular system by reabsorption, most of the wastes that are filtered or secreted are eliminated in the urine. The concentration of these substances in the renal vein leaving the kidneys is therefore lower than their concentrations in the blood entering the kidneys in the renal artery. Some of the blood that passes through the kidneys, in other words, is "cleared" of these waste products; this is known as the **renal plasma clearance.**

The quantity of a substance excreted in the urine within a given period of time depends on the (1) quantity *filtered* through the glomeruli into the tubular fluid; (2) the quantity *secreted* from the unfiltered blood in peritubular capillaries by active transport into the tubules; and (3) the quantity *reabsorbed* by transport from the tubules into the peritubular blood. Filtration and secretion increase renal plasma clearance; reabsorption decreases the amount excreted and thus decreases the clearance rate (table 16.4).

Renal Clearance of Inulin: Measurement of GFR

If a substance is neither reabsorbed nor secreted by the tubules, the amount excreted per minute in the urine will be equal to the amount that is filtered out of the glomeruli. There does not seem to be a single substance produced by the body, however, that is not reabsorbed or secreted to some degree. Plants such as artichokes, dahlias, onions, and garlic, fortunately, do produce such a compound. This compound, a polymer of the monosac-

Figure 16.18. The renal clearance of inulin. (*a*) Inulin is present in the blood entering the glomeruli, and (*b*) some of this blood, together with its dissolved inulin, is filtered. All of this filtered inulin enters the urine, whereas most of the filtered water is returned to the vascular system (is reabsorbed). (*c*) The blood leaving the kidneys in the renal vein, therefore contains less inulin than the blood that entered the kidneys in the renal artery. Since inulin is filtered but neither reabsorbed nor secreted, the inulin clearance rate equals the glomerular filtration rate (GFR).

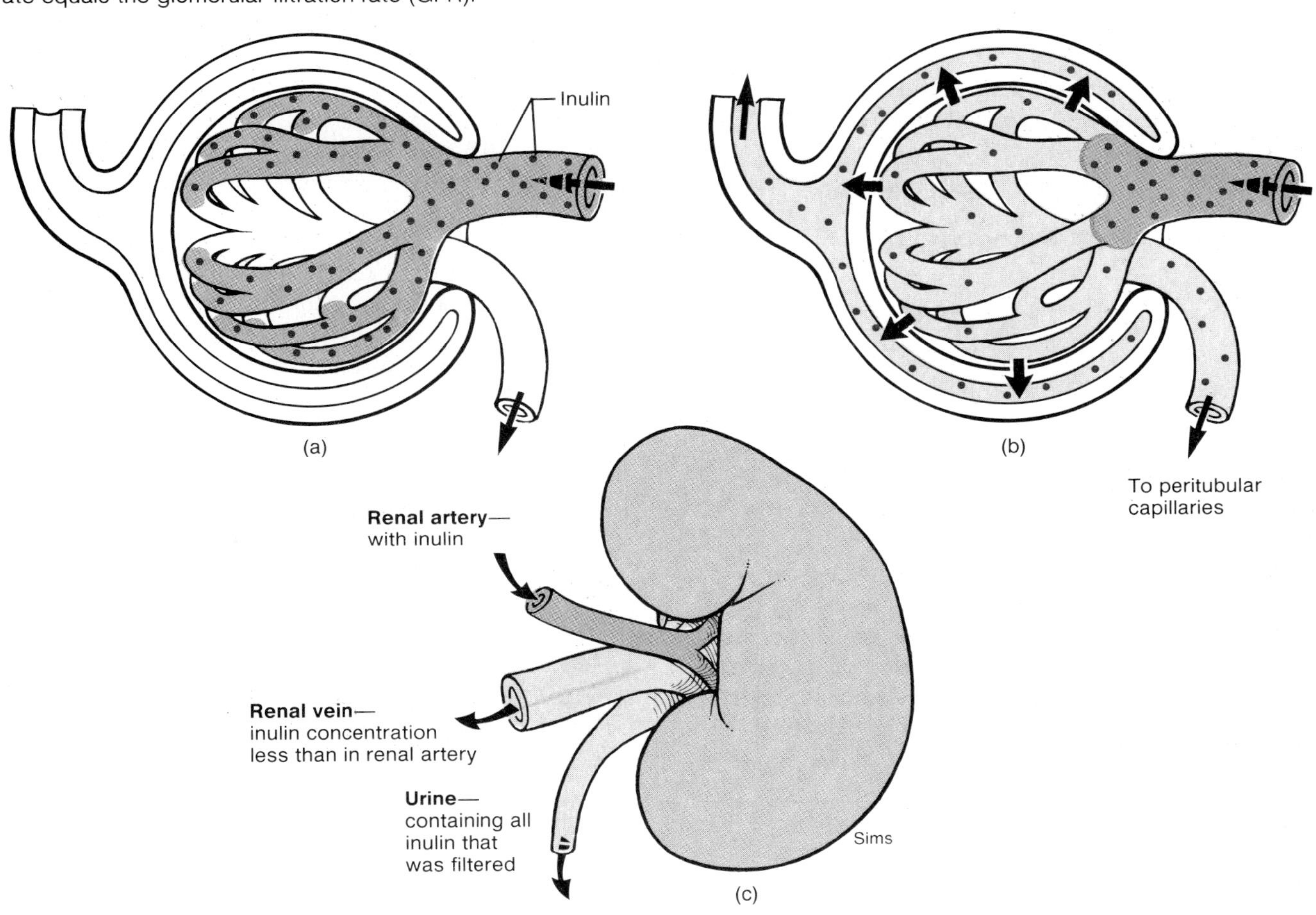

charide fructose, is *inulin.* Once injected into the blood, inulin is filtered by the glomeruli, and the amount of inulin excreted per minute is exactly equal to the amount that was filtered per minute (fig. 16.18).

If the concentration of inulin in urine is measured and the rate of urine formation is determined, the rate of inulin excretion can easily be calculated:

$$\underset{(mg/min)}{\textit{Quantity excreted per minute}} = \underset{\left(\frac{ml}{min}\right)}{V} \times \underset{\left(\frac{mg}{ml}\right)}{U}$$

Where V = rate of urine formation
U = inulin concentration in urine

In order to calculate the rate at which a substance is filtered by the glomeruli (in mg per minute), multiply the ml per minute of plasma that is filtered (the **glomerular filtration rate,** or **GFR**) by the concentration of that substance in the plasma. This is shown in the following equation:

$$\underset{(mg/min)}{\textit{Quantity filtered per minute}} = \underset{\left(\frac{ml}{min}\right)}{GFR} \times \underset{\left(\frac{mg}{ml}\right)}{P}$$

Where P = inulin concentration in plasma

Table 16.5 Renal "handling" of different plasma molecules.

If Substance is	Example	Concentration in Renal Vein	Renal Clearance Rate
Not filtered	Proteins	Same as in renal artery	Zero
Filtered, not reabsorbed nor secreted	Inulin	Less than in renal artery	Equal to GFR (115–125 ml/min)
Filtered, partially reabsorbed	Urea	Less than in renal artery	Less than GFR
Filtered, completely reabsorbed	Glucose	Same as in renal artery	Zero
Filtered and secreted	PAH	Less than in renal artery; approaches zero	Greater than GFR; equal to total plasma flow rate (~625 ml/min)
Filtered, reabsorbed, and secreted	K^+	Variable	Variable

Since inulin is neither reabsorbed nor secreted, the amount filtered equals the amount excreted:

$$\underset{\text{(amount filtered)}}{GFR \times P} = \underset{\text{(amount excreted)}}{V \times U}$$

If the last equation is now solved for the glomerular filtration rate,

$$GFR_{(ml/min)} = \frac{V_{(ml/min)} \times U_{(mg/ml)}}{P_{(mg/ml)}}$$

Suppose, for example, that inulin is infused into a vein and its concentration in the urine and plasma are found to be 30 mg per ml and 0.5 mg per ml, respectively. If the rate of urine formation is 2 ml per minute, the GFR can be calculated as follows:

$$GFR = \frac{2\ \text{ml/min} \times 30\ \text{mg/ml}}{0.5\ \text{mg/ml}} = 120\ \text{ml/min}$$

This equation states that, at a plasma inulin concentration of 0.5 mg per ml, 120 ml of plasma must have been filtered in order to excrete the measured amount of 60 mg that appears in the urine per minute. The glomerular filtration rate is thus 120 ml per minute in this example.

Measurements of the plasma concentration of **creatinine** are often used clinically as an index of kidney function. Creatinine, produced as a waste product of muscle creatine, is secreted to a slight degree by the renal tubules so that its excretion rate is a little above that of inulin. Since it is released into the blood at a constant rate and since its excretion is closely matched to the GFR, an abnormal decrease in GFR causes the plasma creatinine concentration to rise. A simple measurement of blood creatinine concentration can thus provide information about the health of the kidneys.

Clearance Rates. If both the amount of inulin excreted per minute and the plasma inulin concentration are known, the volume of plasma that was filtered per minute (the GFR) can be calculated. Since 120 ml per minute are filtered (enter the glomerular capsules) in the previous example, the amount of inulin contained in 120 ml of plasma is excreted per minute. Since creatinine is filtered like inulin but is also secreted to a slight degree, the excretion rate of creatinine is slightly greater than the filtration rate. The amount of creatinine excreted per minute, in other words, is greater than the amount contained in the 120 ml of plasma that was filtered.

If a substance is reabsorbed to some degree, conversely, the excretion rate will be less than the amount contained in the 120 ml of plasma that was filtered. In order to compare the renal "handling" of various substances in terms of their reabsorption or secretion, the **renal plasma clearance rate** of these substances can be calculated using the same formula used for the determination of the GFR:

$$\textit{Clearance Rate} = \frac{V \times U}{P}$$

Where V = urine volume per minute
U = concentration of substance in urine
P = concentration of substance in plasma

In the case of inulin, the clearance rate is equal to the glomerular filtration rate. If a substance is secreted by the tubules, its clearance rate is greater than the GFR; if a substance is reabsorbed, its clearance rate is less than the GFR (table 16.5). In each of these cases, the clearance rate, in ml per minute, indicates the amount of plasma that originally contained the amount of the substance excreted in a minute's time.

Clearance of Urea

Urea is a waste product of amino acid metabolism that is secreted by the liver into the blood. Despite the fact that it is a waste product, a significant proportion of the filtered urea (40%–60%) is reabsorbed passively, as a result of the high permeability of the tubule membranes to this compound. Not all of the urea contained in the filtered plasma, therefore, is cleared. Using the formula for renal clearance rate previously described,

If $V = 2$ ml/min
$U = 7.5$ mg/ml of urea
$P = 0.2$ mg/ml of urea

$$\text{urea clearance rate} = \frac{2\ \text{ml/min} \times 7.5\ \text{mg/ml}}{0.2\ \text{mg/ml}} = 75\ \text{ml/min}$$

The amount of urea excreted per minute, in other words, was originally contained in 75 ml of plasma. Since 120 ml of plasma were filtered, as determined by the inulin clearance rate, this indicates that approximately 40% of the filtered urea must have been reabsorbed in this example.

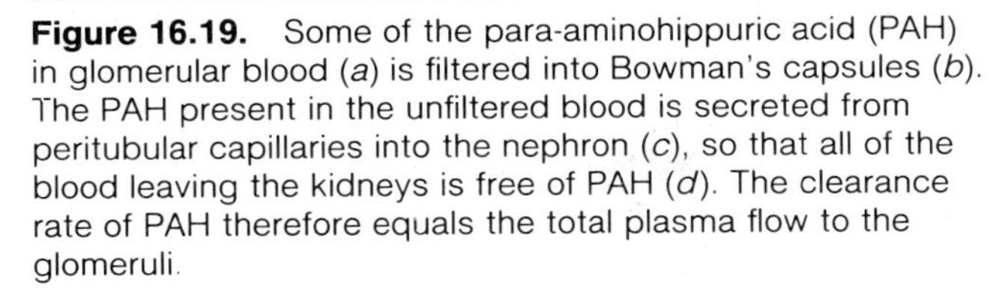

Figure 16.19. Some of the para-aminohippuric acid (PAH) in glomerular blood (*a*) is filtered into Bowman's capsules (*b*). The PAH present in the unfiltered blood is secreted from peritubular capillaries into the nephron (*c*), so that all of the blood leaving the kidneys is free of PAH (*d*). The clearance rate of PAH therefore equals the total plasma flow to the glomeruli.

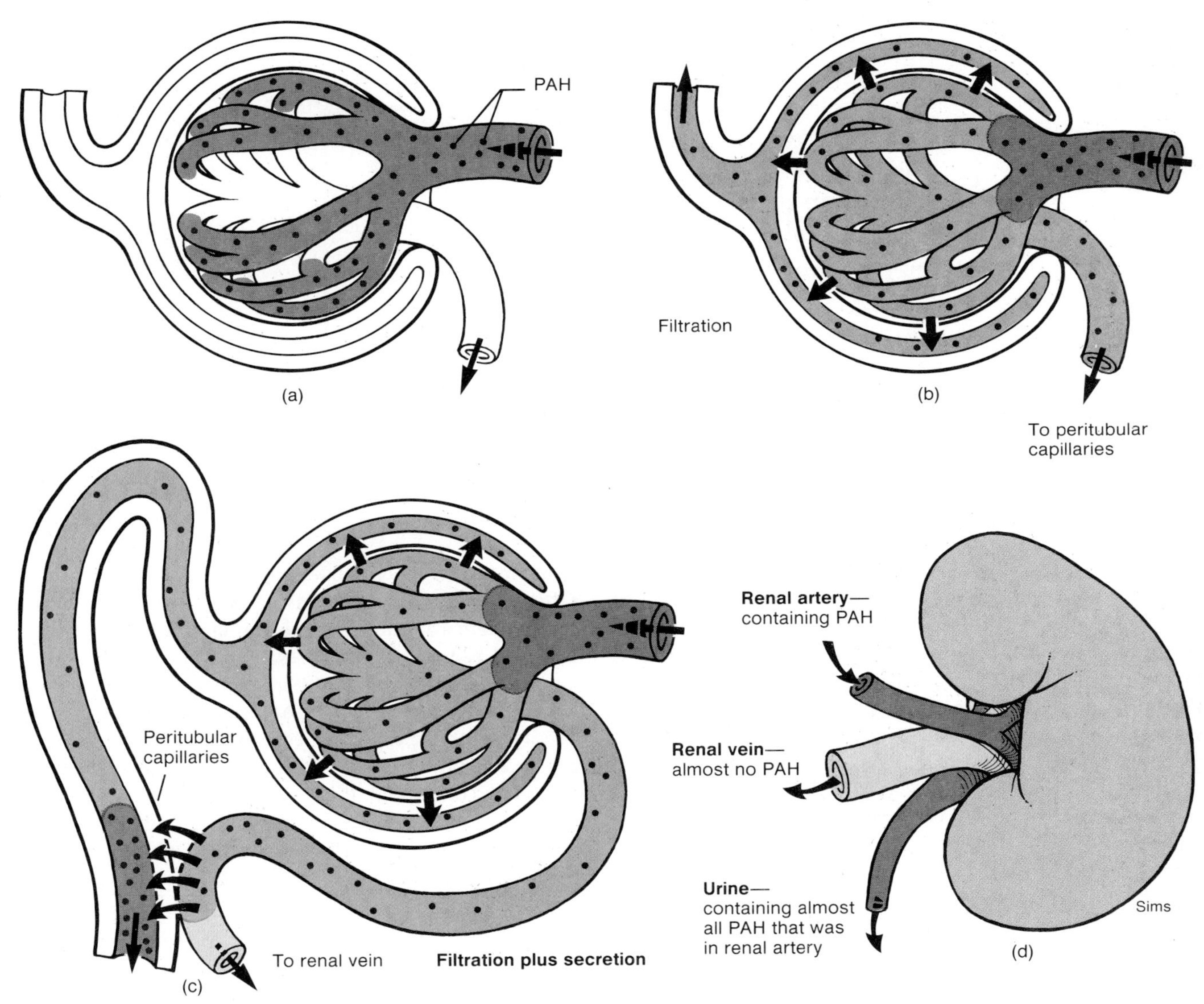

Clearance of PAH: Measurement of Renal Blood Flow

Not all of the blood delivered to the glomeruli is filtered into the glomerular capsules; most of the glomerular blood passes through to the efferent arterioles and peritubular capillaries. The inulin and urea in this unfiltered blood are not excreted but instead return to the general circulation. Blood must thus make many passes through the kidneys before it can be completely cleared of a given amount of inulin or urea.

In order for compounds in the unfiltered renal blood to be cleared, they must be *secreted* into the tubules by active transport from the peritubular capillaries. In this way all of the blood going to the kidneys can potentially be cleared of a secreted compound in a single pass. This is the case for a molecule called **para-aminohippuric acid,** or **PAH,** (fig. 16.19). The clearance rate (in ml/min) of PAH can be used to measure the *total renal blood flow* (in ml/min). The normal PAH clearance rate has been found to average 625 ml/min. Since the glomerular filtration rate averages about 120 ml/min, this indicates that only about 120/625, or roughly 20%, of the renal blood flow is filtered. The remaining 80% passes on to the efferent arterioles.

Figure 16.20. The reabsorption of glucose in the proximal tubule. Glucose is reabsorbed by cotransport with Na^+ from the tubular fluid and then transported by Na^+-independent carriers through the basolateral membrane into the peritubular blood. By this means normally all of the filtered glucose is reabsorbed.

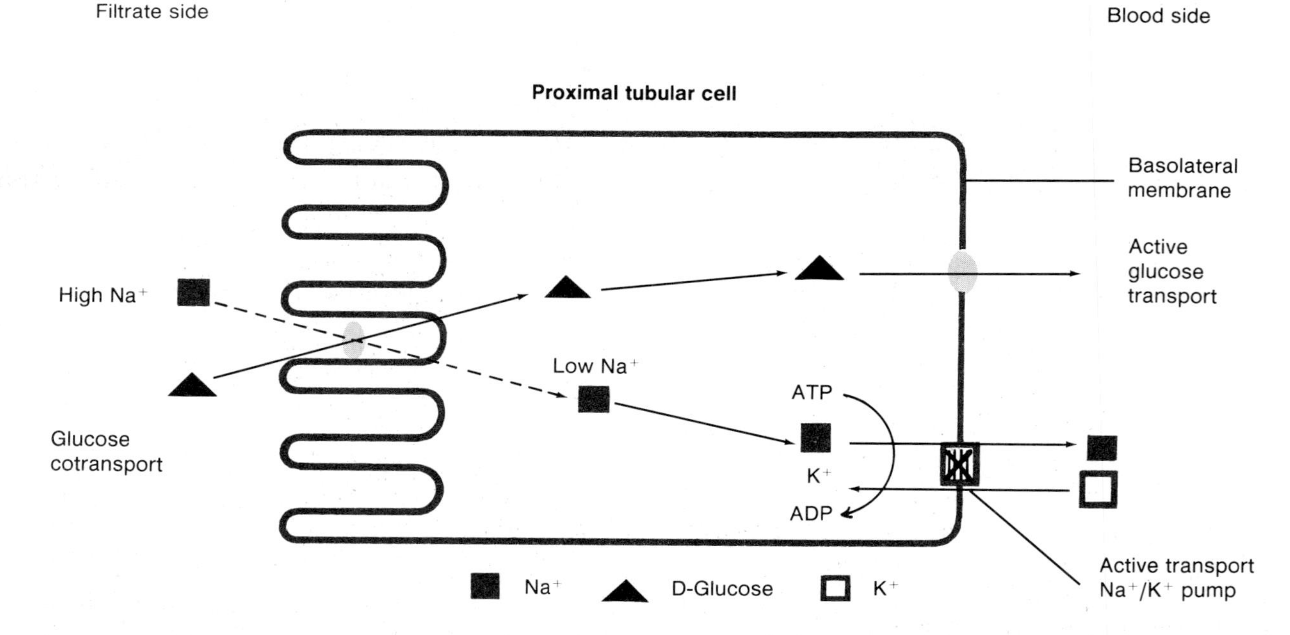

Since filtration and secretion clear only the molecules dissolved in plasma, the PAH clearance rate measures the renal plasma flow rate. In order to convert this to the total renal blood flow, the volume of blood occupied by erythrocytes must be taken into account. If the hematocrit is 45, for example, erythrocytes occupy 45% of the blood volume and plasma accounts for the remaining 55% of the blood volume. The **total renal blood flow** is calculated by dividing the PAH clearance rate by the fractional blood volume occupied by plasma (0.55, in this example). The total renal blood flow in this example is thus 625 ml/min divided by 0.55, or 1.1 L/min.

Many antibiotics, like penicillin, are secreted by the renal tubules and thus have clearance rates greater than the glomerular filtration rate. Because penicillin is rapidly removed from the blood by renal clearance, large amounts must be administered to be effective. The ability of the kidneys to be visualized in radiographs is improved by the injection of Diodrast, a material that is secreted into the tubules and improves contrast by absorbing X rays. Many drugs and some hormones are inactivated in the liver by chemical transformations and are rapidly cleared from the blood by active secretion in the nephrons.

Reabsorption of Glucose and Amino Acids

Glucose and amino acids in the blood are easily filtered by the glomeruli into the renal tubules. These molecules, however, are usually not present in the urine. It can be concluded, therefore, that filtered glucose and amino acids are normally completely reabsorbed by the nephrons.

The reabsorption of glucose and amino acids is an energy-requiring process, which occurs primarily in the proximal convoluted tubules. The energy required for movement of these compounds from the filtrate into the tubule cells is provided by cotransport with Na^+, which diffuses down its electrochemical gradient when it enters the cells. The glucose and amino acids appear to share carriers with Na^+ in the apical membrane. The extrusion of glucose and amino acids from the other side of the cell (across the basolateral membranes) occurs by means of a Na^+-independent active transport carrier (fig. 16.20).

Carrier-mediated transport displays the property of *saturation.* This means that when the transported molecule (such as glucose) is present in sufficiently high concentrations, all of the carriers become "busy," and the transport rate reaches a maximal value. The concentration of transported molecules needed to just saturate the carriers and to just achieve the maximal transport rate is called the **transport maximum** (abbreviated T_m).

Table 16.6 Inherited diseases associated with the presence of specific amino acids in the urine.

Disease	Cause of Disease	Effect of Defect	Treatment
Cystinuria	Renal carriers for cystine and related amino acids are defective.	Kidney stones	Bicarbonate and diuretic administration
Hartnup disease	Renal carriers for tryptophane are defective.	Decreased NAD and NADP within body cells	Nicotinamide administration
Homocystinuria	Enzyme defect results in excessive blood levels of homocystine.	Speech defects, mental retardation	Diet low in methionine, high in cystine
Phenylketonuria	Enzyme defect results in excessive accumulation of phenylalanine and in urinary excretion of phenylpyruvic acid.	Severe mental retardation	Diet low in phenylalanine

The carriers for glucose and amino acids in the renal tubules are not normally saturated and so are able to remove the filtered molecules completely. The T_m for glucose, for example, averages 375 mg per minute, which is well above the rate at which glucose is delivered to the tubules. The rate of glucose delivery can be calculated by multiplying the plasma glucose concentration (about 1 mg per ml) by the GFR (about 125 ml per minute). Approximately 125 mg per minute are thus delivered to the tubules, whereas a rate of 375 mg per minute are required to reach saturation.

Glycosuria. Glucose appears in the urine—a condition called glycosuria—when more glucose passes through the tubules than can be reabsorbed. This occurs when the plasma glucose concentration reaches 180–200 mg per 100 ml. Since the rate of glucose delivery under these conditions is still below the average T_m for glucose, one must conclude that some nephrons have considerably lower T_m values than the average.

The **renal plasma threshold** is the minimum plasma concentration of a substance that results in the excretion of that substance in the urine. The renal plasma threshold for glucose, for example, is 180–200 mg per 100 ml. Glucose is normally absent from urine because plasma glucose concentrations normally remain below this threshold value. The appearance of glucose in the urine (glycosuria) thus occurs only when the plasma concentration of glucose is abnormally high (hyperglycemia).

Fasting hyperglycemia is caused by the inadequate secretion or action of insulin. When this hyperglycemia results in glycosuria, the disease is called **diabetes mellitus.** A person with uncontrolled diabetes mellitus also excretes a large volume of urine, because the excreted glucose carries water with it as a result of the osmotic pressure it generates in the tubules. This condition should not be confused with diabetes insipidus, in which a large volume of dilute urine is excreted as a result of inadequate ADH secretion.

The excretion of amino acids in the urine is not usually due to the presence of a high amino acid concentration in the blood. Rather, specific amino acids are excreted when their carriers are missing or defective due to a genetic disease. Since different classes of amino acids are reabsorbed by different carriers, the types of amino acids that "spill" into the urine are characteristic of the genetic defect (table 16.6).

---

1. *Define clearance rate, and describe how it is measured. Explain why the glomerular filtration rate is equal to the clearance rate of inulin.*
2. *Define the terms* reabsorption *and* secretion. *Describe how the clearance rate is affected by the processes of reabsorption and secretion, and give examples.*
3. *Explain why the total renal blood flow can be measured by the clearance rate of PAH.*
4. *Define transport maximum and renal plasma threshold. Explain why people with diabetes mellitus have glycosuria.*

Renal Control of Electrolyte Balance

The kidneys help regulate the concentrations of plasma electrolytes—sodium, potassium, chloride, bicarbonate, and phosphate—by matching the urinary excretion of these compounds to the amounts ingested. The control of plasma Na^+ is important in the regulation of blood volume and pressure; the control of plasma K^+ is required to maintain proper function of cardiac and skeletal muscles. The regulation of Na^+/K^+ balance is also intimately related to renal control of acid-base balance.

Role of Aldosterone in Na^+/K^+ Balance

Approximately 90% of the filtered Na^+ and K^+ is reabsorbed in the early part of the nephron before the filtrate reaches the distal tubule. This reabsorption occurs at a constant rate and is not subject to hormonal regulation. The final concentration of Na^+ and K^+ in the urine is varied according to the needs of the body by processes that occur in the distal tubule. These processes are regulated by **aldosterone,** a steroid hormone secreted by the adrenal cortex.

Sodium Reabsorption. Although 90% of the filtered sodium is reabsorbed in the early region of the nephron, the amount left in the filtrate delivered to the distal convoluted tubule is still quite large. In the absence of aldosterone, 80% of this amount is automatically reabsorbed through the wall of the distal tubule into the peritubular blood; this is 8% of the amount filtered. The amount of sodium excreted without aldosterone is thus 2% of the amount filtered. Although this percentage seems small, the actual amount of sodium this represents is an impressive 30 g per day excreted in the urine. When aldosterone is secreted in maximal amounts, in contrast, all of the sodium delivered to the distal tubule is reabsorbed. Under these conditions urine contains no Na^+ at all.

Potassium Secretion. About 90% of the filtered K^+ is reabsorbed in the early regions of the nephron (mainly in the proximal tubule). In the absence of aldosterone, all of the filtered K^+ that remains is reabsorbed in the distal tubule. In the absence of aldosterone, therefore, no K^+ is excreted in the urine. The presence of aldosterone stimulates the *secretion of K^+* from the peritubular blood into the distal tubule (fig. 16.21). This secretion is the only means by which K^+ can be eliminated in the urine. When aldosterone secretion is maximal, as much as fifty times more K^+ is excreted in the urine, because of secretion into the distal tubule, than was originally filtered through the glomeruli.

Figure 16.21. Potassium is almost completely reabsorbed in the proximal tubule, but under aldosterone stimulation, it is secreted into the distal tubule. All of the K^+ in urine is derived from secretion rather than from filtration.

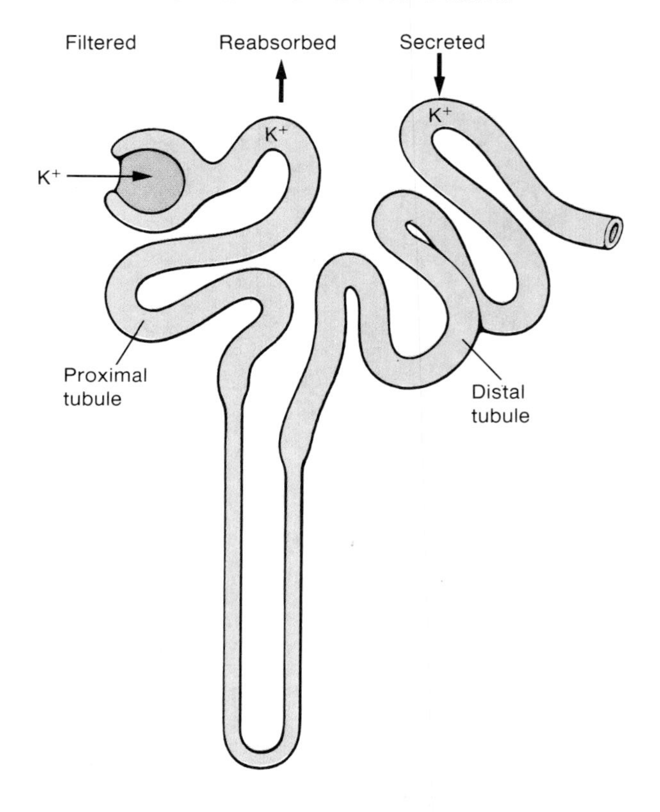

In summary, aldosterone promotes sodium retention and potassium loss from the blood by stimulating the reabsorption of Na^+ and the secretion of K^+ across the wall of the distal convoluted tubules. Since aldosterone promotes the retention of Na^+, it contributes to an increased blood volume and pressure.

The body cannot get rid of excess K^+ in the absence of aldosterone-stimulated secretion of K^+ into the distal tubules. Indeed, when both adrenal glands are removed from an experimental animal, the **hyperkalemia** (high blood K^+) that results can produce fatal cardiac arrhythmias. Abnormally low plasma K^+ concentrations, as might result from excessive aldosterone secretion, can also produce arrhythmias as well as muscle weakness and cramps.

Control of Aldosterone Secretion

Since aldosterone promotes Na^+ retention and K^+ loss, one might predict (on the basis of negative feedback) that aldosterone secretion will be increased when there is a low Na^+ or a high K^+ concentration in the blood. This indeed

Figure 16.22. The juxtaglomerular apparatus (*a*) includes the region of contact of the afferent arteriole with the distal tubule. The afferent arterioles in this region contain granular cells with renin, and the distal tubule cells in contact with the granular cells form an area called the macula densa (*b*). The granular cells of the afferent arteriole are innervated by renal sympathetic nerve fibers.

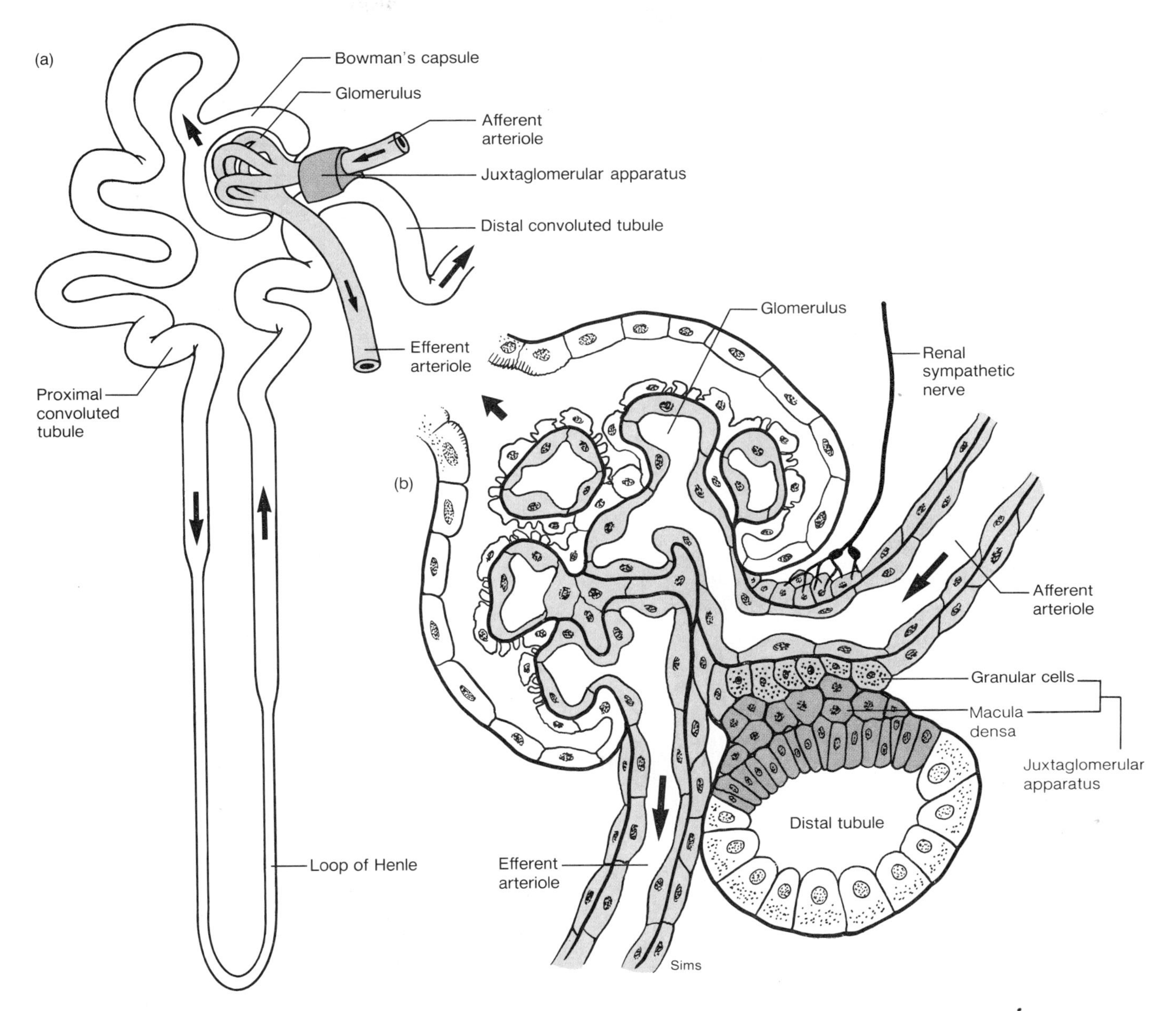

is the case. A rise in blood K^+ *directly* stimulates the secretion of aldosterone from the adrenal cortex. Decreases in plasma Na^+ concentrations also promote aldosterone secretion, but they do so indirectly.

Juxtaglomerular Apparatus. The juxtaglomerular apparatus is the region in each nephron where the afferent arteriole and distal tubule come into contact (fig. 16.22). The microscopic appearance of the afferent arteriole and distal tubule in this small region differs from the appearance in other regions. *Granular cells* within the afferent arteriole secrete the enzyme **renin** into the blood; this enzyme catalyzes the conversion of angiotensinogen (a protein) into angiotensin I (a ten-amino-acid polypeptide).

Secretion of renin into the blood thus results in the formation of angiotensin I, which is then converted to **angiotensin II** by a *converting enzyme* as blood passes through the lungs and other organs. Angiotensin II, in addition to other effects, stimulates the adrenal cortex to secrete aldosterone. Secretion of renin from the granular

Figure 16.23. The sequence of events by which a low sodium (salt) intake leads to increased sodium reabsorption by the kidneys. The dotted arrow and negative sign show the completion of the negative feedback loop.

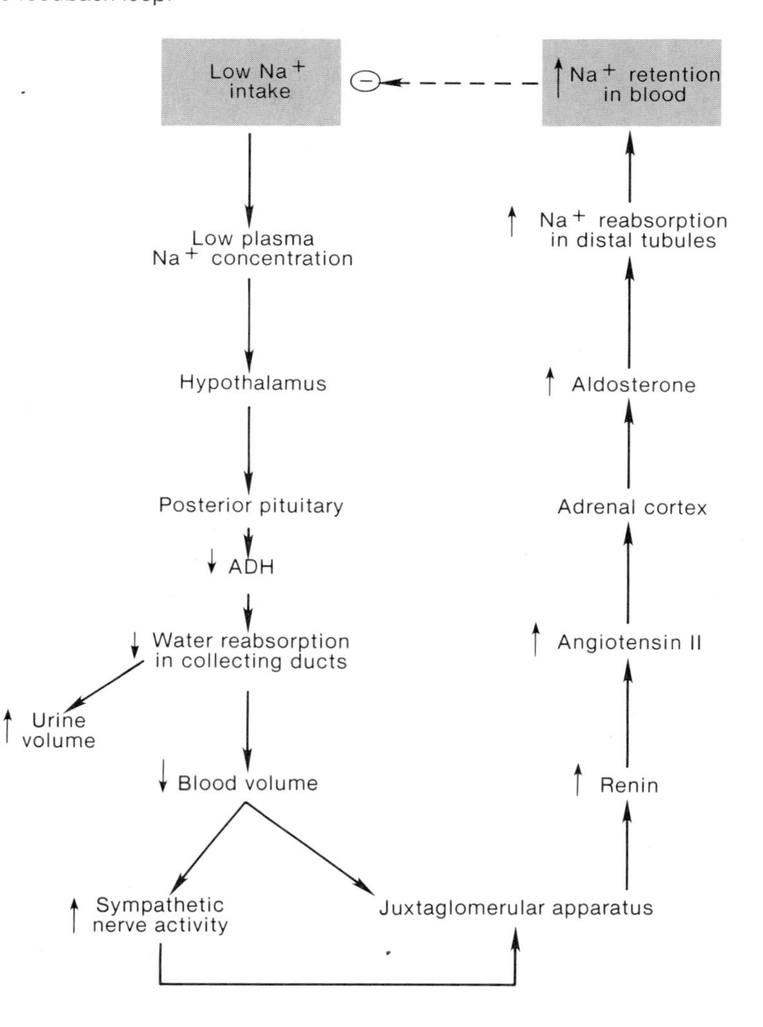

cells of the juxtaglomerular apparatus is thus said to initiate the **renin-angiotensin-aldosterone system.** Conditions that result in renin secretion thus cause increased aldosterone secretion and, by this means, promote the reabsorption of Na^+ in the distal convoluted tubules.

Regulation of Renin Secretion. A fall in plasma Na^+ concentration is always accompanied by a fall in blood volume. This is because ADH secretion is inhibited by the decreased plasma concentration (osmolality); with less ADH, less water is reabsorbed through the collecting ducts and more is excreted in the urine. The fall in blood volume and the fall in renal blood flow that results causes increased renin secretion. Renin secretion is believed to be due in part to the direct effect of blood flow on the granular cells, which may function as baroreceptors in the afferent arterioles. Renin secretion is also stimulated by sympathetic nerve activity, which is increased when the blood volume and pressure fall.

If there is inadequate sodium intake, therefore, the fall in plasma volume that results acts to increase renin secretion. The increased renin secretion acts, via the increased production of angiotensin II, to stimulate aldosterone secretion. In consequence, there is less Na^+ excreted in the urine. This negative feedback system is illustrated in figure 16.23.

Role of the Macula Densa. The region of the distal tubule in contact with the granular cells of the afferent arteriole is called the **macula densa** (fig. 16.22). There is evidence that this region helps to inhibit renin secretion when the blood Na^+ concentration is raised.

Table 16.7 Regulation of renin and aldosterone secretion.

Stimulus	Effect on Renin Secretion	Mechanisms	Angiotensin II Production	Aldosterone Secretion
↓Na^+	Increased	Low blood volume stimulates renal baroreceptors; granular cells release renin.	Increased	Increased
↑Na^+	Decreased	Increased blood volume inhibits baroreceptors; increased Na^+ in distal tubule acts via macula densa to inhibit release of renin from granular cells.	Decreased	Decreased
↑K^+	None	Not applicable	Not changed	Increased
↑Sympathetic nerve activity	Increased	α-adrenergic effect stimulates constriction of afferent arterioles; β-adrenergic effect stimulates renin secretion directly.	Increased	Increased

According to the proposed mechanism, the cells of the macula densa respond to Na^+ within the filtrate delivered to the distal tubule. When the plasma Na^+ concentration is raised, the rate of Na^+ delivered to the distal tubule is also increased. Through an effect on the macula densa, this increase in filtered Na^+ may inhibit the granular cells from secreting renin. Aldosterone secretion thus decreases, and since less Na^+ is reabsorbed in the distal tubule, more Na^+ is excreted in the urine. The regulation of renin and aldosterone secretion is summarized in table 16.7.

Natriuretic Hormone. Expansion of the blood volume causes increased salt and water excretion in the urine. This is due in part to an inhibition of aldosterone secretion, as previously described. There is much experimental evidence, however, that the increased salt excretion that occurs under these conditions is due not only to the inhibition of aldosterone secretion, but also to the increased secretion of another substance with hormone properties. This other substance is called **natriuretic hormone** and is so named because it stimulates salt excretion (the opposite of aldosterone's action). The source and chemical nature of natriuretic hormone remained elusive for many years, but recent evidence has shown that the atria of the heart produce a polypeptide that appears to fit the description of the natriuretic hormone proposed by renal physiologists. This polypeptide is currently known as *atrial natriuretic factor*.

Relationship between Na^+, K^+, and H^+

Hormones, as a general rule, alter the rate of already existing processes. In the absence of aldosterone, the distal tubule reabsorbs 80% of the Na^+ delivered to it; aldosterone can increase this reabsorption to 100%. At a given level of aldosterone, the amount of Na^+ reabsorbed in the distal tubule is a given proportion of the Na^+ delivered to it. Some diuretic drugs inhibit Na^+ reabsorption in the loop of Henle and, therefore, increase the delivery of Na^+ to the distal tubule. As a result, there is an increased reabsorption of Na^+ in the distal convoluted tubule when a person takes these types of diuretics.

Relationship between Na^+ and K^+. The reabsorption of Na^+ in the distal convoluted tubules occurs together with K^+ secretion. This occurs because the aldosterone-stimulated reabsorption of Na^+ creates a large potential difference between the two sides of the tubular wall, with the lumen side very negative (-50 mV) in comparison to the basolateral side. The secretion of K^+ into the tubular fluid is driven by this electrical gradient. Because of the Na^+/K^+ exchange in the distal tubule, an increase in Na^+ reabsorption in the distal tubule results in an increase in K^+ secretion. People who take diuretics that inhibit Na^+ reabsorption in the loop of Henle, for these reasons, tend to have excessive K^+ secretion into the distal tubules and, therefore, excessive K^+ loss in the urine. The actions of different diuretics and their side-effects on blood K^+ are discussed in the last section of this chapter.

The K^+ loss that occurs with many diuretics may present serious side effects to these medications. If K^+ secretion into the distal convoluted tubules is significantly increased, a condition of **hypokalemia** (low blood K^+) may be produced, which must be compensated for by the increased ingestion of potassium. People who take diuretics for the treatment of high blood pressure are usually on a low-sodium diet and often must supplement their meals with potassium chloride (KCl).

Relationship between K^+ and H^+. The plasma K^+ concentration indirectly affects the plasma H^+ concentration (pH). Changes in plasma pH likewise affect the K^+ concentration of the blood. These effects serve to stabilize the *ratio* of K^+ to H^+. When the extracellular H^+ concentration increases, for example, some of the H^+ moves into the tissue cells and causes cellular K^+ to diffuse outward into the extracellular fluid. The plasma concentration of H^+ is thus decreased while the K^+ increases, helping to restabilize the ratio of these ions in the extracellular fluid.

Figure 16.24. In the distal tubule, K^+ and H^+ are secreted in exchange for Na^+. High concentrations of H^+ may therefore decrease K^+ secretion, and vice versa.

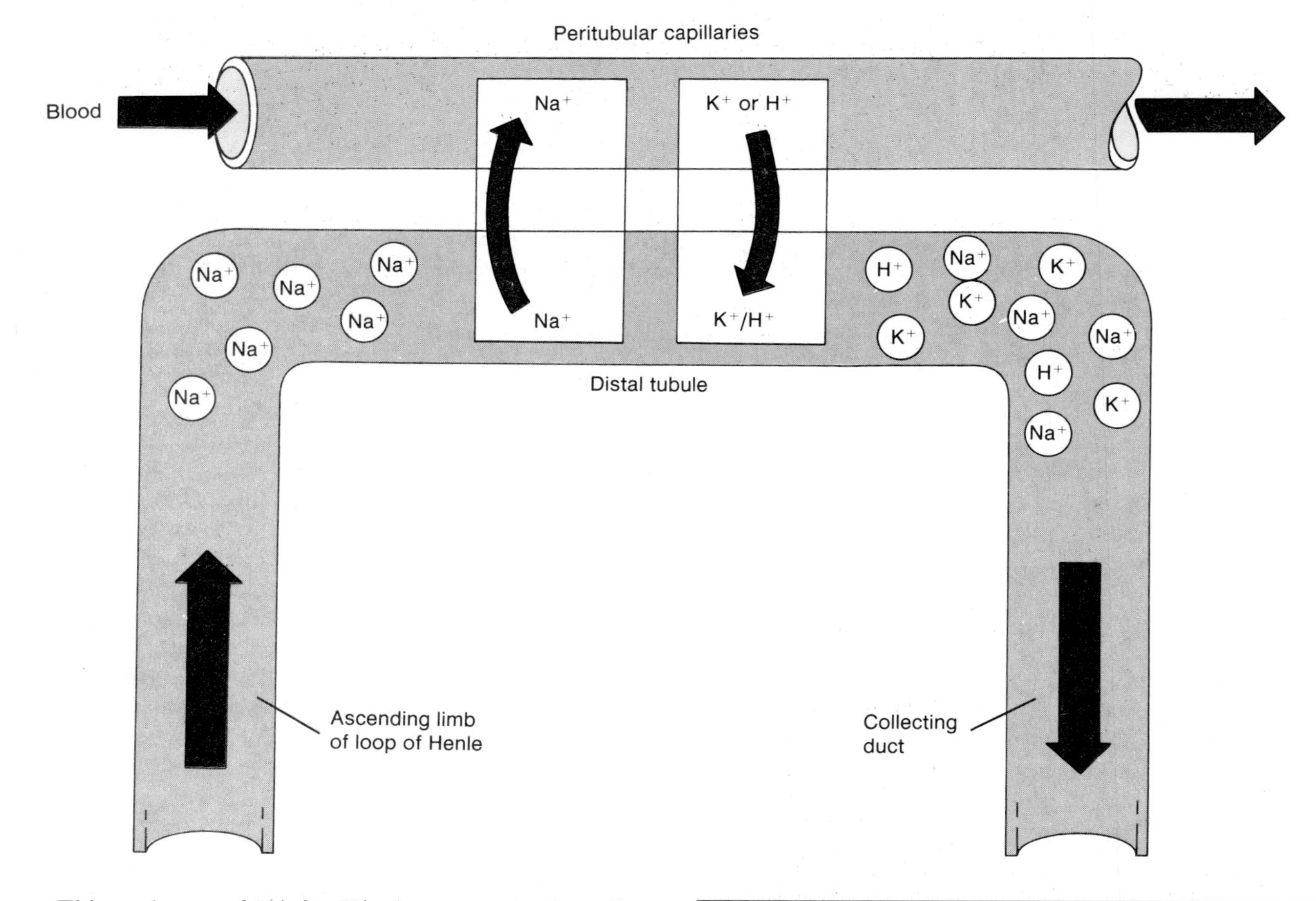

This exchange of H^+ for K^+ also occurs in the cells of the distal tubule when the plasma pH is lowered (fig. 16.24). Under these conditions the tubule cells contain more H^+ and less K^+ than before. Whenever the tubule cells reabsorb Na^+ from the filtrate, they can replace it by secreting either K^+ or H^+ (they usually secrete both ions). When the tubule cells contain increased amounts of H^+ as a result of the fall in plasma pH, however, they secrete increased amounts of H^+ at the expense of K^+, which helps to lower the plasma H^+ concentration and make the urine more acidic. As a side effect, less K^+ is secreted; the plasma concentration of K^+ may thus be increased under these conditions.

Hyperkalemia (high blood K^+) may thus occur in metabolic acidosis. Metabolic alkalosis can have the opposite effect, resulting in lowered plasma K^+ concentrations. This lowering of plasma K^+ is sometimes induced intentionally by the infusion of bicarbonate in patients with heart or renal failure, who are susceptible to hyperkalemia as a result of decreased glomerular filtration.

Aldosterone appears to stimulate the secretion of H^+ as well as K^+ into the distal tubules. Abnormally high aldosterone secretion, in **primary aldosteronism,** or **Conn's syndrome,** therefore, results in both hypokalemia and metabolic alkalosis. Conversely, abnormally low aldosterone secretion, as occurs in **Addison's disease,** can produce hyperkalemia and metabolic acidosis.

---

1. *Describe the effects of aldosterone on the renal nephrons, and explain how aldosterone secretion is regulated.*
2. *Describe how changes in plasma Na^+ concentrations regulate renin secretion, and explain how the secretion of renin acts to help regulate the plasma Na^+ concentration.*
3. *Explain why people who take some diuretic drugs may suffer from hypokalemia and why they must supplement their diets with potassium.*
4. *Explain the mechanisms involved in the lowering of plasma K^+ concentrations by the intravenous infusion of bicarbonate.*

Table 16.8 Classification of metabolic and respiratory components of acidosis and alkalosis.

P_{CO_2}	HCO_3^-	Condition	Causes
Normal	Low	Metabolic acidosis	Increased production of "nonvolatile" acids (lactic acid, ketone bodies, and others), or loss of HCO_3^- in diarrhea
Normal	High	Metabolic alkalosis	Vomiting of gastric acid; hypokalemia; excessive steroid administration
Low	Normal	Respiratory alkalosis	Hyperventilation
High	Normal	Respiratory acidosis	Hypoventilation

Table 16.9 The effect of changes in P_{CO_2} on the blood pH.

P_{CO_2} (mm Hg)	H_2CO_3 (mEq/L)	HCO_3^- (mEq/L)	HCO_3^-/H_2CO_3 ratio	pH	Condition
20	0.6	24	40/1	7.70	Respiratory alkalosis
30	0.9	24	26.7/1	7.53	Respiratory alkalosis
40	1.2	24	20/1	7.40	Normal
50	1.5	24	16/1	7.30	Respiratory acidosis
60	1.8	24	13.3/1	7.22	Respiratory acidosis

Note: A blood pH of less than 7.35 due to high P_{CO_2} is called respiratory acidosis. A blood pH greater than 7.45 due to low P_{CO_2} is called respiratory alkalosis.

Renal Control of Acid-Base Balance

The kidneys help regulate the acid-base balance of the blood through their excretion of H^+ and their reabsorption of HCO_3^-. By means of these processes, the kidneys are responsible for the metabolic component of acid-base balance.

Normal arterial blood has a pH of 7.37–7.45. It should be recalled that the pH number is *inversely* related to the H^+ concentration. An increase in H^+, derived from carbonic acid or the nonvolatile metabolic acids, can lower blood pH. An increase in H^+ derived from metabolic acids, however, is normally prevented from changing the blood pH by **bicarbonate buffer,** which combines with and thus removes from solution the excess H^+.

Respiratory and Metabolic Components of Acid-Base Balance

Acid-base balance has both respiratory and metabolic components. The respiratory component describes the effect of ventilation on arterial P_{CO_2} and thus on the production of carbonic acid (H_2CO_3). The metabolic component describes the effect of nonvolatile metabolic acids—lactic acid, fatty acids, and ketone bodies—on blood pH. Since these acids are normally buffered by bicarbonate (HCO_3^-), the metabolic component can be described in terms of the free HCO_3^- concentration. An increase in metabolic acids "uses up" free bicarbonate, as HCO_3^- is converted to H_2CO_3, and is thus associated with a fall in plasma HCO_3^- concentrations. A decrease in metabolic acids, conversely, is associated with a rise in free HCO_3^-.

Since the respiratory component of acid-base balance is represented by the plasma P_{CO_2} and carbonic acid concentrations and the metabolic component is represented by the free bicarbonate concentrations, the study of acid-base balance can be considerably simplified. A normal plasma pH of 7.40 is obtained when the ratio of bicarbonate to carbonic acid concentrations is twenty to one:

$$\text{pH} \propto \frac{HCO_3^-\text{ concentration}}{H_2CO_3\text{ concentration}} = \frac{20}{1} = 7.40$$

($\propto$ means "is proportional to")

A change in this ratio results in an abnormal blood pH. Pure respiratory acidosis or alkalosis occurs when the HCO_3^- concentration is normal but the P_{CO_2} and H_2CO_3 concentrations are altered. Pure metabolic acidosis or alkalosis occurs when the P_{CO_2} and H_2CO_3 are normal but the HCO_3^- concentration is abnormal. This classification is summarized in table 16.8.

A normal blood pH is produced when the P_{CO_2} is 40 mm Hg and the bicarbonate concentration is 24 milliequivalents (mEq) per liter (one mEq is equal to one millimole times the valence of the ion; in this case, since the valence is 1, mEq = mmole). Table 16.9 shows how changes in P_{CO_2} alter the ratio of HCO_3^- to H_2CO_3 and thus affect blood pH. For the sake of simplicity, this table shows the effect of changing P_{CO_2} when the bicarbonate concentration remains at a constant, normal value. Such changes in P_{CO_2} are produced by hyperventilation and hypoventilation and result in respiratory alkalosis and acidosis, respectively.

A normal plasma P_{CO_2} is maintained by proper lung function. A normal plasma bicarbonate concentration of 24 mEq per L is maintained, despite the continuous production of metabolic acids, by proper kidney function. The kidneys normally reabsorb all of the filtered bicarbonate.

Figure 16.25. Mechanisms of bicarbonate reabsorption.

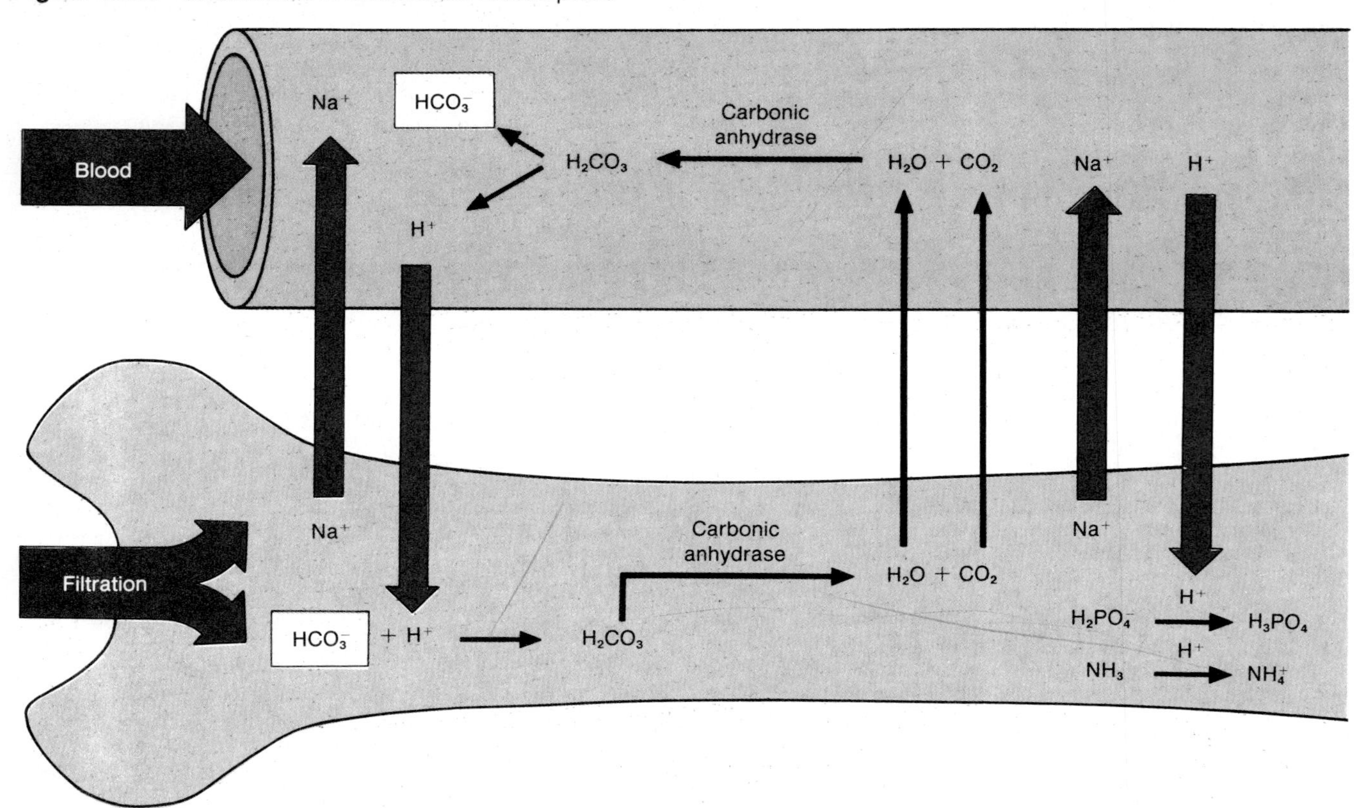

During alkalosis, however, the kidneys can reabsorb bicarbonate in excess of H^+, which is excreted in the urine. In summary, the plasma bicarbonate concentration is the responsibility of the kidneys, whereas the plasma carbonic acid concentration is regulated by the lungs:

$$pH \propto \frac{HCO_3^-}{H_2CO_3} \text{ or } \frac{\text{Kidney function}}{\text{Lung function}}$$

Mechanisms of Renal Acid-Base Regulation

The kidneys help regulate the blood pH by excreting H^+ in the urine and by the reabsorption of bicarbonate. These two mechanisms are interdependent—the reabsorption of bicarbonate occurs as a result of the filtration and secretion of H^+. The kidneys normally reabsorb all of the filtered bicarbonate and excrete about 40–80 mEq of H^+ per day. Normal urine, therefore, is free of bicarbonate and is slightly acidic (with a pH range between 5 and 7).

The apical membranes of the tubule cells are impermeable to bicarbonate. The reabsorption of bicarbonate must therefore occur indirectly. When the urine is acidic, HCO_3^- combines with H^+ to form carbonic acid. Carbonic acid in the filtrate is then converted to CO_2 and H_2O by the action of **carbonic anhydrase.** This enzyme is located in the apical cell membrane that faces the filtrate. Notice that the reaction that occurs in the tubule is the same one that occurs within the red blood cells in pulmonary capillaries.

Carbon dioxide, unlike bicarbonate, can easily pass from the filtrate, through the tubule cells, and enter the blood. Once inside the red blood cells the CO_2 combines with water to form carbonic acid. This reaction is also catalyzed by carbonic anhydrase and results in the addition of equal amounts of H^+ and HCO_3^- to the blood (fig. 16.25).

Table 16.10 Categories of disturbances in acid-base balance, including those that involve mixtures of respiratory and metabolic components.

P_{CO_2} (mmHg)	Bicarbonate (mEq/L)		
	Less than 21	21–26	More than 26
More than 45	Combined metabolic and respiratory acidosis	Respiratory acidosis	Metabolic alkalosis and respiratory acidosis
35–45	Metabolic acidosis	Normal	Metabolic alkalosis
Less than 35	Metabolic acidosis and respiratory alkalosis	Respiratory alkalosis	Combined metabolic and respiratory alkalosis

Production of Bicarbonate in the Tubules. The tubule cell cytoplasm also contains carbonic anhydrase. Some of the CO_2 that enters the tubule cells can thus be converted to carbonic acid, which can in turn dissociate to HCO_3^- and H^+ within the tubule cells. Under acidotic conditions, the H^+ produced in this way is secreted back into the filtrate, whereas the bicarbonate diffuses across the basolateral membranes and enters the blood. More bicarbonate than H^+ is thus returned to the blood as a partial compensation for the state of acidosis.

During alkalosis there is less H^+ secreted into the filtrate. Since the reabsorption of filtered bicarbonate requires the combination of HCO_3^- with H^+ to form carbonic acid, less bicarbonate is reabsorbed. This results in urinary excretion of bicarbonate, which helps to partially compensate for the alkalosis.

In this way, disturbances in acid-base balance caused by respiratory problems can be partially compensated by changes in plasma bicarbonate concentrations. Metabolic acidosis or alkalosis—in which changes in bicarbonate concentrations occur as the primary disturbance—can be similarly compensated in part by changes in ventilation. These interactions of the respiratory and metabolic components of acid-base balance are shown in table 16.10.

Urinary Buffers. In order for H^+ to be excreted in the urine without making the urine damagingly (and perhaps painfully) acidic, the acid must be buffered. Bicarbonate cannot serve this function because it is normally completely reabsorbed. Instead, the urine pH is usually prevented from falling below a value of about 4.6 by the buffering action of phosphates (mainly $H_2PO_4^-$) and ammonia (NH_3). Phosphate enters the urine by filtration. Ammonia (whose presence is strongly evident in a diaper pail or kitty litter box) is produced in the tubule cells by deamination of amino acids. These molecules buffer H^+ as described in the following equations:

$$NH_3 + H^+ \rightarrow NH_4^+ \text{ (ammonium ion)}$$

$$H_2PO_4^- + H^+ \rightarrow H_3PO_4$$

1. *Describe the respiratory and metabolic components of acid-base balance, and explain how these components are represented by the carbonic acid and bicarbonate concentration of blood.*
2. *Explain how the kidneys reabsorb filtered bicarbonate and how this process is affected by acidosis and alkalosis.*
3. *Suppose a person with diabetes mellitus has an arterial pH of 7.30, an abnormally low arterial P_{CO_2}, and an abnormally low bicarbonate concentration. Identify the type of acid-base disturbance, and explain how these values might have been produced.*

Clinical Applications

The importance of kidney function in maintaining homeostasis and the ease with which urine can be collected and used as a mirror of the plasma's chemical composition make the clinical study of renal function and urine composition particularly significant. Further, the ability of the kidneys to regulate blood volume is exploited clinically in the management of high blood pressure.

Table 16.11 Actions of different classes of diuretics.

Category of Diuretic	Example	Mechanism of Action	Major Site of Action
Carbonic anhydrase inhibitors	Acetazolamide	Inhibits reabsorption of bicarbonate	Proximal tubule
Loop diuretics	Furosemide	Inhibits sodium transport	Thick segments of ascending limbs
Thiazides	Hydrochlorothiazide	Inhibits sodium transport	Distal convoluted tubule
Potassium-sparing diuretics	Spironolactone	Inhibits action of aldosterone	Distal convoluted tubule
	Triamterene	Inhibits Na^+/K^+ exchange	Distal convoluted tubule

Use of Diuretics

People who need to lower their blood volume because of hypertension, congestive heart failure, or edema take medications that increase the volume of urine excreted. Such medications are called **diuretics.** There are a number of diuretic drugs in clinical use that act on the renal nephron in different ways (table 16.11). Based on their chemical structure or aspects of their actions, the commonly used diuretics are categorized as carbonic acid inhibitors, loop diuretics, thiazides, osmotic diuretics, and potassium-sparing diuretics.

The most powerful diuretics, inhibiting salt and water reabsorption by as much as 25%, are the drugs that act to inhibit active salt transport out of the ascending limb of the loop of Henle. Examples of these loop diuretics include *furosemide* and *ethacrynic acid.* The thiazide diuretics, like *hydrochlorothiazide,* inhibit salt and water reabsorption by as much as 8% through inhibition of salt transport by the first segment of the distal convoluted tubule. The carbonic anhydrase inhibitors (*acetazolamide*) are much weaker diuretics and act, primarily in the proximal tubule, to prevent the water reabsorption that occurs when bicarbonate is reabsorbed.

When extra solutes are present in the filtrate, they increase the osmotic pressure of the filtrate and in this way decrease the osmotic reabsorption of water throughout the nephron. *Mannitol* is sometimes used clinically for this purpose. Osmotic diuresis can occur in diabetes mellitus due to the presence of glucose in the filtrate and urine; this extra solute causes the excretion of excessive amounts of water in the urine and can result in severe dehydration of the person with uncontrolled diabetes.

The previously mentioned diuretics can, as discussed in an earlier section, result in the excessive secretion of K^+ into the filtrate and its excessive elimination in the urine. For this reason, potassium-sparing diuretics are sometimes used. *Spironolactones* are aldosterone antagonists, which compete with aldosterone for cytoplasmic receptor proteins in the cells of the distal tubule. These drugs, therefore, block the aldosterone stimulation of Na^+ reabsorption and K^+ secretion. *Triamterene* is a different type of potassium-sparing diuretic, which appears to act more directly on the Na^+/K^+ pumps in the distal tubule.

Renal Function Tests and Kidney Disease

Renal function can be tested by techniques that include the PAH clearance rate, which measures total blood flow to the kidneys, and the measurement of GFR by the inulin clearance rate. The plasma creatinine concentration, as previously described, also provides an index of renal function. These tests aid the diagnosis of kidney diseases such as glomerulonephritis and renal insufficiency.

Glomerulonephritis. Inflammation of the glomeruli, or glomerulonephritis, is currently believed to be an *autoimmune disease* which involves the person's own antibodies (as described in chapter 17). These antibodies may have been raised against the basement membrane of the glomerular capillaries, but more commonly, they appear to have been produced in response to streptococcus infections (such as strep throat). A variable number of glomeruli are destroyed in this condition, and the remaining glomeruli become more permeable to plasma proteins. Leakage of proteins into the urine results in decreased plasma colloid osmotic pressure and can therefore lead to edema.

Renal Insufficiency. When nephrons are destroyed—as in chronic glomerulonephritis, infection of the renal pelvis *(pyelonephritis),* and loss of a kidney—or when kidney function is reduced by either arteriosclerosis or blockage by kidney stones, a condition of *renal insufficiency* may develop. This can cause hypertension, due primarily to the retention of salt and water, and *uremia* (high plasma urea concentrations). The inability to excrete urea is accompanied by elevated plasma H^+ (acidosis) and elevated K^+, which are more immediately dangerous than the high urea. Uremic coma appears to result from these associated changes.

Patients with uremia or the potential for developing uremia are often placed on *dialysis* machines. The term *dialysis* refers to the separation of molecules on the basis of size by their ability to diffuse through an artificial semipermeable membrane. Urea and other wastes can easily pass through the membrane pores, whereas plasma proteins are left behind (just as occurs across glomerular capillaries). The plasma is thus cleansed of these wastes as they pass from the blood into the dialyzing fluid. Unlike the tubules, however, the dialysis membrane cannot reabsorb Na^+, K^+, glucose, and other needed molecules. These substances are prevented from diffusing through the membrane by including them in the dialysis fluid. More recent techniques include the use of the patient's own peritoneal membranes (which line the body cavity) for dialysis.

The difficulties encountered in attempting to compensate for renal insufficiency and the many dangers presented by this condition are stark reminders of the importance of kidney function in maintaining homeostasis. The ability of the kidneys to regulate blood volume and chemical composition in a way that can be adjusted according to the body's needs requires great complexity of function. Homeostasis is maintained in large part by coordination of these functions with those of the cardiovascular and pulmonary systems, as described in the preceding chapters.

---

1. *List the different categories of clinical diuretics, and explain how each exerts its diuretic effect.*
2. *Explain why most diuretics can cause excessive loss of K^+, and explain how this is prevented by the potassium-sparing diuretics.*
3. *Define uremia, describe its dangers, and explain how this condition may be corrected by the use of renal dialysis.*

Summary

Structure and Function of the Kidneys p. 478

I. The kidney is divided into an outer cortex and inner medulla.
   A. The medulla is composed of renal pyramids, separated by renal columns.
   B. The renal pyramids empty urine into the calyces that drain into the renal pelvis, and from there to the ureter.

II. Each kidney contains more than a million microscopic functional units called nephrons, which consist of vascular and tubular components.
   A. Filtration occurs in the glomerulus, which receives blood from an afferent arteriole.
   B. Glomerular blood is drained by an efferent arteriole, which delivers blood to peritubular capillaries that surround the nephron tubules.
   C. The glomerular capsule and the proximal and distal convoluted tubules are located in the cortex.
   D. The loop of Henle is located in the medulla.
   E. Filtrate from the distal convoluted tubule is drained into collecting ducts, which plunge through the medulla to empty urine into the calyces.

Glomerular Filtration p. 483

I. A filtrate derived from plasma in the glomerulus must pass through a basement membrane of the glomerular capillaries and through slits in the processes of the podocytes, which comprise the inner layer of the glomerular capsule.
   A. The glomerular ultrafiltrate is formed under the force of blood pressure and has a low protein concentration.
   B. The glomerular filtration rate is 115–125 ml/min.

II. The glomerular filtration rate can be regulated by constriction or dilation of the afferent arterioles.
   A. Sympathetic innervation causes constriction of the afferent arterioles.
   B. Intrinsic mechanisms help to autoregulate the rate of renal blood flow and the glomerular filtration rate.

Reabsorption of Salt and Water p. 485

I. Approximately 65% of the filtered salt and water is reabsorbed across the proximal convoluted tubules.
   A. Sodium is actively transported, chloride follows passively by electrical attraction, and water follows the salt out of the proximal tubule.
   B. Salt transport in the proximal tubules is not under hormonal regulation.

II. The reabsorption of most of the remaining water occurs as a result of the action of the countercurrent multiplier system.
   A. Sodium is actively extruded from the ascending limb, followed passively by chloride.
   B. Since the ascending limb is impermeable to water, the remaining filtrate becomes hypotonic.
   C. Because of this salt transport and because of countercurrent exchange in the vasa recta, the tissue fluid of the medulla becomes hypertonic.
   D. The hypertonicity of the medulla is multiplied by a positive feedback mechanism involving the descending limb, which is passively permeable to water and perhaps to salt.

III. The collecting duct is permeable to water but not to salt.
   A. As the collecting ducts pass through the hypertonic renal medulla, water leaves by osmosis and is carried away in surrounding capillaries.
   B. The permeability of the collecting ducts to water is stimulated by antidiuretic hormone (ADH).

Renal Clearance Rates p. 494

I. Inulin is filtered but neither reabsorbed nor secreted; its clearance rate is thus equal to the glomerular filtration rate.
II. Some of the filtered urea is reabsorbed; its clearance rate is therefore less than the glomerular filtration rate.
III. Since almost all the PAH in blood going through the kidneys is cleared by filtration and secretion, the PAH clearance rate is a measure of the total renal blood flow.
IV. Normally all of the filtered glucose and amino acids are reabsorbed; glycosuria occurs when the transport carriers for glucose become saturated due to hyperglycemia.

Renal Control of Electrolyte Balance p. 500

I. Aldosterone stimulates sodium reabsorption and potassium secretion in the distal convoluted tubule.
II. Aldosterone secretion is stimulated directly by a rise in blood potassium and indirectly by a fall in blood sodium.
   A. Decreased blood flow through the kidneys stimulates the secretion of the enzyme renin from the juxtaglomerular apparatus.
   B. Renin catalyzes the formation of angiotensin I, which is then converted to angiotensin II.
   C. Angiotensin II stimulates the adrenal cortex to secrete aldosterone.
III. Aldosterone stimulates the secretion of H^+ as well as potassium into the filtrate in exchange for sodium.

Renal Control of Acid-Base Balance p. 505

I. The lungs regulate the P_{CO_2} and carbonic acid concentration of the blood, whereas the kidneys regulate the bicarbonate concentration.
II. Filtered bicarbonate combines with H^+ to form carbonic acid in the filtrate.
   A. Carbonic anhydrase in the membranes of microvilli in the tubules catalyzes the conversion of carbonic acid to carbon dioxide and water.
   B. Carbon dioxide is reabsorbed and converted in either the tubule cells or the red blood cells to carbonic acid, which dissociates to bicarbonate and H^+.
   C. In addition to reabsorbing bicarbonate, the kidneys excrete H^+, which is buffered by ammonium and phosphate buffers.

Clinical Applications p. 507

I. Diuretic drugs are used clinically to increase the urine volume and thus to lower the blood volume and pressure.
   A. Loop diuretics and the thiazides inhibit active Na^+ transport in the ascending limb and early portion of the distal tubule, respectively.
   B. Osmotic diuretics are extra solutes in the filtrate that increase the osmotic pressure of the filtrate and inhibit the osmotic reabsorption of water.
   C. The potassium-sparing diuretics act on the distal tubule to inhibit the reabsorption of Na^+ and secretion of K^+.
II. In glomerulonephritis the glomeruli can permit the leakage of plasma proteins into the urine.
III. People with renal insufficiency may be treated by the technique of renal dialysis.

Review Activities

Objective Questions

1. Which of the following statements about the renal pyramids is *false?*
   (a) They are located in the medulla.
   (b) They contain glomeruli.
   (c) They contain collecting ducts.
   (d) They empty urine into the calyces.

Match the following:

2. Active transport of sodium; water follows passively
3. Active transport of sodium; impermeable to water
4. Passively permeable to water and maybe salt
5. Passively permeable to water only

(a) proximal tubule
(b) descending limb
(c) ascending limb
(d) distal tubule
(e) collecting duct

6. Antidiuretic hormone promotes the retention of water by stimulating the
   (a) active transport of water
   (b) active transport of chloride
   (c) active transport of sodium
   (d) permeability of the collecting duct to water
7. Aldosterone stimulates sodium reabsorption and potassium secretion in the
   (a) proximal convoluted tubule
   (b) descending limb of the loop
   (c) ascending limb of the loop
   (d) distal convoluted tubule
   (e) collecting duct
8. Substance *X* has a clearance rate greater than zero but less than that of inulin. What can you conclude about substance *X?*
   (a) It is not filtered.
   (b) It is filtered, but neither reabsorbed nor secreted.
   (c) It is filtered and partially reabsorbed.
   (d) It is filtered and secreted.
9. Substance *Y* has a clearance rate greater than that of inulin. What can you conclude about *Y?*
   (a) It is not filtered.
   (b) It is filtered, but neither reabsorbed nor secreted.
   (c) It is filtered and partially reabsorbed.
   (d) It is filtered and secreted.
10. About 65% of the glomerular ultrafiltrate is reabsorbed in the
   (a) proximal tubule
   (b) distal tubule
   (c) loop of Henle
   (d) collecting duct

11. Diuretic drugs that act in the loop of Henle
    (a) inhibit active sodium transport
    (b) result in the increased flow of filtrate to the distal convoluted tubule
    (c) cause the increased secretion of potassium into the tubule
    (d) promote the excretion of salt and water
    (e) all of the above
12. The appearance of glucose in the urine
    (a) occurs normally
    (b) indicates the presence of kidney disease
    (c) occurs only when the transport carriers for glucose become saturated
    (d) is a result of hypoglycemia
13. Reabsorption of water through the tubules occurs by
    (a) osmosis
    (b) active transport
    (c) facilitated diffusion
    (d) all of the above

Essay Questions

1. Explain how glomerular ultrafiltrate is produced and why it has a low protein concentration.
2. Explain how the countercurrent multiplier system works, and describe its functional significance.
3. Explain how countercurrent exchange occurs in the vasa recta, and describe its functional significance.
4. Explain the mechanisms whereby diuretic drugs may cause an excessive loss of potassium; also explain how the potassium-sparing diuretics work.
5. Explain how the structure of the epithelial wall of the proximal tubule and the distribution of Na^+/K^+ pumps in the epithelial cell membranes contribute to the ability of the proximal tubule to reabsorb salt and water.

Selected Readings

Alexander, E. 1986. Metabolic acidosis: Recognition and etiologic diagnosis. *Hospital Practice* 21:100E.

Anderson, B. 1977. Regulation of body fluids. *Annual Review of Physiology* 39:185.

Bauman, J. W., and F. P. Chinard. 1975. *Renal function: physiological and medical aspects.* St. Louis: C. V. Mosby Co.

Beeuwkes, R., III. 1980. The vascular organization of the kidney. *Annual Review of Physiology* 42:531.

Brenner, B. M., and R. Beeuwkes, III. 1978. The renal circulation. *Hospital Practice* 13:35.

Brenner, B. M., T. H. Hostetter, and H. D. Humes. 1978. Molecular basis of proteinuria of glomerular origin. *New England Journal of Medicine* 298:826.

Buckalew, V. M., Jr., and K. A. Gruber. 1984. Natriuretic hormone. *Annual Review of Physiology* 46:343.

deBold, A. 1985. Atrial natriuretic factor: A hormone produced by the heart. *Science* 230:767.

Giebisch, G. H., and B. Stanton. 1979. Potassium transport in the nephron. *Annual Review of Physiology* 41:241.

Hays, R. M. 1978. Principles of ion and water transport in the kidneys. *Hospital Practice* 13:79.

Hollenberg, N. K. 1986. The kidney in heart failure. *Hospital Practice* 21:81.

Kokko, J. S. 1979. Renal concentrating and diluting mechanisms. *Hospital Practice* 14:110.

Peart, W. S. 1975. Renin-angiotensin system. *New England Journal of Medicine* 292:302.

Rector, F. C., Jr., and M. G. Cogan. 1980. The renal acidoses. *Hospital Practice* 15:99.

Reid, I. A., B. J. Morris, and W. F. Ganong. 1978. The renin-angiotensin system. *Annual Review of Physiology* 40:377.

Renkin, E. M., and R. R. Robinson. 1974. Glomerular filtration. *New England Journal of Medicine* 290:79.

Steinmetz, P. R., and B. M. Koeppen. 1984. Cellular mechanisms of diuretic action along the nephron. *Hospital Practice* 19:125.

Tanner, R. L. 1980. Control of acid excretion by the kidney. *Annual Review of Medicine* 31:35.

Vander, A. J. 1980. *Renal physiology.* 2d ed. New York: McGraw-Hill Book Company.

Walker, L. A., and H. Valtin. 1982. Biological importance of nephron heterogeneity. *Annual Review of Physiology* 44:203.

Warnock, D. G., and F. C. Rector, Jr. 1979. Proton secretion by the kidney. *Annual Review of Physiology* 41:197.

17 The Immune System

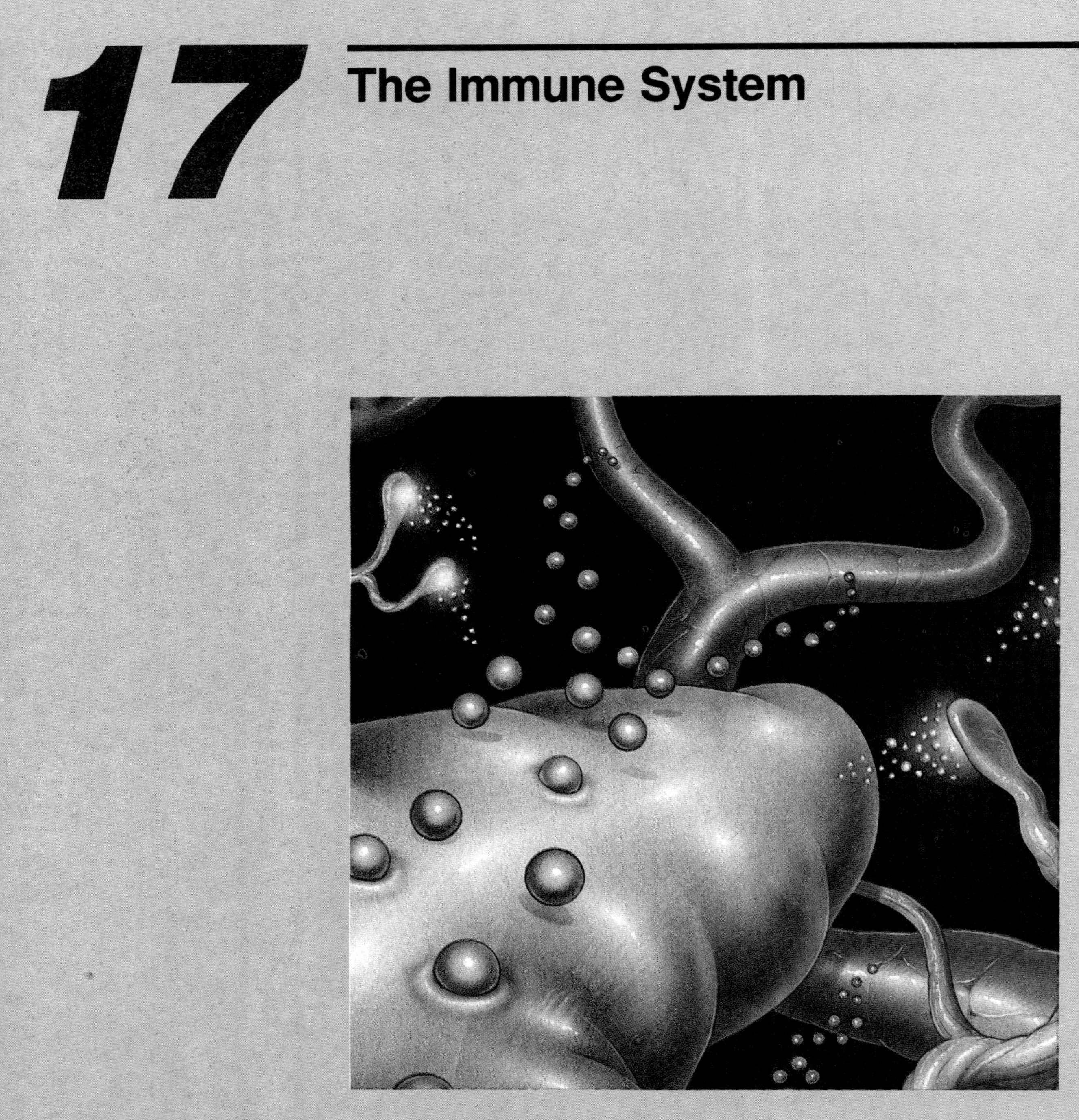

Objectives

By studying this chapter, you should be able to

1. describe some of the mechanisms of nonspecific immunity and distinguish between nonspecific and specific immune defenses
2. describe how B lymphocytes respond to antigens and define the terms *memory cell* and *plasma cell*
3. describe the structure and classifications of antibodies and the nature of antigens
4. describe the complement system and explain how antigen-antibody reactions lead to the destruction of an invading pathogen
5. describe the events that occur during a local inflammation
6. explain how T lymphocytes are classified and describe the function of the thymus gland
7. define the term *lymphokines* and list some of these molecules and their functions
8. describe the histocompatibility antigens and explain the importance of these antigens in the function of the T cell receptor proteins
9. describe the interaction between macrophages and helper T lymphocytes and explain how the helper T cells affect immunological defense by cytotoxic T cells and B cells
10. explain the possible role of suppressor T lymphocytes in the negative feedback control of the immune response
11. explain the possible mechanisms responsible for tolerance of self-antigens
12. describe the ABO and Rh systems of red blood cell antigens, explain how agglutination reactions are produced, and explain the etiology of erythroblastosis fetalis
13. describe the process of active immunity and explain how the clonal selection theory may account for this process
14. describe the mechanisms of passive immunity and give natural and clinical examples of this form of immunization
15. explain the meaning of the term *monoclonal antibodies* and describe some of their clinical uses
16. describe some of the characteristics of cancer and explain how natural killer cells and cytotoxic T lymphocytes provide immunological surveillance against cancer
17. define the term *autoimmune disease,* provide examples of different autoimmune diseases, and explain some of the mechanisms that may be responsible for autoimmune diseases
18. explain how immune complex diseases may be produced and give examples of such diseases
19. distinguish between immediate hypersensitivity and delayed hypersensitivity and describe the mechanisms responsible for each form of allergy

Outline

Table 17.1 Structures and defense mechanisms of nonspecific immunity.

	Structure	Mechanisms
External	Skin	Anatomic barrier to penetration by pathogens; secretions have lysozyme (enzyme that destroys bacteria)
	Digestive tract	High acidity of stomach Protection by normal bacterial population of colon
	Respiratory tract	Secretion of mucus; movement of mucus by cilia; alveolar macrophages
	Genitourinary tract	Acidity of urine Vaginal lactic acid
Internal	Phagocytic cells	Ingest and destroy bacteria, cellular debris, denatured proteins, and toxins
	Interferons	Inhibit replication of viruses
	Complement proteins	Promote destruction of bacteria and other effects of inflammation
	Endogenous pyrogen	Secreted by leukocytes and other cells; produces fever

Table 17.2 Phagocytic cells and their location.

Phagocyte	Location
Neutrophils	Blood and all tissues
Monocytes	Blood and all tissues
Tissue macrophages (histiocytes)	All tissues (including spleen, lymph nodes, bone marrow)
Kupffer cells	Liver
Alveolar macrophages	Lungs
Microglia	Central nervous system

Defense Mechanisms

The immune system includes all of the structures and processes that provide a defense against potential pathogens. These defenses can be grouped into *nonspecific* and *specific* categories.

Nonspecific defense mechanisms are inherited as part of the structure of each organism. Epithelial membranes that cover the body surfaces, for example, restrict infection by most pathogens. The strong acidity of gastric juice (pH 1–2) also helps to kill many microorganisms before they can invade the body. These external defenses are backed by internal defenses, such as phagocytosis, which function in both a specific and nonspecific manner (table 17.1).

Each individual can acquire the ability to defend against specific pathogens by a prior exposure to those pathogens. This specific immune response is a function of lymphocytes. Internal specific and nonspecific defense mechanisms function together to combat infection, with lymphocytes interacting in a coordinated effort with phagocytic cells.

Nonspecific Immunity

Invading pathogens that have crossed epithelial barriers enter connective tissues. These invaders—or chemicals, called *toxins,* secreted from them—may enter blood or lymphatic capillaries and be carried to other areas of the body. The invasion and spread of infection is fought in two stages: (1) nonspecific immunological defenses are employed; if these are sufficiently effective, the pathogens may be destroyed without progression to the next step; (2) lymphocytes may be recruited, and their specific actions used to reinforce the previously nonspecific immune defenses.

Figure 17.1. Diapedesis. White blood cells squeeze through openings between capillary endothelial cells to enter underlying connective tissues.

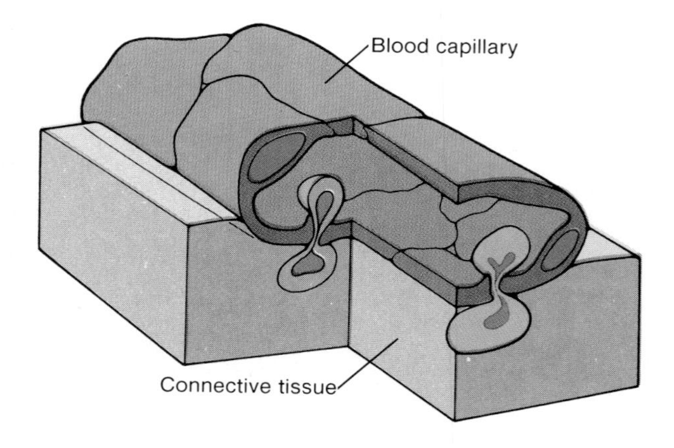

Phagocytosis. There are two major groups of phagocytic cells: (1) neutrophils; and (2) the cells of the *mononuclear phagocyte system.* This latter category includes *monocytes* in the blood, *macrophages* (derived from monocytes) in the connective tissues, and *organ-specific phagocytes* in the liver, spleen, lymph nodes, lungs, and brain (table 17.2).

Connective tissues contain a resident population of all leukocyte types. Neutrophils and monocytes in particular can be highly mobile within connective tissues as they scavenge for invaders and cellular debris. These leukocytes are recruited to the site of an infection by a process known as **chemotaxis**—movement toward chemical attractants. Neutrophils are the first to arrive at the site of an infection; monocytes arrive later and can be transformed into macrophages as the battle progresses.

Figure 17.2. Phagocytosis by a neutrophil or macrophage. A phagocytic cell extends its pseudopods around the object to be engulfed (such as a bacterium). Dots represent lysosomal enzymes (*L* = lysosomes). If the pseudopods fuse to form a complete food vacuole (*1*), lysosomal enzymes are restricted to the organelle formed by the lysosome and food vacuole. If the lysosome fuses with the vacuole before fusion of the pseudopods is complete (*2*), lysosomal enzymes are released into the infected area of tissue.

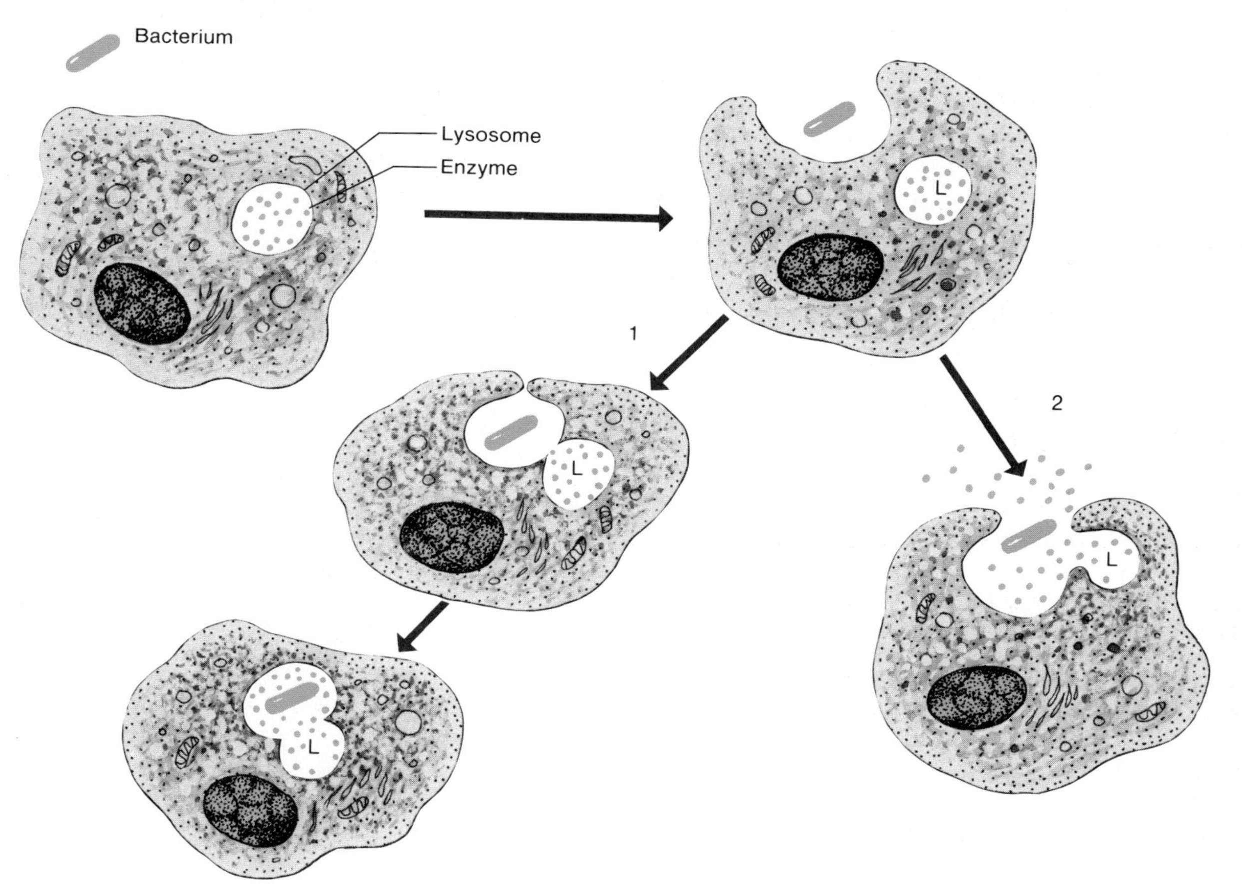

If the infection is sufficiently large, new phagocytic cells from the blood may join those already in the connective tissue. These new neutrophils and monocytes are able to squeeze through the tiny gaps between adjacent endothelial cells in the capillary wall and enter the connective tissues. This process, called **diapedesis,** is illustrated in figure 17.1.

Phagocytic cells engulf particles in a manner similar to the way an amoeba eats. The particle becomes surrounded by cytoplasmic extensions called pseudopods, which ultimately fuse together. The particle thus becomes surrounded by a membrane derived from the plasma membrane (fig. 17.2) and contained within an organelle analogous to a food vacuole in an amoeba. This vacuole then fuses with lysosomes (organelles that contain digestive enzymes), so that the ingested particle and the digestive enzymes remain separated from the cytoplasm by a continuous membrane. Often, however, lysosomal enzymes are released before the food vacuole has completely formed. When this occurs, free lysosomal enzymes may be released into the infected area and contribute to inflammation.

The *Kupffer cells* in the liver, together with phagocytic cells in the spleen and lymph nodes, are **fixed phagocytes.** This term refers to the fact that these cells are immobile ("fixed") in the channels within these organs. As blood flows through the liver and spleen and as lymph percolates through the lymph nodes, foreign chemicals and debris are removed by phagocytosis and chemically inactivated within the phagocytic cells. Invading pathogens are very effectively removed in this manner, so that blood is usually sterile after a few passes through the liver and spleen.

Fever. Fever may be a component of the nonspecific defense system. Body temperature is regulated by the hypothalamus, which contains a thermoregulatory control center (a "thermostat") that coordinates skeletal muscle shivering and the activity of the sympathoadrenal system to maintain body temperature at about 37° C. This thermostat is reset upwards in response to a chemical called **endogenous pyrogen,** secreted by leukocytes. Endogenous pyrogen secretion is stimulated by a chemical called *endotoxin,* which is released by certain bacteria.

Although high fevers are definitely dangerous, many believe that a mild to moderate fever may be a beneficial response that aids recovery from bacterial infections. There is some evidence to support this view, but the mechanisms involved are not clearly understood. One theory is that elevated body temperature may interfere with the nutritional requirements of some bacteria.

Interferons. In 1957, researchers demonstrated that cells infected with a virus produced polypeptides that "interfered with" the ability of a second, unrelated strain of virus to infect other cells in the same culture. These **interferons,** as they were called, thus produced a nonspecific, short-acting resistance to viral infection. This discovery produced a great deal of excitement, but further research in this area was hindered by the fact that human interferons could only be obtained in very small quantities and that animal interferons had little effect in humans. In 1980, however, technological breakthroughs allowed researchers to introduce human interferon genes into bacteria—through a technique called *genetic recombination*—so that bacteria could act as interferon factories.

Leukocytes, fibroblasts, and probably many other cells make their own characteristic types of interferons. These polypeptides act as messengers that protect other cells in the vicinity from viral infection. The viruses are still able to penetrate these other cells, but the ability of the viruses to replicate and assemble new virus particles (fig. 17.3) is inhibited.

Lymphocytes called *T lymphocytes* release interferons in response to viral infections and perhaps as part of their immunological surveillance against cancer. Interferons may destroy cancer cells directly and indirectly by the activation of T lymphocytes and *natural killer cells.* Some of the proposed effects of interferons are summarized in table 17.3.

Recent clinical trials have suggested that interferons may be effective in the treatment of such viral infections as viral hepatitis and herpes-induced "cold sores." Current evidence suggests that interferons may not prove to be a "magic bullet" against cancer but may, when combined with other methods, significantly augment the treatment of some forms of cancer. The emphasis is on treatment rather than prevention here, because the protective effects of interferons are of short duration.

Figure 17.3. The sequence of events that can occur when human cells are infected with virus particles.

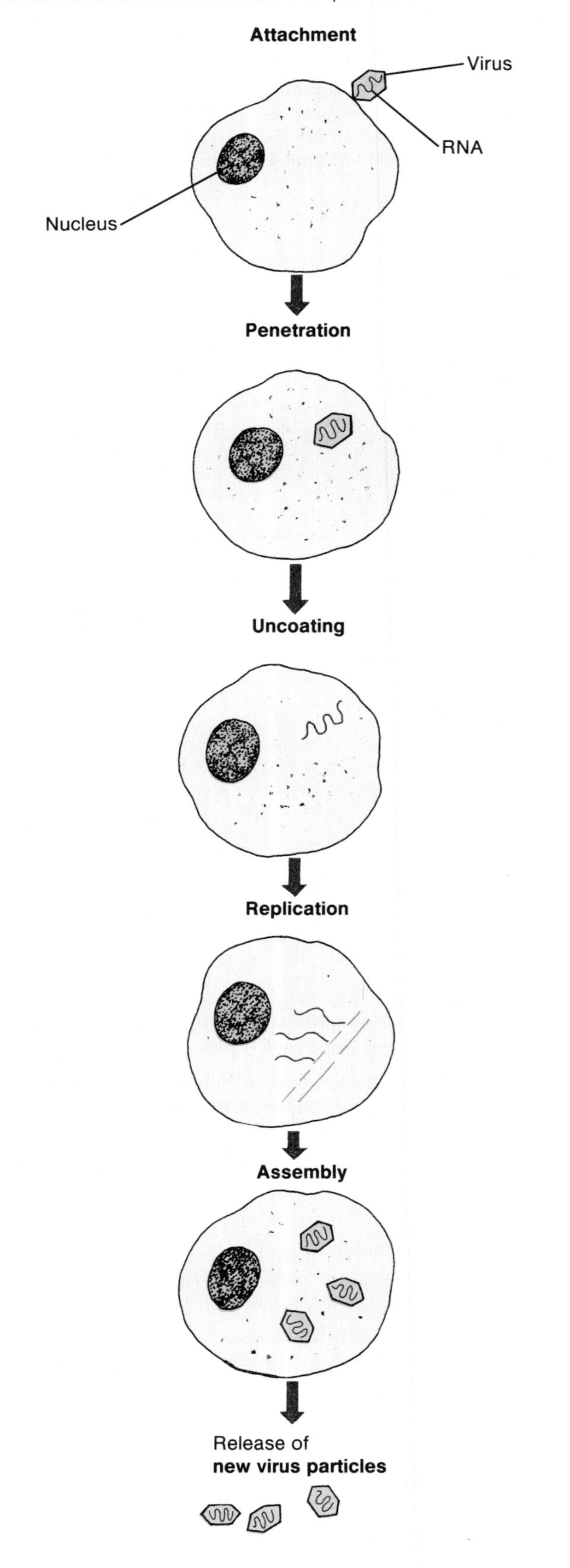

Table 17.3 Some of the proposed effects of interferons.

Stimulation	Inhibition
Macrophage phagocytosis	Cell division
Activity of cytotoxic ("killer") T cells	Tumor growth
Activity of natural killer cells	Maturation of adipose cells
Production of antibodies	Maturation of erythrocytes

Specific Immunity

In 1890, von Behring demonstrated that a guinea pig which had been previously injected with a sublethal dose of diphtheria toxin could survive subsequent injections of otherwise lethal doses of that toxin. Further, von Behring showed that this immunity could be transferred to a second, nonexposed animal by injections of serum from the immunized guinea pig. He concluded that the immunized animal had chemicals in its serum—which he called **antibodies**—that were responsible for the immunity. He also showed that these antibodies conferred immunity only to subsequent diphtheria infections; the antibodies were *specific* in their actions. It was later learned that antibodies are proteins produced by a particular type of lymphocyte.

Antigens. Antigens are molecules that stimulate antibody production and combine with specific antibodies. Most antigens are large molecules (such as proteins) with a molecular weight greater than about 10,000, and they are foreign to the blood and other body fluids (although there are exceptions to both descriptions). The ability of a molecule to function as an antigen depends not only on its size but also on the complexity of its structure. Proteins are therefore more antigenic than polysaccharides, which have a simpler structure. Plastics used in artificial implants are composed of large molecules but are not very antigenic because of their simple, repeating structures.

A large, complex, foreign molecule can have a number of different **antigenic determinant sites,** which are areas of the molecule that stimulate production of and combine with different antibodies. Most naturally occurring antigens have many antigenic determinant sites and stimulate the production of different antibodies with specificities for these sites.

Haptens. Many small organic molecules are not antigenic by themselves but can become antigens if they bond to proteins (and thus become antigenic determinant sites on the proteins). This was discovered by Karl Landsteiner, the same man who discovered the ABO blood groups. By bonding these small molecules—which Landsteiner called

Figure 17.4. Immunoassay using the agglutination technique. Antibodies against a particular antigen are adsorbed to red blood cells or latex particles. When these are mixed with a solution that contains the appropriate antigen, the formation of the antigen-antibody complexes produces clumping (agglutination) that can be seen with the unaided eye.

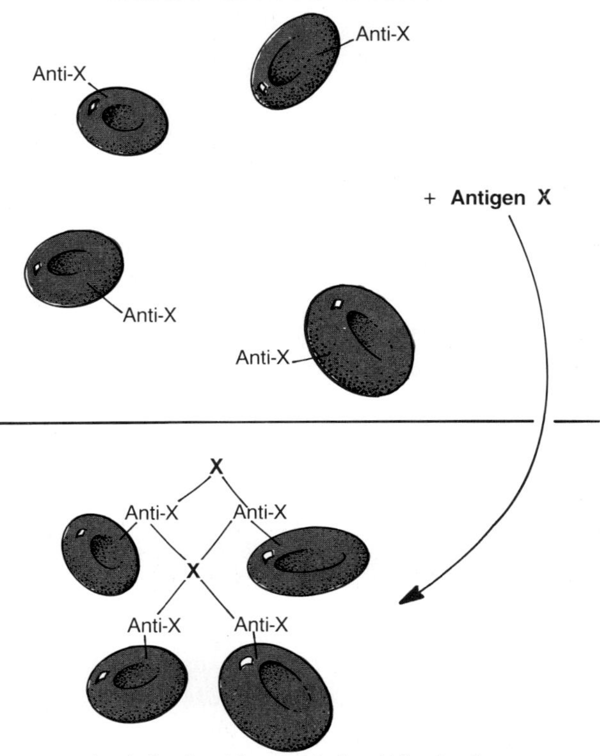

haptens—to proteins in the laboratory, new antigens could be created for research or diagnostic purposes. The bonding of foreign haptens to a person's own proteins can also occur in the body; by this means, derivatives of penicillin, for example, that would otherwise be harmless can produce fatal allergic reactions in susceptible people.

Immunoassays. When the antigen or antibody is attached to the surface of a cell or to particles of latex rubber (in commercial diagnostic tests), the antigen-antibody reaction becomes visible because the particles *agglutinate* (clump) as a result of antigen-antibody bonding (fig. 17.4). These agglutinated particles can be used to assay a variety of antigens, and tests that utilize this procedure are called *immunoassays.* Blood typing and modern pregnancy tests are examples of such immunoassays.

Table 17.4 Comparison of B and T lymphocytes.

Characteristic	B Lymphocytes	T Lymphocytes
Site where processed	Bone marrow	Thymus
Type of immunity	Humoral (secretes antibodies)	Cell-mediated
Subpopulations	Memory cells and plasma cells	Cytotoxic (killer) T cells, helper cells, suppressor cells
Presence of surface antibodies	Yes—IgM or IgD	Not detectable
Receptors for antigens	Present—are surface antibodies	Present—are related to immunoglobulins
Life span	Short	Long
Tissue distribution	High in spleen, low in blood	High in blood and lymph
Percent of blood lymphocytes	10%–15%	75%–80%
Transformed by antigens to	Plasma cells	Small lymphocytes
Secretory product	Antibodies	Lymphokines
Immunity to viral infections	Enteroviruses, poliomyelitis	Most others
Immunity to bacterial infections	Streptococcus, staphylococcus, many others	Tuberculosis, leprosy
Immunity to fungal infections	None known	Many
Immunity to parasitic infections	Trypanosomiasis, maybe to malaria	Most others

Lymphocytes

Leukocytes, erythrocytes, and blood platelets are all ultimately derived from ("stem from") unspecialized cells in the bone marrow. These *stem cells* produce the specialized blood cells, and they replace themselves by cell division so that the stem cell population is not exhausted. Lymphocytes produced in this manner seed the thymus, spleen, and lymph nodes, producing self-replacing lymphocyte colonies in these organs.

The lymphocytes that become seeded in the thymus become **T lymphocytes.** These cells have surface characteristics and an immunological function that is different from other lymphocytes. The thymus, in turn, seeds other organs; about 65% to 85% of the lymphocytes in blood and most of the lymphocytes in lymph nodes are T lymphocytes. T lymphocytes, therefore, come from or had an ancestor that came from the thymus gland.

Most of the lymphocytes that are not T lymphocytes are called **B lymphocytes.** The letter *B* is derived from immunological research performed in chickens. Chickens have an organ called the *bursa of Fabricius* at the hind end of their digestive tracts, which is the equivalent of the thymus located at the anterior end. The thymus processes T lymphocytes; the bursa, in a chicken, processes B lymphocytes. Since mammals do not have a bursa, the *B* is often translated as the "bursa equivalent" for humans and other mammals. It is currently believed that the B lymphocytes in mammals are processed in the bone marrow, which conveniently also begins with the letter *B*.

Both B and T lymphocytes function in specific immunity. The B lymphocytes combat bacterial and some viral infections by secreting antibodies into the blood and lymph. They are therefore said to provide **humoral immunity.** T lymphocytes attack host cells that have become infected with viruses or fungi, transplanted human cells, and cancerous cells. The T lymphocytes do not secrete antibodies; they must come into close proximity or have actual physical contact with the victim cell in order to destroy it. T lymphocytes are therefore said to provide **cell-mediated immunity** (table 17.4).

1. *List the phagocytic cells in blood and lymph, and indicate which organs contain fixed phagocytes.*
2. *Describe the actions of interferons.*
3. *List four properties that antigens usually have. Explain why proteins are more antigenic than polysaccharides.*
4. *Describe the meaning of the term* hapten, *and give an example.*
5. *Distinguish between B and T lymphocytes in terms of their origins and immune functions.*

Figure 17.5. B lymphocytes have antibodies on their surface that function as receptors for specific antigens. The interaction of antigens and antibodies on the surface stimulates cell division and the maturation of the B cell progeny into memory cells and plasma cells. Plasma cells produce and secrete large amounts of the antibody (note the extensive rough endoplasmic reticulum in these cells).

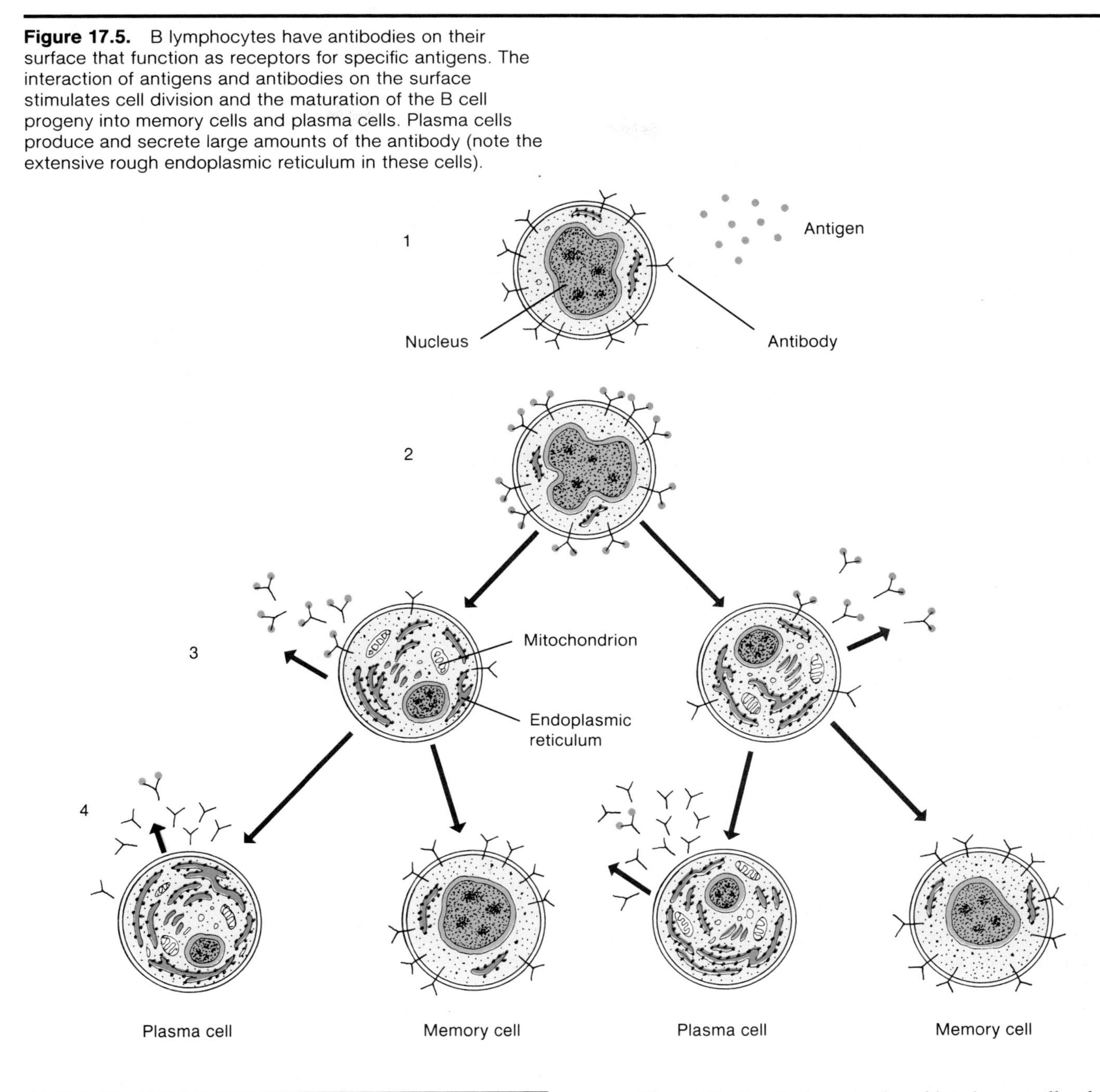

Functions of B Lymphocytes

Exposure of a B lymphocyte to the appropriate antigens results in cell growth followed by many cell divisions. Some of the progeny become **memory cells,** which are indistinguishable from the original cell; others are transformed into **plasma cells** (fig. 17.5). Plasma cells are protein factories that produce about two thousand antibody proteins per second in their brief (five to seven day) life span.

The antibodies that are produced by plasma cells when B lymphocytes are exposed to a particular antigen react specifically with that antigen. Such antigens may be isolated molecules, or they may be molecules at the surface of an invading foreign cell. The specific bonding of antibodies to antigens serves to identify the enemy and to activate defense mechanisms that lead to the invader's destruction.

Figure 17.6. The separation of serum protein by electrophoresis. A = albumin, alpha-1 (α_1), alpha-2 (α_2), beta (β), and gamma (γ) globulin.

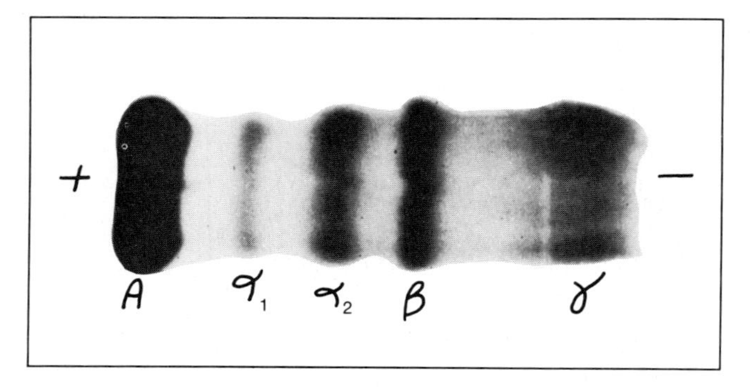

Antibodies

Antibody proteins are also known as **immunoglobulins.** These are found in the gamma globulin class of plasma proteins, as identified by a technique called *electrophoresis,* in which classes of plasma proteins are separated by their movement in an electric field (fig. 17.6). With this technique, five distinct bands of proteins appear: albumin, alpha-1-globulin, alpha-2-globulin, beta globulin, and gamma globulin.

The gamma globulin band is wide and diffuse because it represents a heterogenous class of molecules. Since antibodies are specific in their actions, it could be predicted that different types of antibodies have different structures. An antibody against smallpox, for example, does not confer immunity to poliomyelitis and, therefore, must have a slightly different structure than an antibody against polio. Despite these differences, antibodies are structurally related and form only a few subclasses.

There are five subclasses of immunoglobulins (abbreviated Ig): *IgG, IgA, IgM, IgD,* and *IgE.* Most of the antibodies in serum are in the IgG subclass, whereas most of the antibodies in external secretions (saliva and milk) are IgA (table 17.5). Antibodies in the IgE subclass are involved in allergic reactions.

Table 17.5 The immunoglobulins.

Immunoglobulin	Examples of Functions
IgG	Main form of antibodies in circulation: production increased after immunization
IgA	Main antibody type in external secretions, such as saliva and mother's milk
IgE	Responsible for allergic symptoms in immediate hypersensitivity reactions
IgM	Function as antigen receptors on lymphocyte surface prior to immunization; secreted during primary response
IgD	Function as antigen receptors on lymphocyte surface prior to immunization; other functions unknown

Antibody Structure. All antibody molecules consist of four interconnected polypeptide chains. Two longer, higher molecular weight chains (the *H chains*) are joined to two shorter, lighter *L chains.* Research has shown that these four chains are arranged in the form of a Y. The stalk of the Y has been called the "crystallizable fragment" (abbreviated F_c), whereas the top of the Y is the "antigen-binding fragment" (F_{ab}). This is shown in figure 17.7.

The amino acid sequences of some antibodies have been determined by using antibodies derived from people with multiple myelomas. These lymphocyte tumors are derived from the division of a single B lymphocyte, forming a population of genetically identical cells that secrete identical antibodies. These populations and the antibodies they secrete are different, however, in different patients. Analyses of these antibodies have shown that the F_c regions of different antibodies are the same (are *constant*), whereas the F_{ab} regions are *variable.* Variability of the antigen-binding regions is required for the specificity of antibodies for antigens.

Each of the five immunoglobulin subclasses is characterized by its own unique polypeptide in the constant region of the heavy chains. Substituting one of these polypeptides for another, therefore, changes the immunoglobulin subclass. Before a B lymphocyte is stimulated by an antigen, for example, it has IgM antibodies on its surface that function as *receptors* for the antigen. After stimulation by antigens, the B lymphocyte becomes transformed into a plasma cell that secretes IgG antibodies instead of IgM antibodies. This replacement involves a change only in the constant part of the heavy chains; since the antigen-binding parts are unaltered, the antibody specificity remains the same when IgM is replaced by IgG.

Figure 17.7. Antibodies are composed of four polypeptide chains—two are heavy (*H*) and two are light (*L*). (*a*) A computer-generated model of antibody structure. (*b*) A simplified diagram showing the constant and variable regions. The variable regions are abbreviated V, and the constant regions are abbreviated C. Antigens combine with the variable regions. Each antibody molecule is divided into an F_{ab} (antigen-binding) fragment and an F_c (crystallizable) fragment.

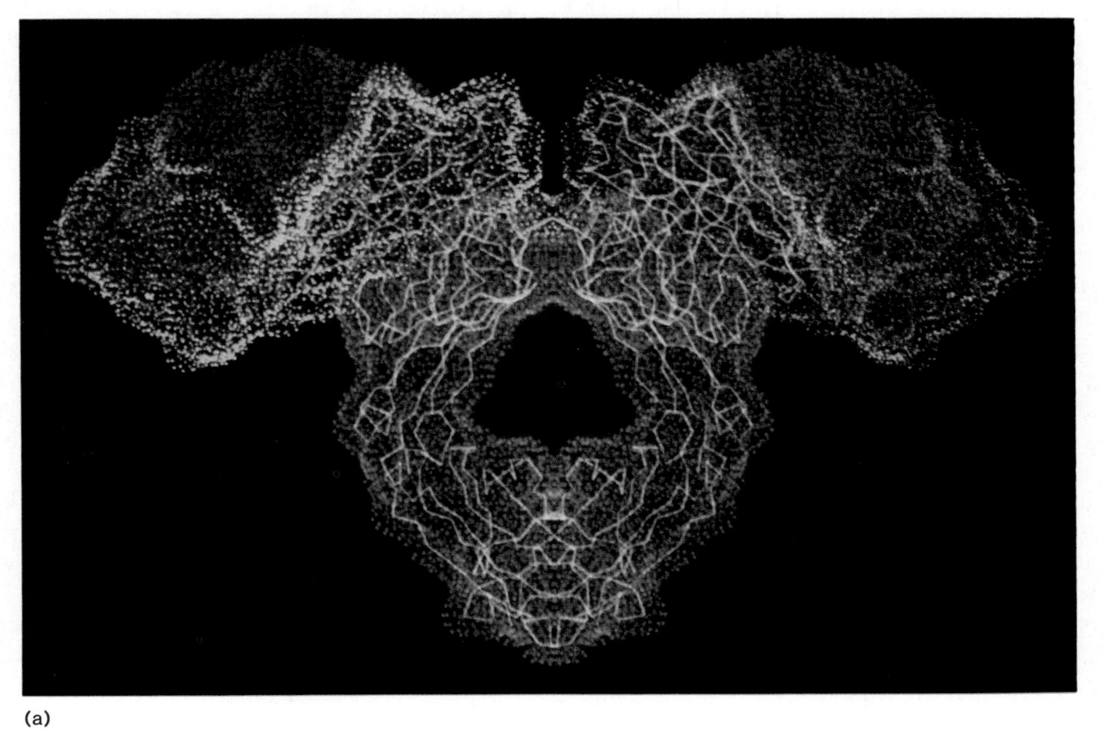

(a)

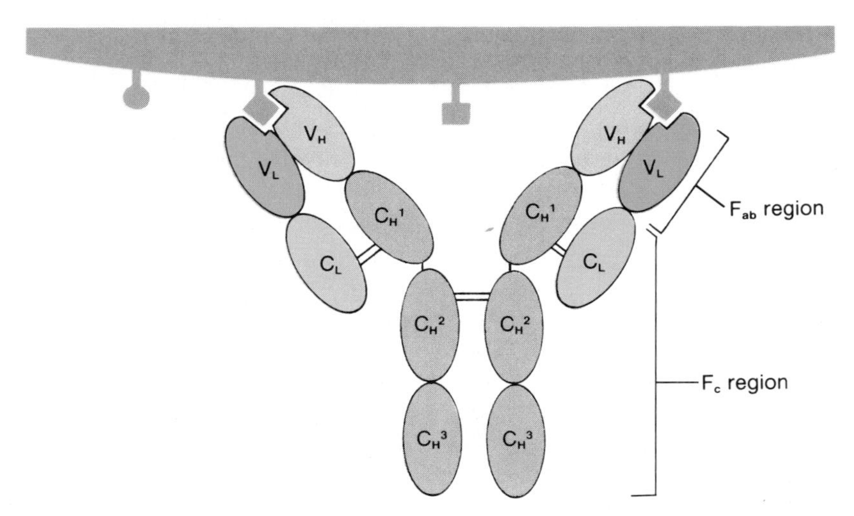

(b)

Diversity of Antibodies. It is estimated that there are about 100 million trillion (10^{20}) antibody molecules in each individual, representing a few million different specificities for different antigens. Considering that antibodies to particular antigens can cross-react to some degree with closely related antigens, this tremendous antibody diversity usually insures that there are some antibodies that can combine with any antigen the person may encounter. These observations evoke a question that has long fascinated scientists: How can a few million different antibodies be produced? A person cannot possibly inherit a correspondingly large number of genes devoted to antibody production.

Two mechanisms have been proposed to explain antibody diversity. First, since different combinations of heavy and light chains can produce different antibody specificities, a person does not have to inherit a million different genes to code for a million different antibodies. If a few hundred genes code for different H chains and a few hundred code for different L chains, different combinations of these polypeptide chains could produce millions of different antibodies. Second, the diversity of antibodies could increase during development if, when some lymphocytes divided, the progeny received antibody genes that have been slightly altered by mutations. This is called *somatic mutation* since the mutation occurs in a body cell rather than in a sperm or ovum. The diversity of antibodies would thus increase as the lymphocyte population increases.

Opsonization. The combination of antibodies with antigens does not itself produce destruction of the antigens or of the pathogenic organisms that contain these antigens. Antibodies, rather, serve to identify the targets for immunological attack and to activate nonspecific immune processes that destroy the invader. Bacteria that are buttered with antibodies, for example, are better targets for phagocytosis by neutrophils and macrophages. The ability of antibodies to stimulate phagocytosis is termed **opsonization.** Immune destruction of bacteria is also promoted by antibody-induced activation of a system of serum proteins known as *complement.*

The Complement System

It was learned in the early part of the twentieth century that rabbit antibodies to sheep red blood cell antigens could not lyse (destroy) these cells unless certain protein components of serum were present. These proteins, called **complement,** are a nonspecific defense system that is activated by the bonding of antibodies to antigens and by this means is directed against specific invaders that have been identified by antibodies.

There are eleven complement proteins, designated C1 (which has three protein components) through C9. These proteins are present in an inactive state within plasma and other body fluids and become activated by the attachment of antibodies to antigens. In terms of their functions, the complement proteins can be subdivided into three components: (1) recognition (C1); (2) activation (C4, C2, and C3, in that order); and (3) attack (C5–C9). The attack phase consists of **complement fixation,** in which complement proteins attach to the cell membrane and destroy the victim cell.

Antibodies of the IgG and IgM subclasses attach to antigens on the invading cell's membrane, bond to C1, and by this means activate its enzyme activity. Activated C1 catalyzes the hydrolysis of C4 into two fragments (fig. 17.8), designated C4a and C4b. The C4b fragment bonds to the cell membrane (is "fixed") and becomes an active enzyme that splits C2 into two fragments, C2a and C2b. The C2a becomes attached to C4b and cleaves C3 into C3a and C3b. Fragment C3b becomes attached to the growing complex of complement proteins on the cell membrane. The C3b converts C5 to C5a and C5b. The C5b and eventually C6 through C9 become fixed to the cell membrane.

Complement proteins C5 through C9 create large pores in the membrane (fig. 17.9). These pores allow the osmotic influx of water, so that the victim cell swells and bursts. Notice that the complement proteins, not the antibodies directly, kill the cell; antibodies only serve as activators of this process. Other molecules can also activate the complement system in an alternate nonspecific pathway that bypasses the early phases of the specific pathway described here.

Complement fragments that are not fixed but instead are liberated into the surrounding fluid have a number of effects. These effects include (1) *chemotaxis*—the liberated complement fragments attract phagocytic cells to the site of complement activation; (2) *opsonization*—phagocytic cells have receptors for C3b, so that this fragment may form bridges between the phagocyte and the victim cell that facilitates phagocytosis; and (3) fragments C3a and C5a *stimulate the release of histamine* from mast cells (a connective tissue cell type) and basophils. As a result of histamine release, there is increased blood flow to the infected area due to vasodilation and increased capillary permeability. The latter effect can result in the leakage of plasma proteins into the surrounding tissue fluid, producing local edema.

Figure 17.8. The fixation of complement proteins. The formation of an antibody-antigen complex causes complement protein C4 to be split into two subunits—C4a and C4b. The C4b subunit attaches (is fixed) to the membrane of the cell to be destroyed (such as a bacterium). This event triggers the activation of other complement proteins, some of which attach (are fixed) to the C4b on the membrane surface.

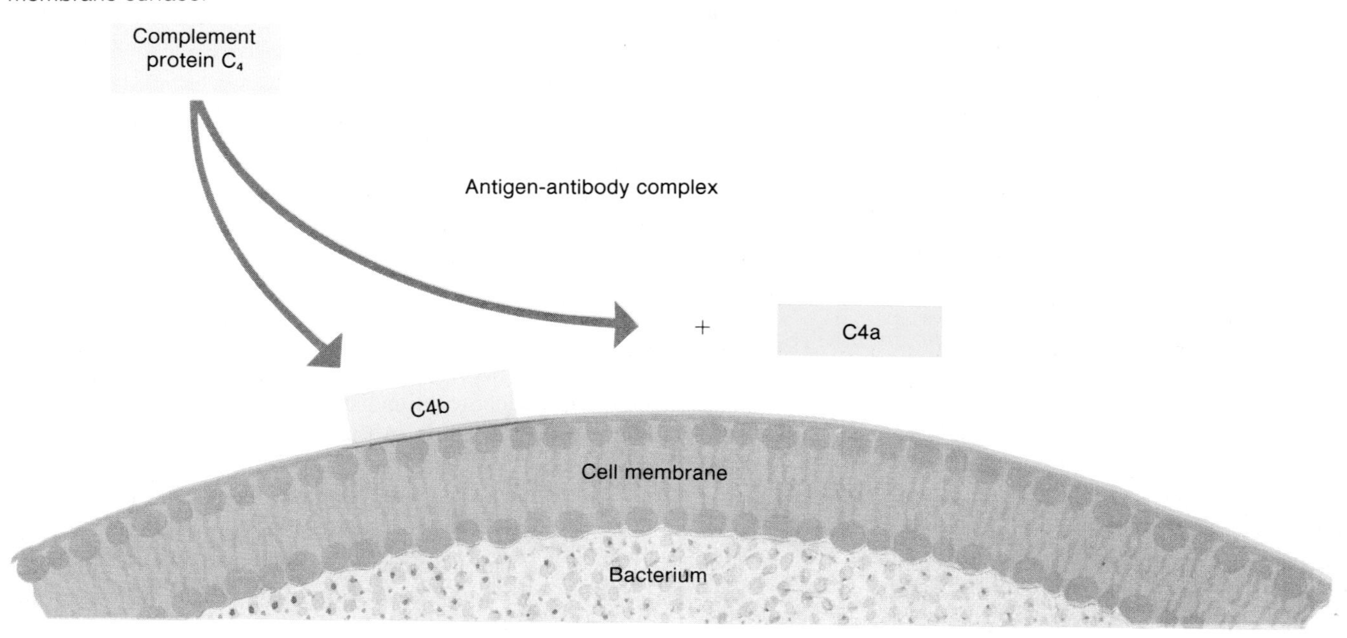

Figure 17.9. Complement proteins C5 through C9 (illustrated as a doughnut-shaped ring) puncture the membrane of the cell to which they are attached (fixed). This aids destruction of the cell.

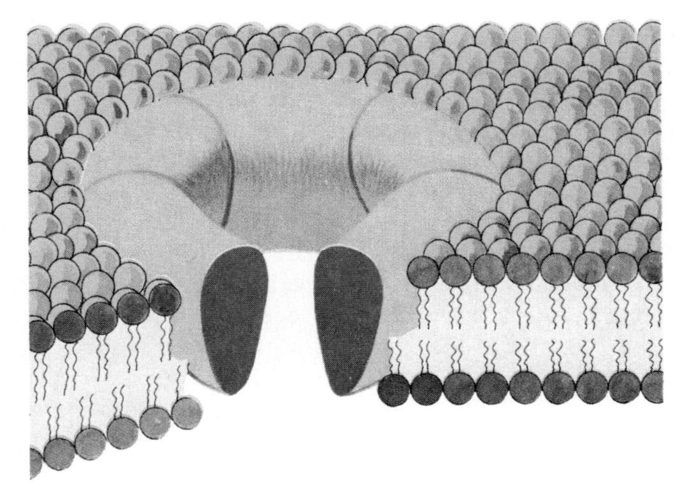

Table 17.6 Summary of events that occur in a local inflammation when a break in the skin permits entry of bacteria.

Category	Events
Nonspecific immunity	Bacteria enter through break in anatomic barrier of skin. Resident phagocytic cells—neutrophils and macrophages—engulf bacteria. Nonspecific activation of complement proteins occurs.
Specific immunity	B cells are stimulated to produce specific antibodies. Phagocytosis is enhanced by antibodies attached to bacterial surface antigens. Specific activation of complement proteins occurs, which stimulates phagocytosis, chemotaxis of new phagocytes to the infected area, and secretion of histamine from tissue mast cells. Diapedesis allows new phagocytic leukocytes (neutrophils and monocytes) to invade the infected area. Vasodilation and increased capillary permeability (as a result of histamine secretion) produce redness and edema.

Local Inflammation

Aspects of the nonspecific and specific immune responses and their interactions are well illustrated by the events that occur when bacteria enter a break in the skin and produce a local inflammation (table 17.6). The inflammatory reaction is initiated by the nonspecific mechanisms of phagocytosis and complement activation. Activated complement further increases this nonspecific response by attracting new phagocytes to the area and by increasing the activity of these phagocytic cells.

After some time, B lymphocytes are stimulated to produce antibodies against specific antigens that are part of the invading bacteria. Attachment of these antibodies to antigens in the bacteria greatly increases the previously nonspecific response. This occurs because of greater activation of complement, which directly destroys the bacteria and which—together with the antibodies themselves—promotes the phagocytic activity of neutrophils, macrophages, and monocytes (fig. 17.10).

As inflammation progresses, the release of lysosomal enzymes from macrophages causes the destruction of leukocytes and other tissue cells. These effects, together with those produced by histamine and other chemicals released from mast cells, produce the characteristic symptoms of a local inflammation: *redness and warmth* (due to vasodilation); *swelling* (edema); and *pus* (the accumulation of dead leukocytes). If the infection continues, the release of endogenous pyrogen from leukocytes and macrophages may produce a fever.

1. *Illustrate the structure of an antibody molecule. Label the constant and variable regions, the F_c and F_{ab} parts, and the heavy and light chains.*
2. *Define opsonization, and name two types of molecules that promote this process.*
3. *Describe complement fixation, and explain the roles of complement fragments that do not become fixed.*
4. *Explain how nonspecific and specific immune mechanisms cooperate during a local inflammation.*

Functions of T Lymphocytes

The thymus processes lymphocytes in such a way that their functions become quite distinct from those of B cells; lymphocytes that reside in the thymus or which came from the thymus or which are derived from cells that came from the thymus are all T lymphocytes that can be distinguished by special techniques from B cells. The T lymphocytes provide specific immune protection without secreting antibodies. This is accomplished in different ways by the three subpopulations of T lymphocytes, which are designated as cytotoxic (killer), helper, and suppressor cells.

Thymus Gland

The thymus gland extends from below the thyroid in the neck into the thoracic cavity. This organ grows during childhood but gradually regresses after puberty. Lymphocytes from the fetal liver and spleen in prenatal life and from the bone marrow postnatally seed the thymus and become transformed into T cells. These lymphocytes in turn enter the blood and seed lymph nodes and other organs, where they divide to produce new T cells when stimulated by antigens.

Small T lymphocytes that have not yet been stimulated by antigens have very long life spans—months or perhaps years. Still, new T cells must be continuously produced to provide efficient cell-mediated immunity. Since the thymus atrophies after puberty, this organ may not be able to provide new T cells in later life. Colonies of T cells in the lymph nodes and other organs are apparently able to produce new T cells under the stimulation of various **thymus hormones.**

Two hormones that are believed to be secreted by the thymus—*thymopoietin I* and *thymopoietin II*—may promote the transformation of lymphocytes into T cells. Another thymus hormone, called *thymosin,* may promote the maturation of T lymphocytes.

There is some experimental evidence supporting the notion that the administration of these thymus hormones may be able to restore cell-mediated immunity in some

Figure 17.10. The entry of bacteria through a cut in the skin produces a local inflammatory reaction. In this reaction, antigens on the bacterial surface are coated with antibodies and ingested by phagocytic cells. Symptoms of inflammation are produced by the release of lysosomal enzymes and by the secretion of histamine and other chemicals from tissue mast cells.

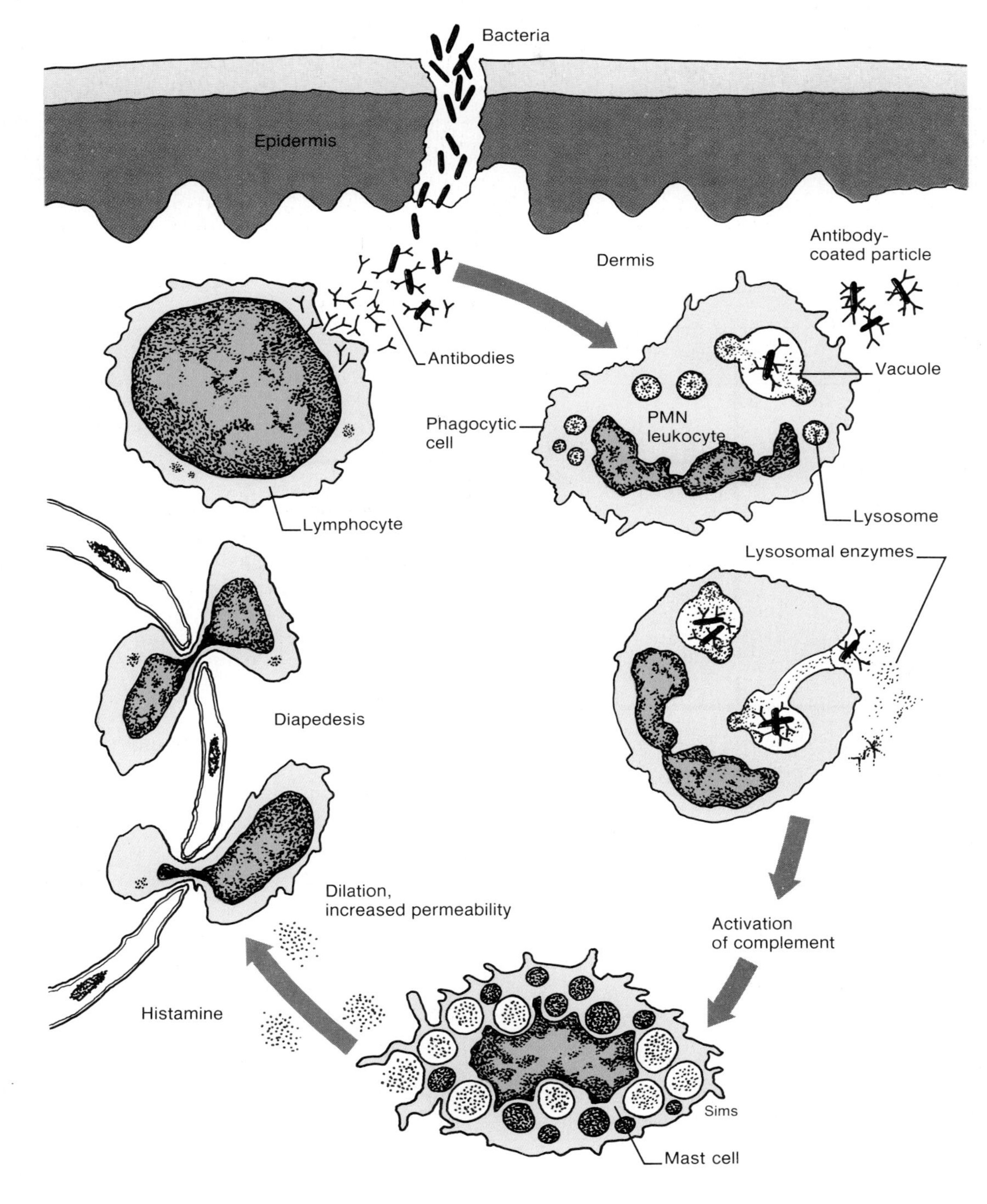

Figure 17.11. A given antigen can stimulate the production of both B and T cell clones. The ability to produce B cell clones, however, is also influenced by the relative effects of helper and suppressor T cells.

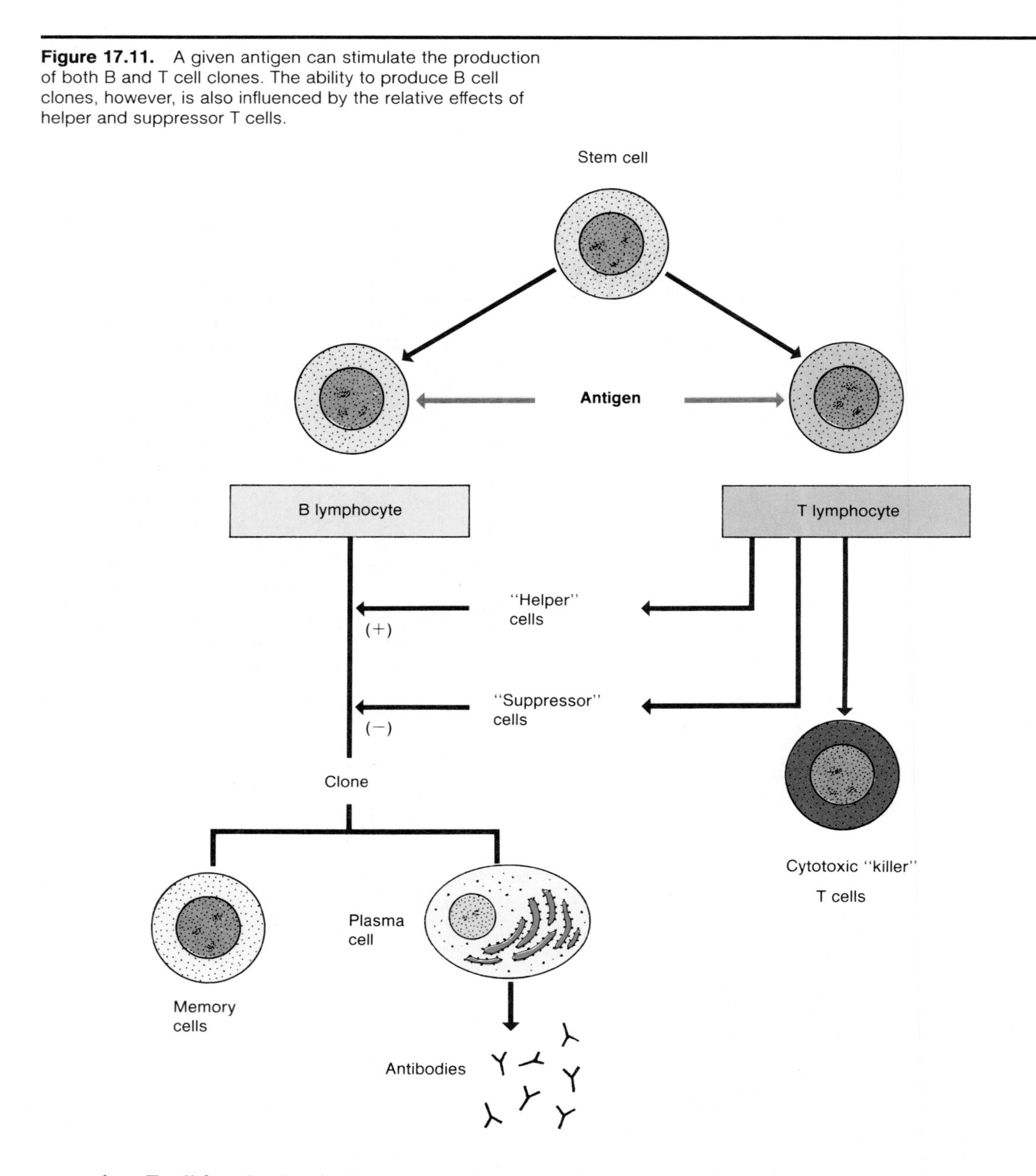

cases where T cell function has declined. This decline occurs in some congenital and acquired diseases as well as naturally in the course of aging in conjunction with an increased susceptibility to viral infections and cancer. Clinical manipulations of T cell function—through the administration of thymus hormones, interferon, interleukin-2, and other techniques—is an exciting possibility that may result from further experimental research.

Cytotoxic, Helper, and Suppressor T Lymphocytes

The **cytotoxic,** or **killer, T lymphocytes** destroy specific victim cells that are identified by specific antigens on their surface. In order to effect this *cell-mediated* destruction, the T lymphocytes must be near or in actual contact with their victim cells (in contrast to B cells, which kill at a distance by secreting antibodies). The mechanisms by which the cytotoxic lymphocytes kill their victim cells are presently not well understood.

The cytotoxic lymphocytes defend against viral and fungal infections and are also responsible for transplant rejection reactions and for immunological surveillance against cancer. Although most bacterial infections are fought by B lymphocytes, some are the targets of cell-mediated attack by cytotoxic T lymphocytes. This is the case with the *tubercle bacilli* that cause tuberculosis. Injections of some of these bacteria under the skin produce inflammation after a latent period of forty-eight to seventy-two hours. This *delayed hypersensitivity* reaction is cell mediated rather than humoral, as shown by the fact that it can be induced in an unexposed guinea pig by an infusion of lymphocytes, but not of serum, from an exposed animal.

The **helper T lymphocytes** and **suppressor T lymphocytes** indirectly participate in the specific immune response by regulating the responses of the B cells (fig. 17.11) and the cytotoxic T cells. The activity of B cells and cytotoxic T cells is increased by helper T lymphocytes and decreased by suppressor T lymphocytes. The amount of antibodies secreted in response to antigens is thus affected by the relative numbers of helper to suppressor T cells that develop in response to a given antigen.

As a result of advances in recombinant DNA technology (genetic engineering) that allow the production of monoclonal antibodies (discussed later), it is now possible for clinical laboratories to distinguish between the different subcategories of lymphocytes by means of antigen "markers" on their surfaces. Counting the lymphocytes in each of these subcategories provides far more information about diseases and their causes than was previously available.

Acquired immune deficiency syndrome (AIDS) is a disease that particularly strikes certain high-risk groups. In terms of the number of cases reported, these are homosexual men (71%), intravenous drug abusers (17%), Haitian immigrants (5%), and hemophiliacs (1%). Through the use of laboratory tests, it has been learned that AIDS is associated with an abnormally low helper T lymphocyte count. This results in decreased immunological function and thus an increased incidence of opportunistic infections. *Pneumocystis carinii pneumonia* and *Kaposi's sarcoma* are two previously rare conditions that account for a high percentage of deaths in AIDS victims. It has been shown that AIDS is caused by a specific strain of virus (HTLV-III) that infects human lymphocytes.

Lymphokines. The T lymphocytes secrete a family of low-molecular weight polypeptides called **lymphokines,** which are believed to mediate the actions of these cells. The lymphokines include (1) *interferon;* (2) *lymphotoxin,* which may be released from cytotoxic T cells and cause destruction of the victim cells; (3) *macrophage-activating factor,* which promotes phagocytic activity; (4) chemicals that act by chemotaxis to attract leukocytes to the infected area; (5) *macrophage-migration-inhibiting factor,* which prevents phagocytic cells from leaving the area; and (6) *interleukin-2,* which is secreted by helper cells and stimulates cell division and proliferation of cytotoxic T cells, suppressor T cells, B cells, and natural killer cells.

Genetic engineering techniques have been used to generate recombinant bacteria that can produce large amounts of interleukin-2 for research and clinical applications. When preliminary research demonstrated that blood incubated *in vitro* with interleukin-2 produced "lymphokine-activated killer cells" that can lyse tumor cells but not normal cells, medical scientists attempted to use this technique in an experimental group of cancer patients. In the first report of this attempt (December, 1985), twenty-five patients with advanced metastatic cancers of different types, in whom all conventional therapy had failed (surgery, chemotherapy, and radiation therapy) were treated with a combination of interleukin-2 and "lymphokine-activated killer cells" from their own peripheral blood. One patient with advanced metastatic melanoma had a complete remission, and ten patients had partial regression of established tumors at the time of the report. Additional research is needed and is ongoing to determine the optimum parameters of this treatment and the effectiveness of different combinations of the methods now available for cancer treatment.

T Cell Receptor Proteins. Unlike B cells, T cells do not make antibodies and thus do not have antibodies on their surface to serve as receptors for antigens. The T cells do, however, have specific receptors for antigens on their membrane surface, and these T cell receptors have recently been identified as molecules closely related to immunoglobulins. The T cell receptors differ from the antibody receptors on B cells in another, and very important, characteristic: they cannot bond to free antigens. In order for a T lymphocyte to respond to a foreign antigen, the antigen must be presented to the T lymphocyte on the membrane of an *antigen-presenting cell.* The chief antigen-presenting cells are macrophages, which present the foreign antigen together with other surface antigens, called *histocompatibility antigens,* to the T lymphocytes. Some knowledge of the histocompatibility antigens is thus required before T cell–macrophage interactions and T cell functions can be understood.

Histocompatibility Antigens. Tissue that is transplanted from one person to another contains antigens that are foreign to the host. This is because all tissue cells, with the exception of mature red blood cells, are genetically marked with a characteristic combination of **histocompatibility antigens** on the membrane surface. The greater the difference in these antigens between the donor and the recipient in a transplant, the greater will be the chance of

Figure 17.12. There are four human histocompatibility antigens (or human leukocyte antigens—HLA). Each of these antigens is coded by a gene located on chromosome number 6.

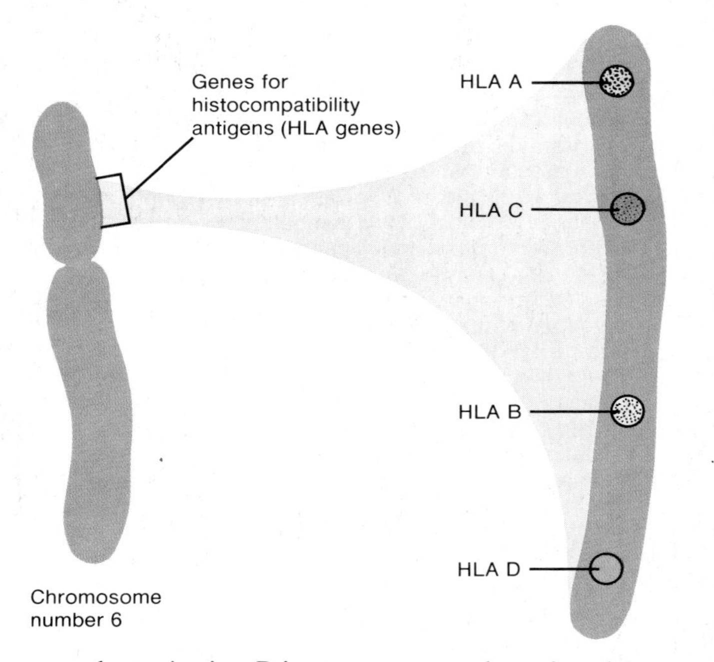

transplant rejection. Prior to organ transplantation, therefore, the "tissue type" of the recipient is matched to that of potential donors. Since the person's white blood cells are used for this purpose, an alternate name for histocompatibility antigens in humans is **human leukocyte antigens,** abbreviated **HLA.**

The histocompatibility antigens (HLA) are proteins that are coded by a group of genes, called the *major histocompatibility complex* (*MHC*), located on chromosome number 6 (fig. 17.12). These four genes are labeled A, B, C, and D. Each of these genes can code for only one protein in a given individual, but this protein can be different in different people. Two people, for example, may both have antigen A3, but one might have antigen B17 while the other has antigen B21. The closer two people are related, the more similar their histocompatibility antigens will be.

Clinical interest has been generated by the observation that certain diseases are much more common in people who have particular histocompatibility antigens (table 17.7). Ankylosing spondylitis (a type of rheumatoid arthritis), for example, is much more common in people who have antigen B27, and psoriasis (a skin disorder) is three times more common in people with antigen B17 than in the general population. Some other diseases that have a high correlation with particular histocompatibility antigens include Hodgkin's disease (a cancer of the lymph nodes), myasthenia gravis, Graves' disease, and juvenile-onset diabetes.

Table 17.7 Association between particular human histocompatibility antigens and diseases.

Disease	HLA Antigen	Frequency in Patients (%)	Frequency in Controls (%)
Ankylosing spondylitis	B27 B13	90 18	7 4
Psoriasis	B17	29	8
	B16	15	5
Graves' disease	B8	47	21
Coeliac disease	B8	78	24
Dermatitis herpetiformis	B8	62	27
Myasthenia gravis	B8	52	24
SLE (systemic lupus erythematosus)	B15	33	8
Multiple sclerosis	A3	36	25
	B7	36	25
Acute lymphatic leukemia	A2 B35	63 25	37 16
Hodgkin's disease	A1	39	32
	B8	26	22
Chronic hepatitis	B8	68	18
Ragweed hay fever	B7	50	19

From H. McDervitt and W. Bodmer, *Lancet 1,* p. 1269, 1974.

Interactions between Macrophages and T Lymphocytes

The major histocompatibility complex of genes produces two classes of HLA antigens, designated *class 1* and *class 2.* The class-1 molecules, which are made by all cells in the body except red blood cells, comprise the HLA type A, B, and C. Class-2 molecules comprise the D group of HLA antigens and are produced only by macrophages and B lymphocytes.

When a foreign particle, such as a virus, infects the body, it is taken into macrophages by phagocytosis and partially digested. The foreign antigens are then moved to the surface of the macrophage, where they are presented in close association with the class-2 HLA antigens. This combination of foreign and HLA antigens is needed for recognition by the receptor protein on the surface of helper T lymphocytes and is thus required for activation of helper T cells (fig. 17.13). The helper T lymphocytes that are activated in this way proliferate and secrete the lymphokines, including interleukin-2, which greatly enhance other aspects of the immune response.

Figure 17.13. (*a*) An electron micrograph showing contact between a macrophage (*left*) and a lymphocyte (*right*). As illustrated in (*b*), such contact between a macrophage and a T cell requires that the helper T cell interact with both the foreign antigen and the HLA class-2 antigen on the surface of the macrophage.

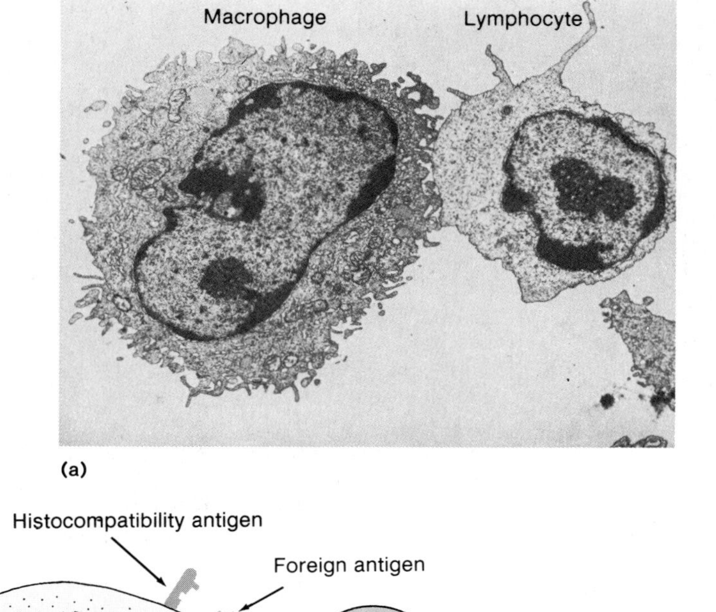

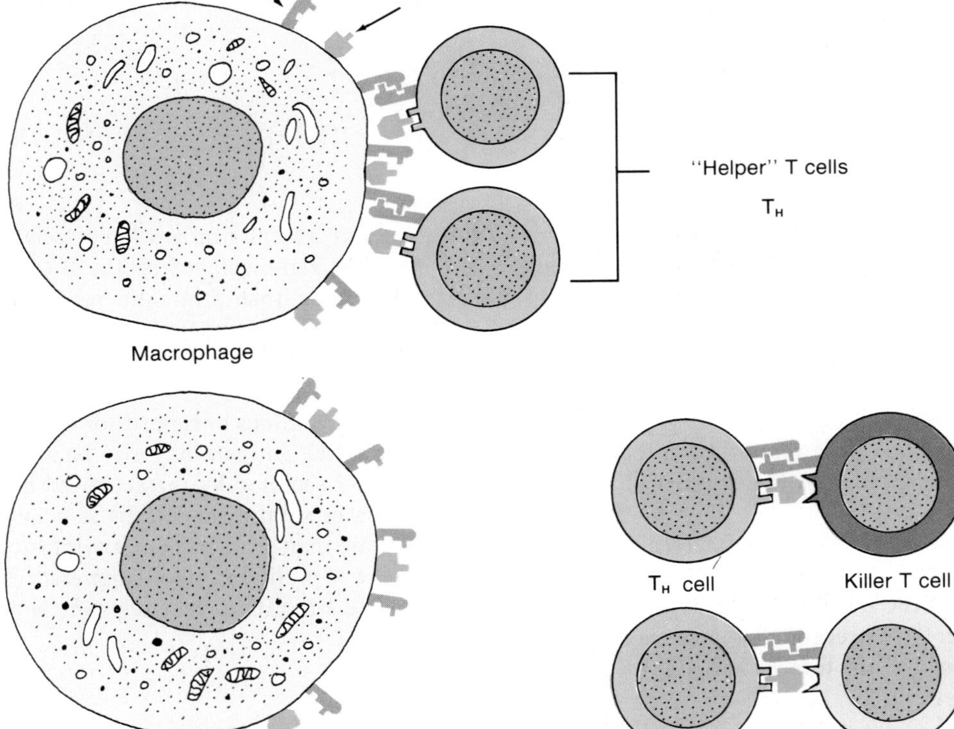

Figure 17.14. In order for a killer T cell to destroy a tissue cell infected with viruses, the T cell must interact with both the foreign antigen and the class-1 HLA antigen on the surface of the infected cell.

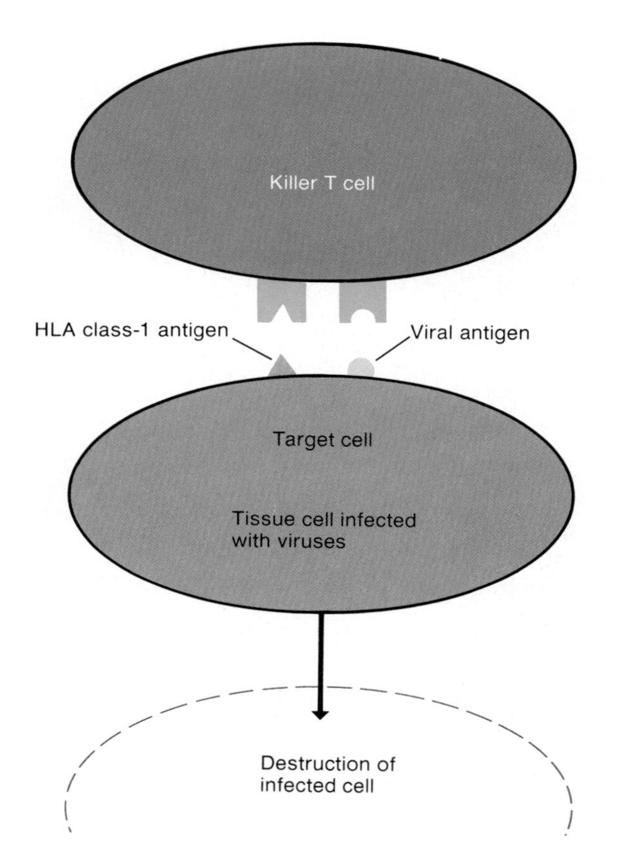

The cytotoxic T lymphocytes are only able to destroy infected cells if those infected cells display the foreign antigens together with the class-1 HLA antigens (fig. 17.14). Those cytotoxic T cells that are able to recognize the foreign antigen are stimulated to divide and proliferate by the interleukin-2 secreted by activated helper T lymphocytes. In this way, activation of helper T lymphocytes by macrophages results in and is required for the full immunological defense provided by the cytotoxic T lymphocytes (fig. 17.15).

Macrophage-induced activation of helper T cells and subsequent secretion of interleukin-2 by these cells is also believed to stimulate the B cell response to an antigen. In this case, the helper T cell receptor protein may combine in a specific fashion with the foreign antigen (attached to the antibody receptor protein) in association with the class-2 HLA antigen displayed on the B cell surface. This interaction may then stimulate proliferation of the activated B cells, their conversion to plasma cells, and the secretion of antibodies directed against the foreign antigen (fig. 17.16).

Subpopulations of suppressor T lymphocytes that are specific to the antigen may also be stimulated by the interleukin-2 secreted by the activated helper T cells. Proliferation of suppressor T cells is believed to occur at a slower rate than proliferation of cytotoxic T cells or B cells. This slower proliferation of suppressor T cells is believed to provide a negative feedback control of the immune response.

Tolerance

The ability to produce antibodies against **non-self-**antigens while tolerating (not producing antibodies against) **self-**antigens occurs early in life when immunological competence is established (the first month or so of postnatal life). If a fetal mouse of one strain receives transplanted antigens from a different strain, therefore, it

Figure 17.15. Interaction between macrophages, helper T lymphocytes, cytotoxic T lymphocytes, and infected cells in the immunological defense against viral infections.

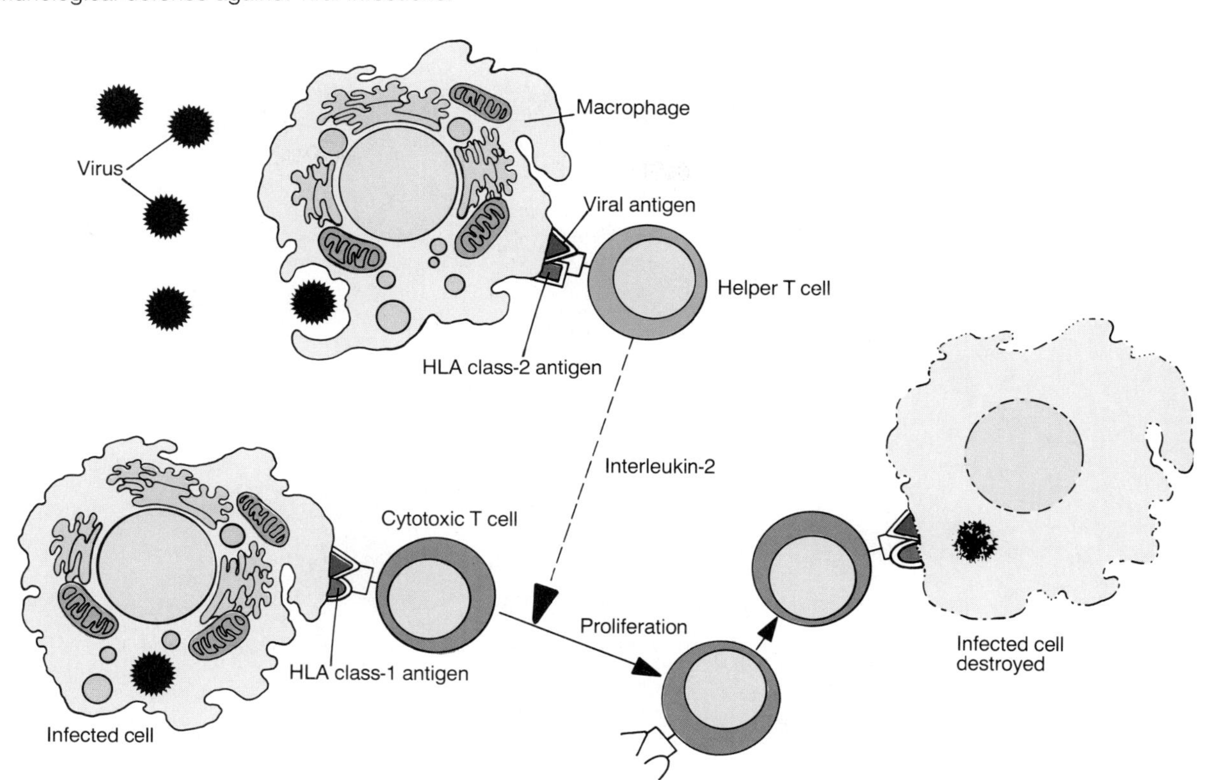

will not recognize tissue transplanted later in life from the other strain as foreign and, as a result, will not immunologically reject the transplant.

The ability of an individual's immune system to recognize and tolerate self-antigens requires continuous exposure of the immune system to those antigens. If this exposure begins when the immune system is weak—such as in fetal and early postnatal life—tolerance is more complete and long lasting than when exposure occurs later in life. Some self-antigens, however, are normally hidden from the blood, such as thyroglobulin within the thyroid gland and lens protein in the eye. An exposure to these self-antigens results in antibody production just as if these proteins were foreign. Antibodies made against self-antigens are called **autoantibodies.**

Two major theories have been proposed to account for immunological tolerance: (1) *clone deletion;* and (2) *immunological suppression.* According to the clone deletion theory, tolerance to self-antigens is achieved by destruction of the lymphocytes that inherit the ability to make autoantibodies. It is not known how this might occur.

According to the immunological suppression theory, the lymphocytes that make autoantibodies are present throughout life but are normally inhibited from attacking self-antigens. This is presumably due to the effects of suppressor T lymphocytes. An alteration in the ratio of suppressor to helper T lymphocytes in later life, therefore, might result in the production of autoantibodies.

1. *Describe the role of the thymus in cell-mediated immunity.*
2. *Define the term* lymphokines, *identify their origin, and list the different functions of these molecules.*
3. *Define the term* histocompatibility antigens, *and explain the importance of class-1 and class-2 HLA antigens in the function of T cells.*
4. *Describe the requirements for activation of helper T cells by macrophages, and explain how helper T cells promote the immunological defenses provided by cytotoxic T cells and by B cells.*
5. *Explain how suppressor T cells can provide a negative feedback control of the immune response and how a defect in suppressor cell function might explain the origin of autoantibodies.*

Figure 17.16. Schematic diagram of the events that are believed to occur in the interactions of macrophages, helper T lymphocytes, and B lymphocytes.

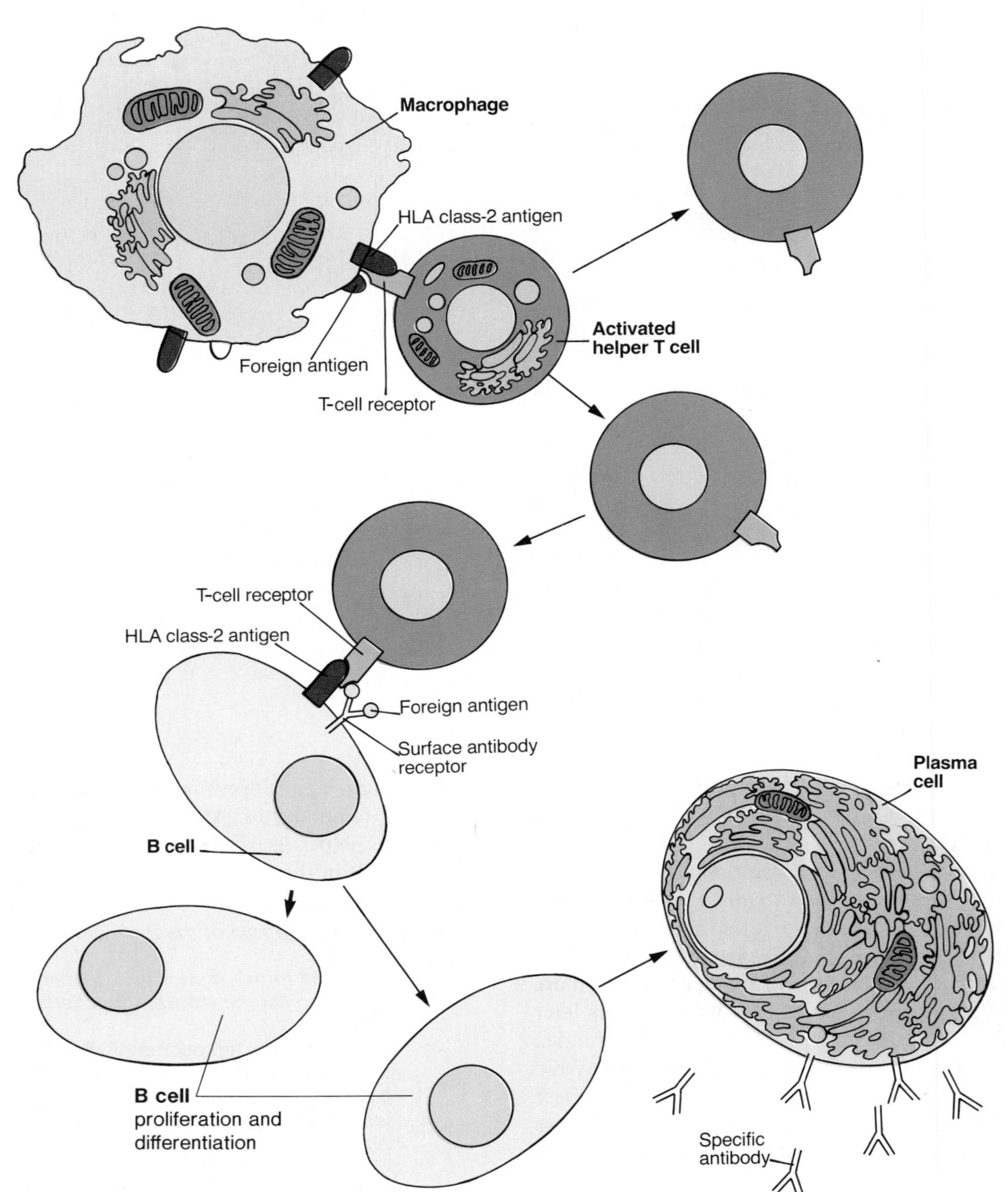

Red Blood Cell Antigens and Blood Typing

The histocompatibility antigens found on the surface of tissue cells are far more varied than the antigens on red blood cells. There are several groups of red blood cell antigens, but the major group is known as the **ABO system.** In terms of the antigens present on the red blood cell surface, a person may be *type A* (with only A antigens), *type B* (with only B antigens), *type AB* (with both A and B antigens), or *type O* (with neither A nor B antigens).

The ABO System

Each person inherits two genes (one from each parent) that control the production of the ABO antigens. The gene for A or B antigens is dominant to the gene for O, since the latter simply means the absence of A or B. A person who is type A, therefore, may have inherited the A gene from each parent (may have the genotype AA) or the A gene from one parent and the O gene from the other parent (and have the genotype AO). Likewise, a person who is type B may have the genotype BB or BO. It follows that a type O person inherited the O gene from each parent (has genotype OO), whereas a type AB person inherited the A gene from one parent and the B gene from the other (there is no dominance-recessive relationship between A and B).

The immune system is tolerant of its own red blood cell antigens. A person who is type A, for example, does not produce anti-A antibodies. Surprisingly, however, people with type A blood do make antibodies against the B antigen, and conversely, people with type B blood make antibodies against the A antigen. This is believed to result from the fact that antibodies made in response to some common bacteria can cross-react with the A or B antigens. A person who is type A, therefore, acquires antibodies that can react with B antigens by exposure to these bacteria but does not develop antibodies that can react with A antigens because this is prevented by tolerance mechanisms.

People who are type AB develop tolerance to both of these antigens and thus do not produce either anti-A or anti-B antibodies. Those who are type O, in contrast, do not develop tolerance to either antigen and, therefore, have both anti-A and anti-B antibodies in their plasma (table 17.8).

Table 17.8 The ABO system of antigens on red blood cells.

Antigen on RBCs	Antibody in Plasma
A	Anti-B
B	Anti-A
O	Anti-A and anti-B
AB	Neither anti-A nor anti-B

Transfusion Reactions. Before transfusions are performed, a *major crossmatch* is made by mixing serum from the recipient with blood cells from the donors. If the types do not match—if the donor is type A, for example, and the recipient is type B—the recipient's antibodies attach to the donor's red blood cells and form bridges that cause the cells to clump together, or **agglutinate** (fig. 17.17). Because of this agglutination reaction, the A and B antigens are sometimes called *agglutinogens* and the antibodies against them are called *agglutinins.* Transfusion errors that result in such agglutination in the blood can produce a blockage of small blood vessels and organ damage.

In emergencies, type O blood has been given to people who are type A, B, AB, or O. Since type O red blood cells lack A or B antigens, the recipient's antibodies cannot cause agglutination of the donor red blood cells. Type O is, therefore, a *universal donor,* but only as long as the volume of plasma donated is small because plasma from a type O person would agglutinate type A, type B, and type AB red blood cells. Likewise, type AB people are *universal recipients* because they lack anti-A and anti-B antibodies and thus cannot agglutinate donor red blood cells. (Donor plasma could agglutinate recipient red blood cells if the transfusion volume were too large). Because of the dangers involved, use of the universal donor and recipient concept in blood transfusions is strongly discouraged.

Rh Factor

Another important group of antigens found in most red blood cells is the *Rh factor* (Rh stands for Rhesus monkey, in which these antigens were first discovered). People who have these antigens are said to be **Rh positive,** whereas those who do not are **Rh negative.** There are fewer Rh negative people because this condition is recessive to Rh positive. The Rh factor is of particular significance when Rh negative mothers give birth to Rh positive babies.

Figure 17.17. The agglutination (clumping) of red blood cells occurs when cells with A-type antigens are mixed with anti-A antibodies and when cells with B-type antigens are mixed with anti-B antibodies. No agglutination would occur with type O blood (not shown).

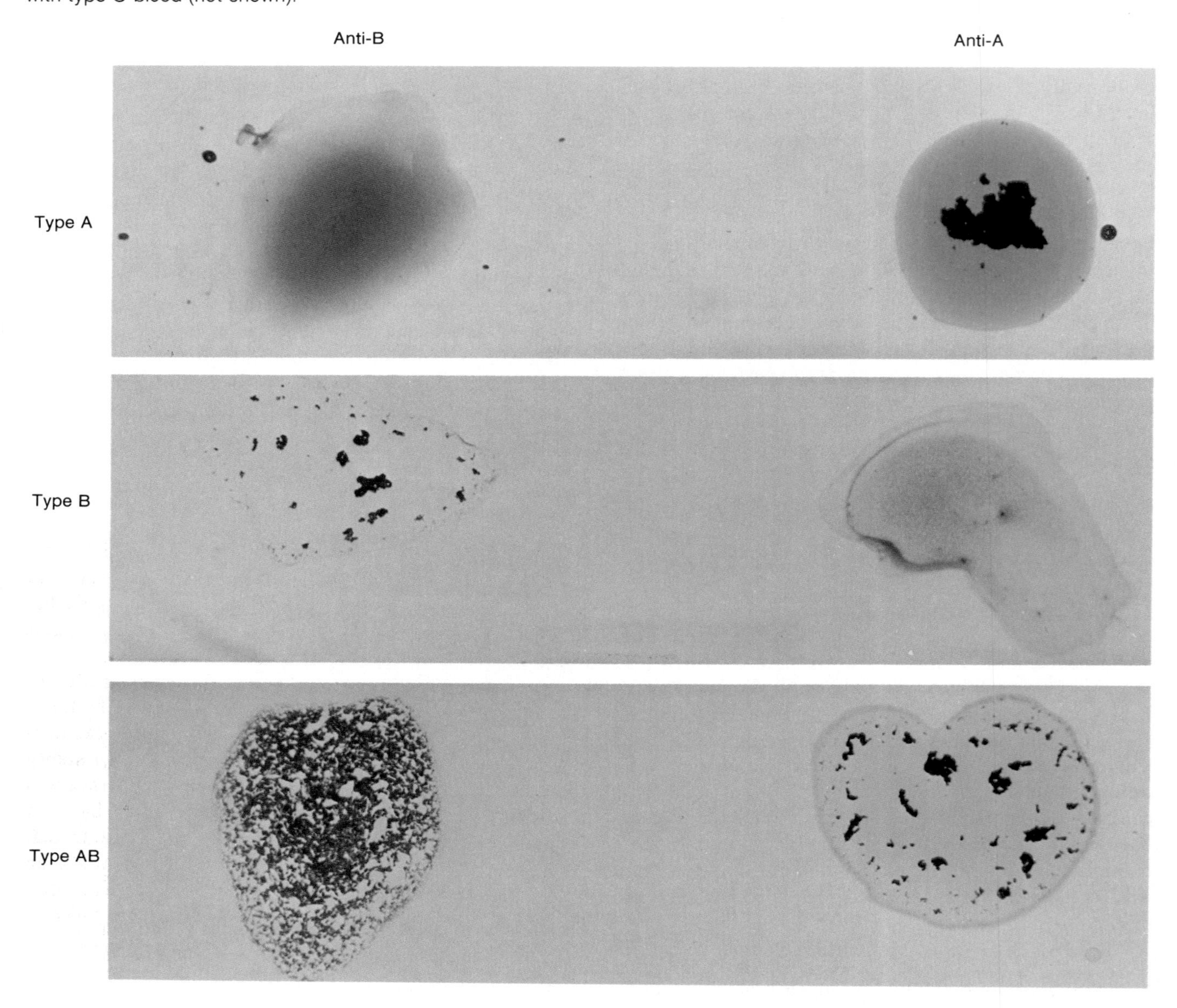

At the time she gives birth, the immune system of an Rh negative mother may become exposed to Rh antigens from her baby. Since these antigens are foreign to her, she can become sensitized and build up lymphocyte clones against the Rh antigen. This does not affect the first baby, which is already born, but can affect subsequent Rh positive fetuses because antibodies against the Rh factor can cross the placenta. These antibodies cause the destruction of the fetus's erythrocytes, so that the baby is born anemic, with a condition called **erythroblastosis fetalis.**

Erythroblastosis fetalis can be prevented by injecting the Rh negative mother with *antibodies against the Rh factor* (one trade name for this is RhoGAM) within seventy-two hours after the birth of each Rh positive baby. This is a type of passive immunization in which the injected antibodies inactivate the Rh antigens and thus prevent the mother from becoming actively immunized to them.

Figure 17.18. Active immunity to a pathogen can be gained by exposure to the fully virulent form or by inoculation with a pathogen whose virulence (ability to cause disease) has been attenuated (reduced) but whose antigens are the same as in the virulent form.

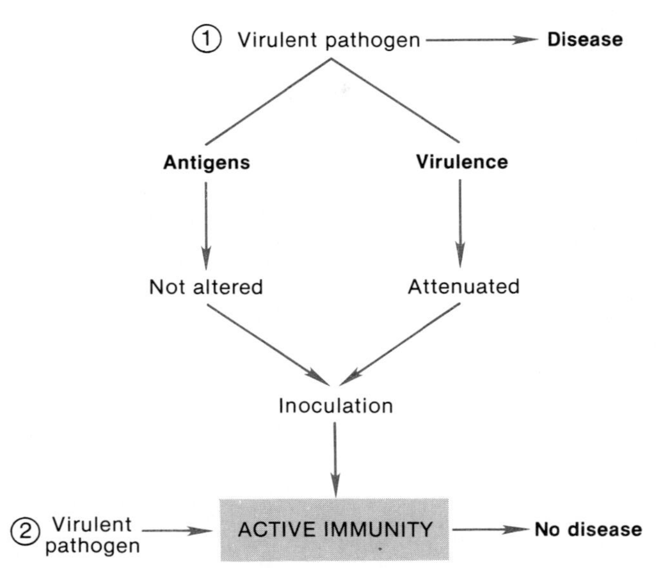

1. *List the red blood cell antigens and the antibodies against red blood cell antigens that are present in each of the ABO blood types.*
2. *Explain how agglutination reactions are produced when different blood types are mixed together.*
3. *Explain how the disease* erythroblastosis fetalis *is produced and how this disease can be prevented.*

Immunizations

It was first known in the mid-eighteenth century that the fatal effects of smallpox could be prevented by inducing mild cases of the disease. This was accomplished at that time by rubbing needles into the pustules of people who had mild forms of smallpox and injecting these needles into healthy people. Understandably, this method of immunization was not widely popular.

Acting on the observation that milkmaids who contracted cowpox—a disease similar to smallpox but less *virulent* (pathogenic)—were immune to smallpox, an English physician, named Edward Jenner, inoculated a healthy boy with cowpox. When the boy recovered, Jenner inoculated him with an otherwise deadly amount of smallpox, from which he also proved to be immune. (This was fortunate for both the boy—who was an orphan—and Jenner; Jenner's fame spread, and as the boy grew into manhood he proudly gave testimonials on Jenner's behalf). This experiment, performed in 1796, began the first widespread immunization program.

A similar, but more sophisticated, demonstration of the effectiveness of immunizations was performed by Louis Pasteur almost a century later. Pasteur isolated the bacteria that cause anthrax and heated them until their ability to cause disease was greatly reduced (their virulence was *attenuated*). He then injected these attenuated bacteria into twenty-five sheep, leaving twenty-five unimmunized. Several weeks later, before a gathering of scientists, he injected all fifty sheep with the completely active anthrax bacteria. All twenty-five of the unimmunized sheep died—all twenty-five of the immunized sheep survived (fig. 17.18).

Active Immunity and the Clonal Selection Theory

When a person is exposed to a particular pathogen for the first time, there is a latent period of five to ten days before measurable amounts of specific antibodies appear in the blood. This sluggish **primary response** may not be sufficient to protect the individual against the disease caused by the pathogen. Antibody concentrations in the blood during this primary response reach a plateau in a few days and decline after a few weeks.

Figure 17.19. A comparison of antibody production in the primary response (upon first exposure to an antigen) to antibody production in the secondary response (upon subsequent exposure to the antigen). The greater secondary response is believed to be due to the development of lymphocyte clones produced during the primary response.

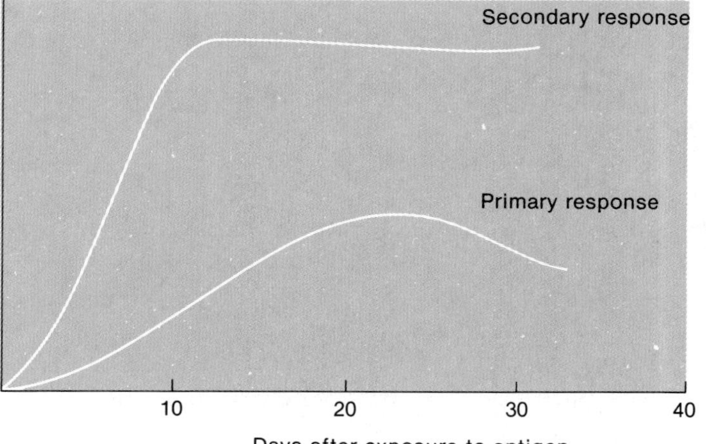

A subsequent exposure of the same individual to the same antigen results in a **secondary response** (fig. 17.19). Compared to the primary response, antibody production during the secondary response is much more rapid. Maximum antibody concentrations in the blood are reached in less than two hours and are maintained for a longer time than in the primary response. This rapid rise in antibody production is usually sufficient to prevent the disease.

Before stimulation by a particular antigen, the B cells that make the appropriate antibodies produce mainly IgM antibodies. These IgM antibodies therefore account for a high proportion of the antibodies made during the primary response. In contrast, most antibodies made during the secondary response are in the IgG subclass. This is shown in figure 17.20.

Clonal Selection Theory. The immunization procedures of Jenner and Pasteur were effective because the people who were inoculated produced a secondary rather than a primary response when exposed to the virulent pathogens. This protection is not simply due to accumulations of antibodies in the blood, because secondary responses occur even after antibodies produced by the primary response have disappeared. Immunizations, therefore, seem to produce a type of "learning" in which the ability of the immune system to combat a particular pathogen is improved by prior exposure.

Figure 17.20. The greater antibody production in the secondary response compared to the primary response is due mainly to increased amounts of IgG antibodies. Notice that the amount of IgM antibodies is about the same in both primary and secondary responses.

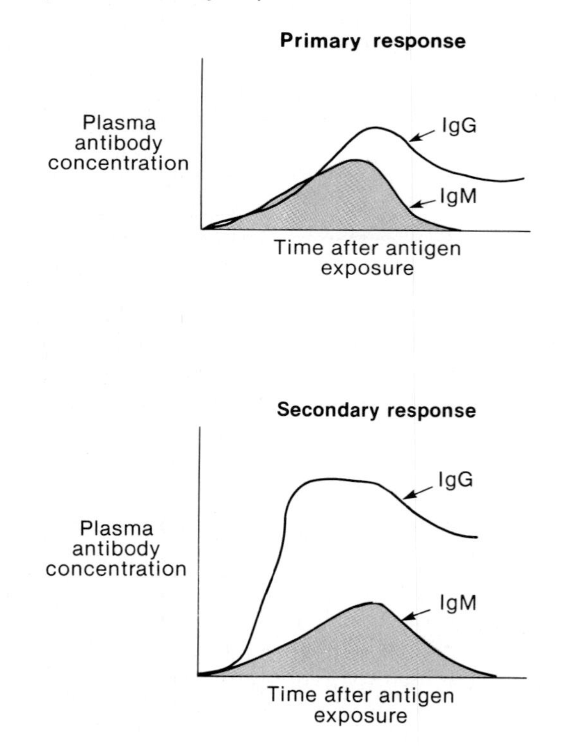

Figure 17.21. According to the clonal selection theory, exposure to an antigen stimulates the production of lymphocyte clones and the maturation of some members of B cell clones into antibody-secreting plasma cells.

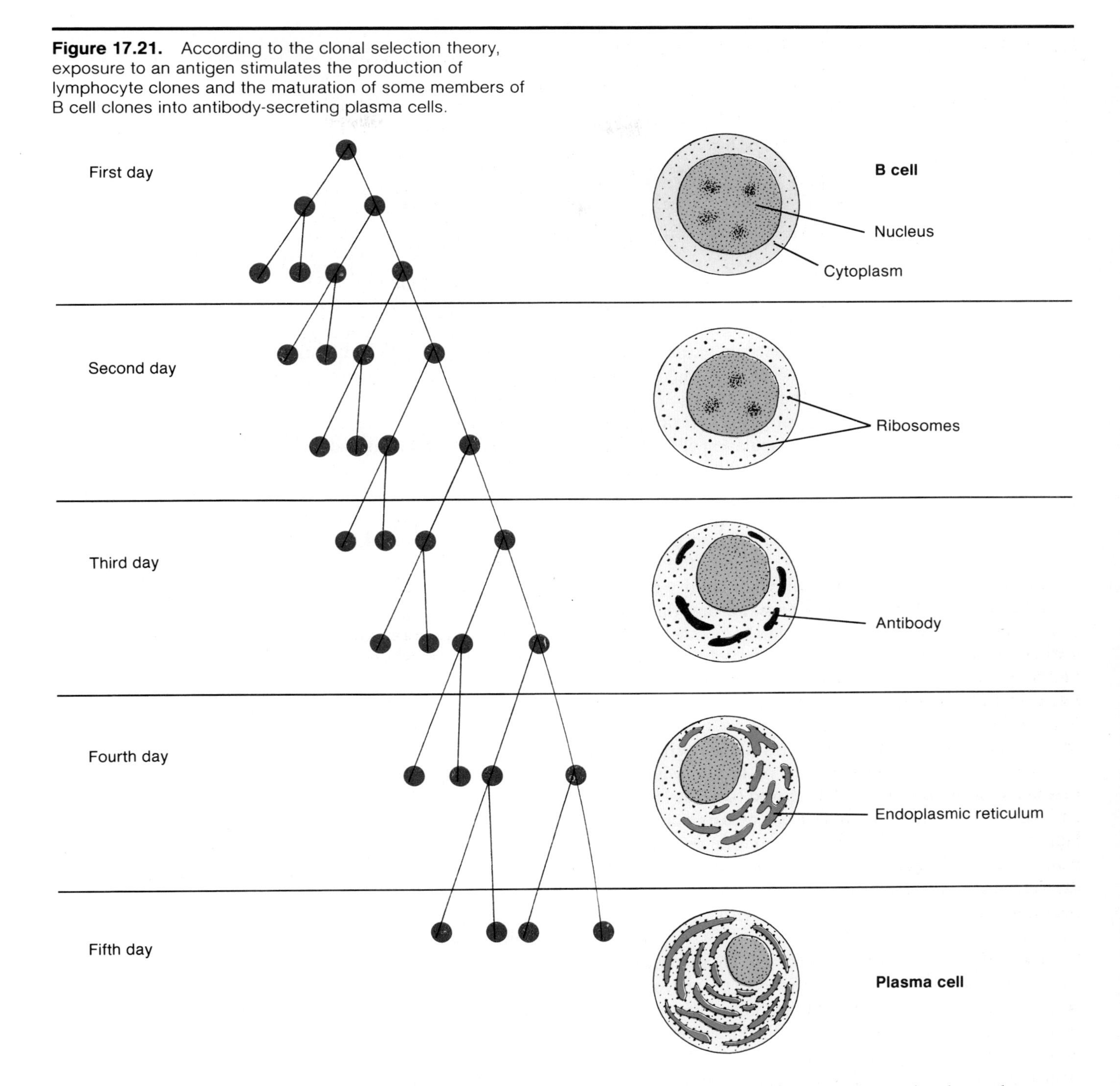

The mechanisms by which secondary responses are produced are not completely understood; the **clonal selection theory,** however, appears to account for most of the evidence. According to this theory, B lymphocytes *inherit* the ability to produce particular antibodies (and T lymphocytes inherit the ability to respond to particular antigens). Some lymphocytes, therefore, can respond to smallpox and produce antibodies against it even if the person has never been previously exposed to this disease.

Exposure to smallpox antigens stimulates these specific lymphocytes to divide many times until a large population of genetically identical cells—a clone—is produced. Some of these cells become plasma cells that secrete antibodies for the primary response; others become memory cells that can be stimulated to secrete antibodies during the secondary response (fig. 17.21).

Table 17.9 Summary of the clonal selection theory (with regard to B cells).

Process	Results
Lymphocytes inherit the ability to produce specific antibodies.	Prior to antigen exposure, lymphocytes are already present in the body that can make the appropriate antibodies.
Antigens interact with antibody receptors on the lymphocyte surface.	Antigen-antibody interaction stimulates cell division and the development of lymphocyte clones containing memory cells and plasma cells that secrete antibodies.
Subsequent exposure to the specific antigens produce a more efficient response.	Exposure of lymphocyte clones to specific antigens results in greater and more rapid production of specific antibodies.

Notice that according to the clonal selection theory (table 17.9), antigens do not induce lymphocytes to make the appropriate antibodies. Rather, antigens select lymphocytes (through interaction with surface receptors) that are already able to make antibodies against that antigen. This is analogous to evolution by natural selection. An environmental agent (in this case, antigens) acts on the genetic diversity already present in a population of organisms (lymphocytes) to cause increasing numbers of the individuals that are selected.

Active Immunity. The development of a secondary response provides **active immunity** against the specific pathogens. The development of active immunity requires prior exposure to the specific antigens, during which time a primary response is produced and the person may get sick. Some parents, for example, deliberately expose their children to others who have measles, chicken pox, and mumps so that their children will be immune to these diseases in later life.

Clinical immunization programs induce primary responses by inoculating people with pathogens whose virulence has been attenuated or destroyed (such as Pasteur's heat-inactivated anthrax bacteria) or by using closely related strains of microorganisms that are antigenically similar but less pathogenic (such as Jenner's cowpox inoculations). These procedures cause the development of lymphocyte clones that can combat the virulent pathogens by producing secondary responses.

The first successful polio vaccine (the Salk vaccine) was composed of viruses that had been inactivated by treatment with formaldehyde. These "killed" viruses were injected into the body, in contrast to the currently used oral (Sabin) vaccine. The oral vaccine contains "living" viruses that have attenuated virulence. These viruses invade the epithelial lining of the intestine and multiply but do not invade nerve tissue. The immune system can, therefore, become sensitized to polio antigens and produce a secondary response if polio viruses that attack the nervous system are later encountered.

Passive Immunity

The ability to mount an immune response—called **immunological competence**—does not develop until about a month after birth. The fetus, therefore, cannot immunologically reject its mother. The immune system of the mother is fully competent but does not usually respond to fetal antigens for reasons that are not completely understood. Some IgG antibodies from the mother do cross the placenta and enter the fetal circulation, however, and these serve to confer **passive immunity** to the fetus.

The fetus and the newborn baby are, therefore, immune to the same antigens as the mother. Since the baby did not itself produce the lymphocyte clones needed to form these antibodies, such passive immunity disappears when the infant is about one month old. If the baby is breast-fed it can receive additional antibodies of the IgA subclass in its mother's first milk (the *colostrum*).

Passive immunizations are used clinically to protect people who have been exposed to extremely virulent infections or toxins, such as snake venom, tetanus, and others. In these cases the affected person is injected with *antiserum* (serum-containing antibodies), also called *antitoxin,* from an animal that has been previously exposed to the pathogen. The animal develops the lymphocyte clones and active immunity and thus has a high concentration of antibodies in its blood. Since the person who is injected with these antibodies does not develop active immunity, he or she must again be injected with antitoxin upon subsequent exposures. A comparison of active and passive immunity is shown in table 17.10.

Monoclonal Antibodies

In addition to their use in passive immunity, antibodies are also commercially prepared for use in research and clinical laboratory tests. In the past, antibodies were obtained by first chemically purifying a specific antigen and then injecting this antigen into animals. Since an antigen typically has many different antigenic determinant sites, however, the antibodies obtained by this method were

Table 17.10 Comparison of active and passive immunity.

Characteristic	Active Immunity	Passive Immunity
Injection of person with	Antigens	Antibodies
Source of antibodies	The person inoculated	An animal that is inoculated with the antigen; or the mother
Method	Injection with killed or attenuated pathogens or their toxins	Natural—transfer of antibodies across the placenta; artificial—injection with antibodies
Time to develop resistance	5 to 14 days	Immediately after injection
Duration of resistance	Long (perhaps years)	Short (days to weeks)
When used	Before exposure to pathogen	Before or after exposure to pathogen

polyclonal; they had different specificities. This decreased their sensitivity to a particular antigenic site and resulted in some degree of cross-reaction with closely related antigen molecules.

In the preparation of monoclonal antibodies, an animal is injected with an antigen and then subsequently killed. B lymphocytes are then obtained from its spleen and placed in thousands of different *in vitro* incubation vessels. These cells soon die, however, unless they are hybridized with cancerous multiple myeloma cells. Cell fusion is promoted by a chemical, polyethylene glycol. The fusion of a B lymphocyte with a cancerous cell produces a hybrid that undergoes cell division and produces a clone, called a *hybridoma.* Each hybridoma secretes large amounts of identical, **monoclonal antibodies.**

The availability of large quantities of pure monoclonal antibodies has resulted in the development of much more sensitive clinical laboratory tests (of pregnancy, for example). These pure antibodies have also been used to pick one molecule (the specific antigen interferon, for example) out of a solution of many molecules and thus isolate and concentrate it. In the future, monoclonal antibodies against specific tumor antigens may aid the diagnosis of cancer. Even more exciting, cytotoxic drugs that can kill normal as well as cancerous cells might be aimed directly at a tumor by combining these drugs with monoclonal antibodies against specific tumor antigens.

1. *Describe two methods used to induce active immunity.*
2. *Explain the characteristics of the primary and secondary immune responses, and draw graphs to illustrate your discussion.*
3. *Explain the clonal selection theory and how this theory accounts for the secondary response.*
4. *Describe passive immunity, and give natural and clinical examples of this type of immunization.*

Tumor Immunology

Oncology (the study of tumors) has revealed that tumor biology is similar to and interrelated with the functions of the immune system. Most tumors appear to be clones of single cells that have become transformed. This is similar to the development of lymphocyte clones in response to specific antigens. Lymphocyte clones, however, are under complex inhibitory control systems—such as those exerted by suppressor T lymphocytes and negative feedback by antibodies. The division of tumor cells, in contrast, is not effectively controlled by normal inhibitory mechanisms. Tumor cells are also relatively unspecialized—they *dedifferentiate,* which means that they become similar to the less-specialized cells of an embryo.

Tumors are described as *benign* when they are relatively slow growing and limited to a specific location (warts, for example). Benign tumors do not undergo **metastasis,** a term that refers to the dispersion of tumor cells and the resultant seeding of new tumors in different locations. *Malignant* tumors grow more rapidly and do metastasize. The term **cancer,** as it is usually used, refers to malignant, life-threatening tumors.

As tumors dedifferentiate, they reveal surface antigens that can stimulate the immune destruction of the tumor cells. Consistent with the concept of dedifferentiation, some of these antigens are proteins produced in embryonic or fetal life that are not normally produced postnatally. Since they are absent at the time immunological competence is established they are treated as foreign and fit subjects for immunological attack when they are produced by cancerous cells. The release of two such antigens into the blood has provided the basis for a laboratory diagnosis of some cancers. *Carcinoembryonic antigen* tests are useful in the diagnosis of colon cancer, for example, and tests for *alpha-fetoprotein* (normally produced only by the fetal liver) help in the diagnosis of liver cancer.

Figure 17.22. A killer T cell (*a*) contacts a cancer cell (the larger cell), in a manner that requires specific interaction with antigens on the cancer cell. The killer T cell releases lymphokines, including toxins that cause the death of the cancer cell (*b*). (Scanning electron micrographs © Andrejs Liepens.)

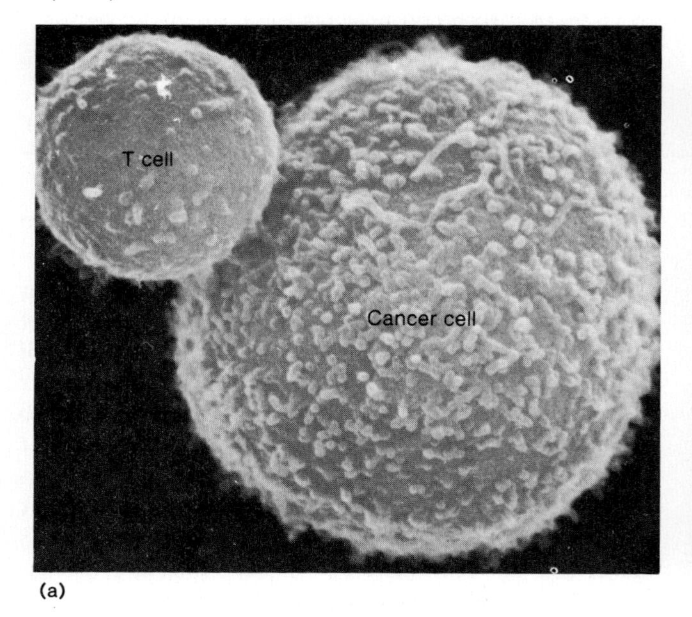

(a)

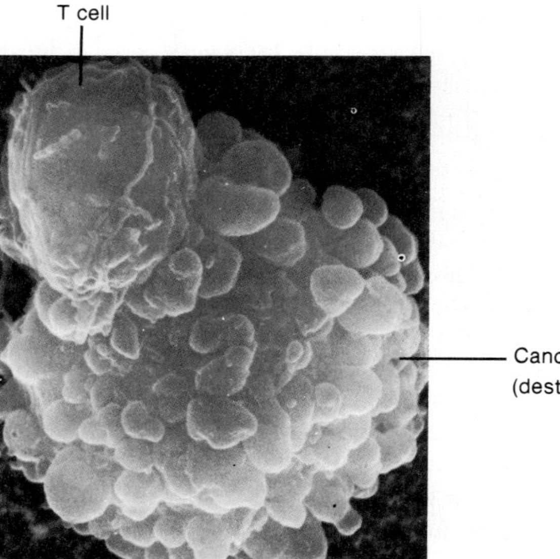

(b)

Tumors are attacked by the cell-mediated immune system, although humoral immunity (antibodies) may have a supportive role. Killer T lymphocytes (fig. 17.22) and natural killer cells recognize antigens on the tumor cell surface and normally destroy such cells before a recognizable tumor develops. This function is known as **immunological surveillance** against cancer.

Natural Killer Cells

Researchers observed that a strain of hairless mice, which genetically lacks a thymus and T lymphocytes, does not suffer from a particularly high incidence of tumor production. This surprising observation led to the discovery of **natural killer (NK) cells,** which are lymphocytes that are related to, but distinct from, T lymphocytes. Unlike killer T cells, NK cells destroy tumors in a nonspecific fashion and do not require prior exposure for sensitization to the tumor antigens. The NK cells thus provide a first line of cell-mediated defense, which is subsequently backed up by a specific response mediated by killer T cells. These two cell types interact, however; the activity of NK cells is stimulated by interferon, released as one of the lymphokines from T lymphocytes.

Effects of Aging and Stress

Susceptibility to cancer varies greatly. The Epstein-Barr virus that causes Burkitt's lymphoma in a few individuals in Africa, for example, can also be found in healthy people throughout the world. In most cases the virus is harmless; in some cases mononucleosis (involving a limited proliferation of white blood cells) is produced. Only rarely does this virus cause the uncontrolled proliferation of leukocytes occurring in Burkitt's lymphoma. The reasons for these different responses to the Epstein-Barr virus and indeed for the different susceptibilities of people to other forms of cancer are not well understood.

It is known that cancer risk increases with age. According to one theory, this is due to the fact that aging lymphocytes accumulate genetic errors over the years that decrease their effectiveness. The secretion of thymus hormones also decreases with age in parallel with a decrease in cell-mediated immune competence. Both of these changes and perhaps others not yet discovered could increase susceptibility to cancer.

Numerous experiments have demonstrated that tumors grow faster in experimental animals subject to stress than in unstressed control animals. This is generally believed to result from the fact that stressed animals, including humans, have increased secretion of corticosteroid

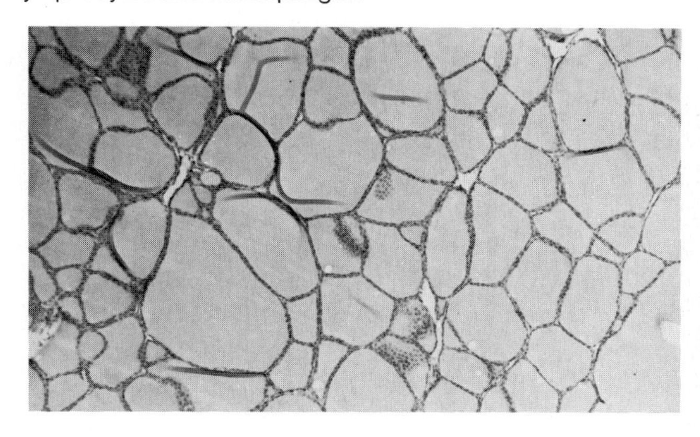

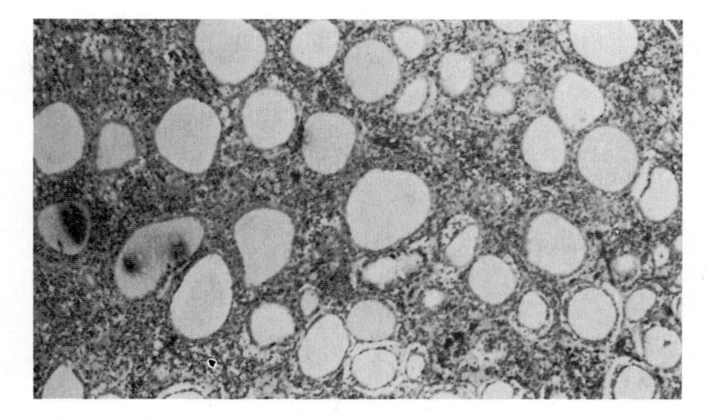

Figure 17.23. Autoimmune thyroiditis in a rabbit, induced experimentally by injection with thyroglobulin. Compare the picture of a normal thyroid (*left*) with that of the diseased thyroid (*right*). The grainy appearance of the diseased thyroid is due to the infiltration of large numbers of lymphocytes and macrophages.

hormones, which act to suppress the immune system (this is why cortisone is given to people who receive organ transplants and to people with chronic inflammatory diseases). Some recent experiments, however, suggest that the stress-induced suppression of the immune system may also be due to other factors that do not involve the adrenal cortex. Future advances in cancer therapy may incorporate methods of strengthening the immune system together with methods that directly destroy tumors.

1. *Explain why cancer cells are believed to be dedifferentiated, and describe some of the clinical applications of this concept.*
2. *Describe what the term "immunological surveillance" against cancer refers to, and identify the cells involved in this function.*
3. *Explain the possible relationships between stress and cancer susceptibility.*

Diseases Caused by the Immune System

The ability of the normal immune system to tolerate self-antigens while it identifies and attacks foreign antigens provides a specific defense against invading pathogens. In every individual, however, this system of defense against invaders at times commits domestic offenses. This can result in diseases that range in severity from the sniffles to sudden death.

Diseases caused by the immune system can be grouped into three interrelated categories: (1) *autoimmune diseases;* (2) *immune complex diseases;* and (3) *allergy,* or *hypersensitivity.* It is important to remember that these diseases are not caused by foreign pathogens but by abnormal responses of the immune system.

Autoimmunity

Autoimmune diseases are those produced by failure of the immune system to recognize and tolerate self-antigens. This failure results in the production of autoantibodies that can cause inflammation and organ damage. Such autoimmune destruction may occur as a result of the following mechanisms.

1. **An antigen that does not normally circulate in the blood may become exposed to the immune system.** Thyroglobulin protein that is normally trapped within the thyroid follicles, for example, can stimulate the production of autoantibodies that cause the destruction of the thyroid (fig. 17.23); this occurs in *Hashimoto's thyroiditis.* Similarly, autoantibodies developed against lens protein in a damaged eye may cause the destruction of a healthy eye (in *sympathetic ophthalmia.*)

Figure 17.24. One proposed mechanism for the development of rheumatoid arthritis. An initial joint inflammation (*1*) results in the release of lysosomal enzymes (*2*) that cause damage to IgG antibodies (*3*). These damaged IgG antibodies act as antigens (*4*) to stimulate B lymphocytes, resulting in the production of IgM antibodies (*5*) against those damaged IgG antibodies/antigens. This, in turn, produces further inflammation.

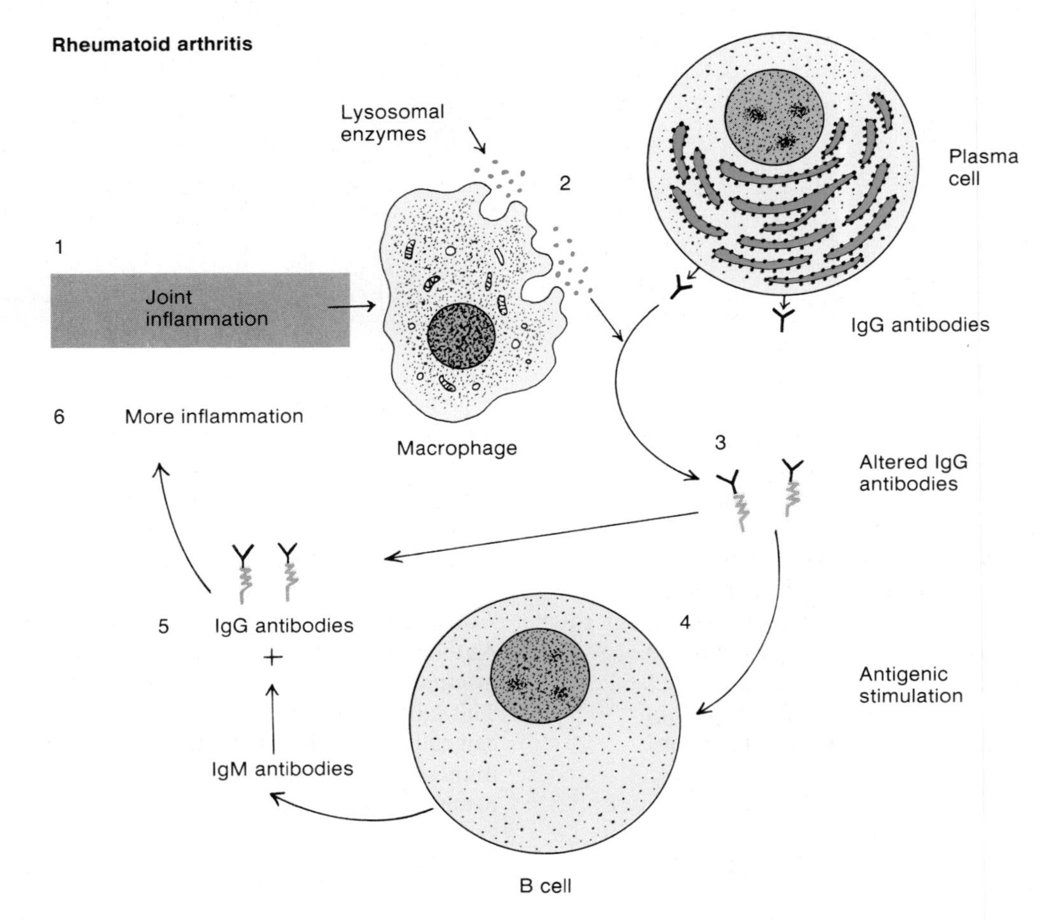

2. **A self-antigen, which is otherwise tolerated, may be altered by combining with a foreign hapten.** The disease *thrombocytopenia* (low platelet count), for example, can be caused by the autoimmune destruction of thrombocytes (platelets). This occurs when drugs such as aspirin, sulfonamide, antihistamines, digoxin, and others combine with platelet proteins to produce new antigens. The symptoms of this disease usually stop when the person stops taking these drugs.

3. **A self-antigen may become damaged and, as a result, may expose new antigenic sites.** This may occur in *rheumatoid arthritis.* An initial inflammation of the joints (perhaps through viral infection) may result in the release of digestive enzymes from lysosomes in phagocytic cells. When these enzymes digest part of IgG antibodies they expose new antigenic sites. This stimulates the production of IgM antibodies directed against the altered IgG proteins (fig. 17.24). The IgM antibodies that a person develops against his own IgG antibodies are called *rheumatoid factor* and can be used to diagnose this condition.

Table 17.11 Examples of autoimmune diseases.

Disease	Antigen	Ig and/or T Cell Response
Postvaccinal and postinfectious encephalomyelitis	Myelin, cross-reactive	T cell
Aspermatogenesis	Sperm	T cell
Sympathetic ophthalmia	Uvea	T cell
Hashimoto's disease	Thyroglobulin	IgG and T cell
Graves' disease	Receptor proteins for TSH	Thyroid-stimulating antibody (TSAb)
Autoimmune hemolytic disease	I, Rh, and others on surface of RBCs	IgM and IgG
Thrombocytopenic purpura	Hapten-platelet or hapten-adsorbed antigen complex	IgG
Myasthenia gravis	Myosin	IgG
Rheumatic fever	Streptococcal cross-reactive with heart	IgG and IgM
Glomerulonephritis	Streptococcal cross-reactive with kidney	IgG and IgM
Rheumatoid arthritis	IgG	IgM to Fc(γ)
Systemic lupus erythematosus	DNA, nucleoprotein, RNA, etc.	IgG

Source: From Barrett, James T.: Textbook of immunology, ed. 4, St. Louis, 1983, The C. V. Mosby Co.

4. **Antibodies produced against foreign antigens may cross-react with self-antigens.** Autoimmune diseases of this sort can occur, for example, as a result of *streptococcus* bacterial infections. Antibodies produced in response to antigens in this bacterium may cross-react with self-antigens in the heart and kidneys. The inflammation induced by such autoantibodies can produce heart damage (including the valve defects characteristic of *rheumatic fever*) and damage to the glomerular capillaries in the kidneys (*glomerulonephritis*).
5. **Self-antigens, such as receptor proteins, may be presented to the helper T lymphocytes together with class-2 HLA antigens.** Normally, only macrophages and B lymphocytes produce class-2 HLA antigens, which are associated with foreign antigens and are recognized by helper T cells. Perhaps as a result of viral infection, however, cells that do not normally produce class-2 HLA antigens may start to do so and, in this way, present a self-antigen to the helper T cells. In *Graves' disease,* for example, the thyroid cells produce class-2 HLA antigens, and the immune system produces autoantibodies against the TSH receptor proteins in the thyroid cells. These autoantibodies, called *TSAb* for "thyroid-stimulating antibody," interact with the TSH receptors and overstimulate the thyroid gland.
6. **Genes that code for non-self-antibodies may mutate to produce autoantibodies.** This would not cause particular diseases but would lead to the increased frequency of autoimmune diseases in general. An increased frequency of autoimmune diseases does in fact occur with age, as predicted by this mutation theory. Table 17.11 provides some examples of autoimmune diseases.

Immune Complex Diseases

The term *immune complexes* refers to combinations of antibodies with antigens that are free rather than attached to bacterial or other cells. The formation of such complexes activates complement proteins and promotes inflammation. This inflammation normally is self-limiting because the immune complexes are removed by phagocytic cells. When large numbers of immune complexes are continuously formed, however, the inflammation may be prolonged. Also, the dispersion of immune complexes to other sites can lead to widespread inflammations and organ damage. The damage produced by the inflammatory response to antigens is called **immune complex disease.**

Immune complex diseases can result from infections by bacteria, parasites, and viruses. In viral hepatitis B, for example, an immune complex that consists of viral antigens and antibodies can cause widespread inflammation of arteries *(periarteritis).* Note that the arterial damage is not caused by the hepatitis virus itself but by the inflammatory process.

Immune complex diseases can also result from the formation of complexes between self-antigens and autoantibodies. This is the case in rheumatoid arthritis, where the inflammation is produced by complexes of altered IgG antibodies (the antigens in this case) and IgM antibodies. Another immune complex disease that has an autoimmune basis is *systemic lupus erythematosus (SLE).* People with SLE produce antibodies against their own DNA and nuclear proteins. This can result in the formation of immune complexes throughout the body, including the glomerular capillaries where glomerulonephritis may be produced.

Table 17.12 Allergy: comparison between immediate and delayed hypersensitivity reaction.

Characteristic	Immediate Reaction	Delayed Reaction
Time for onset of symptoms	Within several minutes	Within one to three days
Lymphocytes involved	B cells	T cells
Immune effector	IgE antibodies	Cell-mediated immunity
Allergies most commonly produced	Hay fever, asthma, and most other allergic conditions	Contact dermatitis (such as to poison ivy and poison oak)
Therapy	Antihistamines and adrenergic drugs	Corticosteroids (such as cortisone)

Allergy

The term *allergy,* usually used synonymously with *hypersensitivity,* refers to particular types of abnormal immune responses to antigens, which are called *allergens* in these cases. There are two major forms of allergy: (1) **immediate hypersensitivity,** which is due to an abnormal B lymphocyte response to an allergen that produces symptoms within seconds or minutes; and (2) **delayed hypersensitivity,** which is an abnormal T cell response that produces symptoms within about forty-eight hours after exposure to an allergen. A comparison between these two types of hypersensitivity is provided in table 17.12.

Immediate Hypersensitivity. Immediate hypersensitivity can produce such symptoms as allergic rhinitis (chronic runny or stuffy nose), conjunctivitis (red eyes), allergic asthma, atopic dermatitis (urticaria, or hives), and others. These symptoms result from the production of antibodies of the IgE subclass, instead of the normal IgG antibodies.

Unlike IgG antibodies, IgE antibodies do not circulate in the blood but instead attach to tissue mast cells (which have membrane receptors for these antibodies). When the person is again exposed to the same allergen, the allergen bonds to the antibodies attached to the mast cells. This stimulates the mast cells to secrete various chemicals, including **histamine** (fig. 17.25). During this

Figure 17.25. Allergy (immediate hypersensitivity) is produced when antibodies of the IgE subclass attach to tissue mast cells. The combination of these antibodies with allergens (antigens that provoke an allergic reaction) cause the mast cell to secrete histamine and other chemicals that produce the symptoms of allergy.

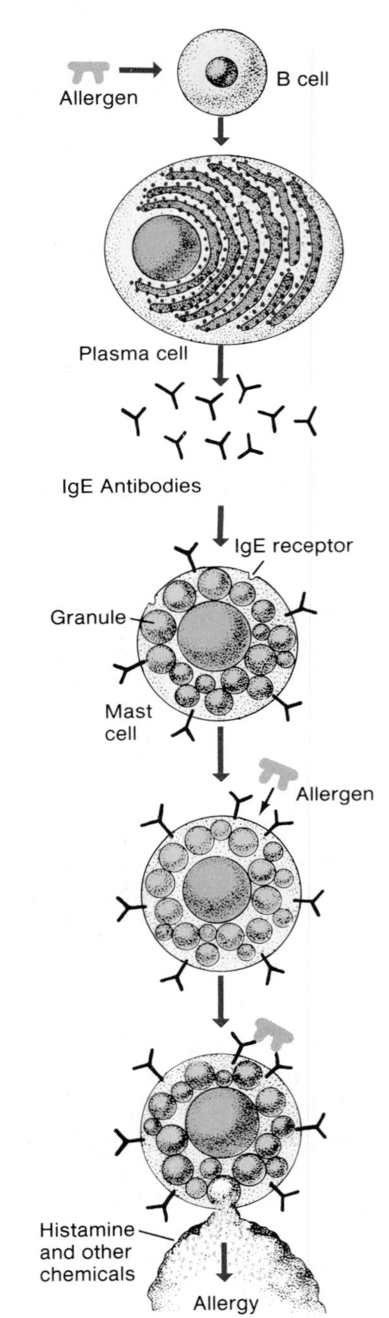

Figure 18.11. The duodenum and associated structures.

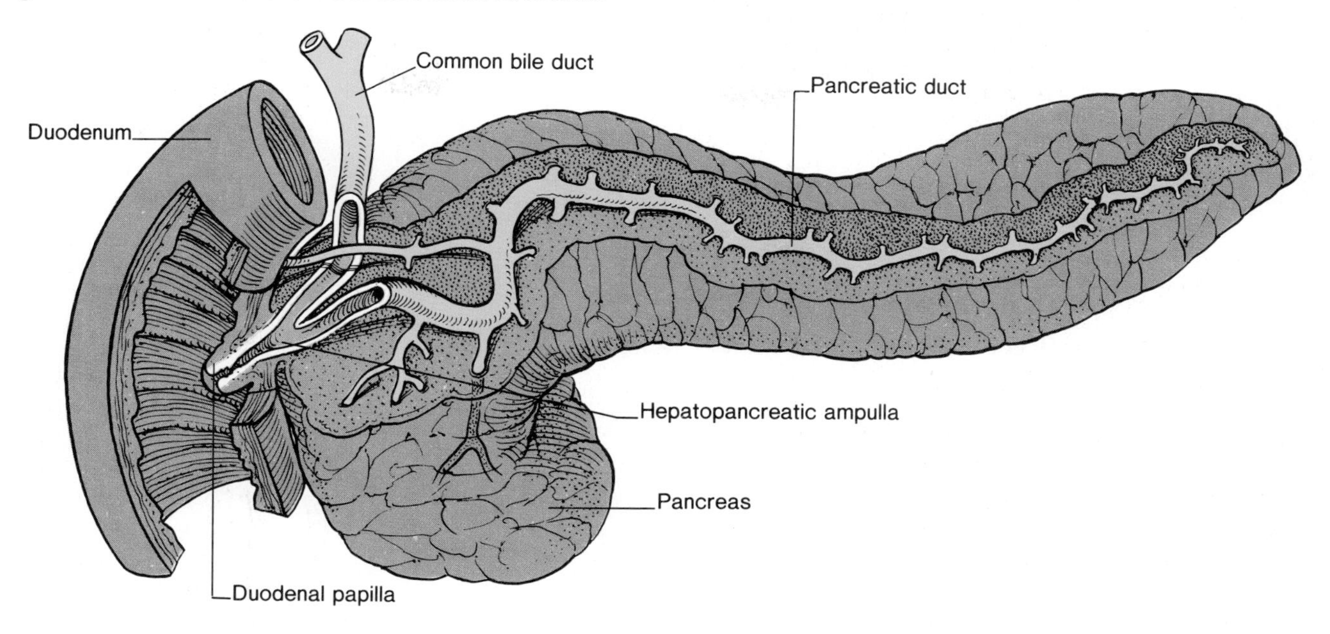

Figure 18.12. The microscopic structure of the duodenum.

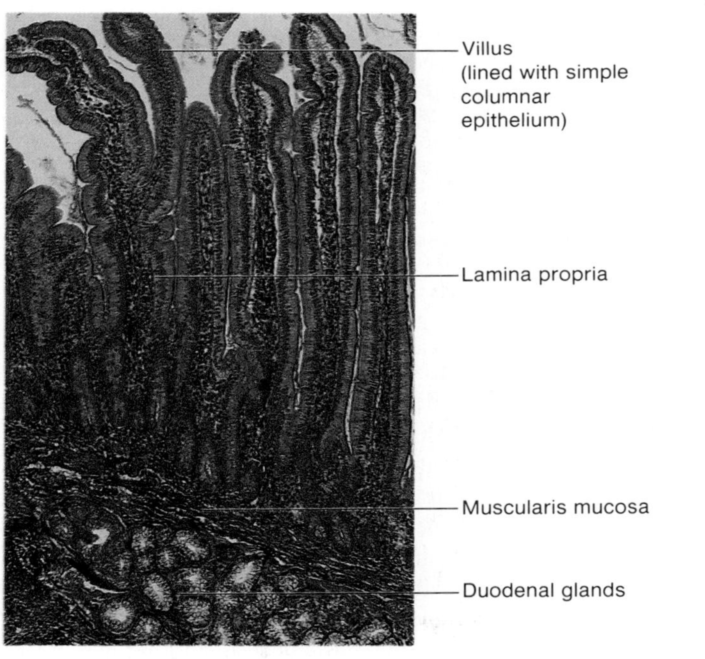

Figure 18.13. A diagram of the structure of an intestinal villus.

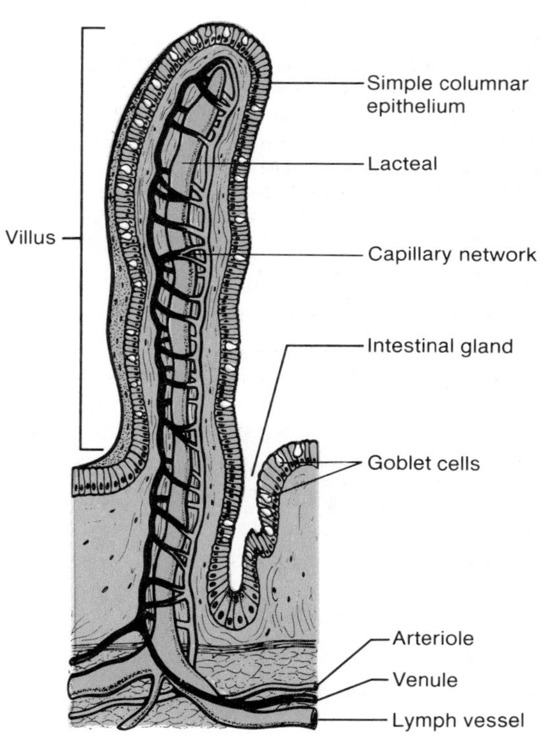

Figure 18.14. (*a*) Intestinal villi and crypts of Lieberkühn. The crypts serve as sites for production of new epithelial cells. The time required for migration of these new cells to the tip of the villi is shown in (*b*). Epithelial cells are exfoliated from the tips of the villi.

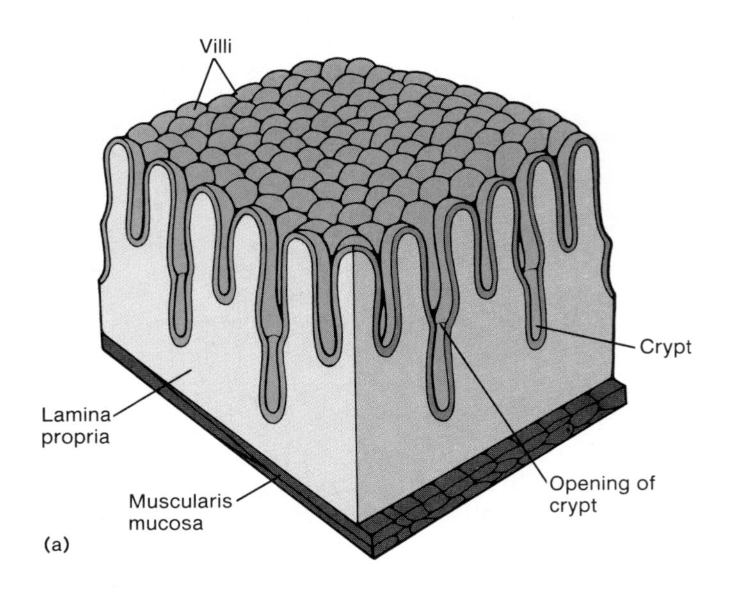

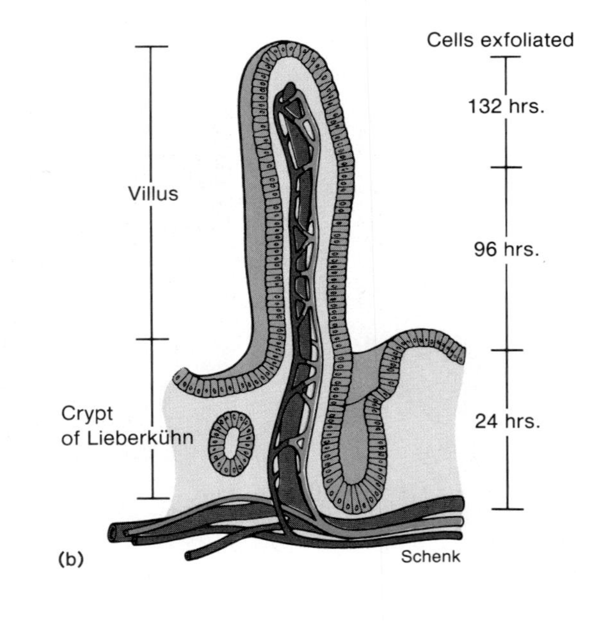

the villi invaginates downwards at various points to form narrow pouches that open through pores to the intestinal lumen. These structures are called the **crypts of Lieberkühn** (fig. 18.14).

Microvilli are fingerlike projections formed by foldings of the cell membrane, which can only be clearly seen in an electron microscope. In a light microscope, the microvilli produce a somewhat vague **brush border** on the edges of the columnar epithelial cells. The term *brush border* is thus often used synonymously with microvilli in descriptions of the intestine (fig. 18.15).

Intestinal Enzymes

In addition to providing a large surface area for absorption, the cell membranes of the microvilli contain digestive enzymes. These enzymes are not secreted into the lumen, but instead remain attached to the cell membrane with their active sites exposed to the chyme. These **brush border enzymes** hydrolyze disaccharides, polypeptides, and other substrates (table 18.3). One brush border enzyme, *enterokinase,* is required for activation of the protein-digesting enzyme *trypsin,* which enters the intestine in pancreatic juice.

The ability to digest milk sugar, or lactose, depends on the presence of a brush border enzyme called lactase. This enzyme is present in all children under the age of four but becomes inactive in most adults. This can result in **lactose intolerance.** The presence of large amounts of undigested lactose in the intestine causes diarrhea, gas, cramps, and other unpleasant symptoms. Yogurt is better tolerated than milk because it contains lactase produced by the yogurt bacteria, which becomes activated in the duodenum and digests lactose.

Intestinal Contractions and Motility

Like cardiac muscle, intestinal smooth muscle is capable of spontaneous electrical activity and automatic, rhythmic contractions. Spontaneous depolarization begins in the longitudinal smooth muscle and is conducted to the circular smooth muscle layer across *nexuses.* The term *nexus* is used here to indicate an electrical synapse between smooth muscle cells. The spontaneous depolarizations, called **pacesetter potentials,** decrease in amplitude as they are conducted from one muscle cell to another, much like excitatory postsynaptic potentials (EPSPs). Also like EPSPs, pacesetter potentials stimulate the production of action potentials in the smooth muscle cells through which they are conducted (fig. 18.16).

Figure 18.15. (*a*) The cell membrane of epithelial cells that line the small intestine are folded into microvilli. (*b*) Glycoproteins within the membranes of the microvilli contain polysaccharides that extend out into the lumen and form the glycocalyx, which covers the brush border epithelium. (*c*) A diagram of an intestinal epithelial cell. (*d*) The microvilli contain actin filaments attached to a terminal web of myosin filaments. This allows the microvilli to shorten.

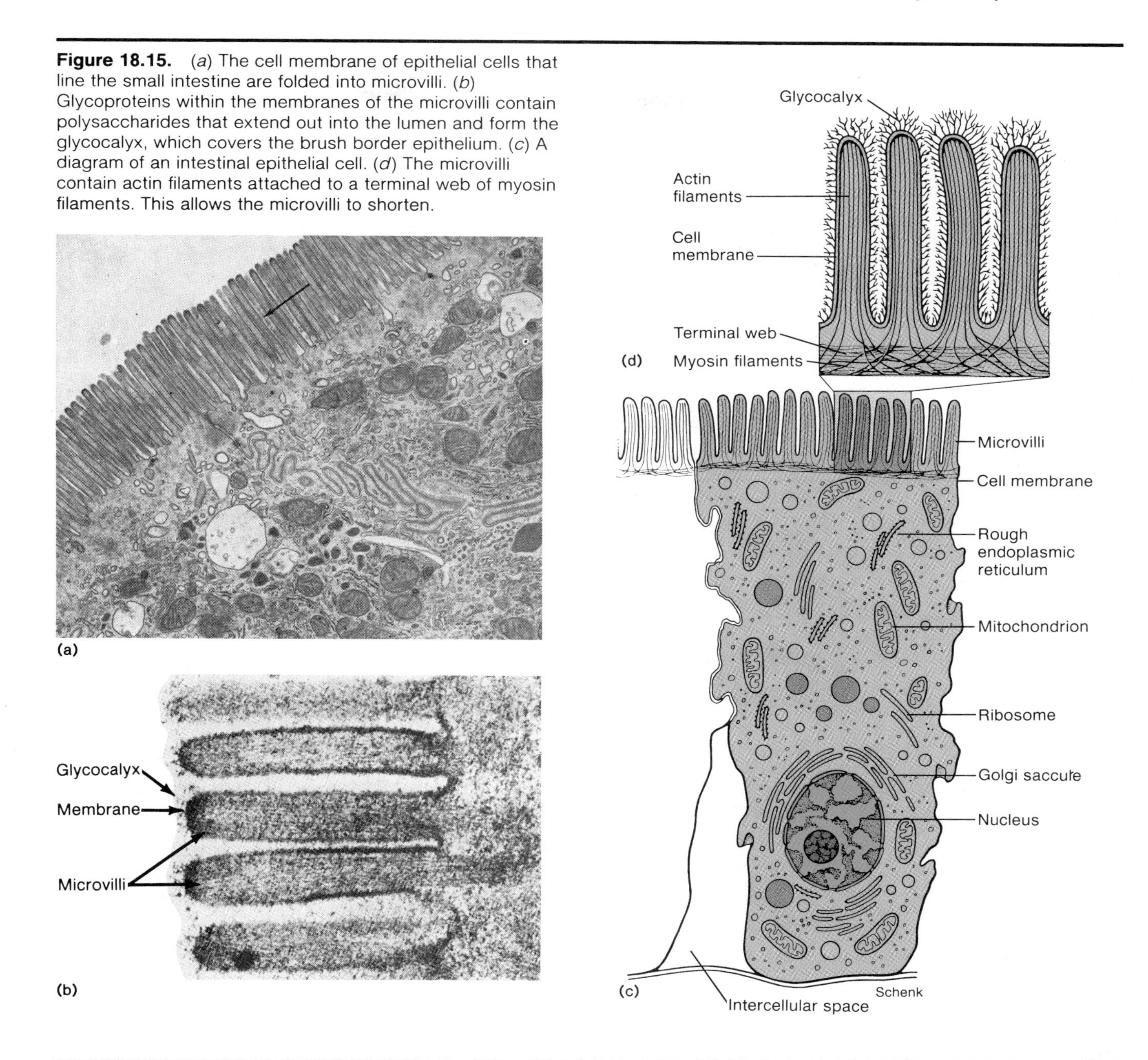

Table 18.3 Brush border enzymes attached to the cell membrane of microvilli in the small intestine.

Category	Enzyme	Comments
Disaccharidase	Sucrase	Digests sucrose to glucose and fructose; deficiency produces gastrointestinal disturbances
	Maltase	Digests maltose to glucose
	Lactase	Digests lactose to glucose and galactose; deficiency produces gastrointestinal disturbances (lactose intolerance)
Peptidase	Aminopeptidase	Produces free amino acids, dipeptides, and tripeptides
	Enterokinase	Activates trypsin (and indirectly other pancreatic juice enzymes); deficiency results in protein malnutrition
Phosphatase	Ca^{++}, Mg^{++}—ATPase	Needed for absorption of dietary calcium; enzyme activity regulated by vitamin D
	Alkaline phosphatase	Removes phosphate groups from organic molecules; enzyme activity may be regulated by vitamin D

Figure 18.16. The smooth muscle of the gastrointestinal tract produces and conducts spontaneous pacesetter potentials. As these potential changes reach a threshold level of depolarization, they stimulate the production of action potentials, which in turn stimulate smooth muscle contraction.

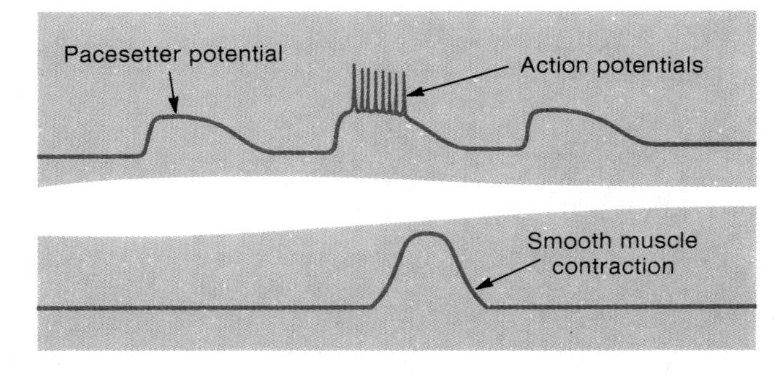

Figure 18.17. Segmentation of the small intestine. Simultaneous contractions of many segments of the intestine help to mix the chyme with digestive enzymes and mucus.

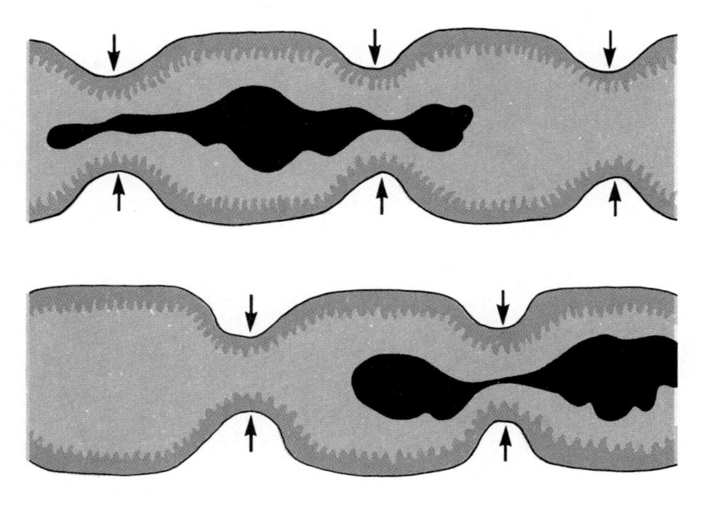

The nexuses conduct the pacesetter potentials, not the action potentials. Action potentials are, therefore, limited to those smooth muscle cells that are depolarized to threshold by the spreading pacesetter potentials. When action potentials are produced, they stimulate smooth muscle contraction. The rate at which this automatic activity occurs is influenced by autonomic nerves. Contraction is stimulated by parasympathetic (vagus nerve) innervation and is reduced by sympathetic nerve activity.

The small intestine has two major types of contractions: peristalsis and segmentation. Peristalsis is much weaker in the small intestine than in the esophagus and stomach. **Intestinal motility**—the movement of chyme through the intestine—is relatively slow and is due primarily to the fact that the pressure at the pyloric end of the small intestine is greater than at the distal end.

The major contractile activity of the small intestine is **segmentation.** This term refers to muscular constrictions of the lumen, which occur simultaneously at different intestinal segments (fig. 18.17). This action serves to mix the chyme more thoroughly.

Figure 18.18. The large intestine.

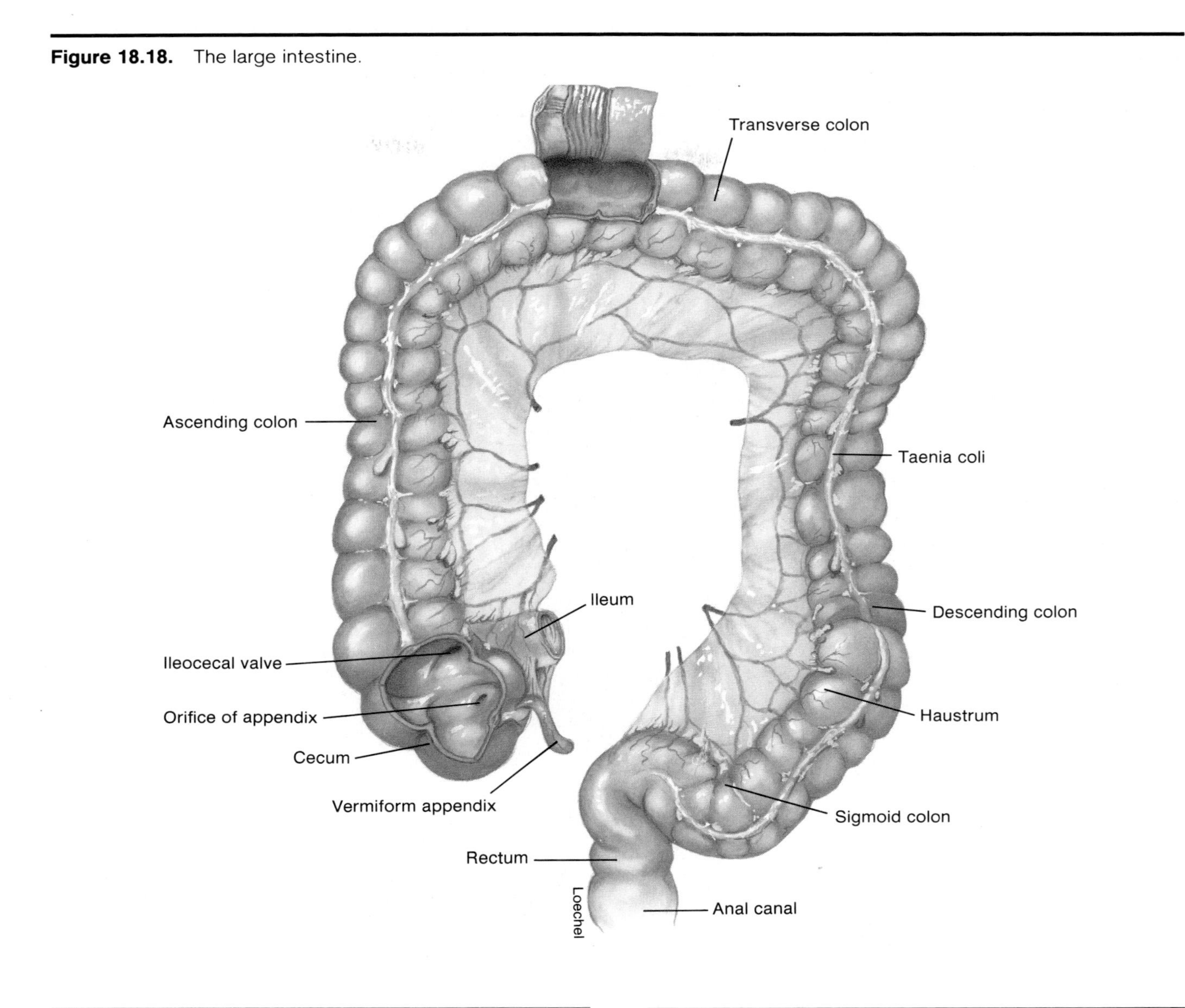

1. *Describe the intestinal structures that increase surface area, and explain the function of the crypts of Lieberkühn.*
2. *Define the term* brush border enzymes, *and list examples. Explain why many adults cannot tolerate milk.*
3. *Describe how smooth muscle contraction in the small intestine is regulated, and explain the function of segmentation.*

Large Intestine

Chyme from the ileum passes into the *cecum,* which is a blind-ending pouch at the beginning of the large intestine, or *colon.* Waste material then passes in sequence through the ascending colon, transverse colon, descending colon, sigmoid colon, rectum, and anal canal (fig. 18.18). Waste material (feces) is excreted through the anus.

Figure 18.19. A radiograph after a barium enema showing the haustra of the large intestine.

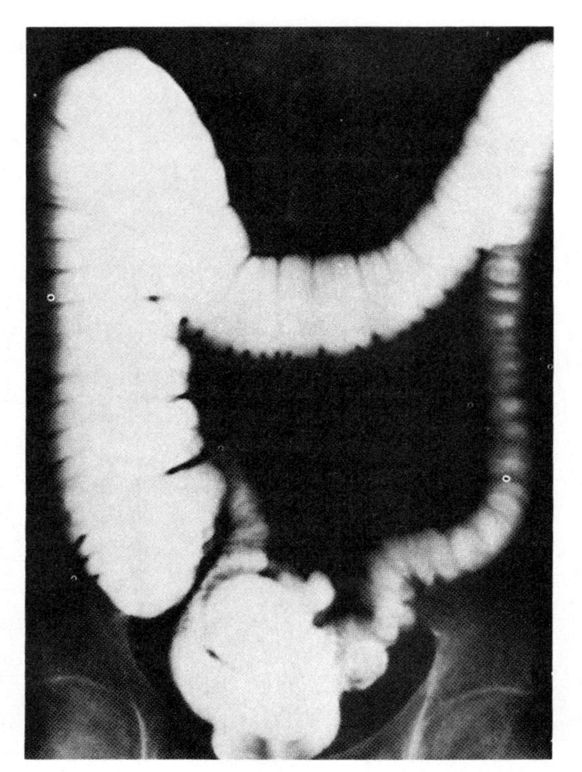

As in the small intestine, the mucosa of the large intestine contains many scattered lymphocytes and lymphatic nodules and is covered by columnar epithelial cells and mucous-secreting goblet cells. Although this epithelium does form crypts of Lieberkühn, there are no villi in the large intestine—the intestinal mucosa therefore presents a flat appearance. The outer surface of the colon bulges outwards to form pouches, or **haustra** (fig. 18.19). Occasionally, the muscularis externa of the haustra may become so weakened that the wall forms a more elongated outpouching, or diverticulum (*divert* = turned aside). Inflammation of these structures is called *diverticulitis*.

The *vermiform appendix* is a short, thin outpouching of the cecum. It does not function in digestion, but like the tonsils, it contains numerous lymphatic nodules (fig. 18.20) and is subject to inflammation—a condition called **appendicitis.** This is commonly detected in its later stages by pain in the lower right quadrant of the abdomen. Rupture of the appendix can cause inflammation of the surrounding body cavity—*peritonitis.* This dangerous event may be prevented by surgical removal of the inflamed appendix (appendectomy).

Fluid and Electrolyte Absorption in the Intestine

Most of the fluid and electrolytes in the lumen of the digestive tract are absorbed by the small intestine. Although a person may only drink about 1.5 L/day of water, the small intestine receives 7–9 L/day as a result of the fluid secreted into the digestive tract by the salivary glands,

Figure 18.20. The microscopic appearance of a cross section of the human appendix.

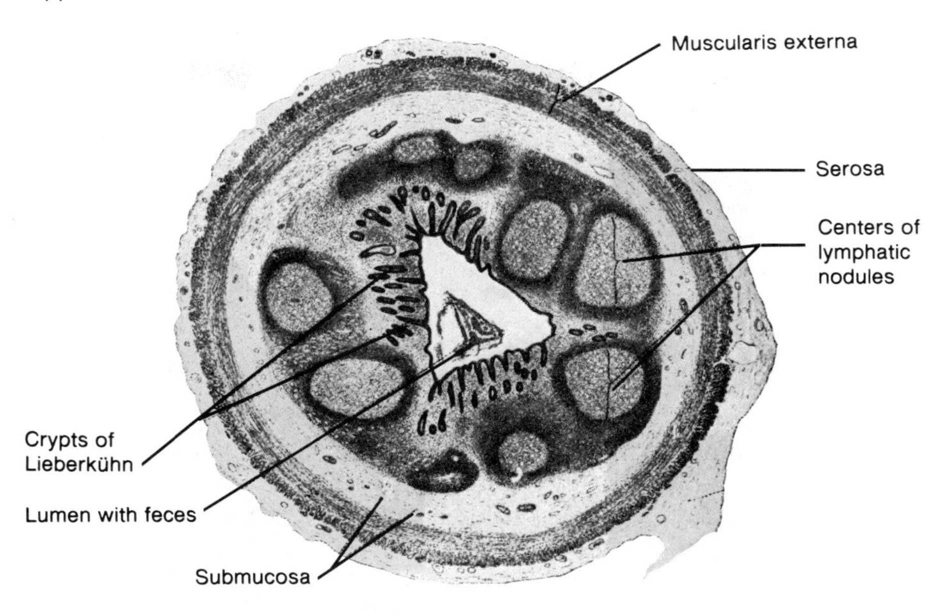

stomach, pancreas, liver, and gallbladder. The small intestine absorbs most of this fluid and passes about 1.5–2.0 L/day of fluid to the large intestine. The large intestine absorbs about 90% of this remaining volume, leaving less than 200 ml/day of fluid to be excreted in the feces.

Absorption of water in the intestine occurs passively as a result of the osmotic gradient created by the active transport of ions. The epithelial cells of the intestinal mucosa are joined together much like those of the kidney tubules and, like the kidney tubules, contain Na^+/K^+ pumps in the basolateral membrane. The analogy with kidney tubules is emphasized by the observation that aldosterone, which stimulates salt and water reabsorption in the renal tubules, also appears to stimulate salt and water absorption in the ileum.

The handling of salt and water transport in the large intestine is made more complex by the fact that the large intestine can secrete, as well as absorb, water. The secretion of water by the mucosa of the large intestine occurs passively as a result of the active transport of Cl^- out of the epithelial cells into the intestinal lumen. Normally the absorption of NaCl and water far exceeds their secretion, but this balance may be altered in some disease states.

Diarrhea is characterized by excessive fluid excretion in the feces. There are three different mechanisms, illustrated by three different diseases, which can cause diarrhea. In *cholera,* severe diarrhea results from a chemical called *enterotoxin,* released from the infecting bacteria. Enterotoxin, acting via stimulation of adenylate cyclase and cAMP production, stimulates active Cl^- and water secretion. In *celiac sprue,* a disease produced in susceptible people by eating foods that contain gluten (proteins from grains such as wheat), diarrhea results from inadequate absorption of fluid due to damage to the intestinal mucosa. In *lactose intolerance,* diarrhea is produced by the increased osmolarity of the contents of the intestinal lumen as a result of the presence of undigested lactose.

Defecation

After electrolytes and water have been absorbed, the waste material that is left passes to the rectum, leading to an increase in rectal pressure and the urge to defecate. If the urge to defecate is denied, feces are prevented from entering the anal canal by the internal anal sphincter. In this case the feces remain in the rectum and may even back

Figure 18.21. The structure of the liver. (*a*) A scanning electron micrograph of the liver. Hepatocytes are arranged in plates so that blood that passes through sinusoids (*b*) will be in contact with each liver cell. (Photo from: *Tissues and Organs: A Text Atlas of Scanning Electron Microscopy* by R. G. Kessel and R. Kardon. W. H. Freeman and Company. © 1979.)

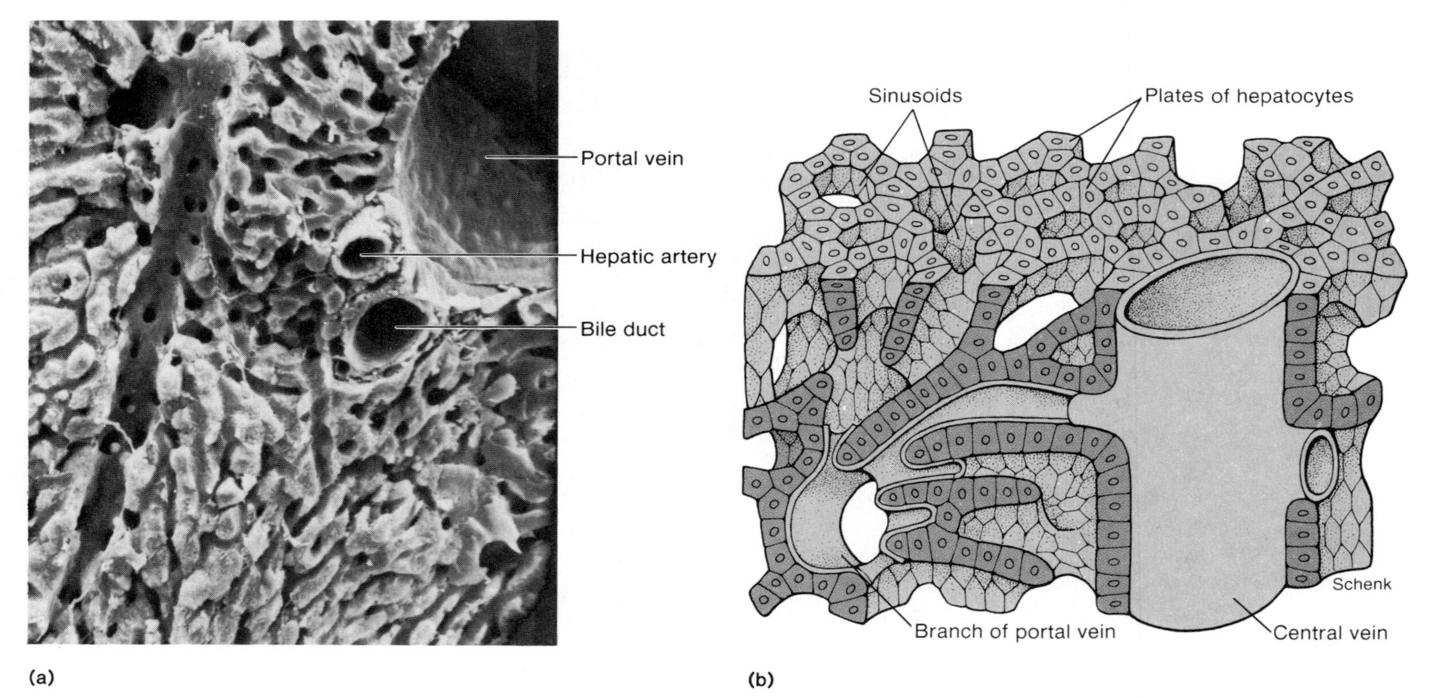

up into the sigmoid colon. The **defecation reflex** normally occurs when the rectal pressure rises to a particular level that is determined, to a large degree, by habit. At this point the internal anal sphincter relaxes to admit feces into the anal canal.

During the act of defecation the longitudinal rectal muscles contract to increase rectal pressure and the internal and external anal sphincter muscles relax. Excretion is aided by contractions of abdominal and pelvic skeletal muscles, which raise the intra-abdominal pressure and help push the feces from the rectum through the anal canal and out the anus.

1. *Describe how electrolytes and water are absorbed in the large intestine, and explain how diarrhea may be produced.*
2. *Describe the structures and mechanisms involved in defecation.*

Liver, Gallbladder, and Pancreas

The liver, which is the largest internal organ, lies immediately beneath the diaphragm in the abdominal cavity. Attached to the inferior surface of the liver, between the right and quadrate lobes, is the pear-shaped gallbladder. This organ is approximately 10 cm long by 3.5 cm wide. The pancreas is located behind the stomach, along the posterior abdominal wall.

Structure of the Liver

Although the liver is the largest internal organ, it is, in a sense, only one to two cells thick. This is because the liver cells, or **hepatocytes,** form **plates** that are one to two cells thick and separated from each other by large capillary spaces, called **sinusoids** (fig. 18.21). The sinusoids are lined with phagocytic **Kupffer cells,** but the large intercellular gaps between adjacent Kupffer cells make these sinusoids more highly permeable than other capillaries. The plate structure of the liver and the high permeability of the sinusoids allow each hepatocyte to have direct contact with the blood.

Figure 18.22. The flow of blood and bile in a liver lobule. Blood flows within sinusoids from a portal vein to the central vein (from the periphery to the center of a lobule). Bile flows within hepatic plates from the center to bile ducts at the periphery of a lobule.

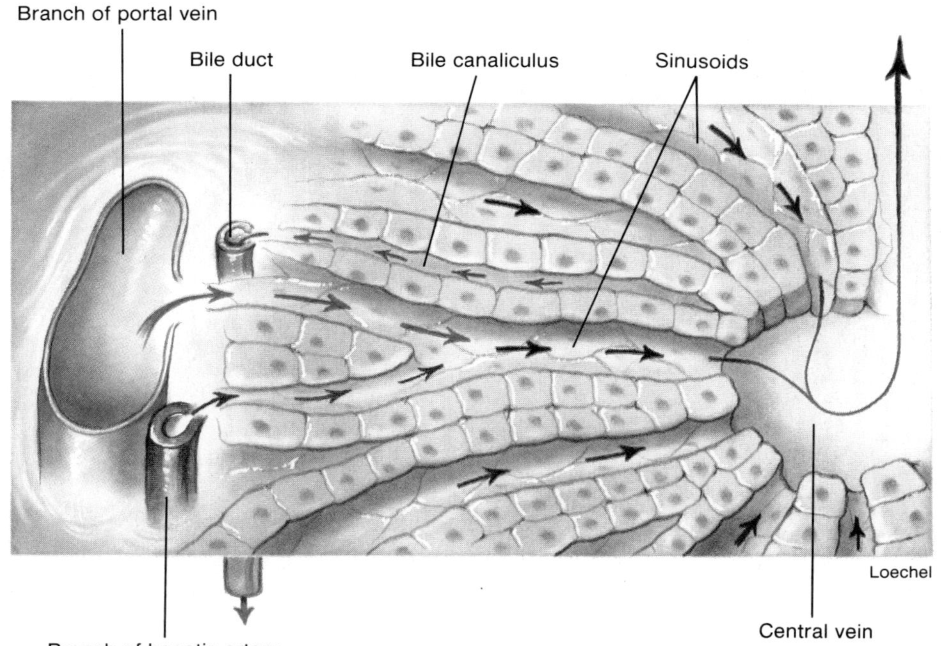

In **cirrhosis,** large numbers of liver lobules are destroyed and replaced with permanent connective tissue and "regenerative nodules" of hepatocytes. These regenerative nodules don't have the platelike structure of normal liver tissue and are therefore less functional. One indication of this decreased function is the entry of ammonia from the hepatic portal blood into the general circulation. Cirrhosis may be caused by chronic alcohol abuse, viral hepatitis, and other agents that attack liver cells.

Portal System. The products of digestion that are absorbed into blood capillaries in the intestine do not directly enter the general circulation. Instead, this blood is delivered first to the liver. Capillaries in the digestive tract drain into the *hepatic portal vein,* which carries this blood to capillaries in the liver; it is not until the blood has passed through this second capillary bed that it enters the general circulation through the *hepatic vein* that drains the liver. The term **portal system** is used to describe this unique pattern of circulation: capillaries → vein → capillaries → vein. In addition to receiving venous blood from the intestine, the liver also receives arterial blood via the *hepatic artery.*

Liver Lobules. The hepatic plates are arranged into functional units called **liver lobules** (fig. 18.22). In the middle of each lobule is a *central vein,* and at the periphery of each lobule are branches of the hepatic portal vein and of the hepatic artery, which open into the spaces *between* hepatic plates. Arterial blood and portal venous blood, containing molecules absorbed in the GI tract, thus mix as the blood flows within the sinusoids from the periphery of the lobule to the central vein. The central veins of different liver lobules converge to form the hepatic vein, which carries blood from the liver to the inferior vena cava.

Bile is produced by the hepatocytes and secreted into thin channels called **bile canaliculi,** located *within* each hepatic plate (fig. 18.22). These bile canaliculi are drained at the periphery of each lobule by *bile ducts,* which in turn drain into *hepatic ducts* that carry bile away from the liver. Since blood travels in the sinusoids and bile travels in the opposite direction within the hepatic plates, blood and bile do not mix in the liver lobules.

Figure 18.23. Enterohepatic circulation. Substances secreted in the bile may be absorbed by the intestinal epithelium and recycled to the liver via the hepatic portal vein.

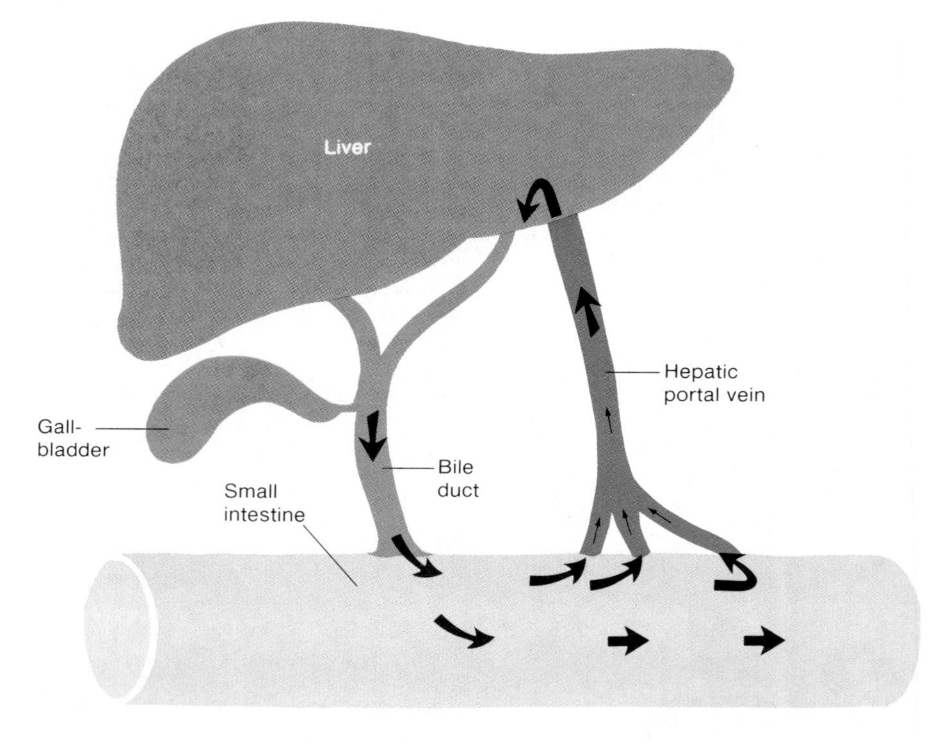

Enterohepatic Circulation. In addition to the normal constituents of bile, a wide variety of exogenous compounds (drugs) are secreted by the liver into the bile ducts (table 18.4). The liver can thus "clear" the blood of particular compounds by removing them from the blood and excreting them into the intestine with the bile. (The liver can also clear the blood by other mechanisms that will be described in a later section.)

Many compounds that are released with the bile into the intestine are not excreted with the feces, however. Some of these can be absorbed through the small intestine and enter the hepatic portal blood. These absorbed molecules are thus carried back to the liver, where they can be again secreted by hepatocytes into the bile ducts. Compounds that recirculate between the liver and intestine in this way are said to have an **enterohepatic circulation** (fig. 18.23).

Table 18.4 Compounds that the liver excretes into the bile.

	Compound	Comments
Endogenous (naturally occurring)	Bile salts, urobilinogen, cholesterol	High percentage is absorbed and has an enterohepatic circulation[1]
	Lecithin	Small percentage is absorbed and has an enterohepatic circulation
	Bilirubin	No enterohepatic circulation
Exogenous (drugs)	Ampicillin, streptomycin, tetracycline	High percentage is absorbed and has an enterohepatic circulation
	Sulfonamides, penicillin	Small percentage is absorbed and has an enterohepatic circulation

[1]Compounds with an enterohepatic circulation are absorbed to some degree by the intestine and are returned to the liver in the hepatic portal vein.

Table 18.5 Summary of the major categories of liver functions.

Functional Category	Actions
Detoxification of blood	Phagocytosis by Kupffer cells
	Chemical alteration of biologically active molecules (hormones and drugs)
	Production of urea, uric acid, and other molecules that are less toxic than parent compounds
	Excretion of molecules in bile
Carbohydrate metabolism	Conversion of blood glucose to glycogen and fat
	Production of glucose from liver glycogen and from other molecules (amino acids, lactic acid) by gluconeogenesis
	Secretion of glucose into the blood
Lipid metabolism	Synthesis of triglyceride and cholesterol
	Excretion of cholesterol in bile
	Production of ketone bodies from fatty acids
Protein synthesis	Production of albumin
	Production of plasma transport proteins
	Production of clotting factors (fibrinogen, prothrombin, and others)
Secretion of bile	Synthesis of bile salts
	Conjugation and excretion of bile pigment (bilirubin)

Functions of the Liver

As a result of its very large and diverse enzymatic content, its unique structure, and the fact that it receives venous blood from the intestine, the liver has a wider variety of functions than any other organ in the body. A summary of the major categories of liver function is presented in table 18.5.

Bile Production and Secretion. The liver produces and secretes 250–1,500 ml of bile per day. The major constituents of bile include bile salts, bile pigment (bilirubin), phospholipids (mainly lecithin), cholesterol, and inorganic ions (table 18.6).

Table 18.6 Composition of the bile.

Component	Concentration
pH	5.7–8.6
Bile salts	140–2230 mg/100 ml
Lecithin	140–810 mg/100 ml
Cholesterol	97–320 mg/100 ml
Bilirubin	12–70 mg/100 ml
Urobilinogen	5–45 mg/100 ml
Sodium	145–165 mEq/L
Potassium	2.7–4.9 mEq/L
Chloride	88–115 mEq/L
Bicarbonate	27–55 mEq/L

Bile pigment, or bilirubin, is produced in the spleen, liver, and bone marrow from heme groups (minus the iron), derived from hemoglobin. Without the protein part of hemoglobin, the **free bilirubin** is not very water-soluble and thus must be carried in the blood attached to albumin proteins. This protein-bound bilirubin can neither be filtered by the kidneys into the urine, nor can it be directly excreted by the liver into the bile.

The liver can take some of the free bilirubin out of the blood and conjugate (combine) it with glucuronic acid. This **conjugated bilirubin** is water-soluble and is secreted into the bile. Once the conjugated bilirubin enters the intestine it is converted by bacteria into another pigment—**urobilinogen**—which is partially responsible for the color of the feces. About 30% to 50% of the urobilinogen, however, is absorbed by the intestine and enters the portal blood. Some of this is returned to the intestine in an enterohepatic circulation; the rest enters the general circulation (fig. 18.24). The urobilinogen in plasma, unlike free bilirubin, is not attached to albumin and, therefore, is easily filtered by the kidneys into the urine, giving urine its characteristic yellow color.

The **bile salts** are derivatives of cholesterol that have two to four polar groups on each molecule. The principal bile salts in humans are *cholic acid* and *chenodeoxycholic acid* (fig. 18.25). In aqueous solutions these molecules "huddle" together to form aggregates known as **micelles.** The nonpolar parts are located in the central region of the micelle (away from water), whereas the polar groups face water around the periphery of the micelle. Lecithin, cholesterol, and other lipids enter these micelles in a process that aids the digestion and absorption of fats.

Jaundice is a yellow staining of the tissues produced by high blood concentrations of either free or conjugated bilirubin. Jaundice due to high blood levels of conjugated bilirubin in adults may result when bile excretion is blocked by gallstones. Since free bilirubin is derived from heme, jaundice due to high blood levels of free bilirubin is usually caused by an excessively high rate of red blood cell destruction. This is the cause of jaundice in newborn babies who suffer from erythroblastosis fetalis (described in chapter 17). *Physiological jaundice of the newborn* is due to high levels of free bilirubin in otherwise healthy neonates. This type of jaundice may be caused by the rapid fall in blood hemoglobin concentrations that normally occurs at birth, or in premature infants it may be caused by inadequate amounts of hepatic enzymes that are needed to conjugate bilirubin and thus excrete it in the bile.

Newborn infants with jaundice are usually treated by *phototherapy,* in which they are placed under blue light in the 400–500 nm wavelength range. This light is absorbed by bilirubin in cutaneous vessels and results in the conversion of bilirubin to a more water-soluble isomer, which is soluble in plasma without having to be conjugated with glucuronic acid. The more water-soluble photoisomer of bilirubin can then be excreted in the bile.

Detoxification of the Blood. The liver can remove biologically active molecules such as hormones and drugs from the blood by (1) excretion of these compounds in the bile; (2) phagocytosis by Kupffer cells, which line the sinusoids; and (3) chemical alteration of these molecules within the hepatocytes.

Ammonia, for example, is a very toxic molecule produced by the action of bacteria in the intestine. The observation that portal blood has an ammonia concentration that is four to fifty times greater than in the hepatic vein means that this compound is removed by the liver. The liver has the enzymes needed to convert ammonia into less toxic *urea* molecules, which are secreted by the liver into the blood and excreted by the kidneys in the urine. Similarly, the liver converts toxic porphyrins into *bilirubin* and toxic purines into *uric acid.*

Steroid hormones and other nonpolar compounds, such as many drugs, are inactivated in their passage through the liver by modifications of their chemical structures. The liver has enzymes that convert these molecules into more polar (more water-soluble) forms by **hydroxylation** (the addition of OH^- groups) and by **conjugation** with highly polar groups such as sulfate and glucuronic acid. Polar derivatives of steroid hormones and drugs are less biologically active and, because of their increased water solubility, are more easily excreted by the kidneys into the urine.

Secretion of Glucose, Triglycerides, and Ketone Bodies. The liver helps to regulate the blood glucose concentration by either removing glucose from or adding glucose to the blood, according to the needs of the body. After a carbohydrate-rich meal, the liver can remove some glucose from the portal blood and convert it into glycogen and triglycerides (**glycogenesis** and **lipogenesis,** respectively). During fasting, the liver secretes glucose into the blood. This glucose can be derived from the breakdown of stored glycogen in a process called **glycogenolysis,** or it can be produced by the conversion of noncarbohydrate molecules (such as amino acids) into glucose in a process called **gluconeogenesis.** The liver also contains the enzymes required to convert free fatty acids into ketone bodies (**ketogenesis**), which are secreted into the blood in large amounts during fasting. These processes are controlled by hormones.

Production of Plasma Proteins. Plasma albumin and most of the plasma globulins (with the exception of immunoglobulins) are produced by the liver. Albumin comprises about 70% of the total plasma protein and contributes most to the colloid osmotic pressure of the blood. The globulins produced by the liver have a wide variety of functions, including the transport of cholesterol and triglycerides, transport of steroid and thyroid hormones, inhibition of trypsin activity, and blood clotting. Clotting factors I (fibrinogen), II (prothrombin), III, V, VII, IX, and XI are all produced by the liver.

Gallbladder

The gallbladder is a saclike organ attached to the inferior surface of the liver. This organ stores and concentrates bile, which drains to it from the liver by way of the hepatic ducts, bile duct, and *cystic duct,* respectively. A sphincter valve at the neck of the gallbladder allows a storage capacity of about 35 to 50 ml. The inner mucosal layer of the gallbladder is arranged in rugae similar to those of the stomach. When the gallbladder fills with bile, it expands to the size and shape of a small pear. Bile is a yellowish-green fluid containing bile salts, bilirubin, cholesterol, and other compounds as previously discussed. Contraction of the muscularis ejects bile from the cystic duct into the *common bile duct,* which conveys bile into the duodenum (fig. 18.26).

Bile is continuously produced by the liver and drains through the hepatic and common bile ducts to the duodenum. When the small intestine is empty of food, the *sphincter of Oddi* closes, and bile is forced up the cystic duct to the gallbladder for storage.

Figure 18.26. The pancreatic duct joins the common bile duct to empty its secretions through the duodenal papilla into the duodenum. The release of bile and pancreatic juice into the duodenum is controlled by the sphincter of Oddi.

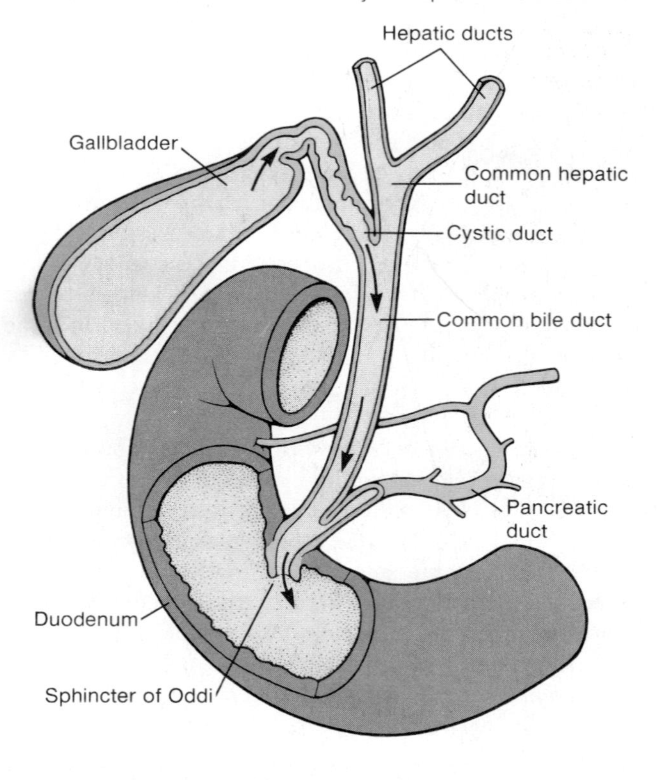

Approximately twenty million Americans have **gallstones,** which can produce painful symptoms by obstructing the cystic or common bile ducts. Gallstones commonly contain cholesterol as their major component. Cholesterol normally has an extremely low water solubility (20 μg/L), but it can be present in bile at two million times its water solubility (40 g/L) because it enters the hydrophobic centers of mixed micelles of bile salts and lecithin. In order for gallstones to be produced, the liver must secrete enough cholesterol to create a supersaturated solution, and some substance must serve as a nucleus for the formation of cholesterol crystals. The gallstone is formed from cholesterol crystals that become hardened by the precipitation of inorganic salts (fig. 18.27). There is evidence that the gallbladder of people with gallstones secretes a glycoprotein that may serve as the nucleating factor in the formation of cholesterol crystals. Gallstones may sometimes be dissolved by treatment with the bile salt chenodeoxycholic acid, or they may have to be removed surgically.

Figure 18.27. (*a*) An X ray of a gallbladder that contains gallstones. (*b*) A posterior view of a gallbladder that has been removed (cholecystectomy) and cut open to reveal its gallstones (bilary calculi). A dime is placed in the photo to show relative size.

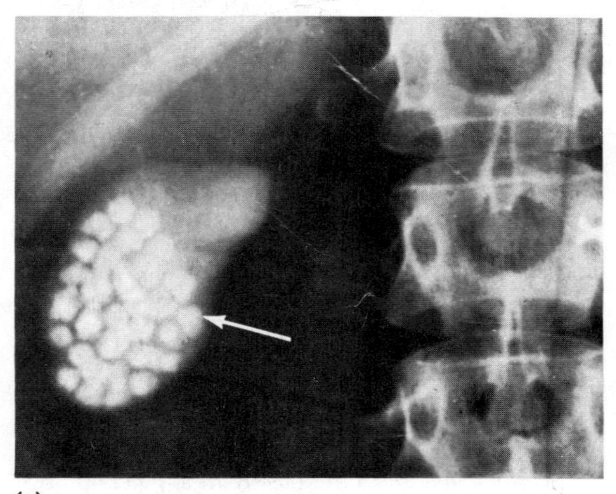

(a)

(b)

Pancreas

The pancreas is a soft, lobulated, glandular organ that has both exocrine and endocrine functions. The endocrine function is performed by clusters of cells, called the **islets of Langerhans,** that secrete the hormones insulin and glucagon into the blood. As an exocrine gland, the pancreas secretes pancreatic juice through the pancreatic duct (fig. 18.28) into the duodenum. Within the lobules of the pancreas are the exocrine secretory units, called **acini.** Each acinus consists of a single layer of epithelial cells surrounding a lumen, into which the constituents of pancreatic juice are secreted.

Figure 18.28. The pancreas is both an exocrine and an endocrine gland. Pancreatic juice—the exocrine product—is secreted by acinar cells into the pancreatic duct. Scattered "islands" of cells, called the islets of Langerhans, secrete the hormones insulin and glucagon into the blood.

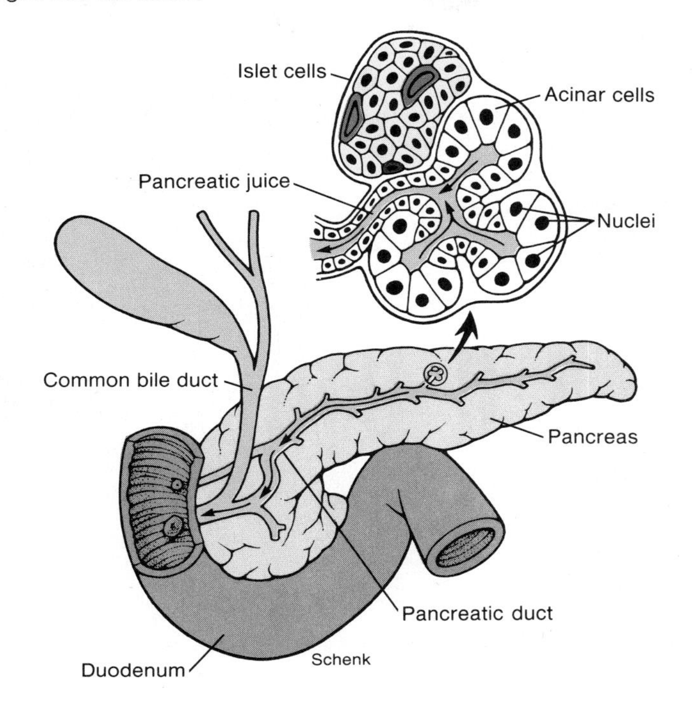

Table 18.7 Enzymes in pancreatic juice.

Enzyme	Zymogen	Activator	Action
Trypsin	Trypsinogen	Enterokinase	Cleaves internal peptide bonds
Chymotrypsin	Chymotrypsinogen	Trypsin	Cleaves internal peptide bonds
Elastase	Proelastase	Trypsin	Cleaves internal peptide bonds
Carboxypeptidase	Procarboxypeptidase	Trypsin	Cleaves last amino acid from carboxyl-terminal end of polypeptide
Phospholipase	Prophospholipase	Trypsin	Cleaves fatty acids from phospholipids such as lecithin
Lipase	None	None	Cleaves fatty acids from glycerol
Amylase	None	None	Digests starch to maltose and short chains of glucose molecules
Cholesterolesterase	None	None	Releases cholesterol from its bonds with other molecules
Ribonuclease	None	None	Cleaves RNA to form short chains
Deoxyribonuclease	None	None	Cleaves DNA to form short chains

Pancreatic Juice. Pancreatic juice contains water, bicarbonate, and a wide variety of digestive enzymes that are secreted into the duodenum. These enzymes include (1) **amylase,** which digests starch; (2) **trypsin,** which digests protein; and (3) **lipase,** which digests triglycerides; other pancreatic enzymes are indicated in table 18.7. It should be noted that the complete digestion of food molecules in the small intestine requires the action of both pancreatic enzymes and brush border enzymes.

Most pancreatic enzymes are produced as inactive molecules, or *zymogens,* which help to minimize the risk of self-digestion within the pancreas. The inactive form of trypsin, called trypsinogen, is activated within the small intestine by the catalytic action of the brush border enzyme *enterokinase.* Enterokinase converts trypsinogen to active trypsin. Trypsin, in turn, activates the other zymogens of pancreatic juice (fig. 18.29) by cleaving off polypeptide sequences that inhibit the activity of these enzymes.

Figure 18.29. The pancreatic protein-digesting enzyme *trypsin* is secreted in an inactive form known as trypsinogen. This inactive enzyme (zymogen) is activated by a brush border enzyme, enterokinase (*EN*), located in the cell membrane of microvilli. Active trypsin in turn activates other zymogens in pancreatic juice.

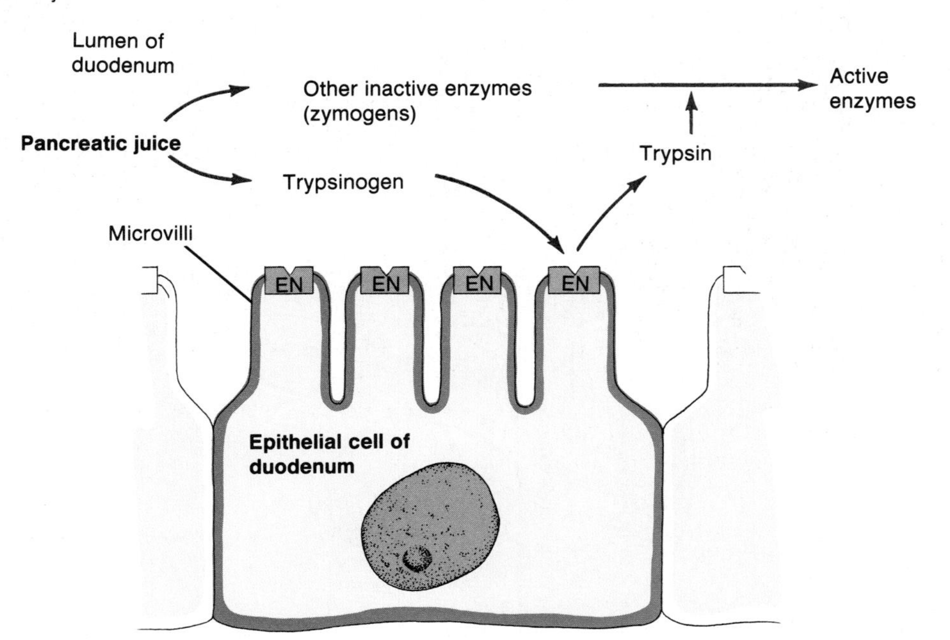

The activation of trypsin is, therefore, the triggering event for the activation of other pancreatic enzymes. Actually, the pancreas does produce small amounts of active trypsin, yet the other enzymes don't become active until pancreatic juice enters the duodenum. This is because pancreatic juice also contains a small protein called *pancreatic trypsin inhibitor,* which attaches to trypsin and inactivates it in the pancreas.

Inflammation of the pancreas may result when the various safeguards against self-digestion are insufficient. **Acute pancreatitis** is believed to be caused by the reflux of pancreatic juice and bile from the duodenum into the pancreatic duct. The leakage of trypsin into the blood also occurs, but trypsin is inactive in the blood because of the inhibitory action of two plasma proteins, alpha-1-antitrypsin and alpha-2-macroglobulin. Pancreatic amylase may also leak into the blood, but it is not active because its substrate (glycogen) is not present in blood. Pancreatic amylase activity can be measured *in vitro,* however, and these measurements are commonly performed to assess the health of the pancreas.

1. *Describe the structure of liver lobules, and trace the pathways for the flow of blood and bile in the lobules.*
2. *Describe the composition and function of bile, and trace the flow of bile from the liver and gallbladder to the duodenum.*
3. *Explain how the liver inactivates and excretes compounds such as hormones and drugs.*
4. *Describe the enterohepatic circulation of bilirubin and urobilinogen.*
5. *Explain how the liver helps maintain a constant blood glucose concentration and how the pattern of venous blood flow permits this task to be performed.*
6. *Describe the endocrine and exocrine structures and functions of the pancreas, and explain why the pancreas does not normally digest itself.*

Digestion and Absorption of Carbohydrates, Lipids, and Proteins

The caloric (energy) value of food is found predominantly in its content of carbohydrates, lipids, and proteins. In the average American diet, carbohydrates account for approximately 50% of the total calories, protein accounts for 11% to 14%, and lipids make up the balance. These food molecules consist primarily of long combinations of subunits (monomers), which must be digested by hydrolysis reactions into the free monomers before absorption can occur. The characteristics of the major digestive enzymes are summarized in table 18.8.

Table 18.8 Summary of the sources and activities of the major digestive enzymes.

Region or Source					
Organ	Source	Substrate	Enzymes	Optimum pH	Products
Mouth	Saliva	Starch	Salivary amylase	6.7	Maltose
Stomach	Gastric glands	Protein	Pepsin	1.6–2.4	Shorter polypeptides
Duodenum	Pancreatic juice	Starch	Pancreatic amylase	6.7–7.0	Maltose, maltriose, and oligosaccharides
		Polypeptides	Trypsin, chymotrypsin, carboxypeptidase	8.0	Amino acids, dipeptides, and tripeptides
		Triglycerides	Pancreatic lipase	8.0	Fatty acids and monoglycerides
	Epithelial membranes	Maltose	Maltase	5.0–7.0	Glucose
		Sucrose	Sucrase	5.0–7.0	Glucose + fructose
		Lactose	Lactase	5.8–6.2	Glucose + galactose
		Polypeptides	Aminopeptidase	8.0	Amino acids, dipeptides, tripeptides

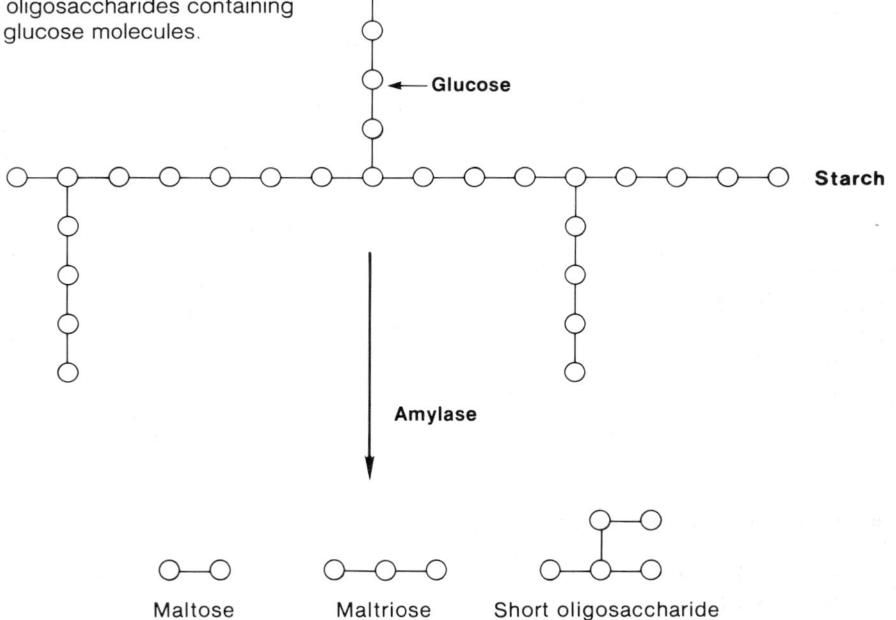

Figure 18.30. Pancreatic amylase digests starch into maltose, maltriose, and short oligosaccharides containing branch points in the chain of glucose molecules.

Digestion and Absorption of Carbohydrates

Most of the ingested carbohydrates are in the form of starch, which is a long polysaccharide of glucose in the form of straight chains with occasional branchings. The most commonly ingested sugars are the disaccharides sucrose (table sugar, consisting of glucose and fructose) and lactose (milk sugar, consisting of glucose and galactose). The digestion of starch begins in the mouth with the action of **salivary amylase,** or **ptyalin.** This enzyme cleaves some of the bonds between adjacent glucose molecules, but most people don't chew their food long enough for sufficient digestion to occur in the mouth. The digestive action of salivary amylase stops when the bolus enters the stomach because this enzyme is inactivated at the low pH of gastric juice.

The digestion of starch, therefore, occurs mainly in the duodenum as a result of the action of **pancreatic amylase.** This enzyme cleaves the straight chains of starch to produce the disaccharide *maltose* and the trisaccharide *maltriose.* Pancreatic amylase, however, cannot hydrolyze the bond between glucose molecules at the branch points in the starch. As a result, short, branched chains of glucose molecules, called *oligosaccharides,* are released together with maltose and maltriose by the activity of this enzyme (fig. 18.30).

Figure 18.31. Polypeptide chains are digested into free amino acids, dipeptides, and tripeptides by the action of pancreatic juice enzymes and brush border enzymes. The amino acids, dipeptides, and tripeptides enter duodenal epithelial cells. Dipeptides and tripeptides are hydrolyzed into free amino acids within the epithelial cells, and these products are secreted into capillaries that carry them to the hepatic portal vein.

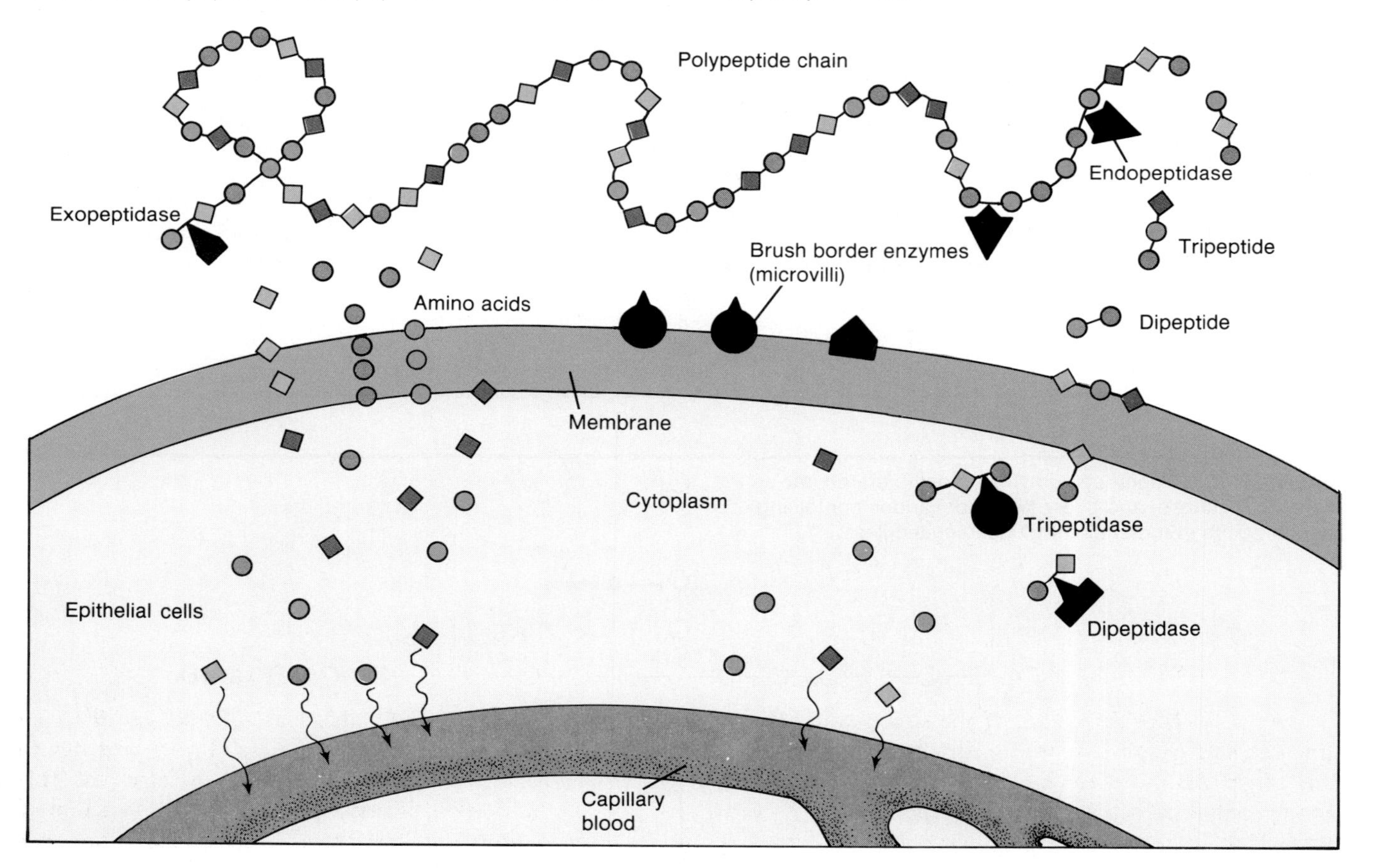

Maltose, maltriose, and oligosaccharides of glucose released from partially digested starch, together with the disaccharides sucrose and lactose, are hydrolyzed to their monosaccharides by brush border enzymes, located on the microvilli of the epithelial cells in the small intestine. The absorption of these monosaccharides across the membrane of the microvilli occurs by means of **coupled transport.** In this process, glucose binds to the same carrier as Na^+ and enters the epithelial cell as Na^+ diffuses down its electrochemical gradient. This is a type of active transport, because energy from ATP is needed to maintain the Na^+ gradient. Glucose is then secreted from the epithelial cells into capillaries within the villi.

Digestion and Absorption of Proteins

Protein digestion begins in the stomach with the action of pepsin. Pepsin results in the liberation of some amino acids, but the major products of pepsin digestion are short-chain polypeptides. This activity helps to produce a more homogenous chyme, but it is not essential for the complete digestion of protein that occurs—even in people with total gastrectomies—in the small intestine.

Most protein digestion occurs in the duodenum and jejunum. The pancreatic juice enzymes **trypsin, chymotrypsin,** and **elastase** cleave peptide bonds within the interior of the polypeptide chains. These enzymes are thus grouped together as *endopeptidases.* Enzymes that remove amino acids from the ends of polypeptide chains, in contrast, are *exopeptidases.* These include the pancreatic juice enzyme **carboxypeptidase,** which removes amino acids from the carboxyl-terminal end of polypeptide chains, and the brush border enzyme **aminopeptidase.** Aminopeptidase cleaves amino acids from the amino-terminal end of polypeptide chains.

Figure 18.32. Pancreatic lipase digests fat (triglycerides) by cleaving off the first and third fatty acids. This produces free fatty acids and monoglycerides. Sawtooth structures indicate hydrocarbon chains in the fatty acids.

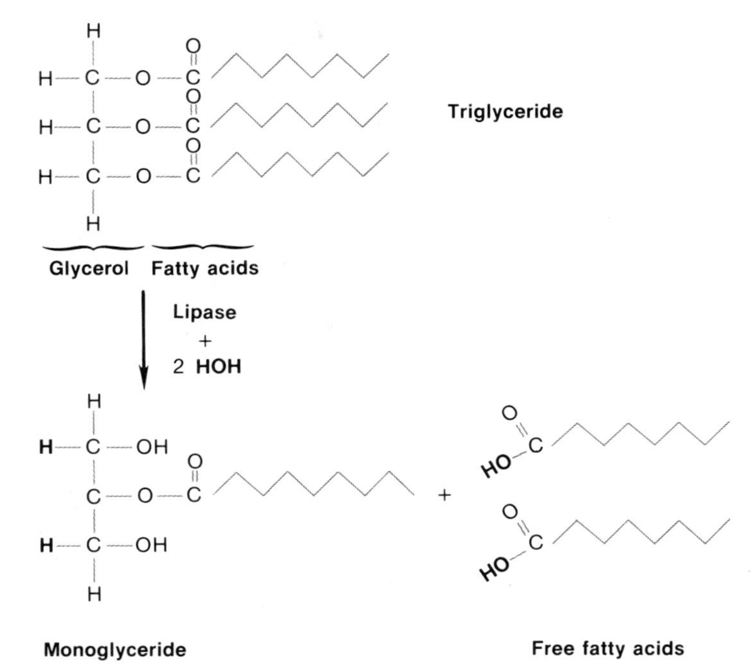

As a result of the action of these enzymes, polypeptide chains are digested into free amino acids, dipeptides, and tripeptides. The free amino acids are absorbed through the epithelial cells of the intestinal mucosa and secreted into blood capillaries. This absorption is carrier-mediated, involving the coupled transport of free amino acids with Na^+, and uses four different carrier systems for different classes of amino acids. The dipeptides and tripeptides may enter epithelial cells by a different carrier system, but they are then digested within these cells into amino acids, which are secreted into the blood (fig. 18.31). Newborn babies appear to be capable of absorbing a substantial amount of undigested proteins (hence they can absorb antibodies from their mother's first milk); in adults, however, only the free amino acids enter the portal vein. Foreign food protein, which would be very antigenic, does not normally enter the blood.

Digestion and Absorption of Fats

Although the salivary glands and stomach produce lipases, there is very little fat digestion until the fat in chyme arrives in the duodenum in the form of fat globules. Through mechanisms described in the next section, the arrival of fat in the duodenum serves as a stimulus for the secretion of bile. Mixed micelles of bile salts, lecithin, and cholesterol are secreted into the duodenum and act to break up the fat droplets into much finer droplets. This process, called **emulsification,** results in the formation of tiny *emulsification droplets* of triglycerides. Note that emulsification is not chemical digestion—the bonds joining glycerol and fatty acids are not hydrolyzed by this process.

Digestion of Lipids. The emulsification of fat aids digestion because the smaller and more numerous emulsification droplets present a greater surface area than the unemulsified fat droplets that originally entered the duodenum. Fat digestion occurs at the surface of the droplets through the enzymatic action of **pancreatic lipase,** which is aided in its action by a protein called *colipase,* also secreted by the pancreas, which coats the emulsification droplets and "anchors" the lipase enzyme to the droplets. Through hydrolysis, lipase removes two of the three fatty acids from each triglyceride molecule and thus liberates *free fatty acids* and *monoglycerides* (fig. 18.32). **Phospholipase A** likewise digests phospholipids such as lecithin into fatty acids and lysolecithin (the remainder of the lecithin molecule after two fatty acids are removed).

Figure 18.33. Steps in the digestion of fat (triglycerides) and the entry of fat digestion products (fatty acids and monoglycerides) into micelles of bile salts secreted by the liver.

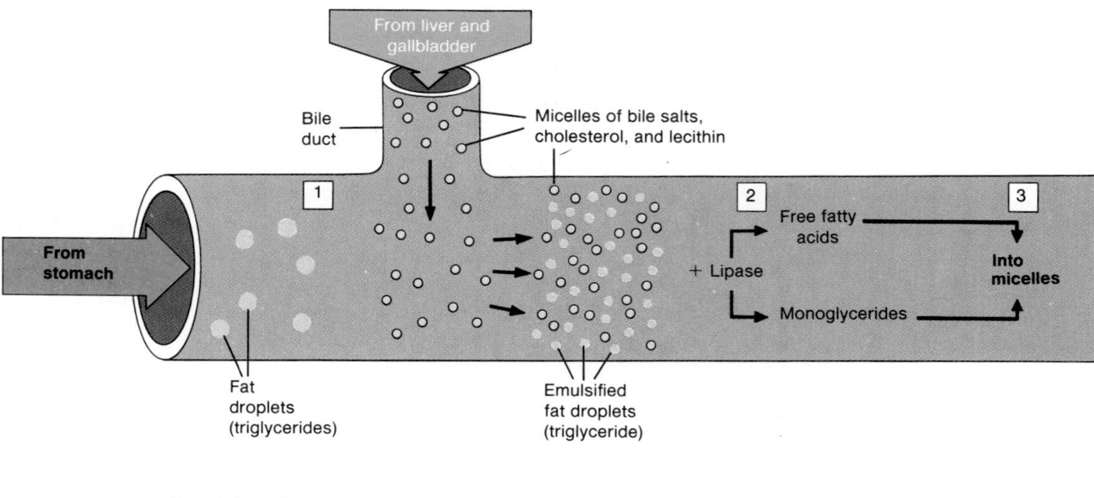

Free fatty acids, monoglycerides, and lysolecithin are more polar than the undigested lipids and are able to move more easily into the mixed micelles of bile salts, lecithin, and cholesterol (fig. 18.33). These micelles then move to the brush border of the intestinal epithelium where absorption occurs.

Absorption of Lipids. Free fatty acids, monoglycerides, and lysolecithin can leave the micelles and pass through the membrane of the microvilli to enter the intestinal epithelial cells. There is also some evidence that the micelles may be transported intact into the epithelial cells and that the lipid digestion products may be removed intracellularly from the micelles. In either event, these products are used to *resynthesize* triglycerides and phospholipids within the epithelial cells. This is different from the absorption of amino acids and monosaccharides, which pass through the epithelial cells without being altered.

Triglycerides, phospholipids, and cholesterol are then combined with protein inside the epithelial cells to form small particles called **chylomicrons.** These tiny combinations of lipid and protein are secreted into the lymphatic capillaries of the intestinal villi (fig. 18.34). Absorbed lipids thus pass through the lymphatic system, eventually entering the venous blood by way of the thoracic duct. The absorption of lipids is thus significantly different from that of amino acids and monosaccharides, which enter the portal blood.

Once the chylomicrons are in the blood, their triglyceride content is removed by the enzyme **lipoprotein lipase,** which is attached to the endothelium of blood vessels. This enzyme hydrolyzes triglycerides and thus provides free fatty acids and glycerol for use by the tissue cells. The remaining *remnant particles,* containing cholesterol, are taken up by the liver; this is a process of endocytosis, which requires membrane receptors for the protein part (or *apoprotein*) of the remnant particle. Cholesterol and triglycerides produced by the liver are combined with other apoproteins and secreted into the blood as *very-low-density lipoproteins (VLDL),* which serve to deliver cholesterol to different organs. Excess cholesterol is returned from these organs to the liver attached to *high-density lipoproteins (HDL),* as described in chapter 13.

1. *List the enzymes involved in carbohydrate digestion, indicating their origin, sites of action, substrates, and products.*
2. *List the enzymes involved in protein digestion, indicating their origins, sites of action, and mode of action (endopeptidase or exopeptidase). Compare the characteristics of pepsin and trypsin.*
3. *Describe how bile aids both the digestion and absorption of fats. Explain how the absorption of fat differs from the absorption of amino acids and monosaccharides.*
4. *Trace the pathway and fate of a molecule of triglyceride and cholesterol in a chylomicron within an intestinal epithelial cell.*

Figure 18.34. Fatty acids and monoglycerides from the micelles within the small intestine are absorbed by epithelial cells and converted intracellularly into triglycerides. These are then combined with protein to form chylomicrons, which enter the lymphatic vessels (lacteals) of the villi. These lymphatic vessels transport the chylomicrons to the thoracic duct, which empties them into the venous blood.

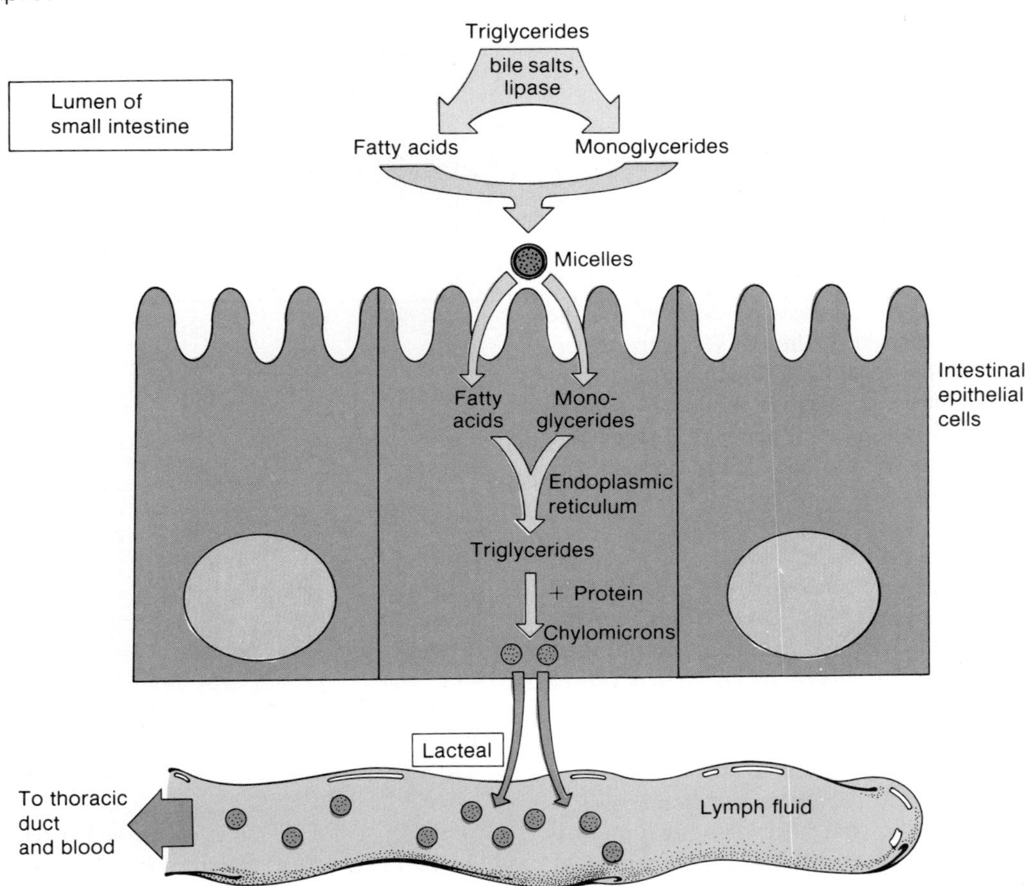

Neural and Endocrine Regulation of the Digestive System

The motility and glandular secretions of the GI tract are, to a large degree, automatic. Neural and endocrine control mechanisms, however, can stimulate or inhibit these automatic functions to help coordinate the different stages of digestion. The sight, smell, or taste of food, for example, can stimulate salivary and gastric secretions via activation of the vagus nerve, which helps to "prime" the digestive system in preparation for a meal. Stimulation of the vagus, in this case, originates in the brain and is a conditioned reflex (as Pavlov demonstrated by training dogs to salivate in response to a bell). The vagus nerve is also involved in the reflex control of one part of the digestive system by another—these are "short reflexes," which don't involve the brain.

The GI tract is both an endocrine gland and a target for the action of various hormones. Indeed, the first hormones to be discovered were gastrointestinal hormones. In 1902, Bayliss and Starling discovered that the duodenum produced a chemical regulator, which they named **secretin;** in 1905, these scientists proposed that secretin was but one of many yet undiscovered chemical regulators produced by the body. They coined the term *hormones* for this new class of regulators. Other investigators in 1905 discovered that an extract from the stomach antrum (pyloric region) stimulated gastric secretion. The hormone **gastrin** was thus the second hormone to be discovered.

Table 18.9 Summary of the physiological effects of gastrointestinal hormones.

Secreted by	Hormone	Effects
Stomach	Gastrin	Stimulates parietal cells to secrete HCl Stimulates chief cells to secrete pepsinogen Maintains structure of gastric mucosa
Small intestine	Secretin	Stimulates water and bicarbonate secretion in pancreatic juice Potentiates actions of cholecystokinin on pancreas
Small intestine	Cholecystokinin (CCK)	Stimulates contraction of the gallbladder Stimulates secretion of pancreatic juice enzymes Potentiates action of secretin on pancreas Maintains structure of exocrine pancreas (acini)
Small intestine	Gastric inhibitory peptide (GIP)	Inhibits gastric emptying Inhibits gastric acid secretion Stimulates secretion of insulin from endocrine pancreas (islets of Langerhans)

The chemical structures of gastrin, secretin, and the duodenal hormone **cholecystokinin** (*CCK*) were determined in the 1960s. More recently, a fourth hormone produced by the small intestine, **gastric inhibitory peptide** *(GIP),* has been added to the list of proven GI tract hormones. The effects of these hormones are summarized in table 18.9.

Regulation of Gastric Function

Gastric motility and secretion are, to some extent, automatic. Waves of contraction that serve to push chyme through the pyloric sphincter, for example, are initiated spontaneously by pacesetter cells in the greater curvature of the stomach. The secretion of hydrochloric acid (HCl) and pepsinogen, likewise, can be stimulated in the absence of neural and hormonal influences by the presence of cooked or partially digested protein in the stomach. The effects of autonomic nerves and hormones are superimposed on this automatic activity. The extrinsic control of gastric function is conveniently divided into three phases: (1) the cephalic phase; (2) the gastric phase; and (3) the intestinal phase. These are summarized in table 18.10.

Cephalic Phase. The cephalic phase of gastric regulation refers to control by the brain via the vagus nerve. As previously discussed, various conditioned stimuli can evoke gastric secretion. Activation of the vagus nerve can stimulate HCl and pepsinogen secretion by three mechanisms: (1) direct vagal stimulation of the gastric parietal and chief cells (the primary mechanism); (2) vagal stimulation of gastrin secretion by the G cells, which in turn stimulates the parietal and chief cells to secrete HCl and pepsinogen, respectively; and (3) stimulation of secretion by increases in gastric blood flow.

Table 18.10 The cephalic, gastric, and intestinal phases in the regulation of gastric acid secretion.

Phase of Regulation	Description
Cephalic phase	1. Sight, smell, and taste of food cause stimulation of vagus nuclei in brain 2. Vagus stimulates acid secretion a) Direct stimulation of parietal cells (major effect) b) Stimulation of gastrin secretion; gastrin stimulates acid secretion (lesser effect)
Gastric phase	1. Distension of stomach stimulates vagus nerve; vagus stimulates acid secretion 2. Amino acids and peptides in stomach lumen stimulate acid secretion a) Direct stimulation of parietal cells (lesser effect) b) Stimulation of gastrin secretion; gastrin stimulates acid secretion (major effect) 3. Gastrin secretion inhibited when pH of gastric juice falls below 2.5
Intestinal phase	1. Neural inhibition of gastric emptying and acid secretion a) Arrival of chyme in duodenum causes distension, increase in osmotic pressure b) These stimuli activate a neural reflex that inhibits gastric activity 2. Gastric inhibitory peptide (GIP) secreted by duodenum in response to fat in chyme; GIP inhibits gastric acid secretion

Gastric Phase. The presence of short polypeptides and amino acids in the stomach stimulates the G cells to secrete gastrin and the parietal and chief cells to secrete HCl and pepsinogen; this is called the gastric phase of regulation. Since gastrin also stimulates HCl and pepsinogen secretion, a *positive feedback mechanism* develops: as more HCl and pepsinogen are secreted, more short polypeptides and amino acids are released from the ingested protein, thus stimulating more secretion of gastrin and, therefore, more secretion of HCl and pepsinogen (fig. 18.35).

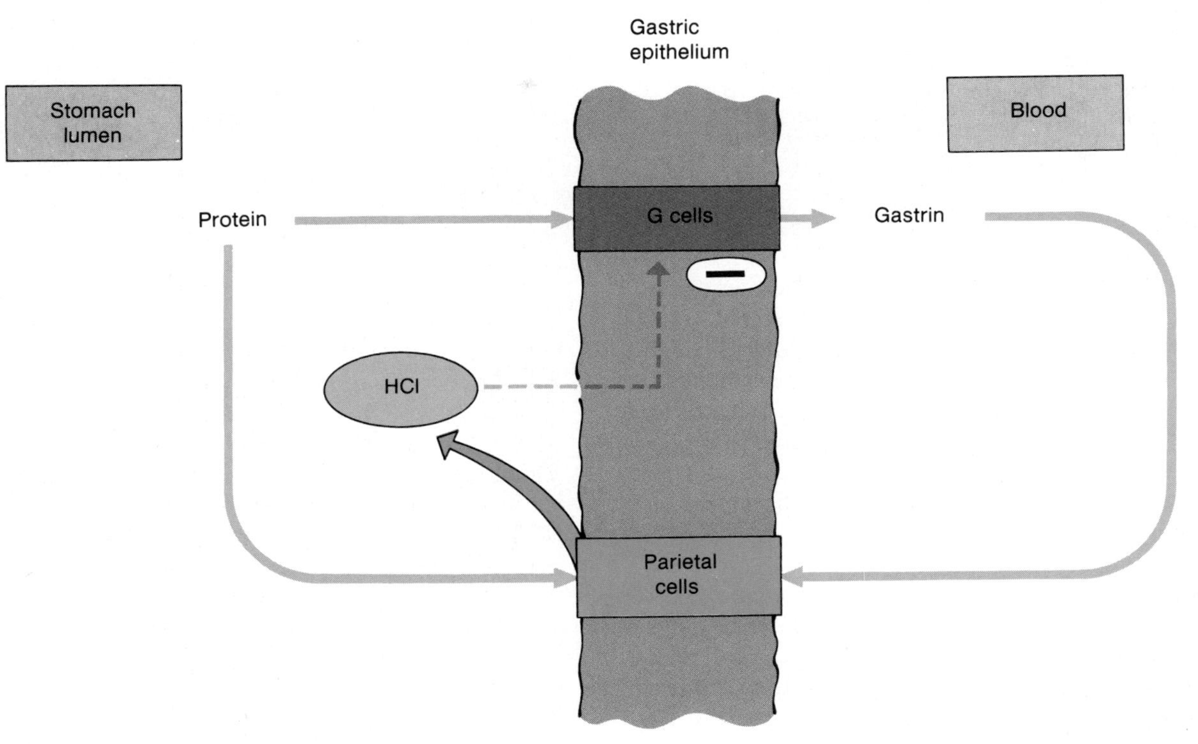

Figure 18.35. The stimulation of gastric acid (HCl) secretion by the presence of proteins in the stomach lumen and by the hormone gastrin. The secretion of gastrin is inhibited by gastric acidity. This forms a negative feedback loop.

Secretion of HCl during the gastric phase is also regulated by a *negative feedback mechanism.* As the pH of gastric juice drops, so does the secretion of gastrin—at a pH of 2.5 gastrin secretion is reduced, and at a pH of 1.0 gastrin secretion is totally abolished. The secretion of HCl thus declines accordingly. The presence of proteins and polypeptides in the stomach help to buffer the acid and thus to prevent a rapid fall in gastric pH; more acid can thus be secreted when proteins are present than when they are absent. The arrival of protein into the stomach thus stimulates acid secretion two ways—by the positive feedback mechanism previously discussed and by lessening of the negative feedback control of acid secretion. The amount of acid secreted is, by means of these effects, closely matched to the amount of protein ingested. As the stomach is emptied, the protein buffers leave, the pH thus falls, and the secretion of gastrin and HCl is accordingly inhibited.

Intestinal Phase. The intestinal phase of gastric regulation refers to the inhibition of gastric activity when chyme enters the small intestine. Investigators in 1886 demonstrated that the addition of olive oil to a meal inhibits gastric emptying, and in 1929 it was shown that the presence of fat inhibits gastric juice secretion. This inhibitory intestinal phase of gastric regulation is due to both a neural reflex originating from the duodenum and to a chemical hormone secreted by the duodenum.

The arrival of chyme into the duodenum increases its osmolality. This stimulus, together with stretch of the duodenum and possibly other stimuli, produces a neural reflex that results in the inhibition of gastric motility and secretion. The presence of fat in the chyme also stimulates the duodenum to secrete a hormone that inhibits gastric function. This inhibitory hormone is believed to be gastric inhibitory peptide (GIP).

The inhibitory neural and endocrine mechanisms during the intestinal phase prevent the further passage of chyme from the stomach to the duodenum. This gives the duodenum time to process the load of chyme that it has previously received. Since GIP is stimulated by fat in the chyme, a breakfast of bacon and eggs takes longer to pass through the stomach—and makes one feel "fuller" for a longer time—than does a breakfast of pancakes and syrup.

Table 18.11 Regulation of pancreatic juice and bile secretion by the hormones *secretin* and *cholecystokinin* (CCK) and by the neurotransmitter *acetylcholine,* released from parasympathetic nerve endings.

Description	Secretin	CCK	Acetylcholine (Vagus Nerve)
Stimulus for release	Acidity of chyme decreases duodenal pH below 4.5	Fat and protein in chyme	Sight, smell of food; distension of stomach
Second messenger	Cyclic AMP	Ca^{++}	Ca^{++}
Effect on pancreatic juice	Stimulates water and bicarbonate secretion; potentiates action of CCK	Stimulates enzyme secretion; potentiates action of secretin	Stimulates enzyme secretion
Effect on bile	Stimulates secretion	Potentiates action of secretin; stimulates contraction of gallbladder	Stimulates contraction of gallbladder

Regulation of Pancreatic Juice and Bile Secretion

The arrival of chyme into the duodenum stimulates the intestinal phase of gastric regulation and, at the same time, stimulates reflex secretion of pancreatic juice and bile. The entry of new chyme is thus retarded as the previous load is digested. The secretion of pancreatic juice and bile is stimulated by both neural reflexes initiated in the duodenum and by secretion of the duodenal hormones cholecystokinin (CCK) and secretin.

Pancreatic Juice. The secretion of pancreatic juice is stimulated by both secretin and CCK. The release of secretin occurs in response to a fall in duodenal pH below 4.5; this pH fall occurs for only a short time, however, because the acidic chyme is rapidly neutralized by alkaline pancreatic juice. The secretion of CCK occurs in response to the fat content of chyme in the duodenum.

Secretin stimulates the production of bicarbonate by the pancreas. Since bicarbonate neutralizes the acidic chyme and since secretin is released in response to the low pH of chyme, this completes a negative feedback loop where the effects of secretin inhibit its secretion. Cholecystokinin, in contrast, stimulates the production of pancreatic enzymes such as trypsin, lipase, and amylase. Secretin and CCK can have different effects on the same cells (the pancreatic acinar cells) because their actions are mediated by different intracellular compounds that act as second messengers. The second messenger of secretin action is cyclic AMP, whereas the second messenger for CCK is Ca^{++} (table 18.11).

Secretion of Bile. The liver secretes bile continuously, but this secretion is greatly augmented following a meal. This increased secretion is due to the release of secretin and CCK from the duodenum. Secretin is the major stimulator of bile secretion by the liver, and CCK enhances this effect. The arrival of chyme in the duodenum also causes the gallbladder to contract and eject bile. Contraction of the gallbladder occurs in response to neural reflexes from the duodenum and in response to stimulation by CCK.

Trophic Effects of Gastrointestinal Hormones

Patients with tumors of the stomach pylorus have high acid secretion and hyperplasia (growth) of the gastric mucosa. Surgical removal of the pylorus reduces gastric secretion and prevents growth of the gastric mucosa. Patients with peptic ulcers are sometimes treated by vagotomy—cutting of the vagus nerve. Vagotomy also reduces acid secretion but has no effect on the gastric mucosa. These observations suggest that the hormone gastrin, secreted by the pyloric mucosa, may exert stimulatory, or *trophic,* effects on the gastric mucosa. The structure of the gastric mucosa, in other words, is dependent on the effects of gastrin.

In the same way, the structure of the acinar (exocrine) cells of the pancreas is dependent upon the trophic effects of CCK. Perhaps this explains why the pancreas, as well as the stomach, atrophies during starvation. Since neural reflexes appear to be capable of regulating digestion, perhaps the primary function of the GI hormones is trophic—that is, maintenance of the structure of their target organs.

---

1. *Describe the positive and negative feedback mechanisms that operate during the gastric phase of HCl and pepsinogen secretion.*
2. *Describe the mechanisms involved in the intestinal phase of gastric regulation, and explain why a fatty meal takes longer to leave the stomach than a meal low in fat.*
3. *Explain the hormonal mechanisms involved in the production and release of pancreatic juice and bile.*

Summary

Introduction to the Digestive System p. 552

I. The digestion of food molecules involves the hydrolysis of these molecules into their subunits.
   A. The digestion of food occurs in the lumen of the GI tract and is catalyzed by specific enzymes.
   B. The digestion products are absorbed through the intestinal mucosa and enter the blood or lymph.

II. The layers (tunics) of the GI tract are, from the inside outward, mucosa, submucosa, muscularis, and serosa.
   A. The mucosa consists of a simple columnar epithelium, a thin layer of connective tissue, called the lamina propria, and a thin layer of smooth muscle, called muscularis mucosa.
   B. The submucosa is composed of connective tissue, the muscularis consists of layers of smooth muscles, and the serosa is connective tissue covered with the visceral peritoneum.
   C. The submucosa contains the submucosal plexus, and the muscularis contains the myenteric plexus of autonomic nerves.

Esophagus and Stomach p. 555

I. Peristaltic waves of contraction push food through the lower esophageal sphincter into the stomach.

II. Swallowing, or deglutition, occurs in three phases and involves structures of the buccal cavity, pharynx, and esophagus.

III. The stomach consists of a cardia, fundus, body, and pyloris (antrum), which ends with the pyloric sphincter.
   A. The lining of the stomach is thrown into folds, or rugae, and the mucosa is formed into gastric pits and gastric glands.
   B. The parietal cells of the gastric glands secrete HCl, and the chief cells secrete pepsinogen.
   C. In the acidic environment of gastric juice, pepsinogen is converted into the active protein-digesting enzyme called pepsin.
   D. Some digestion of protein occurs in the stomach; the most important function of the stomach is the secretion of intrinsic factor, which is needed for the absorption of vitamin B_{12} in the intestine.

Small Intestine p. 560

I. Regions of the small intestine include the duodenum, jejunum, and ileum; the common bile duct and pancreatic duct empty into the duodenum.

II. Fingerlike extensions of mucosa called villi project into the lumen, and at the bases of the villi the mucosa forms narrow pouches called the crypts of Lieberkühn.
   A. New epithelial cells are formed in the crypts.
   B. The membrane of intestinal epithelial cells is folded to form microvilli; this is called the brush border of the mucosa and serves to increase surface area.

III. Digestive enzymes, called brush border enzymes, are located in the membranes of the microvilli.

IV. The small intestine exhibits two major types of movements—peristalsis and segmentation.

Large Intestine p. 565

I. The large intestine is divided into the cecum, colon, rectum, and anal canal.
   A. The vermiform appendix is attached to the inferior medial margin of the cecum.
   B. The colon consists of ascending, transverse, descending, and sigmoid portions.
   C. Bulges in the walls of the large intestine are called haustra.

II. Three types of movements occur in the large intestine: peristalsis, haustral churning, and mass movement.

III. The large intestine absorbs water and electrolytes.
   A. Although most of the water that enters the GI tract is absorbed in the small intestine, about 1–1.5 L/day pass to the large intestine, which absorbs about 90% of this amount.
   B. Na^+ is actively absorbed and water follows passively, in a manner analagous to the reabsorption of NaCl and water in the renal tubules.

IV. Defecation occurs when the anal sphincters relax and contraction of other muscles raises the rectal pressure.

Liver, Gallbladder, and Pancreas p. 568

I. The liver, the largest internal organ, is composed of functional units called lobules.
   A. Liver lobules consist of plates of hepatic cells separated by capillary sinusoids.
   B. Blood flows from the periphery of each lobule, where the hepatic artery and portal vein empty through the sinusoids and out the central vein.
   C. Bile flows within the hepatocyte plates, in canaliculi, to the bile ducts.
   D. Substances excreted in the bile can be returned to the liver in the hepatic portal blood; this is called an enterohepatic circulation.
   E. Bile consists of a pigment called bilirubin, bile salts, cholesterol, and other molecules.
   F. The liver detoxifies the blood by excreting substances in the bile, by phagocytosis, and by chemical inactivation.
   G. The liver modifies the plasma concentrations of proteins, glucose, triglycerides, and ketone bodies.

II. The gallbladder serves to store and concentrate the bile, and it releases bile through the cystic duct and common bile duct to the duodenum.

III. The pancreas is both an exocrine and an endocrine gland.
   A. The endocrine portion is known as the islets of Langerhans and secretes the hormones insulin and glucagon.
   B. The exocrine acini of the pancreas produce pancreatic juice, which contains various digestive enzymes and bicarbonate.

Digestion and Absorption of Carbohydrates, Lipids, and Proteins p. 576

I. The digestion of starch begins in the mouth through the action of salivary amylase.
   A. Pancreatic amylase digests starch into disaccharides and short-chain oligosaccharides.
   B. Complete digestion into monosaccharides is accomplished by brush border enzymes.

II. Protein digestion begins in the stomach by the action of pepsin.
   A. Pancreatic juice contains protein-digesting enzymes, including trypsin, chymotrypsin, and others.
   B. The brush border contains digestive enzymes that help to complete the digestion of proteins into amino acids.
   C. Amino acids, like monosaccharides, are absorbed and secreted into capillary blood entering the portal vein.
III. Lipids are digested in the small intestine after being emulsified by bile salts.
   A. Free fatty acids and monoglycerides enter particles called micelles, formed in large part by bile salts, and in this form, or as free molecules, they are absorbed.
   B. Once inside the mucosal epithelial cells, these subunits are used to resynthesize triglycerides.
   C. Triglycerides in the epithelial cells, together with proteins, form chylomicrons, which are secreted into the central lacteals of the villi.
   D. Chylomicrons are transported by lymph to the thoracic duct and there enter the blood.

Neural and Endocrine Regulation of the Digestive System p. 581

I. The regulation of gastric function occurs in three phases.
   A. In the cephalic phase, the activity of higher brain centers, acting via the vagus nerve, stimulates gastric juice secretion.
   B. In the gastric phase, the secretion of HCl and pepsin is controlled by the gastric contents and by the hormone gastrin, secreted by the gastric mucosa.
   C. In the intestinal phase, the activity of the stomach is inhibited by neural reflexes from the duodenum and by gastric inhibitory peptide (GIP), secreted by the duodenum.
II. The secretion of the hormones secretin and cholecystokinin (CCK) regulate pancreatic juice and bile secretion.
   A. Secretin secretion is stimulated by the arrival of acidic chyme into the duodenum.
   B. CCK secretion is stimulated by the presence of fat in the chyme arriving in the duodenum.
   C. Contraction of the gallbladder occurs in response to a neural reflex and to the secretion of CCK by the duodenum.
III. Gastrointestinal hormones may be needed for the maintenance of the GI tract and accessory digestive organs.

Review Activities

Objective Questions

1. Intrinsic factor
   (a) is secreted by the stomach
   (b) is a polypeptide
   (c) promotes absorption of vitamin B_{12} in the intestine
   (d) helps prevent pernicious anemia
   (e) all of the above
2. Intestinal enzymes such as lactase are
   (a) secreted by the intestine into the chyme
   (b) produced by the crypts of Lieberkühn
   (c) produced by the pancreas
   (d) attached to the cell membrane of microvilli in the epithelial cells of the mucosa
3. Which of the following statements about gastric secretion of HCl is *false?*
   (a) HCl is secreted by parietal cells.
   (b) HCl hydrolyzes peptide bonds.
   (c) HCl is needed for the conversion of pepsinogen to pepsin.
   (d) HCl is needed for maximum activity of pepsin.
4. Most digestion occurs in the
   (a) mouth
   (b) stomach
   (c) small intestine
   (d) large intestine
5. Which of the following statements about trypsin is true?
   (a) Trypsin is derived from trypsinogen by the digestive action of pepsin.
   (b) Active trypsin is secreted into the pancreatic acini.
   (c) Trypsin is produced in the crypts of Lieberkühn.
   (d) Trypsinogen is converted to trypsin by the brush border enzyme enterokinase.
6. During the gastric phase, the secretion of HCl and pepsinogen is stimulated by
   (a) vagus nerve stimulation that originates in the brain
   (b) polypeptides in the gastric lumen and by gastrin secretion
   (c) secretin and cholecystokinin from the duodenum
   (d) all of the above
7. The secretion of HCl by the stomach mucosa is inhibited by
   (a) neural reflexes from the duodenum
   (b) the secretion of gastric inhibitory peptide from the duodenum
   (c) the lowering of gastric pH
   (d) all of the above
8. The first organ to receive the blood-borne products of digestion is the
   (a) liver
   (b) pancreas
   (c) heart
   (d) brain
9. Which of the following statements about hepatic portal blood is true?
   (a) It contains absorbed fat.
   (b) It contains ingested proteins.
   (c) It is mixed with bile in the liver.
   (d) It is mixed with blood from the hepatic artery in the liver.

Essay Questions

1. Explain how the gastric secretion of HCl and pepsin is regulated during the cephalic, gastric, and intestinal phases.
2. Describe how pancreatic enzymes become activated in the lumen of the intestine, and explain the need for these mechanisms.
3. What is the function of bicarbonate in pancreatic juice? Explain why ulcers are more likely to be located in the duodenum than in the stomach.
4. Explain why the pancreas is considered to be both an exocrine and an endocrine gland. Given this information, predict what effects tying of the pancreatic duct would have on pancreatic structure and function.
5. Explain how jaundice is produced when (*a*) the person has gallstones, (*b*) the person has a high rate of red blood cell destruction, and (*c*) the person has liver disease. In which of these cases would phototherapy for the jaundice be effective? Explain.

Selected Readings

Binder, H. J. 1984. The pathophysiology of diarrhea. *Hospital Practice* 19:107.

Bleich, H. L., and E. S. Boro. 1979. Protein digestion and absorption. *New England Journal of Medicine* 300:659.

Bortoff, A. 1972. Digestion. *Annual Review of Physiology* 28:201.

Carey, M. C., D. M. Small, and C. M. Bliss. 1983. Lipid digestion and absorption. *Annual Review of Physiology* 45:651.

Chou, C. C. 1982. Relationship between intestinal blood flow and motility. *Annual Review of Physiology* 44:29.

Christensen, R. R. 1971. The controls of gastrointestinal movements; Some old and new views. *New England Journal of Medicine* 285:483.

Cohen, S. 1983. Neuromuscular disorders of the gastrointestinal tract. *Hospital Practice* 18:121.

Davenport, H. W. 1982. *Physiology of the digestive tract.* 5th ed. Chicago: Year Book Medical Publishers.

Dockray, G. J. 1979. Comparative biochemistry and physiology of gut hormones. *Annual Review of Physiology* 41:83.

Freeman, H. J., and Y. S. Kim. 1978. Digestion and absorption of proteins. *Annual Review of Physiology* 29:99.

Gardner, J. D., and R. T. Jensen. 1986. Receptors and cell activation associated with pancreatic enzyme secretion. *Annual Review of Physiology* 48:103.

Gray, G. M. 1975. Carbohydrate digestion and absorption: Role of the small intestine. *New England Journal of Medicine* 292:1225.

Gollan, J. L., and A. B. Knapp. 1985. Bilirubin metabolism and congenital jaundice. *Hospital Practice* 20:83.

Grossman, M. I. 1979. Neural and hormonal regulation of gastrointestinal function: An overview. *Annual Review of Physiology* 41:27.

Guth, P. H. 1982. Stomach blood flow and acid secretion. *Annual Review of Physiology* 44:3.

Hersey, S. J., S. H. Norris, and A. J. Gilbert. 1984. Cellular control of pepsinogen secretion. *Annual Review of Physiology* 46:393.

Holt, K. M., and J. I. Isenberg. 1985. Peptic ulcer disease: Physiology and pathophysiology. *Hospital Practice* 20:89.

Kappas, A. and A. P. Alvarez. June 1975. How the liver metabolizes foreign substances. *Scientific American.*

McGuigan, J. E. 1978. Gastrointestinal hormones. *Annual Review of Physiology* 29:99.

Moog, F. November 1981. The lining of the small intestine. *Scientific American.*

Salen, G. and S. Shefer. 1983. Bile acid synthesis. *Annual Review of Physiology* 45:679.

Sanders, M. J., and A. H. Soll. 1986. Characterization of receptors regulating secretory function in the fundic mucosa. *Annual Review of Physiology* 48:89.

Smith, B. F., and T. Lamont. 1984. The pathogenesis of gallstones. *Hospital Practice* 19:93.

Soll, A. and J. H. Walsh. 1979. Regulation of gastric acid secretion. *Annual Review of Physiology* 41:35.

Walsh, J. H. and M. I. Grossman. 1975. Gastrin. *New England Journal of Medicine* 292: first part, p. 1324; second part, p. 1377.

Weisbrodt, N. W. 1981. Patterns of intestinal motility. *Annual Review of Physiology* 43:21.

Williams, J. A. 1984. Regulatory mechanisms in pancreas and salivary acini. *Annual Review of Physiology* 46:361.

19 Regulation of Metabolism

Objectives

By studying this chapter, you should be able to

1. explain the requirements for adequate amounts of vitamins, minerals, essential amino acids and fatty acids, and calories in the diet
2. define the terms *energy reserves* and *circulating energy substrates* and describe how these interact during anabolism and catabolism
3. describe the regulation of eating and describe in general terms the endocrine control of metabolism
4. describe the actions of insulin and glucagon and explain how the secretion of these hormones is regulated
5. explain how insulin and glucagon regulate metabolism during feeding and fasting
6. explain the causes and symptoms of type I and type II diabetes mellitus and of reactive hypoglycemia
7. describe the metabolic effects of epinephrine and the glucocorticoids
8. describe the effects of thyroxine on cell respiration and the relationship between thyroxine levels and the basal metabolic rate
9. explain the causes and symptoms of hypothyroidism and hyperthyroidism in terms of the actions of thyroid hormones
10. describe the metabolic effects of growth hormone and the body's requirements for growth hormone and thyroxine for proper growth
11. describe the actions of parathyroid hormone, 1,25-dihydroxyvitamin D_3, and calcitonin and explain how the secretion of these hormones is regulated
12. describe how 1,25-dihydroxyvitamin D_3 is produced and explain why this compound is needed to prevent osteomalacia and rickets

Outline

Table 19.1 Recommended daily allowances for vitamins and elements.

	Infants	Children	Adolescents (15–18 yrs)		Adults (23–50 yrs)	
	0–6 mos	4–6 yrs	Males	Females	Males	Females
Weight, kg (lb.)	6 (13)	20 (44)	66 (145)	55 (120)	70 (154)	55 (120)
Height, cm (in.)	60 (24)	112 (44)	176 (69)	163 (64)	178 (70)	163 (64)
Protein, g	kg × 2.2	30	56	46	56	44
Fat-soluble vitamins						
Vitamin A, μg	420	500	1000	800	1000	800
Vitamin D, μg	10	10	10	10	5	5
Vitamin E activity, mg	3	6	10	8	10	8
Water-soluble vitamins						
Ascorbic acid, mg	35	45	60	60	60	60
Folic acid, μg	30	200	400	400	400	400
Niacin, mg	6	11	18	14	18	13
Riboflavin, mg	0.4	1.0	1.7	1.3	1.6	1.2
Thiamine, mg	0.3	0.9	1.4	1.1	1.4	1.0
Vitamin B_6, mg	0.3	1.3	2.0	2.0	2.2	2.0
Vitamin B_{12}, μg	0.5	2.5	3.0	3.0	3.0	3.0
Elements						
Calcium, mg	360	800	1200	1200	800	800
Phosphorus, mg	240	800	1200	1200	800	800
Iodine, μg	40	90	150	150	150	150
Iron, mg	10	10	18	18	10	18
Magnesium, mg	50	200	400	300	350	300
Zinc, mg	3	10	15	15	15	15

Source: Food and Nutrition Board, *Recommended Dietary Allowances,* 9th ed., National Academy of Sciences—National Research Council, Washington, D.C., 1980.

Nutritional Requirements

Vitamins and minerals are needed in the diet for enzymatic reactions to proceed, and the essential amino acids and fatty acids are required for the metabolism of protein and fat. The energy needs of the body are met by the caloric value of food.

Living tissue is maintained by the constant expenditure of energy. This energy is obtained directly from ATP and indirectly from the cell respiration of glucose, fatty acids, ketone bodies, amino acids, and other organic molecules. These molecules are ultimately obtained from food, but they can also be obtained from the glycogen, fat, and protein stored in the body. The energy value of food is commonly measured in *kilocalories,* which are also called "big calories" and spelled with a capital letter (*C*alories).

In addition to its caloric value, food also supplies the essential amino acids and fatty acids. The eight **essential amino acids** are lysine, methionine, valine, leucine, isoleucine, tryptophane, phenylalanine, and threonine. The **essential fatty acids** are linoleic acid and linolenic acid. These molecules are termed *essential* because the body cannot make them and thus is forced to obtain them in the diet for proper protein and fat synthesis.

Vitamins and Elements

Vitamins are small organic molecules that cannot be made by the body and that serve as coenzymes in metabolic reactions. There are two groups of vitamins—the fat-soluble vitamins (A, D, E, and K) and the water-soluble vitamins. Water-soluble vitamins include thiamine (B_1), riboflavin (B_2), niacin (B_3), pyridoxine (B_6), pantothenic acid, biotin, folic acid, B_{12}, and vitamin C (ascorbic acid). Recommended daily allowances for these vitamins are shown in table 19.1.

Many of the water-soluble vitamins serve as coenzymes in the metabolism of carbohydrates, lipids, and proteins. Thiamine, for example, is needed for the activity of the enzyme that converts pyruvic acid to acetyl coenzyme A. Riboflavin and niacin are needed for the production of FAD and NAD, respectively; these latter compounds serve as coenzymes that transfer hydrogens during cell respiration. Pyridoxine is a cofactor for the enzymes involved in amino acid metabolism. Deficiencies of the water-soluble vitamins can, for obvious reasons, have widespread effects in the body.

Many fat-soluble vitamins have highly specialized functions. Vitamin K, for example, is required for the production of prothrombin and for clotting factors VII, IX, and X. Vitamin D is converted into a hormone that participates in the regulation of calcium balance. The visual

Table 19.2 Recommended daily intake of trace elements.

Element	Safe and Adequate Intake (mg/day)
Iron (males)	10
Iron (females)	18
Zinc	15
Manganese	2.5 to 5.0
Fluorine	1.5 to 4.0
Copper	2.0 to 3.0
Molybdenum	0.15 to 0.5
Chromium	0.05 to 0.2
Selenium	0.05 to 0.2
Iodine	0.15

Source: From the Food and Nutrition Board of the National Academy of Science, 1980.

pigments in the rods and cones of the retina are derived from vitamin A. Vitamin A and related compounds, called *retinoids,* also have effects on genetic expression in epithelial cells; these compounds are now used clinically in the treatment of some skin conditions, and researchers are attempting to derive related compounds that may aid the treatment of some cancers.

Elements are needed as cofactors for specific enzymes and for a wide variety of other critical functions. Elements that are required in relatively large amounts per day include sodium, potassium, magnesium, calcium, phosphorus, and chlorine (table 19.1). In addition, the following **trace elements** are recognized as essential: iron, zinc, manganese, fluorine, copper, molybdenum, chromium, and selenium. These must be ingested in amounts ranging from 50 micrograms to 18 milligrams per day (table 19.2).

Caloric Requirements

There are tremendous variations in the energy requirements of people. These are partly due to differences in "fuel efficiency," so that some people consume more calories than others do at comparable levels of physical activity. This "fuel efficiency" is regulated in part by the thyroid gland. The differences in daily energy requirements among people, however, are largely due to differences in physical activity.

Average daily energy expenditures may range from about 1,300 to 5,000 kilocalories per day. The average values for people not engaged in heavy manual labor but who are active during their leisure time is about 2,900 kilocalories per day for a man and 2,100 kilocalories per day for a woman. People engaged in office work, the professions, sales, and comparable occupations consume approximately 5 kilocalories per minute during work. More physically demanding occupations may require energy expenditures of 7.5 to 10 kilocalories per minute.

When the caloric intake is greater than the energy expenditures, excess calories are stored primarily as fat. This is true regardless of the source of the calories—carbohydrates, protein, or fat—because these molecules can be converted to fat by the metabolic pathways described in chapter 5. Appropriate body weights are indicated in table 19.3.

Table 19.3 Desirable body weights, according to sex, age, height, and body frame.

	Height Feet	Height Inches	Small Frame	Medium Frame	Large Frame
Men	5	2	128–134	131–141	138–150
	5	3	130–136	133–143	140–153
	5	4	132–138	135–145	142–156
	5	5	134–140	137–148	144–160
	5	6	136–142	139–151	146–164
	5	7	138–145	142–154	149–168
	5	8	140–148	145–157	152–172
	5	9	142–151	148–160	155–176
	5	10	144–154	151–163	158–180
	5	11	146–157	154–166	161–184
	6	0	149–160	157–170	164–188
	6	1	152–164	160–174	168–192
	6	2	155–168	164–178	172–197
	6	3	158–172	167–182	176–202
	6	4	162–176	171–187	181–207

Weights at ages 25–59 based on lowest mortality. Weight in pounds according to frame (in indoor clothing weighing 5 lbs., shoes with 1" heels).

	Height Feet	Height Inches	Small Frame	Medium Frame	Large Frame
Women	4	10	102–111	109–121	118–131
	4	11	103–113	111–123	120–134
	5	0	104–115	113–126	122–137
	5	1	106–118	115–129	125–140
	5	2	108–121	118–132	128–143
	5	3	111–124	121–135	131–147
	5	4	114–127	124–138	134–151
	5	5	117–130	127–141	137–155
	5	6	120–133	130–144	140–159
	5	7	123–136	133–147	143–163
	5	8	126–139	136–150	146–167
	5	9	129–142	139–153	149–170
	5	10	132–145	142–156	152–173
	5	11	135–148	145–159	155–176
	6	0	138–151	148–162	158–179

Weights at ages 25–59 based on lowest mortality. Weight in pounds according to frame (in indoor clothing weighing 3 lbs., shoes with 1" heels).

Source: Metropolitan Insurance Companies.

Obesity is a risk factor for cardiovascular diseases, renal disease, diabetes mellitus, gallbladder disease, the development of kidney stones, and some malignancies (particularly endometrial and breast cancer). Obesity in childhood is due to an increase in both the size and number of adipose cells; weight gain in adulthood is due mainly to an increase in adipose cell size, although the number of these cells may also increase in extreme weight gains. When weight is lost, the size of the adipose cells decreases, but the number of adipose cells does not decrease. It is thus important to prevent further increases in weight in all overweight people but particularly so in children.

The degree of obesity can be most conveniently determined by the **body mass index.** This is equal to the nude body weight (in kilograms) divided by the square of the barefoot height (in meters). For example, a person who weighs 170 pounds (77.1 kg) and is five-feet-nine-inches tall (1.75 m), has a body mass index of $77.1 \div 1.75^2 = 25.1$. The normal body mass index is 20–25 for males and 19–24 for females. Statistically, the morbidity and mortality rates are significantly increased when the body mass index is greater than 30. When the body mass index is over 40, the risk factor for cardiovascular disease due to obesity is comparable to risk factors due to smoking, hypertension, or hyperlipidemia.

1. *List the fat-soluble vitamins, and describe some of their functions.*
2. *Explain the roles of vitamins B_1, B_2, and B_3 in energy metabolism.*
3. *List the essential amino acids and fatty acids, and explain their significance.*
4. *In a single sentence, completely explain how a person can gain fat. Describe the effects of weight gain and loss on adipose cells.*

Regulation of Energy Metabolism

Food and stored energy reserves in the body can provide circulating energy substrates for the tissue cells. The utilization and storage of energy is regulated by a variety of hormones.

The term **metabolism** refers to all of the chemical changes that occur within the cells of the body. On the basis of energy flow, these reactions comprise two categories: **anabolism** and **catabolism.** Anabolic reactions require the input of energy (obtained from the hydrolysis of ATP) and result in the formation of large, energy-rich molecules such as triglycerides, glycogen, and protein. Catabolism refers to the hydrolysis of these molecules into their subunits and to the use of these subunits in cell respiration for the generation of energy used to make ATP.

The molecules that can be oxidized for energy by the processes of cell respiration may be derived from the **energy reserves** of glycogen, fat, or protein in the body, or they can be derived from the products of digestion that are absorbed through the small intestine. Since these molecules—glucose, fatty acids, amino acids, and others—are carried by the blood to the tissue cells for use in cell respiration, they can be called **circulating energy substrates** (fig. 19.1).

Because of differences in cellular enzyme content, different organs have different *preferred energy sources.* The brain has an almost absolute requirement for blood glucose as its energy source, for example. A fall in the plasma concentration of glucose below about 50 mg per 100 ml can thus "starve" the brain and have disastrous consequences. Resting skeletal muscles, in contrast, use fatty acids as their preferred energy source. Similarly, ketone bodies, lactic acid, and amino acids can be used to different degrees as energy sources by various organs. The plasma normally contains adequate concentrations of all of these circulating energy substrates to meet the energy needs of the body.

Eating

Ideally, one should eat the kinds and amounts of foods that provide adequate vitamins, minerals, essential amino acids and fatty acids, and calories. Proper caloric intake is that which maintains energy reserves (primarily fat and glycogen) and results in a body weight within an optimum range for health.

There is a tendency for body weight to be stable despite short-term changes in caloric intake. It has thus been proposed that there may be some mechanism that is sensitive to the amount of body fat. Although this mechanism is not known, it is clear that there is a relationship between body fat and endocrine function. The secretion of anterior pituitary hormones is affected in a variety of ways. Obese women, for example, may experience menstrual cycle abnormalities and hirsutism (hairiness), whereas the sudden weight loss seen in women with *anorexia nervosa* may produce amenorrhea (cessation of the menstrual cycle). Abnormalities in growth hormone, ACTH, and prolactin secretion have also been observed in obese people.

Eating behavior appears to be at least partially controlled by areas of the hypothalamus. Lesions (destruction) in the ventromedial area of the hypothalamus produce *hyperphagia,* or overeating, and obesity in experimental animals. Lesions of the lateral hypothalamus, in contrast, produce *hypophagia* and weight loss.

The chemical neurotransmitters that may be involved in neural pathways mediating eating behavior are being investigated. There is evidence, for example, that endorphins may be involved, because injections of naloxone (a morphine-blocking drug) suppresses overeating in rats.

Figure 19.1. A schematic flow chart of energy pathways in the body.

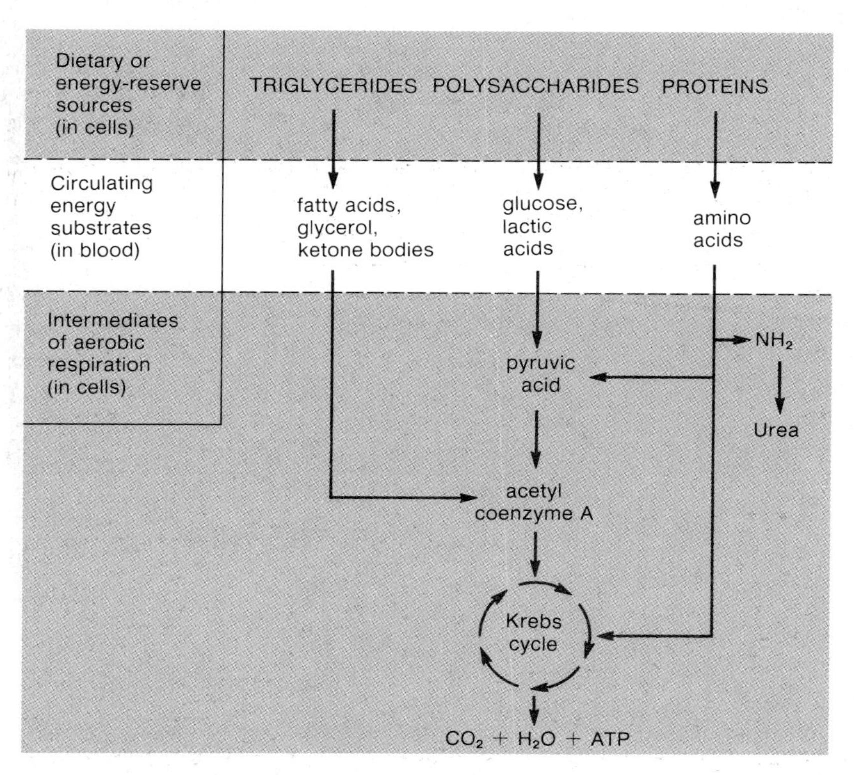

There is also evidence that the neurotransmitters norepinephrine and serotonin may be involved; injections of norepinephrine into the brain of rats cause overeating, whereas injections of serotonin have the opposite effect. Interestingly, the intestinal hormone cholecystokinin (CCK) also appears to function as a neurotransmitter in the brain, and it has been shown that injections of CCK cause experimental animals to stop eating. More research is needed to elucidate the structure and processes involved in the regulation of eating.

Hormonal Regulation of Metabolism

The absorption of energy carriers from the intestine is not continuous; it rises to high levels following meals and tapers toward zero between meals. Despite this, the plasma concentration of glucose and other energy substrates does not remain high during periods of absorption and does not normally fall below a certain level during periods of fasting. During the absorption of digestion products from the intestine, energy substrates are removed from the blood and deposited as energy reserves from which withdrawals can be made during times of fasting (fig. 19.2). This assures that there will be an adequate plasma concentration of energy substrates to sustain tissue metabolism at all times.

Figure 19.2. The balance of metabolism can be tilted toward anabolism (synthesis of energy reserves) or catabolism (utilization of energy reserves) by the combined actions of various hormones. Growth hormone and thyroxine have both anabolic and catabolic effects.

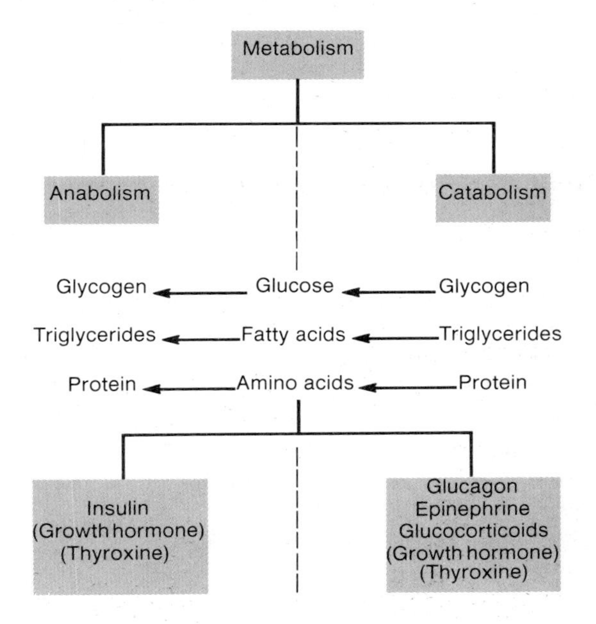

Table 19.4 Summary of the endocrine regulation of metabolism.

Hormone	Blood Glucose	Carbohydrate Metabolism	Protein Metabolism	Lipid Metabolism
Insulin	Decreased	↑Glycogen formation ↓Glycogenolysis ↓Gluconeogenesis	↑Amino acid transport	↑Lipogenesis ↓Lipolysis ↓Ketogenesis
Glucagon	Increased	↓Glycogen formation ↑Glycogenolysis ↑Gluconeogenesis	No direct effect	↑Lipolysis ↑Ketogenesis
Growth hormone	Increased	↑Glycogen formation ↑Gluconeogenesis ↓Glucose utilization	↑Protein synthesis	↓Lipogenesis ↑Lipolysis ↑Ketogenesis
Glucocorticoids	Increased	↑Glycogen formation ↑Gluconeogenesis	↓Protein synthesis	↓Lipogenesis ↑Lipolysis ↑Ketogenesis
Epinephrine	Increased	↓Glycogen formation ↑Glycogenolysis ↑Gluconeogenesis	No direct effect	↑Lipolysis ↑Ketogenesis
Thyroxine	No effect	↑Glucose utilization	↑Protein synthesis	No direct effect

The rate of deposit and withdrawal of energy substrates into and from the energy reserves and the conversion of one type of energy substrate into another are regulated by the actions of hormones. The balance between anabolism and catabolism is determined by the antagonistic effects of hormones such as insulin, glucagon, growth hormone, thyroxine, and others (fig. 19.2). The specific metabolic effects of these hormones are summarized in table 19.4.

1. *Define the terms* energy reserves *and* circulating energy carriers, *and give examples.*
2. *Describe the structures and neurotransmitters that may be involved in the regulation of eating.*
3. *Which hormones promote an increase in blood glucose? Which promote a decrease? List the hormones that stimulate fat synthesis (lipogenesis) and fat breakdown (lipolysis).*

Energy Regulation by the Islets of Langerhans

Scattered within a "sea" of pancreatic exocrine tissue (the acini) are islands of hormone-secreting cells. These **islets of Langerhans** contain three distinct cell types that secrete different hormones (fig. 19.3). The most numerous are the *beta cells;* these cells comprise 60% of each islet and secrete the hormone **insulin.** The *alpha cells* comprise about 25% of each islet and secrete the hormone **glucagon.** The least numerous cell type, the *delta cells,* produce **somatostatin.** The latter hormone is identical to the somatostatin produced by the hypothalamus, which acts to inhibit growth-hormone secretion from the pituitary.

Figure 19.3. The cellular composition of a normal islet of Langerhans.

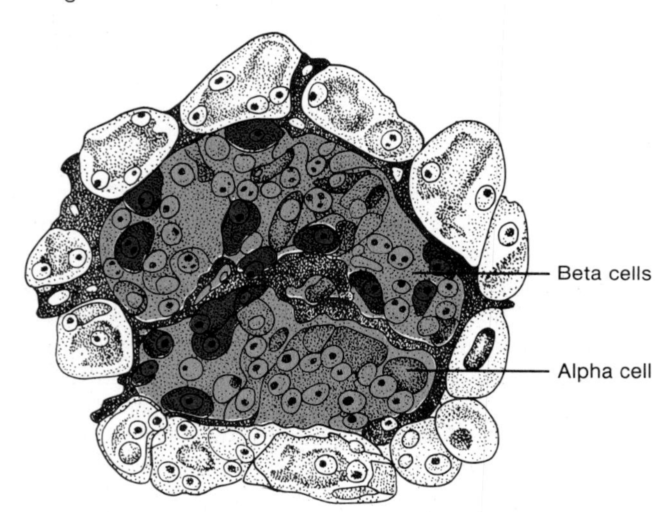

All three hormones are polypeptides. Insulin consists of two polypeptide chains—one that is twenty-one amino acids long and another that is thirty amino acids long—joined together by disulfide bonds. Glucagon is a twenty-one-amino-acid polypeptide, and somatostatin contains fourteen amino acids. Insulin was the first of these hormones to be discovered (in 1921). The importance of insulin in diabetes mellitus was immediately appreciated, and clinical use of insulin in the treatment of this disease began almost immediately after its discovery. The physiological role of glucagon was discovered later, and the importance of glucagon in the development of diabetes has only recently been suspected. The physiological significance of islet-secreted somatostatin is not currently known.

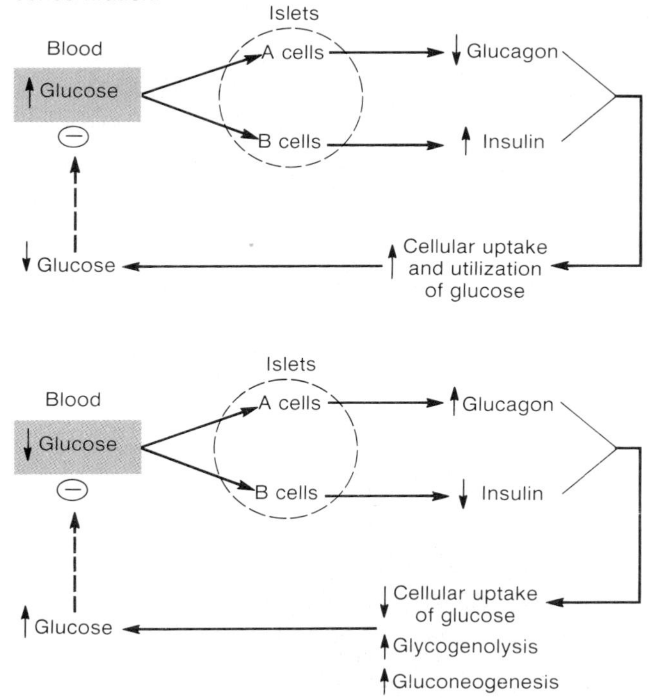

Figure 19.4. The secretion from the B (beta) cells and A (alpha) cells of the islets of Langerhans is regulated to a large degree by the blood glucose concentration. A high blood glucose concentration stimulates insulin and inhibits glucagon secretion. A low blood glucose concentration, conversely, stimulates glucagon and inhibits insulin secretion. The actions of insulin and glucagon provide negative feedback control of the blood glucose concentration.

Regulation of Insulin and Glucagon Secretion

Insulin and glucagon secretion is largely regulated by the plasma concentrations of glucose and, to a lesser degree, of amino acids. The alpha and beta cells, therefore, act as both the sensors and effectors in this control system. Since the plasma concentration of glucose and amino acids rises during the absorption of a meal and falls during fasting, the secretion of insulin and glucagon likewise changes between periods of absorption and periods of fasting. These changes in insulin and glucagon secretion, in turn, cause changes in plasma glucose and amino acid concentrations and thus help to maintain homeostasis via negative feedback loops (fig. 19.4).

Effects of Glucose. During the absorption of a carbohydrate meal, the plasma glucose concentration rises. This rise in plasma glucose (1) stimulates the beta cells to secrete insulin and (2) inhibits the secretion of glucagon from the alpha cells. Insulin acts to *stimulate the cellular uptake of plasma glucose.* A rise in insulin secretion therefore lowers the plasma glucose concentration. Since glucagon has the antagonistic effect of raising the plasma glucose concentration, the inhibition of glucagon secretion complements the effect of increased insulin during the absorption of a carbohydrate meal. A rise in insulin and a fall in glucagon secretion thus help to lower the high plasma glucose concentration that occurs during periods of absorption.

Figure 19.5. Changes in blood glucose and plasma insulin concentrations after the ingestion of 100 grams of glucose in an oral glucose tolerance test.

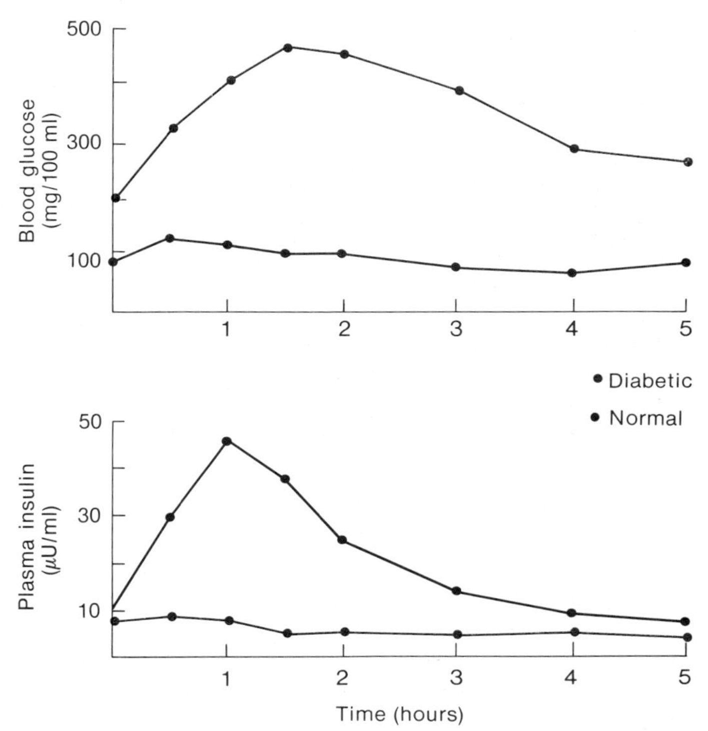

During fasting, the plasma glucose concentration falls. During fasting, therefore, (1) insulin secretion decreases and (2) glucagon secretion increases. These changes in hormone secretion prevent the cellular uptake of blood glucose into organs such as the muscles, liver, and adipose tissue and promote the release of glucose from the liver (through the actions of glucagon). A negative feedback loop is therefore completed (fig. 19.4), which helps to retard the fall in plasma glucose concentration that occurs during fasting.

The ability of the beta cells to secrete insulin, as well as the ability of insulin to lower blood glucose, is measured clinically by the **oral glucose tolerance test** (fig. 19.5). In this procedure, a person drinks a glucose solution and blood samples are taken periodically for plasma glucose measurements. In a normal person the rise in blood glucose produced by drinking this solution is reversed to normal levels within two hours following glucose ingestion.

People with **diabetes mellitus**—due to the inadequate secretion or action of insulin—maintain a state of high plasma glucose concentration (hyperglycemia) during the oral glucose tolerance test (fig. 19.5). People who have **reactive hypoglycemia** (low plasma glucose concentration due to excessive insulin secretion) have lower-than-normal blood glucose concentrations five hours following glucose ingestion.

Effects of Ingested Protein. Although the ingestion of carbohydrates causes a rise in insulin and a fall in glucagon secretion, the ingestion of proteins stimulates the secretion of both hormones. This rise in insulin secretion helps to complete a negative feedback loop because insulin promotes the uptake of amino acids into the tissues and the incorporation of these amino acids into proteins. The importance of a simultaneous rise in glucagon and insulin secretions can be understood in terms of the need to maintain a constant blood glucose concentration. A rise in insulin secretion, if not accompanied by a simultaneous rise in glucagon secretion, would cause hypoglycemia in the absence of ingested carbohydrates. A rise in both insulin and glucagon secretions thus promotes lowering of the amino acid concentration of the blood, while the blood glucose concentration remains relatively constant due to the antagonistic effects of insulin and glucagon.

Since most meals contain both carbohydrates and proteins, insulin secretion is normally raised following meals, but the effects on glucagon secretion are variable. In an average meal that is rich in carbohydrates, the suppressive effects of high plasma glucose are more potent than the stimulatory effects of amino acids on glucagon secretion. In general, therefore, the insulin secretion is increased and the glucagon secretion is decreased during the period of absorption.

Effects of Autonomic Nerves. The islets of Langerhans receive both parasympathetic and sympathetic innervation. The activation of the parasympathetic system during meals stimulates insulin secretion at the same time that gastrointestinal function is stimulated. The activation of the sympathetic system, in contrast, stimulates glucagon secretion and inhibits insulin secretion. The effects of glucagon, together with those of epinephrine, produce a "stress hyperglycemia" when the sympathoadrenal system is activated.

Effects of Gastric Inhibitory Peptide (GIP). Surprisingly, insulin secretion increases more rapidly following glucose ingestion than it does following an intravenous injection of glucose. This is due to the fact that the intestine, in response to glucose ingestion, secretes a hormone that stimulates insulin secretion before the glucose is absorbed. Insulin secretion thus begins to rise in anticipation of a rise in blood glucose. The intestinal hormone that mediates this effect is believed to be gastric inhibitory peptide (GIP).

Table 19.5 summarizes the effects of various factors that regulate insulin and glucagon secretion.

Table 19.5 Regulation of insulin and glucagon secretion.

Regulator	Effect on Insulin Secretion	Effect on Glucagon Secretion
Hyperglycemia	Stimulates	Inhibits
Hypoglycemia	Inhibits	Stimulates
Gastric inhibitory peptide	Stimulates	Stimulates (?)
Sympathetic nerve impulses	Inhibits	Stimulates
Parasympathetic nerve impulses	Stimulates	Inhibits
Amino acids	Stimulates	Stimulates
Somatostatin	Inhibits*	Inhibits*

*Inhibitory effects of somatostatin on insulin and glucagon secretion may not occur under normal conditions.

The mechanisms that regulate insulin and glucagon secretion and the actions of these hormones normally prevent the plasma glucose concentration from rising above 170 mg per 100 ml after a meal or from falling below about 50 mg per 100 ml between meals. This regulation is important because abnormally high blood glucose can damage tissue cells (as may occur in diabetes mellitus), and abnormally low blood glucose can damage the brain. The later effect results from the fact that glucose enters the brain by facilitated diffusion; when the rate of this diffusion is too low, due to low plasma glucose concentrations, the supply of metabolic energy for the brain may become inadequate. This can result in weakness, dizziness, personality changes, and ultimately in coma and death.

Insulin and Glucagon: Period of Absorption

The lowering of plasma glucose by insulin is, in a sense, a side effect of the primary action of this hormone. Insulin is the major hormone that promotes anabolism in the body. During absorption of the products of digestion and the subsequent rise in the plasma concentrations of circulating energy substrates, insulin promotes the cellular uptake of plasma glucose and its incorporation into glycogen and fat. As previously described, insulin also promotes the cellular uptake of amino acids and their incorporation into proteins. The stores of large, energy reserve molecules are thus increased while the plasma concentrations of glucose and amino acids are decreased.

The synthesis of triglycerides (fat) within adipose cells depends upon insulin-stimulated glucose uptake from plasma. Once inside the adipose cells, glucose can be converted into α-glycerol phosphate and acetyl coenzyme A (acetyl CoA). The formation of fatty acids from acetyl CoA and condensation of three fatty acids with glycerol yields triglycerides. Although the adipose cells can also

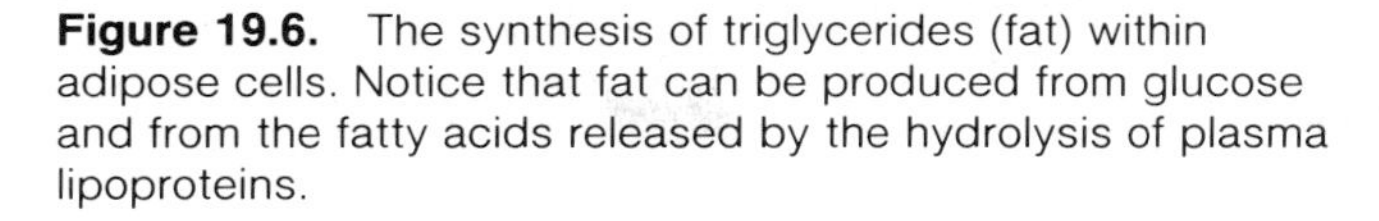

Figure 19.6. The synthesis of triglycerides (fat) within adipose cells. Notice that fat can be produced from glucose and from the fatty acids released by the hydrolysis of plasma lipoproteins.

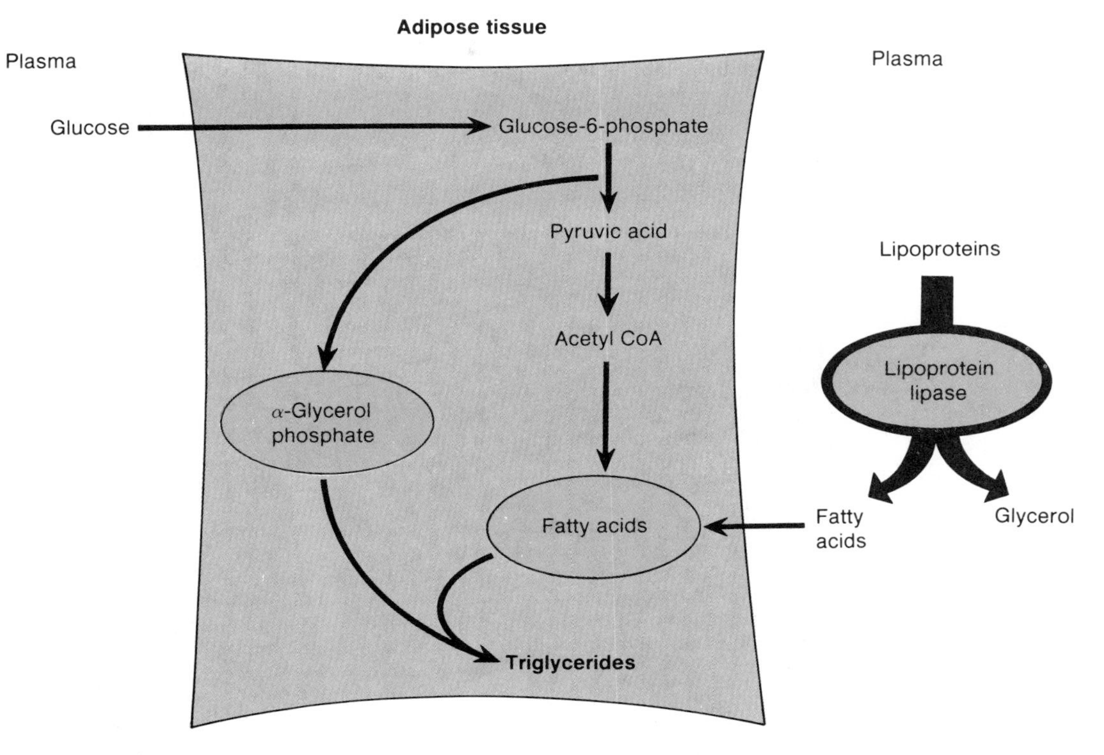

utilize free fatty acids from the blood, they cannot incorporate glycerol from the blood into triglycerides. This is because adipose cells lack the enzymes needed to convert glycerol to α-glycerol phosphate, which is the precursor needed in triglyceride synthesis (fig. 19.6). Entry of blood glucose into adipose cells—which is directly dependent on insulin—thus determines the rate at which fat is produced.

A nonobese 70 kg man has approximately 10 kg of stored fat. Since 250 g of fat can supply the energy requirements for one day, this reserve fuel is sufficient for about forty days. Glycogen is less efficient as an energy reserve, and less is stored in the body; there is about 100 g of glycogen stored in the liver and about 200 g in skeletal muscles. Insulin promotes the cellular uptake of glucose into the liver and muscles and the conversion of glucose into glucose-6-phosphate. In the liver and muscles, this can be changed into glucose-1-phosphate, which is used as the precursor of glycogen. Once the stores of glycogen are filled, the continued ingestion of excess calories results in the continued production of fat rather than of glycogen.

Insulin and Glucagon: Period of Fasting

Glucagon stimulates and insulin suppresses the hydrolysis of liver glycogen, or **glycogenolysis.** During times of fasting, when glucagon secretion is high and insulin secretion is low, therefore, liver glycogen is used as a source of additional blood glucose. This process is essentially the reverse of that which formed glycogen and results in the liberation of free glucose from glucose-6-phosphate by the action of an enzyme called *glucose-6-phosphatase.* Only the liver has this enzyme and therefore only the liver can use its stored glycogen as a source of additional blood glucose. Since muscles lack glucose-6-phosphatase, the glucose-6-phosphate produced from muscle glycogen can only be used for glycolysis by the muscle cells themselves.

Since there are only about 100 g of stored glycogen in the liver, adequate blood glucose levels could not be maintained for very long during fasting using this source alone. The low levels of insulin secretion during fasting, together with elevated glucagon secretion, however, promote **gluconeogenesis:** the formation of glucose from noncarbohydrate molecules. Low insulin allows the release of

Figure 19.7. Increased glucagon secretion and decreased insulin secretion during fasting favors catabolism. These hormonal changes result in elevated release of glucose, fatty acids, ketone bodies, and amino acids into the blood. Notice that the liver secretes glucose that is derived both from the breakdown of liver glycogen and from the conversion of amino acids in gluconeogenesis.

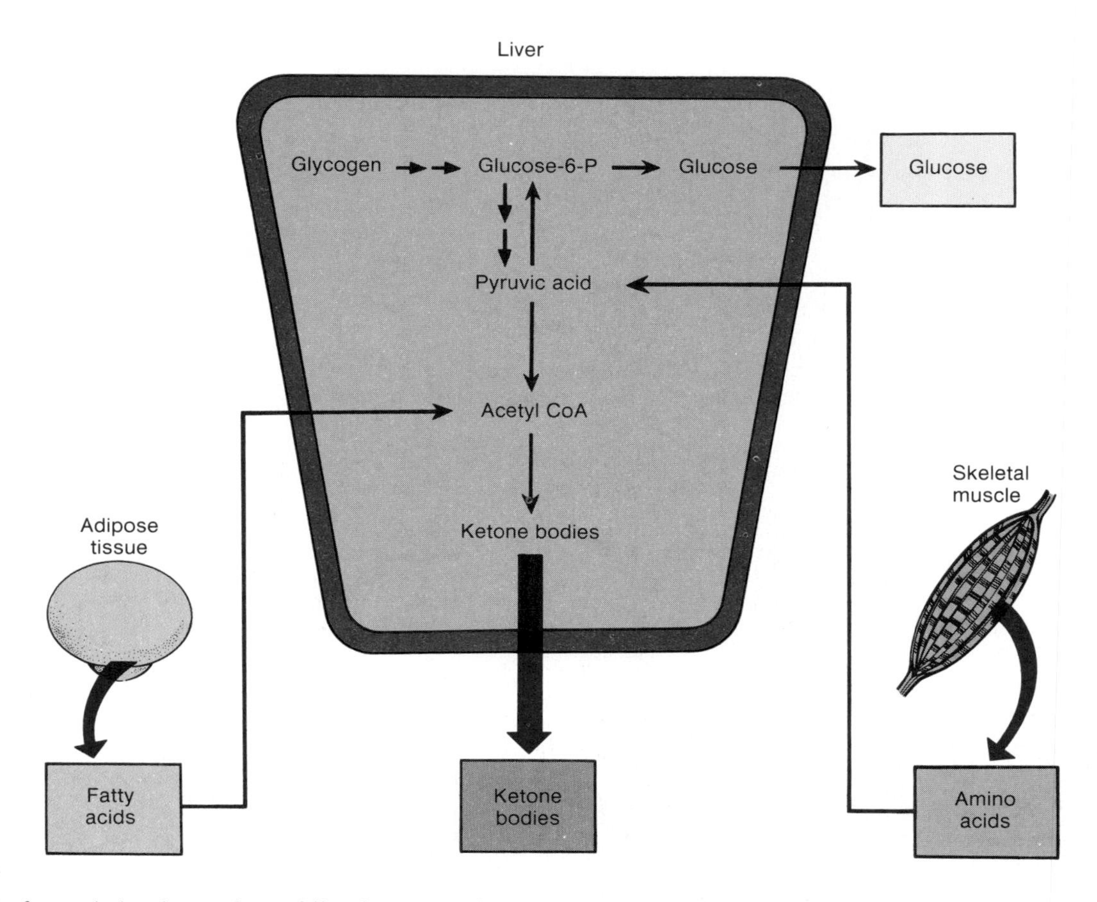

amino acids from skeletal muscles, while glucagon stimulates the production of enzymes in the liver that convert amino acids to pyruvic acid and pyruvic acid to glucose. During prolonged fasting and exercise, gluconeogenesis in the liver using amino acids from muscles may be the only source of blood glucose.

The secretion of glucose from the liver during fasting thus compensates for the low plasma glucose concentrations and helps to provide the glucose needed by the brain. In the presence of low insulin secretion, however, other organs cannot utilize blood glucose as an energy source. This helps to spare glucose for the brain, but alternative energy sources are needed for these other organs. The major alternative energy sources that are used by the liver and skeletal muscles are free fatty acids and ketone bodies.

Glucagon, in the presence of low insulin levels, stimulates in adipose cells an enzyme called hormone-sensitive lipase. This enzyme catalyzes the hydrolysis of stored triglycerides and the release of free fatty acids and glycerol into the blood. Glucagon, in the presence of low insulin levels, also stimulates enzymes in the liver that convert some of these fatty acids into ketone bodies. These derivatives of fatty acids—including acetoacetic acid, β-hydroxybutyric acid, and acetone—can also be released into the blood (fig. 19.7). Several organs in the body can use ketone bodies, like fatty acids, as a source of acetyl CoA in aerobic respiration.

Through the stimulation of **lipolysis** (the breakdown of fat) and **ketogenesis** (the formation of ketone bodies) the high glucagon and low insulin levels that are found during fasting provide circulating energy substrates for use by the muscles, liver, and other organs. Through liver glycogenolysis and gluconeogenesis, these hormonal changes help to provide adequate levels of blood glucose to sustain the metabolism of the brain. The antagonistic action of insulin and glucagon (fig. 19.8) thus promotes appropriate metabolic responses during periods of fasting and periods of absorption. The actions of insulin and glucagon are summarized in table 19.6.

Figure 19.8. The inverse relationship between insulin and glucagon secretion during the absorption of a meal and during fasting. Changes in the insulin : glucagon ratio tilts metabolism toward anabolism during the absorption of food and toward catabolism during fasting.

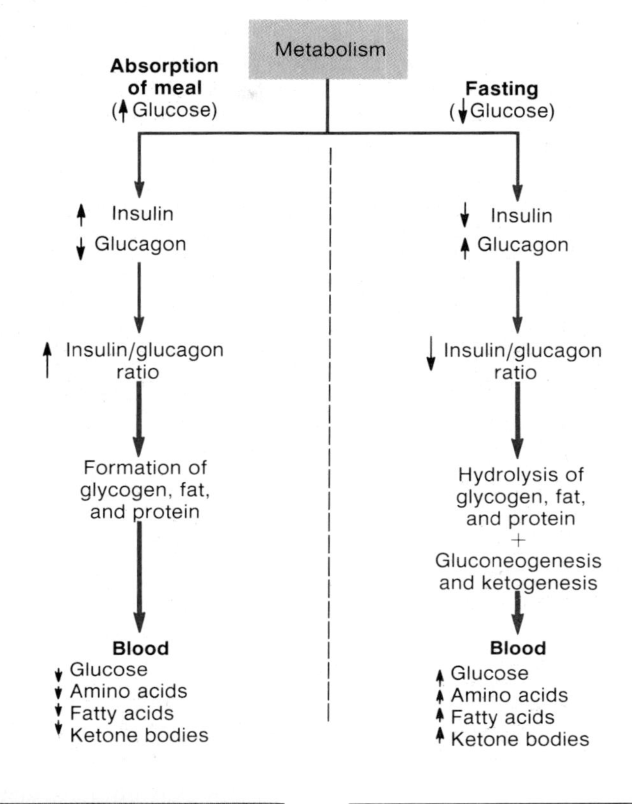

Table 19.6 Comparison of the metabolic effects of insulin and glucagon.

Actions	Insulin	Glucagon
Cellular glucose transport	↑	0
Glycogen synthesis	↑	↓
Glycogenolysis in liver	↓	↑
Gluconeogenesis	↓	↑
Amino acid uptake; protein synthesis	↑	0
Inhibition of amino acid release; protein degradation	↓	0
Lipogenesis	↑	0
Lipolysis	↓	↑
Ketogenesis	↓	↑

Code: ↑ = Increased
↓ = Decreased
0 = No direct effect

1. *Describe how the secretions of insulin and glucagon change during periods of absorption and periods of fasting. Explain how these changes in hormone secretion are produced.*
2. *Describe how the synthesis of fat in adipose cells is regulated by insulin. Explain how fat metabolism is regulated by insulin and glucagon during periods of absorption and fasting.*
3. *Define the following terms:* glycogenolysis, gluconeogenesis, *and* ketogenesis. *Explain how insulin and glucagon affect the processes these terms describe during periods of absorption and fasting.*
4. *Explain why the liver, but not skeletal muscles, can secrete glucose into the blood. Describe two pathways that contribute to the hepatic secretion of glucose.*

Table 19.7 Comparison of juvenile-onset and maturity-onset diabetes mellitus.

Characteristics	Juvenile-Onset (Type I)	Maturity-Onset (Type II)
Usual age at onset	Under 20 years	Over 40 years
Development of symptoms	Rapid	Slow
Percent of diabetic population	About 10%	About 90%
Development of ketoacidosis	Common	Rare
Association with obesity	Rare	Common
Beta cells of islets	Destroyed	Usually not destroyed
Insulin secretion	Decreased	Normal or increased
Autoantibodies to islet cells	Present	Absent
Associated with particular HLA antigens	Yes	No
Usual treatment	Insulin injections	Diet; oral stimulators of insulin secretion

Diabetes Mellitus and Hypoglycemia

Chronic high blood glucose, or hyperglycemia, is the hallmark of the disease **diabetes mellitus.** The name of this disease is derived from the fact that glucose "spills over" into the urine when the blood glucose concentration is too high. The hyperglycemia of diabetes mellitus results from either the insufficient secretion of insulin by the beta cells of the islets of Langerhans or the inability of secreted insulin to stimulate the cellular uptake of glucose from the blood. Diabetes mellitus, in short, results from the inadequate secretion or action of insulin.

There are two forms of diabetes mellitus. In **type I,** or **insulin-dependent,** diabetes the beta cells are destroyed and secrete little or no insulin. This form of the disease accounts for only about 10% of the cases of diabetes in the country. About 90% of the people who have diabetes have **type II,** or **non-insulin-dependent,** diabetes mellitus. Type I diabetes is also called *juvenile-onset diabetes,* because this condition is usually diagnosed in people under the age of thirty. Type II, or *maturity-onset, diabetes* is usually diagnosed in people over the age of thirty. Some comparisons of these two forms of diabetes mellitus are shown in table 19.7.

Type I Diabetes Mellitus

Type I diabetes mellitus results when the beta cells of the islets of Langerhans are destroyed by the effects of a virus or another environmental agent or by the action of autoantibodies. Removal of the insulin-secreting cells in this way causes hyperglycemia and the appearance of glucose in the urine. Without insulin, glucose cannot enter the adipose cells; the rate of fat synthesis thus lags behind the rate of fat breakdown and large amounts of free fatty acids are released from the adipose cells.

In a person with uncontrolled type I diabetes, many of the fatty acids released from adipose cells are converted into ketone bodies in the liver. This may result in an elevated ketone body concentration in the blood (ketosis), and if the buffer reserve of bicarbonate is neutralized, it may also result in *ketoacidosis.* During this time, the glucose and excess ketone bodies that are excreted in the urine act as osmotic diuretics and cause the excessive excretion of water in the urine. This can produce severe dehydration, which, together with ketoacidosis and associated disturbances in electrolyte balance, may lead to coma and death (fig. 19.9).

In addition to the lack of insulin, people with type I diabetes have an abnormally high secretion of glucagon from the alpha cells of the islets. Glucagon stimulates glycogenolysis in the liver and thus helps to raise the blood glucose concentration. Glucagon also stimulates the production of enzymes in the liver that convert fatty acids to ketone bodies. Some researchers believe that the full symptoms of diabetes result from high glucagon secretion as well as from the absence of insulin. The lack of insulin may be largely responsible for hyperglycemia and for the release of large amounts of fatty acids into the blood. The high glucagon secretion may contribute to the hyperglycemia and be largely responsible for the development of ketoacidosis.

Insulin Resistance: Obesity and Type II Diabetes

The effects produced by insulin, or any hormone, depend on the concentration of that hormone in the blood and on the sensitivity of the target tissue to given amounts of the hormone. Tissue responsiveness to insulin, for example, varies under normal conditions. For reasons that are incompletely understood, exercise increases insulin sensitivity and obesity decreases insulin sensitivity of the target tissues. The islets of a nondiabetic obese person, therefore, must secrete high amounts of insulin to maintain the blood glucose concentration in the normal range.

Figure 19.9. The sequence of events by which an insulin deficiency may lead to coma and death.

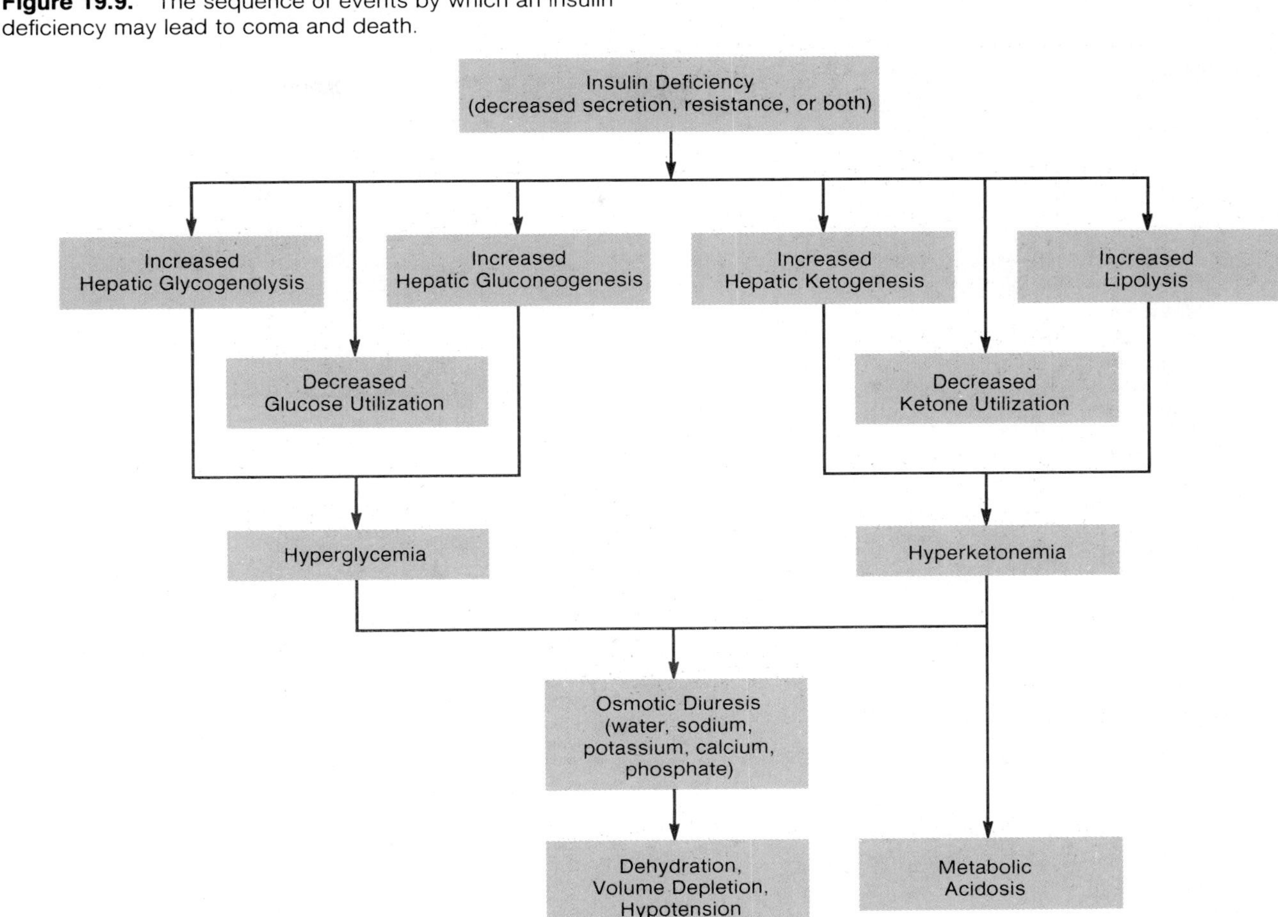

Type II diabetes is usually slow to develop, is hereditary, and occurs most often in people who are overweight. Unlike type I diabetes mellitus, people who have type II diabetes have normal or even elevated levels of insulin in their blood. Despite this, people with type II diabetes have hyperglycemia if untreated. This must mean that, even though the insulin levels may be in the normal range, the amount of insulin secreted is inadequate because there is a decreased tissue sensitivity to the effects of insulin.

Since obesity decreases insulin sensitivity, people who are genetically predisposed to insulin resistance may develop symptoms of diabetes when they gain weight. Conversely, this type of diabetes mellitus can usually be controlled by increasing tissue sensitivity to insulin through diet and exercise. If this is not sufficient, oral drugs are available that increase insulin secretion and also stimulate tissue responsiveness to insulin.

People with type II diabetes don't usually develop ketoacidosis. The hyperglycemia itself, however, can be dangerous on a long-term basis. Diabetes is the second leading cause of blindness in the United States, and people with diabetes frequently have circulatory problems that increase the tendency to get gangrene and increase the risk of atherosclerosis. The causes of damage to the retina and lens of the eyes and to blood vessels are not well understood. It is believed, however, that these problems may result from a long-term exposure to high blood glucose.

Table 19.8 Comparison of coma due to diabetic ketoacidosis and to hypoglycemia.

	Diabetic Ketoacidosis	Hypoglycemia
Onset	Hours to days	Minutes
Causes	Insufficient insulin; other diseases	Excess insulin; insufficient food; excessive exercise
Symptoms	Excessive urination and thirst; headache, nausea, and vomiting	Hunger, headache, confusion, stupor
Physical Findings	Deep, labored breathing; breath has acetone odor; blood pressure decreased, pulse weak; skin is dry	Pulse, blood pressure, and respiration are normal; skin is pale and moist
Laboratory Findings	Urine: glucose present, ketone bodies increased Plasma: glucose and ketone bodies increased, bicarbonate decreased	Urine: no glucose, ketone bodies at normal concentration Plasma: glucose concentration low, bicarbonate normal

Hypoglycemia

People with type I diabetes mellitus are dependent upon insulin injections to prevent hyperglycemia and ketoacidosis. If inadequate insulin is injected, the person may enter a coma as a result of the ketoacidosis, electrolyte imbalance, and dehydration that develop. An overdose of insulin, however, can also produce a coma as a result of the hypoglycemia (abnormally low blood glucose levels) produced. The physical signs and symptoms of diabetic and hypoglycemic coma are sufficiently different (table 19.8) to allow hospital personnel to distinguish between these two types.

Less severe symptoms of hypoglycemia are usually produced by an oversecretion of insulin from the islets of Langerhans after a carbohydrate meal. This **reactive hypoglycemia** is caused by an exaggerated response of the beta cells to a rise in blood glucose and is most commonly seen in adults who are genetically predisposed to type II diabetes. People with reactive hypoglycemia must, therefore, limit their intake of carbohydrates and eat small meals at frequent intervals, rather than two or three meals per day.

The symptoms of reactive hypoglycemia include tremor, hunger, weakness, hypoglycemia, blurred vision, and impaired mental ability. The appearance of some of these symptoms, however, does not necessarily indicate reactive hypoglycemia and a given level of blood glucose does not always produce these symptoms. For these reasons a number of tests must be performed, including the oral glucose tolerance test, to confirm the diagnosis of reactive hypoglycemia. In the glucose tolerance test, reactive hypoglycemia is shown when the initial rise in blood glucose produced by the ingestion of a glucose solution triggers excessive insulin secretion, so that the blood glucose levels fall below normal within five hours (fig. 19.10).

Figure 19.10. An idealized oral glucose tolerance test in a person with reactive hypoglycemia. The blood glucose concentration falls below the normal range within five hours of glucose ingestion as a result of excessive insulin secretion.

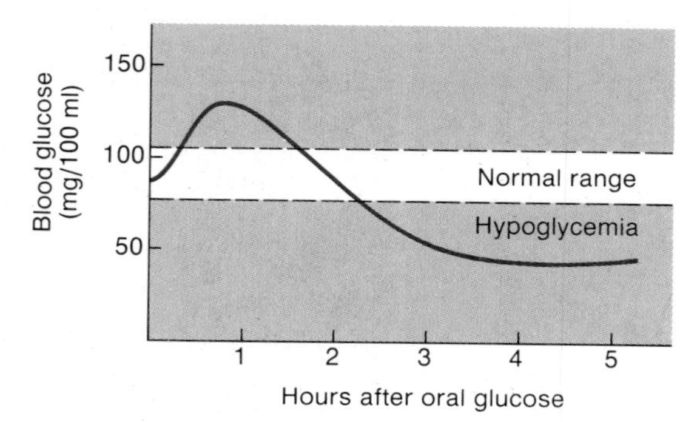

1. *Explain how ketoacidosis and dehydration are produced in a person with type I diabetes mellitus.*
2. *Describe the causes of hyperglycemia in a person with type II diabetes, and explain how weight loss may help to control this condition.*
3. *Explain how reactive hypoglycemia is produced, and describe the dangers of this condition.*

Metabolic Regulation by Adrenal Hormones, Thyroxine, and Growth Hormone

The anabolic effects of insulin are antagonized by glucagon, as previously described, and by the actions of a variety of other hormones. The hormones of the adrenals, thyroid, and anterior pituitary (specifically growth hormone) antagonize the action of insulin on carbohydrate and lipid metabolism. The actions of insulin, thyroxine, and growth hormone, however, can act synergistically in the stimulation of protein synthesis.

Adrenal Hormones

The adrenal gland consists of two different parts that have different embryonic origins, that secrete different hormones, and that are regulated by different control

Figure 19.11. Cyclic AMP (cAMP) serves as a second messenger in the actions of epinephrine (and glucagon) on liver and adipose tissue metabolism.

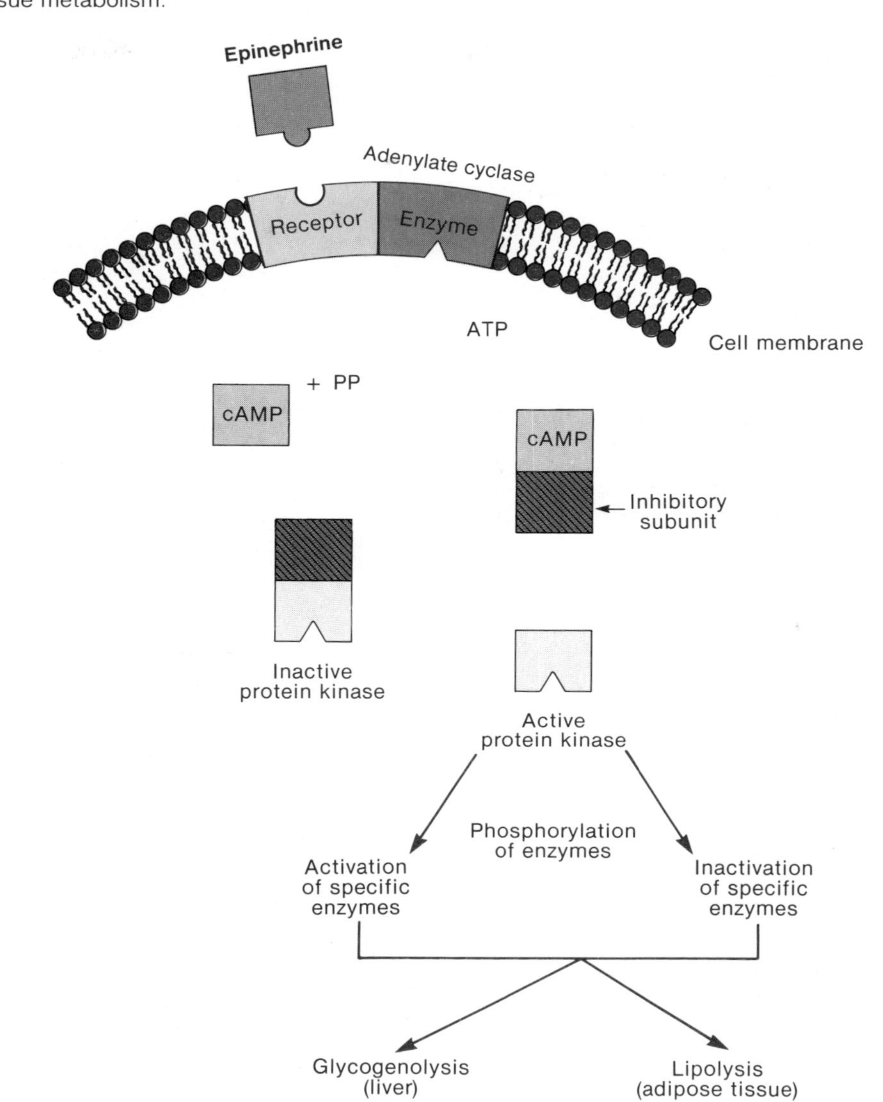

systems. The **adrenal medulla** secretes catecholamine hormones—epinephrine and lesser amounts of norepinephrine—in response to sympathetic nerve stimulation. The **adrenal cortex** secretes corticosteroid hormones. These are grouped into two functional categories: **mineralocorticoids,** such as aldosterone, which regulate Na^+ and K^+ balance, and **glucocorticoids,** such as hydrocortisone (cortisol), which participate in metabolic regulation. The secretion of glucocorticoids is stimulated by the anterior pituitary hormone ACTH.

Metabolic Effects of Epinephrine. The metabolic effects of epinephrine are similar to those of glucagon. Both stimulate glycogenolysis and the release of glucose from the liver and lipolysis and the release of fatty acids from adipose tissue. These actions occur in response to glucagon during fasting, when low blood glucose stimulates glucagon secretion, and in response to epinephrine during the "fight-or-flight" reaction to stress. The latter effect provides circulating energy substrates in anticipation of the need for intense physical activity. Glucagon and epinephrine have similar mechanisms of action; the actions of both are mediated by cyclic AMP (fig. 19.11).

Metabolic Effects of Glucocorticoids. Hydrocortisone and other glucocorticoids are secreted by the adrenal cortex in response to ACTH stimulation. The secretion of ACTH from the anterior pituitary occurs as part of the

Figure 19.12. The catabolic actions of glucocorticoids help raise the blood concentration of glucose and other energy-carrier molecules.

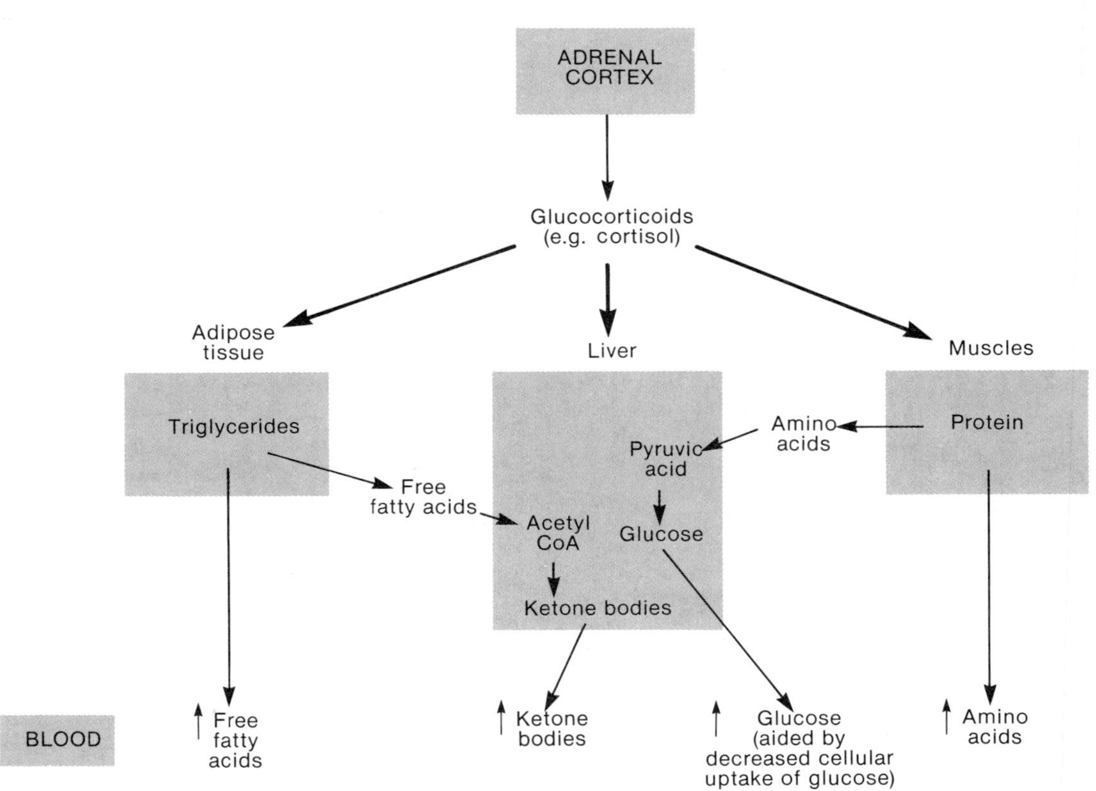

general adaptation syndrome in response to stress. Since prolonged fasting or prolonged exercise certainly qualify as stressors, ACTH—and thus glucocorticoid secretion—is stimulated under these conditions. The increased secretion of glucocorticoids during prolonged fasting or exercise supports the effects of increased glucagon and decreased insulin secretion from the islets.

Like glucagon, hydrocortisone promotes lipolysis and ketogenesis; it also stimulates the synthesis of hepatic enzymes that promote gluconeogenesis. Although it stimulates enzyme (protein) synthesis in the liver, hydrocortisone promotes protein breakdown in the muscles. This latter effect increases the blood levels of amino acids and thus provides the substrates needed by the liver for gluconeogenesis. The release of circulating energy substrates—amino acids, glucose, fatty acids, and ketone bodies—into the blood in response to hydrocortisone (fig. 19.12) helps to compensate for a state of prolonged fasting or exercise. Whether these metabolic responses are beneficial in other stressful states is open to question.

Thyroxine

The thyroid follicles secrete thyroxine, also called tetraiodothyronine (T_4), in response to stimulation by TSH from the anterior pituitary. Almost all organs in the body are targets of thyroxine action. Thyroxine itself, however, is not the active form of the hormone within the target cells; thyroxine is a prehormone that must first be converted to triiodothyronine (T_3) within the target cells to be active. Acting via its conversion to T_3, thyroxine (1) regulates the rate of cell respiration and (2) contributes to proper growth and development, particularly during early childhood.

Thyroxine and Cell Respiration. Thyroxine stimulates the rate of cell respiration in almost all cells in the body. This is believed to be due to thyroxine-induced lowering of cellular ATP concentrations. ATP exerts an end-product inhibition of cell respiration so that when ATP concentrations increase, the rate of cell respiration decreases. Conversely, a lowering of ATP concentrations, as may occur in response to thyroxine, stimulates cell respiration.

Figure 19.13. A mechanism proposed to explain the effects of thyroid hormones on basal metabolic rate. Through the activation of genes and the stimulation of protein synthesis, thyroid hormones increase the activity of the Na^+/K^+ pump. This active transport carrier accounts for a large percentage of the energy expenditures in the cell. The concentration of ATP, therefore, declines as a result of this increased energy usage, and the decreased ATP concentrations stimulate increased cellular respiration.

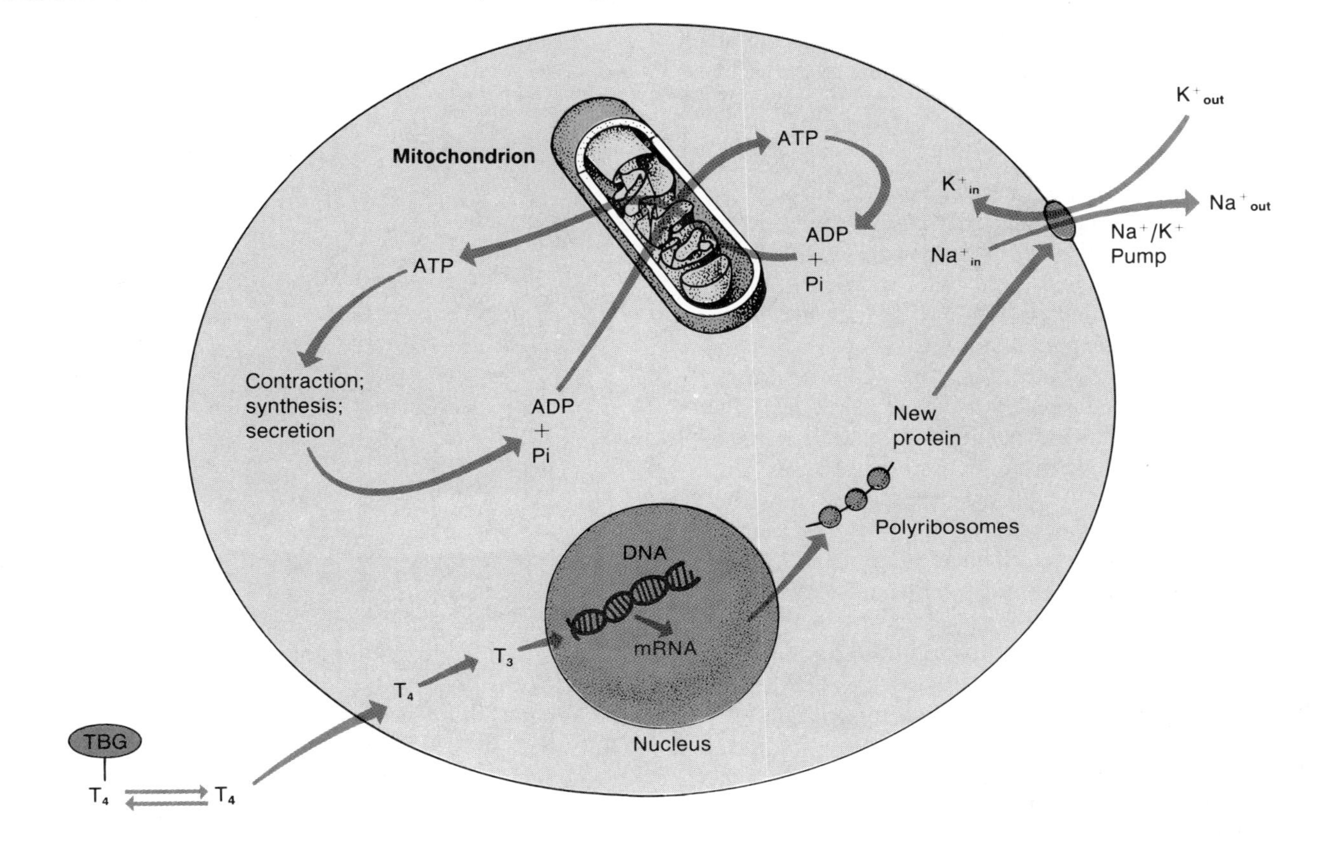

The mechanisms of thyroxine action are not entirely understood. One effect of thyroxine, however—the stimulation of Na^+/K^+ pump activity in cell membranes—could account for the thyroxine-induced lowering of cell ATP concentrations. The active transport of Na^+ and K^+ represents a significant energy "sink" in the cell, accounting for about 12% of the calories consumed at rest. Through stimulation of Na^+/K^+ pumps, therefore, thyroxine could significantly decrease ATP concentrations and thus stimulate the rate of cell respiration (fig. 19.13).

When a person wakes up in the morning, the total energy consumption of the body is at its lowest or basal level, known as the **basal metabolic rate (BMR).** Most commonly measured by the rate of oxygen consumption, the BMR indicates the "idling speed" of the body. Since the activity of the Na^+/K^+ pumps contributes significantly to the energy consumed in the basal state and since the activity of these pumps is set by thyroxine secretion, the BMR can be used as an index of thyroid function. Indeed, such measurements were used clinically to evaluate thyroid function prior to the development of direct chemical determinations of T_4 and T_3 in the blood.

The coupling of energy-releasing reactions to energy-requiring reactions is never 100% efficient; a proportion of the energy is always lost as heat. Much of the energy liberated during cell respiration and much of the energy released by the hydrolysis of ATP escapes as heat. Since thyroxine stimulates both ATP consumption and cell respiration, the actions of thyroxine result in the production of metabolic heat.

The heat-producing, or *calorigenic* (*calor* = heat), *effects* of thyroxine are required for cold adaptation. This does not mean that people who are cold-adapted have high levels of thyroxine secretion. Rather, thyroxine levels in the normal range coupled with the increased activity of the sympathoadrenal system are responsible for cold adaptation. Thyroxine exerts a permissive effect on the ability of the sympathoadrenal system to increase heat production in response to cold stress.

Thyroxine in Growth and Development. Through its stimulation of cell respiration, thyroxine stimulates the increased consumption of circulating energy substrates such as glucose, fatty acids, and other molecules. These effects, however, are mediated at least in part by the activation of genes; thyroxine thus stimulates both RNA and protein synthesis. As a result of its stimulation of protein synthesis throughout the body, thyroxine is considered to be an anabolic hormone like insulin and growth hormone.

Because of its stimulation of protein synthesis, thyroxine is needed for growth of the skeleton and, most importantly, for the proper development of the central nervous system. This latter effect is particularly significant during prenatal development and the first two years after birth. Hypothyroidism during this time may result in **cretinism** (fig. 19.14). Unlike dwarfs, who have normal thyroxine secretion but a low secretion of growth hormone, cretins suffer from severe mental retardation.

Hypothyroidism and Hyperthyroidism. As might be predicted from the effects of thyroxine, people who are hypothyroid have an abnormally low basal metabolic rate (BMR) and experience weight gain and lethargy. There is also a decreased ability to adapt to cold stress when there is a thyroxine deficiency. Another symptom of hypothyroidism is **myxedema**—accumulation of mucoproteins in subcutaneous connective tissues. Hypothyroidism can be produced by a variety of causes, including insufficient TRH secretion from the hypothalamus, insufficient TSH secretion from the pituitary, or insufficient iodine in the diet. Hypothyroidism due to lack of iodine is accompanied by excessive TSH secretion, which stimulates growth of the thyroid and the production of a goiter.

A goiter can also be produced by another mechanism. In **Graves's disease,** autoantibodies are produced that have TSH-like effects on the thyroid. Since the production of these autoantibodies is not controlled by negative feedback, this results in the excessive stimulation of the thyroid. A goiter is thus produced that is associated with a hyperthyroid state. Hyperthyroidism produces a high BMR accompanied by weight loss, nervousness, irritability, and an intolerance to heat. The symptoms of hypothyroidism and hyperthyroidism are compared in table 19.9.

Figure 19.14. Cretinism is a disease of infancy caused by an underactive thyroid gland.

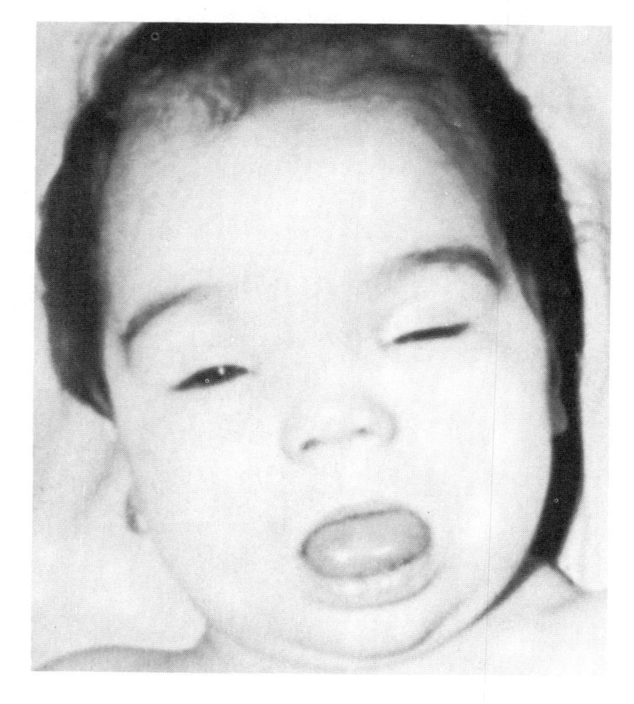

Table 19.9 Comparison of hypothyroidism and hyperthyroidism.

	Hypothyroid	Hyperthyroid
Growth and Development	Impaired growth	Accelerated growth
Activity and Sleep	Decreased activity; increased sleep	Increased activity; decreased sleep
Temperature Tolerance	Intolerance to cold	Intolerance to heat
Skin Characteristics	Coarse, dry skin	Smooth skin
Perspiration	Absent	Excessive
Pulse	Slow	Rapid
Gastrointestinal Symptoms	Constipation; decreased appetite; increased weight	Frequent bowel movements; increased appetite; decreased weight
Reflexes	Slow	Rapid
Psychological Aspects	Depression and apathy	Nervous, "emotional"
Plasma T_4 Levels	Decreased	Increased

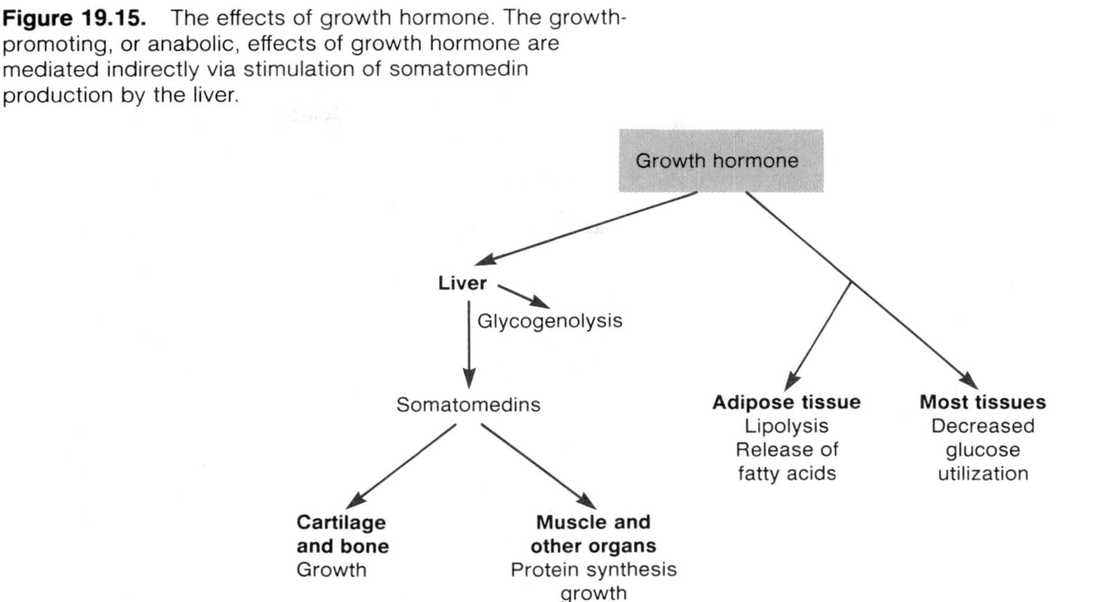

Figure 19.15. The effects of growth hormone. The growth-promoting, or anabolic, effects of growth hormone are mediated indirectly via stimulation of somatomedin production by the liver.

Growth Hormone

The anterior pituitary secretes growth hormone, also called somatotropic hormone, in larger amounts than any other of its hormones. As its name implies, growth hormone stimulates growth in children and adolescents. The continued high secretion of growth hormone in adults, particularly under the conditions of fasting and other forms of stress, implies that this hormone can have important metabolic effects even after the growing years have ended.

Regulation of Growth Hormone Secretion. The secretion of growth hormone is inhibited by somatostatin, which is produced by the hypothalamus and secreted into the hypothalamo-hypophyseal portal system. In addition, a newly discovered hypothalamic-releasing hormone appears to stimulate growth hormone secretion. Growth hormone thus appears to be unique among the anterior pituitary hormones in that its secretion is controlled by both a releasing and an inhibiting hormone from the hypothalamus. The secretion of growth hormone follows a circadian ("about a day") pattern, increasing during sleep and decreasing during periods of wakefulness.

Growth hormone secretion is stimulated by an increase in the plasma concentrations of amino acids and by a decrease in the plasma glucose concentration. The secretion of growth hormone is, therefore, stimulated during absorption of a high protein meal, when amino acids are absorbed. The secretion of growth hormone is also stimulated during prolonged fasting when plasma glucose is low and plasma amino acid concentration is raised by the breakdown of muscle protein.

Effects of Growth Hormone on Metabolism. The fact that growth hormone secretion is increased during fasting and also during absorption of a protein meal reflects the complex nature of this hormone's action. Growth hormone has both anabolic and catabolic effects; it promotes protein synthesis (anabolism) and it also stimulates the catabolism of fat and release of fatty acids from adipose tissue. Growth hormone is similar to insulin in its former effect and similar to glucagon in its latter effect.

In terms of its action on lipid and carbohydrate metabolism, growth hormone is said to have an anti-insulin effect. A rise in the plasma fatty acid concentration induced by growth hormone results in decreased rates of glycolysis in many organs. This inhibition of glycolysis by fatty acids, perhaps together with a more direct action of growth hormone, results in decreased glucose utilization by the tissues. As a consequence, experimental animals that have had their pituitary removed (and thus lack growth hormone) experience hypoglycemia and are more sensitive to the effects of insulin than intact animals.

Growth hormone stimulates the cellular uptake of amino acids and protein synthesis in many organs of the body. These actions are useful during a protein-rich meal; amino acids are removed from the blood and used to form proteins, and the plasma concentration of glucose and fatty acids is increased to provide alternate energy sources (fig. 19.15). The anabolic effect of growth hormone on protein synthesis is particularly important during the growing years, when it contributes to increases in bone length and in the mass of many soft tissues.

Figure 19.16. The hepatic production of polypeptides called *somatomedins* is stimulated by growth hormone. These compounds in turn produce the anabolic effects that are characteristic of the actions of growth hormone in the body.

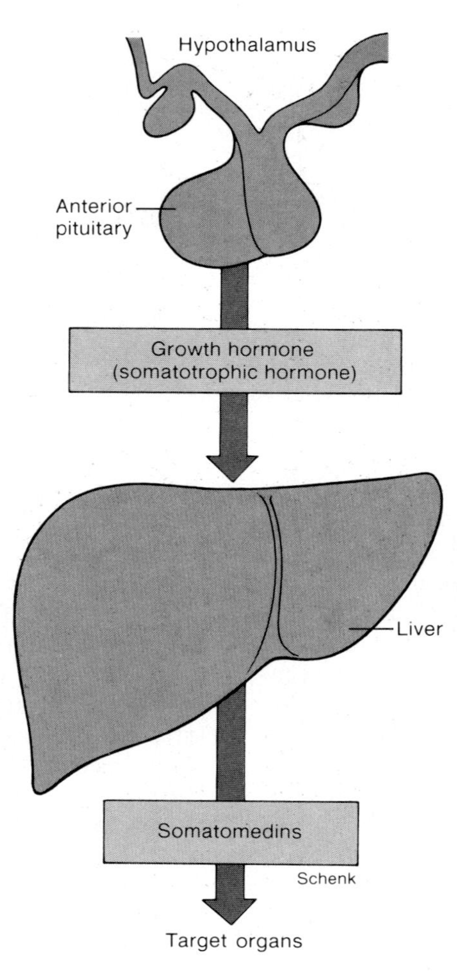

Effects of Growth Hormone on Body Growth. The anabolic effects of growth hormone and their resultant stimulation of skeletal growth are believed to be produced indirectly. This was demonstrated by the observation that growth hormone by itself may not stimulate bone growth *in vitro* (outside the body); the plasma from growth hormone-stimulated animals must be present. This is because the growth-promoting effects are stimulated indirectly by polypeptides called **somatomedins,** which are produced by the liver and secreted into the blood in response to growth hormone stimulation. The somatomedins are therefore responsible for, or mediate, the stimulation of skeletal growth induced by growth hormone (fig. 19.16). Skeletal growth is produced by the stimulation of chondrocytes in the epiphyseal disc cartilages of bones in growing children and adolescents. This skeletal growth stops when the epiphyseal discs are converted to bone after the growth spurt during puberty, despite the fact that growth hormone secretion remains as high in adults as in children.

An excessive secretion of growth hormone in children can produce **gigantism** in which people may grow up to eight feet tall. An excessive growth hormone secretion that occurs after the epiphyseal discs have sealed, however, cannot produce increases in height. The oversecretion of growth hormone in adults results in an elongation of the jaw and deformities in the bones of the face, hands, and feet. This condition, called **acromegaly,** is accompanied by the growth of soft tissues and coarsening of the skin (fig. 19.17).

Figure 19.17. The progression of acromegaly in one individual, from age nine (*a*), sixteen (*b*), thirty-three (*c*), and fifty-two (*d*) years. The coarsening of features and disfigurement are evident by age thirty-three and severe at age fifty-two.

(a)

(b)

(c)

(d)

An inadequate secretion of growth hormone during the growing years results in **dwarfism.** An interesting variant of this is *Laron dwarfism,* in which there is a genetic insensitivity to the effects of growth hormone. It is also believed that the short stature of African pygmies may be due to a genetically low sensitivity to the effects of growth hormone.

An adequate diet, particularly of proteins, is required for the production of somatomedins. This helps to explain the common observation that many children are significantly taller than their immigrant parents, who may not have had an adequate diet in their youth. Children with protein malnutrition **(kwashiorkor)** have low growth rates and low somatomedin levels, despite the fact that their growth hormone levels are abnormally elevated. When these children eat an adequate diet, somatomedin levels and growth rates increase.

1. *Describe the effects of epinephrine and of the glucocorticoids on the metabolism of carbohydrates and lipids, and explain the significance of these effects as a response to stress.*
2. *Explain the actions of thyroxine on the basal metabolic rate, and explain why people who are hypothyroid have a tendency to gain weight and are less resistant to cold stress.*
3. *Describe the effects of growth hormone on the metabolism of lipids, glucose, and amino acids.*
4. *Explain how growth hormone stimulates skeletal growth.*

Table 19.10 The endocrine regulation of calcium and phosphate balance.

Hormone	Effect on Intestine	Effect on Kidneys	Effect on Bone	Associated Diseases
Parathyroid hormone (PTH)	No direct effect	Stimulates Ca^{++} reabsorption Inhibits PO_4^{-3} reabsorption	Stimulates resorption	Osteitis fibrosa cystica with hypercalcemia due to excess PTH
1,25-dihydroxyvitamin D_3	Stimulates absorption of Ca^{++} and PO_4^{-3}	Stimulates reabsorption of Ca^{++} and PO_4^{-3}	Stimulates resorption	Osteomalacia (adults) and rickets (children) due to deficiency of 1,25-dihydroxyvitamin D_3
Calcitonin	None	Inhibits resorption of Ca^{++} and PO_4^{-3}	Stimulates deposition	None

Regulation of Calcium and Phosphate Balance

The calcium and phosphate concentrations of plasma are affected by bone formation and resorption, intestinal absorption, and urinary excretion of these ions. These processes are regulated by parathyroid hormone, 1,25-dihydroxyvitamin D_3, and calcitonin, as summarized in table 19.10.

The skeleton, in addition to providing support for the body, serves as a large store of calcium and phosphate in the form of *hydroxyapatite* crystals. The calcium phosphate in these hydroxyapatite crystals is derived from the blood by the action of bone-forming cells, or **osteoblasts.** Bone resorption (dissolution), produced by the action of **osteoclasts** (fig. 19.18), results in the return of bone calcium and phosphate to the blood.

The formation and resorption of bone occur constantly at rates determined by the hormonal balance. Body growth during the first two decades of life occurs because bone formation proceeds at a faster rate than bone resorption. By age fifty or sixty, the rate of bone resorption often exceeds the rate of bone formation. The constant activity of osteoblasts and osteoclasts allows bone to be remodeled throughout life. The position of the teeth, for example, can be changed by orthodontic appliances (braces), which cause bone resorption on the pressure-bearing side and bone formation on the opposite side of the alveolar sockets.

The plasma concentrations of calcium and phosphate are maintained, despite the changing rates of bone formation and resorption, by hormonal control of the intestinal absorption and urinary excretion of these ions. These hormonal control mechanisms are very effective at maintaining the plasma calcium and phosphate concentrations within narrow limits. Plasma calcium, for example, is normally maintained at about 2.5 millimolar, or 5 milliequivalents per liter (a milliequivalent equals a millimole times the valence of the ion, in this case, times two).

The maintenance of normal plasma calcium concentrations is important because of the wide variety of effects that calcium has in the body. In addition to its role in bone formation, excitation-contraction coupling in muscles, and as a second messenger in the action of some hormones, calcium is needed to maintain proper membrane permeability. An abnormally low plasma calcium concentration increases the permeability of the cell membranes to Na^+ and other ions. Hypocalcemia, therefore, enhances the excitability of nerves and muscles and can result in muscle spasm (tetany).

There are a variety of bone disorders associated with abnormal calcium and phosphate balance. In **osteomalacia** (in adults) and **rickets,** (in children) bone demineralization and softening occur as a result of inadequate vitamin D. Excessive parathyroid hormone secretion, resulting in excessive demineralization of bone and high levels of blood calcium, produces **osteitis fibrosa cystica.** The most common bone disorder is **osteoporosis,** in which decreased bone mass per volume results in fractures in response to ordinary bone stress. This condition occurs in both sexes but advances more rapidly in women after menopause. The causes of osteoporosis are not known, but it may be associated with age-related decreases in the intestinal absorption of calcium. The treatment of this condition includes estrogen therapy (for postmenopausal women), increased dietary calcium, and sometimes small amounts of vitamin D or its derivatives.

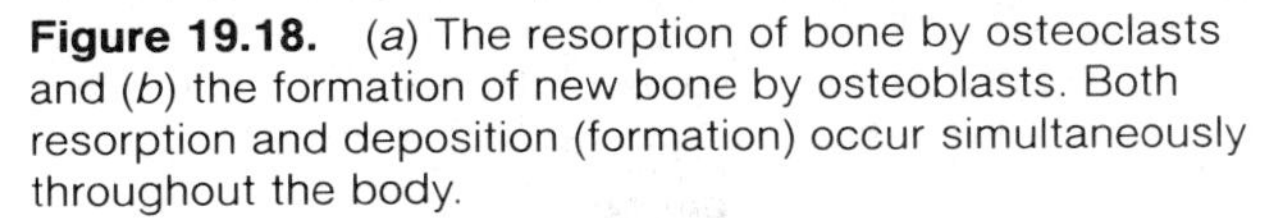

Figure 19.18. (*a*) The resorption of bone by osteoclasts and (*b*) the formation of new bone by osteoblasts. Both resorption and deposition (formation) occur simultaneously throughout the body.

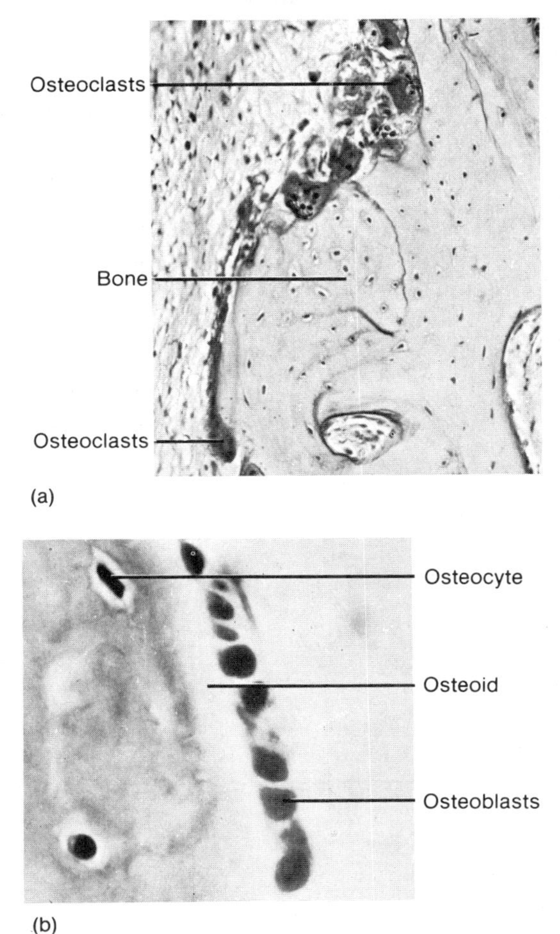

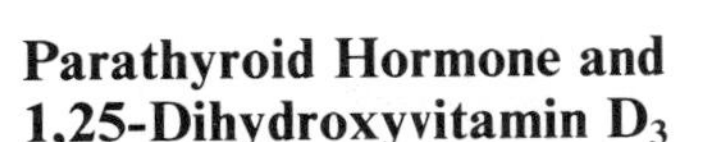

Parathyroid Hormone and 1,25-Dihydroxyvitamin D_3

Parathyroid hormone, an eighty-four-amino-acid polypeptide secreted by the parathyroid glands, acts to help raise the plasma concentration of calcium. The effects of parathyroid hormone include (1) the stimulation of bone resorption; (2) the stimulation of calcium reabsorption in the kidneys, a process that returns calcium to the blood, so that less calcium is excreted in the urine; and (3) the inhibition of renal phosphate reabsorption, so that more phosphate is excreted in the urine. Parathyroid hormone also promotes the formation of 1,25-dihydroxyvitamin D_3, and so it indirectly helps to raise plasma calcium levels via the effects of this other hormone.

The production of **1,25-dihydroxyvitamin D_3** begins in the skin, where vitamin D_3 is produced from its precursor molecule (7-dehydrocholesterol) under the influence of sunlight. When the skin does not make sufficient vitamin D_3 because of insufficient exposure to sunlight, this compound must be ingested in the diet—that is why it is called a vitamin. Whether this compound is secreted into the blood from the skin or enters the blood after being absorbed from the intestine, vitamin D_3 functions as a *prehormone,* which must be chemically changed in order to be biologically active.

Figure 19.19. The pathway for the production of the hormone 1,25-dihydroxyvitamin D_3. This hormone is produced in the kidneys from the inactive precursor, 25-hydroxyvitamin D_3 (formed in the liver). This latter molecule is produced from vitamin D_3 secreted by the skin.

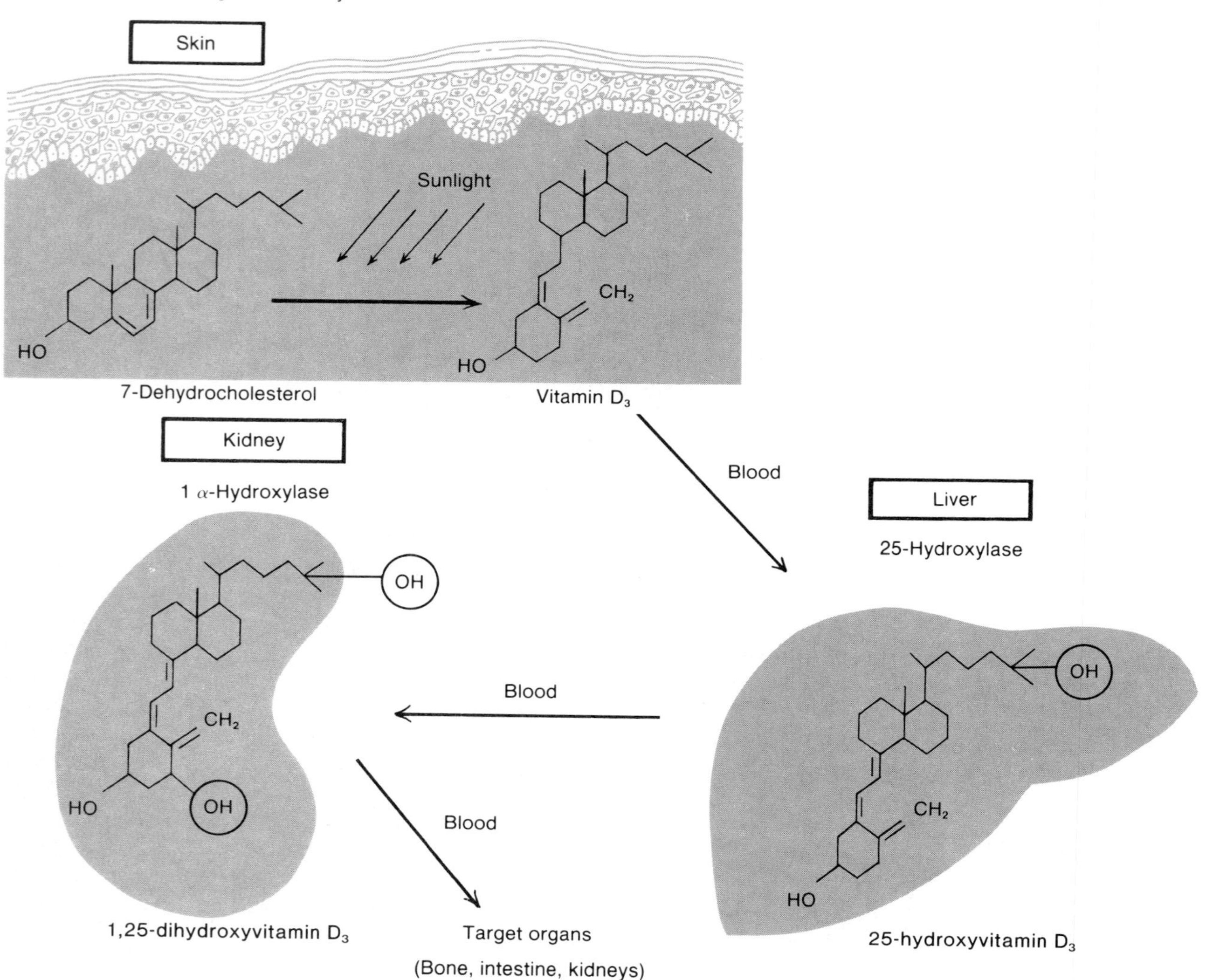

An enzyme in the liver adds a hydroxyl group (OH) to carbon 25, which converts vitamin D_3 into 25-hydroxyvitamin D_3. In order to be active, however, another hydroxyl group must be added to the first carbon. Hydroxylation of the first carbon is accomplished by an enzyme in the kidneys, which converts the molecule to 1,25-dihydroxyvitamin D_3 (fig. 19.19). The activity of this enzyme in the kidneys is stimulated by parathyroid hormone (fig. 19.20).

The hormone 1,25-dihydroxyvitamin D_3 helps to raise the plasma concentrations of calcium and phosphate by stimulating (1) the intestinal absorption of calcium and phosphate, (2) the resorption of bones, and (3) the renal reabsorption of calcium and phosphate so that less is excreted in the urine. Notice that 1,25-dihydroxyvitamin D_3, but not parathyroid hormone, directly stimulates intestinal absorption of calcium and phosphate and promotes the reabsorption of phosphate in the kidneys.

Figure 19.20. A decrease in plasma Ca^{++} directly stimulates the secretion of parathyroid hormone (*PTH*). The production of 1,25-dihydroxyvitamin D_3 also rises when Ca^{++} is low because PTH stimulates the final hydroxylation step in the formation of this compound in the kidneys.

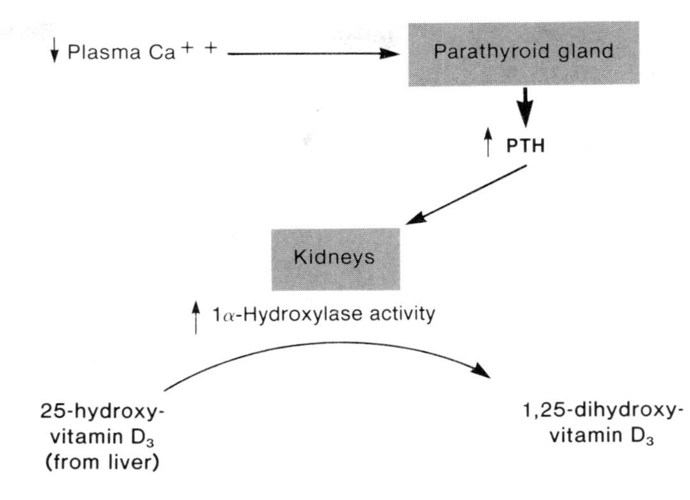

Since 1,25-dihydroxyvitamin D_3 directly stimulates bone resorption, it seems paradoxical that this hormone is needed for proper bone deposition and, in fact, that inadequate amounts of 1,25-dihydroxyvitamin D_3 result in the bone demineralization of osteomalacia and rickets. This apparent paradox may be explained logically by the fact that *the primary function of 1,25-dihydroxyvitamin D_3 is stimulation of intestinal Ca^{++} and PO_4^{-3} absorption.* When calcium intake is adequate, the major result of 1,25-dihydroxyvitamin D_3 action is the availability of Ca^{++} and PO_4^{-3} in sufficient amounts to promote bone deposition. Only when calcium intake is inadequate does the direct effect of 1,25-dihydroxyvitamin D_3 on bone resorption become significant, acting to assure proper blood Ca^{++} levels.

Another possible explanation of the paradoxical effects of vitamin D is that a different metabolite of this vitamin may act as a hormone and stimulate bone deposition. There is evidence that *24,25-dihydroxyvitamin D_3* may have this function. Further research is required to determine the precise mechanism by which vitamin D functions to promote bone deposition indirectly and to prevent osteomalacia and rickets when there is adequate calcium in the diet.

Negative Feedback Control of Calcium and Phosphate Balance

The secretion of parathyroid hormone is controlled by the plasma calcium concentrations. Its secretion is stimulated by low-calcium and inhibited by high-calcium concentrations. Since parathyroid hormone stimulates the final hydroxylation step in the formation of 1,25-dihydroxyvitamin D_3, a rise in parathyroid hormone results in an increase in production of 1,25-dihydroxyvitamin D_3. Low blood calcium can thus be corrected by the effects of increased parathyroid hormone and 1,25-dihydroxyvitamin D_3 (fig. 19.21).

It is possible that plasma calcium levels might fall while phosphate levels remain normal. In this case, the increased secretion of parathyroid hormone and the production of 1,25-dihydroxyvitamin D_3 that result could abnormally raise phosphate levels while acting to restore normal calcium levels. This is prevented by the fact that parathyroid hormone inhibits renal reabsorption in the kidneys, so that more phosphate is excreted in the urine (fig. 19.21). In this way blood calcium levels can be raised to normal without excessively raising blood phosphate concentrations.

Figure 19.21. The negative feedback loop that returns low blood Ca^{++} concentrations to normal without simultaneously raising blood phosphate levels above normal.

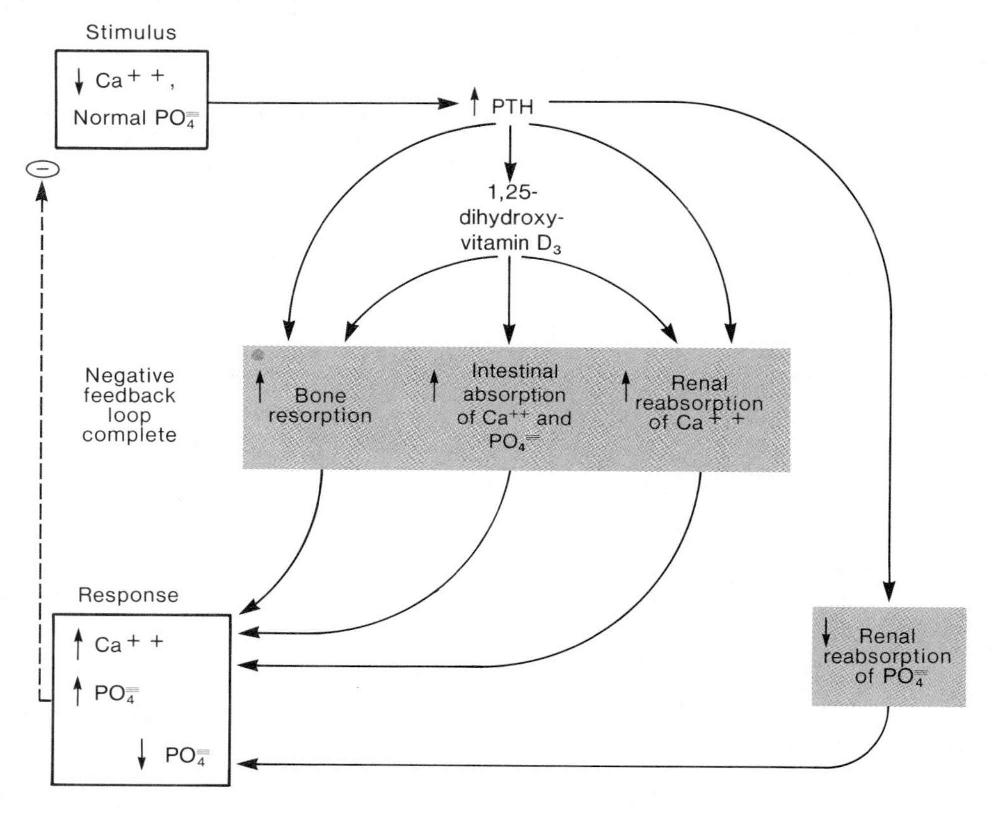

Calcitonin

Experiments in the 1960s revealed that high blood calcium in dogs may be lowered by a hormone secreted from the thyroid gland. This hormone thus has an effect opposite to that of parathyroid hormone and 1,25-dihydroxyvitamin D_3. The calcium-lowering hormone, called **calcitonin,** was found to be a thirty-two-amino-acid polypeptide secreted by *parafollicular cells,* or *C cells,* in the thyroid, which are distinct from the follicular cells that secreted thyroxine.

The secretion of calcitonin is stimulated by high plasma calcium levels and acts to lower calcium levels by (1) inhibiting the activity of osteoclasts, thus reducing bone resorption and (2) stimulating the urinary excretion of calcium and phosphate by inhibiting their reabsorption in the kidneys (fig. 19.22).

Figure 19.22. Negative feedback control of calcitonin secretion.

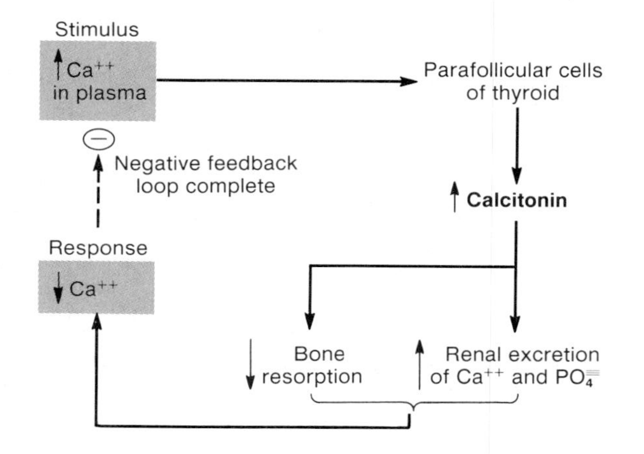

Although it is attractive to think that calcium balance is regulated by the effects of antagonistic hormones, the significance of calcitonin in human physiology remains unclear. Patients with their thyroid gland surgically removed (as for thyroid cancer) are *not* hypercalcemic, as would have been expected if calcitonin were needed to lower blood calcium levels. Similarly, patients who receive injections of many times the normal amounts of calcitonin do not become hypocalcemic. The ability of very large, pharmacological doses of calcitonin to inhibit osteoclast activity and bone resorption, however, is clinically useful in the treatment of *Paget's disease,* in which osteoclast activity causes softening of bone.

1. *Describe how the secretion of parathyroid hormone and of calcitonin is regulated.*
2. *List the steps involved in the formation of 1,25-dihydroxyvitamin D_3, and indicate how this formation is influenced by parathyroid hormone.*
3. *Describe the actions of parathyroid hormone, 1,25-dihydroxyvitamin D_3, and calcitonin on the intestine, skeletal system, and kidneys, and explain how these actions affect the blood levels of calcium.*
4. *Explain how the* in vivo *effects of 1,25-dihydroxyvitamin D_3 differ when calcium intake is adequate or inadequate.*

Summary

Nutritional Requirements p. 590

I. Vitamins and elements serve as cofactors and coenzymes.
   A. Vitamins are divided into those that are fat-soluble (A, D, E, and K) and those that are water-soluble.
   B. Many water-soluble vitamins are needed for the activity of the enzymes involved in cell respiration.

II. The caloric intake must be sufficient to meet the energy expenditures of the body; an excessive caloric intake results in obesity.

Regulation of Energy Metabolism p. 592

I. The body tissues can use circulating energy substrates, including glucose, fatty acids, ketone bodies, lactic acid, amino acid, and others, for cell respiration.
   A. Different organs have different preferred energy sources.
   B. Circulating energy substrates can be obtained from food or from the energy reserves of glycogen, fat, and protein in the body.

II. Eating behavior is regulated, at least in part, by the hypothalamus.
   A. Lesions of the ventromedial area of the hypothalamus produce hyperphagia, whereas lesions of the lateral hypothalamus produce hypophagia.
   B. A variety of neurotransmitters have been implicated in the control of eating behavior; these include the endorphins, norepinephrine, serotonin, and cholecystokinin.

III. The control of energy balance in the body is regulated by the anabolic and catabolic effects of a variety of hormones.

Energy Regulation by the Islets of Langerhans p. 594

I. A rise in plasma glucose concentration stimulates insulin and inhibits glucagon secretion.
   A. Amino acids stimulate the secretion of both insulin and glucagon.
   B. Insulin secretion is also stimulated by parasympathetic innervation of the islets and by the action of gastric inhibitory peptide (GIP), secreted by the intestine.

II. During the intestinal absorption of a meal, insulin promotes the uptake of blood glucose into tissue cells.
   A. This lowers the blood glucose concentration and increases the energy reserves of glycogen, fat, and protein.
   B. Insulin is required for the production of fat by adipose cells.

III. During periods of fasting, insulin secretion decreases and glucagon secretion increases.
   A. Glucagon stimulates glycogenolysis in the liver, gluconeogenesis, lipolysis, and ketogenesis.
   B. These effects help maintain adequate levels of blood glucose for the brain and provide alternate energy sources for other organs.

Diabetes Mellitus and Hypoglycemia p. 600

I. Diabetes mellitus and reactive hypoglycemia represent disorders of the islets of Langerhans.
   A. Type I, or juvenile-onset, diabetes occurs when the beta cells are destroyed; the resulting lack of insulin and excessive glucagon secretion produce the symptoms of this disease.
   B. Type II, or maturity-onset, diabetes occurs as a result of a relative tissue insensitivity to insulin and inadequate insulin secretion; this condition is aggravated by obesity.
   C. Reactive hypoglycemia occurs when the islets secrete excessive amounts of insulin in response to a rise in blood glucose concentration.

Metabolic Regulation by Adrenal Hormones, Thyroxine, and Growth Hormone p. 602

I. The adrenal hormones involved in energy regulation include epinephrine from the adrenal medulla and glucocorticoids (mainly hydrocortisone) from the adrenal cortex.
   A. The effects of epinephrine are similar to those of glucagon.
   B. Glucocorticoids promote the breakdown of muscle protein and the conversion of amino acids to glucose in the liver.

II. Thyroxine stimulates the rate of cell respiration in almost all cells in the body.
   A. Thyroxine thus sets the basal metabolic rate (BMR), which is the rate at which energy is consumed by the body under resting conditions.
   B. Thyroxine also promotes protein synthesis and by this means is needed for proper body growth and development, particularly of the central nervous system.

III. The secretion of growth hormone is regulated by releasing and inhibiting hormones from the hypothalamus.
   A. The secretion of growth hormone is stimulated by a protein meal and by a fall in glucose, as occurs during fasting.
   B. Growth hormone stimulates catabolism of lipids and inhibits glucose utilization.
   C. Growth hormone also stimulates protein synthesis and thus promotes body growth.
   D. The anabolic effects of growth hormone, including the stimulation of bone growth in childhood, is believed to be produced indirectly via polypeptides called somatomedins.

Regulation of Calcium and Phosphate Balance p. 610

I. Bone contains calcium and phosphate in the form of hydroxyapatite crystals; this serves as a reserve supply of calcium and phosphate for the blood.
   A. The formation and resorption of bone is produced by the action of osteoblasts and osteoclasts, respectively.
   B. The plasma concentrations of calcium and phosphate are also affected by absorption from the intestine and by the urinary excretion of these ions.

II. Parathyroid hormone stimulates bone resorption and calcium reabsorption in the kidneys; this hormone thus acts to raise the blood calcium concentration.
   A. The secretion of parathyroid hormone is stimulated by a fall in blood calcium levels.
   B. Parathyroid hormone also inhibits reabsorption of phosphate in the kidneys, so that more phosphate is excreted in the urine.

III. 1,25-dihydroxyvitamin D_3 is derived from vitamin D by hydroxylation reactions in the liver and kidneys.
   A. The last hydroxylation step is stimulated by parathyroid hormone.
   B. 1,25-dihydroxyvitamin D_3 stimulates the intestinal absorption of calcium and phosphate, resorption of bone, and renal reabsorption of phosphate.

IV. A rise in parathyroid hormone, accompanied by the increased production of 1,25-dihydroxyvitamin D_3, helps to maintain proper blood levels of calcium and phosphate in response to a fall in calcium levels.

V. Calcitonin is secreted by the parafollicular cells of the thyroid gland.
   A. Calcitonin secretion is stimulated by a rise in blood calcium levels.
   B. Calcitonin, at least at pharmacological levels, acts to lower blood calcium by inhibiting bone resorption and stimulating the urinary excretion of calcium and phosphate.

Review Activities

Objective Questions

Match the following:

1. Absorption of carbohydrate meal
2. Absorption of protein meal
3. Fasting

(a) rise in insulin, rise in glucagon
(b) fall in insulin, rise in glucagon
(c) rise in insulin, fall in glucagon
(d) fall in insulin, fall in glucagon

Match the following:

4. Growth hormone
5. Thyroxine
6. Hydrocortisone

(a) increased protein synthesis, increased cell respiration
(b) protein catabolism in muscles; gluconeo-genesis in liver
(c) protein synthesis in muscles, decreased glucose utilization
(d) fall in blood glucose, increased fat synthesis

7. A lowering of blood glucose concentration promotes
   (a) decreased lipogenesis
   (b) increased lipolysis
   (c) glycogenolysis
   (d) all of the above
8. Glucose can be secreted into the blood by the
   (a) liver
   (b) muscles
   (c) liver and muscles
   (d) liver, muscles, and brain
9. The basal metabolic rate is determined primarily by
   (a) hydrocortisone
   (b) insulin
   (c) growth hormone
   (d) thyroxine

10. Somatomedins are required for the anabolic effects of
    (a) hydrocortisone
    (b) insulin
    (c) growth hormone
    (d) thyroxine

11. The increased intestinal absorption of calcium is stimulated directly by
    (a) parathyroid hormone
    (b) 1,25-dihydroxyvitamin D_3
    (c) calcitonin
    (d) all of the above

12. A rise in blood calcium levels directly stimulates
    (a) parathyroid hormone secretion
    (b) calcitonin secretion
    (c) 1,25-dihydroxyvitamin D_3 formation
    (d) all of the above

Essay Questions

1. Compare the metabolic effects of fasting to the state of uncontrolled type I diabetes. Explain the hormonal similarities of these conditions.
2. Glucocorticoids stimulate the breakdown of protein in muscles but the synthesis of protein in the liver. Explain the significance of these differences.
3. Describe how thyroxine affects cell respiration, and explain why a hypothyroid person has a tendency to gain weight and has a reduced tolerance to cold.
4. Compare and contrast the metabolic effects of thyroxine and growth hormone.
5. Why is vitamin D considered to be both a vitamin and a prehormone? Explain why people with osteoporosis might be helped by taking controlled amounts of vitamin D.

Selected Readings

Austin, L. A., and H. Heath, III. 1981. Calcitonin: Physiology and pathophysiology. *New England Journal of Medicine* 304:269.

Barret, E. J., and R. A. DeFronzo. 1984. Diabetic ketoacidosis: Diagnosis and treatment. *Hospital Practice* 19:89.

Cahill, G. F., and H. O. McDevitt. 1981. Insulin-dependent diabetes mellitus: The initial lesion. *New England Journal of Medicine* 304:454.

Cheng, K., and J. Larner. 1985. Intracellular mediators of insulin action. *Annual Review of Physiology* 47:405.

DeLuca, H. F. 1980. The vitamin D hormonal system: Implications for bone disease. *Hospital Practice* 15:57.

Eisenbarth, G. S. 1986. Type 1 diabetes mellitus: A chronic autoimmune disease. *New England Journal of Medicine* 314:1360.

Gardner, L. I. July 1972. Deprivation dwarfism. *Scientific American.*

Goodman, D. S. 1984. Vitamin A and retinoids in health and disease. *New England Journal of Medicine* 310:1023.

Habener, J. F., and J. E. Mahaffey. 1978. Osteomalacia and disorders of vitamin D metabolism. *Annual Review of Medicine* 29:327.

Hahn, T. J. 1986. Physiology of bone: Mechanisms of osteopenic disorders. *Hospital Practice* 21:73.

Haussler, M. R., and T. A. McCain. 1977. Basic and clinical concepts related to vitamin D metabolism and action. *New England Journal of Medicine* 297: first part, p. 974; second part, p. 1041.

Hirsch, J. 1984. Hypothalamic control of appetite. *Hospital Practice* 19:131.

Isaksson, O. G. P., S. Edén, and J.-O. Jansson. 1985. Mode of action of pituitary growth hormone on target cells. *Annual Review of Physiology* 47:483.

Levine, M. 1986. New concepts in the biology and biochemistry of ascorbic acid. *New England Journal of Medicine* 314:892.

Notkins, A. L. November 1979. The cause of diabetes. *Scientific American.*

Oppenheimer, J. H. 1979. Thyroid hormone action at the cellular level. *Science* 203:971.

Phillips, L. S., and R. Vassilopoulou-Sellin. 1980. Somatomedins. *New England Journal of Medicine* 302: first part, p. 371; second part, p. 438.

Raisz, L. G., and B. E. Kream. 1981. Hormonal control of skeletal growth. *Annual Review of Physiology* 43:225.

Siperstein, M. D. 1985. Type II diabetes: Some problems in diagnosis and treatment. *Hospital Practice* 20:55.

Sterling, S. 1979. Thyroid hormone action at the cellular level. *New England Journal of Medicine* 300: first part, p. 117; second part, p. 173.

Tepperman, J. 1980. *Metabolic and Endocrine Physiology.* 4th ed. Chicago: Year Book Medical Publishers.

Unger, R. H., R. E. Dobbs, and L. Orci. 1979. Insulin, glucagon, and somatostatin secretion in the regulation of metabolism. *Annual Review of Physiology* 40:307.

Unger, R. H., and L. Orci. 1981. Glucagon and the A cell: Physiology and pathophysiology. *New England Journal of Medicine* 304: first part, p. 1518; second part, p. 1575.

Van Wyk, J., and L. E. Underwood. 1978. Growth hormone, somatomedins, and growth failure. *Hospital Practice* 13:57.

Wynder, E. L., and D. P. Rose. 1984. Diet and breast cancer. *Hospital Practice* 19:73.

20 Reproduction

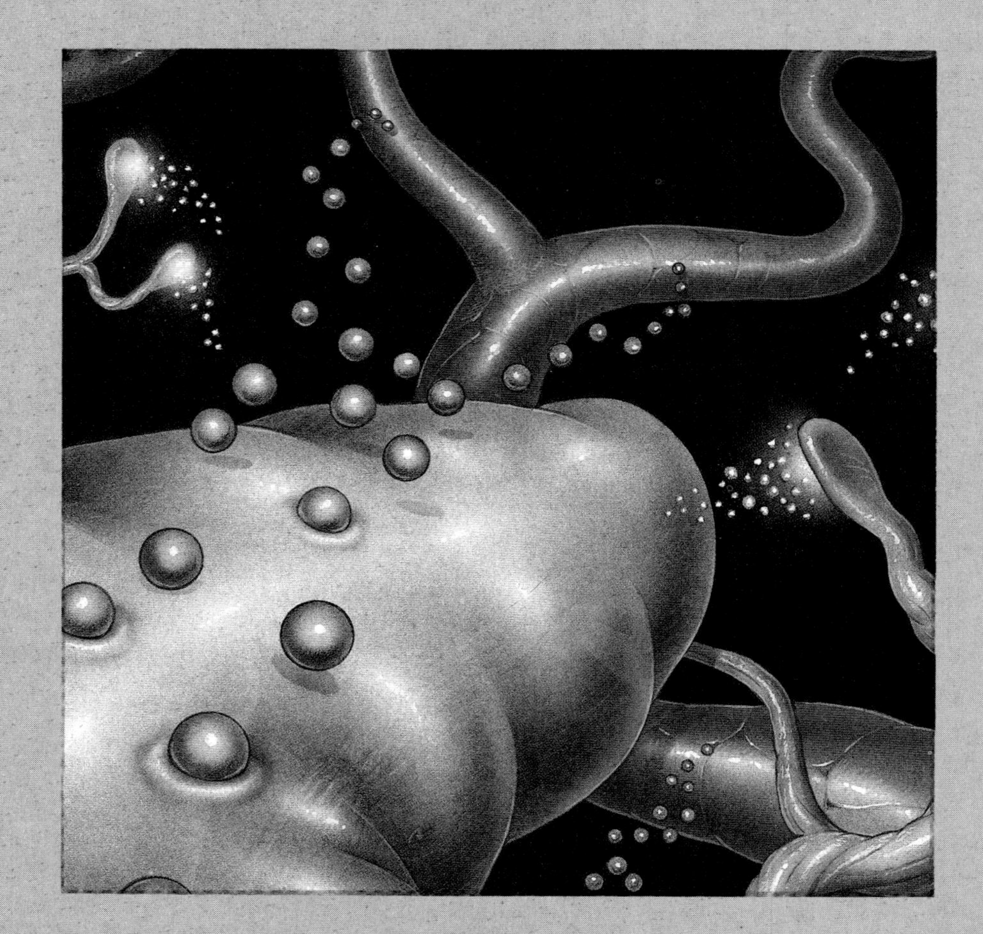

Objectives

By studying this chapter, you should be able to

1. describe how the chromosomal sex of an embryo is determined by the chromosomal content and how this relates to the development of testes or ovaries
2. explain how the development of sex accessory organs and external genitalia is affected by the type of gonads present in the embryo
3. describe the hormonal changes that occur during puberty, the mechanisms that might control the onset of puberty, and the changes in secondary sexual characteristics that result during puberty
4. explain how the secretions of pituitary gonadotrophic hormones (FSH and LH) are regulated in the male and describe the actions of FSH and LH on the testis
5. describe the structure of the testis and how the interstitial Leydig cells and seminiferous tubules may interact
6. describe the stages of spermatogenesis and the roles of Sertoli cells in spermatogenesis
7. describe the hormonal control of spermatogenesis and the effects of androgens on the male accessory sexual organs
8. explain the physiology of erection and ejaculation and describe the composition of semen and the requirements for male fertility
9. describe oogenesis and the stages of ovarian follicle development through ovulation and the formation of a corpus luteum
10. explain the hormonal interactions involved in the control of ovulation
11. describe the changes that occur in ovarian sex steroid secretion during a nonfertile cycle and the function and fate of a corpus luteum during a nonfertile cycle
12. explain how the secretion of FSH and LH is controlled through negative and positive feedback mechanisms during a menstrual cycle
13. explain how the contraceptive pill functions to prevent ovulation
14. describe the cyclic changes that occur in the endometrium and the hormonal mechanisms that cause these changes
15. describe the acrosomal reaction and the events that occur at fertilization, blastocyst formation, and implantation
16. explain how menstruation and further ovulation are normally prevented during pregnancy
17. describe the structure and functions of the placenta
18. list the hormones secreted by the placenta and describe their actions
19. describe the factors that stimulate uterine contractions during labor and parturition and explain how the onset of labor may be regulated
20. explain the hormonal requirements for development of the mammary glands during pregnancy and why lactation does not normally occur during pregnancy
21. describe the milk-ejection reflex

Outline

Figure 20.1. The human life cycle. Numbers in parentheses indicate haploid state (23 chromosomes) and diploid state (46 chromosomes).

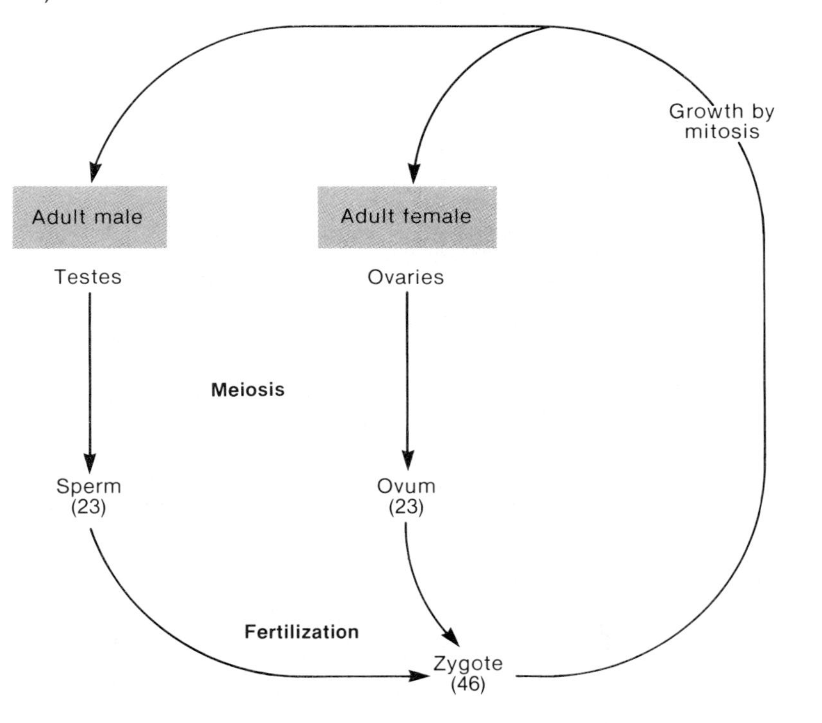

Sexual Reproduction

"A chicken is an egg's way of making another egg." Phrased in more modern terms, genes are "selfish." Genes, according to this view, do not exist in order to make a well-functioning chicken (or other organism). The organism, rather, exists and functions so that the genes can survive beyond the mortal life of individual members of a species. Whether or not one accepts this rather cynical view, it is clear that reproduction is an essential function of life. The incredible complexity of structure and function in living organisms could not be produced in successive generations by chance; mechanisms must exist to transmit the blueprint (genetic code) from one generation to the next. *Sexual reproduction,* in which genes from two individuals are combined in random and novel ways with each new generation, offers the further advantage of introducing great variability into a population. This variability of genetic constitution helps to insure that some members of a population will survive changes in the environment over evolutionary time.

In sexual reproduction, **germ cells,** or **gametes** (sperm and ova), are formed within the *gonads* (testes and ovaries) by a process of reduction division, or *meiosis.* During this type of cell division, the normal number of chromosomes in most human cells—forty-six—is halved, so that each gamete receives twenty-three chromosomes. Fusion of a sperm and egg cell (ovum) in the act of **fertilization** results in restoration of the original chromosome number of forty-six in the fertilized egg (the *zygote*). Growth of the zygote into an adult member of the next generation occurs by means of mitotic cell divisions, as described in chapter 3. When this individual reaches puberty, mature sperm or ova will be formed by meiosis within the gonads so that the life cycle can be continued (fig. 20.1).

Sex Determination

Each zygote inherits twenty-three chromosomes from its mother and twenty-three chromosomes from its father. This does not produce forty-six different chromosomes, but rather, twenty-three pairs of *homologous chromosomes.* Each member of a homologous pair, with the important exception of the sex chromosomes, looks like the other and contains similar genes (such as those coding for eye color,

Figure 20.20. The endocrine control of spermatogenesis. During puberty both testosterone and FSH are required to initiate spermatogenesis. In the adult, however, testosterone alone can maintain spermatogenesis.

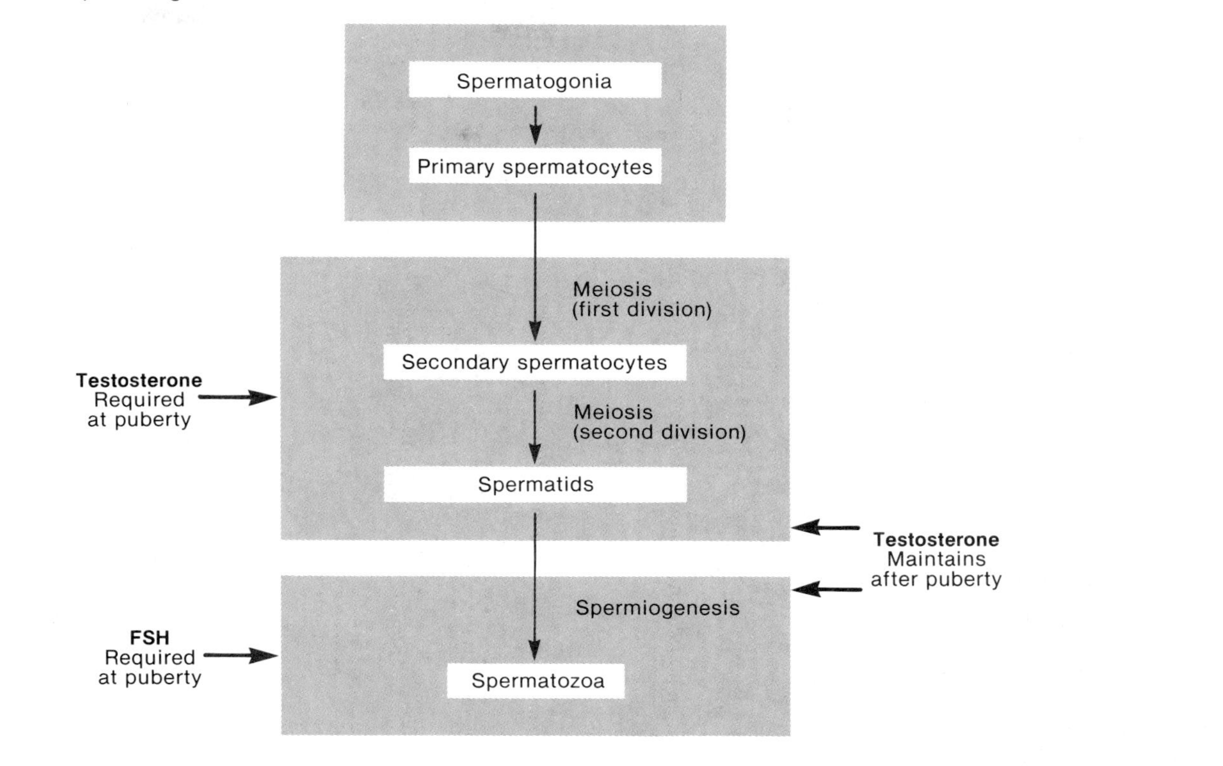

Figure 20.21. A human spermatozoan: (*a*) a diagrammatic representation; (*b*) a scanning electron micrograph. (From: *Tissues and Organs: A Text Atlas of Scanning Electron Microscopy* by R. G. Kessel and R. Kardon. W. H. Freeman and Company. © 1979.)

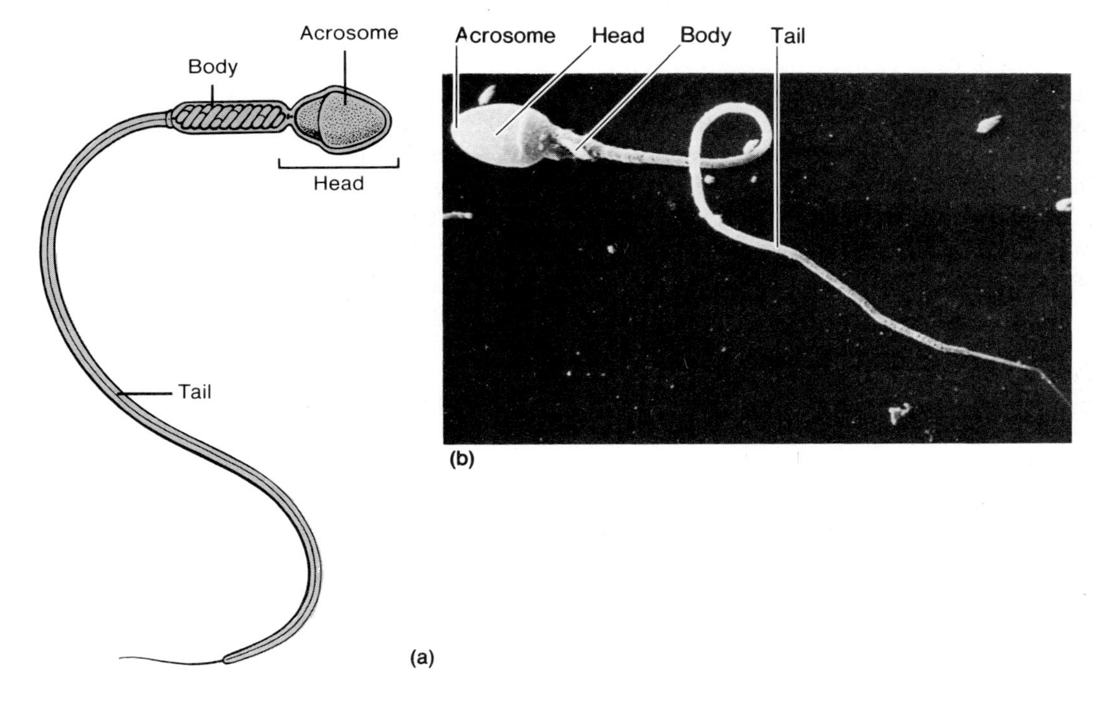

Figure 20.22. Spermatozoa travel from seminiferous tubules to the rete testis, the ductuli efferentes, and the epididymis. Spermatozoa leave the scrotum and enter the body cavity in a tube called the vas deferens.

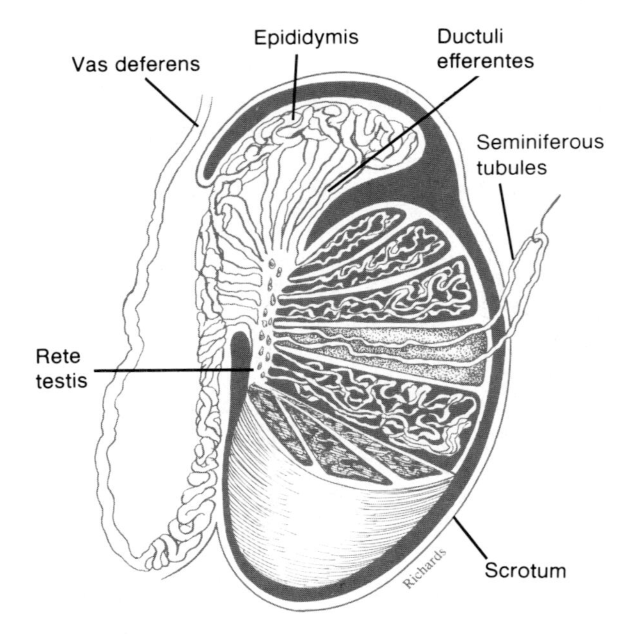

Male Sex Accessory Organs

The seminiferous tubules are connected at both ends to the *rete testis* (fig. 20.22). Spermatozoa and tubular secretions are moved to this area of the testis and are drained via the *ductuli efferentes* into the **epididymis.** The epididymis is a single-coiled tube, four to five meters long if stretched out, that receives the tubular products. Spermatozoa enter at the "head" of the epididymis and are drained from its "tail" by a single tube, the **vas deferens.**

Spermatozoa that enter the head of the epididymis are nonmotile. During their passage through the epididymis, spermatozoa gain motility and undergo other maturational changes—sperm that leave the epididymis are more resistant to changes in pH and temperature, are motile, and can become able to fertilize an ovum. Sperm obtained from the seminiferous tubules, in contrast, cannot fertilize an ovum. The epididymis serves as a site for sperm maturation and for the storage of sperm between ejaculations.

The vas deferens carries sperm from the epididymis out of the scrotum and into the body cavity. In its passage, the vas deferens obtains fluid secretions of the **seminal vesicles** and **prostate** gland. This fluid, now called *semen,* is carried by the ejaculatory duct to the *urethra* (fig. 20.23). Sperm can be stored within a widened area of the vas deferens known as the *ampulla.*

The seminal vesicles and prostate are androgen-dependent accessory sexual organs—they atrophy if androgen is withdrawn by castration. The seminal vesicles secrete fluid containing fructose (which serves as an energy source for the spermatozoa), citric acid, coagulation proteins, and prostaglandins. The prostate secretes a liquefying agent and the enzyme *acid phosphatase,* which is often measured clinically to assess prostate function. Abnormal growth of the prostate often occurs in older men—a condition called **benign prostatic hyperplasia (BPH).** The causes of this condition are not well understood.

Figure 20.23. The male accessory glands and ducts associated with the urethra and urinary bladder.

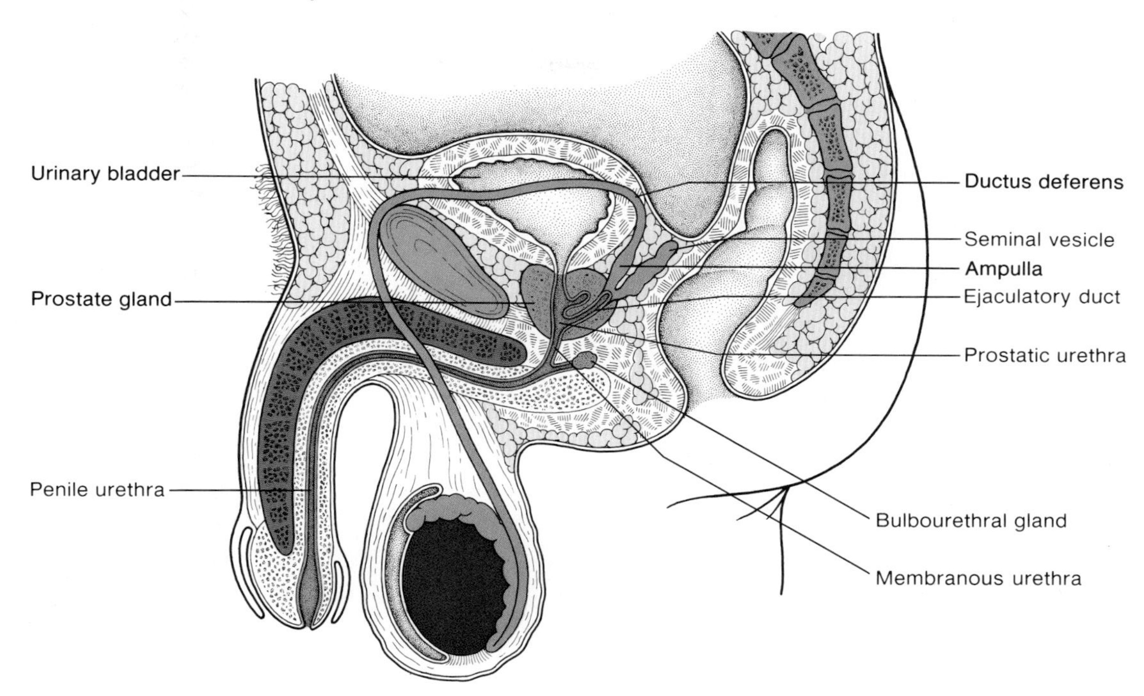

Table 20.6 Control of erection and ejaculation.

Regulation	Effect	Result
Parasympathetic nerves	Vasodilation produces increased blood flow into erectile tissues—corpora cavernosa and corpus spongiosum. As tissues become turgid, venous outflow is partially occluded, further increasing accumulation of blood.	Erection
Sympathetic nerves	Peristaltic waves of contraction in tubular system—primarily epididymis, vas deferens, and ejaculatory ducts, and contraction of prostate and seminal vesicles.	Ejaculation

Erection, Emission, and Ejaculation

Erection, accompanied by increases in the length and width of the penis, is achieved as a result of blood flow into the "erectile tissues" of the penis. These erectile tissues include two paired structures—the *corpora cavernosa*—located on the dorsal side of the penis, and one unpaired *corpus spongiosum* on the ventral side (fig. 20.24). The urethra runs through the center of the corpus spongiosum. The erectile tissue forms columns extending the length of the penis, although the corpora cavernosa do not extend all the way to the tip.

Erection is achieved as a result of parasympathetic nerve-induced vasodilation of arterioles that allows blood to flow into the penis (table 20.6). As the erectile tissues become engorged with blood and the penis becomes turgid, venous outflow of blood is partially occluded, thus aiding erection. The term *emission* refers to the movement of semen into the urethra, and *ejaculation* refers to the forcible expulsion of semen from the urethra out of the penis. Emission and ejaculation are stimulated by sympathetic nerves, which cause peristaltic contractions of the tubular system, contractions of the seminal vesicles and prostate, and contractions of muscles at the base of the penis. Sexual function in the male thus requires the synergistic action (rather than antagonistic action) of the parasympathetic and sympathetic systems.

Figure 20.24. The structure of the penis showing the attachment, blood and nerve supply, and the arrangement of the erectile tissue.

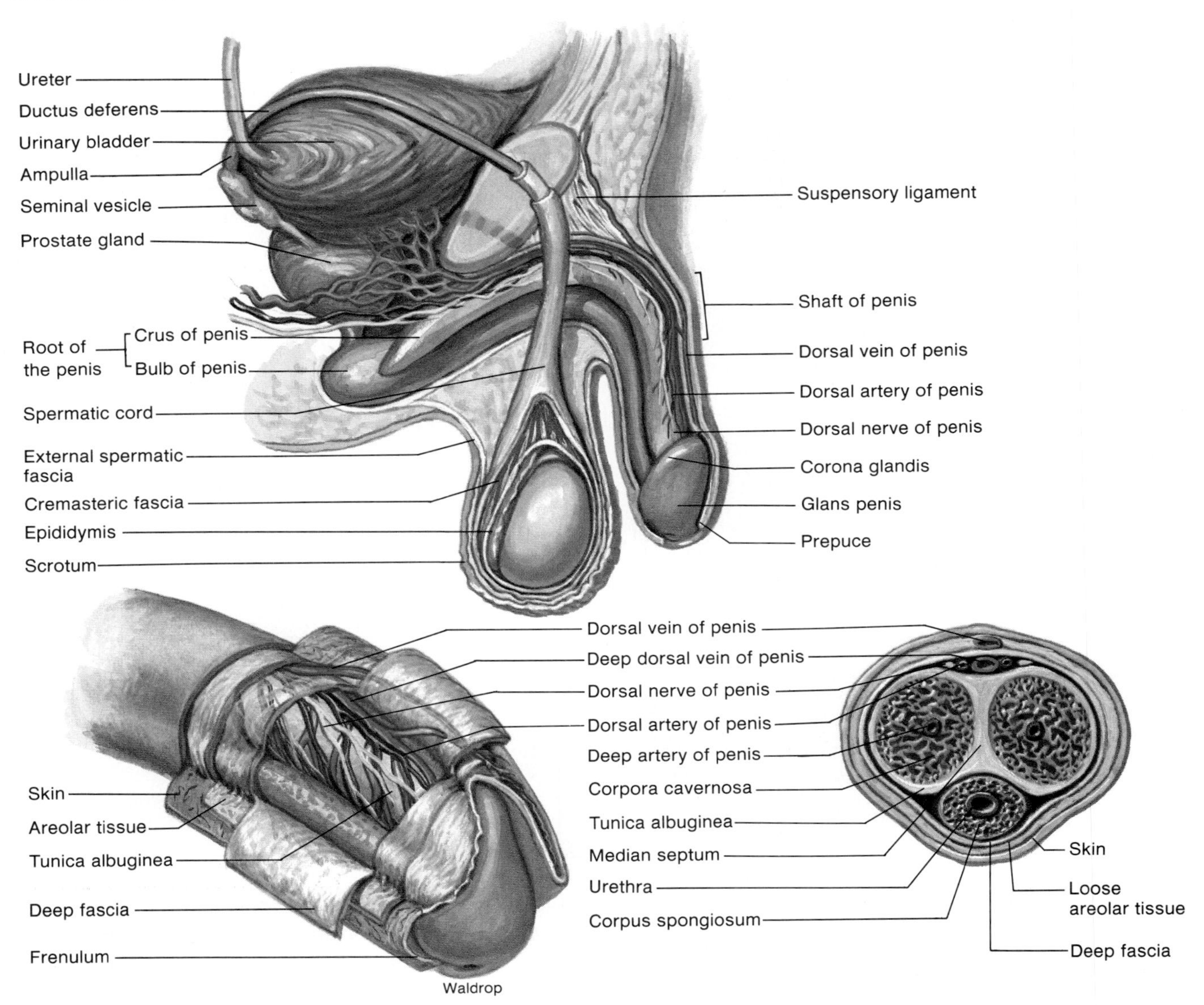

Table 20.7 Some characteristics and reference values used in the clinical examination of semen.

Characteristic	Reference Value
Volume of ejaculate	1.5–5.0 ml
Sperm count	40–250 million/ml
Sperm motility	
Percent of motile forms:	
1 hour after ejaculation	70% or more
3 hours after ejaculation	60% or more
Leukocyte count	0–2000/ml
pH	7.2–7.8
Fructose concentration	150–600 mg/100 ml

Source: Modified from Glasser, L. Seminal fluid and subfertility. Diagnostic Medicine, July/August 1981, p. 28.

Male Fertility

The male ejaculates about 1.5–5.0 ml of semen. The bulk of this fluid (45% to 80%) is produced by the seminal vesicles, and about 15% to 30% is contributed by the prostate. The sperm content in human males averages 60–150 million per milliliter in the ejaculated semen. Some of the values of normal human semen are summarized in table 20.7.

A sperm concentration below about twenty million per milliliter is termed *oligospermia,* and is associated with decreased fertility. A total sperm count below about fifty million per ejaculation is clinically significant in male infertility. In addition to low sperm counts as a cause of infertility, some men have antibodies against their own sperm (this is very common in men with vasectomies). Some women also produce antibodies against sperm.

Figure 20.25. A simplified illustration of a vasectomy, in which a segment of the ductus deferentia is removed through an incision in the scrotum.

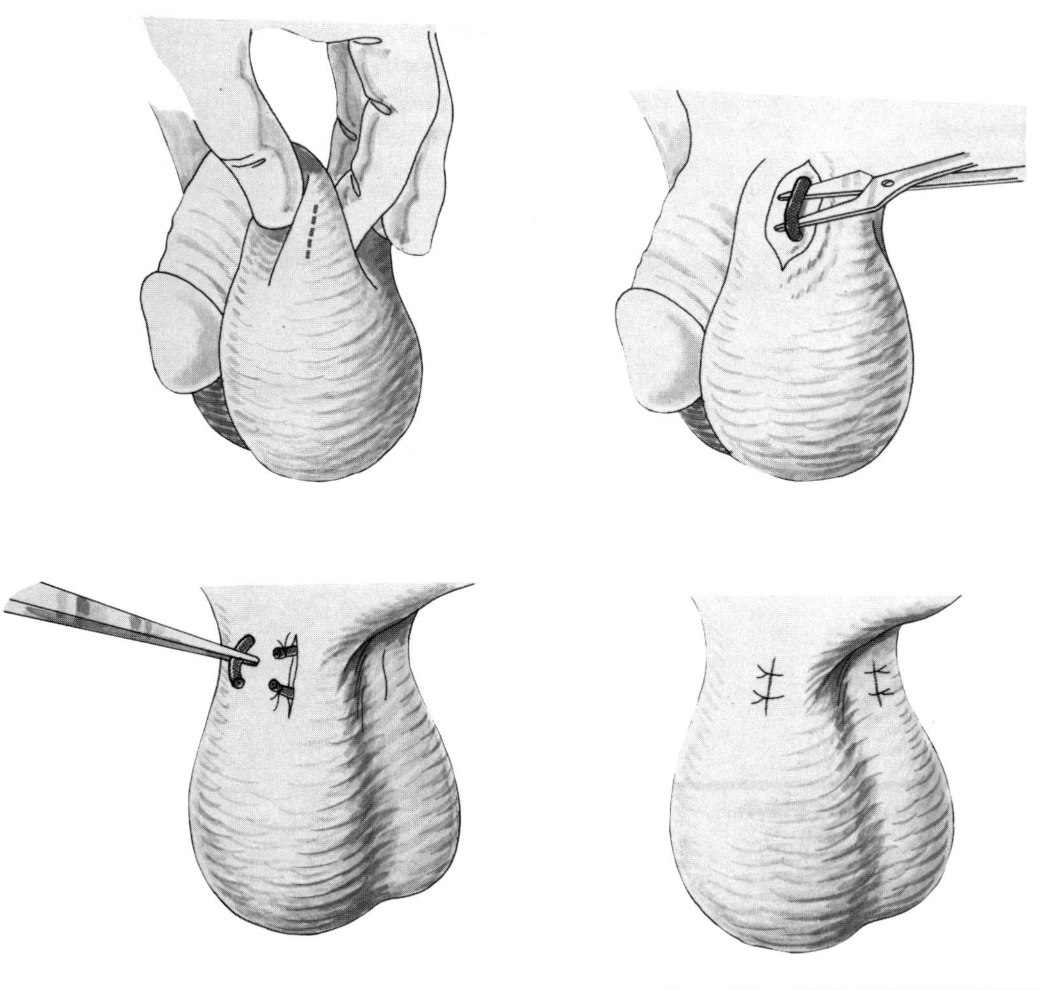

Vasectomy (fig. 20.25) is commonly performed as a contraceptive method. In this procedure the vas deferens is cut and tied or, in some cases, a valve or similar device is inserted. This procedure interferes with sperm transport but does not directly affect the secretion of androgens from Leydig cells in the interstitial tissue. Since spermatogenesis continues, the sperm produced cannot be drained from the testes and instead accumulates in "crypts" that form in the seminiferous tubules and vas deferens. These crypts present sites of inflammatory reactions in which spermatozoa are phagocytosed and destroyed by the immune system.

Although androgen secretion from the Leydig cells is not directly affected by a vasectomy, it is clear that tubular function may be affected by the inflammatory reactions that result. Since the tubules have been shown to secrete 5α-reduced androgens and perhaps other hormones, and since the physiological importance of this endocrine function is not presently understood, it is clear that more research is needed to establish the long-term safety of this contraceptive procedure.

1. *Describe the effects of castration on FSH and LH secretion in the male, and explain the experimental evidence that suggests that the testes produce a polypeptide that specifically inhibits FSH secretion.*
2. *Describe the two compartments of the testes with respect to their (a) structure, (b) function, and (c) response to gonadotrophin stimulation. Describe two ways that these compartments appear to interact.*
3. *Using a diagram, describe the stages of spermatogenesis. Explain why spermatogenesis can continue throughout life without using up all of the spermatogonia.*
4. *Describe the structure and proposed functions of the Sertoli cells in the seminiferous tubules.*
5. *Explain how FSH and androgens synergize to stimulate sperm production at puberty. Describe the hormonal requirements for spermatogenesis after puberty.*

Figure 20.26. A dorsal view of the female reproductive organs showing the relationship of the ovaries, uterine tubes, uterus, cervix, and vagina.

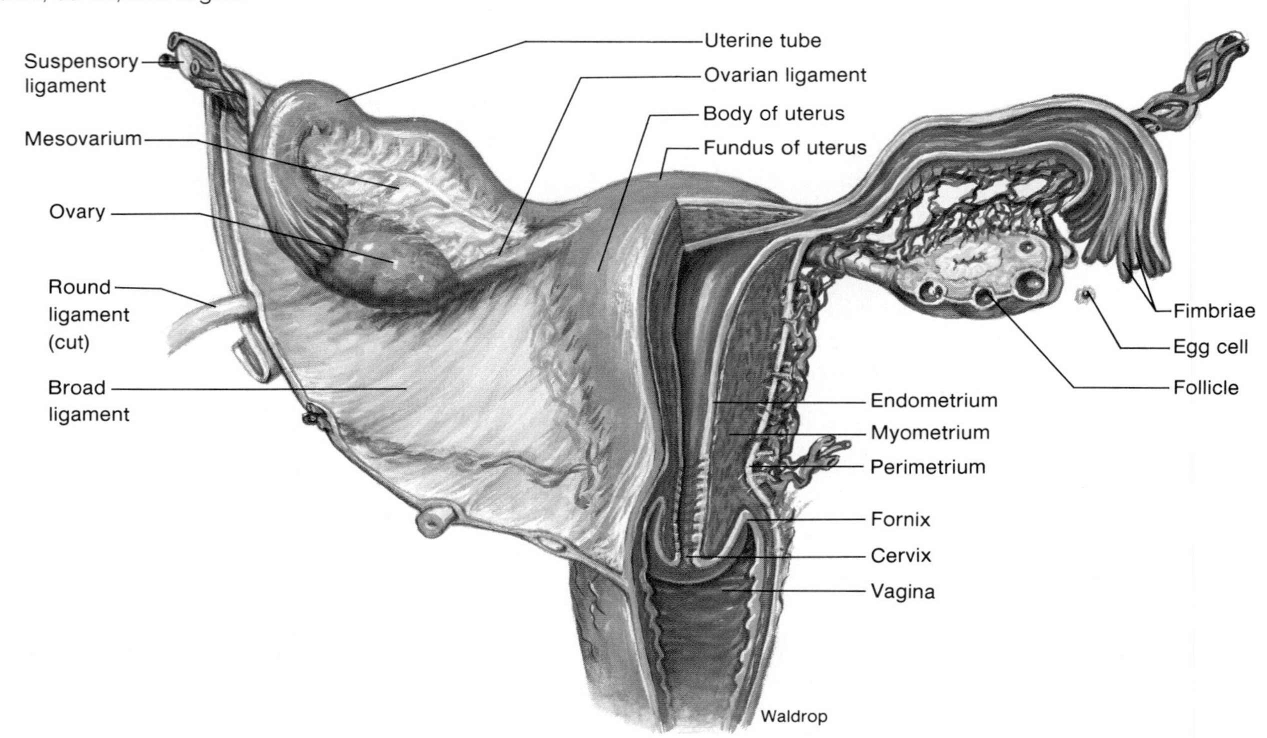

Female Reproductive System

The two ovaries (fig. 20.26) are located within the body cavity, suspended by means of ligaments from the pelvic girdle. Extensions, or *fimbriae,* of the **uterine,** or **fallopian, tubes** partially cover each ovary. Ova that are released from the ovary—in a process called *ovulation*—are normally drawn into the fallopian tubes by the action of the ciliated epithelial lining of the tubes. The lumen of each fallopian tube is continuous with the **uterus** (or womb), a pear-shaped muscular organ also suspended within the pelvic girdle by ligaments.

The uterus narrows to form the *cervix* (*cervi* = neck), which opens to the **vagina.** The only physical barrier between the vagina and uterus is a plug of *cervical mucus.* These structures—the vagina, uterus, and fallopian tubes—constitute the sex accessory organs of the female (fig. 20.27). Like the sex accessory organs of the male, the female genital tract is affected by gonadal steroid hormones. Cyclic changes in ovarian secretion, as will be described in the next section, cause cyclic changes in the epithelial lining of the genital tract. The epithelial lining of the uterus, known as the *endometrium,* is most dramatically altered during the ovarian cycle.

The vaginal opening is located immediately posterior to the opening of the urethra. Both openings are covered by inner *labia minora* and outer *labia majora* (fig. 20.28). The *clitoris,* which is homologous to the penis, as previously described, is located at the anterior margin of the labia minora.

Ovarian Cycle

The germ cells that migrate into the ovaries during early embryonic development multiply, so that by about five months of gestation (prenatal life) the ovaries contain approximately six to seven million oogonia. The production of new oogonia stops at this point and never resumes again. Towards the end of gestation the oogonia begin meiosis, at which time they are called **primary oocytes.** Like spermatogenesis in the prenatal male, oogenesis is arrested at prophase I of the first meiotic division. The number of primary oocytes decreases throughout a woman's reproductive years. The ovaries of a newborn girl contain about two million oocytes, but this number is reduced to about 300,000–400,000 by the time the girl enters puberty. Oogenesis ceases entirely at menopause (the time menstruation stops).

Figure 20.27. Organs of the female reproductive system seen in sagittal section.

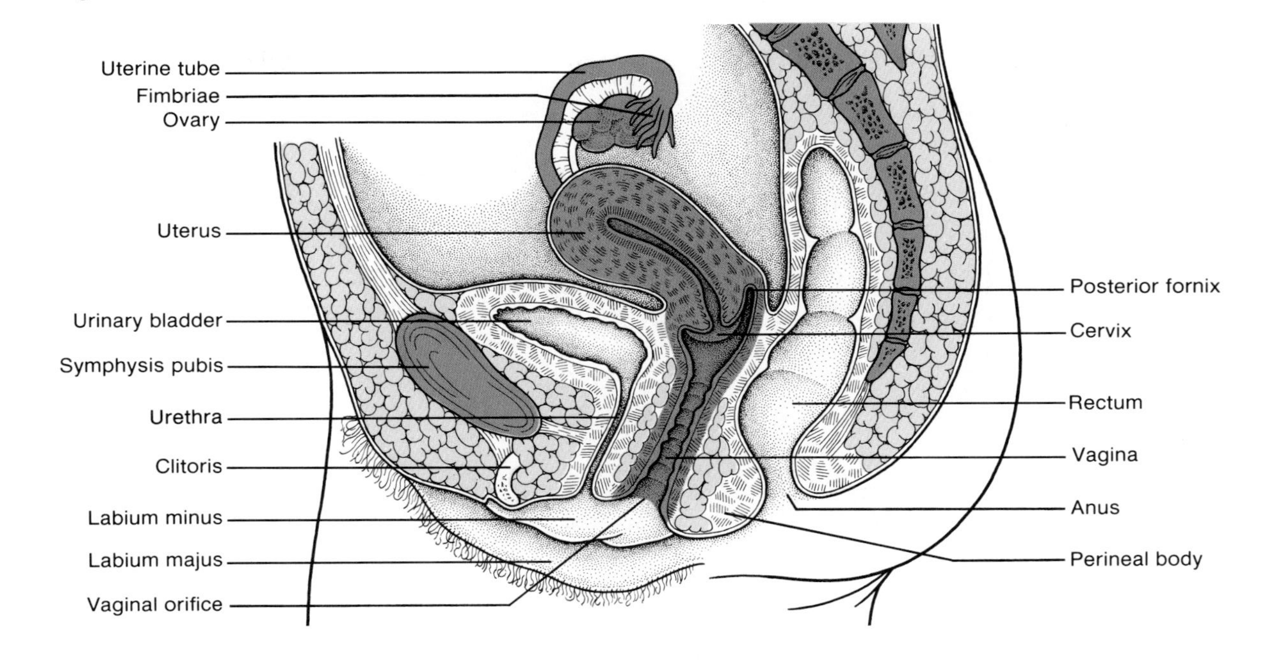

Figure 20.28. The external female genitalia.

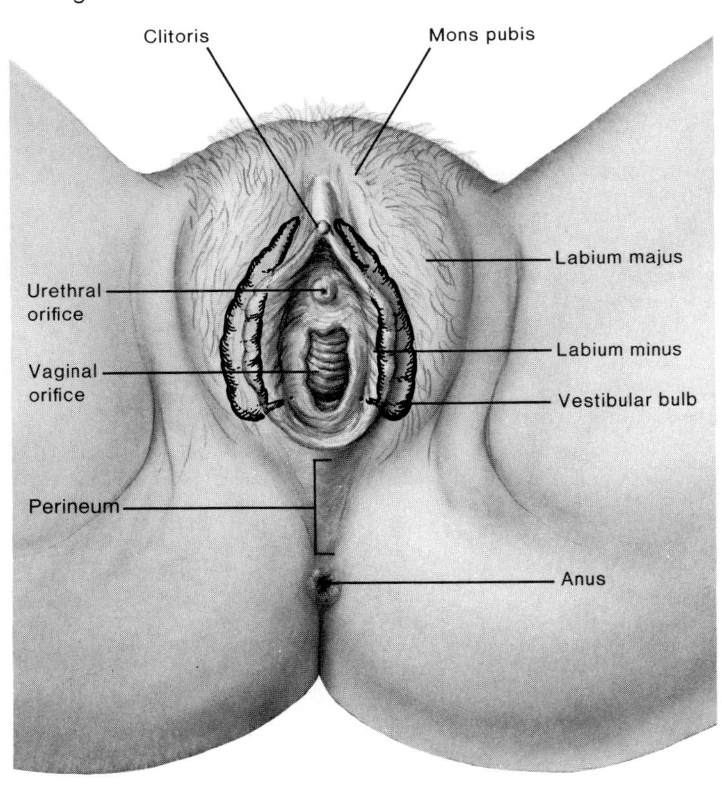

Figure 20.29. (*a*) A photomicrograph of primordial and secondary follicles. (*b*) A diagram of the parts of a secondary follicle.

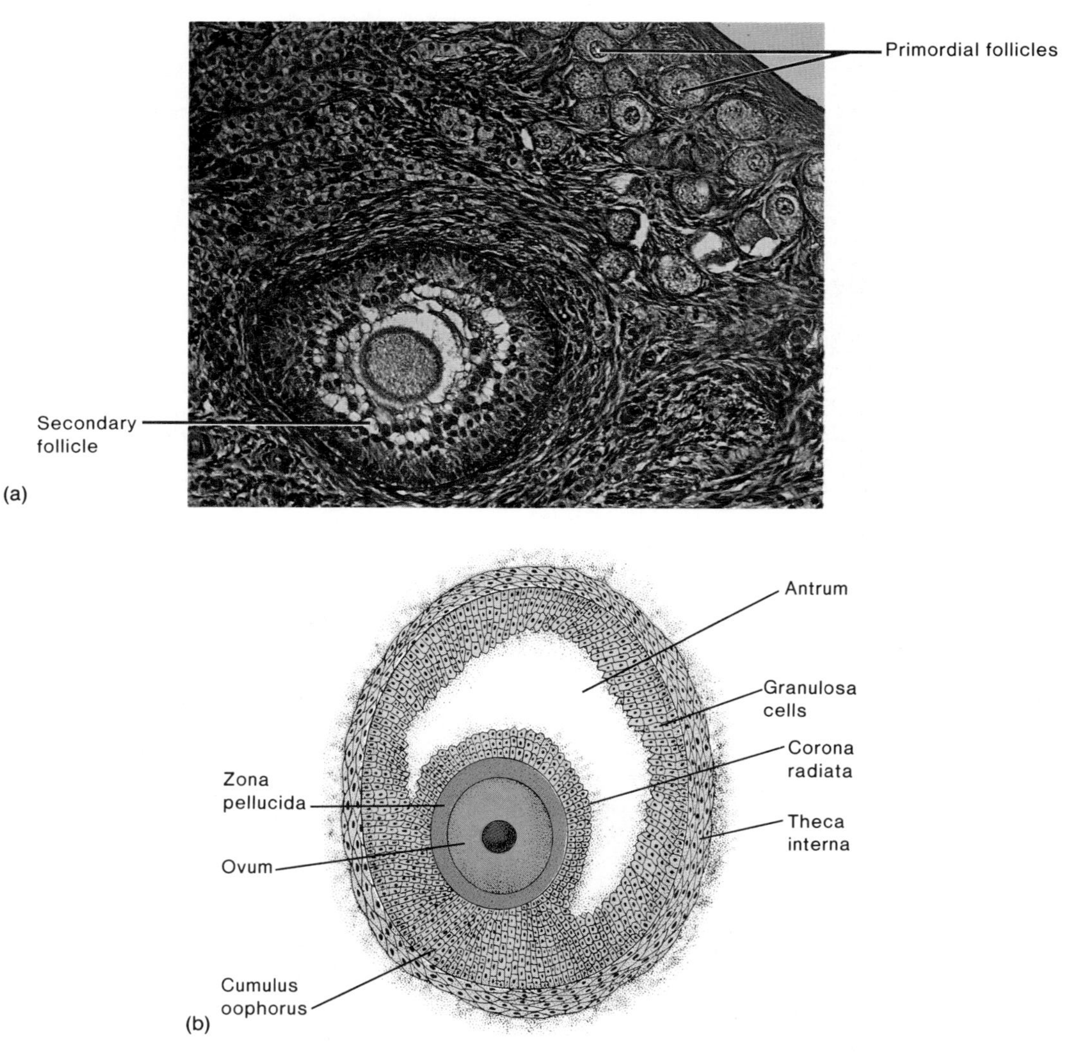

Primary oocytes that are not stimulated to complete the first meiotic division are contained within tiny follicles, called **primordial follicles.** In response to gonadotrophin stimulation some of these oocytes and follicles get larger, and the follicular cells divide to produce numerous small **granulosa cells** that surround the oocyte and fill the follicle. A follicle at this stage in development is called a **primary follicle.**

Some primary follicles will be stimulated to grow still bigger and develop a fluid-filled cavity, called an *antrum,* at which time they are called **secondary follicles** (fig. 20.29). The granulosa cells of secondary follicles form a ring around the circumference of the follicle and form a mound that supports the ovum. This mound is called the *cumulus oophorous.* Some granulosa cells also encircle the oocyte, forming a *corona radiata.* Between the oocyte and the corona radiata is a thin gel-like layer of proteins and polysaccharides, called the *zona pellucida.* Under the stimulation of FSH from the anterior pituitary, the granulosa cells secrete increasing amounts of estrogen as the follicles grow. Interestingly, the granulosa cells produce estrogen from its precursor testosterone, which is supplied by cells of the *theca interna* layer immediately outside the follicle.

As the follicle develops, the primary oocyte completes its first meiotic division. This does not form two complete cells, however, because only one cell—the **secondary oocyte**—gets all the cytoplasm. The other cell formed at this time becomes a small *polar body* (fig. 20.30), which eventually fragments and disappears. The secondary oocyte enters the second meiotic division, but meiosis is arrested at metaphase II and is never completed unless fertilization occurs.

Figure 20.30. (*a*) A primary oocyte at metaphase I of meiosis. Note the alignment of chromosomes (arrow). (*b*) A human secondary oocyte formed at the end of the first meiotic division and the first polar body (arrow).

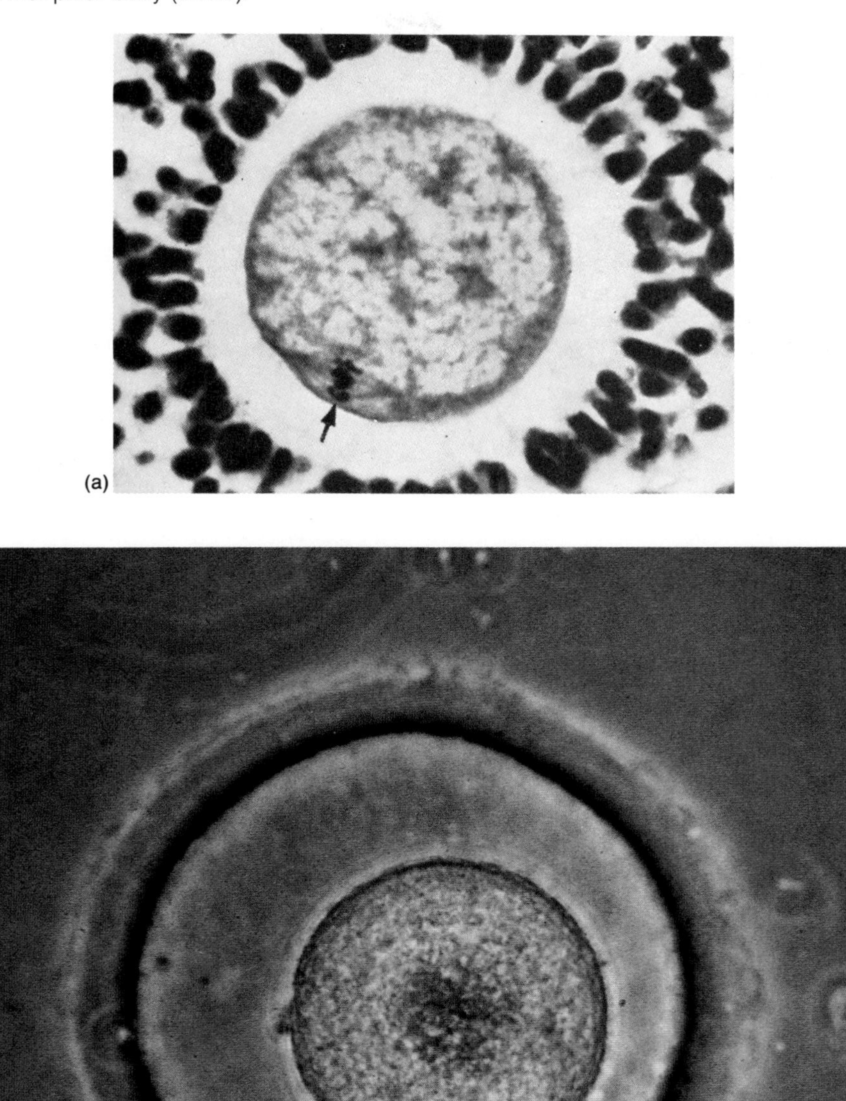

Figure 20.31. A graafian follicle within the ovary of a monkey.

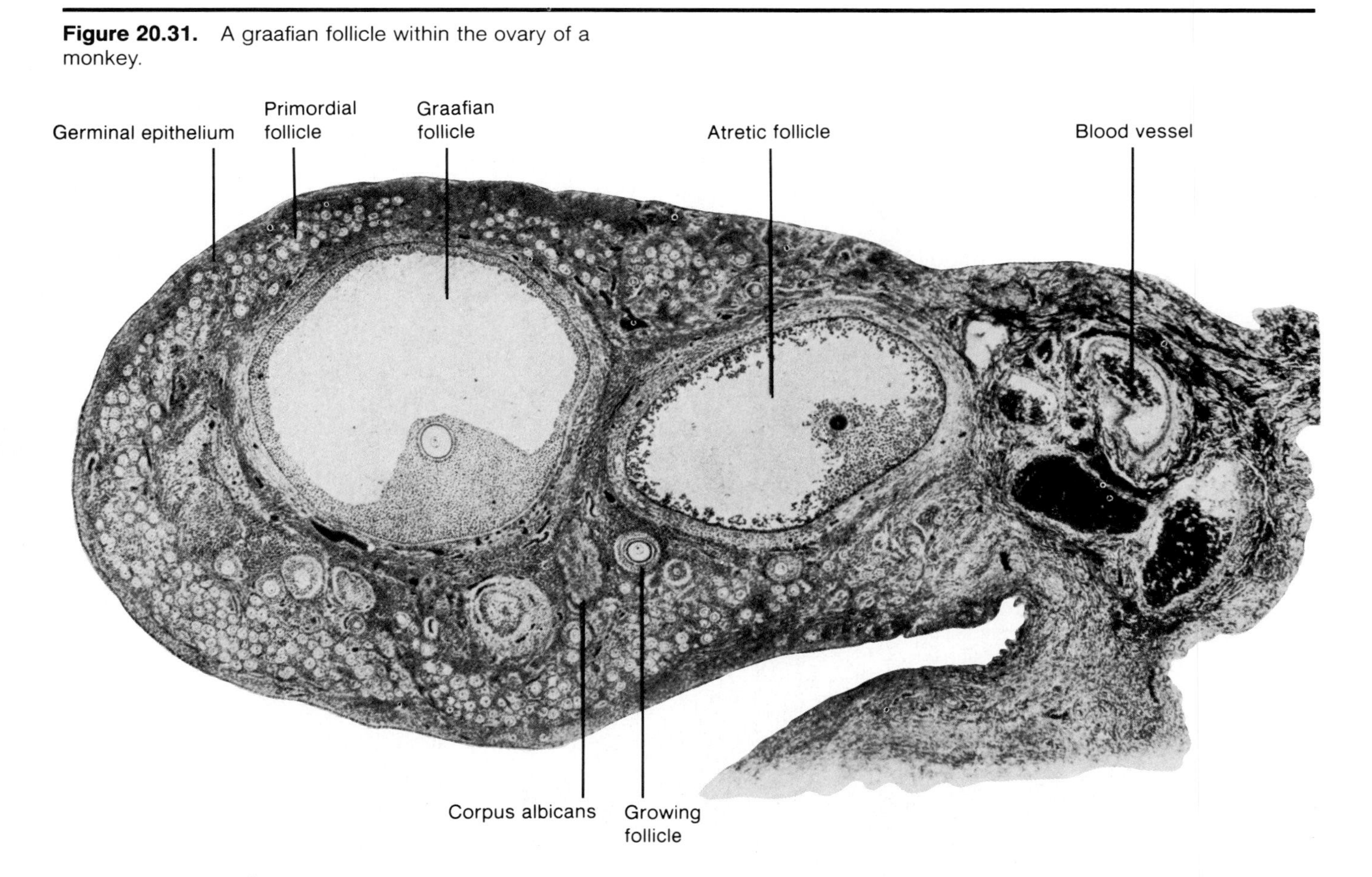

Ovulation

Usually, by about ten to fourteen days after the first day of menstruation, only one follicle has continued its growth to become a mature **graafian follicle** (fig. 20.31); other secondary follicles during that cycle regress and become *atretic.* The graafian follicle is so large that it forms a bulge on the surface of the ovary. Under proper hormonal stimulation this follicle will rupture—much like the popping of a blister—and extrude its oocyte into the uterine tube in the process of **ovulation** (fig. 20.32).

The released cell is a secondary oocyte, surrounded by the zona pellucida and corona radiata. If it is not fertilized, it disintegrates in a couple of days. If a sperm passes through the corona radiata and zona pellucida and enters the cytoplasm of the secondary oocyte, the oocyte completes the second meiotic division. In this process the cytoplasm is again not divided equally; most of the cytoplasm remains in the zygote (fertilized egg), leaving another polar body which, like the first, disintegrates (fig. 20.33).

Changes continue in the ovary following ovulation. The empty follicle, under the influence of luteinizing hormone from the anterior pituitary, undergoes structural and biochemical changes to become a **corpus luteum.** Unlike the ovarian follicles, which secrete only estrogen, the corpus luteum secretes two sex steroid hormones: estrogen and progesterone. Toward the end of a nonfertile cycle the corpus luteum regresses and is changed into a nonfunctional *corpus albicans.* These cyclic changes in the ovary are summarized in figure 20.34.

1. *Compare the structure and contents of a primordial follicle, primary follicle, secondary follicle, and graafian follicle.*
2. *Define ovulation, and describe the changes that occur in the ovary following ovulation in a nonfertile cycle.*
3. *Describe oogenesis, and explain why only one mature ovum is produced by this process.*
4. *Compare the hormonal secretions of the ovarian follicles with those of a corpus luteum.*

Figure 20.32. Ovulation from a human ovary.

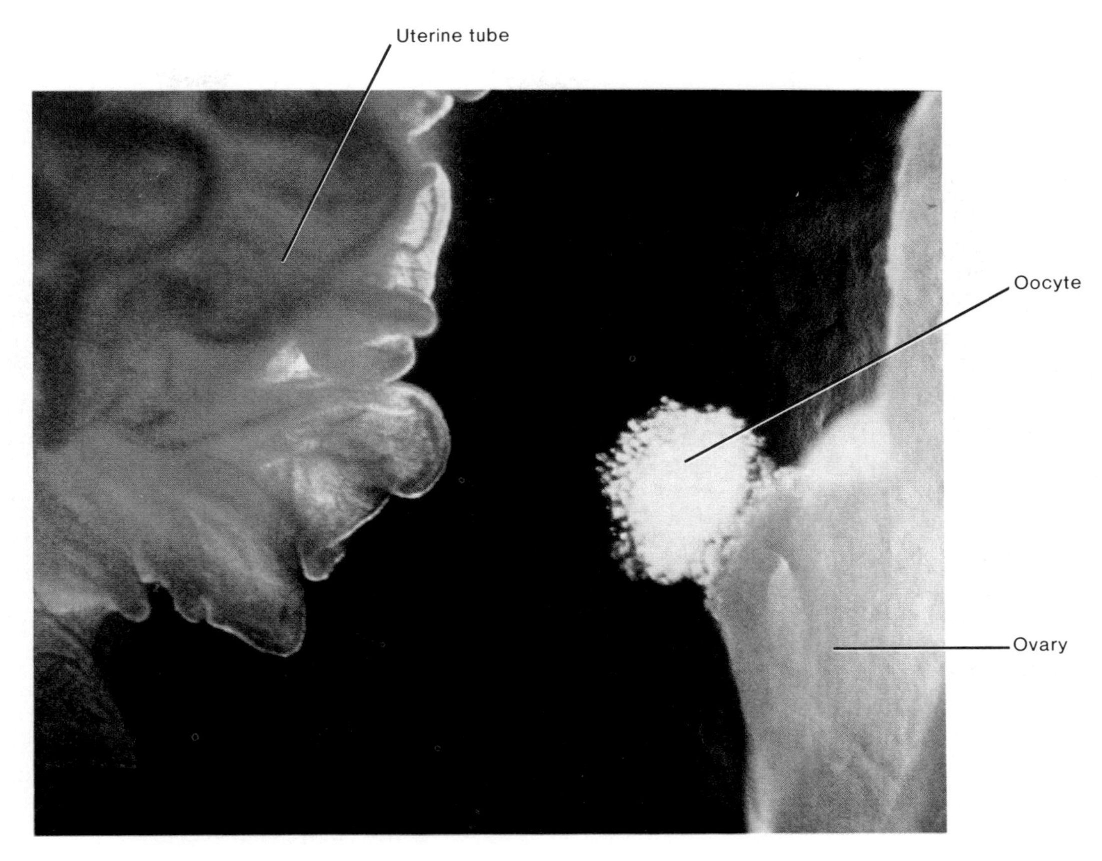

Figure 20.33. A schematic diagram of the process of oogenesis. During meiosis, each primary oocyte produces a single haploid gamete. If the secondary oocyte is fertilized, it forms a secondary polar body and becomes a zygote.

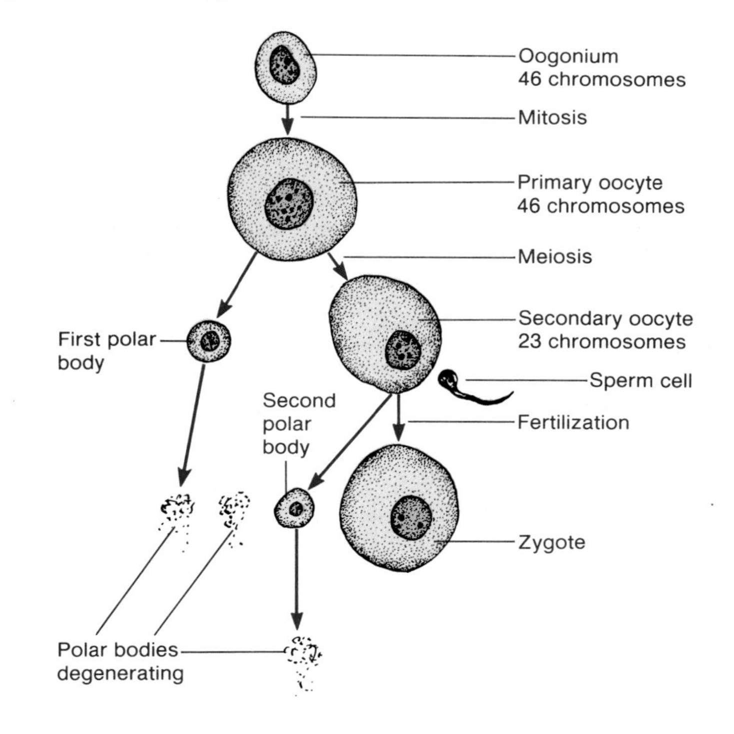

Figure 20.34. A schematic diagram of an ovary showing the various stages of ovum and follicle development.

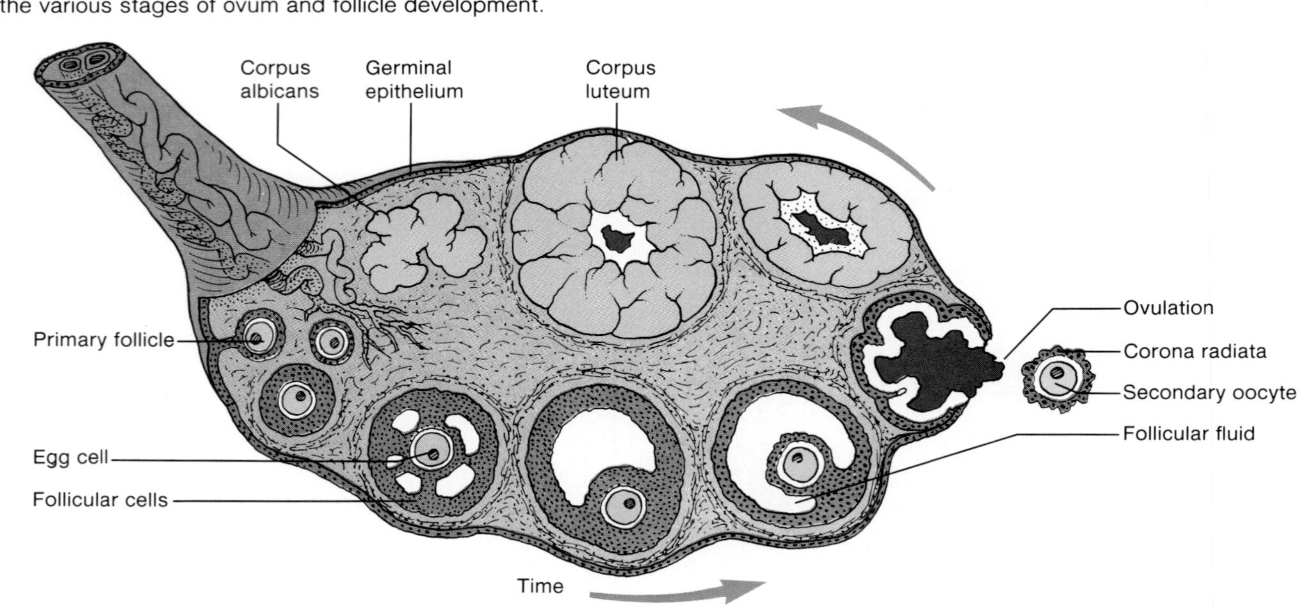

Menstrual Cycle

Humans, apes, and old-world monkeys have cycles of ovarian activity that repeat at approximately one-month intervals; hence the name **menstrual cycle** (*menstru* = monthly). The term *menstruation* is used to indicate the periodic shedding of the stratum functionale of the endometrium (epithelial lining of the uterus), which becomes thickened prior to menstruation under stimulation by ovarian steroid hormones. In primates (other than new-world monkeys) this shedding of the endometrium is accompanied by bleeding. There is no bleeding, in contrast, when other mammals shed the endometrium; their cycles, therefore, are not called menstrual cycles (even though the cycle of a cow, for example, is also about a month long).

Humans and other primates that have menstrual cycles may permit copulation at any time of the cycle. Nonprimate mammals, in contrast, are sexually receptive only at a particular time in their cycles (shortly before or shortly after ovulation). These animals are therefore said to have *estrous cycles*. Bleeding occurs in some animals (such as dogs and cats) that have estrous cycles shortly before they permit copulation. This bleeding is a result of high estrogen secretion and is not associated with shedding of the endometrium.

The bleeding that accompanies menstruation, in contrast, can be experimentally induced by removing the ovaries. This demonstrates that menstrual bleeding is caused by the withdrawal of ovarian hormones. During the normal menstrual cycle, the secretion of ovarian hormones rises and falls in a regular fashion, causing cyclic changes in the endometrium and other sex steroid-dependent tissues.

Phases of the Menstrual Cycle: Pituitary and Ovary

The average menstrual cycle has a duration of about twenty-eight days. Since it is a cycle, there is no beginning or end, and the changes that occur are generally gradual. It is convenient, however, to call the first day of menstruation "day one" of the cycle, because menstrual blood flow is the most apparent of the changes that occur. It is also convenient to divide the cycle into four phases based on changes that occur in the ovary and in the endometrium. The ovaries are in the **follicular phase** starting on the first day of menstruation and ending on the day of ovulation. After ovulation, the ovaries are in the **luteal phase** until the first day of menstruation. The cyclic changes that occur in the endometrium are called the *menstrual, proliferative,* and *secretory phases* and will be discussed separately.

Menstrual and Follicular Phases. The menstrual phase lasts from day one to day four or five of the average cycle. During this time, the secretions of ovarian steroid hormones are at their lowest ebb, and the ovaries contain only primordial and primary follicles. During the *follicular phase,* which lasts from day one to about day thirteen of the cycle (this is highly variable), some of the primary

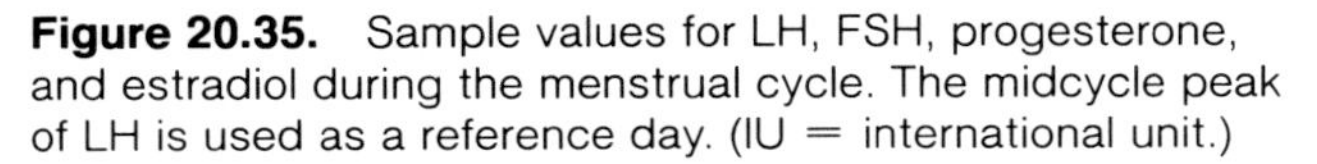

Figure 20.35. Sample values for LH, FSH, progesterone, and estradiol during the menstrual cycle. The midcycle peak of LH is used as a reference day. (IU = international unit.)

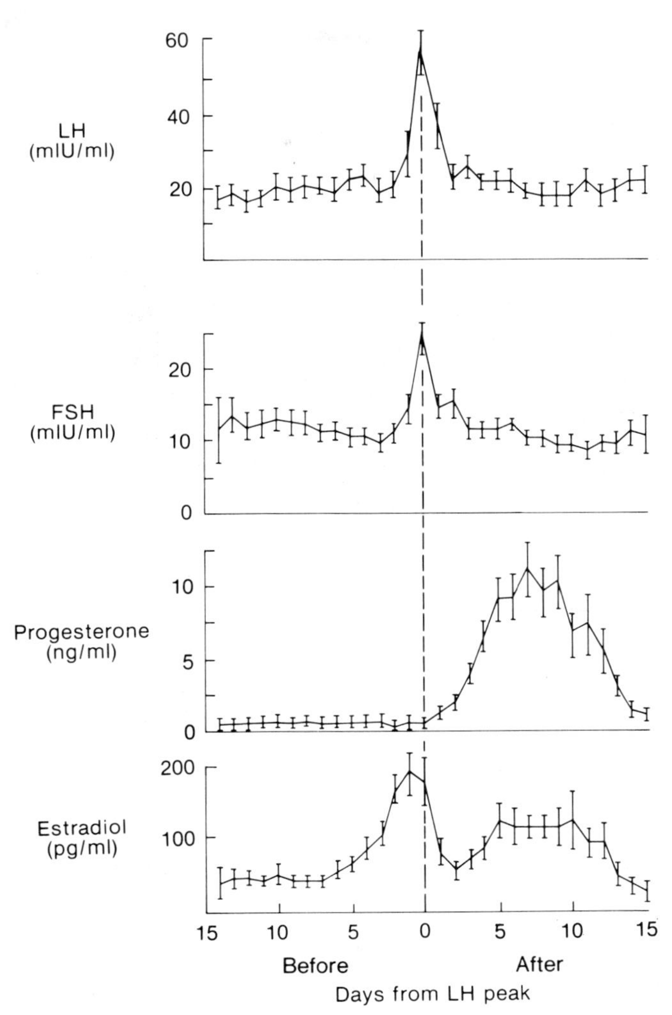

follicles grow, form an antrum, and become secondary follicles. Toward the end of the follicular phase one follicle in one ovary reaches maturity and becomes a graafian follicle. As follicles grow, the granulosa cells secrete an increasing amount of **estradiol** (the principal estrogen), which reaches its highest concentration in the blood at about day twelve of the cycle.

The growth and development of the follicles and the secretion of estradiol are stimulated by **FSH (follicle-stimulating hormone).** Although it has often been stated that FSH levels rise during the follicular phase, most studies fail to show significant changes in FSH concentration during this time of the cycle (fig. 20.35). There are, however, increasing numbers of FSH receptor proteins in the granulosa cells at this time, so perhaps the increased growth and secretion of the follicles is due to the increased sensitivity of the follicles to FSH.

The rapid rise in estradiol secretion from the follicles during the follicular phase stimulates a "surge" of **LH (luteinizing hormone)** secretion from the anterior pituitary. This *LH surge* occurs at the last day of the follicular phase as a result of the **positive feedback** of estradiol on the pituitary and hypothalamus.

The LH surge begins about twenty-four hours before ovulation and reaches its peak about sixteen hours before ovulation. Since GnRH (gonadotrophin-releasing hormone) from the hypothalamus stimulates the pituitary to secrete both FSH and LH, there is a simultaneous, though smaller, surge in FSH secretion. Since this is the only time of the cycle that FSH secretion significantly increases, many investigators believe that this midcycle peak in FSH acts as a stimulus for the development of new follicles for the next month's cycle.

Figure 20.36. Phases of the menstrual cycle in relation to ovarian changes and hormone secretion, and the relationship between changes in the ovaries and the endometrium of the uterus during different phases of the menstrual cycle.

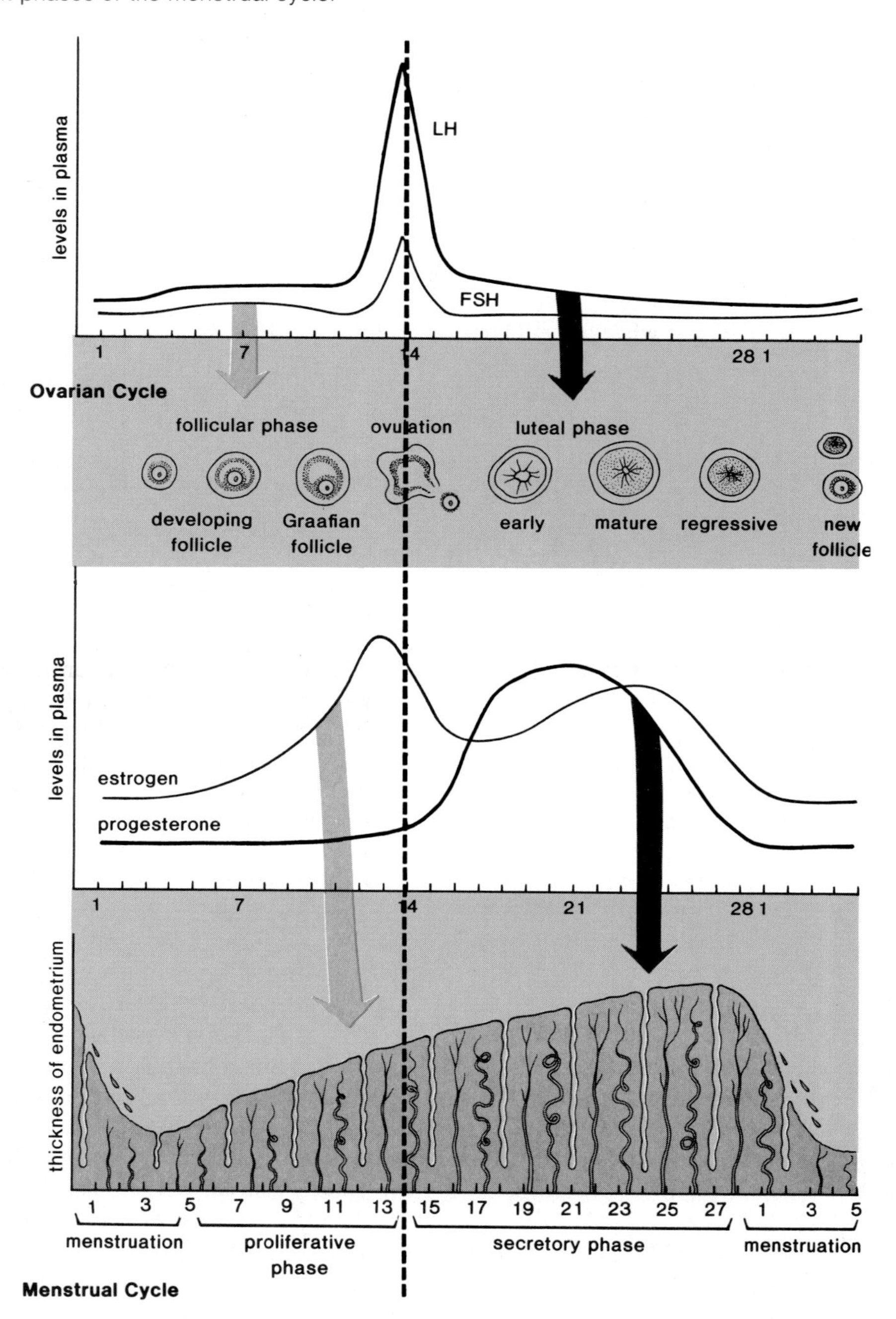

Ovulation. Under the influence of FSH stimulation, the graafian follicle grows so large that it becomes a thin-walled "blister" on the surface of the ovary. The growth of the follicle is accompanied by a rapid rate of increase in estradiol secretion. This rapid increase in estradiol, in turn, triggers the LH surge. Finally, the surge in LH secretion causes the wall of the graafian follicle to rupture (fig. 20.36 *top*). Ovulation occurs, therefore, as a result of the sequential effects of FSH followed by LH on the ovarian follicles.

By means of the positive feedback effect of estradiol on LH secretion, the follicle in a sense sets the time for its own ovulation. This is because ovulation is triggered by an LH surge, and the LH surge is triggered by increased estradiol secretion, which occurs while the follicle grows. In this way the graafian follicle does not normally ovulate until it has reached the proper size and degree of maturation.

In ovulation, a secondary oocyte, arrested at metaphase II of meiosis, is released into a uterine tube. This oocyte is still surrounded by a zona pellucida and corona radiata as it begins its journey to the uterus. Normally only one ovary ovulates per cycle, with the left and right ovary alternating in successive cycles. Interestingly, if one ovary is removed the remaining ovary does not skip cycles, but ovulates every month. The mechanisms by which this regulation is achieved are not understood.

Luteal Phase. After ovulation, the empty follicle is stimulated by LH to become a new structure, the **corpus luteum** (fig. 20.37). This change in structure is accompanied by a change in function. Whereas the developing follicles secrete only estradiol, the corpus luteum secretes both estradiol and **progesterone.** Progesterone levels in the blood are negligible before ovulation but rise rapidly to reach a peak during the luteal phase at approximately one week after ovulation.

The combined high levels of estradiol and progesterone during the luteal phase exert a **negative feedback inhibition** of FSH and LH secretion. This serves to retard development of new follicles, so that further ovulation does not normally occur during that cycle. Although this might seem like locking the barn door after the horse (ovum) has escaped, it does prevent more horses (ova) from escaping. In this way multiple ovulations (and possible pregnancies) on succeeding days of the cycle are prevented.

High levels of estrogen and progesterone during the nonfertile cycle do not persist for very long, however, and new follicles do start to develop towards the end of one cycle, in preparation for the next cycle. Estrogen and progesterone levels fall during the late luteal phase (starting about day twenty-two), because the corpus luteum regresses and stops functioning. In lower mammals, the decline in corpus luteum function is caused by a hormone secreted by the uterus called *luteolysin.* A similar hormone has not yet been identified in humans, and the cause of corpus luteum regression in humans is not well understood. Luteolysis (breakdown of the corpus luteum) can be prevented by high LH secretion, but LH levels remain low during the luteal phase as a result of negative feedback inhibition by ovarian steroids.

Figure 20.37. A corpus luteum in a human ovary.

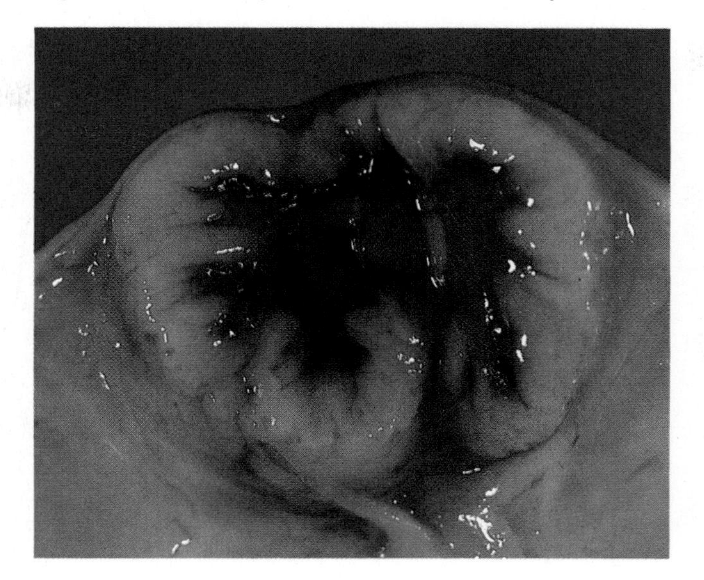

With the declining function of the corpus luteum, estrogen and progesterone fall to very low levels by day twenty-eight of the cycle. The withdrawal of ovarian steroids causes menstruation and permits a new cycle of ovarian follicle development to progress.

Cyclic Changes in the Endometrium

In addition to a description of the female cycle in terms of the phases of ovarian function, the cycle can also be described in terms of the changes that occur in the endometrium. Three phases can be identified on this basis (fig. 20.36 *bottom*): (1) the proliferative phase; (2) the secretory phase; and (3) the menstrual phase.

Table 20.8 Phases of the menstrual cycle.

Phase of Cycle		Hormonal Changes		Tissue Changes	
Ovarian	Endometrial	Pituitary	Ovary	Ovarian	Endometrial
Follicular (days 5–13)	Proliferative	No apparent change	Estradiol secretion rises (due to FSH stimulation of follicles)	Follicles grow; graafian follicle develops (due to FSH stimulation)	Mitotic division increases thickness of endometrium; spiral arteries develop (due to estradiol stimulation)
Ovulatory (day 14)	Proliferative	LH surge (and increased FSH) stimulated by positive feedback from estradiol	Fall in estradiol secretion	Graafian follicle is ruptured and ovum is extruded into fallopian tube	No change
Luteal (days 15–28)	Secretory	LH and FSH decrease (due to negative feedback of steroids)	Progesterone and estrogen secretion increase, then fall	Development of corpus luteum (due to LH stimulation); regression of corpus luteum	Glandular development in endometrium (due to progesterone stimulation)
Follicular (days 1–4)	Menstrual	FSH and LH remain low	Estradiol and progesterone remain low	Primary follicles grow	Outer two-thirds of endometrium is shed with accompanying bleeding

The **proliferative phase** of the endometrium occurs while the ovary is in its follicular phase. The increasing amounts of estradiol secreted by the developing follicles stimulates growth (proliferation) of the stratum functionale of the endometrium. In humans and other primates, spiral arteries develop in the endometrium during this phase. Estradiol may also stimulate the production of receptor proteins for progesterone at this time, in preparation for the next phase of the cycle.

The **secretory phase** of the endometrium occurs when the ovary is in its luteal phase. In this phase, increased progesterone secretion stimulates the development of mucus glands. As a result of the combined actions of estradiol and progesterone, the endometrium becomes thick, vascular, and "spongy" in appearance during the time of the cycle following ovulation. It is therefore well prepared to accept and nourish an embryo if fertilization occurs.

The **menstrual phase** occurs as a result of the fall in ovarian hormone secretion during the late luteal phase. Necrosis (cellular death) and sloughing of the stratum functionale of the endometrium may be produced by constriction of the spiral arteries. The spiral arteries appear to be responsible for bleeding during menstruation, because lower animals that lack spiral arteries don't bleed when they shed their endometrium. The phases of the menstrual cycle are summarized in table 20.8.

The cyclic changes in ovarian secretion cause other cyclic changes in the female genital ducts. High levels of estradiol secretion, for example, cause cornification of the vaginal epithelium (the upper cells die and become filled with keratin). High levels of estradiol also cause the production of a thin, watery cervical mucus, which can be easily penetrated by spermatozoa. During the luteal phase of the cycle, the high levels of progesterone cause the cervical mucus to become thick and sticky after ovulation has occurred.

Cyclic changes in ovarian hormone secretion also cause cyclic changes in *basal body temperature.* In the **rhythm method** of birth control, a woman measures her oral basal body temperature upon waking to determine when ovulation has occurred. On the day of the LH peak, when estradiol secretion begins to decline, there is a slight drop in basal body temperature. Starting about one day after the LH peak, the basal body temperature sharply rises as a result of progesterone secretion and remains elevated throughout the luteal phase of the cycle (fig. 20.38). The day of ovulation can be accurately determined by this method, making the method useful in increasing fertility if conception is desired. Since the day of the cycle in which ovulation occurs is quite variable in many women, however, the rhythm method is not very reliable for contraception by predicting when the next ovulation will occur. Much more effective means of contraception—including the contraceptive pill and intrauterine device—are currently available.

Figure 20.38. Changes in basal body temperature during the menstrual cycle.

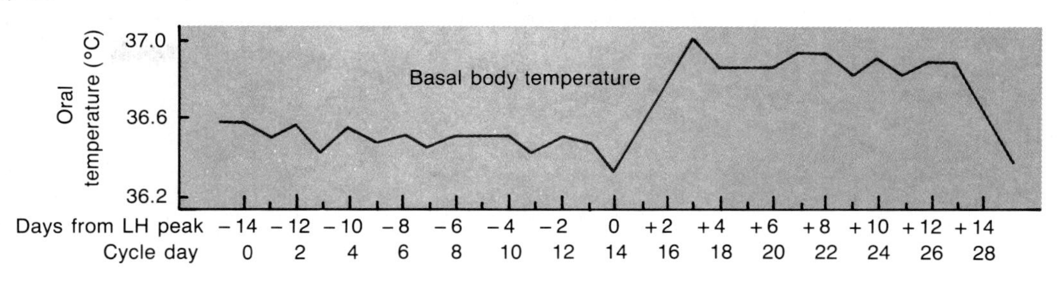

Contraceptive Pill

About ten million women in the United States and sixty million women in the world are currently using **oral steroid contraceptives.** These contraceptives usually consist of a synthetic estrogen combined with a synthetic progesterone in the form of pills that are taken once each day for three weeks after the last day of a menstrual period. This procedure causes an immediate increase in blood levels of ovarian steroids (from the pill), which is maintained for the normal duration of a monthly cycle. As a result of *negative feedback inhibition* of gonadotrophin secretion, *ovulation never occurs.* The entire cycle is like a false luteal phase, with high levels of progesterone and estrogen and low levels of gonadotrophins.

Since the contraceptive pills contain ovarian steroid hormones, the endometrium proliferates and becomes secretory just as it does during a normal cycle. In order to prevent an abnormal growth of the endometrium, women stop taking the pill after three weeks. This causes estrogen and progesterone levels to fall, and permits menstruation to occur. The contraceptive pill is an extremely effective method of birth control, but it does have potentially serious side effects—including an increased incidence of thromboembolism, cardiovascular disorders, and endometrial and breast cancer. It has been pointed out, however, that the mortality risk of contraceptive pills is still much lower than the risk of death from the complications of pregnancy—or from automobile accidents.

The Menopause

The term *menopause* means literally "pause in the menses" and refers to the cessation of ovarian activity that occurs at about the age of fifty. During the postmenopausal years, which account for about a third of a woman's life-span, no new ovarian follicles develop and the ovaries cease secreting estradiol. This fall in estradiol is due to defects in the ovaries, not in the pituitary; indeed, FSH and LH secretion by the pituitary is elevated due to the absence of negative feedback inhibition from estradiol. Like prepubertal boys and girls, the only estrogen found in the blood of postmenopausal women is that formed by aromatization of the weak androgen androstenedione, secreted principally by the adrenal cortex, into a weak estrogen called estrone.

It is the withdrawal of estradiol secretion from the ovaries that is most responsible for the many debilitating symptoms of menopause. These include vasomotor disturbances (which produce "hot flashes"), urogenital atrophy, and the increased development of osteoporosis (see chapter 19). Estrogen replacement therapy, often in combination with progesterone, helps to alleviate these symptoms, although this treatment is given cautiously because of the increased risks associated with taking estrogen pills.

1. *Describe the changes that occur in the ovary and endometrium during the follicular phase, and explain the hormonal control of these changes.*
2. *Describe the hormonal regulation of ovulation.*
3. *Describe the formation, function, and fate of the corpus luteum, and describe the changes that occur in the endometrium during the luteal phase.*
4. *Explain the significance of negative feedback inhibition during the luteal phase, and explain the hormonal control of menstruation.*

Figure 20.39. The process of fertilization. (*a, b*) Diagrammatic representations. (*c*) A scanning electron micrograph of sperm bound to the egg surface. As the head of the sperm encounters the gelatinous corona radiata of the egg (*b2*), the acrosomal vesicle ruptures and the sperm digests a path for itself by the action of enzymes released from the acrosome (*b3, b4*). When the cell membrane of the sperm contacts the cell membrane of the egg (*b5*), they become continuous, and the sperm nucleus and other contents move into the egg cytoplasm.

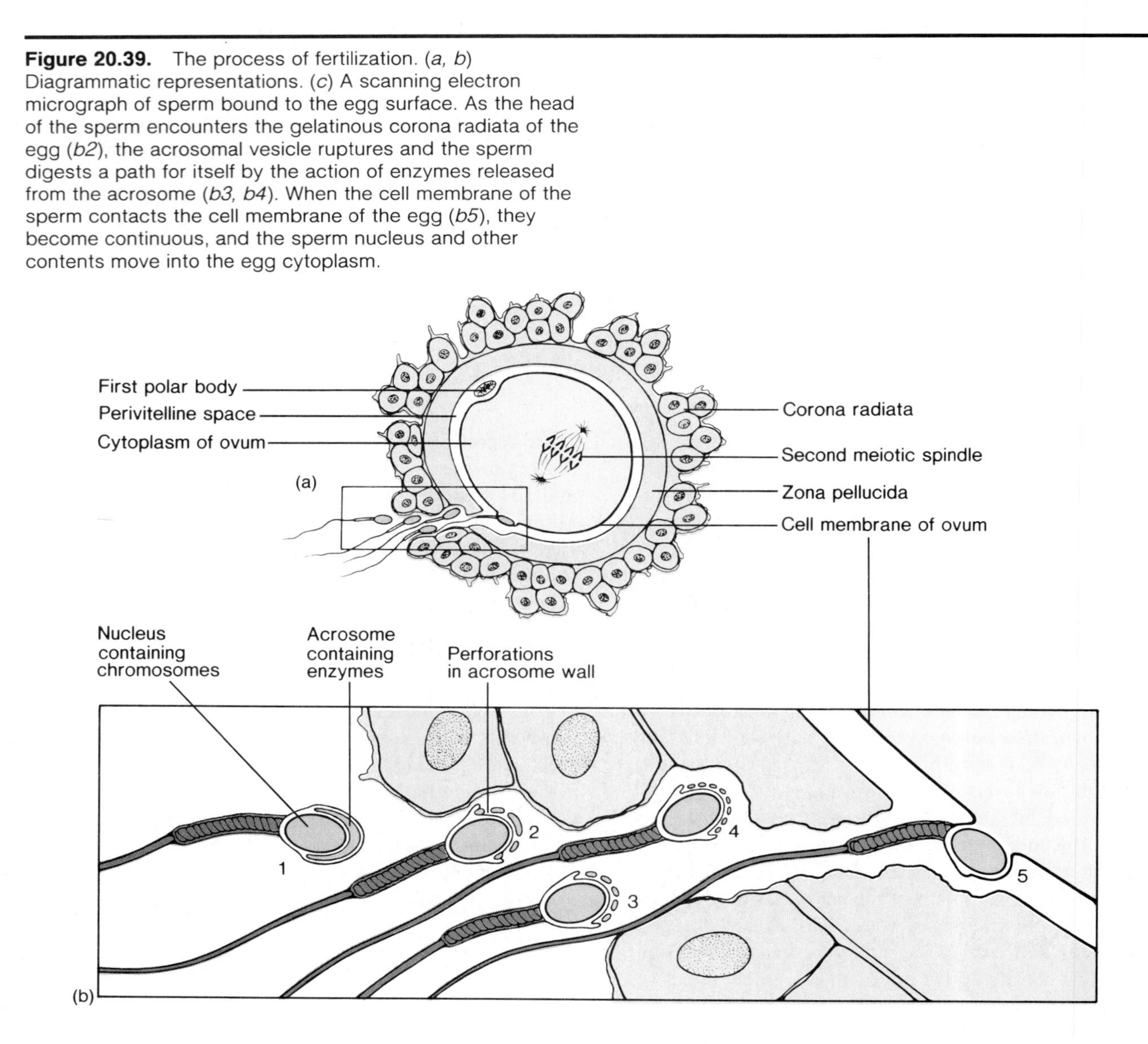

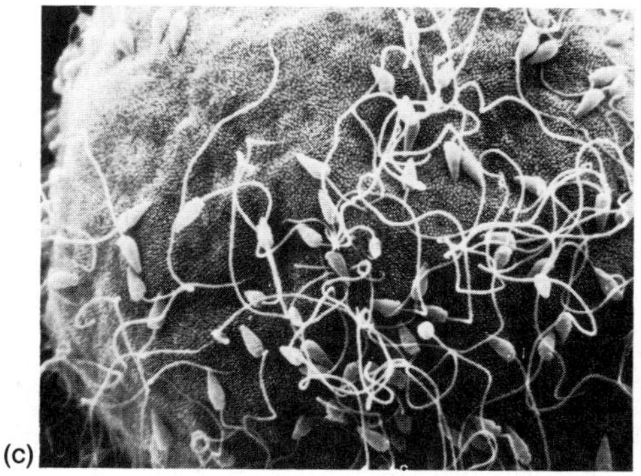

Figure 20.40. An electron micrograph showing the head of a human sperm with its nucleus and acrosomal cap.

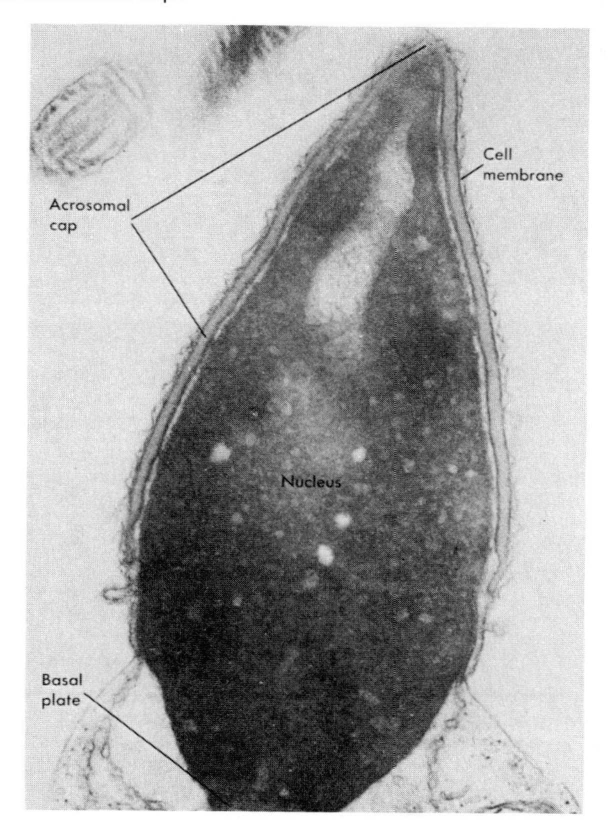

Fertilization, Pregnancy, and Parturition

During the act of sexual intercourse a man ejaculates an average of 300 million sperm into the vagina. This tremendous number is needed because of the high sperm fatality rate—only about 100 survive to enter each fallopian tube. During their passage through the female reproductive tract the sperm gain the ability to fertilize an ovum. This process is called **capacitation.** The changes that occur in capacitation are not known. Experiments have shown, however, that freshly ejaculated sperm are infertile; they must be present in the female tract for at least seven hours before they can fertilize an ovum.

A woman usually ovulates only one ovum a month, producing a total number of less than 450 during her reproductive years. Each ovulation releases a secondary oocyte arrested at metaphase of the second meiotic division. The secondary oocyte, as previously described, enters the fallopian tube surrounded by its zona pellucida (a thin transparent layer of protein and polysaccharides) and corona radiata of granulosa cells (fig. 20.39).

The head of each spermatozoan is capped by an organelle called an *acrosome* (fig. 20.40). The acrosome contains digestive enzymes—a trypsinlike protein-digesting enzyme and hyaluronidase (which digests hyaluronic acid, an important constituent of connective tissues). Fertilization normally occurs in the fallopian tubes. When sperm meets ovum, an **acrosomal reaction** occurs that exposes the acrosome's digestive enzymes and allows the sperm to penetrate the corona radiata and the zona pellucida. The acrosomal enzymes are not released in this process; rather, the sperm tunnels its way through these barriers by digestion reactions that are localized to its acrosomal cap.

As the first sperm tunnels its way through the zona pellucida, a chemical change in the zona occurs that prevents other sperm from entering. Only one sperm, therefore, is allowed to fertilize one ovum. As fertilization occurs, the secondary oocyte is stimulated to complete its

Figure 20.41. A secondary oocyte, arrested at metaphase II of meiosis, is released at ovulation. If this cell is fertilized, it completes its second meiotic division and produces a second polar body.

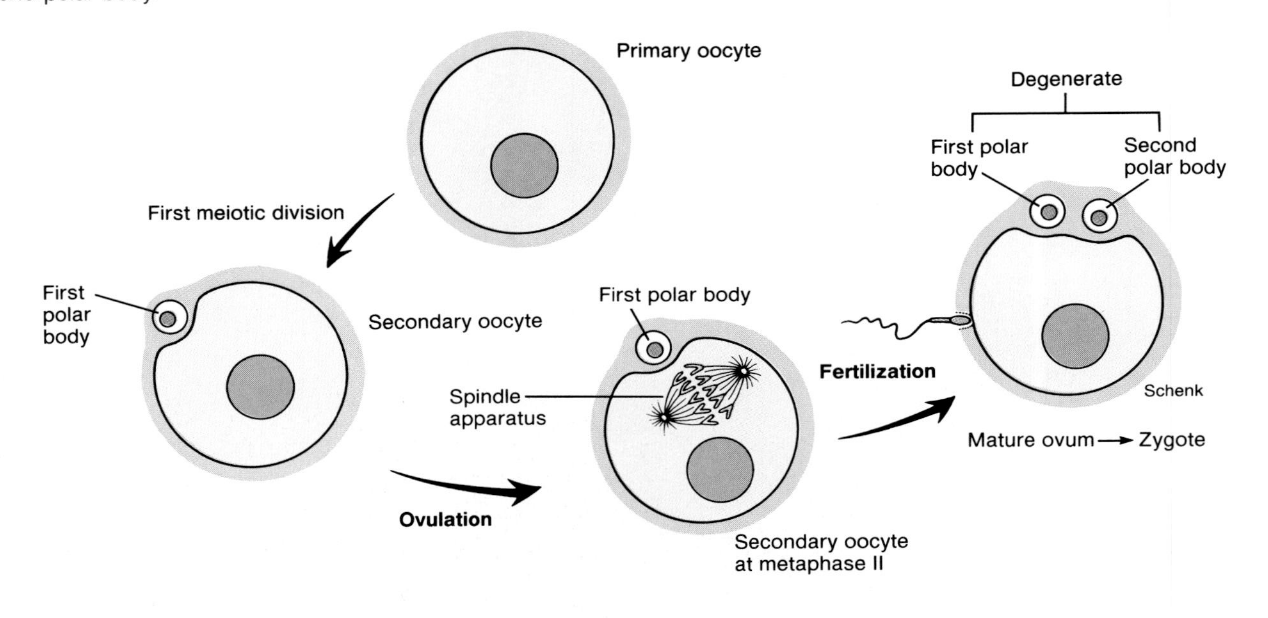

Figure 20.42. Fertilization and the union of chromosomes from the sperm and ovum to form the zygote.

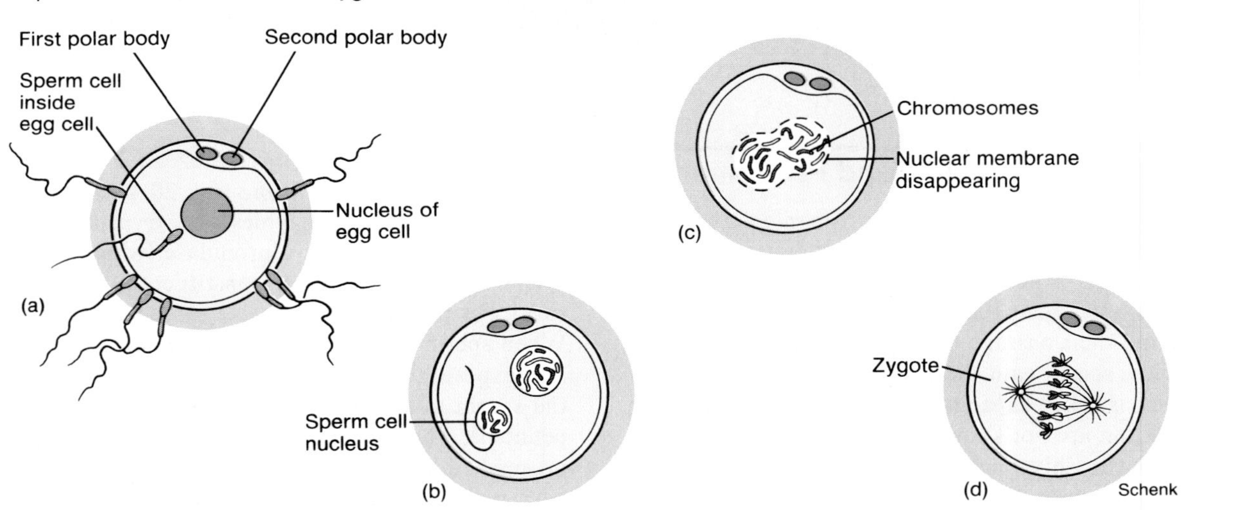

second meiotic division (fig. 20.41). Like the first meiotic division, the second produces one cell that contains all of the cytoplasm—the mature ovum or egg cell—and one polar body. The second polar body, like the first, ultimately fragments and disintegrates.

At fertilization, the entire sperm enters the cytoplasm of the much larger egg cell. Within twelve hours the nuclear membrane in the ovum disappears, and the haploid number of chromosomes (twenty-three) in the ovum is joined by the haploid number of chromosomes from the sperm. A fertilized egg, or *zygote,* containing the diploid number of chromosomes (forty-six) is thus formed (fig. 20.42).

Figure 20.43. A diagrammatic representation of the ovarian cycle, fertilization, and the morphogenic events of the first week. The numbers indicate the days after fertilization. Implantation of the blastocyst begins between the fifth and seventh day and is generally completed by the tenth day.

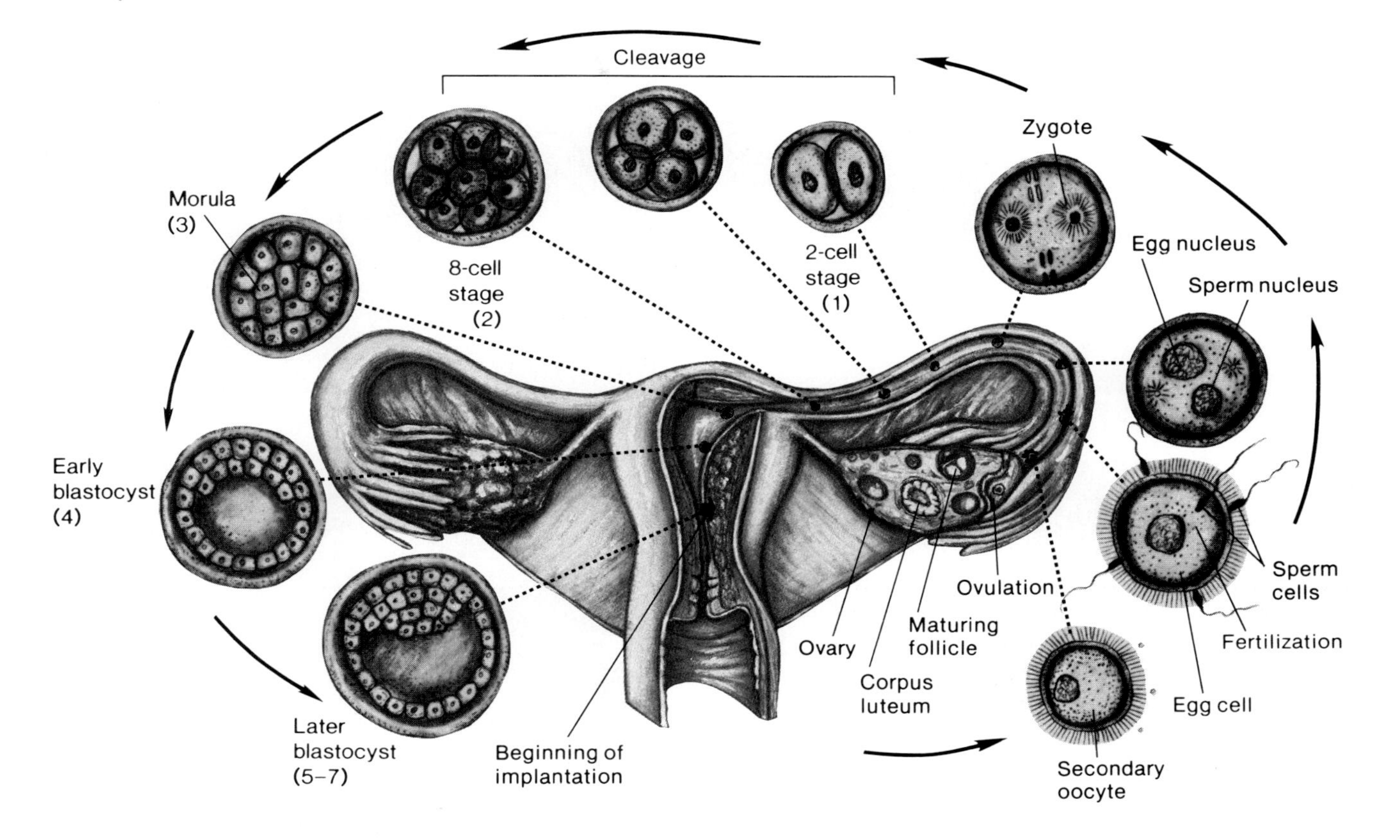

A secondary oocyte that is ovulated but not fertilized does not complete its second meiotic division, but instead disintegrates twelve to twenty-four hours after ovulation. Fertilization therefore cannot occur if intercourse takes place beyond one day following ovulation. Sperm, in contrast, can survive up to three days in the female reproductive tract. Fertilization therefore can occur if intercourse is performed within three days prior to the day of ovulation.

Cleavage and Formation of a Blastocyst

At about thirty to thirty-six hours after fertilization the zygote divides by mitosis—a process called *cleavage*—into two smaller cells. The rate of cleavage is thereafter accelerated. A second cleavage, performed about forty hours after fertilization, produces four cells. By about fifty to sixty hours after fertilization a third cleavage occurs, producing a ball of eight cells called a **morula** (= mulberry). This very early embryo enters the uterus three days after ovulation has occurred (fig. 20.43).

Cleavage continues so that a morula consisting of thirty-two to sixty-four cells is produced by the fourth day after fertilization. The embryo remains unattached to the uterine wall for the next two days, during which time it undergoes changes that convert it into a hollow structure called a **blastocyst** (fig. 20.44). The blastocyst consists of two parts: (1) an *inner cell mass,* which will become the body of the fetus; and (2) a surrounding *chorion,* which will become part of the placenta. The cells that form the chorion are called *trophoblast cells.*

On the sixth day following fertilization, the blastocyst attaches to the uterine wall, with the side containing the inner cell mass against the endometrium. The trophoblast cells produce enzymes that allow the blastocyst to

Figure 20.44. Stages of pre-embryonic development of a human ovum fertilized in a laboratory (*in vitro*) as seen in scanning electron micrographs. (*a*) 4-cell stage; (*b*) cleavage at the 16-cell stage; (*c*) a morula; and (*d*) a blastocyst.

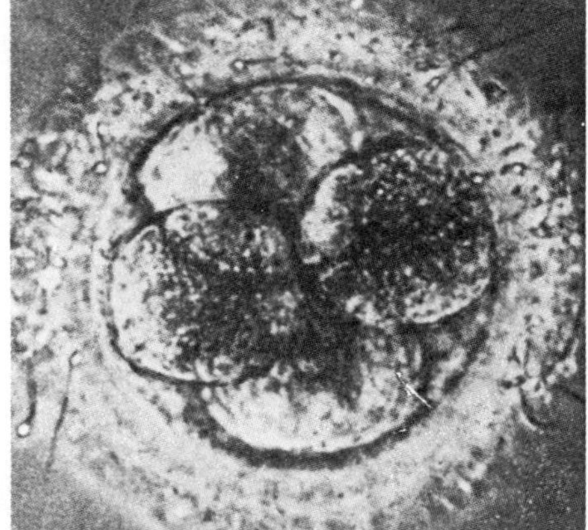

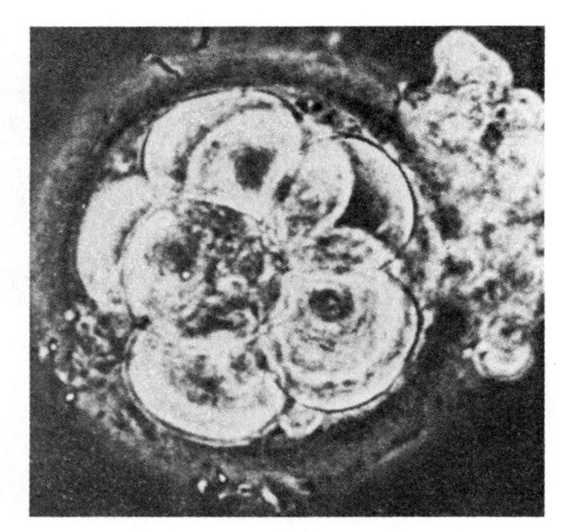

(b)

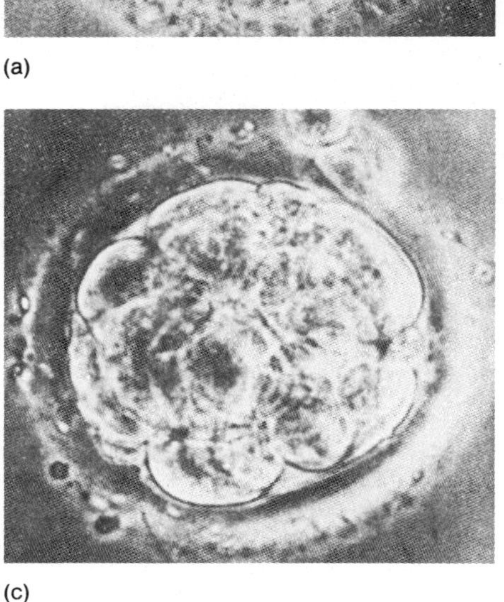

(c)

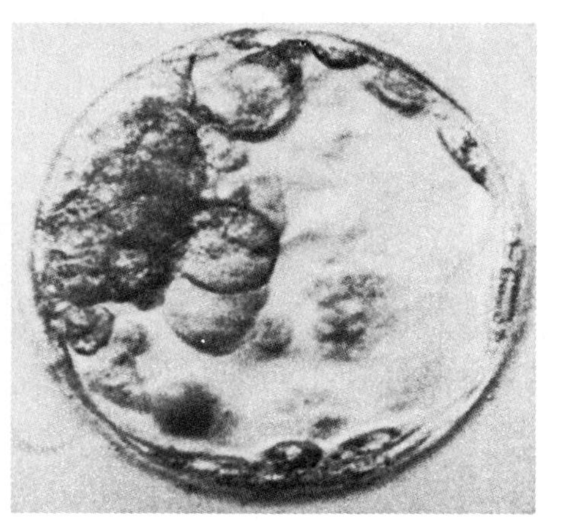

"eat its way" into the thick endometrium. This begins the process of **implantation,** or **nidation,** and by the seventh day the blastocyst is usually completely buried in the endometrium (fig. 20.45).

Implantation and Formation of a Placenta

In nonfertile cycles the corpus luteum begins to decrease its secretion of steroids about ten days after ovulation. This withdrawal of steroids, as previously described, causes necrosis and sloughing of the endometrium following day twenty-eight of the nonfertile cycle. If fertilization and implantation have occurred, however, these events must obviously be prevented to maintain the pregnancy.

Chorionic Gonadotrophin. The blastocyst saves itself from being eliminated with the endometrium by secreting a hormone that indirectly prevents menstruation. Even before the sixth day when implantation occurs, the trophoblast cells of the chorion secrete **chorionic gonadotrophin (hCG**—the *h* stands for *human*). This hormone is identical to LH in its effects and therefore is able to maintain the corpus luteum past the time when it would otherwise regress. The secretion of estradiol and progesterone is thus maintained and menstruation is normally prevented.

The secretion of hCG declines by the tenth week of pregnancy (fig. 20.46). Actually, this hormone is only required for the first five to six weeks of pregnancy, because the placenta itself becomes an active steroid hormone-secreting gland. At the fifth to sixth week the mother's corpus luteum begins to regress (even in the presence of hCG), but the placenta secretes more than sufficient amounts of steroids to maintain the endometrium and prevent menstruation.

Figure 20.45. The blastocyst adheres to the endometrium on about the sixth day as shown in diagram (a). (b) is a scanning electron micrograph showing the surface of the endometrium and implantation at twelve days following fertilization.

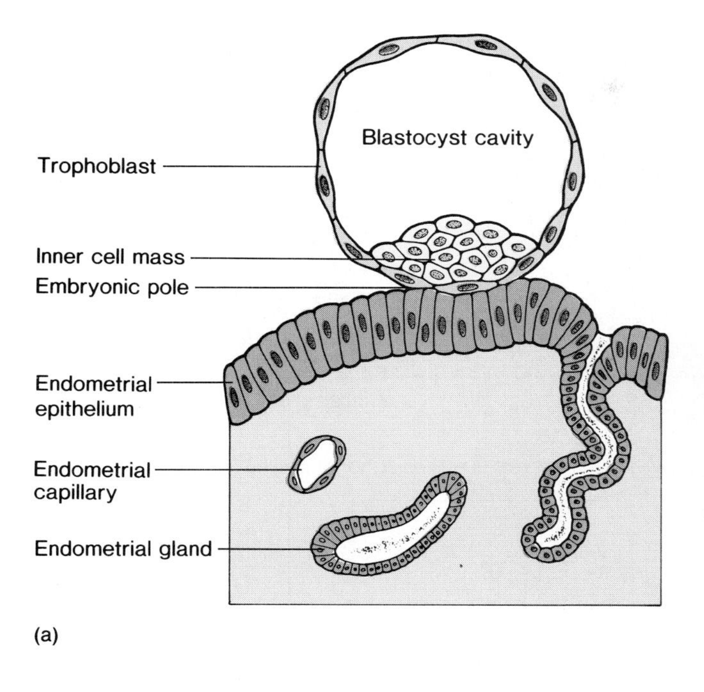

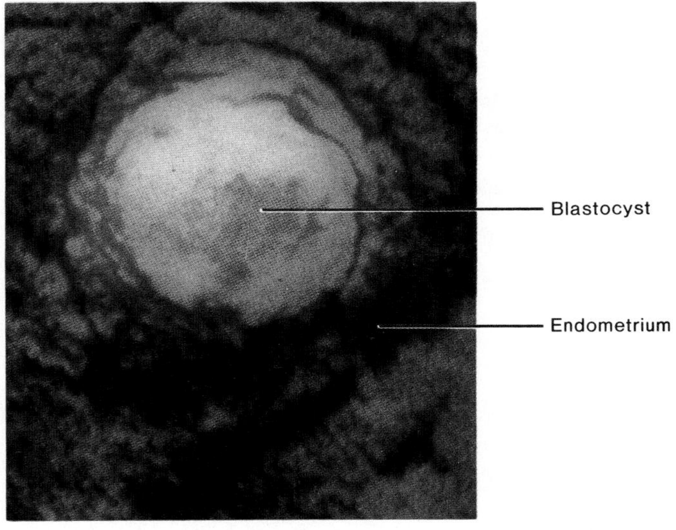

Figure 20.46. Human chorionic gonadotrophin (hCG) is secreted by trophoblast cells during the first trimester of pregnancy. This hormone maintains the mother's corpus luteum for the first five and a half weeks. After that time the placenta becomes the major sex-hormone-producing gland, secreting increasing amounts of estrogen and progesterone throughout pregnancy.

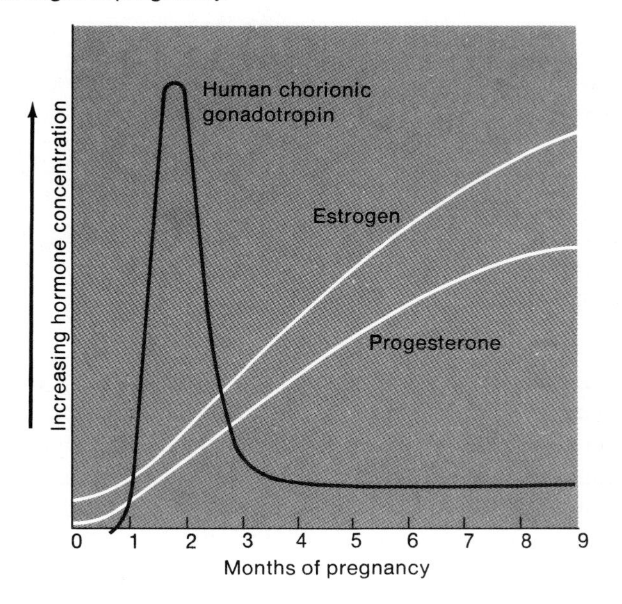

All **pregnancy tests** assay for the presence of hCG in blood or urine, because this is the only hormone that is secreted by the blastocyst but not by the mother's endocrine glands. Modern pregnancy tests detect the presence of hCG by the use of antibodies against hCG or by use of cellular receptor proteins for hCG. Extremely sensitive techniques utilizing monoclonal antibodies against a subunit of hCG and radioimmunoassay (chapter 17) permit pregnancy to be detected in a clinical laboratory as early as seven to ten days after conception.

Chorionic Membranes. Between days seven and twelve, as the blastocyst becomes completely embedded in the endometrium, the chorion becomes a two-cell-thick structure, consisting of an inner *cytotrophoblast* layer and an outer *syncytiotrophoblast* layer. The inner cell mass (which will become the body of the fetus), meanwhile, also develops two cell layers. These are the *ectoderm* (which will form such organs as the nervous system and skin) and the *endoderm* (which will eventually form the gut and its derivatives). A third, middle embryonic layer—the *mesoderm*—is not yet seen at this stage. The embryo at this stage is a two-layer-thick disc separated from the cytotrophoblast of the chorion by an *amniotic cavity.*

Figure 20.47. After the syncytiotrophoblast has created blood-filled cavities in the endometrium, these cavities are invaded by extensions of the cytotrophoblast (*a*). These extensions, called villi, branch extensively to produce the chorion frondosum (*b*). The developing embryo is surrounded by a membrane known as the amnion.

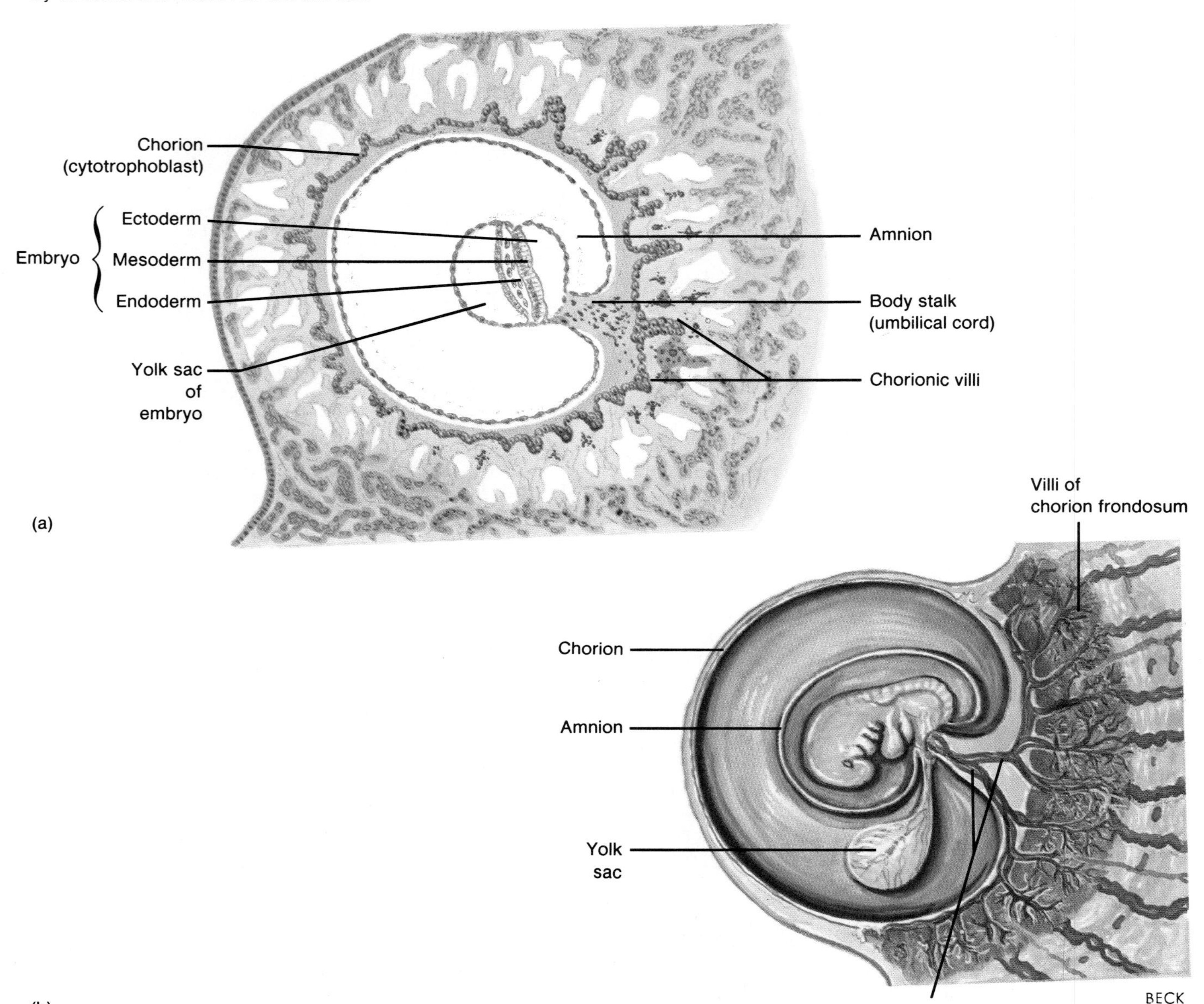

As the syncytiotrophoblast invades the endometrium, it secretes protein-digesting enzymes that create many small, blood-filled cavities in the maternal tissue. The cytotrophoblast then forms projections, or *villi,* (fig. 20.47) that grow into these pools of venous blood, producing a leafy-appearing structure called the *chorion frondosum* (*frond* = leaf). This occurs only on the side of the chorion that faces the uterine wall. As the embryonic structures grow, the other side of the chorion bulges into the cavity of the uterus, loses its villi, and becomes smooth in appearance.

Formation of the Placenta and Amniotic Sac. As the blastocyst is implanted in the endometrium and the chorion develops, the cells of the endometrium also undergo changes. These changes, including cellular growth and the accumulation of glycogen, are called the **decidual reaction.** The maternal tissue in contact with the chorion frondosum is called the *decidua basalis.* These two structures—chorion frondosum (fetal tissue) and decidua basalis (maternal tissue)—together form the functional unit known as the **placenta.**

Figure 20.48. Blood from the fetus is carried to and from the chorion frondosum by umbilical arteries and veins. The maternal tissue between the chorionic villi is known as the decidua basalis, and this tissue, together with the villi, form the functioning placenta. The space between chorion and amnion is obliterated, and the fetus lies within the fluid-filled amniotic sac.

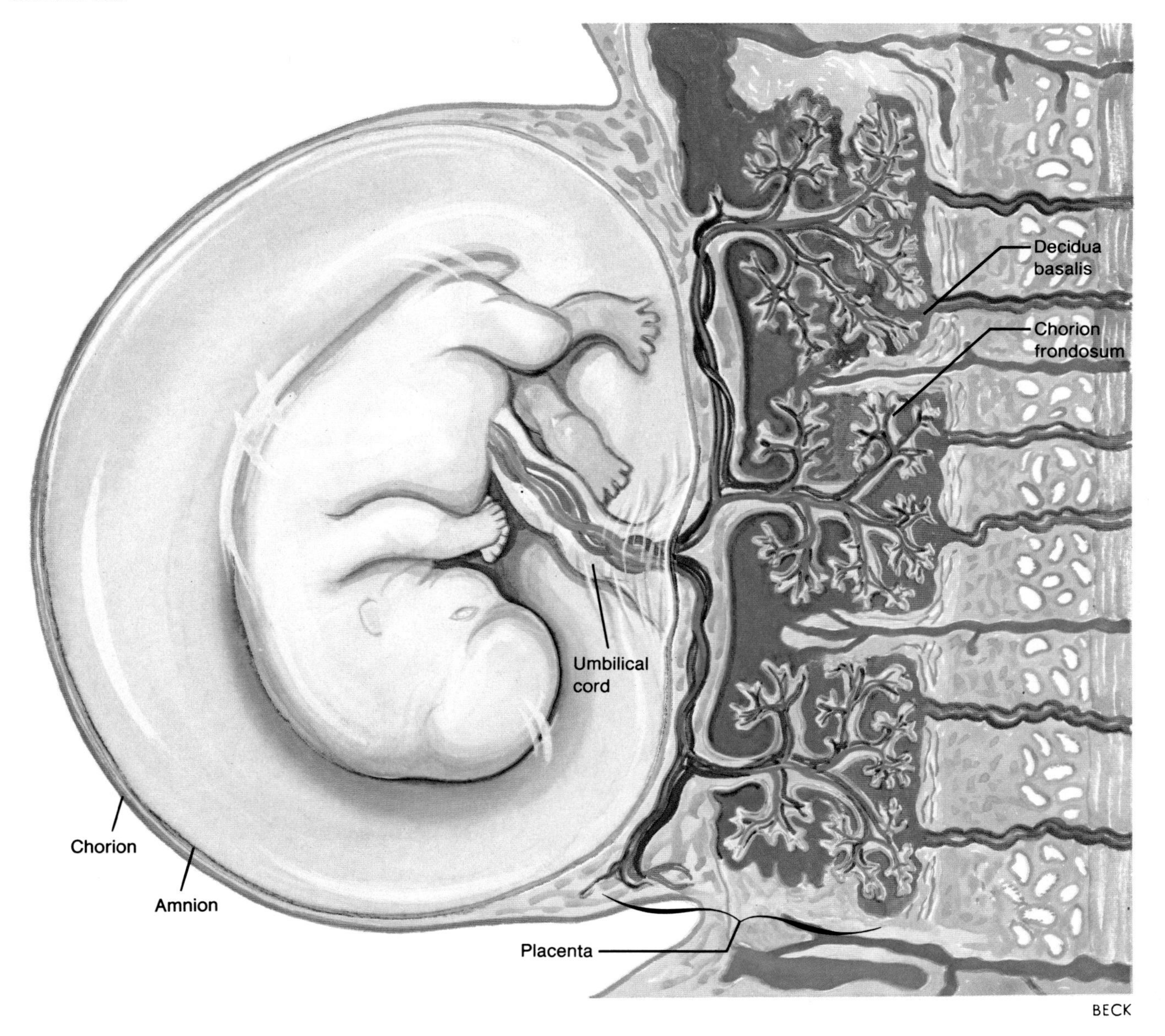

The human placenta is a disc-shaped structure that is continuous at its outer surface with the smooth part of the chorion, which bulges into the uterine cavity. Immediately beneath the chorionic membrane is the amnion, which has grown to envelop the entire fetus (fig. 20.48). The fetus, together with its umbilical cord, is therefore located within the fluid-filled *amniotic sac.*

Amniotic fluid is formed initially as an isotonic secretion, which is later increased in volume and decreased in concentration by urine from the fetus. Amniotic fluid also contains cells that are sloughed off from the fetus, placenta, and amniotic sac. Since all of these cells are derived from the same fertilized egg, all have the same genetic composition. Many genetic abnormalities can be

Figure 20.49. Amniocentesis. In this procedure amniotic fluid, together with suspended cells, is withdrawn for examination. Various genetic diseases can be detected prenatally by this means.

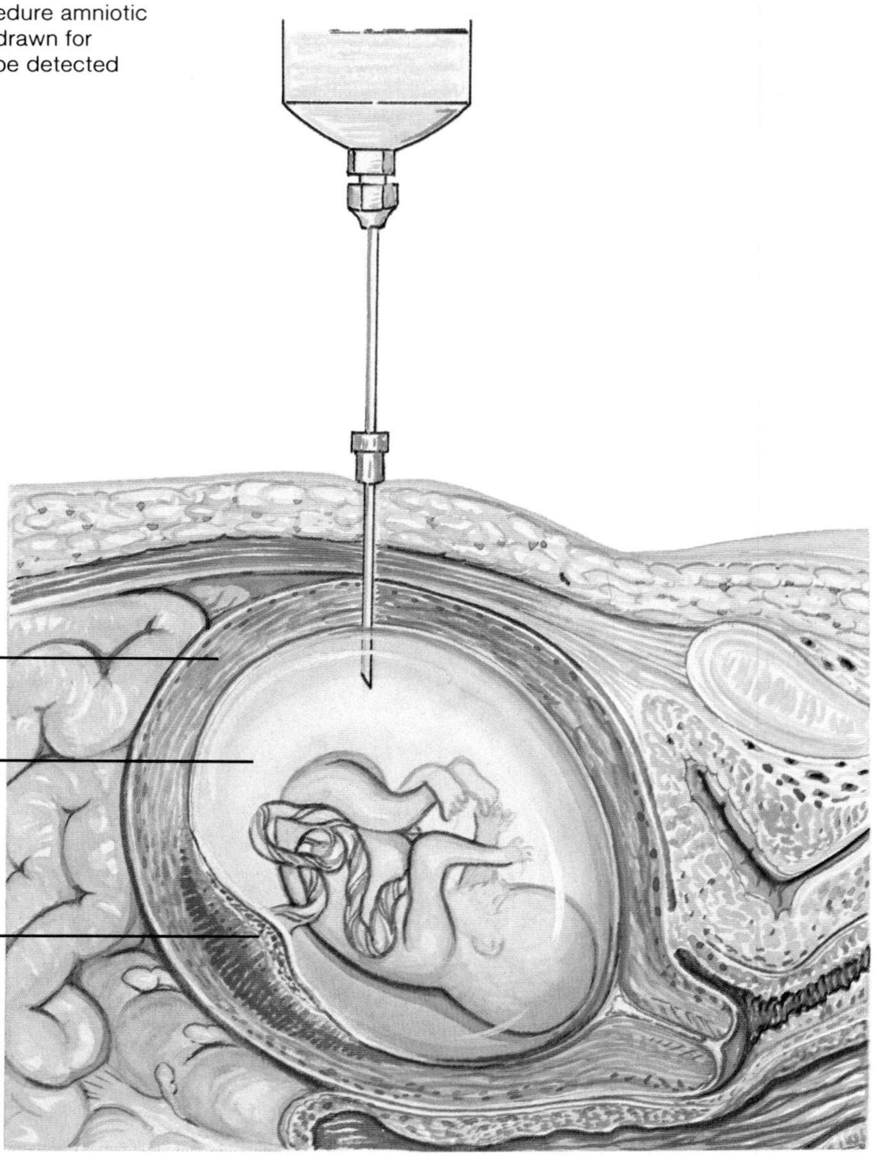

detected by aspiration of this fluid and examination of the cells thus obtained. This procedure is called **amniocentesis** (fig. 20.49).

Amniocentesis is usually performed at the fourteenth or fifteenth week of pregnancy, when the amniotic sac contains 175–225 ml of fluid. Genetic diseases such as Down's syndrome (where there are three instead of two chromosomes number 21) can be detected by examining chromosomes; diseases such as Tay-Sachs disease, in which there is a defective enzyme involved in the formation of myelin sheaths, can be detected by biochemical techniques.

The amniotic fluid that is withdrawn contains fetal cells at a concentration that is too low to permit direct determination of genetic or chromosomal disorders. These cells must therefore be cultured *in vitro* for at least two weeks before they are present in sufficient numbers for the laboratory tests required. A newer method, called **chorionic villus biopsy,** is now available to detect genetic disorders much earlier than permitted by amniocentesis. In chorionic villus biopsy, a catheter is inserted through the cervix to the chorion, and a sample of a chorionic villus is obtained by suction or cutting. Genetic tests can be performed directly on the villus sample, since this sample contains much larger numbers of fetal cells than does a sample of amniotic fluid. Chorionic villus biopsy can provide genetic information at ten to twelve weeks gestation; such information obtained by amniocentesis, in comparison, is not generally available before twenty weeks.

Figure 20.50. The circulation of blood within the placenta. Maternal blood is delivered to and drained from the spaces between the chorionic villi. Fetal blood is brought to blood vessels within the villi by branches of the umbilical artery and is drained by branches of the umbilical vein.

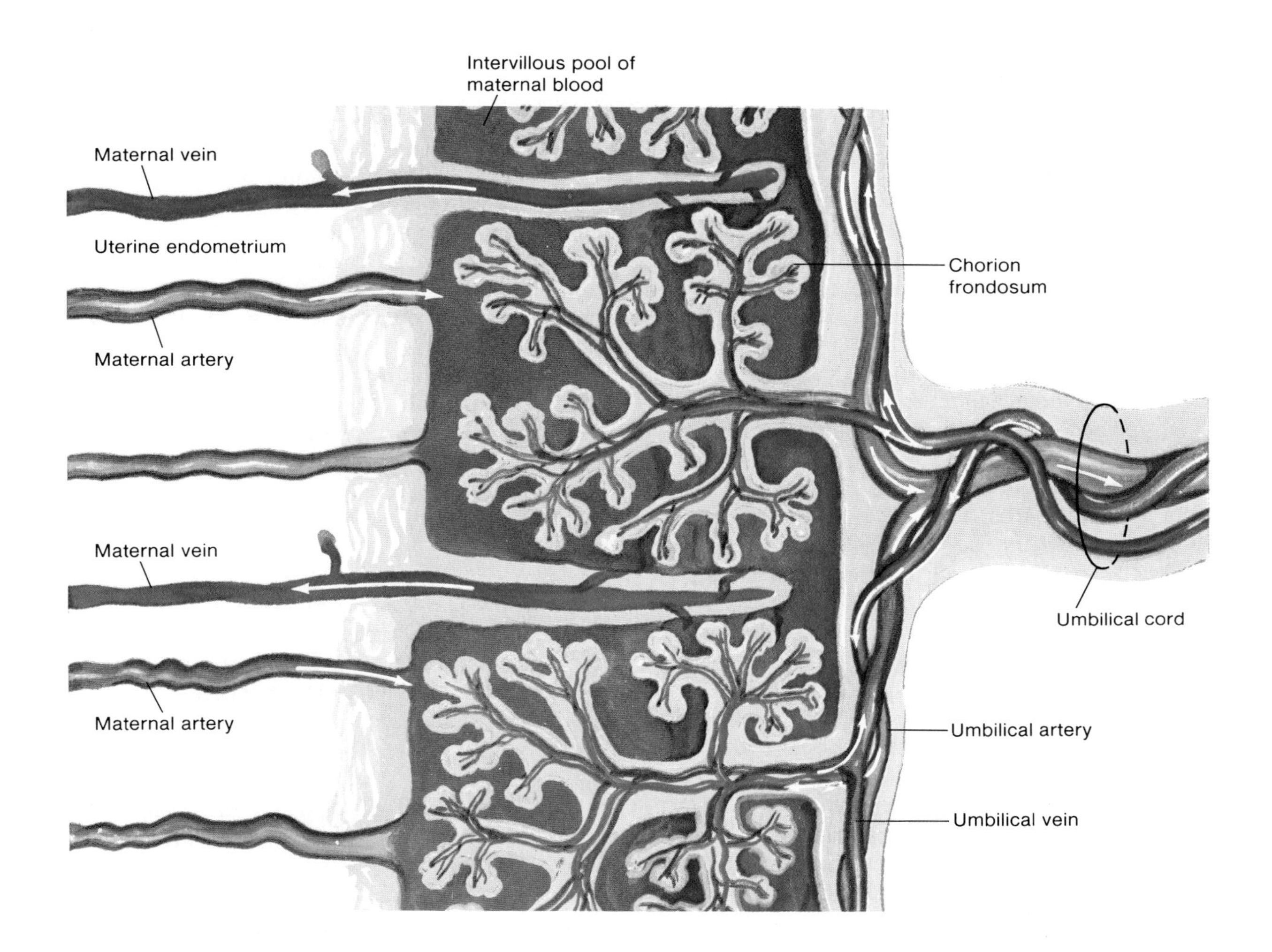

Major structural abnormalities, which may not be predictable from genetic analysis, can often be detected by *ultrasound.* Sound wave vibrations are reflected from the interface of tissues with different densities—such as the interface between the fetus and amniotic fluid—and used to produce an image. This technique is so sensitive that it can be used to detect a fetal heartbeat several weeks before it can be detected by a stethoscope.

Exchange of Molecules across the Placenta

The *umbilical artery* delivers fetal blood to vessels within the villi of the chorion frondosum of the placenta. This blood circulates within the villi and returns to the fetus via the *umbilical vein.* Maternal blood is delivered to and drained from the cavities within the decidua basalis that are located between the chorionic villi (fig. 20.50). In this way, maternal and fetal blood are brought close together but never mix within the placenta.

mono- (Gr.) One, single.
monoclonal antibodies Identical antibodies derived from a clone of genetically identical plasma cells.
monocyte A mononuclear, nongranular leukocyte that is phagocytic and able to be transformed into a macrophage.
monomers A single molecular unit of a longer, more complex molecule; monomers are joined together to form dimers, trimers, and polymers; the hydrolysis of polymers eventually yields separate monomers.
monosaccharides Also called simple sugars; the monomers of more complex carbohydrates. Examples include glucose, fructose, and galactose.
-morph, morpho- (Gr.) Form, shape.
motile Capable of self-propelled movement.
motor neuron An efferent neuron that conducts action potentials away from the central nervous system and innervates effector organs (muscles and glands). It forms the ventral roots of spinal nerves.
motor unit A lower motor neuron and all of the skeletal muscle fibers stimulated by branches of its axon. Larger motor units (more muscle fibers per neuron) produce more power when the unit is activated, but smaller motor units afford a finer degree of neural control over muscle contraction.
mucous membrane The layers of visceral organs that include the lining epithelium, submucosal connective tissue, and (in some cases) a thin layer of smooth muscle (the muscularis mucosa).
muscle spindles Sensory organs within skeletal muscles that are composed of intrafusal fibers and are sensitive to muscle stretch; they provide a length detector within muscles.
myelin sheath A sheath surrounding axons, which is formed by successive wrappings of a neuroglial cell membrane. Myelin sheaths are formed by Schwann cells in the peripheral nervous system and by oligodendrocytes within the central nervous system.
myocardial infarction An area of necrotic tissue in the myocardium that is filled in by scar (connective) tissue.
myofibrils Subunits of striated muscle fibers that consist of successive sarcomeres; myofibrils run parallel to the long axis of the muscle fiber, and the pattern of their filaments provides the striations characteristic of striated muscle cells.
myogenic Originating within muscle cells; this term is used to describe self-excitation by cardiac and smooth muscle cells.
myoglobin A molecule composed of globin protein and heme pigment, related to hemoglobin, but containing only one subunit (instead of the four in hemoglobin), and found in striated muscles. Myoglobin serves to store oxygen in skeletal and cardiac muscle cells.
myoneural junction Also called the neuromuscular junction; a synapse between a motor neuron and the muscle cell that it innervates.
myosin The protein that forms the A bands of striated muscle cells; together with the protein actin, myosin provides the basis for muscle contraction.
myxedema A type of edema associated with hypothyroidism; it is characterized by accumulation of mucoproteins in tissue fluid.

N

NAD Nicotinamide adenine dinucleotide; a coenzyme derived from niacin that functions to transport electrons in oxidation-reduction reactions; it helps to transport electrons to the electron transport chain within mitochondria.
naloxone A drug that antagonizes the effects of morphine and endorphins.
necrosis Cellular death within tissues and organs.
negative feedback Mechanisms in the body that act to maintain a state of internal constancy, or homeostasis; effectors are activated by changes in the internal environment, and the actions of the effectors serve to counteract these changes and maintain a state of balance.
neoplasm A new, abnormal growth of tissue, as in a tumor.
nephron The functional unit of the kidneys, consisting of a system of renal tubules and a vascular component that includes capillaries of the glomerulus and the peritubular capillaries.
neuroglia The supporting tissue of the nervous system, consisting of neuroglial, or glial, cells. In addition to providing support, the neuroglial cells participate in the metabolic and bioelectrical processes of the nervous system.
neurons Nerve cells, consisting of a cell body that contains the nucleus, short branching processes, called dendrites, that carry electrical charges to the cell body, and a single fiber, or axon, that conducts nerve impulses away from the cell body.
neurotransmitter A chemical contained in synaptic vesicles in nerve endings, which is released into the synaptic cleft and stimulates the production of either excitatory or inhibitory postsynaptic potentials.
neutrons Electrically neutral particles that exist together with positively charged protons in the nucleus of atoms.
nexus A bond between members of a group; the type of bonds present in single-unit smooth muscles.
nidation Implantation of the blastocyst into the endometrium of the uterus.
Nissl bodies Granular-appearing structures in the cell bodies of neurons that have an affinity for basic stain; they correspond to ribonucleoprotein.
nodes of Ranvier Gaps in the myelin sheath of myelated axons, located approximately 1 mm apart. Action potentials are produced only at the nodes of Ranvier in myelinated axons.
norepinephrine A catecholamine released as a neurotransmitter from postganglionic sympathetic nerve endings and as a hormone (together with epinephrine) by the adrenal medulla.
nucleolus A dark-staining area within a cell nucleus; the site where ribosomal RNA is produced.
nucleoplasm The protoplasm of a nucleus.
nucleotide The subunit of DNA and RNA macromolecules; each nucleotide is composed of a nitrogenous base (adenine, guanine, cytosine, and thymine or uracil), a sugar (deoxyribose or ribose), and a phosphate group.
nucleus The organelle, surrounded by a double saclike membrane, called the nuclear envelope, that contains the DNA and genetic information of the cell.
nystagmus Involuntary, oscillatory movements of the eye.

O

obese Excessively fat.
oligo- (Gr.) Few, small.
oligodendrocytes A type of neuroglial cell; it forms myelin sheaths around axons in the central nervous system.
oncology The study of tumors.
oncotic pressure The colloid osmotic pressure of solutions produced by proteins; in plasma, it serves to counterbalance the outward filtration of fluid from capillaries due to hydrostatic pressure.
oo- (Gr.) Pertaining to an egg.
oogenesis The formation of ova in the ovaries.
opsonization The role of antibodies in enhancing the ability of phagocytic cells to attack bacteria.
optic disc The area of the retina where axons from ganglion cells gather to form the optic nerve and where blood vessels enter and leave the eye; it corresponds to the blind spot in the visual field due to the absence of photoreceptors.
organ A structure in the body composed of a number of primary tissues that perform particular functions.
organelle A membrane-enclosed structure within cells that performs specialized tasks. The term includes mitochondria, Golgi apparatus, endoplasmic reticulum, nuclei, and lysosomes; it is also used for some structures not enclosed by a membrane, such as ribosomes and centrioles.
organ of Corti The structure within the cochlea responsible for hearing. It consists of hair cells and supporting cells on the basilar membrane that help transduce sound waves into nerve impulses.
osmolality A measure of the total concentration of a solution; the number of moles of solute per kilogram of solvent.
osmoreceptors Sensory neurons that respond to changes in the osmotic pressure of the surrounding fluid.
osmosis The passage of solvent (water) from a more dilute to a more concentrated solution through a membrane that is more permeable to water than to the solute.

osmotic pressure A measure of the tendency for a solution to gain water by osmosis when separated by a membrane from pure water; directly related to the osmolality of the solution, it is the pressure required to just prevent osmosis.
osteo- (G.) Pertaining to bone.
osteoblasts Cells that produce bone.
osteocytes Bone-forming cells that have become entrapped within a matrix of bone; these cells remain alive due to nourishment supplied by canaliculi within the extracellular material of bone.
osteomalacia Softening of bones due to a deficiency of vitamin D and calcium.
osteoporosis Demineralization of bone, seen most commonly in the elderly. It may be accompanied by pain, loss of stature, and other deformities and fractures.
ovaries The gonads of a female that produce ova and secrete female sex steroids.
ovi- (La.) Pertaining to egg.
oviduct The part of the female reproductive tract that transports ova from the ovaries to the uterus. Also called the uterine or fallopian tube.
ovulation The extrusion of a secondary oocyte out of the ovary.
oxidative phosphorylation The formation of ATP by using energy derived from electron transport to oxygen; this occurs in the mitochondria.
oxidizing agent An atom that accepts electrons in an oxidation-reduction reaction.
oxyhemoglobin A compound formed by the bonding of molecular oxygen to hemoglobin.
oxyhemoglobin saturation The ratio, expressed as a percentage, of the amount of oxyhemoglobin compared to the total amount of hemoglobin in blood.
oxytocin One of the two hormones produced in the hypothalamus and secreted by the posterior pituitary (the other hormone is vasopressin); oxytocin stimulates the contraction of uterine smooth muscles and promotes milk ejection in females.

P

pacemaker A group of cells that has the fastest spontaneous rate of depolarization and contraction in a mass of electrically coupled cells; in the heart, this is the sinoatrial, or SA, node.
PAH Para-aminohippuric acid; a substance used to measure total renal plasma flow because its clearance rate is equal to the total rate of plasma flow to the kidneys; PAH is filtered and secreted but not reabsorbed by the renal nephrons.
parathyroid hormone (PTH) A polypeptide hormone secreted by the parathyroid glands, PTH acts to raise the blood Ca^{++} levels primarily by stimulating reabsorption of bone.
Parkinson's disease A tremor of the resting muscles and other symptoms caused by inadequate dopamine-producing neurons in the basal ganglia of the cerebrum. Also called paralysis agitans.
parturition Birth.
passive immunity Specific immunity granted by the administration of antibodies made by another organism.
Pasteur effect A decrease in the rate of glucose utilization and lactic acid production in tissues or organisms by their exposure to oxygen.
pathogen Any disease-producing microorganism or substance.
pepsin The protein-digesting enzyme secreted in gastric juice.
peptic ulcer An injury to the mucosa of the esophagus, stomach, or small intestine caused by acidic gastric juice.
peri- (Gr.) Around, surrounding.
perilymph The fluid between the membranous and bony labyrinth of the inner ear.
perimysium The connective tissue surrounding a fascicle of skeletal muscle fibers.
periosteum Connective tissue covering bones; it contains osteoblasts and is therefore capable of forming new bone.
peristalsis Waves of smooth muscle contraction in smooth muscles of the tubular digestive tract, involving circular and longitudinal muscle fibers at successive locations along the tract; it serves to propel the contents of the tract in one direction.
permease A term used to indicate membrane transport carriers and to emphasize the similarity of specificity and other properties that transport carriers have with enzymes.
pH The pH of a solution is equal to the logarithm of 1 over the hydrogen ion concentration. The pH scale goes from zero to 14; a pH of 7.0 is neutral, whereas solutions with lower pH are acidic and solutions with higher pH are basic.
phagocytosis Cellular eating; the ability of some cells (such as white blood cells) to engulf large particles (such as bacteria) and digest these particles by merging the food vacuole containing these particles with a lysosome containing digestive enzymes.
phonocardiogram A visual display of the heart sounds.
phospodiesterase An enzyme that cleaves cyclic AMP into inactive products, thus inhibiting the action of cyclic AMP as a second messenger.
photoreceptors Sensory cells (rods and cones) that respond electrically to light; they are located in the retinas of the eyes.
pineal gland A gland within the brain that secretes the hormone melatonin and is affected by sensory input from the photoreceptors of the eyes.
pinocytosis Cell drinking; invagination of the cell membrane to form narrow channels that pinch off into vacuoles; it provides cellular intake of extracellular fluid and dissolved molecules.
plasma The fluid portion of the blood. Unlike serum (which lacks fibrinogen), plasma is capable of forming insoluble fibrin threads when in contact with test tubes.
plasma cells Cells derived from B lymphocytes that produce and secrete large amounts of antibodies; they are responsible for humoral immunity.
platelets Disc-shaped structures, 2 to 4 micrometers in diameter, that are derived from bone marrow cells called megakaryocytes. Platelets circulate in the blood and participate (together with fibrin) in forming blood clots.
pluripotent A property of early embryonic cells, which are able to specialize in a number of ways to produce tissues characteristic of different organs.
pneumotaxic center A neural center in the pons that rhythmically inhibits inspiration in a manner independent of sensory input.
-pod, -podium (Gr.) Foot, leg, extension.
polar body A small daughter cell formed by meiosis that degenerates in the process of oocyte production.
polar molecule A molecule in which the shared electrons are not evenly distributed, so that one side of the molecule is relatively negatively (or positively) charged in comparison with the other side; polar molecules are soluble in polar solvents such as water.
poly- (Gr.) Many.
polydipsia Excessive thirst.
polymer A large molecule formed by the combination of smaller subunits, or monomers.
polymorphonuclear leukocyte A granular leukocyte containing a nucleus with a number of lobes connected by thin, cytoplasmic strands; this term includes neutrophils, eosinophils, and basophils.
polypeptide A chain of amino acids connected by covalent bonds called peptide bonds. A very large polypeptide is called a protein.
polyphagia Excessive eating.
polysaccharide A carbohydrate formed by covalent bonding of numerous monosaccharides; examples include glycogen and starch.
polyuria Excretion of an excessively large volume of urine in a given period.
portal system Two capillary beds in series, where blood from the first is drained by veins into a second capillary bed, which in turn is drained by veins that return blood to the heart. The two major portal systems in the body are the hepatic portal system and the hypothalamo-hypophyseal portal system.
posterior Anatomical term denoting a backside position.
posterior pituitary The part of the pituitary gland that is derived from the brain; it secretes vasopressin (ADH) and oxytocin, produced in the hypothalamus. Also called the neurohypophysis.
postsynaptic inhibition The inhibition of a postsynaptic neuron by axon endings that release a neurotransmitter that induces hyperpolarization (inhibitory postsynaptic potentials).

presynaptic inhibition Neural inhibition in which axoaxonic synapses inhibit the release of neurotransmitter chemicals from the presynaptic axon.
pro- (Gr.) Before, in front of, forward.
prolactin A hormone secreted by the anterior pituitary that stimulates lactation (acting together with other hormones) in the postpartum female. It may also participate (along with the gonadotrophins) in regulating gonadal function in some mammals.
prophylaxis Prevention or protection.
proprioceptor A sensory receptor that provides information about body position and movement; examples include receptors in muscles, tendons, and joints as well as the sense of equilibrium provided by the semicircular canals of the inner ear.
proto- (Gr.) First, original.
proton A unit of positive charge in the nucleus of atoms.
protoplasm A general term that includes cytoplasm and nucleoplasm.
pseudo- (Gr.) False.
pseudohermaphrodite An individual who has some of the physical characteristics of both sexes, but lacks functioning gonads of both sexes; a true hermaphrodite has both testes and ovaries.
pseudopods Footlike extensions of the cytoplasm that enable some cells (with amoeboid motion) to move across a substrate; pseudopods also are used to surround food particles in the process of phagocytosis.
ptyalin Also called salivary amylase; an enzyme in saliva that catalyzes the hydrolysis of starch into smaller molecules.
puberty The period of time in an individual's life-span when secondary sexual characteristics and fertility develop.
pulmonary circulation The part of the vascular system which includes the pulmonary artery and pulmonary veins; it transports blood from the right ventricle of the heart, through the lungs, and back to the left atrium of the heart.
pupil The opening at the center of the iris of the eye.
pyramidal tracts Also called corticospinal tracts, these motor tracts descend without synaptic interruption from the cerebrum to the spinal cord. In the spinal cord, these fibers synapse either directly or indirectly (via spinal interneurons) with the lower motor neurons of the spinal cord.
pyrogen A fever-producing substance.

Q

QRS complex The part of an electrocardiogram that is produced by depolarization of the ventricles.

R

recruitment In terms of muscle contraction, the successive stimulation of more and larger motor units in order to produce increasing strengths of muscle contraction.
reduced hemoglobin Hemoglobin with iron in the reduced ferrous state, which is able to bond with oxygen but is not combined with oxygen. Also called deoxyhemoglobin.
reducing agent An electron donor in a coupled oxidation-reduction reaction.
releasing hormones Polypeptide hormones secreted by neurons in the hypothalamus that travel in the hypothalamo-hypophyseal portal system to the anterior pituitary gland and stimulate the anterior pituitary to secrete specific hormones.
REM sleep The stage of sleep in which dreaming occurs; it is associated with rapid eye movements (REM). REM sleep occurs three to four times each night and lasts from a few minutes to over an hour.
renal Pertaining to the kidneys.
renal plasma clearance rate The milliliters of plasma that are cleared of a particular solute per minute by the excretion of that solute in the urine; if there is no reabsorption or secretion of that solute by the nephron tubules, this is equal to the glomerular filtration rate.
renal pyramids The medulla of the human kidney.
repolarization The reestablishment of the resting membrane potential after depolarization has occurred.
respiratory acidosis A lowering of the blood pH below 7.35 due to the accumulation of CO_2 as a result of hypoventilation.
respiratory alkalosis A rise in blood pH above 7.45 due to the excessive elimination of blood CO_2 as a result of hyperventilation.
respiratory distress syndrome Also called hyaline membrane disease; most frequently occurring in premature infants, this syndrome is caused by abnormally high alveolar surface tension as a result of a deficiency in lung surfactant.
resting potential The potential difference across a cell membrane when the cell is in an unstimulated state. The resting potential is always negatively charged on the inside of the membrane compared to the outside.
reticular activating system (RAS) A complex network of nuclei and fiber tracts within the brain stem that produces nonspecific arousal of the cerebrum to incoming sensory information. The RAS thus maintains a state of alert consciousness, and must be depressed during sleep.
retina The layer of the eye that contains neurons and photoreceptors (rods and cones).
rhodopsin Visual purple; a pigment in rod cells that undergoes a photochemical dissociation in response to light and, in so doing, stimulates electrical activity in the photoreceptors.
ribosomes Particles of protein and ribosomal RNA that form the organelles responsible for the translation of messenger RNA and protein synthesis.
rickets A condition caused by a deficiency of vitamin D and associated with interference of the normal ossification of bone.
rigor mortis The stiffening of a dead body, due to the depletion of ATP and the production of rigor complexes between actin and myosin in muscles.
RNA Ribonucleic acid; a nucleic acid consisting of the nitrogenous bases adenine, guanine, cytosine, and uracil; the sugar ribose; and phosphate groups. There are three types of RNA found in cytoplasm: messenger RNA (mRNA), transfer RNA (tRNA), and ribosomal RNA (rRNA).
rods One of the two categories of photoreceptors (along with cones) in the retina of the eye; rods are responsible for black-and-white vision under low illumination.

S

saccadic eye movements Very rapid eye movements that occur constantly and that change the focus on the retina from one point to another.
saltatory conduction The rapid passage of action potentials from one node of Ranvier to another in myelinated axons.
sarcomere The structural subunit of a myofibril in a striated muscle, equal to the distance between two successive Z lines.
sarcoplasm The cytoplasm of striated muscle cells.
sarcoplasmic reticulum The smooth or agranular endoplasmic reticulum of skeletal muscle cells; it surrounds each myofibril and serves to store Ca^{++} when the muscle is at rest.
Schwann cell A neuroglial cell of the peripheral nervous system that forms sheaths around peripheral nerve fibers. Schwann cells also direct regeneration of peripheral nerve fibers to their target cells.
sclera The tough white outer coat of the eyeball that is continuous anteriorly with the clear cornea.
second messenger A molecule or ion whose concentration within a target cell is increased by the action of a regulator compound (e.g., hormone or neurotransmitter) and which stimulates the metabolism of that target cell in a way characteristic of the actions of that regulator molecule—that is, in a way that mediates the intracellular effects of that regulatory compound.
secretin A polypeptide hormone secreted by the small intestine in response to acidity of the intestinal lumen; along with cholecystokinin, secretin stimulates the secretion of pancreatic juice into the small intestine.
semicircular canals Three canals of the bony labyrinth that contain endolymph, which is continuous with the endolymph of the membranous labyrinth of the cochlea; the semicircular canals provide a sense of equilibrium.
semilunar valves The valve flaps of the aorta and pulmonary artery at their juncture with the ventricles.

seminal vesicles The paired organs located on the posterior border of the urinary bladder that empty their contents into the vas deferens and thus contribute to the semen.

seminiferous tubules The tubules within the testes that produce spermatozoa by meiotic division of their germinal epithelium.

semipermeable membrane A membrane with pores of a size that permits the passage of solvent and some solute molecules but restricts the passage of other solute molecules.

sensory neuron An afferent neuron that conducts impulses from peripheral sensory organs into the central nervous system.

serosa An outer epithelial membrane that covers the surface of a visceral organ.

Sertoli cells Nongerminal, supporting cells in the seminiferous tubules. Sertoli cells envelop spermatids and appear to participate in the transformation of spermatids into spermatozoa.

serum The fluid squeezed out of a clot as it retracts; the supernatant when a sample of blood clots in a test tube and is centrifuged; serum is plasma without fibrinogen (which has been converted to fibrin in clot formation).

sex chromosomes The X and Y chromosomes; the unequal pairs of chromosomes involved in sex determination (which is due to the presence or absence of a Y chromosome). Females lack a Y chromosome and normally have the genotype XX; males have a Y chromosome and normally have the genotype XY.

shock As it relates to the cardiovascular system, this refers to a rapid, uncontrolled fall in blood pressure, which in some cases becomes irreversible and leads to death.

sickle-cell anemia A hereditary, autosomal recessive trait that occurs primarily in people of African ancestry, in which it evolved apparently as a protection (in the carrier state) against malaria. In the homozygous state, hemoglobin S is made instead of hemoglobin A; this leads to the characteristic sickling of red blood cells, hemolytic anemia, and organ damage.

sinus A cavity.

sinusoids Blood channels that appear as cavitylike in the surrounding tissue; they function as a type of capillary space that is relatively large and lined (in the liver sinusoids) by phagocytic cells of the reticuloendothelial system.

sleep apnea A temporary cessation of breathing during sleep, usually lasting for several seconds.

sliding filament theory The theory that the thick and thin filaments of a myofibril slide past each other, while maintaining their initial length, during muscle contraction.

smooth muscle Nonstriated, spindle-shaped muscle cells with a single nucleus in the center; involuntary muscle in visceral organs that is innervated by autonomic nerve fibers.

sodium/potassium pump An active transport carrier, with ATPase enzymatic activity, that acts to accumulate K^+ within cells and extrude Na^+ from cells, thus maintaining gradients for these ions across the cell membrane.

soma-, somato-, -some (Gr.) Body, unit.

somatomedins A group of small polypeptides that are believed to be produced in the liver in response to growth hormone stimulation and to mediate the actions of growth hormone on the skeleton and other tissues.

somatostatin A polypeptide produced in the hypothalamus that acts to inhibit the secretion of growth hormone from the anterior pituitary; somatostatin is also produced in the islets of Langerhans of the pancreas, but its function there has not been established.

somatotrophic hormone Growth hormone; an anabolic hormone secreted by the anterior pituitary that stimulates skeletal growth and protein synthesis in many organs.

sounds of Korotkoff The sounds heard when blood pressure measurements are taken. These sounds are produced by the turbulent flow of blood through an artery that has been partially constricted by a pressure cuff.

spermatogenesis The formation of spermatozoa, including meiosis and maturational processes in the seminiferous tubules.

spermiogenesis The maturational changes that transform spermatids into spermatozoa.

sphygmo- (Gr.) The pulse.

sphygmomanometer A manometer (pressure transducer) used to measure the blood pressure.

spindle fibers Filaments that extend from the poles of a cell to its equator and attach to chromosomes during the metaphase stage of cell division. Contraction of the spindle fibers pulls the chromosomes to opposite poles of the cell.

spironolactones Diuretic drugs that act as an aldosterone antagonist.

steroid A lipid, derived from cholesterol, that has three six-sided carbon rings and one five-sided carbon ring. These form the steroid hormones of the adrenal cortex and gonads.

striated muscle Skeletal and cardiac muscle, the cells of which exhibit cross-banding, or striations, due to arrangement of thin and thick filaments into sarcomeres.

stroke volume The amount of blood ejected from each ventricle at each heartbeat.

sub- (La.) Under, below.

substrate In enzymatic reactions, the molecules that combine with the active sites of an enzyme and are converted to products by catalysis of the enzyme.

sulcus A groove or furrow; a depression in the cerebrum that separates folds, or gyri, of the cerebral cortex.

super-, supra- (La.) Above, over.

suppressor T cells A subpopulation of T lymphocytes that acts to inhibit the production of antibodies against specific antigens by B lymphocytes.

surfactant In the lungs, a mixture of phospholipids and proteins produced by alveolar cells that reduces the surface tension of the alveoli and contributes to the elastic properties of the lungs.

sym-, syn- (Gr.) With, together.

synapse A region where a nerve fiber comes into close or actual contact with another cell and across which nerve impulses are transmitted either directly or indirectly (via the release of chemical neurotransmitters).

synergistic Pertaining to regulatory processes or molecules (such as hormones) that have complementary or additive effects.

systemic circulation The circulation that carries oxygenated blood from the left ventricle in arteries to the tissue cells and that carries blood depleted in oxygen via veins to the right atrium; the general circulation, as compared to the pulmonary circulation.

systole The phase of contraction in the cardiac cycle. When unmodified, this term refers to contraction of the ventricles; the term *atrial systole* refers to contraction of the atria.

T

tachycardia Excessively rapid heart rate, usually applied to rates in excess of 100 beats per minute. In contrast to an excessively slow heart rate (below 60 beats per minute), which is termed *bradycardia.*

target organ The organ that is specifically affected by the action of a hormone or other regulatory process.

T cell A type of lymphocyte that provides cell-mediated immunity, in contrast to B lymphocytes that provide humoral immunity through the secretion of antibodies. There are three subpopulations of T cells: cytotoxic, helper, and suppressor.

telo- (Gr.) An end; complete; final.

tendon The dense regular connective tissue that attaches a muscle to the bones of its origin and insertion.

testosterone The major androgenic steroid secreted by the Leydig cells of the testes after puberty.

tetanus A term used to mean either a smooth contraction of a muscle (as opposed to muscle twitching) or a state of maintained contracture of high tension.

thalassemia A group of hemolytic anemias caused by the hereditary inability to produce either the alpha or beta chain of hemoglobin. It is found primarily among Mediterranean people.

thorax The part of the body cavity above the diaphragm; the chest.

threshold The minimum stimulus that just produces a response.

thrombocytes Blood platelets; disc-shaped structures in blood that participate in clot formation.

thrombus A blood clot, produced by the formation of fibrin threads around a platelet plug.

thyroxine Also called tetraiodothyronine, or T_4. The major hormone secreted by the thyroid gland, which regulates the basal metabolic rate and stimulates protein synthesis in many organs; a deficiency of this hormone in early childhood produces cretinism.
tinnitus A ringing sound or other noise that is heard but is not related to external sounds.
toxin A poison.
toxoid A modified bacterial endotoxin that has lost toxicity but retains its ability to act as an antigen and stimulate antibody production.
tracts A collection of axons within the central nervous system, forming the white matter of the CNS.
trans- (La.) Across, through.
transamination The transfer of an amino group from an amino acid to an alpha-keto acid, forming a new keto acid and a new amino acid, without the appearance of free ammonia.
transpulmonary pressure The pressure difference across the wall of the lung, equal to the difference between intrapulmonary pressure and intrapleural pressure.
triiodothyronine Abbreviated T_3; a hormone secreted in small amounts by the thyroid; the active hormone in target cells formed from thyroxine.
tropomyosin A filamentous protein that attaches to actin in the thin filaments and that acts, together with another protein called troponin, to inhibit and regulate the attachment of myosin cross-bridges to actin.
trypsin A protein-digesting enzyme in pancreatic juice that is released into the small intestine.
tympanic membrane The eardrum; a membrane separating the external from the middle ear that transduces sound waves into movements of the middle ear ossicles.

U

universal donor A person with blood type O, who is able to donate blood to people with other blood types in emergency blood transfusions.
universal recipient A person with blood type AB, who can receive blood of any type in emergency transfusions.
urea The chief nitrogenous waste product of protein catabolism in the urine, formed in the liver from amino acids.
uremia The retention of urea and other products of protein catabolism due to inadequate kidney function.
urobilinogen A compound formed from bilirubin in the intestine; some is excreted in the feces and some is absorbed and enters the enterohepatic circulation, where it may be excreted either in the bile or in the urine.

V

vasa-, vaso- (La.) Pertaining to blood vessels.
vasa vasora Blood vessels that supply blood to the walls of large blood vessels.
vasectomy Surgical removal of a portion of the vas (ductus) deferens to induce infertility.
vasoconstriction Narrowing of the lumen of blood vessels due to contraction of the smooth muscles in their walls.
vasodilation Widening of the lumen of blood vessels due to relaxation of the smooth muscles in their walls.
vein A blood vessel that returns blood to the heart.
ventilation Breathing; the process of moving air into and out of the lungs.
vertigo A feeling of movement or loss of equilibrium.
virulent Pathogenic, or able to cause disease.

Z

zygote A fertilized ovum.
zymogens Inactive enzymes that become active when part of their structure is removed by another enzyme or by some other means.

Credits

Photographs

Chapter 1

1.1, 1.2, 1.3, 1.4a, 1.7, 1.9, 1.10: © Edwin Reschke; **1.12:** © SIU Biomedical Communications/ Journalism Services.

Chapter 2

2.26: © Edwin Reschke.

Chapter 3

3.3: © Sandra L. Wolin; **3.5a–b:** © Kwang W. Jeon; **3.6 (1–4):** © R. H. Albertin, M.A.; **3.7a–b, 3.8:** © Richard Chao; **3.9a:** © K. R. Porter; **3.10:** © E. G. Pollack; **3.11:** © Richard Chao; **3.14:** Margery Shaw, courtesy of Upjohn Company; **3.15a:** © K. R. Porter; **3.16a–e:** © Edwin Reschke: **3.18a:** O. L. Miller, Jr., *Journal of Cell Physiology,* 74 (1969); **3.19:** © Alexander Rich; **3.23a:** © K. R. Porter; **3.25a:** © Daniel S. Friend.

Chapter 5

5.9: © Fernandes-Morán, V. M.D., Ph.D.

Chapter 6

6.4a: © Carolyn Chambers; **6.4b:** Kessel and Kardon/© W. H. Freeman and Company; **6.11:** © Richard Chao.

Chapter 7

7.2: © Dr. Kerry L. Openshaw; **7.4:** H. Webster, from Hubbard, John: *The Vertebrate Peripheral Nervous System.* © Plenum Press, 1974; **7.6:** © Edwin Reschke; **7.12:** © Churchill Livingstone, Inc.; **7.17a–b:** © MacMillan Journals, Ltd.; **7.18:** © John Heuser, Washington University School of Medicine, St. Louis, MO.

Chapter 8

8.1b: © Edwin Reschke; **8.16b:** © Journalism Services; **8.17:** © Martin Rotker/Taurus Photos; **8.19, 8.21:** © Lester Bergman & Associates; **8.25:** © Edwin Reschke.

Chapter 9

9.7a: © Martin Rotker/Taurus; **9.9b:** © Igaku-Shoin, Ltd.

Chapter 10

10.14: © Dean E. Hillman; **10.27:** Kessel and Shih/© Springer-Verlag, Berlin; **10.36a:** © Thomas Sims; **10.42b:** © F. Werblin; **10.50:** © David H. Hubel.

Chapter 11

11.3: © Edwin Reschke; **11.5:** Narco Scientific, Inc.; **11.7a:** © Edwin Reschke; **11.18:** Huxley, H. E., *The Structure and Function of Muscle,* vol. 1, 2d ed. © 1972 Academic Press, Inc.; **11.19a–b, 11.20(top):** © H. E. Huxley; **11.19c:** Kessel and Kardon/© W. H. Freeman and Company; **11.29, 11.31:** Hans Hoppler, *Respiratory Physiology* 44:94(1981).

Chapter 12

12.2: © Edwin Reschke; **12.3(both):** © Avril Somylo, Ph.D.; **12.13a:** © Igaku-Shoin, Ltd.

Chapter 13

13.2b: © Igaku-Shoin, Ltd.; **13.16a–b:** © Richard Menard; **13.17 (all):** © James Shaffer; **13.20:** © Don Fawcett; **13.23:** Historical Pictures Service, Chicago; **13.24a–b:** American Heart Association; **13.29:** © Manfred Kage/Peter Arnold, Inc.

Chapter 14

14.9: Markell, E. K., and Voge, M.: *Medical Parasitology,* 6th ed. © W. B. Saunders Company, 1986; **14.16a–b:** Donald S. Baim, from Hurst et al.: *The Heart,* 5th ed. © McGraw-Hill Book Company, 1982; **14.20a–b:** © Neils A. Lassen, Copenhagen, Denmark.

Chapter 15

15.2: Murray, John F.: *The Normal Lung,* 2nd ed. © W. B. Saunders Company, 1986; **15.3a–b:** American Lung Association; **15.5a:** West, J. B.: *Respiratory Physiology: The Essentials.* © 1979 Williams and Wilkins Company, Baltimore; **15.6:** Yokochi, C., and Rohen, J. W.: *Photographic Anatomy of the Human Body,* 2nd ed. © Igaku-Shoin, Ltd., 1978; **15.7:** American Lung Association; **15.9a–b, 15.11:** Edward C. Vasquez, R.T., C.R.T., Department of Radiologic Technology, Los Angeles City College; **15.15a–b:** Comroe, J. H.: *Physiology of Respiration.* © 1974 Yearbook Medical Publishers, Inc.; **15.18:** Courtesy of Warren E. Collins, Braintree, MA; **15.21a–c:** © Oscar Auerbach; **15.24 (both):** Kessel and Kardon/© W. H. Freeman and Company; **15.38a–c:** McCuray, P. R., *Sickle-Cell Disease.* © Medcom, Inc. 1973.

Chapter 16

16.3b: Kessel and Kardon/© W. H. Freeman and Company; **16.6a:** © Gordon Leedale/BioPhoto Associates; **16.7a:** Daniel Friend, from Bloom, W., and Fawcett, D. W.: *A Textbook of Histology,* 10th ed. © W. B. Saunders Company, 1975.

Chapter 17

17.6: © Stuart I. Fox; **17.7a:** Computer graphic by Arthur J. Olson, Ph.D./© 1985 All rights reserved. Scripps Clinic and Research Foundation, La Jolla, CA 92037; **17.13a:** Rosenthal, Alan S.: *New England Journal of Medicine,* 303 (1980) 1153; **17.17 (all):** © Stuart I. Fox; **17.22a–b:** © Andrejs Liepens, Department of Biology, Memorial University, St. John's, Newfoundland; **17.23 (both):** © Noel R. Rose; **17.26 (both):** Philip S. Norman, from Middleton, E., Reed, C. E., and Ellis, E. F.: *Allergy: Principles and Practice.* © C. V. Mosby Company, 1978; **17.27a:** Kessel and Shih/ © Springer-Verlag, Berlin; **17.27b:** Acarology Laboratory, Ohio State University.

Chapter 18

18.4b, 18.6: Courtesy of Utah Valley Hospital, Department of Radiology; **18.7:** © Edwin Reschke; **18.12:** © Manfred Kage/Peter Arnold; **18.15a:** © Keith Porter; **18.15b:** Alpers, D. H., and Seetharam, B.: *New England Journal of Medicine* 296 (1977) 1047; **18.19:** Sodeman, W. A., and Watson, T. M.: *Pathologic Physiology,* 6th ed. © W. B. Saunders Company, 1969; **18.20:** After Sobotta, from Bloom, W., and Fawcett, D. W.: *A Textbook of Histology,* 10th ed. © W. B. Saunders Company, 1975; **18.21a:** Kessel and Kardon/© W. H. Freeman and Company; **18.27a:** © Carroll Weiss/RBP; **18.27b:** © Sheril D. Burton

Chapter 19

19.14: © Lester V. Bergman and Associates; **19.17a–d:** *American Journal of Medicine* 20 (1956) 133; **19.18a–b:** Bhasker, S. N., ed.: *Orban's Oral Histology and Embryology,* 9th ed. © C. V. Mosby Company, 1980.

Chapter 20

20.2a–b: Wisniewski, L. P., and Hirschhorn, K.: "A Guide to Human Chromosome Defects," 2nd ed., White Plains: The March of Dimes Birth Defects Foundation, BD:OAS 16(6), 1980; **20.3a–d:** Williams, R. H.: *Textbook of Endocrinology,* 6th ed. © W. B. Saunders Company, 1981; **20.5a–b:** From Redding, A., and Hirschhorn, K.: "A Guide to Human Chromosome Defects." D. Bergsma (ed). New York: The National Foundation-March of Dimes, BD:OAS IV(4), 1968; **20.7i–j:** © Landrum Shettles; **20.18a:** © Gordon Leedale/ BioPhoto Associates; **20.21b:** Kessel and Kardon/© W. H. Freeman and Company; **20.29a:** © Edwin Reschke; **20.30a:** Blandau, R. J.: *A Textbook of Histology,* 10th ed. © W. B. Saunders Company, 1975; **20.30b:** © Landrum Shettles; **20.31:** Bloom, W., and Fawcett, D. W.: *A Textbook of Histology,* 10th ed. © W. B. Saunders Company, 1975; **20.32:** © Landrum Shettles; **20.37:** © Martin Rotker/Taurus Photos; **20.39c:** Tegner, M. J., and Epel, D.: "Sea Urchin Sperm-Egg Interactions Studied with the Scanning Electron Microscope," *Science,* vol. 179, pp. 685–688, 16 February 1973/© AAAS; **20.40:** Lucian Zamboni, from Greep, Roy, and Weiss, Leon: *Histology,* 3rd ed. © McGraw-Hill Book Company, 1973; **20.44a–d:** © R. G. Edwards; **20.45b:** © Landrum Shettles.

Line Art

Chapter 1

1.4: From Van De Graaff, Kent M., *Human Anatomy.* © 1984 Wm. C. Brown Publishers, Dubuque, Iowa. All Rights Reserved. Reprinted by permission. **1.13, 1.19:** From Mader, Sylvia S., *Inquiry Into Life,* 4th ed. © 1976, 1979, 1982, 1985 Wm. C. Brown Publishers, Dubuque, Iowa. All Rights Reserved. Reprinted by permission.

Chapter 2

2.5: From Hole, John W., Jr., *Human Anatomy and Physiology,* 3rd ed. © 1978, 1981, 1984 Wm. C. Brown Publishers, Dubuque, Iowa. All Rights Reserved. Reprinted by permission.

Chapter 3

3.1, 3.23b: From Mader, Sylvia S., *Inquiry Into Life,* 4th ed. © 1976, 1979, 1982, 1985 Wm. C. Brown Publishers, Dubuque, Iowa. All Rights Reserved. Reprinted by permission. **3.9, 3.12:** From Mader, Sylvia S., *Biology: Evolution, Diversity, and the Environment.* © 1985 Wm. C. Brown Publishers, Dubuque, Iowa. All Rights Reserved. Reprinted by permission. **3.15:** From Van De Graaff, Kent M., *Human Anatomy.* © 1984 Wm. C. Brown Publishers, Dubuque, Iowa. All Rights Reserved. Reprinted by permission. **3.16, 3.17:** From Van De Graaff, Kent M., and Stuart I. Fox, *Concepts of Human Anatomy and Physiology.* © 1986 Wm. C. Brown Publishers, Dubuque, Iowa. All Rights Reserved. Reprinted by permission. **3.23c:** From Hole, John W., Jr., *Human Anatomy and Physiology,* 3rd ed. © 1978, 1981, 1984 Wm. C. Brown Publishers, Dubuque, Iowa. All Rights Reserved. Reprinted by permission.

Chapter 4

4.5: From Van De Graaff, Kent M., and Stuart I. Fox, *Concepts of Human Anatomy and Physiology.* © 1986 Wm. C. Brown Publishers, Dubuque, Iowa. All Rights Reserved. Reprinted by permission.

Chapter 6

6.14: From Mader, Sylvia S., *Inquiry Into Life,* 4th ed. © 1982, 1985 Wm. C. Brown Publishers, Dubuque, Iowa. All Rights Reserved. Reprinted by permission.

Chapter 7

7.1, 7.5: From Hole, John W., Jr., *Human Anatomy and Physiology,* 3rd ed. © 1978, 1981, 1984 Wm. C. Brown Publishers, Dubuque, Iowa. All Rights Reserved. Reprinted by permission. **7.13:** Modified from Kutchai, H. C., "Generation and Conduction of Action Potentials," in *Physiology,* by Berne, R. M., and M. N. Levy, (eds.). St. Louis: The C. V. Mosby Company, 1983. **7.17c:** From Gilula, Reeves, and Steinbach. *Nature,* 235: 262–265. © MacMillan Journals Limited.

Chapter 8

8.1, 8.8, 8.22, 8.24: From Van De Graaff, Kent M., *Human Anatomy.* © 1984 Wm. C. Brown Publishers, Dubuque, Iowa. All Rights Reserved. Reprinted by permission. **8.9, 8.10, 8.16, 8.20:** From Van De Graaff, Kent M., and Stuart I. Fox, *Concepts of Human Anatomy and Physiology.* © 1986 Wm. C. Brown Publishers, Dubuque, Iowa. All Rights Reserved. Reprinted by permission. **8.13, 8.26:** From Hole, John W., Jr., *Human Anatomy and Physiology,* 3rd ed. © 1978, 1981, 1984 Wm. C. Brown Publishers, Dubuque, Iowa. All Rights Reserved. Reprinted by permission. **8.23:** From Hole, John W., Jr., *Essentials of Human Anatomy and Physiology.* © 1982 Wm. C. Brown Publishers, Dubuque, Iowa. All Rights Reserved. Reprinted by permission. **8.27:** After Demers, L. M., "The Effects of Prostaglandins," in *Diagnostic Medicine,* September 1984, p. 37. Oradell, NJ: Medical Economics Company, Inc.

Chapter 9

9.1, 9.3: From Mader, Sylvia S., *Inquiry Into Life,* 4th ed. © 1976, 1979, 1982, 1985 Wm. C. Brown Publishers, Dubuque, Iowa. All Rights Reserved. Reprinted by permission. **9.2, 9.5, 9.13, 9.14, 9.15:** From Hole, John W., Jr., *Human Anatomy and Physiology,* 3rd ed. © 1978, 1981, 1984 Wm. C. Brown Publishers, Dubuque, Iowa. All Rights Reserved. Reprinted by permission. **9.4a, 9.9:** From Van De Graaff, Kent M., *Human Anatomy Laboratory Textbook,* 2nd ed. © 1981, 1984 Wm. C. Brown Publishers, Dubuque, Iowa. All Rights Reserved. Reprinted by permission. **9.4b, 9.7, 9.12:** From Van De Graaff, Kent M., *Human Anatomy.* © 1984 Wm. C. Brown Publishers, Dubuque, Iowa. All Rights Reserved. Reprinted by permission. **9.6:** From Peele, T. L., *The Neuroanatomic Basis for Clinical Neurology,* 2nd ed. © 1961 by McGraw-Hill Book Company. **9.8, 9.11, 9.16, 9.18, 9.19:** From Van De Graaff, Kent M., and Stuart I. Fox, *Concepts of Human Anatomy and Physiology.* © 1986 Wm. C. Brown Publishers, Dubuque, Iowa. All Rights Reserved. Reprinted by permission. **9.10:** From Noback, Charles R., and Robert J. Demarest, *The Human Nervous System: Basic Principles of Neurobiology,* 2nd ed. © 1975 by McGraw-Hill Book Company.

Chapter 10

10.5, 10.32, 10.41, 10.48: From Van De Graaff, Kent M., *Human Anatomy.* © 1984 Wm. C. Brown Publishers, Dubuque, Iowa. All Rights Reserved. Reprinted by permission. **10.6, 10.20, 10.21, 10.22, 10.28, 10.31, 10.33, 10.34, 10.35, 10.36, 10.39:** From Van De Graaff, Kent M., and Stuart I. Fox, *Concepts of Human Anatomy and Physiology.* © 1986 Wm. C. Brown Publishers, Dubuque, Iowa. All Rights Reserved. Reprinted by permission. **10.9:** From Fox, Stuart I., *A Laboratory Guide to Human Physiology: Concepts and Clinical Applications,* 3rd ed. © 1976, 1980, 1984 Wm. C. Brown Publishers, Dubuque, Iowa. All Rights Reserved. Reprinted by permission. **10.10, 10.11, 10.13, 10.16, 10.17, 10.19, 10.42a, 10.46:** From Hole, John W., Jr., *Human Anatomy and Physiology,* 3rd ed. © 1978, 1981, 1984 Wm. C. Brown Publishers, Dubuque, Iowa. All Rights Reserved. Reprinted by permission. **10.40:** From Mader, Sylvia S., *Inquiry Into Life,* 4th ed. © 1976, 1979, 1982, 1985 Wm. C. Brown Publishers, Dubuque, Iowa. All Rights Reserved. Reprinted by permission.

Chapter 11

11.1, 11.7b, 11.17: From Hole, John W., Jr., *Human Anatomy and Physiology,* 3rd ed. © 1978, 1981, 1984 Wm. C. Brown Publishers, Dubuque, Iowa. All Rights Reserved. Reprinted by permission. **11.2:** From Van De Graaff, Kent M., *Human Anatomy.* © 1984 Wm. C. Brown Publishers, Dubuque, Iowa. All Rights Reserved. Reprinted by permission. **11.4:** From Fox, Stuart I., *A Laboratory Guide to Human Physiology: Concepts and Clinical Applications,* 3rd ed. © 1976, 1980, 1984 Wm. C. Brown Publishers, Dubuque, Iowa. All Rights Reserved. Reprinted by permission. **11.22:** From Morki, E., "Contractile Proteins of the Heart," *Hospital Practice,* June 1983, p. 97. New York: H. P. Publishing Company, Inc.

Chapter 12

12.1, 12.5, 12.7, 12.8, 12.13: From Van De Graaff, Kent M., and Stuart I. Fox, *Concepts of Human Anatomy and Physiology.* © 1986 Wm. C. Brown Publishers, Dubuque, Iowa. All Rights Reserved. Reprinted by permission. **12.4, 12.6, 12.9:** From Van De Graaff, Kent M., *Human Anatomy.* © 1984 Wm. C. Brown Publishers, Dubuque, Iowa. All Rights Reserved. Reprinted by permission.

Chapter 13

13.1, 13.3, 13.27: From Hole, John W., Jr., *Human Anatomy and Physiology,* 3rd ed. © 1978, 1981, 1984 Wm. C. Brown Publishers, Dubuque, Iowa. All Rights Reserved. Reprinted by permission. **13.2, 13.6, 13.18:** From Van De Graaff, Kent M., *Human Anatomy.* © 1984 Wm. C. Brown Publishers, Dubuque, Iowa. All Rights Reserved. Reprinted by permission. **13.9, 13.12, 13.13, 13.19, 13.24, 13.32:** From Van De Graaff, Kent M., and Stuart I. Fox, *Concepts of Human Anatomy and Physiology.* © 1986 Wm. C. Brown Publishers, Dubuque, Iowa. All Rights Reserved. Reprinted by permission. **13.14:** Drawn from *Emergency Medicine,* Volume II, Issue #2, February 15, 1979. **13.28:** From Benson, Harold J., et al., *Anatomy and Physiology Laboratory Textbook, Complete Version,* 3rd ed. © 1970, 1976, 1983 Wm. C. Brown Publishers, Dubuque, Iowa. All Rights Reserved. Reprinted by permission. **13.31:** Adapted from Marchand, A., "Case of the Month, Circulating Anticoagulants: Chasing the Diagnosis," *Diagnostic Medicine,* June 1983, p. 14.

Chapter 14

14.8: From Hole, John W., Jr., *Human Anatomy and Physiology,* 3rd ed. © 1978, 1981, 1984 Wm. C. Brown Publishers, Dubuque, Iowa. All Rights Reserved. Reprinted by permission. **14.17:** Reprinted by permission of *Practical Cardiology.* **14.18:** From Astrand, P., and K. Rodahl, *Textbook of Work Physiology.* © 1977 McGraw-Hill, Inc. Reprinted by permission of McGraw-Hill Book Company. **14.21, 14.25, 14.31:** From Van De Graaff, Kent M., and Stuart I. Fox, *Concepts of Human Anatomy and Physiology.* © 1986 Wm. C. Brown Publishers, Dubuque, Iowa. All Rights Reserved. Reprinted by permission. **14.23:** From Feigal, E. O., "Physics in the Cardiovascular System," in *Physiology and Biophysics,* Volume II, T. C. Ruch and H. D. Patton, editors. © 1974 W. B. Saunders, Philadelphia, PA. **14.27:** From Fox, Stuart I., *A Laboratory Guide to Human Physiology: Concepts and Clinical Applications,* 3rd ed.

© 1976, 1980, 1984 Wm. C. Brown Publishers, Dubuque, Iowa. All Rights Reserved. Reprinted by permission. **14.30:** Modified from Geddes, L. A., *The Direct and Indirect Measurement of Blood Pressure.* © 1970 by Year Book Medical Publishers, Inc., Chicago.

Chapter 15

15.4: From Weibel, E. R., *Morphometry of the Human Lung,* Springer-Verlag OHG, Heidelberg, 1963. Reprinted by permission. **15.5, 15.8, 15.16, 15.19, 15.29, 15.39, 15.40:** From Hole, John W., Jr., *Human Anatomy and Physiology,* 3rd ed. © 1978, 1981, 1984 Wm. C. Brown Publishers, Dubuque, Iowa. All Rights Reserved. Reprinted by permission. **15.14, 15.28, 15.32:** From Van De Graaff, Kent M., and Stuart I. Fox, *Concepts of Human Anatomy and Physiology.* © 1986 Wm. C. Brown Publishers, Dubuque, Iowa. All Rights Reserved. Reprinted by permission. **15.17:** From Van De Graaff, Kent M., *Human Anatomy.* © 1984 Wm. C. Brown Publishers, Dubuque, Iowa. All Rights Reserved. Reprinted by permission. **15.42:** From Wasserman, K., A. L. Van Kessel, and G. G. Burton, in *Journal of Applied Physiology,* Vol. 22, pp. 71–85. © 1967 American Physiological Society.

Chapter 16

16.1, 16.3, 16.4, 16.21: From Mader, Sylvia S., *Inquiry Into Life,* 4th ed. © 1976, 1979, 1982, 1985 Wm. C. Brown Publishers, Dubuque, Iowa. All Rights Reserved. Reprinted by permission. **16.2, 16.5, 16.6b:** From Hole, John W., Jr., *Human Anatomy and Physiology,* 3rd ed. © 1978, 1981, 1984 Wm. C. Brown Publishers, Dubuque, Iowa. All Rights Reserved. Reprinted by permission. **16.7, 16.11, 16.15, 16.18, 16.19, 16.22:** From Van De Graaff, Kent M., and Stuart I. Fox, *Concepts of Human Anatomy and Physiology.* © 1986 Wm. C. Brown Publishers, Dubuque, Iowa. All Rights Reserved. Reprinted by permission.

Chapter 17

17.1, 17.10: From Van De Graaff, Kent M., and Stuart I. Fox, *Concepts of Human Anatomy and Physiology.* © 1986 Wm. C. Brown Publishers, Dubuque, Iowa. All Rights Reserved. Reprinted by permission. **17.15:** Modified from "The Immune System in AIDS," by Jeffrey Lawrence. Copyright © 1985 by Scientific American, Inc. All rights reserved. **17.16:** From McDevitt, Hugh O., "The HLA System and Its Relation to Disease," *Hospital Practice,* July 1985, p. 57. New York: H. P. Publishing Company, Inc.

Chapter 18

18.1, 18.5, 18.8, 18.26: From Hole, John W., Jr., *Human Anatomy and Physiology,* 3rd ed. © 1978, 1981, 1984 Wm. C. Brown Publishers, Dubuque, Iowa. All Rights Reserved. Reprinted by permission. **18.2, 18.4, 18.11:** From Van De Graaff, Kent M., *Human Anatomy.* © 1984 Wm. C. Brown Publishers, Dubuque, Iowa. All Rights Reserved. Reprinted by permission. **18.3, 18.9, 18.14, 18.15, 18.16, 18.18, 18.21, 18.22, 18.28:** From Van De Graaff, Kent M., and Stuart I. Fox, *Concepts of Human Anatomy and Physiology.* © 1986 Wm. C. Brown Publishers, Dubuque, Iowa. All Rights Reserved. Reprinted by permission. **18.13:** From Mader, Sylvia S., *Inquiry Into Life,* 4th ed. © 1976, 1979, 1982, 1985 Wm. C. Brown Publishers, Dubuque, Iowa. All Rights Reserved. Reprinted by permission.

Chapter 19

19.1, 19.16: From Van De Graaff, Kent M., and Stuart I. Fox, *Concepts of Human Anatomy and Physiology.* © 1986 Wm. C. Brown Publishers, Dubuque, Iowa. All Rights Reserved. Reprinted by permission. **19.4, 19.5:** From Parker, Mary L., Plides, Rosita S., Chao, Kuen-Lan, Cornblath, Marvin, and Kipnis, David M., "Juvenile Diabetes Mellitus, A Deficiency in Insulin," *Diabetes* 1968; 17: 27–32. Reproduced with permission from the American Diabetes Association, Inc. **19.9:** From Barrett, E. J., and R. A. DeFronao, "Diabetic Ketoacidosis: Diagnosis and Management," *Hospital Practice,* April 1984, p. 89.

Chapter 20

20.7, 20.12, 20.18, 20.34, 20.39, 20.45: From Van De Graaff, Kent M., *Human Anatomy.* © 1984 Wm. C. Brown Publishers, Dubuque, Iowa. All Rights Reserved. Reprinted by permission. **20.16, 20.21, 20.23, 20.27, 20.36:** From Mader, Sylvia S., *Inquiry Into Life,* 4th ed. © 1976, 1979, 1982, 1985 Wm. C. Brown Publishers, Dubuque, Iowa. All Rights Reserved. Reprinted by permission. **20.17, 20.24, 20.26, 20.41, 20.42, 20.52:** From Van De Graaff, Kent M., and Stuart I. Fox, *Concepts of Human Anatomy and Physiology.* © 1986 Wm. C. Brown Publishers, Dubuque, Iowa. All Rights Reserved. Reprinted by permission. **20.19, 20.25:** From Fawcett, D. W., chapter 2, p. 43, in *Handbook of Physiology,* section 7, Vol. V, "Endocrinology: Male Reproductive System." American Physiological Society, Washington, DC. **20.22, 20.28, 20.33, 20.43, 20.46:** From Hole, John W., Jr., *Human Anatomy and Physiology,* 3rd ed. © 1978, 1981, 1984 Wm. C. Brown Publishers, Dubuque, Iowa. All Rights Reserved. Reprinted by permission. **20.35:** From Thorneycroft, Ian H., et. al., *American Journal of Obstetrics and Gynecology,* Vol. 3, © 1971 The C. V. Mosby Company.

Index

References in italics refer to illustrations.

B

C

D

E

F

G

H

I

J

K

L

M

N

Q

R

S

T

U

V

W

Z